Extremo

From Biology to Biotechnology

LIST OF BOOKS EDITED BY THE AUTHORS

Ben-Amotz, A., J. E. W. Polle, and D. V. Subba Rao, eds. 2009. *The Alga Dunaliella: Biodiversity, Physiology, Genomics and Biotechnology*. Enfield, NH: Science Publishers.

Satoskar, A., and R. Durvasula, eds. *Pathogenesis of Leishmaniasis: New Developments in Research*. Berlin: Springer, 2014.

Sree Hari Rao, V., and R. Durvasula, eds. 2013. *Dynamic Models of Infectious Diseases*. Berlin: Springer.

Subba Rao, D. V., ed. 1996. *Present and Future of Oceanographic Programs in Developing Countries, Vienna and Honolulu*. IAPSO Publication Scientifique No. 36 and Andhra University Memoirs No. 3. Visakhapatnam, India: Andhra University.

Subba Rao, D. V., ed. 2002. *Pelagic Ecology Methodology*. Rotterdam, Netherlands: Swets & Zeitlinger and Balkema Publishers.

Subba Rao, D. V., ed. 2006. *Algal Cultures, Analogues of Blooms and Applications*. Vol. 1. Enfield, NH: Science Publishers.

Subba Rao, D. V., ed. 2006. *Algal Cultures, Analogues of Blooms and Applications*. Vol. 2. Enfield, NH: Science Publishers.

(a)

(b)

(c)

(d)

(e)

(f)

Sampling in the Antarctic for extremophiles; Dr. Amber G. Teufel and Dr. Rachael M. Morgan-Kiss.

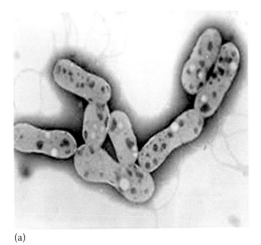

(a)

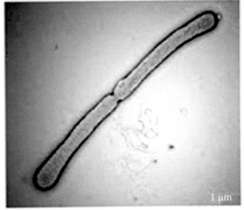

(b)

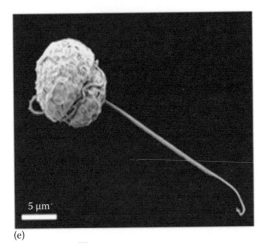

(c)

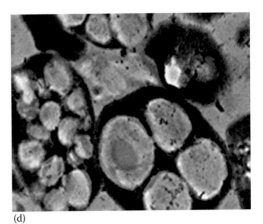

(d)

(e)

(f)

(a) *Baslnearium* and (b) *Thermotomaculam*—hydrothermal vents; Dr. Mino Sayaka and Dr. Satoshi Nakagawa. (c) *Haloferax mediterranei*. Colonies of halophilic microorganisms isolated from salted water (Santa Pola); Dr. Rosa María Martínez-Espinosa. (d) *Haloferax mediterranei* cells (PHA production); Dr. Vanesa Bautista Saiz. SEM images: (e) *Gymnodinium* sp. and (f) *Brachiomonas* sp. isolated from Lake Bonney within the McCurdo Dry Valleys, Antarctica; Dr. Wei Li, Montana University, USA.

(a)

(b)

(c)

(d)

(e)

(f)

Location of the desert Sahara in the northern part of the African continent (a–e) and distribution of Tunisian hypersaline systems and oases, where studies were conducted. Examples of (d) hypersaline environments and (e) desert plants in southern Tunisia (details in their chapter; Dr. Ameur Cherif). (f) Coastal saltern ponds, Spain; Dr. Rosa María Martínez-Espinosa.

Extremophiles
From Biology to Biotechnology

Edited by
Ravi Durvasula
D. V. Subba Rao

CRC Press
Taylor & Francis Group
Boca Raton London New York

CRC Press is an imprint of the
Taylor & Francis Group, an **informa** business

CRC Press
Taylor & Francis Group
6000 Broken Sound Parkway NW, Suite 300
Boca Raton, FL 33487-2742

First issued in paperback 2020

ISBN-13: 978-0-367-57232-7 (pbk)
ISBN-13: 978-1-4987-7492-5 (hbk)

Library of Congress Cataloging-in-Publication Data

Names: Durvasula, Ravi, editor.
Title: Extremophiles : from biology to biotechnology / edited by Ravi Durvasula and D.V. Subba Rao.
Description: Boca Raton : Taylor & Francis, a CRC title, part of the Taylor & Francis imprint, a member of the Taylor & Francis Group, the academic division of T&F Informa plc, 2018. | Includes bibliographical references and index.
Identifiers: LCCN 2017031972 | ISBN 9781498774925 (hardback : alk. paper)
Subjects: LCSH: Extreme environments--Microbiology. | Extreme environments--Biotechnology.
Classification: LCC QR100.9 .E953 2018 | DDC 578.75/8--dc23
LC record available at https://lccn.loc.gov/2017031972

Visit the Taylor & Francis Web site at
http://www.taylorandfrancis.com

and the CRC Press Web site at
http://www.crcpress.com

To all researchers on extremophiles,
in particular the best,
for their pioneering contributions.

Contents

Chapter 16

Ravi Durvasula and D. V. Subba Rao

Preface

The Romans used the term *extremus* to describe something outside the norm, to which MacElroy (1974) conjoined *philos* and offered the term *extremophile* to define organisms—from bacteria to fishes—that inhabit extreme environments, ranging from the frozen Antarctic to abyssal hot hydrothermal vents. Extremophiles are nature's ultimate extreme survivors that live in challenging habitats under multiple stresses of temperature, light, salinity, and pH. The quote "That which does not kill us makes us stronger" by the German philosopher Friedrich Nietzsche (1844–1900) is applicable to extremophiles. Studies on aquatic extremophiles have increased exponentially from 21 in 1997 to the current 494; likewise, the numbers of identified extremophiles from all habitats increased from 279 to 2860 in the same period. Descriptions of extremophiles, including "microbes that thrive under conditions that would kill other creatures" (Madigan and Marr 1997), "the unseen majority" (Whitman et al. 1998), "the indestructible" (Copley 1999), "ultimate extreme survivors" (Rothschild and Mancinelli 2001), "unique microbes that are found in unique environments" (Brock 2012), and "tough microbes produce tough molecules" (Anitori 2012), influenced us most to compile this collection. An earlier study on the halophiles *Chlamydomonas plethora* and *Nitzschia frustule*, isolated from Kuwait Bay, Arabian Gulf (Subba Rao et al. 2005), and more recent studies on a thermophile microalga, *Scenedesmus* species, from Soda Dam, New Mexico (Durvasula et al. 2015) have been the motivation behind this book.

We are inspired by two more recent publications in *Nature* (Hashimoto et al. 2016; Bittel 2016) on proteins from an extremotolerant Tardigrade that help human DNA withstand radiation, leading us to think, "Somewhere, something incredible is waiting to be known" (credited to Carl Sagan). Of interest is the discovery of rich soil microbial communities dominated by Actinobacteria, Chloroflexi, Proteobacteria, Acidobacteria, and two new phyla AD3 and WPS-2 by Dr. Belinda C. Ferrari's team in the frozen expanses of Antarctica. These communities aerobically scavenged atmospheric H2 and CO at rates sufficient to provide sources of energy and carbon to support their ecosystem functioning (Ji et al. 2017). This stunning finding "opens up the possibility of atmospheric gases supporting life on other planets."

This book presents an overview of the recent developments and components of the current understanding of extremophiles. Deliberately, we have steered away from descriptive taxonomic studies of these biota. Understanding extremophiles and their utility in biotechnology involves studying their habitat and their physiological and biochemical adaptations. Mechanisms for the production of biocatalysts that are functional under extreme conditions remain largely unknown. There is an unimagined profusion of microbial life with an extraordinary amount of genetic diversity, particularly the Archaea.

Several specialists were invited to present a current review of their understanding of extremophiles and to discuss the utility of extremophiles in biotechnology. Each chapter is authored by an expert or group of experts who combine their own research with literature reviews on adaptation mechanisms of extremophiles. These chapters include research on extremophiles from the Antarctic, the Bay of Bengal, Saudi Arabia, Spain, Tunisia, and deep-sea vents, and on the commercially important extremozymes, carotenoids, bioactive compounds, and secondary metabolites of medicinal value. The chapter on *Dunaliella salina* from the Bay of Bengal differs from the rest because of the nature of its new data. One chapter addresses recent advances in the genomics of microbiota, gene sequencing, and potential biotechnological applications of extremophiles.

We realize that some topics could not be covered, mainly because the potential contributors were busy with fieldwork in far-flung Antarctic research stations or otherwise busy with teaching and applying for research grants. To the authors, we are most grateful for their understanding, immense patience, high level of professional and scholarly efforts, cordial and prompt cooperation, and valuable contributions to the overall project. Each chapter was reviewed by an internationally renowned scientist whose comments enhanced the quality of presentation, and we remain most grateful to them.

This book is quite technical but hopefully quite readable. We believe this book will provide some insights for undergraduate and graduate students and researchers interested in the structure and functioning of extremophiles. The much diffused information on culturing extremophiles is collated and presented. Additional references and a list of books and conference proceedings on extremophiles are provided. A glossary of specialized terms and a subject index are given at the end.

We thank Bala T. Durvasula for constant help and patient handling of the manuscripts.

Ravi V. Durvasula
Albuquerque, New Mexico

D. V. Subba Rao
Riverview, Florida

REFERENCES

Anitori, R.P., ed. 2012. *Extremophiles, Microbiology and Biotechnology.* Norfolk, UK: Caister Academic Press.

Brock, T.D. 2012. *Thermophilic Microorganisms and Life at High Temperatures.* New York: Springer.

Copley, J. 1999. Indestructible. *New Sci* 164: 45–46.

Durvasula, R., Hurwitz, I., Fieck, A. et al. 2015. Culture, growth, pigments and lipid content of *Scenedesmus* species, an extremophile microalga from Soda Dam, New Mexico in wastewater. *Algal Res* 10: 128–133.

Hashimoto, T., Horikawa, D.D., and Kunieda, T. 2016. Extremotolerant tardigrade genome and improved radiotolerance of human cultured cells by tardigrade-unique protein. *Nat Commun* 7: 12808.

Ji, M., Greening, C., Vanwonterghem, I. et al. 2017. Atmospheric trace gases support primary production in Antarctic desert surface soil. *Nature* doi:10.1038/nature25014.

MacElroy, R.D. 1974. Some comments on the evolution of extremophiles. *Biosystems* 5: 74–75.

Madigan, M.T. and Marr, B.L. 1997. Extremophiles, Sci. American 276: 66–71.

Rothschild, L.J. and Mancinelli, R.L. 2001. Life in extreme environments. *Nature* 409: 1092–1101.

Subba Rao, D.V., Pan, Y. and Al-Yamani, F. 2005. Growth, and photosynthetic rates of *Chalmydomonas plethora* and *Nitzschia frustula* cultures isolated from Kuwait Bay, Arabian Gulf and their potential as live algal food for tropical mariculture. *Mar Ecol* 26: 63–71.

Whitman, W.B., Coleman, D.C. and Wiebe, W.J. 1998. Prokaryotes: The unseen majority. *Proc Natl Acad Sci USA* 95: 6578–6583.

Acknowledgments

The editors are most grateful to the contributors of the chapters and for meeting requirements from time to time despite their heavy teaching and research workload. The chapters were first reviewed by the editors and then submitted to external reviewers. We remain most grateful to the following conscientious reviewers, who have selflessly given of their time by providing constructive comments on the various chapters, and thus helped in producing this book.

To the authors, we are most grateful for their understanding, immense patience, high level of professional and scholarly efforts, cordial and prompt cooperation, and valuable contributions to the overall project. Each chapter was reviewed by an internationally renowned scientist whose comments enhanced the quality of presentation, and we remain most grateful to them.

For their invitation to us to prepare this volume, and for their excellent cooperation and immense patience throughout the production of this book, we thank John Sulzycki and Jennifer Blaise, CRC Publisher and his editorial assistant in Boca Raton, Florida.

With pleasure, we thank Adel Rosario for her meticulous proof corrections.

Dr. Elizaveta Bonch-Osmolovskaya
Vinogradsky Institute of Microbiology
Russian Academy of Sciences
Moscow, Russia

Dr. Daniele Daffonchio
Biological and Environmental Sciences and Engineering Division
KAUST, King Abdullah University of Science and Technology
Thuwal, Kingdom of Saudi Arabia

Dr. Subrata K. Das
Functional Genomics of Extremophiles
Institute of Life Sciences
Nalco Square, Bhubaneswar, India

Dr. Kevin G. Devine
Pharmaceutical Sciences
London Metropolitan University
London, United Kingdom

Dr. Ivy Hurwitz
Department of Internal Medicine
Center for Global Health
University of New Mexico
Albuquerque, New Mexico

Dr. Krishnamurthi Kannan
Environmental Health Division
CSIR–National Environmental Engineering Research Institute (CSIR-NEERI)
Nehru Marg, Nagpur, India

Dr. Balasaheb Kapadnis
Functional Genomics of Extremophiles
Institute of Life Sciences
Nalco Square, Bhubaneswar, India

Dr. Sunil K. Khare
Biochemistry Enzyme and Microbial Biochemistry
Indian Institute of Technology, Delhi
Hauz-Khas, New Delhi, India

Dr. Bhavdish Narain Johri
Department of Biotechnology
Barkatullah University
Bhopal, India

Dr. Izabela Michalak
Wrocław University of Technology
Department of Advanced Material Technologies
Wrocław, Poland

Dr. Noura Raddadi
Industrial and Environmental Biotechnologies Research Unit
University of Bologna
Bologna, Italy

Dr. Sarada Ravi
Plant Cell Biotechnology Department
CSIR–Central Food Technological Research Institute
Mysore, India

Dr. Marlis Reich
Molecular Ecology Group
University of Bremen
Bremen, Germany

Dr. Tulasi Satyanarayana
Department of Microbiology
University of Delhi South Campus
New Delhi, India

Dr. G. Tsiamis
Environmental Microbiology
Department of Environmental and Natural Resources Management
University of Patras
Agrinio, Greece

Dr. José Maria Vega
Plant Biochemistry and Molecular Biology
Universidad de Sevilla
Sevilla, Spain

Dr. Kenneth Wilson
Department of Biology
University of Saskatchewan
Saskatoon, Saskatchewan, Canada

PHOTO CREDITS

Drs. Amber G. Teufel and Rachael M. Morgan-Kiss: Sampling in the Antarctic.

Frontispiece: Biota

Dr. Mino Sayaka and Dr. Satoshi Nakagawa: (a) *Baslnearium* and (b) *Thermotomaculam*—hydrothermal vents.

Dr. Rosa María Martínez-Espinosa: (c) *Haloferax mediterranei* and colonies of halophilic micro-organisms isolated from salt water (Santa Pola).

Dr. Vanesa Bautista Saiz: (d) *Haloferax mediterranei* cells.

Dr. Wei Li, Montana University: Scanning electron microscopy images of (e) *Gymnodinium* sp. and (f) *Brachiomonas* sp. isolated from Lake Bonney within the McMurdo Dry Valleys, Antarctica.

Drs. Sylvia Herter and David E. Gilbert, DOE Joint Genome Institute, Walnut Creek, California, for *Chloroflexux aurantiacus.*

Dr. Ameur Cherif: (a–c) The desert terrain of Sahara in the northern part of the African continent. (d) Example of hypersaline environments in southern Tunisia. (e) Example of desert plants in southern Tunisia (details in their chapter; Dr. Amur Cherif). (f) Coastal saltern ponds, Spain; Dr. Rosa Maria Martinez-Espinosa.

Examples of hypersaline environments in southern Tunisia.
Examples of desert plants in southern Tunisia.
Examples of microorganisms isolated from different saline systems of Tunisia.
Examples of desert insects recovered from desert areas in southern Tunisia.

Dr. Suman Keerthi: *Dunaliella salina* from Bay of Bengal salterns.

About the Editors

Dr. Ravi Durvasula, a physician trained in infectious diseases at Yale University, is a professor of medicine and the director of the Global Health Center at the University of New Mexico School of Medicine. While at Yale, he developed a novel molecular approach, termed paratransgenesis, which serves as a "Trojan horse" approach to control disease transmission. Dr. Durvasula has been applying paratrangenic strategies with the goal of reducing the transmission of a spectrum of arthropod-borne diseases, including leishmaniasis, Chagas disease, and agricultural diseases, such as Pierce's disease. He has worked collaboratively in many regions of the world, including Tunisia, India, Brazil, Argentina, the United Kingdom, and Guatemala. He has published more than 60 peer-reviewed scientific publications and reports and coedited two books, *Dynamic Models of Infectious Diseases* (Springer 2013) and *Pathogenesis of Leishmaniasis* (Springer 2014). Dr. Durvasula has applied the paratransgenesis strategy to commercial aquaculture to deploy molecules that interfere with the transmission cycles of infectious pathogens in shrimp. Through applications that involve genetically engineered cyanobacteria and other unicellular green algae, Dr. Durvasula has been engaged in the study of extremophiles from various locations in the world.

Dr. D.V. Subba Rao is a biological oceanographer and formerly an emeritus scientist at the Bedford Institute of Oceanography, Dartmouth, Nova Scotia. He is currently with the Center for Global Health, Department of Medicine, University of New Mexico, Albuquerque, NM. His research and teaching activities have engaged him at the Commonwealth Scientific and Industrial Research Organisation, Australia; Johns Hopkins University, Baltimore, Maryland; the University of New Mexico, Albuquerque; and the Kuwait Institute for Scientific Research, Kuwait and India. He has published 130 papers in peer-reviewed journals on phytoplankton ecology; the physiological ecology of primary production; picoplankton; the recurrence of red tides; phycotoxin episodes; the impact of ballast water introductions, tsunamis, and mega engineering projects on coastal marine environments; and the biotechnological utility of microalgae. Besides publishing 40 technical reports, Dr. Rao has edited two volumes of *Oceanographic Science in Developing Countries* (1990), *Present and Future of Oceanographic Programs in Developing Countries* (1996), *Pelagic Ecology Methodology* (2002), two volumes of *Algal Cultures, Analogues of Blooms and Applications* (2006), and *The Alga Dunaliella: Biodiversity, Physiology, Genomics and Biotechnology* (2009).

About the Contributors

Dr. Alan Barozzi earned a master's degree in biology from the University of Milan–Bicocca, Italy, in 2014. His thesis work focused on the characterization of the intestinal bacterial community associated with the beetle *Psacothea hilaris hilaris*, a pest insect native to China, now spreading also in Europe, in order to identify a strategy to counteract the parasitic activity of this insect affecting his primary symbionts. Since January 2015, he has been a PhD student in marine science at the King Abdullah University of Science and Technology, Kingdom of Saudi Arabia. He is currently with the Department of Food, Environmental and Nutritional Sciences, University of Milan, Milan, Italy. His research is focused on the investigation of microbial communities that thrive in extreme environments, with particular interest in the Red Sea deep hypersaline anoxic basins, with the main aim of investigating the microbial diversity harbored in the brine–seawater interfaces and along the brine bodies.

Hitarth B. Bhatt is currently pursuing his doctoral research in microbiology under the supervision of Prof. Satya P. Singh at UGC-CAS Department of Biosciences, Saurashtra University, Rajkot, India. He has been awarded a meritorious fellowship from the University Grants Commission, India, under the Basic Science Research Scheme. He is working on molecular and enzymatic diversity of haloalkaliphilic bacteria from the saline desert of Little Rann of Kutch. His research area also covers cloning, overexpression, and characterization of recombinant enzymes from haloalkaliphilic bacteria. He has expertise in polyphasic taxonomy and microbial systematics of bacteria. He has published one research paper in a refereed journal and contributed one invited chapter in an edited book of a reputed publisher.

Dr. María José Bonete is a biochemist. She received a licenciado degree (equivalent to the MS degree) in chemistry from the Universidad de Valencia, Spain, in 1977 and a PhD degree in science from the Universidad de Murcia, Spain, in 1982. She spent a research period at the University of Bath, United Kingdom. She is a professor of biochemistry and molecular biology in the Departamento de Agroquímica y Bioquímica, Facultad de Ciencias, Universidad de Alicante, Spain. Her main research activities include halophilic proteins, the molecular biology of halophiles, and nitrogen metabolism and its regulation. She is the leader of the group "Biotechnology of Extremophiles." The main topic of the group is the nitrogen cycle in haloarchaea and its application for the bioremediation and production of biomolecules. She is a member of the Biochemical Society, Spanish Society for Biochemistry and Molecular Biology, Sociedad Española de Biotecnología, and small and medium enterprises (SMEs), and an editorial board member for Archaea. She also serves as a reviewer internationally for 6 international granting agencies and 25 journals.

Dr. Sara Borin graduated with a degree in food science in 2001 and obtained a PhD in pesticide chemistry, biochemistry, and ecology from the University of Milan, Italy, completing her scientific training in several European laboratories as a leader in the field of environmental microbiology. Since 2008 Sara Borin has been an associate professor of agriculture microbiology and biotechnology at the Department of Food, Environmental and Nutritional Sciences (DeFENS), University of Milan. She is studying the molecular ecology of bacteria applied to the study of environmental bacterial communities, how they contribute to ecosystem functioning, and how they can be exploited in sustainable environmental biotechnologies, emphasizing: (1) bioremediation and energy, (2) plant microbiota and plant growth promotion, (3) ecosystem functioning in extreme environments, and (4) insect microbiota.

Himadri Bose, after completing his MSc in microbiology from Vellore Institute of Technology, Tamil Nadu, India, joined the Department of Microbiology, University of Delhi for his PhD to work on the utility of carbonic anhydrase of the bacterium *Aeribacillus pallidus* in carbon sequestration in 2013 under the supervision of Prof. T. Satyanarayana. During this period, he isolated alkaliphilic and moderately thermophilic bacteria from Indian hot water springs and sediment samples, screened for carbonic anhydrases, and selected *A. pallidus*, isolated from hot water springs of Pepariya of Madhya Pradesh state of India, for detailed investigation. Carbonic anhydrase was produced by this bacterium, which was purified and characterized. He has also attempted to clone and express the gene encoding γ-carbonic anhydrase in *Escherichia coli*. There was good expression, but the activity could not be detected. He is currently making an effort to clone and express carbonic anhydrase in active form. He has published three papers in *Bioprocess and Biosystems Engineering, Energy & Fuels*, and *Environmental Science & Pollution Research*; a review in *Frontiers in Microbiology*; and an article in Botanica. He received his PhD in October 2017 for his work on carbon sequestration. He has finished his PhD in 2017 and is now pursuing his postdoctoral research at Department of Biotechnology IIT Kharagpur, India.

Dr. Abdellatif Boudabous earned a third cycle doctorate (PhD in 1978) and a state doctorate (1983) on the taxonomy, phylogeny, and ecology of marine and terrestrial *Bacillus* species, both from Provence University, Luminy, Marseille, France. He has developed a team working on microbial ecology, taxonomy, phylogeny, and bacterial biochemistry. Several groups and bacteriological genera and species of *Bacillus*, *Frankia*, *Geodermatophilus*, *Lactobacillus*, and *Lactococcus* are studied with members of the Laboratory of Microorganisms and Active Biomolecules (LMBA), Faculty of Sciences, University of Tunis El Manar, Tunisia. For 17 years, the team has also developed studies on the antibiotic resistance of several pathogens, *Escherichia coli*, *Staphylococcus aureus*, *Enterococcus*, *Salmonella*, and *Pseudomonas*. In the last 20 years, the team has been interested in microbes in extreme environments, with European partners and universities. A large and long cooperation has been developed with several universities, Milan (Italy), Lyon 1 (France), La Rioja (Spain), New Hampshire (United States), Ioannina and Patras (Greece), and the University of Stif (Algeria). These studies were published in about 170 articles with Professor Boudabous, and up to 400 papers were realized by others members of the LMBA, Faculty of Sciences, University of Tunis El Manar. In 1995, Professor Boudabous also created and coordinated, with different institutions and members of the laboratory, a third cycle formation (master's and PhD) in microbiology and molecular epidemiology at Tunis El Manar University (1995–2017). Besides participating in several scientific committees, in 2013 Professor Boudabous was appointed president of the Tunisian National Commission for the Evaluation of Scientific Activities.

Dr. Mónica L. Camacho Carrasco is senior lecturer of biochemistry and molecular biology, Faculty of Sciences, University of Tunis, Tunisia. She is also currently affiliated with the Departamento de Agroquímica y Bioquímica, Facultad de Ciencias, Universidad de Alicante, Alicante, Spain. She has participated in many research projects related to halophilic archaea since 1984. Her two present projects, granted by the Ministry of Economy and Competitiveness, are BIO2013-42921-P (Excellence) and CTM2013-43147-R (Challenges). She has published some papers in specialized international journals since 1986, and she has participated in national and international congresses since 1985, within subjects associated with her research activity. She had a postdoctoral stay at the University of Bath, United Kingdom, for a full year (1992), working with proteins of archaea. She has directed several doctoral theses and research works. She has participated in the cooperation with various national and international research groups.

Dr. Vikas Singh Chauhan is working as a principal scientist at Plant Cell Biotechnology, Council of Scientific and Industrial Research–Central Food Technological Research Institute, Mysore,

India. He possesses a doctoral degree in bioscience, and microalgal biotechnology is his primary area of research interest. He was involved in setting up one of India's largest *Spirulina* production plants at Nanjangud. He has also been involved in the commercialization of the knowledge base of the institute. He is actively involved in guiding postgraduate and doctoral students. He has published 24 research papers.

Dr. Ameur Cherif graduated with a degree in natural sciences from the Faculty of Sciences of Tunis, University of Tunis El Manar, Tunisia, in June 1995. Following his master's degree in genetics and molecular biology in 1997, he obtained a PhD degree in microbiology in 2001. He was recruited as assistant in 2001 and then maître assistant. He obtained his Habilitation Universitaire, HDR, in 2007, and was appointed associate professor in microbiology at the Higher Institute for Biotechnology of Sidi Thabet (ISBST), University of Manouba, Tunisia, where he was elected twice as director of the institute (2010–2014). He is the head of the research laboratory of Biotechnology and Bio-Geo Resources Valorization at the ISBST. In December 2012, he was promoted as full professor in the ISBST. His current research activities focus on several aspects of molecular microbial ecology and biotechnology; these include (1) the phylogeny and ecology of extremophilic microbes, particularly from deserts, marine sediments, and saline systems (Chott and Sebkha); (2) the microbial ecology of symbionts and commensal and pathogen microbiota in different hosts, and development of the symbiotic biological control approach (arthropods); and (3) biotechnological applications of active biomolecules, such as bacteriocins, antibiotics, esterase, several enzymes, and biosurfactants, and microbial resource management for plant growth promotion and bioremediation. Professor Cherif represented Tunisia in European Cost Action FA 0701; was a partner member in several EU FP7, Tempus, Erasmus, H2020, and bilateral projects; and serves on various review and editorial committees. In his research career, Professor Cherif has published more than 70 peer-reviewed articles and several book chapters.

Dr. Hanene Cherif graduated with a degree in natural sciences from the Faculty of Sciences of Tunis (FST), University of Tunis El Manar, Tunisia, in June 1998. Following her master's degree in microbiology in 2001, she obtained her PhD degree in microbiology in 2009. She joined the Laboratory of Microorganisms and Active Biomolecules at the FST as a postdoctoral researcher for three years in the ambit of EU-FP7 project BIODESERT (grant no. 245746) (2010–2012). She continued her work in the Laboratory of Biotechnology and Bio-Geo Resources Valorization at the Higher Institute for Biotechnology of Sidi Thabet, University of Manouba, Tunisia, as a postdoc before she won a second postdoctoral position in the H2020 MADFORWATER project in April 2017. Her current work focuses on two aspects: (1) microbial diversity in extreme environments and (2) microbial resource management for plant growth promotion.

Dr. Habib Chouchane graduated with a degree in natural sciences from the Faculty of Sciences of Tunis, University of Tunis El Manar, Tunisia, in June 1994. From 1994 to 2007, he was recruited as a secondary school teacher of natural sciences. He obtained his master's degree in 2005 and his PhD degree in 2010 in biology from the University of Toulouse. He was recruited as an assistant in 2007 at the University of Gafsa, and then assistant professor at the Higher Institute for Biotechnology of Sidi Thabet, University of Manouba, Tunisia, in 2013, where he joined the research laboratory Biotechnology and Bio-Geo Resources Valorization. His main research activities focus on microbial biopolymers with valuable applications in bioremediation and health. His current collaborations include the WE-MET project (Sustainable Wastewater Treatment Coupled to Energy Recovery with Microbial Electrochemical Technologies) and the MADFORWATER project (Development and Application of Integrated Technological and Management Solutions for Wastewater Treatment and Efficient Reuse in Agriculture Tailored to the Needs of Mediterranean African Countries).

Dr. Elena Crotti has been an assistant professor at the Department of Food, Environmental and Nutritional Sciences, University of Milan, Italy, since December 2013. Crotti earned her PhD in chemistry, biochemistry, and ecology of pesticides from the University of Milan in 2009. After obtaining her PhD, she worked as a postdoctoral fellow at the University of Milan. Her research interests are in the fields of microbial ecology and biotechnology, applied to marine ecosystems and microbe–insect associations.

Dr. Daniele Daffonchio has been a professor of bioscience at the King Abdullah University of Science and Technology (KAUST), Biological and Environmental Sciences and Engineering Division, Kingdom of Saudi Arabia, since April 2014. Before KAUST, Daffonchio was a professor of microbial systems biotechnology at the Department of Food, Environmental and Nutritional Sciences (November 2010–March 2014) and associate professor at DiSTAM (March 2002–October 2010), University of Milan, Italy. From November 1998 to October 1999, he was an adjunct professor of industrial microbiology at the University of Urbino, Fano, Italy, and from November 1995 to February 2002, he was an assistant professor at DiSTAM, University of Milan. Daffonchio initiated his research career in 1994 as a postdoctoral scientist at the LabMet, University of Gent, Belgium, after getting a PhD in Chemistry, biochemistry, and ecology of pesticides at the University of Milan in 1993. His research interests are in the microbial ecology and biotechnology of complex ecosystems in conventional and extreme aquatic and terrestrial habitats. The actual research deals with the study and exploitation of extremophile microorganisms and microbial communities along the water stress continuity from the Arabian Desert to the depth of the brine pools in the Red Sea.

Kruti G. Dangar is pursuing her PhD in biotechnology in the UGC-CAS Department of Biosciences, Saurashtra University, Rajkot, Gujarat, India, under the supervision of Prof. Satya P. Singh. She has been awarded a research project under the Women Scientist Scheme A by the Department of Sciences and Technology, New Delhi, India. She is working on the molecular diversity, metagenomics, and biotechnological aspects of the haloalkaliphilic actinomycetes of the coastal regions of the Gujarat.

Dr. Ravi Durvasula, a physician trained in infectious diseases at Yale University, is a professor of medicine and the director of the Global Health Center, Department of Medicine at the University of New Mexico School of Medicine. While at Yale, he developed a novel molecular approach, termed paratransgenesis, which serves as a "Trojan horse" approach to control disease transmission. Dr. Durvasula has been applying paratrangenic strategies with the goal of reducing the transmission of a spectrum of arthropod-borne diseases, including leishmaniasis, Chagas, disease, and agricultural diseases, such as Pierce's disease. He has worked collaboratively in many regions of the world, including Tunisia, India, Brazil, Argentina, the United Kingdom, and Guatemala. He has published more than 60 peer-reviewed scientific publications and reports and coedited two books, *Dynamic Models of Infectious Diseases* (Springer 2013) and *Pathogenesis of Leishmaniasis* (Springer 2014). Dr. Durvasula has applied the paratransgenesis strategy to commercial aquaculture to deploy molecules that interfere with the transmission cycles of infectious pathogens in shrimp. Through applications that involve genetically engineered cyanobacteria and other unicellular green algae, Dr. Durvasula has been engaged in the study of extremophiles from various locations in the world.

Darine El Hidri is a PhD student in microbiology in the Laboratory of Biotechnology and Bio-Geo Resources Valorization at the Higher Institute of Biotechnology of Sidi Thabet, Tunisia. She obtained a four-year license in life and earth sciences (2005) and a master's degree of hydrobiology (2009) from the Faculty of Sciences of Bizerte, University of Carthage, Tunisia. Her current research activities focus on (1) the diversity and phylogeny of extremophilic bacteria from arid (oasis) and saline systems (Chott and Sebkha), and (2) the adaptation and ecology of the haloalkaliphilic bacterium suspected through genome analyses.

Khaled Elmnasri graduated with a degree in life and earth sciences from the Faculty of Sciences of Bizerte, University of Carthage, Tunisia, in August 2005. In 2011, he obtained a master's degree in hydrobiology devoted to the microbial community of scorpions. His current research activities focus on arthropod–bacteria symbiosis (commensal and pathogen) and biological pest control, at the Biotechnology and Bio-Geo Resources Valorization Laboratory in the Higher Institute of Biotechnology of Sidi Thabet, Tunisia.

Dr. Niels Thomas Eriksen has an MSc in biology from Odense University, Denmark (1995), and a PhD in biochemistry from the University of Southern Denmark (1998). Since 1999, he has been employed at Aalborg University, Denmark, and is presently an associate professor in bioprocess technology at the Department of Chemistry and Bioscience and head of the study board in chemistry, biotechnology, and environmental engineering. His research interests include bioprocess technology and applied microbiology in general, and he has worked with the cultivation and utilization of bacteria, cyanobacteria, yeast, filamentous fungi, eukaryote microalgae, and invertebrates. He has always had a particular interest in the cultivation of phototrophic and heterotrophic microalgae.

Dr. Julia María Esclapez Espliego works in the area of biochemistry at the Departamento de Agroquímica y Bioquímica, Facultad de Ciencias, University of Alicante, Spain. She obtained her degree in biology in 1999, achieving two academic achievement awards. In 2004, she obtained a PhD in biology with a European Doctor Mention and Extraordinary Award Doctorate. She has achieved predoctoral and postdoctoral short-term fellowships. Her main areas of research are based on the homologous and heterologous expression of halophilic proteins, the purification and characterization of novel halophilic proteins, the crystallization and resolution of protein structures, and transcriptional studies of genes involved in the metabolism of nitrogen and carbon. She has published several papers in journals with a high impact factor, such as the *Proceedings of the National Academy of Sciences of the United States of America*, and has much experience in teaching biochemistry.

Dr. Besma Ettoumi graduated with a degree in natural sciences from the Faculty of Sciences of Tunis, University of Tunis El Manar, Tunisia, in 2001. She obtained her master's degree (in April 2004) and her PhD (in December 2009) working on molecular and enzymatic typing of species of the *Bacillus cereus* group and phylogenetic diversity and activities of marine bacteria, respectively. She obtained a six-month postdoctoral position in the European project BIODESERT, a two-year postdoctoral position in the MetalBiorec Italian Project, and a one-year position in the Research and Technological Innovation Department of ENI—a company involved in exploration and production, gas, and power in Tunisia. She is currently with the Higher Institute of Biotechnology, Biotechnology and Bio-Geo Resources Valorization, University of Manouba, Biotechnopole Sidi Thabet, Ariana, Tunisia. Dr. Ettoumi has collaborated on several research topics, such as the phylogeny of marine bacteria and their biotechnological potential as biosurfactant producers, and the diversity of bacteria involved in growth promotion under stress conditions.

Dr. Raoudha Ferjani is a postdoctoral researcher within the frame of the MADFORWATER project (Development and Application of Integrated Technological and Management Solutions for Wastewater Treatment and Efficient Reuse in Agriculture Tailored to the Needs of Mediterranean African Countries). She obtained a four-year license in life sciences in June 2009, a master's degree of microbiology in 2011, and a PhD degree in microbiology in 2016, from the Faculty of Sciences of Tunis, University of Tunis El Manar, Tunisia. She is currently with the Laboratory of Microorganisms and Active Biomolecules, Faculty of Sciences of Tunis, University of Tunis

El Manar. In her master's and PhD degrees, she worked on the framework of the BIODESERT project (Biotechnology from Desert Microbial Extremophiles for Supporting Agriculture Research Potential in Tunisia and Southern Europe). The current research activities of Dr. Ferjani focus on microbial resources management for plant growth promotion and phytoremediation.

Dr. Imen Fhoula is a Tunisian microbiologist. She graduated with a degree in natural sciences from the Faculty of Sciences of Tunis, University of Tunis El Manar, Tunisia, in June 2005. She is currently with the Laboratory of Microorganisms and Active Biomolecules, Faculty of Sciences of Tunis, University of Tunis El Manar. She received her master's degree in microbiology in 2008 and obtained her PhD degree in microbiology in 2014. Her scientific interests focus on several aspects of molecular microbial ecology; systematic, technological, and probiotic properties; and their potential applications for organisms (human, animals, and plants). This includes the phylogeny and ecology of lactic acid bacteria, particularly from rhizospheres, plants, feces, wheat, and insects of hard environments (saline systems); the microbial ecology of arthropod symbionts; prevalence and acquired antibiotic resistance; the biotechnological applications of active biomolecules, such as bacteriocins, esterase, autolysine, and several enzymes; quality sourdough bread; and antiproliferative activity against several types of cancer cells. Dr. Fhoula has published several peer-reviewed papers.

Dr. Andrew J. Gates joined the School of Biological Sciences, Faculty of Sciences, University of East Anglia, Norwich, UK, as a lecturer in 2011. He is a member of the molecular microbiology theme and the Centre for Molecular and Structural Biochemistry. Gates holds a BSc (Hons) degree in molecular biology from the Faculty of Medicine at the University of Newcastle upon Tyne, United Kingdom, and a PhD in chemistry from the University of East Anglia. Prior to his appointment within the School of Biological Sciences, he worked with Prof. David Richardson as a senior postdoctoral research associate, which initiated his interest in the genetics and biochemistry of assimilatory nitrate reduction, and nitrogen and iron respiratory systems in bacteria.

Dr. Sangeeta D. Gohel is an assistant professor in the UGC-CAS Department of Biosciences, Saurashtra University, Rajkot, Gujarat, India. She completed her MSc and PhD degrees in microbiology from the same department. She received the UGC Research Fellowship in Sciences for Meritorious Students during her PhD. She is a recipient of a CSIR-SRF (direct) and UGC Research Associateship. Dr. Gohel has completed an UGC-sponsored start-up research project on molecular cloning, expression, and characterization of extracellular serine proteases. Currently, she is working on the DST-SERB project Rhizospheric Microbes and Soil Metagenome, and another major research project sponsored by the Internal Quality Assurance Cell, Saurashtra University, Rajkot, India. She has published her work in five reputed international journals as a first author and contributed seven book chapters. She has presented her work in 30 international and national conferences, and many of her presentations were adjudged the best. Her h-index and citations are 6 and 91, respectively, with a cumulative impact factor of 12. Dr. Gohel has been working on the molecular phylogeny and diversity of haloalkaliphilic actinomycetes; extremozyme purification and characterization; the cloning, expression, and structure function analysis of enzymes; antimicrobial activity profiling; and the screening of rhizospheric microbes and metagenomics from various saline habitats of the Gujarat Coast, India.

Amel Guesmi is a PhD student in microbiology in the Laboratory of Biotechnology and Bio-Geo Resources Valorization at the Higher Institute of Biotechnology of Sidi Thabet, University of Manouba, Tunisia. She graduated with a degree in life and earth sciences from the Faculty of Sciences of Tunis, University of Tunis El Manar, Tunisia, in 2003. She is currently with the Laboratory of Microorganisms and Active Biomolecules, Faculty of Sciences of Tunis, University of Tunis El Manar. She obtained her master's degree in microbiology in 2006 from the University of

Tunis El Manar. Her current research activities focus on (1) the study of the ecology and diversity of aerobic spore-forming bacteria and microbial community composition in deserts and saline systems (Chott and Sabkha), and (2) the exploration of these organisms for biotechnological applications in industrial and environmental domains.

Dr. Chadlia Hamdi is a postdoctoral researcher in the Laboratory of Microorganisms and Active Biomolecules, Faculty of Sciences of Tunis, University of Tunis El Manar, Tunisia. She is also currently affiliated with the Higher Institute of Biotechnology, Biotechnology and Bio-Geo Resources Valorization, University of Manouba, Biotechnopole, Sidi Thabet, Ariana, Tunisia. She graduated with a degree in life and earth sciences in 2003 from the Faculty of Sciences of Tunis. She obtained her master's and PhD degrees in microbiology in 2006 and 2014, respectively, from the University of Tunis El Manar. Her scientific interests are in pathogens of the honeybee, *Apis mellifera*, and in the development of a symbiotic biological approach for controlling American foulbrood disease and other honeybee pests. Her current research activity focuses on bacterial enzymes and their biotechnological applications in industrial and environmental domains.

Dr. Jane A. Irwin obtained her BA (Mod) in biochemistry in 1987, followed by a PhD in biochemistry from Trinity College, Dublin, Ireland in 1994. She subsequently carried out postdoctoral research in University College Dublin on enzymes from psychrophilic bacteria from marine environments. She was appointed College Lecturer in the School of Veterinary Medicine, University College Dublin in 1999 and now works in the general area of glycobiology, with emphasis on glycan-degrading enzymes, in particular fucosidase, at different mucosal surfaces, and also on microbial adhesion. Her recent work concerns gene expression in the female reproductive tract of cattle and horses. She has published 34 papers, and 4 book chapters in a variety of areas, including mucosal biology, antimicrobials, and veterinary education, and supervised five PhD students.

Dr. Atef Jaouani graduated with a degree in biological engineering in 1997 from the National School for Engineers of Sfax (ENIS), Tunisia. After a first DEA (master's degree) in biological engineering fron ENIS (1999), he obtained a second DEA of sciences from the Free University of Brussels (2001). In 2004, he obtained his PhD degree from the same university, with the main subject dealing with the depollution of polyphenol-rich effluents by physical-chemical and biological means, a case study of olive mill wastewaters. Currently, Dr. Jaouani is an associate professor at the Higher Institute for Applied Biological Sciences, Tunis, University of Tunis El Manar, Tunisia, and a senior researcher at the Laboratory of Microorganisms and Active Biomolecules. He is also currently with the Higher Institute of Biotechnology, Biotechnology and Bio-Geo Resources Valorization, University of Manouba, Biotechnopole Sidi Thabet, Ariana, Tunisia. He participates in several international projects financed by the European Union, North Atlantic Treaty Organization, and Gates Foundation (United States), where he coordinates several work package in research projects (WPs). He is a senior expert at the German Agency for International Cooperation (GIZ) in the field of bioresource management and sustainable development. He is a scientific advisor for several Tunisian small and medium enterprises working in the field of water and wastewater treatment. His major research interests are environmental biotechnology, bioprocesses, food and environmental microbiology, cell and enzyme immobilization, mycoremediation, and wastewater treatment. Dr. Jaouani has published more than 30 peer-reviewed articles and book chapters and serves on various review and editorial committees.

Dr. Suman Keerthi obtained an MSc in biotechnology from Sri Venkateswara University, Tirupati, India, in 2006 and a PhD from Andhra University, India, in 2012. For his doctoral thesis, he isolated and established algal cultures and investigated the biotechnologically important carotenogenic chlorophyte *Dunaliella* sp. (~60 strains), diatoms *Cylindrotheca closterium* (pennate) and *Odontella aurita* (centric), and a cyanophte *Synecochococcus* sp. Strains were confirmed utilizing their morphology and their molecular typing of ribosomal DNA. Of the *Dunaliella*, 16 strains were carotenogenic. Dr. Keerthi raised *Dunaliella* in autotrophic and mixotrophic modes and investigated variations in cellular carotene and lipids in relation to nitrogen resources. He has published nine papers in international journals. Dr. Keerthi has presented his findings at several workshops, conferences, and symposia. Dr. Keerthi also studied carbon dioxide sequestration through culturing medically useful microalgae in photobioreactors linked to gas outlets of industries at Visakhapatnam. As a project scientist with the Centre for Marine Living Resources and Ecology, Ministry of Earth Sciences, Cochin, India, he has participated in several cruises and applied CHEMTAX pigment analyses on samples from the Arabian Sea and the Indian Ocean and related their variations to structural and functional microalgal groups. Presently, Dr. Keerthi is in charge of the research and development of large-scale microalgal cultures in mariculture operations at BMR Groups, BMR Industries Pvt LTD, Chennai.

Dr. Adinarayana Kunamneni is a staff scientist in the Department of Internal Medicine, University of New Mexico Health Sciences, Albuquerque. He earned a PhD in pharmaceutical sciences from Andhra University, Andhra Pradesh, India, and spent his postdoctoral years at the Durban University of Technology, South Africa; the Spanish National Research Council; the University of Kentucky, Lexington; and the University of New Mexico and New Mexico VA Health Care System. Dr. Kunamneni is a recipient of a CSIR-SRF fellowship (direct), an NRF/DOL Scarce Skills fellowship, a postdoctoral fellowship from the Spanish Ministry of Science and Education, and a European Community Marie Curie Fellowship. Dr. Kunamneni authored 50 peer-reviewed papers (3050 citations), with 37 peer-reviewed primary research articles, 8 reviews, and 5 book chapters. His research interests include recombinant and nonrecombinant fermentation technology, bioactive marine natural products, biocatalysis, protein engineering, and drug discovery and development.

Dr. Antje Labes is a trained microbiologist active in the field of the biotechnology of marine natural products. Her major research fields are the biotechnology of marine microbes and marine natural products. In recent years, Antje has worked as a senior scientist at GEOMAR-Biotech, a center dedicated to natural product biotechnology using marine resources. The research focuses on the identification of new natural products from various natural sources, including marine macro- and microorganisms, and the evaluation of their potential for applications in medicine, cosmetics, agriculture, and food products. GEOMAR-Biotech is a part of the Helmholtz Centre for Ocean Research Kiel (GEOMAR). GEOMAR is one of the leading institutes in marine sciences in Europe, performing biological, geological, physical, and theoretical oceanography research. She now holds a professorship for microbiology and molecular biology at the Flensburg University of Applied Science, Kiel, Germany, in the Department of Bioprocess Engineering. During her academic career, she trained and mentored more than 50 graduate students and 8 PhD students. Labes started and was involved in many research projects associated with the field of blue biotechnology that laid the basis for a strong network of cooperation partners throughout Europe and internationally, including interactions with small and medium enterprises and industry, leading to the foundation of a regional network on marine biotechnology in northern Germany. Antje has authored more than 30 publications in peer-reviewed journals and 2 patents.

Anna Vegara Luque is a PhD student at the University of Alicante, Spain. She earned a degree in biology in 2010. She obtained a master's degree in biotechnology in 2011 from the University of Alicante. Her first publication is Esclapez et al., *Journal of Biotechnology* 193, 100–107, 2015. She has had three oral presentations at conferences and meetings, and she has participated in congresses with posters since 2010. She collaborates in the project BIO2013-42921-P, granted by the Ministry of Economy and Competitiveness. Her areas of research are protein interactions, heterologous expression, and the purification and characterization of halophilic proteins of nitrogen metabolism from *Haloferax mediterranei*. She has been an honorary collaborator since 2012.

Mouna Mahjoubi is a PhD student in microbiology in the Laboratory of Biotechnology and Bio-Geo Resources Valorization at the Higher Institute of Biotechnology of Sidi Thabet, University of Manouba, Ariana, Tunisia. She graduated with a degree in life sciences from the Faculty of Sciences of Bizerte, University of Carthage, Tunisia, in June 2009. She obtained a master's degree in microbiology in 2011 from the Faculty of Sciences of Tunisia, University of Tunis El Manar, Tunisia. Her current research activities focus on the characterization of the microbial community dynamics and bioremediation potential of hydrocarbonoclastic bacteria in marine polluted ecosystems in the ambit of the European project ULIXES (Unravelling and Exploiting Mediterranean Microbial Diversity and Ecology for Xenobiotics and Pollutants Clean Up).

Dr. Francesca Mapelli discussed her PhD thesis in agricultural ecology at the University of Milan, Italy, in 2012. Since the same year, she has been a postdoc fellow at the University of Milan. The scientific and research activity covers several areas under the field of microbial ecology of extreme ecosystems, including the study of prokaryotic diversity and system functioning in the Mediterranean Sea, with a focus on eastern Mediterranean deep hypersaline anoxic basins. Dr. Mapelli has extensive experience in sampling activities of extreme ecosystems, and she has participated in several oceanographic cruises onboard research vessels, collecting sediment and water samples from mud volcanoes, deep hypersaline anoxic basins, and the water column in the Mediterranean Sea.

Dr. Rosa María Martínez-Espinosa works in the areas of biochemistry and molecular biology at the Departamento de Agroquímica y Bioquímica, Facultad de Ciencias, University of Alicante, Spain. She obtained her degree in biology in 1998, and a PhD in biology in 2003. She has achieved postdoctoral short-term fellowships to carry out research on the N cycle in the United Kingdom between 2004 and 2008. Her interest lies in carotenoid production by haloarchaea and in the characterization of the molecular biology and biochemistry of the N cycle in haloarchaea. Her studies have related to electron transfer between proteins, and the purification and characterization of the enzymes involved using spectroscopic, molecular biology, and biochemistry techniques. She has published more than 30 book chapters and several papers in journals with a high impact factor. She is a member of the Spanish Society for Biochemistry and Molecular Biology, Sociedad Española de Biotecnología, and SEM; an editorial board member for *Bioscience and Bioengineering*, and serves as an international reviewer for more than two international granting agencies and five journals.

Dr. Ahmed Slaheddine Masmoudi graduated with a degree in molecular and cellular biology from the Louis Pasteur University of Strasbourg, Paris, in June 1985. He obtained his PhD degree in molecular and cellular biology in 1988. He was recruited as an assistant professor in 1992. Then he obtained his state doctorate in 1993. He was promoted to professor in biochemistry and molecular biology at the Faculty of Sciences of Monastir in Tunisia in 1997, and then as full professor at the Institute of Biotechnology in Monastir in 2003. In September 2012, he was elected the director of the Higher Institute for Biotechnology of Sidi Thabet, Biotechnology and Bio-Geo Resources Valorization, University of Manouba, Ariana, Tunisia. His main research activities are in several disciplines, including biochemistry and cellular and molecular biology, with a specific interest in (1) the biotechnological valorization of microbial resources, (2) the study of protein posttranslational modification, and (3) aspects of the polymorphism of lipid-associated proteins involved in diabetes.

Dr. Giuseppe Merlino has been a postdoctoral scientist at King Abdullah University of Science and Technology, Biological and Environmental Sciences and Engineering Division, Kingdom of Saudi Arabia, since July 2014. Dr. Merlino started his research activity as a postdoc at the Department of Food, Environmental and Nutritional Sciences (January 2012–June 2014). During his working period in Milan, the research activity was focused on the microbial ecology of anaerobic metabolic processes and on the biotechnological application of technologies based on such processes for the production of renewable energy and for the bioremediation of organochloride-impacted environments. The actual research interest is still focused on the microbial ecology and biotechnology of microbial processes, including, in particular, the hydrocarbon degradation potential and the biodegradation of the chlorinated compounds by the microbial communities harbored in the Red Sea. The actual work includes the elucidation of the microbial diversity along the entire water column, as well as in sub-sea-floor sediments and brine pools; the dynamics and processes shaping these communities; and the isolation and characterization of novel bacteria, with the final aim of exploiting them for biotechnological applications.

Dr. Grégoire Michoud has been a postdoctoral scientist at King Abdullah University of Science and Technology, Biological and Environmental Sciences and Engineering Division, Kingdom of Saudi Arabia, since April 2015. He recently obtained his PhD from the Université de Bretagne Occidentale, Brest, France, after a master's degree in microbiology from the Universite Pierre et Marie Curie, Paris. He is interested in the characterization of the coping mechanisms of prokaryotes in extreme environments by functional genomics and culture approaches. He is also interested in high hydrostatic pressure experiments concerning either cultural, biochemical, physical, or genomic standpoints.

Dr. Sayaka Mino received a PhD in marine life science from Hokkaido University, Japan, in 2015. Her research interest is in the area of microbial population genetics, dealing with genetic diversity, the mechanisms responsible for genetic and metabolic differentiation, and the distribution pattern of microbes from geographically separated deep-sea hydrothermal systems. She has primarily worked on the cultivation of chemolithoautotrophs (e.g., Epsilonproteobacteria and Aquificales) thriving at deep-sea hydrothermal vents around the globe, and has tried to understand genetic differentiations between them. She is currently an assistant professor in the Laboratory of Microbiology, Division of Marine Life Science, Faculty of Fisheries Sciences, Hokkaido University, Minato-cho, Hakodate, Japan.

Dr. Francesco Molinari studied chemistry as an undergraduate student at the University of Milan, Italy, where he was also a PhD student under the supervision of Prof. Cesare Gennari, working on supramolecular and stereoselective chemistry. He was a postdoc student in Lisbon with Prof. Joaquim Cabral working on membrane bioreactors, and from 1992 a lecturer at the University of Milan. From 2000 to 2015, he was associate professor at the Department of Food, Environmental and Nutritional Sciences (DeFENS) of the University of Milan. From July 2015, he has been a full professor at DeFENS. He is the coauthor of more than 100 papers in international journals (peer reviewed), 1 world patent, and 4 European patents. His main research interests are focused on (1) microbial and enzymatic biotransformations in the food, pharma, and environmental fields; (2) stereoselective biocatalysis, and (3) biotechnology of fermentation.

Dr. Rachael Marie Morgan-Kiss is an associate professor of microbial ecophysiology in the Department of Microbiology, Miami University. She grew up in a small town located on Vancouver Island, British Columbia, Canada. She received a BSc from the University of Victoria, British Columbia, in 1995 and a PhD from the University of Western Ontario, Ontario, in 2000. Her PhD dissertation focused on the adaptation of the photosynthetic apparatus in a psychrophilic green alga isolated from an ice-covered lake in the McMurdo Dry Valleys, Antarctica. Her postdoctoral work focused on bacterial fatty acid synthase and oxidation, as well as anoxygenic photosynthesis. Her current research program focuses on polar microbiology and specifically on the diversity and function of microbial eukaryotes residing in ice-covered Antarctic lakes. Research projects in her laboratory combine field studies in Antarctica with physiological studies on a large collection of polar photosynthetic and eukaryotic microorganisms. Dr. Morgan-Kiss has presented more than

50 talks and posters at scientific meetings and has published more than 30 scientific articles and book chapters that reflect her broad interests in lipid biochemistry, photobiology, and environmental adaptation to extreme environments.

Dr. Amor Mosbah graduated with a degree in natural sciences from the Faculty of Sciences of Tunis, University of Tunis El Manar, Tunisia, in June 1997. From 1997 to June 2002, he was a PhD scholar at the AFMB UMR 7257, CNRS, Aix Marseille. From June 2002 to 2003, he was recruited as a researcher at the AFMB UMR 7257, CNRS, Aix Marseille. In 2003, he was recruited by Cellpep S.A. (a French pharmaceutical company), and in 2007 by Ambrillia Biopharma Inc. (a Canadian pharmaceutical company). In 2011, he worked at the Institut de Chimie de Rennes, France. In 2015, he was recruited as assistant professor at the Higher Institute for Biotechnology of Sidi Thabet of the University of Manouba, Tunisia, where he joined the research laboratory Biotechnology and Bio-Geo Resources Valorization. His main research activities focus on biochemistry, bioinformatics, and structural studies of natural and synthetic biomolecules; this includes peptide and protein synthesis, purification and structural characterization, organic compound purification and characterization, and biomass valorization, with valuable applications in agro-alimentary and health. His current collaborations include many local and international industry and academic collaborations (Latoxan, University of Aix Marseille 2, and University of Rennes 1, France).

Dr. Sandeep Narayan Mudliar is working as a principal scientist at Plant Cell Biotechnology, Council of Scientific and Industrial Research–Central Food Technological Research Institute, Mysore, India. He obtained his M.Tech. and PhD in chemical engineering. His areas of interest include algal bioengineering for integrated CO_2 capture and conversion to value-added products, microbial and plant cell culture fermentation, agro-residues to value-added products, resource recovery from food wastes, and modeling and scale-up of bioprocesses.

Dr. Afef Najjari has been an assistant professor in bioinformatics in the Laboratory of Microorganisms and Active Biomolecules, Faculty of Sciences of Tunisia, University of Tunis El Manar, Tunisia, since 2015. She obtained a four-year license in life sciences (June 2002), a master's degree of microbiology (March 2004), and a PhD in microbiology (February 2010) from the same faculty. In her master's and PhD degrees, she worked on the genetic diversity and enzymatic activities of lactic acid bacteria. Najjari has won two postdoctoral positions: The first (January 2010–December 2012) was financed by the European Commission in the ambit of the EU-FP7 project BIODESERT (grant no. 245746). The second one was in the frame of a Fulbright scholarship program (January 2014–June 2014) at the Microbial Ecology and Environmental Genomics Laboratory of Oklahoma State University, Stillwater. She has also realized several second periods in the laboratory of project partners at the University of Ioannina, Greece. As a postdoc, Najjari has worked on microbial diversity in extreme environments and mainly on archaeal groups. Currently, besides her other topics of interest, she is working on functional genomics, high-throughput data analysis (pyrosequencing and metagenomics), gene expression, database development, and transcriptome analysis (microarrays and quantitative polymerase chain reaction).

Dr. Satoshi Nakagawa is an associate professor of the Laboratory of Marine Environmental Microbiology, Division of Applied Biosciences, Graduate School of Agriculture, Kyoto University, Japan. He is also a visiting scientist of the Japan Agency for Marine-Earth Science and Technology (JAMSTEC). He received a PhD from Kyoto University in 2005. He worked as a research scientist at JAMSTEC, and became an associate professor of the Faculty of Fisheries Sciences at Hokkaido University, Japan, in 2009, and in 2015 became an associate professor of the Graduate School of Agriculture, Kyoto University. Dr. Nakagawa's research focus is the ecology, physiology, and evolution of microorganisms that live in extreme marine environments, including deep-sea hydrothermal fields. His current projects include (1) microbial genomics, transcriptomics, and gly-comics to better understand how marine microbes respond to their surrounding environments and how they interact with other organisms, and (2) ecophysiological and population genetic analyses of marine extremophiles. He is a member of the International Society for Microbial Ecology, the American Geophysical Union, the Japanese Society of Microbial Ecology, and the Japan Society for Bioscience, Biotechnology, and Agrochemistry.

Dr. Anila Narayan obtained her PhD degree in biotechnology from Mysore University, India. She is currently working as assistant professor in the Department of Botany at St. Xavier's College for Women, Aluva, Kerala, India. She is also currently affiliated with Plant Cell Biotechnology (PCBT) Department, CSIR-Central Food Technological Research Institute (CFTRI), Mysore, India. She developed an *Agrobacterium*-mediated transformation protocol for *Dunaliella*, a marine micro-alga. The carotenoid biosynthetic pathway of *Dunaliella* was further metabolically engineered by expressing the β-carotene ketolase gene from *Haematococcus pluvialis* in *Dunaliella* using a rubisco-driven chloroplast-targeted bkt gene. This resulted in the formation of ketocarotenoids, including astaxanthin from β-carotene in *Dunaliella*. The genetic transformation method has been stable for generations and has opened up the feasibility of using transgenic *Dunaliella* for cell factory applications.

Dr. Mohamed Neifar is a Tunisian biochemist and enzymologist. He received his ED, MD, and PhD in biology engineering from the National Engineering School of Sfax, University of Sfax, Tunisia. From 2010 to 2012, he conducted postdoctoral research in the laboratories of Prof. Daniele Daffonchio and Prof. Abdellatif Boudabous in the European project BIODESERT (Biotechnology from Desert Microbial Extremophiles for Supporting Agriculture Research Potential in Tunisia and Southern Europe). He joined the research laboratory Higher Institute of Biotechnology, Biotechnology and Bio-Geo Resources Valorization in the Biotechnology Department of the Institute of Biotechnology of Sidi Thabet, University of Manouba, Ariana, Tunisia, as an assistant professor in 2013. His scientific interests focus on microbial catalysts and their applications in bioremediation, bioenergy, and sustainable agriculture. His current collaborations include the MADFORWATER project (Development and Application of Integrated Technological and Management Solutions for Wastewater Treatment and Efficient Reuse in Agriculture Tailored to the Needs of Mediterranean African Countries), the AADMEN project (An African Desert Soil Microbial Ecology Research Network), and an innovative project in green chemistry (Development of Biofertilizers for Sustainable Agriculture in Tunisia) supported by a partnership between PhosAgro; the United Nations Educational, Scientific, and Cultural Organization; and the International Union of Pure and Applied Chemistry. Dr. Neifar has published more than 30 peer-reviewed papers and book chapters.

Dr. Hadda Imen Ouzari graduated with a degree in natural sciences from the Faculty of Sciences of Tunis (FST) University of Tunis El Manar, Tunis, Tunisia, in 1995. After completing a PhD in microbiology (2002) at the FST, she was recruited as an assistant professor at the Research Center of Biotechnology of Borj-Cedria, Tunisia. Following a postdoctoral qualification (habilita-tion) in 2008, she was promoted as an associate professor and later as professor, in 2013, at the FST. Professor Ouzari is teaching different courses in molecular biology, food microbiology, and micro-bial ecology and currently coordinating research on microbiology and molecular epidemiology in the FST as a principal investigator in the Laboratory of Microorganisms and Active Biomolecules. The main research activities of Professor Ouzari are on several aspects of microbial ecology and food microbiology, including (1) plant–microbe interactions and plant growth-promoting rhizobac-teria selection, (2) biotechnological potentialities and microbial application in bioremediation, and (3) lactic acid bacteria characterization and microbial food security. She has published about 50 peer-reviewed papers and more than 600 nucleotide sequences. She is a referee for eight scientific journals and has supervised the research of many PhD and master's students. Professor Ouzari has contributed to different national and international projects, as well as EU FP7, Erasmus, and H2020 programs. As president of the Tunisian Society of Microbiology, she contributes to research promotion, scientific interaction between researchers from different countries, and the discussion and dissemination of research data.

Dr. Carmen Pire obtained her chemistry degree in 1994 and her PhD in 1998. Since 2008, she has been a biochemistry with the Departamento de Agroquímica y Bioquímica, Facultad de Ciencias, Universidad de Alicante, Alicante, Spain. Her main areas of research are carbon and nitrogen metabolism in halophilic archaea, the purification and characterization of halophilic enzymes and homologous and heterologous overexpression, protein structure studies and protein design by site-directed mutagenesis, and regulation studies of nitrogen metabolism by transcriptional analysis. She has 25 papers published in these topics, and has participated in several projects, with PI being one of them. Now she is a researcher for two projects granted by the Ministry of Economy and Competitiveness: BIO2013-42921-P (Excellence) and CTM2013-43147-R (Challenges).

Dr. Vikram H. Raval is currently working as a DST-SERB young scientist in the UGC-CAS Department of Biosciences, Saurashtra University, Rajkot, Gujarat, India, with a major research project on the metagenomics of saline habitats of the Gujarat region by the government of India. Earlier, he worked as an assistant professor in this department. Dr. Raval has completed his master's and doctorate in biotechnology from Saurashtra University. During his doctoral research, he was awarded junior and senior research fellowships under the DBT Multi-Institutional Research Project. During predoctoral research at Nicholas Piramal India Limited, Mumbai, as a research associate, he worked on the isolation, identification, and maintenance of fungal cultures; metabolite library preparation; and screening of anti-infective agents. Dr. Raval pursued his doctoral research under the guidance of Prof. S. P. Singh at Saurashtra University. He has 10 years of research experience in microbes from the saline habitats of Gujarat Coast, India. He has worked on the biochemical and molecular properties of extracellular enzymes of halophilic and haloalkaliphilic bacteria, and the metagenomics of various saline ecosystems. He has published six research papers in referred journals and contributed six invited chapters in edited books of reputed publishers.

Dr. Sarada Ravi works as a senior principal scientist and head of the Plant Cell Biotechnology Department, Council of Scientific and Industrial Research–Central Food Technological Research Institute, Mysore, India. She holds a PhD in biochemistry from the University of Mysore, India. She has handled several projects in the field of microalgae biotechnology, focusing on carotenoids, lipids, and algal genetic transformation for food and fuel applications. She has authored more than 75 publications in peer-reviewed journals and has more than 14 patents to her credit. She is an elected member of the National Academy of Sciences, India, and has won best-paper awards at several national and international seminars. She was also a recipient of the Central Food Technological Research Institute Award as best scientist in 2004. She has guided a number of students toward doctoral and master's degrees.

David J. Richardson joined the School of Biological Sciences, University of East Anglia, Norwich, United Kingdom, as a lecturer in 1988. He soon became a member of the Centre for Metalloprotein Spectroscopy and Biology, which encompasses colleagues from both the Schools of Chemical and Biological Sciences. As a bacterial biochemist, he has used a range of disciplines to unravel the respiratory processes of anaerobic bacteria from soils, marine environments, and the human gastrointestinal tract. His work has revealed the great adaptive flexibility of highly branched pathways of respiratory chains by the isolation, structural characterization, and functional analysis of their enzymes and electron transfer proteins. Work on the nitrogen cycle, as well as the mechanisms by which bacteria use insoluble extracellular minerals as respiratory electron acceptors, is now having a major impact on environmental and climate science, industrial biotechnology, and synthetic biology. Richardson's contributions have been recognized by the award of the Society for General Microbiology Fleming Medal (1999), a Royal Society Wolfson merit fellowship (2008), and appointment to the Biotechnology and Biological Sciences Research Council (2012). He has published more than 200 peer-reviewed papers that have been cited around 12,000 times. His research currently has an h-index of 61 (Google Scholar).

Dr. Vanesa Bautista Saiz earned a degree in biology from the University of Alicante, Spain (1999). She obtained her PhD degree at the same university in 2010, and in 2012 she received the PhD Extraordinary Award. Since February 2007, she has been a technical specialist at the Agrochemistry and Biochemistry Department, University of Alicante. From September 2011, she has also held a position as associate teacher in the same department, where she concentrates her teaching and research work. Her main areas of research are based on the genomic and proteomic study of halophilic enzymes, the characterization and heterologous expression of proteins involved in the metabolism of nitrogen and carbon, and the study of transcriptional regulators involved in nitrogen metabolism and the genomics of *Haloferax mediterranei*.

Dr. Tulasi Satyanarayana, after obtaining a PhD in 1978, had postdoctoral stints at Paul Sabatier University and the Institute of Applied Sciences, Toulouse, France. In 1988, he joined the Department of Microbiology, University of Delhi, South Campus, New Delhi, India. He has been a professor since 1998. His research focus has been on understanding the diversity of extremophilic fungi and bacteria, their enzymes and potential applications, heterotrophic carbon sequestration, metagenomics, and the heterologous expression of industrial enzymes. He has published more than 200 scientific papers and reviews, has edited six books, and has three patents to his credit. He is a fellow of the National Academy of Agricultural Sciences, Association of Microbiologists of India (AMI), Mycological Society of India (MSI), Biotech Research Society of India (BRSI), and AP Academy of Sciences, and a recipient of the Dr. G. B. Manjrekar Award of the AMI in 2003, Dr. V. S. Agnihotrudu Memorial Award of MSI in 2009, and Malaviya Memorial Award of BRSI in 2012 for his distinguished contributions. He is an editor of *Climate Change* and *Sustainable Environment*, and a member of the editorial boards of *Bioresource Technology*, *Indian Journal of Biotechnology*, and *Indian Journal of Experimental Biology*. He reviews papers for several national and international journals. During 40 years of research and teaching, he has mentored 26 scholars for their PhD, completed several major research projects, and visited several countries. He is now president of the Association of Microbiologists of India.

Amit K. Sharma is currently pursuing his doctoral research, under the supervision of Prof. Satya P. Singh, in microbiology in the UGC-CAS Department of Biosciences, Saurashtra University, Rajkot, India. His current interests are the molecular diversity and extracellular proteases of actinomycetes from seawater. At present, he is a senior research fellow of the University Grants Commission, New Delhi, India, under the UGC-BSR scheme. He was earlier awarded a dissertation fellowship by the Internal Quality Assurance Cell of Saurashtra University during the academic year 2009–2010. He has published three research papers and contributed two book chapters.

Daris P. Simon has obtained his master's degree in biotechnology from Cochin University of Science and Technology, Kerala, India. He has registered for a PhD in biotechnology from the University of Mysore. He is currently with Plant Cell Biotechnology (PCBT) Department, CSIR-Central Food Technological Research Institute (CFTRI), Mysore, India. He has carried out the genetic transformation of the microalga *Dunaliella salina* for the modification of the carotenoid pathway. The cloning of the β-carotene hydroxylase gene in *D. salina* has resulted in the overproduction of violaxanthin and zeaxanthin. He has also analyzed virulence gene-inducing factors in spent medium of *D. salina*.

Dr. Satya P. Singh is a professor and head of the UGC-CAS Department of Biosciences, Saurashtra University, Rajkot, India, and coordinator of the UGC-CAS program. He has a master's in microbiology from G. B. Pant University, Pantnagar, India, and a doctoral degree from Griffith University, Brisbane, Australia. He has worked at the National Food Research Institute, Tsukuba, Japan, as visiting scientist and visited Yangoon University, Myanmar, as visiting professor. He has 25 years of research experience in the diversity, phylogeny, and enzymatic characteristics of halophilic and haloalkaliphilic bacteria, actinomycetes, and archaea of various saline ecosystems of Gujarat Coast, India. He has published 82 research papers and 21 book chapters. His h-index and citations are 22 and 1490, respectively, with a cumulative impact factor of 135. He is a reviewer of a large number of international peer-reviewed journals, including *Biochimica et Biophysica Acta*, *PLoS One*, *Enzyme and Microbial Technology*, and *Critical Reviews in Biotechnology*. Sixteen PhD and 17 M.Phil. students have completed their research under his supervision. Professor Singh has research collaborations with Dr. Kiyoshi Hayashi (National Food Research Institute, Tsukuba, and Toyo University, Japan), Prof. Peter Rogers (Griffith University, Brisbane, Australia), Prof. S. K. Khare (Indian Institute of Technology, New Delhi, India), and Prof. Sanjay Kapoor (University of Delhi University, South Campus, New Delhi, India).

Asma Soussi is a PhD student in microbiology at the University of Tunis El Manar, Tunisia. She is conducting research studies at the Higher Institute of Biotechnology of Sidi Thabet, Tunisia, in the Laboratory of Biotechnology and Bio-Geo Resources Valorization. She is currently with the Biological and Environmental Sciences and Engineering Division, King Abdullah University of Science and Technology, Thuwal, Kingdom of Saudi Arabia. In 2010, she graduated from the University of Carthage, Tunis, Tunisia, with her engineering bachelor's in industrial biotechnology obtained from the National Institute of Applied Sciences and Technology. She acquired her master's degree in microbiology from the University of Tunis El Manar in 2012, in which she was particularly interested in the study of cyanobacteria biodiversity in the extreme environments of southern Tunisia. Currently, she is focusing on her PhD studies on the role of plant growth promotion bacteria in the biodissolution of phosphates and their direct application in agriculture.

Vidyashankar Srivatsan holds an MSc degree in biosciences from Sri Sathya Sai Institute of Higher Learning, Prasanthi Nilayam, India. He is currently working as senior research fellow in the Plant Cell Biotechnology Department of the Council of Scientific and Industrial Research–Central Food Technological Research Institute, Mysore, India, and has registered for a doctoral degree at the University of Mysore, India. His current areas of interest are microalgal lipids, mass cultivation of microalgae, flocculation of microalgae, extraction of lipids, and use of algal biomass as animal feed. He has five publications in peer-reviewed journals and has participated in several national and international conferences. He is a recipient of an International Travel Grant Award from the Department of Science and Technology, India.

Dr. Amber Grace Teufel (Siebenaler) is a postdoctoral fellow in the Department of Biology at Brookhaven National Laboratory in Upton, New York, under the direction of Dr. Ian Blaby. She was born in Blakeslee, Ohio, but now resides on Long Island with her husband, Philip, and their two dogs. She received her BSc from the University of Toledo, Ohio, in 2008, where she studied maize transcription factors, and she received her PhD from Miami University in Oxford, Ohio, in 2016. Her PhD dissertation focused on the influence of abiotic drivers (light and nutrients) on the photobiology and diversity of Antarctic lake phytoplankton communities. She has given numerous presentations at both domestic and international meetings, explaining her work on algal photobiology and community diversity, and her graduate work has been published. Her current research focuses on a bioinformatic and comparative genomic approach to identifying unstudied proteins known as the GreenCut. By combining high-throughput and low-throughput approachs (bioinformatic and data mining approaches with high-throughput laboratory automation and low-throughput genetic, biochemical, and protein analysis), she hopes her research will contribute to the identification and function of the currently unstudied GreenCut proteins in *Chlamydomonas reinhardtii*.

Foram J. Thakrar is currently working on her PhD under the supervision of Prof. Satya P. Singh at the UGC-CAS Department of Biosciences, Saurashtra University, Rajkot, Gujarat, India. She is working on the molecular diversity of the alkaline proteases of the haloalkaliphilic actinomycetes isolated from the saline habitats of the Gujarat Coast. She is a UGC-BSR meritorious fellow in the Department of Biosciences. She has completed her master's in microbiology from the Department of Biosciences, Saurashtra University, Rajkot, India. Thakrar has also worked as assistant professor in microbiology at Government College Vankal, Veer Narmad South Gujarat University, Surat, India, and as laboratory technician in the Department of Biosciences, Saurashtra University, Rajkot, Gujarat, India.

Javier Torregrosa-Crespo has been a PhD student in the biochemistry and molecular biology area of the Departamento de Agroquímica y Bioquímica, Facultad de Ciencias, Universidad de Alicante, Alicante, Spain, since 2014. He obtained his degree in biology in 2012, earning the Academic Achievement Award. He completed his studies with a master's in biotechnology from the International Business School Aliter, Madrid. He worked for the international company ABENGOA S.A. in Sevilla in 2013, which focuses on cloning and expressing enzymes in fungi for the production of bioethanol. Currently, he is working on biochemical studies of halophilic enzymes and their potential biotechnological applications as part of the research group in charge of the project CTM2013-43147-R, granted by the Spanish Ministry of Economy and Competitiveness.

CHAPTER 1

Extremophiles
Nature's Amazing Adapters

Ravi Durvasula and D. V. Subba Rao

CONTENTS

1.1 INTRODUCTION

1.1.1 Tree of Life: Extremophiles

The tree of life has three main trunks: (1) bacteria, (2) Archaea, and (3) Eukarya, each represented by a large number of extremophiles known for their remarkable diversity of morphology, biochemistry, genomics, and biosynthesis of many adaptive compounds (Table 1.1).

Extremophiles represent a variety of biota from diverse environments. Examples range from an active ecosystem brimming with 4000 species in the subglacial Lake Whillans, 800 m beneath the Antarctic ice sheet (Christner et al. 2014; Fox 2014), to the metazoan pompei worms that live in deep-sea hydrothermal vents at 2 km below the ocean surface (Desbruyères and Laubier 1980), to the commercially important Antarctic krill, *Euphausia superba*, living in the Southern Ocean.

Despite adaptation to wide environments, most extremophiles are simple organisms (MacElroy 1974), either single celled or filamentous, and include a myriad of prokaryotes that do not contain a nucleus or any other membrane-bound organelle. Many of the unique environments inhabited by extremophiles and the physiological adaptations of these organisms will be explored in detail throughout this book. In this introductory chapter, we provide an overview of some of the most important features pertaining to the biology of extremophiles (Figure 1.1).

Table 1.1 Comparison of Bacteria, Archaea, and Eukarya

	Bacteria (Eubacteria)[a]	Archaea[a]	Eukarya
Domain	Large domain of prokaryotes	Domain or kingdom of single-celled biota	Multicellular
Phylogenetically	Different	Different from bacteria but similar to Eukarya	Different
Habitat	Ubiquitous Flourished on earth for >3.5 Ga	Extreme harsh environment	Ubiquitous Appeared in the past 2 Ga
Shape	Larger surface area to volume, leading to higher growth rate, shorter generation time	Similar to bacteria Larger surface area to volume, leading to higher growth rate, shorter generation time	Varied
Organisms	bacteria and microorganisms	Microorganisms	Plants, animals, and microorganisms such as fungi and protists
Cell	Single	Single	Multicellular
Nucleus and organelles	Absent	Absent	True nucleated
Cell	Prokaryote	Prokaryote	Eukaryote
Cell wall	Peptidoglycan/ lipopolysaccharide	Pseudopeptidoglycan Unique membrane lipids with branched fatty chains composed of repeating units of isoprene	Have cell walls, but do not contain peptidoglycan Mitochondria and chloroplasts present
Reproduction	Asexual, binary fission, budding but form spores	Asexual, binary fission, budding	Meiosis
Metabolism		Diverse Do not use glycolysis pathway to break down glucose Functional Krebs cycle pathway absent	
Protein synthesis (first amino acid formed)	Formylmethionine	Methionine	Methionine
DNA	Mostly circular chromosome and plasmids	Circular chromosome and plasmids	Linear chromosome, rarely plasmids DNA wrapped around proteins called histones
DNA replication, transcription, and translation		Similar to eukaryotes	
RNA		Complex, polymerases similar to bacteria and eukaryotes	
Histones	Not present	Present	Present
Organelles	Not present	Not present	Present
Ribosomes	70S	70S	80S
Antibiotic sensitivity	Sensitive	Not sensitive	Resistant to those that affect bacteria

[a] Earlier archaea were classified as bacteria and were called archaebacteria but now are considered distinct.

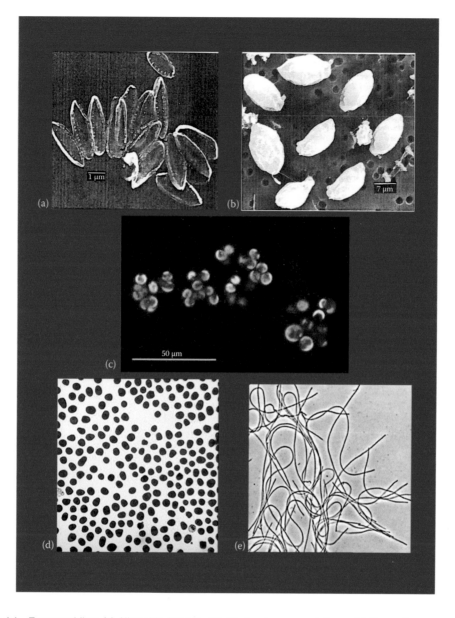

Figure 1.1 Extremophiles: (a) *Nitzschia frustula*, (b) *Chalmydomonas plethora*, (c) *Scenedesmus* sp. nov., (d) *Dunaliella salina*, and (e) *Chloroflexus* aurantiacus. (Courtesy of Dr. Sylvia Herter and Dr. D.E. Gilbert.)

Biota that grow at low temperatures are known as mesophiles, compared with the cold-adapted psychrophilles or psychrotrophs that have slower metabolic rates. In their DNA replication, transcription, and translation, Archaea are similar to nucleated eukaryotes. Unique to Archaea are their membrane lipids, with branched fatty chains linked to head groups via ether linkages (Sarma et al. 2010), which facilitate biotechnological applications such as thermostable DNA polymerases.

Thermophilic extremophiles are physiologically active at temperatures ranging from 0°C to 120°C and pH ranges from 0 to 12 (Antranikian and Egotova 2007). Molecular studies have demonstrated that eukaryotes are more adaptable than prokaryotes, which may explain the greater novel protist diversity in extreme environments (Roberts 1999).

Extremophiles live under harsh environmental conditions and carry out biochemical processes, where survival strategies hinge on the production of extremozymes. The study of these extremozymes is fascinating because they provide data on not only protein folding, stability, structure, and function, but also applications to biotechnology.

The prokaryotes have flourished on earth for more than 3.5 giga-annum (Ga). The unicellular mitotic eukaryotes are autotrophic, appeared around 2.5 Ga (Horneck 2000), and are represented by microalgae, the first organisms to exhibit photosynthesis. Of the $4-6 \times 10^{30}$ prokaryotes that include bacteria (eubacteria), archaea, and cyanobacteria that inhabit the earth (Whitman et al. 1998), about 1 million live in the oceans (Table 1.2). As Gould (2014) noted, "On any possible, reasonable, or fair criterion, bacteria are—and always have been—the dominant forms of life on earth." There are up to 1 million estimated eukaryotes (algal species) (Guiry 2012). Yet, we do not know how many exist as extremophiles and how many are cultured. But we should recollect a quote from Louis Pasteur, "The role of the infinitely small is infinitely large."

Publications on the aquatic extremophiles are growing in number; there were 21 before 1997 and 507 after 1997, mostly on heterotrophs and mixotrophs.

1.1.2 Categories of Extremophiles

Based on the stressor, about 15 categories of extremophiles can be recognized. The stressors include temperature, radiation, pressure, vacuum, desiccation, salinity, pH, oxygen tension, chemicals, and methane. Because of the latitude of stressors, there is no single culture method for all categories of extremophiles, which has limited studies of their growth. Based on their living conditions, six broad categories of extremophiles are recognized:

1. *Thermophile*: Organisms requiring optimum growth temperatures of 50°C or higher. In the case of hyperthermophiles, the optimum may be between 80°C and 110°C.
2. *Halophile*: Organisms requiring at least 0.2 M (3%–30%) salt for growth.
3. *Psychrophile*: Organisms requiring optimum growth temperatures of 15°C or lower and unable to grow above 20°C.
4. *Alkaliphile*: Organisms exhibiting optimal growth at pH values above 10.
5. *Acidophile*: Organisms with optimum growth at or below pH 2.
6. *Piezophile or barophile*: Organisms that live optimally at high hydrostatic pressure.

1.1.3 Extremophile Biota

Extremophiles are capable of sustaining biological life under one extreme environmental stressor, such as temperature or salinity (Horikoshi and Grant 1998), as opposed to polyextremophiles that thrive under conditions of multiple stressors. The archaea *Sulfolobus acidocaldarus*, for example, is a polyextremophile that lives at pH 3 and 80°C (Rothschild and Mancinelli

Table 1.2 Abundance of Biota and Extremophiles

Category	Examples	Number of Species	Global Abundance	Nutrition
Prokaryote 0.5–2.0 μm	Archaea	1,320	$4-6 \times 10^{30}$	Osmotrophy and photosynthesis
	bacteria	20,000[a]	1.37×10^{21}[b]	Heterotrophy?
	Cyanobacteria	834	1×10^{27}	Mixotrophy
Eukaryota 2–200 μm	Diatoms and dinoflagellates Protists	32×10^3 to 1×10^6	6.25×10^{25}	Mostly autotrophic Some mixotrophic

[a] From Mora, C. et al., *PLoS Biol.*, 9 (8), e1001127, 2011; Dolan, J.R., *J. Eukaryot. Microbiol.*, 63, 410, 2016.
[b] Retrieved from http://www.eoearth.org/view/article/154477.

2001). During the KRSE2011 cruise in the atypical Red Sea waters, characterized by high temperature and hypersalinity, a great diversity of either bacterial or archaeal genomes belonging to Thaumarchaeota, Euryarchaeota, Actinobacteria, Cyanobacteria, Bdellovibrionaeota, Proteobacteria, and Marinimicrobia were recorded (Haroon et al. 2016). The biota included asphototrophic *Prochlorococcus* and *Synechococcus*; the common marine bacterium, *Alteromonas macleodii*; an ammonia-oxidizing genus, *Nitrosopelagicus*; euryarchaeotal Marine Group II organisms; and members of the alpha- and gamma-proteobacteria, such as *Aeromicrobium, Erythrobacter, Maritimibacter, Idiomarina, Marinobacter, Candidatus, Thioglobus*, and several unclassified *Gammaproteobacteria*. Additionally, actinobacterial *Acidiimicrobia* and *Nocardioides* were also present.

Physiologically, extremophiles are heterogeneous and include aerobes, anaerobes, chemotrophs, chemorganotrophs, phototrophs, chemolithotrophs, and photoheterotrophs (Table 1.3).

Carnobacterium pleistocenium, a bacterium, remained viably cryopreserved in the ancient ice of Fox Tunnel, Alaska, and the deep Vostak (Hoover and Pikuta 2010). In the Antarctic glacial surfaces, cryoconite sediment surfaces were covered by biofilms and were shown to be "hot spots" of biological activity (Smith et al. 2016). These biofilms were dominated by Cyanobacteria (27%), Actinobacteria (24%), Proteobacteria (22%), and Bacteroidetes (19%). The cyanobacterium Oscillatoria was photoautotrophic and fixed at about 1.60 kg C within cryoconite across the Canada Glacier per season, while Bacteroidetes was heterotrophic.

Enigmatic extremophiles that live under limiting extremes of temperature, pH, salt concentration, and ionizing radiation prompted Brock (1978), Horikoshi and Grant (1998), Oren (1999), Seckbach (1999, 2000; Seckbach et al. 2013), Gerday and Glansdorff (2007), and Anitori (2012) to discuss the wondrous array of life's adaptations by these bizarre biota. Tardigrades, known as "water bears" or "space bears," are the ultimate extremophiles, living in near-deadly environmental conditions. By a shutdown of all metabolic activities, tardigrades survive exposure to lethal doses of UV radiation (5000 Gy gamma-rays) and tolerate extreme desiccation, temperatures from $-273°C$ to $100°C$, high pressure (7.5 GPa), and immersion in organic solvents (Bittel 2016; Hashimoto et al. 2016).

Chemotrophs, mostly represented by prokaryotic acidophiles, have the ability to use inorganic chemicals as electron donors in anoxic environments (Johnson and Hallberg 2008).

A study of the energetics of the thermophilic and hyperthermophilic Archaea and bacteria isolated from marine hydrothermal systems, heated sediments, continental solfataras, hot springs, water heaters, and industrial waste showed that they catalyze widely varying metabolic processes (Amend and Shock 2001). Electron donors in these microbes include H_2; Fe^{2+}; H_2S; S; $S_2O_3^{2-}$; $S_4O_6^{2-}$; sulfide minerals; CH_4; various mono-, di-, and hydroxycarboxylic acids; alcohols; amino acids; and complex organic substrates. Electron acceptors include O_2, Fe^{3+}, CO_2, CO, NO_3^-, NO_2^-, NO, N_2O, SO_4^{2-}, SO_3^{2-}, $S_2O_3^{2-}$, and S.

1.1.4 Growth and Photosynthetic Vigor

Scenedesmus species, a thermophile microalga from Soda Dam, New Mexico, grew under outdoor harsh conditions (6524–7360 µmoL photons m^{-2} s^{-1} and 40°C); biomass peaked at 10.41 × 10^6 cells mL^{-1}, 6.92 µg Chlorophyll *a* L^{-1}, and 4.49 µgcarotene L^{-1} in wastewater (Durvasula et al. 2015). These cells had 0.34–1.08 pg $cell^{-1}$ carotenoids, comparable to the literature values, 0.24–4.75. Indoor cultures had 16.7–36.6 pg $cell^{-1}$, the highest among the microalgae.

The few data on the photosynthetic functioning of the extremophile microalgae suggest a high photosynthetic vigor. Cultures of the halophiles *Chlamydomonas plethora* and *Nitzschia frustula* isolated from Kuwait had maximum division rates (μ_{max}) of 2.5 and 3.4 day^{-1}, respectively, and yielded the highest assimilation numbers $\left(P_m^B: \mu g\,C\,[\mu g\,Chl\,a]^{-1}\,h^{-1}\right)$: 22.8 for *C. plethora* and 18.1 for *N. frustula*, with initial slopes (α^B: ng C [µg Chl *a*]$^{-1}$ h^{-1} [µmol m^{-2} s^{-1}]$^{-1}$) of 79.5 for

Table 1.3 Types of Extremophiles Based on Their Habitat

Environmental Variable	Type	Characteristic	Habitat	Example	Reference
Temperature	Hyperthermophile	Growth >80°C, Upper limit 110°C	Mid-Atlantic Ridge	*Pyrolobus fumari*	Blöchl et al. 1997
	Thermophile	120°C, Upper limit 70°C, 60°C–80°C, Upper limit 70°C	Juan de Fuca Ridge, Hunter's Hot Springs, Oregon	*Geobacterium*, *Synechococuus lividis*	Kashefi and Lovely 2003, Meeks and Castenholtz 1971
	Mesophile	15°C–60°C, Upper limit 45°C		93 thermophiles and hyperthermophiles	Zheng and Wu 2010
	Psychrophile	<15°C, Upper limit 25°C		*Polaribacter*, *Cryptococcus*	Feller and Gerday 2003
Radiation			Core of nuclear reactors	*Deinococcus radiodurans*, *Tardigrades*	Paje et al. 1997, Copley 1999, Bittel 2016, Hashimoto et al. 2016
Pressure	Barophile	Weight loving		*Photobacterium* sp.	MacDonald et al. 2002
	Piezophile	Pressure loving		*Pyrococcus* sp.	Di Giulio 2005
Vacuum		Tolerates vacuum		*Deinococcus radiodurans*	Bauermeister et al. 2011
Desiccation	Xerophiles	Anhydrobiotic, 120 years of desiccation, 6000 atm pressure, Temperature, 272°C and 151°C		*Tardigrades*	Copley 1999, Rothschild and Mancinelli 2001
Salinity	Halophile	2–5.5 M Na Cl	Cabo Rojo, Puerto Rico, Great Salt Lake, Utah	*Haloterrigena thermotolerans*	Montalvo-Rodríguez et al. 2000
pH	Alkaliphile	pH >8.5–11.5	Lake Magadi, Tanzania	*Natronococcus occultus*	Jones et al. 1998, Oren 2002
	Acidophile	pH low, 0.7		*Picrophilus oshimae*	Van de Vossenberg et al. 1998
Oxygen tension	Anaerobe	Cannot tolerate O_2		*Methanocaldococcus jannaschii*	Bult et al. 1996
Chemicals	Metals	Tolerates metals		*Cyanidium caldarium*, *Ferroplasma acidarmanus*	Csatorday et al. 1981, Golyshina et al. 2000
Methane	Methane	Methogens		*Pyrolobus fumarii*, *Methanopyrus kamdleri*	Madigan and Oren 1999

C. plethora and 39.6 for *N. frustula*, the highest observed so far (Subba Rao et al. 2005). Kottmeier and Sullivan (1988) demonstrated high photosynthetic efficiencies (15 mg C [mg Chl *a* h^{-1}] h^{-1}) in sea ice microalgae from McMurdo Sound temperatures of –24.0°C to 0.0°C, and brine salinities from 150‰ to <20‰ have been reported in the upper layers of Antarctic sea ice.

Carbon fixation in the cyanobacterium *Synechococcus* sp., growing at ~70°C in a Yellowstone National Park hot spring, was the greatest at 65°C, but it was supraoptimal at 70°C and suboptimal at 55°C (Miller et al. 1998). Miller et al. (1998) speculated that these cells surviving at lethal temperature and subjected to visible and ultraviolet (UV) radiation are "avoiding competition with other phototrophs under these nonoptimal contions."

While discussing the ecological importance of the Antarctic sea ice diatoms, Thomas and Dieckmann (2002) pointed out that their survival requires a complex suite of physiological and metabolic adaptations, such as the production of highly active antioxidative protective enzymes, that is, catalase, glutathione peroxidase, and glutathione reductase. A chemotrophic ecosystem consisting of a white mat of sulfur-eating microbes, as well as large clams, was reported underneath the calved iceberg Larsen (Marchant 2017).

1.1.5 Mechanisms: Enzymes

Extremophiles serve as model organisms for bioengineering enzymes, biofuel production, and bioremediation (Rossi et al. 2003; Satyanarayana et al. 2005; Kumar et al. 2011). Three main mechanisms centered around enzymes, such as DNA polymerases, lipases, proteases, dehydrogenases, pullulanases, amylases, xylanases, phosphatases, lipases, and cellulases, enable the survival of extremophiles. These enzymes facilitate (1) protein stability, repairing damage and preventing unfolding of proteins; (2) lipid stability, allowing for the production of thermotolerant lipids; and (3) DNA stability, stabilizing DNA via thermostabilizers, reverse gyrase, and DNA-binding proteins.

Furthermore, acidophiles use powerful cytoplasmic proton pumps to maintain a proton concentration gradient of >4 to keep their cytoplasm at near-neutral pH (Rainey and Oren 2006), while alkaliphiles cannot grow at neutral pH but only at pH 9–10.

1.1.6 Evolution of Strategies

Because of their diverse and extreme habitats, extremophiles have developed unique structural and biochemical adaptation strategies to function under environmental stress. In this respect, marked differences exist between the heterotrophs and autotrophs. Archaea are known to perform methanogenesis and fix carbon from organic and inorganic electron donors and acceptors, and thus play crucial roles in the earth's global geochemical cycles (Offre et al. 2013).

Chloroflexus aurantiacus (plate) is an interesting thermophile that occurs in hot springs with a pH above 5.5. This gliding (0.01–0.04 μm s^{-1}) and filamentous thermophile harbors bacteriochlorophyll *c* and bacteriochlorophyll *a*, in addition to β- and γ-carotene and glycopsides (Pierson and Castenholtz 1974). The discovery of *C. aurantiacus* has revolutionized anoxygenic photosynthetic research. During the past 40 years, 18 new species of extremophilic anoxygenic phototrophs were discovered and studied (Nadir et al. 2005). Under anaerobic conditions, *C. auranticus* is photoheterotrophic, while under aerobic conditions it is chemoheterotrophic, in darkness or light. It lives up to 70°C, and the optimum growth is at 52°C–60°C.

Under chemotrophy, *C. aurantiacus* lacks chromosomes, but chromosome synthesis takes place under phototrophic metabolism. Anoxygenic photosynthesis takes place at 0°C–70°C and pH values of 3–11 and salinities of 0% to approximately 32% NaCl. Tang et al. (2011) provided genomic evidence that suggests numerous gene adaptations and replacements in *C. aurantiacus* to facilitate life under both anaerobic and aerobic conditions.

Another fascinating survival strategy in the heterotrophic bacteria includes the production of biocatalysts or enzymes called "extremozymes" and production of antifreeze glycoproteins in the psychrophiles as an adaptation to freezing conditions. The autotrophs have a suite of diverse pigments, including oligosaccharide mycosporine-like amino acids, scytonemins, carotenoids, phycobiliproteins, and chlorophylls, that provide a comprehensive broadband strategy for coping with the multiple stressors of high irradiance, variable salinity, and low temperatures in extreme cryoenvironments. Protection against high radiation damage includes repair mechanisms for damaged DNA (Rainey and Oren 2006). Differences in the hydrophobic core and electrostatic interactions exist in extremophile Archaea, which, in turn, manifest in their protein adaptations (Reid et al. 2013). Investigations are focused on these physiological and biochemical survival strategies to explore the production of bioactive compounds and their utility in industries such as pharmaceuticals, bioengineering, biofuel production, and bioremediation.

1.1.7 Extracellular Polymeric Substances

The production of EPSs is a key adaptive response that facilitates entrapment, retention, and the survival of ice biota, thus defining ice as a habitat for some extremophiles (Ewert and Deming 2013). Three sea ice diatoms, *Synedropsis* sp., *Fragilariopsis curta*, and *F. cylindrus*, showed differences in the production and composition of EPSs (Aslam et al. 2012). *Synedreopsis* sp. produce EPS carbohydrates that are relatively soluble, while both *Fragilariopsis* spp. and *F. curta* produce colloidal EPSs with high concentrations of amino sugars, which allows for their survival under freezing conditions (Aslam et al. 2012). *Melosira*, another sea ice diatom, produces EPSs that allow it to create convoluted ice-pore morphologies in sea ice (Krembs et al. 2011), while *Fragilariopsis cylindrus*, the most dominant species in polar sea ice, produces EPSs in the form of antifreeze proteins (AFPs) that facilitate depression of the freezing point in the presence of salt (Bayer-Giraldi et al. 2011). An analysis of biofilms produced by one or two species of acidic eukaryotes from Rio Tinto, Spain, further yielded a correlation between EPSs and metal concentrations, thus suggesting mineral adsorption in the matrix of EPSs (Aguilera et al. 2008).

Analyses of natural sea ice microbial communities of Arctic and Antarctic sea ice showed an abundance of ice-binding proteins (IBPs). These small proteins, approximately 25 kDa in size, have mRNA transcripts that are similar to those of proteins involved in core cellular processes, such as photosynthesis (Uhlig et al. 2015). Interestingly, 89% of these transcripts show homology to those from diatoms, haptophytes, and crustaceans, confirming IBP's role for survival of sea ice communities.

The Antarctic euryhaline chlorophyte, *Chlamydomonas* sp., produces an EPS to increase its habitability and primary productivity (Raymond 2011). One of the challenges of the cryosphere is the availability of liquid water for relatively short periods each year (Laybourn-Parry et al. 2012). The IBP produced by this euryhaline is thought to facilitate retention of frozen sea water during its life cycle (Raymond 2011).

1.1.8 Extremophiles and Utility: Selected Examples

Extremophiles are considered cell factories (Hamilton et al. 2001), and a few of their applications are shown in Table 1.4.

Extremophile responses are varied and dependent on their biotopes, strategies, metabolism, thermostability, and phylogeny. For example, even in the same extremophile, *Pyrococcus furiosus*, which grows at 50°C–115°C, multiple activities are noticed (Table 1.5). The molecular weight of their metabolites and enzymes varies from 40 to 272 kD based on optimal temperature and pH.

Microfungi living inside the porous rocks in the Antarctic dry valleys tolerate hard desiccation, exposure to high UV, and wide fluctuations in temperature. These conditions are analogous to those

Table 1.4   Utility of Extremophiles

Examples	Application or Utility	Reference
Hyperthermophiles	Enzymes: Thermostable proteases, lipases, esterases, starch- and xylan-degrading enzymes	Bertoldo and Antranikian 2002
Vibrios and *Alteromonas* strains	Microbial exopolysaccharides: Tissue regeneration and cardiovascular diseases (antithrombotic/proangiogenic effects)	Guezennec 2002
Dunaliella	Cosmetics, wellness, nutrition, therapeutics	Von Oppen-Bezalel and Shaish 2009
Syntrophus aciditrophicus	Hydrogen from waste	McInerney et al. 2007
Extremophile bacteria	Bioactive compounds	Giddings and Newman 2015
Thermophilic bacilli	Biodegradation of hydrocarbons	Margesin and Schinner 2001
Geobacillus thermoglucosidasius, *Thermobacterium sacchatolyticum*, *Thermoanerobacter mathranii*	Bioethanol–biofuel production	Taylor et al. 2012

found on Mars, and can be used for understanding the functioning of eukaryotic models (Onofri et al. 2004).

As a strategy to survive extreme conditions, the extremophiles have a "survival kit" of enzymes, as summarized below:

	Stressor	Response	Application	Reference
Thermophile	Radiation resistant at 100°C	Exopolysaccharide thermostable proteins	Skin protection	Gabani et al. 2014
Halophiles	Salinity	Ectoines and glycine betaine	Cosmetics	Lentzen and Schwarz 2006
Psychrophiles and piezophiles	Pressure up to 120 MPa and 108°C temperature	Increase polyunsaturated fatty acids	Membrane fluidity	Zeng et al. 2009

Given the many unique physiological processes that permit survival in harsh conditions, proprietary products and processes have been discovered and patented. There are numerous patents based on marine extremophiles; 50 of them are on hydrothermal vent microorganisms from beyond national jurisdictions (Vierros et al. 2012). Examples include

1. *Pyrolobus fumaria*: US 7781198 patent for a polymerase to be used in biotechnology, Verenium Corporation (2010, 2012). Also, patents US 7056703 (2006), US 7049101 (2000), WO2003023029, and WO2002020735 for a hydrotherm vent, 90°C–113°C.
2. *Pyrodictium abyssi*: WO2005094543 for proteins and nucleic acid encoding, Verenium Corporation (Diversa). Also, US 7459172 patent (2008) for hyperthermophilic archaea whose optimal growth temperature range is 80°C–105°C.
3. *Thermococcus barophilus*: Contributor to US 20110045489 (New England Biolabs) and US 20100311142 (Korea Ocean Research and Development Institute) patents for hydrogen production.
4. *Thermococcus marinus*: KR20100064731 for a heat-resistant DNA polymerase, University Sungkyunkwan, Korea hyperthermophile.
5. *Vibrio diabolicaus*: US 7015206 patent for polysaccharide for use in bone repair and other medical purposes, U.S. patent, Institut Francais de le Recherche pour l'Exploitation de la Mer (2006).

Table 1.5 Multiple Activities of *Pyrococcus furiosus*

Enzyme	MW (kDa)	Optimum Temperature (°C)	Optimum pH	Half-Life Thermostability at °C	Biotechnological Utility	Reference
α-Amylase (intracellular)	76	92	7			Antranikian et al. 2005
α-Amylase (extracellular)	100	100	5.5–6	13h at 98		Antranikian et al. 2005
α-Glucosidase	125	115	5.5			Eichler 2001
Pullulanase	90	105	6			
Pullulan-hydrolase	77	90	5	2 at 95		Yang et al. 2004
β-Glucosidase	232	102	5	13h at 110	Color brightening	Lebbink et al 2001
Endoglucanase	36	100	6	40h at 95		Bauer et al. 1999
β-D-Mannosidase	240	105	7.4	60h at 90		Bauer et al. 1999
Endochitinase	40	90–95	6	1h at 120	Biomass utilization	Gao et al. 2003
Exochitinase	55	90–95	6			Gao et al. 2003
Serine protease	150	115	6–9	0.33h at 105		Antranikian et al. 2005
Metalloprotease	128 and 79	100–75	6.5–7			Ward et al. 2002
DNA polymerase	90	72–78	9	4h at 95		Antranikian et al. 2005
Aldalose	272	50				Lorentzen et al. 2004
Aminoacylase	170	100	6.5		Pharmaceutical	Story et al. 2001
Esterase		100		2h at 120		Sehgal and Kelly 2003

6. *Alcanivorax dieselolei*: CN 1904033 (2007) for environmental remediation—bacteria degrade alkanes (saturated hydrocarbons), No. 3 Institute of Oceanography, China, alkane-degrading bacterium isolated from east Pacific and deep-sea sediment.

7. *Thermodesulfatator indicus*: For a novel polypeptide having thermostable DNA polymerase activity and comprising or consisting of an amino acid sequence. (Anderson et al. 2012).

1.1.9 Genomics and Genetic Engineering

The remarkable adaptability of extremophiles to temperature, salinity, pressure, radiation, and other physical and chemical factors offers an unparalleled opportunity to explore novel genetic structures and regulatory mechanisms. Indeed, large-scale efforts to understand the genomic correlates of extremophile biology are underway across the globe, with some fascinating findings.

Picochlorum SENEW3, a broadly halotolerant green alga that was isolated from mesophilic brackish water lagoons, tolerates extreme changes in temperature, light, and salinity (Foflonker et al. 2016). Despite a greatly streamlined genome of 13.5 Mbp, this alga exhibits rapid recovery from salinity stress through large-scale differential expression, the expression of adaptive genes that have been horizontally transferred, and the colocalization of genes involved in stress adaptation.

The acidophilic *Euglena mutabilis* from Rio Tinto, Spain, exhibits considerable differential gene expression, and its transcription patterns for genes involved in stabilizing photosystem II and repair from UV damage and nonphotochemical quenching and oxidative stress are highly adapted (Puente-Sánchez et al. 2016).

When cultivated in 500 µM copper, the photosynthetic activity of the extremophilic green alga, *Chlamydomonas acidophila*, is not adversely affected. Transcriptome studies show activation of genes that are likely related to heavy metal tolerance (Olsson et al. 2015).

Of interest is a novel pathway exploited by hyperthermophilic microbes to reduce carbon dioxide directly into products (Keller et al. 2013). The extremophile *Pyrococcus furiosus* does not have the capacity to incorporate carbon dioxide naturally at 100°C; this was accomplished by the heterologous expression of five genes of the carbon fixation cycle of another archaeon, *Metallosphaera sedula*, that grows autotrophically at 73°C (Keller et al. 2013). The engineered *P. furiosus* follows a new pathway and uses hydrogen as a reducing agent of sugars to produce acetyl-CoA and 3-hydroxypropionic acid, a mass produced, globally important bulk chemical.

The first genome study of an extremophile was on the autotrophic archaeon *Methanococcus jannaschii*, where 1738 predicted protein-coding genes were identified (Bult et al. 1996). Most of the genes in *M. jannaschii* are similar to those in eukaryotes. The genome sequence of extremophiles is crucial for the discovery of novel bioactive compounds, genes, and gene products through data mining, which in turn can be used in proteomic technologies and the biotechnology industry (Giddings and Newman 2015). The genome sequencing of 7 species of algae, *Chlamydomonas reinhardtii*, *Phaeodactylum tricornutum*, *Thalassiosira pseudonana*, *Cyanidoschyzon merolae*, *Ostreococcus lucimarinus*, *Ostreococcus tauri*, and *Micromonas pusilla*, has since been completed (Radakovits et al. 2010). However, metagenomics (Marco 2011), particularly the sequence-driven analysis that uses conserved DNA sequences to design polymerase chain reaction (PCR) primers to screen clones for the sequence of interest (Schloss and Handelsman 2003), offers a hope to study even the unculturable microbes.

Since extremophiles are known to produce biomolecules and extremozymes, which are useful in industrial bioconversions and applications, large quantities of extremophile biomass are essential for genomic studies. However, less than 1% of the extremophile microbiota have been brought into pure culture. Biotechnological progress has been hampered by the inability to isolate and establish pure cultures and to understand their metabolic potential.

Rapidly evolving molecular strategies aimed at introducing foreign DNA into organisms and editing genomic material have gained traction in the study of extremophile biology, with attendant benefits to biotechnology applications. *Geobacillus* is an aerobic thermophilic bacillus. Of the several species of *Geobacillus*, *G. gargensis* was isolated from the Garga hot spring located in Eastern Siberia, Russia. *Geobacillus vulcani* was isolated from a shallow marine hydrothermal vent in Eolian Island Volcano. Through metabolic engineering of two strains of *Geobacillus thermoglucosidasiusm*, lactate dehyderogenase and pyruvate formate lyase pathways were eliminated to facilitate high levels of ethanol production (Cripps 2009).

Cyanobacteria, the only prokaryotes with characteristics to include (1) multiple chromosomal copies, (2) a high content of photosynthetically active proteins in the thylakoids, (3) the presence of exopolysaccharides and extracellular glycolipids, and (4) the existence of a circadian rhythm, are highly amenable for genetic engineering (Heidorn et al. 2011). As such, engineered cyanobacteria have been demonstrated to generate high-value products (Ducat et al. 2011).

Exopolysaccharides identified from some extremophiles show promise as biosurfactants and emulsifiers and in other polymeric roles (Paniagua-Michel et al. 2014). In a review, Wright et al. (2003) discussed several examples from the Japanese Marine Science and Technology Center (JAMSTEC) and the Institut Francais de Recherche pour l'Exploitation de la Mer (IFREMER) that demonstrated the production of novel exopolysaccharides by high-pressure treatment (greater than tens of megapascals) of piezophile eubacteria and Archae. It may be possible to develop transgenic extremophile algal strains for recombinant protein expression and engineered photosynthesis, similar to the studies of Rosenberg et al. (2008).

1.1.10 Biotechnology

Investigations of extremophile physiology, biochemistry, and genomics, such as isolation of the thermostable DNA polymerase, will advance biomedical applications of these organisms and foster industrial biotransformation processes, such as detergent formulation, food processing, biosensors, and bioremediation, which already have a €5 billion per year market (Antranikian and Egorova 2007). For example, halophilic microbes produce bacteriorhodopsin, used as an optical recording material (Ventosa and Nieto 1995). Recent innovative culturing approaches, environmental genome sequencing, and whole genome sequencing have provided new opportunities for the biotechnological exploration of extremophiles. Still, only a few extremophile algae are utilized in biotechnology for the exploitation of commercially important bioactive compounds. Examples include *Dunaliella*, a unicellular biflagellate, and the cyanobacteria *Synechocystis* and *Synechococcus*. *Dunaliella* is cultivated on a mass scale for the production of high-value carotenoids and zeaxanthin (Jin et al. 2003). Shaish et al. (2009) discussed that this β-carotene can potentially protect low-density lipoprotein (LDL) cholesterol against oxidation, which inhibits atherosclerosis. The heterotrophic marine microalga *Cryptothecodinium cohnii* accumulates lipids that are 30%–50% decosahexaaenoic acid (DHA 22:6) (De Swaaf et al. 2003) and might serve as a valuable resource for biotechnological utilization.

Some of the extremophiles synthesize biotic compounds and extremozymes that can remain stable under exacting bioprocess engineering, and therefore will find utility in biotechnology. Some of the nonmethogenic extremophiles have tolerance mechanisms for thermostable proteins, salt tolerance proteins that have biotechnological utility in the production of thermophilic enzymes, detergent formulation, and mineralization of high-salinity organic waste (Mermelstein and Zeikus 1998). Sequencing of extremophile genomes will provide not only insight into the mechanisms that permit life under stressful conditions, as in *Exiguobcterium povilionensi* (White et al. 2013), but also clues for genetic regulatory machinery that may have far-reaching biotechnological applications.

There is every need to study the genomes of autotrophic, heterotrophic, and mixotroophic extremophiles to investigate their enzymes, which would disclose a variety of biocatalysts. A good beginning has been made with genome analyses of nearly 50 extremophilic bacteria and Archaea (Rossi et al. 2003) and 139 genomes (Haroon et al. 2016). Combined with these, studies on EPSs,

AFPs, and IBPs may shed light on their roles in nutrient cycling and carbon transport through the sea ice cover, and adaptation strategies against prevailing extreme environmental conditions and for bioprospecting novel exoenzymes.

Here, we present a collection of papers that explore some of the myriad facets of extremophile biology, physiology, genetics, and environmental adaptations. In addition to these, in the appendix we have provided the methodology utilized for culturing extremophiles and information on their media. From studies on the biota from semiarid deserts to Antarctic waters, and from the deep-sea hydrothermal vents to the salterns, and from studies on extremozymes for their bioprospecting, this compilation should provide insights into the fascinating world of extremophiles and encourage the reader to pursue much greater study of nature's amazing adapters.

REFERENCES

Aguilera, A., Souza-Egipsy, V., Martin-Uriz, P.S. et al. 2008. Extracellular matrix assembly in extreme acidic eukaryotic biofilms and their possible implications in heavy metal adsorption. *Aquat Toxicol* 88: 257–266.

Amend, J.P., and Shock, E.L. 2001. Energetics of overall metabolic reactions of thermophilic and hyperthermophilic Archaea and Bacteria. *FEMS Microbiol Rev* 25 (2): 175–243.

Anderson, I., Saunders, E., Lapidus, A. et al. 2012. Complete genome sequence of the thermophilic sulfate-reducing ocean bacterium *Thermodesulfatator indicus* type strain (CIR29812(T)). *Stand Genomic Sci* 6 (2): 155–164.

Anitori, R.P. (ed.). 2012. *Extremophiles: Microbiology and Biotechnology.* Norfolk, UK: Caister Academic Press.

Antranikian, G., and Egotova, K. 2007. Extremophiles, a unique resource of biocatalysts for industrial biotechnology. In C. Gerday and N. Glansdorff (eds.), *Physiology and Biochemistry of Extremophiles.* Washington, DC: ASM Press, pp. 361–408.

Antranikian, G., Vorgias, C.E., and Bertoldo, C. 2005. Extreme environments as a source for microorganisms and novel biocatalysts. *Adv Biochem Eng Biotechnol* 96: 219–262.

Aslam, S., Cresswell-Maynard, T., Thomas, D.N. et al. 2012. Production and characterization of the intra- and extracellular carbohydrates and polymeric substances (EPS) of three sea-ice diatom species, and evidence for a cryoprotective role for EPS. *J Phycol* 48: 1494–1509.

Bauer, M.W., Driskilll, L.E., Callen, W. et al. 1999. An endogluconase, EglA, from the hyperthermophilic archaeon *Pyrococcus furiosus* hydrolyzes beta-1,4 bonds in mixed linkage (1→3),(1→4) beta-D-glucans and cellulose. *J Bacteriol* 181: 284–290.

Bauermeister, A., Moeller, R., Reitz, G. et al. 2011. Effect of relative humidity on *Deinococcus radiodurans'* resistance to prolonged desiccation, heat, ionizing, germicidal, and environmentally relevant UV radiation. *Microb Ecol* 61: 715–722.

Bayer-Giraldi, M., Weikusat, I., Besir, H. et al. 2011. Characterization of an antifreeze protein from the polar diatom *Fragilariopsis cylindrus* and its relevance in sea ice. *Cryobiol* 63 (3): 210–219.

Bertoldo, C., and Antranikian, G. 2002. Starch-hydrolyzing enzymes from thermophilic archaea and bacteria. *Curr Opin Chem Biol* 6: 151–160.

Bittel, J. 2016. Tardigrade protein helps human DNA withstand radiation. *Nature.* doi: 10.1038/nature.2016.20648.

Blöchl, E., Rachel, R., Burggraf, S. et al. 1997. *Pyrolobus fumarii*, gen. and sp. nov., represents a novel group of archaea, extending the upper temperature limit for life to 113 degrees C. *Extremophiles* 1: 14–21.

Brock, T.D. 1978. *Thermophilic Microorganisms and Life at High Temperatures.* New York: Springer Verlag.

Bult, C.J, White, O., and Olsen, G.J. 1996. Complete genome sequence of the methanogenic archaeon, *Methanococcus jannaschii. Science* 273: 1058–1073.

Christner, B.C., Priscu, J.C., Achberger, A.M. et al. 2014. A microbial ecosystem beneath the West Antarctic ice sheet. *Nature* 512: 310–313.

Copley, J. 1999. Indestructible. *New Sci* 164: 45–46.

Cripps, R.E. 2009. Metabolic engineering of *Geobacillus thermoglucosidasius* for high yield ethanol production. *Metab Eng* 11: 398–408.

Csatorday, K., Maccoll, R., and Berns, D.S. 1981. Accumulation of protoporphyrn IX and Zn protoporphyrin IX in *Cyanidium caldarium*. *Proc Nati Acad Sci USA* 78 (3): 1700–1702.

Desbruyères, D., and Laubier, L. 1980. *Alvinella pompejana* gen. sp. nov., Ampharetidae aberrant des sources hydrothermales de la ride Est-Pacifique. *Oceanologica Acta* 3 (3): 267–274.

de Swaaf, M.E., Pronk, J.T., and Sijtsma, L. 2003. Fed-batch cultivation of the docosahexaenoic-acid-producing marine alga *Crypthecodinium cohnii* on ethanol. *Appl Microbiol Biotechnol* 61 (1): 40–43.

Di Giulio, M. 2005. A comparison of proteins from *Pyrococcus furiosus* and *Pyrococcus abyssi*: Barophily in the physicochemical properties of amino acids and in the genetic code. *Gene* 346: 1–6.

Dolan, J.R. 2016. Marine protists: Diversity and dynamics. *J Eukaryot Microbiol* 63: 410.

Ducat, D.C., Way, J.C., and Silver, P.A. 2011. Engineering cyanobacteria to generate high-value products. *Trends Biotechnol* 29: 95–103.

Durvasula, R., Hurwitz, I., Fieck, A. et al. 2015. Culture, growth, pigments and lipid content of *Scenedesmus* species, an extremophile microalga from Soda Dam, New Mexico in wastewater. *Algal Res* 10: 128–133.

Eichler, J. 2001. Post-translational modification unrelated to protein glycosylation follows translocation of the S-layer glycoprotein across plasma membrane of the haloarchaeon *Haloferax volcanii*. *Eur J Biochem* 2568: 4366–4373.

Ewert, M., and Deming, J.W. 2013. Sea ice microorganisms: Environmental constraints and extracellular responses. *Biology* 2 (2): 603–628.

Feller, G., and Gerday, V. 2003. Psychrophilic enzymes: Hot topics in cold adaptation. *Nat Microbiol* 1: 200–208.

Foflonker, F., Ananyev, G., Qiu, H. et al. 2016. The unexpected extremophile: Tolerance to fluctuating salinity in the green alga *Picochlorum*. *Algal Res* 16: 465–472.

Fox, D. 2014. Lakes under the ice: Antarctica's secret garden. *Nature* 512: 244–246.

Gabani, P., Prakash, D., and Singh, O.V. 2014. Bio-signature of ultraviolet-radiation-resistant extremophiles from elevated land. *Am J Microbiol Res* 2 (3): 94–104.

Gao, J., Bauer, M.W., Shockley, K.R. et al. 2003. Growth of hyperthermophilic archaeon *Pyrococcus furiosus* on chitin involves two family 18 chitinases. *Appl Environ Microbiol* 69 (6): 3119–3128.

Gerday, C., and Glansdorff, N. (eds.). 2007. *Physiology and Biochemistry of Extremophiles*. Washington, DC: ASM Press.

Giddings, L., and Newman, D.J. 2015. *Bioactive Compounds from Extremophiles: Genomic Studies, Biosynthetic Gene Clusters, and New Dereplication Methods*. Springer Briefs in Microbiology: Extremophilic Bacteria. Berlin: Springer.

Golyshina, O.V., Pivovarova, T.A., Karavaiko, G.I. et al. 2000. *Ferroplasma acidiphilum* gen. nov., sp. nov., an acidophilic, autotrophic, ferrous-iron-oxidizing, cell-wall-lacking, mesophilic member of the Ferroplasmaceae fam. nov., comprising a distinct lineage of the Archaea. *Int J Syst Evol Microbiol* 50 (3): 997–1006.

Gould, S.J. 2014. Bacteria evolve with no limit in sight. *Cell* 156: 1119–1121.

Guezennec, J. 2002. Deep-sea hydrothermal vents: A new source of innovative bacterial exopolysaccharides of biotechnological interest? *J Ind Microbiol Biotechnol* 29: 204–208.

Guiry, M.D. 2012. How many species of algae are there? *J Phycol* 48: 1057–1063.

Hamilton, S.C., Farchaus, J.W., and Davis, M.C. 2001. DNA polymerases as engines for biotechnology. *Biotechniques* 31: 370–376, 378–380, 382–373.

Haroon, M.F., Thompson, L.R., Parks, D.H. et al. 2016. A catalogue of 136 microbial draft genomes from Red Sea metagenomes. *Sci Data* 3: 160050.

Hashimoto, T., Horikawa, D.D., Saito, Y. et al. 2016. Extremotolerant tardigrade genome and improved radiotolerance of human cultured cells by tardigrade-unique protein. *Nat Commun* 7: 12808.

Heidorn, T., Camsund, D., Huang, H.-H. et al. 2011. Synthetic biology in cyanobacteria: Engineering and analyzing novel functions. *Methods Enzymol* 497: 539–579.

Hoover, R.B., and Pikuta, E.V. 2010. Psychrophilic and psychrotolerant microbial extremophiles in polar environments. NASA STI, CRS Press, Document ID 20100002095. Washington, DC: NASA.

Horikoshi, K., and Grant, W.D. (eds.). 1998. *Extremophiles—Microbial Life in Extreme Environments*. Paris: Wiley-Liss.

Horneck, G. 2000. The microbial world and the case for Mars. *Space Sci* 48: 1053–1063.

Jin, E.S., Feth, B., and Melis, A. 2003. A mutant of the green alga *Dunaliella salina* constitutively accumulates zeaxanthin under all growth conditions. *Biotechnol Bioeng* 81: 115–124.

Johnson, D.B., and Hallberg, K.B. 2008. Carbon, iron and sulfur metabolism in acidophilic micro-organisms. *Adv Microb Physiol* 54: 202–256.

Jones B.E., Grant, W. Duckworth, A.W. et al. 1998. Microbial diversity of soda lakes. *Extremophiles* 2: 191–200.

Kashefi, K., and Lovely, D.R. 2003. Extending the upper temperature limit for life. *Science* 301: 934.

Keller, M.W., Schut, G.J., Lipscomb, G.L. et al. 2013. Exploiting microbial hyperthermophilicity to produce an industrial chemical, using hydrogen and carbon dioxide. *Proc Natl Acad Sci USA* 110: 5840–5845.

Kottmeier, S.T., and Sullivan, C.W. 1988. Sea ice microbial communities (SIMCO). IX. Effects of temperature and salinity on rates of metabolism and growth of autotrophs and heterotyrophs. *Polar Biol* 8: 293–304.

Krembs, C., Eicken, H., and Deming, J.W. 2011. Exopolymer alteration of physical properties of sea ice and implications for ice habitability and biogeochemistry in a warmer Arctic. *Proc Natl Acad Sci USA* 108: 3653–3658.

Kumar, L., Awasthi, G., and Singh, B. 2011. Extremophiles: A novel source of industrially important enzymes. *Biotechnology* 10: 121–135.

Laybourn-Parry, J., Tranter, M., and Hodson, A.J. 2012. *The Ecology of Snow and Ice Environments*. Oxford: Oxford University Press.

Lebbink, J.H.G., Kaper, T., Bron, P. et al. 2000. Improving low-temperature catalysis in the hyperthermostable *Pyrococcus furiosus* β-glucosidase CelB by directed evolution. *Biochemistry* 39 (13): 3656–3665.

Lentzen, G., and Schwarz, T. 2006. Extremolytes: Natural compounds from extremophiles for versatile applications. *Appl Microbiol Biotechnol* 72 (4): 623–634.

Lorentzen, E., Hensel, R., Knura, T. et al. 2004. Structural basis of allosteric regulation and substrate specificity of the non-phosphorylating glyceraldehyde 3-phosphate dehydrogenase from *Thermoproteus tenax*. *J Mol Biol* 341: 815–828.

MacDonald, A.G., Martinac, B., and Bartlett, D.H. 2002. High pressure experiments with the porins from the barophile *Photobacterium profundum* SS9. In H. Rikimaru (ed.), *Progress in Biotechnology*. Vol. 19. Amsterdam: Elsevier, pp. 311–316.

MacElroy, R.D. 1974. Some comments on the evolution of extremophiles. *Biosystems* 5: 74–75.

Madigan, M.T., and Oren, A. 1999. Thermophilic and halophilic extremophiles. *Curr Opin Microbiol* 2: 265–269.

Marchant, J. 2017. Giant iceberg's split exposes hidden ecosystem. *Nature* 549(7673): 443. doi: 10.1038/549443a.

Marco, D. (ed.). 2011. *Metagenomics: Current Innovations and Future Trends*. Poole, UK: Caister Academic Press.

Margesin, R., and Schinner, F. 2001. Biodegradation and bioremediation of hydrocarbons in extreme environments. *Appl Microbiol Biotechnol* 56: 650–663.

McInerney, M.J., Rohlin, L., Mouttaki, H. et al. 2007. The genome of *Syntrophus aciditrophicus*: Life at the thermodynamic limit of microbial growth. *Proc Natl Acad Sci USA* 104 (18): 7600–7605.

Meeks, J.C., and Castenholtz, R.W. 1971. Growth and photosynthesis in an extreme thermopile, *Synechococcus lividis* (Cyanophyta). *Arch Mikro* 78: 25–41.

Mermelstein, L.D., and Zeikus, J.G. 1998. Anaerobic nonmethogenic extremophiles. In K. Horokoshi and W.D. Grant (eds.), *Extremophiles: Microbial Life in Extreme Environments*. New York: Wiley-Liss, pp. 255–284.

Miller, S.R., Wingard, C.E., and Castenholtz, R.W. 1998. Effects of visible light and UV radiation on photosynthesis in a population of a hot spring cyanobacterium, *Synechococcus* sp., subjected to high-temperature stress. *Appl Environ Microbiol* 64 (10): 3893–3899.

Montalvo-Rodríguez, R., López-Garriga, J., Vreeland, R.H. et al. 2000. *Haloterrigena thermotolerans* sp. nov., a halophilic archaeon from Puerto Rico. *Int J Syst Evol Microbiol* 50: 1065–1067.

Mora, C., Tittensor, D.P., Adl, S. et al. 2011. How many species are there on earth and in the ocean? *PLoS Biol* 9 (8): e1001127.

Nadir, N., Zer, H., Shochat, S. et al. 2005. *Photoinhibition—A historical perspective*. In Govindjee, J.T. Bealty, H. Gest, and J.F. Allen (eds.), *Advances in Photosynthesis and Respiration*, vol. 20, *Discoveries in Photosynthesis*. Dordrecht: Springer, pp. 931–958.

Offre, P., Spang, A., and Schleper, C. 2013. Archaea in biogeochemical cycles. *Ann Rev Microbiol* 67: 437–457.

Olsson, S., Puente-Sánchez, F., Gómez, M.J. et al. 2015. Transcriptional response to copper excess and identification of genes involved in heavy metal tolerance in the extremophilic microalga *Chlamydomonas acidophila*. *Extremophiles* 19 (3): 657–672.

Onofri, S., Selbmann, L., Zucconi, L. et al. 2004. Antarctic microfungi as models for exobiology. *Planet Space Sci* 52: 229–237.

Oren, A. 1999. *Microbiology and Biogeochemistry of Hypersaline Environments*. Boca Raton, FL: CRC Press.

Oren, A. 2002. Molecular ecology of extremely halophilic Archaea and Bacteria. *FEMS Microbiol Ecol* 39: 1–7.

Paje, M.L.F., Neilan, B.A., and Couperwhite, I. 1997. A *Rhodoccus* species that thrives on medium saturated with liquid benzene. *Microbiology* 143: 2975–2981.

Paniagua-Michel, J. de. J., Olmos-Soto, J., and Morales, E.R. 2014. Algal and microbial exopolysaccharides: New insights as biosurfactants and bioemulsifiers. *Adv Food Nutr Res* 73: 221–257.

Pierson, B.K., and Castenholtz, R.W. 1974. A phototrophic gliding filamentous bacterium of hot springs, *Chloroflexus aurantiacus*, gen. and sp. nov. *Arch Microbiol* 100: 5–24.

Puente-Sánchez, F., Pieper, D.H., and Arce-Rodríguez, A. 2016. Draft genome sequence of the deep-subsurface actinobacterium *Tessaracoccus lapidicaptus* IPBSL-7^T. *Genome Announc* 4 (5): e01078-16.

Radakovits, R., Jinkerson, R.E., Darzins, A. et al. 2010. Genetic engineering algae for enhanced biofuel production. *Eukaryot Cell* 9: 486–501.

Rainey, F.A., and Oren, A. (eds.). 2006. Microorganisms and the methods to handle them. In F. Rainey and A. Oren (eds.), *Methods in Microbiology: Extremophiles*. Vol. 35. San Diego: Elsevier, pp. 1–25.

Raymond, J.A. 2011. Algal ice-binding proteins change the structure of sea ice. *Proc Natl Acad Sci USA* 108 (24): E198.

Reid, J.C., Lewis, H., Trejo, E. et al. 2013. Protein adaptations in archaeal extremophiles. *Archaea* 2013: 373275.

Roberts, D.M.L. 1999. Eukaryotic cells under extreme conditions. In J. Seckbach (ed.), *Enigmatic Microorganisms and Life in Extreme Environments*. London: Kluwer Academic, pp. 165–173.

Rosenberg, J.N., Oyler, G.A., Wilkinson, L. et al. 2008. A green light for engineered algae: Redirecting metabolism to fuel a biotechnology revolution. *Curr Opin Biotechnol* 19: 430–436.

Rossi, M., Ciaramella, M., Cannio, R. et al. 2003. Extremophiles. *J Bacteriol* 185: 3683–3689.

Rothschild, L.J., and Mancinelli, R.L. 2001. Life in extreme environments. *Nature* 409: 1092–1101.

Sarma, D.S., Coker, J.A., and DasSarma, P. 2010. Archaea—Overview. In M. Schaechter (ed.), *Desk Encyclopedia of Microbiology*. 2nd ed. Amsterdam: Academic Press, pp. 118–139.

Satyanarayana, T., Raghukumar, C., and Shivaji, S. 2005. Extremophilic microbes: Diversity and perspective. *Curr Sci* 89: 78–90.

Schloss, P.D., and Handelsman, J. 2003. Biotechnological prospects from metagenomics. *Curr Opin Biotechnol* 14 (3): 303–310.

Seckbach, J. 1999. The Cyanidiophyceae: Hot spring algae. In J. Seckbach (ed.), *Enigmatic Microorganisms and Life in Extreme Environments*. Dordrecht: Kluwer Academic, pp. 425–435.

Seckbach, J. 2000. Acidophilic microorganisms, In J. Seckbach (ed.), *Journey to Diverse Microbial Worlds*. Dordrecht: Kluwer Academic, pp. 107–116.

Seckbach, J., Oren, A., and Stan-Lotter, H. (eds.). 2013. *Polyextremophiles: Life under Multiple Forms of Stress*. Vol. 27, 1st ed. Dordrecht: Springer.

Sehgal, A.C., and Kelly, R.M. 2003. Strategic selection of hyperthermophilic esterases for resolution of 2-arylpropionic esters. *Biotechnol Prog* 19: 1410–1416.

Shaish, A., Harari, A., Kamari, Y. et al. 2009. *Application of Dunaliella in Atherosclerosis*. In A. Ben-Amotz, J.E.W. Polle, and D.V. Subba Rao (eds.), *Alga Dunaliella Biodiversity, Physiology, Genomics and Biotechnology*. Enfield, NH: Science Publishers, pp. 475–494.

Smith, H.J., Schmit, A., Foster, R. et al. 2016. Biofilms on glacial surfaces: Hotspots for biological activity. *Biofilms Microbiomes* 2: 16008.

Story, S.V., Grunden, A.M., and Adams, M.W. 2001. Characterization of an aminoacylase from the hyperthermophilic archaeon *Pyrococcus furiosus*. *J Bacteriol* 183 (14): 4259–4268.

Subba Rao, D.V., Pan, Y., and Al-Yamani, F. 2005. Growth, and photosynthetic rates of *Chalmydomonas plethora* and *Nitzschia frustula* cultures isolated from Kuwait Bay, Arabian Gulf and their potential as live algal food for tropical mariculture. *Mar Ecol* 26: 63–71.

Tang, K.H., Barry, K., Chertkov, O. et al. 2011. Complete genome sequence of the filamentous anoxygenic phototrophic bacterium *Chloroflexus aurantiacus*. *BMC Genomics* 29 (12): 334.

Taylor, M.P., Bauer, R., Mackay, S. et al. 2012. Extremophiles and their application to biofuel research. *Extremophiles* 233–265.

Thomas, D.N., and Dieckmann, G.S. 2002. Antarctic sea ice—A habitat for extremophiles. *Science* 295: 641–644.

Uhlig, C., Kilpert, F., Frickenhaus, S. et al. 2015. In situ expression of eukaryotic ice-binding proteins in microbial communities of Arctic and Antarctic sea ice. *ISME J* 9 (11): 2537–2540.

Van de Vossenberg, J.L., Driessen, A.J., and Konings, W.N. 1998. The essence of being an extremophile: The role of the unique archaeal membrane lipids. *Extremophiles* 2: 163–170.

Ventosa, A., and Nieto, J.J. 1995. Biotechnological applications and potentialities of halophilic microorganisms. *World J Microbiol Biotechnol* 11: 85–94.

Vierros, M., McDonald, A., and Salvatore, A. 2012. Oceans and sustainability: The governance of marine areas beyond national jurisdiction. In *Green Economy and Good Governance for Sustainable Development: Opportunities, Promises and Concerns*. Tokyo: United Nations University Press, pp. 221–244.

von Oppen-Bezalel, L., and Shaish, A. 2009. Application of the colorless carotenoids, phytoene, and phytofluene in cosmetics, wellness, nutrition, and therapeutics. In A. Ben-Amotz, J.E.W. Polle, and D.V. Subba Rao (eds.), *The Alga Dunaliella*. Enfield, NH: Science Publishers, pp. 423–444.

Ward, J.V., Tockner, K., Arscott, D.B. et al. 2002. Riverine landscape diversity. *Freshwater Biol* 47: 517–539.

White, R.A., III, Grassa, C.J., and Suttle, C.A. 2013. Draft genome sequence of *Exiguobacterium pavilionensis* strain RW-2, with wide thermal, salinity, and pH tolerance, isolated from modern freshwater microbialites. *Genome Announc* 1 (4): e00597-13.

Whitman, W.B., Coleman, D.C., and Wiebe, W.J. 1998. Prokaryotes: The unseen majority. *Proc Natl Acad Sci USA* 95: 6578–6583.

Wright, P.C., Westacott, R.E., and Burja, A.M. 2003. Piezotolerance as a metabolic engineering tool for the biosynthesis of natural products. *Biomol Eng* 20: 325–331.

Yang, S.J., Lee, H.-S., Park, C.-S. et al. 2004. Enzymatic analysis of an amylolytic enzyme from the hyperthermophilic archaeon *Pyrococcus furiosus* reveals its novel catalytic properties as both an α-amylase and a cyclodextrin-hydrolyzing enzyme. *Appl Environ Microbiol* 70: 5988–5995.

Zeng, X., Birrien, J.L., Fouquet, Y. et al. 2009. *Pyrococcus* CH1, an obligate piezophilic hyperthermophile: Extending the upper pressure-temperature limits for life. *ISME J* 3 (7): 873–876.

Zheng, H., and Wu, H. 2010. Gene-centric association analysis for the correlation between the guanine-cytosine content levels and temperature range conditions of prokaryotic species. *BMC Bioinformatics* 11 (Suppl. 11): S7.

CHAPTER 2

Microbial Diversity and Biotechnological Potential of Microorganisms Thriving in the Deep-Sea Brine Pools

Alan Barozzi, Francesca Mapelli, Grégoire Michoud, Elena Crotti,
Giuseppe Merlino, Francesco Molinari, Sara Borin, and Daniele Daffonchio

CONTENTS

2.1 INTRODUCTION

In the last few years, extreme environments have attracted the attention of microbiologists since they have been described to host unique microbial communities able to flourish in harsh physio-chemical conditions. Microorganisms adapted to thrive in habitats that are considered hostile for the majority of living forms have been indicated as microbial extremophiles. According to the different and specific stressful conditions present in the extreme environments, extremophiles are classified as acidophiles or alkalophiles, thermophiles or psychrophiles, barophiles and piezophiles, halophiles, metalophiles, and radiophiles (Raddadi et al. 2015). Some authors have highlighted the hypothesis that life in early earth had been arisen in conditions similar to the ones occurring in extreme environments, such as deep-sea hydrothermal vents (DSHVs) or deep hypersaline anoxic basins (DHABs) (Shock 1992; van der Wielen et al. 2005; Cockell 2006), sustaining speculation about the first common ancestor evolved on our planet or about the possibility of life in extraterrestrial bodies. Research on extremophilic isolates is important to define the limit of prokaryotic life on our planet; recently, the upper limits for life detected so far in terms of high temperature (122°C) and pressure (120 MPa) have been registered for two extremophiles isolated from DSHVs, that is, *Methanopyrus kandleri* strain 116 and *Pyrococcus yayanosii* CH1, respectively (Takai et al. 2008;

Zeng et al. 2009). Furthermore, beyond the interest in the investigation of the microbial diversity associated with these peculiar ecosystems, aiming also to shed light on the functioning of the systems themselves, the attention toward extremophiles has grown since they represent a source of new biotechnological potential (Antranikian et al. 2005; Adrio and Demain 2014; Raddadi et al. 2015).

Extremophiles produce extremophilic enzymes, namely, extremozymes, and/or accumulate protective organic biomolecules, that is, extremolytes, that allow them to hinder the extreme physiochemical conditions (Raddadi et al. 2015). Hence, extremophile organisms represent a potential source of novel enzymes or molecules with specific properties: not only have whole microbial cells, purified enzymes, cell extracts, and immobilized cells attracted the attention of scientists (mainly for their use in chemical reactions as biocatalysts), but also extremolytes are of interest. For example, the latter find remarkable applications in the field of cosmetics, being potentially useful for the pharmaceutical sector or the food industry (Raddadi et al. 2015). Nowadays, enzymes are applied in many industrial, pharmaceutical, and biotechnological processes. In a recent report, BBC Research, a market research company that monitors modifications driven by science and technology, published that the global market for industrial enzymes reached approximately US$4.8 billion in 2013, and a forecast of US$7.1 billion has been made for 2018 (BBC Research 2011). Sixty percent of the global market for enzymes is held by proteases, which are used in detergent, pharmaceutical, food, leather, silk, and agrochemical applications (Adrio and Demain 2014). Due to their unique properties, extremophiles can be exploited in diverse sectors, encompassing bioremediation and biofuel applications, as well as pharmaceutical and medical ones (Raddadi et al. 2015). For instance, thermophilic proteases, lipases, and cellulases have been introduced in several industrial applications (Adrio and Demain 2014). Another well-known example of an extremozyme with a market of US$500 million is the Taq DNA polymerase, obtained from the thermophilic archaeal microorganism *Thermus aquaticus*, which revolutionized the molecular biology–related sector (De Carvalho 2011).

Oceans and seas cover nearly three-quarters of our planet surface, containing a huge microbial diversity and hosting many marine extreme environments. As highlighted in the European Science Foundation reports, marine extremophiles are a resource of inestimable biotechnological potential and are thus receiving great attention, from the so-called "blue growth" perspective, to develop novel marine biotechnologies as a strategy to sustain a bio-based economy (Børresen et al. 2010; Dixon et al. 2011). Among the different extreme marine environments, in this chapter we focus on hypersaline habitats located on the seafloor, the abovementioned DHABs. Being ones of the most challenging marine habitats found on earth, their unique features greatly affect the distribution, structure, and richness of the microbial communities living therein. DHABs originate from the exposure of subsurface-stratified evaporitic rocks to the seawater, and due to the difference in water density, the mixing between the brine and the overlying seawater is prevented. This results in the formation of stable brines, characterized by reducing conditions, the absence of light and oxygen, high pressure (up to 38 MPa), and hypersalinity (5–10 times higher than seawater), with the presence of a sharp halocline at the brine–seawater interface (van der Wielen et al. 2005; Daffonchio et al. 2006; Borin et al. 2009). All these features specialize in prokaryotic assemblages, making brine lakes a hot spot of microbial life with variable metabolic activities (Figure 2.1c). For instance, DHABs represent a source of halophile-derived enzymes with interesting applications in stereoselective biocatalysis (De Vitis et al. 2015). Halophiles have developed different strategies to survive under high salt concentrations, such as the ability to synthetize compatible solutes to keep the integrity of the cell structure and function and the capability to accumulate in the cytoplasm inorganic ions, allowing proteins to remain active and stable in the presence of salts. Both osmoprotectants and salt-adapted molecules could find applications in many different industrial sectors (Poli et al. 2010; Dalmaso et al. 2015; Raddadi et al. 2015). From this perspective, such microorganisms can be grown in bioreactors to obtain the required biomass or to produce molecules of interest. However, one of the main constraints for using these extremophiles, and in general many marine microorganisms in fermentation technology, is their general poor cultivability (Joint et al. 2010; Prakash

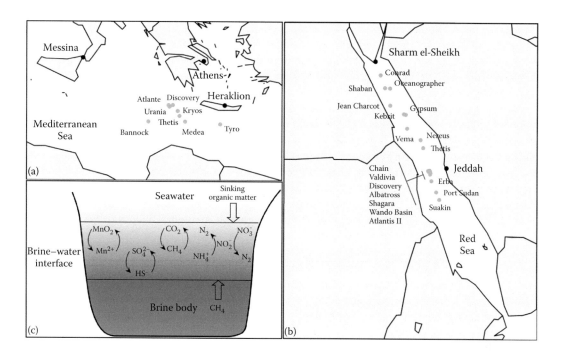

Figure 2.1 Localization of DHABs and general scheme of prokaryotic metabolisms occurring in the seawater–brine transition zones. (a) Locations of DHABs in the Mediterranean Sea. (b) Locations of DHABs in the Red Sea. (c) Simplified scheme showing the most common metabolisms present in the brine pools. From left to right are shown the manganese cycle, the sulfate reduction and sulfide oxidation cycle, the methanogenesis and aerobic (anaerobic) methane oxidation cycle, and the anammox and denitrification cycle. (From Abdallah, R. Z. et al., *Front. Microbiol.*, 5, 487, 2014; Antunes, A. et al., *Environ. Microbiol. Rep.*, 3, 416–433, 2011; Borin, S. et al., *Proc. Natl. Acad. Sci. U.S.A.*, 106, 9151–9156, 2009; Daffonchio, D. et al., *Nature*, 440, 203–207, 2006; Ngugi, D. K. et al., *ISME J.*, 9, 396–411, 2015.)

et al. 2013). Since the vast part of these microorganisms has not yet been cultivated in pure cultures, the characterization of their metabolic properties is limited to the availability of molecular data. Although nowadays the application of omic technologies is receiving great attention, the importance of methods based on microbial cultivation must not be underestimated. Cultivation-dependent techniques are useful to evaluate the microbial metabolic potential and physiology and, indeed, to exploit biotechnologically the microorganisms or the microbial-related biomolecules (Joint et al. 2010; Prakash et al. 2013). For this reason, several cultivation strategies have been developed and/or improved, such as the use of oligotrophic conditions in the media or growth systems that mimic natural environments, the addition of growth factors or other microbial-derived molecules, and *in situ* cultivation approaches (Prakash et al. 2013).

Here, we present an overview of the microbial diversity associated with DHABs, discussing the biotechnological potential of extremophiles in light of the environmental conditions they thrive in. In particular, DHABs placed in two different geographical locations have been considered, the Mediterranean (Figure 2.1a), and the Red Sea (Figure 2.1b). The last part of the chapter discusses the potential of genomics and metagenomics for mining new bioactive molecules for biotechnological applications.

2.2 MEDITERRANEAN DHABS

A large area indicated as "anoxic lake region" is located across the deep Eastern Mediterranean Ridge (Yakimov et al. 2013), an accretionary complex subject to the continental drift of the African,

Eurasian, and Anatolian plates. In this area, DHABs have been discovered since the 1980s. Bannock and Tyro were the first basins described (Jongsma et al. 1983; Cita et al. 1985), and about 10 years later, three additional basins were found in the same area and named Discovery, Urania, and L'Atalante (MEDRIFF Consortium 1995). In recent years, thanks to the intense seafloor exploration activities performed by the Italian National Research Council (CNR), the Medee, Thethis, and Kryos basins were discovered (La Cono et al. 2011; Yakimov et al. 2013, 2015), increasing the number of known Mediterranean DHABs up to eight.

The origin of these basins is due to the dissolution of evaporites, large buried salt deposits formed during the Messinian salinity crisis, about 5 million years ago (Camerlenghi 1990). The dissolution occurred after tectonic events, according to two possible scenarios. The first is the dissolution of evaporites after contact with seawater due to the collision of the African and Eurasian plates, while the second consists of the release of interstitial entrapped evaporated seawater from the subbottom evaporite deposits (Vengosh and Starinsky 1993; Vengosh et al. 1998; Cita 2006; Briand 2008). The geochemistry of the brine strongly differs between the basins. The most clamorous example is the athalassohaline brine of the Discovery basin, which shows a magnesium concentration up to 10 times higher than the concentrations recorded in the other DHAB brines (van der Wielen et al. 2005). Until the recent discovery of the Kryos basin, Discovery was the only athalassohaline DHAB of the Mediterranean Sea, and it has a nearly saturated $MgCl_2$ brine, which makes this basin the marine environment displaying the highest $MgCl_2$ concentration ever detected on earth.

2.2.1 Microbial Diversity

The occurrence in DHAB brines of several multiple extreme conditions (i.e., anoxia, pressure, and salinity) determines the selection of highly adapted microorganisms, very diverse from those inhabiting the overlying seawater column. The brine pools were well separated by a density difference from the overlying seawater, due to a stable interface hosting picno-, oxy-, and chemoclines. Salinity gradually increases along a few meters' depth of the halocline interface (1–3 m, except in the Medee basin), reaching in the brines values up to 7–10 times higher than those in the overlying seawater. Cluster analysis of the chemical characteristics of the Bannock, Discovery, L'Atalante, and Urania brines was perfectly mirrored by the same analysis of the prokaryote dataset (van der Wielen et al. 2005). The presence of specific microbial assemblages in different DHABs highlights a strong selection determined by the geochemical features of the different basins. The interfaces of the Mediterranean brine pools have been finely dissected through an accurate sampling strategy initially set up to study the stratification of the prokaryote network in the Bannock interface (Daffonchio et al. 2006). Across the seawater–brine interface depth profile, gradients of nutrients' concentration correlate with the distribution of specific microbial populations and indicate the microbial production or consumption of the different chemical species used for redox cycling. Due to the changing electron donor and acceptor concentrations, seawater–brine interfaces are hot spots of prokaryote abundance, with a dominance of unique bacterial lineages (Daffonchio et al. 2006; Yakimov et al. 2007). Bannock interface was specifically inhabited by bacterial Mediterranean Sea Brine Lake lineages 2–6 (MSBL2–6 lineages) and the archaeal lineage MSBL1. Among the bacterial lineages, MSBL2 showed high similarity to the SB1 detected in a brine pool of the Red Sea. All the bacterial MSBL lineages were retrieved in the deeper layers of the Bannock interface, in correspondence to the high salinity values to which they are apparently well adapted (Daffonchio et al. 2006). In addition, the anoxic deep layers were largely colonized by sulfate reducers belonging to Deltaproteobacteria, comprising families differently distributed according to salinity values. The Archaea MSBL1 and those belonging to the division ANME-1, responsible for anaerobic oxidation of methane (AOM), were also abundant in anoxic layers (Daffonchio et al. 2006), a result that indicates the importance of sulfur and methane cycles. Sulfur cycling and methanogenesis were also found to be the main drivers of prokaryotic communities in the interface of the Urania basin, one of the most sulfidic marine water bodies on the planet (Borin et al. 2009). Activity measurements and chemical and microbiological data collected in the Urania basin vertical profile identified the archaeal MSBL1 as a group of methanogens highly adapted to the extreme conditions of DHABs (Borin et al. 2009). Delta- and Epsilonproteobacteria, players of the sulfur cycle known as sulfate-reducing (SRB) and sulfide-oxidizing (SOB) bacteria, respectively, were found to be additional important components of the Urania chemocline (Borin et al. 2009). A study of the gene for the dissimilatory sulfite reductase *dsrA* along the Urania and L'Atalante vertical profile showed that SRB comprise highly diverse populations (van der Wielen and Heijs 2007). As observed in Urania and Bannock, and also in the case of L'Atalante, a stratification of Proteobacteria was detected according to the distribution of sulfur chemical species. Deltaproteobacteria were abundant in deep interface layers and brine, while Epsilon- and Gammaproteobacteria dominated the upper and intermediate depths of the interface (Yakimov et al. 2007). Gammaproteobacteria were indicated as the putative responsible for the high level of CO_2 fixation detected in L'Atalante (Yakimov et al. 2007). Archaeal zonation was also described across the L'Atalante basin, where the upper interface was dominated by Crenarchaeota, while Euryarchaeota inhabit the lower layers (Yakimov et al. 2007).

Little information is available about the prokaryotic diversity of Discovery. It was suggested that its brine hosts prokaryotic life, due to the detection of 16S rRNA signatures (van der Wielen et al. 2005). However, the extremely high magnesium concentration in this basin is incompatible with life (Hallsworth et al. 2007). A recent 16S rRNA pyrosequencing study of the Discovery interface that dissected the halocline in three parts according to salinity values (upper, middle, and lower halocline) showed that only few operational taxonomic units (OTUs) are shared between the three levels (Kormas et al. 2015).

The relationship between prokaryotic community composition and the geochemical characteristics of the Mediterranean DHABs was obtained mainly by applying molecular methods, since cultivation-based approaches allowed the isolation of a limited number of bacterial and archaeal strains (Sass et al. 2001; Daffonchio et al. 2006; Borin et al. 2009; Sorokin et al. 2015), yet showed that living cells are thriving in the DHABs.

While most of microbial diversity studies focused on Prokarya, some studies shed light on the diversity and role of Eukarya (Alexander et al. 2009; Filker et al. 2013; Stoeck et al. 2014) and viruses in DHABs. It was recently shown that virus-induced mortality was responsible for 85% of the extracellular DNA detected in DHAB sediments, where it can be preserved despite a high DNAse activity, indicating the possible role of extracellular DNA as a trophic source (Corinaldesi et al. 2014). Extracellular DNA was mainly derived from Proteobacteria and Euryarchaeota, and differed from that extracted from the microbial cells in both L'Atalante and Medee sediments (Corinaldesi et al. 2014). The ability of plasmid DNA to be preserved in DHAB brines and retain transformability potential was demonstrated by transformation experiments of naturally competent *Acinetobacter baylii* BD413 cells, after incubation for up to 32 days in the brines (Borin et al. 2008). The high amount of extracellular DNA detected in DHAB sediments could play an important role as a gene reservoir for horizontal gene transfer (HGT), potentially affecting biotechnological trait dispersal within the DHAB microbiomes and constituting a possible source of genetic information exploitable by heterologous expression studies.

2.2.2 Biotechnological Potential

Despite the advances in high-throughput sequencing technologies that have opened a great perspective in the exploitation of metabolites from the DHAB microbiomes (discussed in Section 2.4), only a few studies have focused on the biotechnological potential of DHAB extremophiles.

Ferrer and colleagues (2005) investigated the seawater–brine interface of the Urania basin to unravel the occurrence of new enzymatic diversity, choosing esterases as model enzymes. After collecting the water samples from the Urania interface, they added crude oil to a subsample and performed a long-term incubation at low temperature, mimicking conditions occurring in Mediterranean DHABs. By screening for esterase activity a metagenomic expression library in *Escherichia coli* obtained from the microcosms' DNA, five novel genes with esterase activity were identified and the relative enzymes were purified. The specificity of these enzymes was assessed using triglycerols and p-nitrophenyl esters (p-NP-esters) with different acyl chain lengths. The decreasing specificity occurring as the chain length increased suggested that the novel enzymes were not lipases. Using p-NP-esters, the authors identified temperature and pH optima for the activity in the ranges of 40°C–60°C and 8–9, respectively. One of the enzymes retained 80% of its maximum activity at pH 12. The authors also tested the activity of the enzymes after exposure to pressure values between 0 and 40 MPa. One of the enzymes showed increasing activity at higher hydrostatic pressure, with a maximum of 20 MPa. All the enzymes were stable in nonpolar and medium polar solvents, while some were also stable in polar solvents. One of the esterases presented high enantioselectivity for the hydrolysis of synthon solketal, an important intermediate for the synthesis of therapeutic drugs. Following structural analysis, one of the novel enzymes resulted in the largest known esterase and presented a high structural complexity with three catalytic sites. Phylogenetic analysis suggested that three of the esterases affiliated with Gram-negative bacteria and one with single-cell eukaryotes, while two of them did not show significant similarity with any known taxa (Ferrer et al. 2005).

In another study, De Vitis and colleagues (2015) screened a collection of 33 strains of 10 Gammaproteobacteria and Firmicutes genera isolated from Bannock, Discovery, L'Atalante, and Urania, aiming to identify strains able to perform the catalytical resolution of racemic propyl ester of anti-2-oxotricyclo[2.2.1.0]heptan-7-carboxylic acid ((R,S)-**1**), a key intermediate for the synthesis of D-cloprostenol. Since for the synthesis of this prostaglandin only the (7*R*)-enantiomer of (R,S)-**1** is of interest, a stereoselective biocatalytic tool could be of great interest. An initial test on marine broth medium using growing cells could resolve (R,S)-**1** through esterase (six strains) or ketoreductase (eight strains) activities. Four strains showed both enantioselective esterase and ketoreductase activity. The experimental conditions were optimized using different media, assessing the optimal salt concentration and the use of resting cells. Higher conversion and enantioselectivity rates were obtained using a richer medium containing 3% NaCl. In the absence of salt, the best-performing strain was *Bacillus horneckiae* 15A, which showed the highest enantiomeric ratio and high rates, reaching 50% conversion in two hours. The study showed that strain 15A provided highly stereoselective reduction of (R,S)-**1**, while *Halomonas aquamarina* 9B enantioselectively hydrolyzed (R,S)-**1**. Given the yield to which (R)-**1** can be produced, easily recovered, and purified, these two bacterial strains further exemplify the potential application of DHAB-derived enzymes for stereoselective biocatalysis (De Vitis et al. 2015).

Overall, the studies described in this section indicate Mediterranean DHABs as a reservoir of novel interesting enzymes capable of high enantioselectivity, and active in a large range of pH, temperature, and pressure conditions.

2.3 RED SEA DHABS

Along the central axis of the Red Sea, between latitudes of 19°N and 27°N, about 25 brine bodies are present in a range of depths between 1196 and 2850 m below sea level (Guan et al. 2015). The first brine pools of the Red Sea were discovered in the 1960s (Swallow and Crease 1965; Miller et al. 1966), and in the following 30 years, further research expeditions revealed the current known number of DHABs (Pautot et al. 1984; Hartmann et al. 1998).

The Red Sea represents an active rift system generated by the divergent movement of the Arabian and African tectonic plates, in a process started approximately 25 million years ago (Schardt 2015). The high number of DHABs is closely related to the tectonic activity present in the region, similar to that observed in the Mediterranean Sea. Coupled with the formation of new oceanic crust, thick evaporitic deposits accumulated during the Late Miocene underwent redissolution in several points of the central Red Sea rift system. In the brine pool trenches, the evaporite dissolution remarkably increased deep water density, determining distinct stratification from the overlying water column. Metal-bearing hydrothermal flows that remain entrapped in the dense layer of the Red Sea brine pools generate highly mineralized brines (Gurvich 2006). A relevant consequence of this formation process is reflected in the different geochemical composition of each basin and is related to the nature of the mineral deposits and their metalliferous compositions (Gurvich 2006) (Figure 2.1c).

The presence of hydrothermal vents along the central axial rift of the Red Sea makes the brine basins rather dynamic and different from the Mediterranean ones, being affected by the hot fluid injections (Schardt 2015). The most remarkable and investigated basin is the Atlantis II Deep, located at a latitude of 21°N. In Atlantis II, the temperature of the brine reaches 68°C, and together with the nearby Discovery Deep, it represents a high temperature pool compared with other Red Sea DHABs (Hartmann et al. 1998; Swift et al. 2012). The brines in the majority of the other pool bodies are warmer than the normal deep water column (around 22°C), but do not exceed 26°C (Hartmann et al. 1998). This anomaly in the recorded temperature had been the trigger for the intensification of the Red Sea bottom investigation (Swallow and Crease 1965).

2.3.1 Microbial Diversity

Despite the presence of DHABs in the Red Sea being proved 50 years ago, a limited number of studies have investigated the microbial communities adapted to live in these challenging environments, and still now, not all the basins have been explored (Antunes et al. 2011). For a long time, the studies on DHABs in the Red Sea were mainly focused on the geochemistry of the pools, looking to evaluate the economic potential of the ores gathered in the mineral deposits, in correspondence with the brines. Atlantis II is the largest known hydrothermal mineral deposit on the Red Sea seafloor, with 20 m thick sedimentary succession, originated from the precipitation of metals from the brine body (Laurila et al. 2014). This large accumulation of minerals gives an idea of the entrapping capacity of metals, but also of the nutrients of this environment. The accumulation of metals and nutrients, especially in the seawater–brine interfaces (Larock et al. 1979; Eder et al. 2002; Guan et al. 2015), supports the presence of different ecological niches exploited by highly diverse microorganisms with specialized metabolisms, which are closely related to the geochemical composition of the basins. The following paragraphs summarize the state of the art of the microbial diversity of the Red Sea brine pools, focusing on Atlantis II, Discovery, Kebrit, Nereus, and Erba.

In most of the interfaces and the DHAB brine bodies of the Red Sea, the available data indicate that microbial communities are dominated by Bacteria rather than Archaea (Bougouffa et al. 2013; Wang et al. 2013; Guan et al. 2015), as previously observed in the Mediterranean (Daffonchio et al. 2006; Yakimov et al. 2015). On the contrary, the interfaces of the Atlantis II and Kebrit Deeps represent an exception, similar to the Urania basin in the Mediterranean Sea (van der Wielen et al. 2005). One hypothesis is that the high concentration of reduced sulfur species in the Kebrit Deep (150 µM H_2S) and the Urania basin has limited the presence of bacterial species. Moreover, the presence of fairly high dissolved oxygen concentrations in Kebrit and Atlantis II (four to nine times higher than other Red Sea hypersaline basins) (Guan et al. 2015) could allow the presence of ammonia-oxidizing archaea (AOA) (Abdallah et al. 2014; Ngugi et al. 2015). The archaeal community associated with the brine interfaces in all the considered basins is dominated by OTUs of the phylum Thaumarchaeota that cover from 64% (Erba Deep) to 99% (Atlantis II and Discovery Deeps) of the total archaeal community. The 16S rRNA gene pyrosequencing showed that only a single AOA phylotype, belonging to the class Shallow Marine Group I (SMGI) (Francis et al. 2005) and to the genus *Nitrosopumilus*, covered 98% of the archaeal diversity in the seawater–brine interfaces. The absence of this phylotype in the water columns and the presence of different genotypes in DHABs with different geochemical characteristics indicate the adaptation of these microbes to the brine environments (Guan et al. 2015; Ngugi et al. 2015).

A cluster analysis of archaeal communities, based on the Jaccard similarity index, has highlighted a different profile for the Kebrit Deep (Guan et al. 2015). This basin presents high salinity (26%), low pH (5.5), and a high concentration of H_2S and CO_2 (Antunes et al. 2011). In 2001, two microorganisms were isolated from the Kebrit Deep, representing the first culturable microorganisms from a DHAB. They were affiliated with the genus *Halanaerobium*, a group of anaerobic, halophilic, and fermenting bacteria (Eder et al. 2001).

The bacterial communities in the brines of all five characterized DHABs are very diverse, mainly represented by the phyla Proteobacteria, Bacteroidetes, Deferribacters, and Chloroflexi. In the Kebrit, Erba, and Nereus Deeps, the class of Deltaproteobacteria is common, while it is less abundant in the hot brines of the Atlantis II and Discovery Deeps. These two pools are mostly inhabited by Nitrospinae-like bacteria in the interface and by Gammaproteobacteria in the lower convective layers (Bougouffa et al. 2013; Guan et al. 2015).

Even though closely located, the Atlantis II and Discovery Deeps showed rather different geochemical profiles and residing microbial communities. Atlantis II is a hydrothermally active system where the temperature is still increasing, whereas Discovery is a hot stable system. As a consequence, Atlantis II presents four convective layers where the microbial communities are diversified (Bougouffa et al. 2013; Abdallah et al. 2014). Alpha diversity was higher in the lower convective layer of Atlantis II, where the temperature (68°C), pressure (20 MPa), and salinity (24.8%) conditions are more extreme than in any other parts of the brine pool water column. The authors attributed such a higher alpha diversity to the extreme conditions, which could prevent the dominance of specific phylotypes, increasing the relative proportion of the "rare biosphere" (Bougouffa et al. 2013).

2.3.2 Biotechnological Potential

As for the Mediterranean DHABs, as well as the Red Sea ones, only a few recent studies have explored the neglected potential of biotechnological resources in brine pools (Mohamed et al. 2013; Sagar et al. 2013a, 2013b).

The potential of novel biocatalysts, such as lipolytic enzymes resistant to environmental stresses such as temperature, salinity, acidic pH, and high concentration of heavy metal inhibitors, has been investigated in the deepest convective layer of Atlantis II (Mohamed et al. 2013). This layer presented very peculiar and extreme physiochemical features, like temperature around 68°C, salinity reaching 27%, a pH of 5.3, and high concentrations of heavy metals (Laurila et al. 2014). The screening of a fosmid library from the total DNA extracted from cells filtered from brines revealed five recombinant clones that formed a clear halo-zone on tributyrin agar, indicative of lipolytic activity (Mohamed et al. 2013). Fosmid insert sequencing revealed a 945 bp open reading frame (ORF) encoding for an esterase or lipase called EstATII and belonging to the hormone-sensitive lipase family. A catalytic domain, typical of the alpha- and beta-hydrolase family, and an esterase and lipase domain (HSL) were identified through bioinformatic analysis. Purification and biochemical characterization of the EstATII enzyme allowed us to obtain a halotolerant esterase with an optimum temperature of 65°C, able to work at a NaCl concentration of up to 4.5 M, in the presence of different metals. Compared with other esterases, such as EstR (Quyen et al. 2007), EstA (Chu et al. 2008), and EstH112 (Oh et al. 2012), the highest activity of EstATII, in different stressful conditions, makes it a very interesting enzyme, considering that the size of the lipolytic enzyme market is in the billion-dollar range (Hasan et al. 2006).

Sagar and colleagues (2013a, 2013b) focused on the screening of active biomolecules against different human cancer cell lines. Until now, more than 3000 natural molecules active on widely different diseases have been extracted from marine microbes and updated on the Antibase database

(Laatsch 2010). The innovative aspect of the works of Sagar and colleagues was the investigation of remote extreme environments, such as the brine pools, as a potential source of new bioactive molecules. The cytotoxic and apoptotic potential of brine bacterial extracts from the interfaces of the Atlantis II, Discovery, Kebrit, Nereus, and Erba Deeps were investigated (Sagar et al. 2013a). The effect of extracts from 12 isolates was tested on three different cancer cell lines in humans: HeLa (cervical carcinoma), MCF-7 (breast adenocarcinoma), and DU145 (prostate carcinoma). The majority of the isolates induced apoptotic and cytotoxic activity, and one of the extracts from a *Halomonas* sp. strain, isolated from the Nereus Deep, had a strong effect on the cervical carcinoma and breast adenocarcinoma cells.

In a further effort (Sagar et al. 2013b), the same group further screened human cancer cell line extracts obtained from 24 strains isolated from the seawater–brine interface (19 strains), brine (1 strain), and sediments (4 strains). Thirteen extracts induced apoptosis, and six of them could determine more than 70% mortality in HeLa cancer cells. The most effective bacterial extracts were derived from isolates belonging to the species *Chromohalobacter salexigens*, *Chromohalobacter israelensis*, *Halomonas meridiana*, and *Idiomarina loihiensis*, highlighting the wide diversity of brine pool microorganisms capable of bioactive molecule production (Sagar et al. 2013b).

2.4 GENOMICS AND METAGENOMICS FOR THE DISCOVERY OF NEW BIOACTIVE MOLECULES

The already limited cultivability of marine microorganisms is further challenged in the brine pools due to the extreme environmental conditions, which complicates the setup of suitable conditions for cultivation in the laboratory. This results in a low number of isolates described with respect to the diversity observed through molecular microbial ecology approaches, in both the Mediterranean and Red Sea brines. Cultivation attempts of the most abundant prokaryote groups were until now unsuccessful, and the isolates retrieved from the brine pools generally represent a minority of the diversity detected, since they are not observed in the cultivation-independent studies. This highlights the inadequacy of the actual cultivation approaches (Antunes et al. 2011).

Genomes and metagenomes may be exploited to discover traits of biotechnological interest, potentially overcoming the cultivation barrier (Owen et al. 2015). Metagenomic analysis can search for either phylogenetic (e.g., 16S rRNA) or functional signatures of specific pathways or enzymes in the total DNA directly extracted from an environmental matrix. Sequence-driven metagenomics is very useful for assessing the diversity of microorganisms in a specific environment, while the search for new enzymes should undergo a function-driven metagenomic approach (Handelsman 2004). The latter approach avoids incorrect annotation by recognizing genes by their function, rather than their sequences (Dalmaso et al. 2015). Single-cell genomics, metatranscriptomics, and metaproteomics can also be used to search for novel enzymes. Metatranscriptomic analyses permit us to separate active and inactive cells, by focusing on the transcriptionally active genes, whereas metaproteomics focuses directly on the enzymes present in the biochemical pathways of interest. Single-cell genomics allows us to study the entire biochemical process of only single uncultured cells, without needing to have them in a high-cell-density culture (Kennedy et al. 2010). As seen earlier in this chapter, the most studied brine pools by the application of the abovementioned methods, which on the whole are referred to as omic tools, are those in the Mediterranean Sea. Besides the previously mentioned functional metagenomics on the Urania DHAB, which discovered and expressed five novel esterases, Ferrer and coworkers (2012) investigated Lake Thetis through metagenomics. Sequence reconstruction revealed important enzymes involved in different pathways, including CO_2 fixation, sulfur reduction and sulfur oxidation pathways, iron storage, and carbon source assimilation. Analyses of the amino acid compositions of the encoded enzymes revealed an overrepresentation of acid residues in the brine proteome. This is typical of halophilic organisms and seems to be one of the main mechanisms adopted by microbes to cope with the extremely high salt concentration in the DHABs. Another interesting aspect in the metagenome obtained from the brine–seawater interface was the presence of three autotrophic carbon dioxide fixation pathways, which probably sustained an active sulfur cycle in the interface. In particular, the enzymes found to be part of the sulfide oxidation pathways, like the multienzymatic *Sox* complex, which catalyzes the complete oxidation of reduced sulfur compounds to sulfate, could be particularly interesting from a biotechnological perspective (Ferrer et al. 2012). Indeed, the environmental conditions of different industrial processes are similar to those of the brine pools (temperature, pH, and salt content), and the relevant enzymes might play interesting roles in bioprocesses such as wastewater treatments. In other cases, enzymatic bioconversion might contribute to keep the environmental conditions of the process within the physiological range of microorganisms. For instance, sulfate reduction consumes protons, and thus results in a pH increase of the fermentation medium (Lens Piet and Kuenen 2001).

Also, the genome sequencing of isolates obtained from brine pools is contributing to the identification of new functions or enzymes. The genome of the extremely halophilic archaeon *Halorhabdus tiamatea*, isolated from the Shaban Deep in the Red Sea, showed that it possesses genes coding for a trehalose synthase and a lactate dehydrogenase, contrary to the other members of this species that were isolated in less harsh environments (Antunes et al. 2011). A proteomic investigation on this strain showed a high number of glycoside hydrolases, the majority of which were expressed (Werner et al. 2014). From the sets of annotated genes, the protein expression profiles, and the measure of glycosidase activity, it was shown that *H. tiamatea* is adapted to degrade hemicellulose of algal and plant origin. These characteristics, together with the presence in the genome of genes coding for proteins typical of thermophiles and for bacteriorhodopsins, suggested that the evolutionary history of this archaea occurred in environments different from those of the brine pools. This strain and its eventual derivatives could be used for the enzymatic conversion of polysaccharides from lignocellulosic biomass, a major area in biorefinery research. Moreover, since *H. tiamatea* enzymes are adapted to resist high temperature and salinity, they might be interesting for industrial applications (Horn et al. 2012).

A single, highly abundant *Nitrosopumilus*-like phylotype has been found to dominate the archaeal community in five different brine–seawater interfaces in the Red Sea (Ngugi et al. 2015). By a combination of metagenomic and single-cell genomic techniques, it was shown that this novel phylotype possesses a putative proline-glutamate "switch," possibly implicated in osmotolerance, a trait of interest to combat desiccation in plants grown in arid conditions or in saline soils.

2.5 CONCLUSION

The microbial diversity of DHABs has received increased attention in recent years. The intrinsically extreme features of the brine pools make these habitats unique environments for highly specialized organisms. We have discussed the prokaryotic diversity in the known DHABs of the Mediterranean and Red Seas. Our analysis showed that each basin has a unique community, where phylogenetic and functional diversity are largely driven by the geochemical conditions of each system. In general, Bacteria prevail over Archaea, although the latter have been shown to have important roles. In all cases, the characterization of the prokaryote microbiomes of DHABs is still in its infancy, with most of the novel divisions detected only by nucleotide markers, without any functional information available.

The rapid advancement in high-throughput sequencing technologies opens a novel perspective for the exploration of a still hidden functional diversity. The improvements of cultivation-based approaches are also contributing to the assessment of the metabolic potential of DHABs. The discovery of new genes, metabolites, and enzymes, all with "extreme" functional characteristics is increasingly attracting the attention of researchers and industries for the development of novel bio-technological applications. Some studies have started to examine the DHABs for the isolation of novel extremozymes with potential industrial applications (Ferrer et al. 2005; Mohamed et al. 2013; De Vitis et al. 2015), but we are only at the beginning of a promising field of investigation.

ACKNOWLEDGMENTS

This work was conducted with the financial support of the Red Sea Research Center at the King Abdullah University of Science and Technology under the CCF program.

REFERENCES

Abdallah, Rehab Z., Mustafa Adel, Amged Ouf et al. 2014. Aerobic methanotrophic communities at the Red Sea brine-seawater interface. *Frontiers in Microbiology* 5: 487.

Adrio, Jose L., and Arnorld L. Demain. 2014. Microbial enzymes: Tools for biotechnological processes. *Biomolecules* 4: 117–139.

Alexander, Eva, Alexandra Stock, Hans-Werner Breiner et al. 2009. Microbial eukaryotes in the hypersaline anoxic L'Atalante deep-sea basin. *Environmental Microbiology* 11: 360–381.

Antranikian, Garabed, Costantinos E. Vorgias, and Costanzo Bertoldo. 2005. Extreme environments as a resource for microorganisms and novel biocatalysts. *Advances in Biochemical Engineering Biotechnology* 96: 219–262.

Antunes, André, David K. Ngugi, and Ulrich Stingl. 2011. Microbiology of the Red Sea (and other) deep-sea anoxic brine lakes. *Environmental Microbiology Reports* 3: 416–433.

BBC Research. 2011. Enzymes in industrial applications: Global markets. Report BIO030 F. Wellesley, MA: BBC Research.

Borin, Sara, Lorenzo Brusetti, Francesca Mapelli et al. 2009. Sulfur cycling and methanogenesis primarily drive microbial colonization of the highly sulfidic Urania deep hypersaline basin. *Proceedings of the National Academy of Science of United States of America* 106: 9151–9156.

Borin, Sara, Elena Crotti, Francesca Mapelli, Isabella Tamagnini, Cesare Corselli, and Daniele Daffonchio. 2008. DNA is preserved and maintains transforming potential after contact with brines of the deep anoxic hypersaline lakes of the eastern Mediterranean Sea. *Saline Systems* 4: 10.

Børresen, Torger, Catherine Boyen, Alan Dobson et al. 2010. Marine biotechnology: A new vision and strategy for Europe. Marine Board-ESF Position Paper 15. Strasbourg: European Science Foundation, pp. 1–94.

Bougouffa, Salim, Jiangke K. Yang, Onon O. Lee et al. 2013. Distinctive microbial community structure in highly stratified deep-sea brine water columns. *Applied and Environmental Microbiology* 79: 3425–3437.

Briand, Frederic, ed. 2008. The Messinian salinity crisis from mega-deposits to microbiology—A consensus report. In *CIESM Workshop Monographs No. 33*, Monte Carlo, Monaco, pp. 1–168.

Camerlenghi, Angelo. 1990. Anoxic basins of the eastern Mediterranean: Geological framework. *Marine Chemistry* 31: 1–19.

Chu, Xinmin, Haoze He, Changquan Guo, and Baolin Sun. 2008. Identification of two novel esterases from a marine metagenomic library derived from South China Sea. *Biotechnologically Relevant Enzymes and Proteins* 80: 615–625.

Cita, Maria B. 2006. Exhumation of Messinian evaporites in the deep-sea and creation of deep anoxic brine-filled collapsed basins. *Sedimentary Geology* 188–189: 357–378.

Cita, Maria B., Fulvia S. Aghib, Alessandra Cambi et al. 1985. Precipitazione attuale di gesso in un bacino anossico profondo; prime osservazioni geologiche, idrologiche, paleontologiche sul Bacino Bannock (Mediterraneo orientale). *Giornale di Geologia* 47: 143–163.

Cockell, Charles S. 2006. The origin and emergence of life under impact bombardment. *Philosophical Transactions of the Royal Society B* 361: 1845–1856.

Corinaldesi, Cinzia, Michael Tangherlini, Gian M. Luna, and Antonio Dell'Anno. 2014. Extracellular DNA can preserve the genetic signatures of present and past viral infection events in deep hypersaline anoxic basins. *Proceedings of the Royal Society B—Biological Sciences* 281: 20133299.

Daffonchio, Daniele, Sara Borin, Tullio Brusa et al. 2006. Stratified prokaryote network in the oxic-anoxic transition of a deep sea halocline. *Nature* 440: 203–207.

Dalmaso, Gabriel Z. L., Davis Ferreira, and Alane B. Vermelho. 2015. Marine extremophiles: A source of hydrolases for biotechnological applications. *Marine Drugs* 13: 1925–1965.

De Carvalho, Carla C. C. R. 2011. Enzymatic and whole cell catalysis: Finding new strategies for old processes. *Biotechnology Advances* 29: 75–83.

De Vitis, Valerio, Benedetta Guidi, Martina L. Contente et al. 2015. Marine microorganisms as source of stereoselective esterases and ketoreductases: Kinetic resolution of a prostaglandin intermediate. *Marine Biotechnology* 17: 144–152.

Dixon, Brian, Marcel Jaspars, Niall McDonough, and Jan-Bart. 2011. A new dawn for marine biotechnology in Europe. *Biotechnology Advances* 29: 453–456.

Eder, Wolfgang, Linda L. Jahnke, Mark Schmidt, and Robert Huber. 2001. Microbial diversity of the brine-seawater interface of the Kebrit Deep, Red Sea, studied via 16S rRNA gene sequences and cultivation methods. *Environmental Microbiology* 67: 3077–3085.

Eder, Wolfgang, Mark Schmidt, Marcus Koch, Dieter Garbe-Schönberg, and Robert Huber. 2002. Prokaryotic phylogenetic diversity and corresponding geochemical data of the brine-seawater interface of the Shaban Deep, Red Sea. *Environmental Microbiology* 4: 758–763.

Ferrer, Manuel, Olga V. Golyshina, Tatyana N. Chernikova et al. 2005. Microbial enzymes mined from the Urania deep-sea hypersaline anoxic basin. *Chemistry & Biology* 12: 895–904.

Ferrer, Manuel, Johannes Werner, Tatyana N. Chernikova et al. 2012. Unveiling microbial life in the new deep-sea hypersaline Lake Thetis. Part II: A metagenomic study. *Environmental Microbiology* 14: 268–281.

Filker, Sabine, Alexandra Stock, Hans-Werner Breiner et al. 2013. Environmental selection of protistan plankton communities in hypersaline anoxic deep-sea basins, Eastern Mediterranean Sea. *Microbiology Open* 2: 54–63.

Francis, Christopher A., Kathryn J. Roberts, Michael J. Beman, Alyson E. Santoro, and Brian B. Oakley. 2005. Ubiquity and diversity of ammonia-oxidizing archaea in water columns and sediments of the oceans. *Proceedings of the National Academy of Sciences of the United States of America* 102: 14683–14688.

Guan, Yue, Tyas Hikmawan, André Antunes, David Ngugi, and Ulrich Stingl. 2015. Diversity of methanogens and sulfate-reducing bacteria in the interfaces of five deep-sea anoxic brines of the Red Sea. *Research in Microbiology* 1–12.

Gurvich, Evgeny G. 2006. *Metalliferous Sediments of the World Ocean: Fundamental Theory of Deep-Sea Hydrothermal Sedimentation.* Berlin: Springer.

Hallsworth, John E., Michail M. Yakimov, Peter N. Golyshin et al. 2007. Limits of life in MgCl2-containing environments: Chaotropicity defines the window. *Environmental Microbiology* 9: 801–813.

Handelsman, Jo. 2004. Metagenomics: Application of genomics to uncultured microorganisms. *Microbiology and Molecular Biology Reviews* 68: 669–685.

Hartmann, Martin, Jan C. Scholten, Peter Stoffers, and F. Wehner. 1998. Hydrographic structure of brine-filled deeps in the Red Sea—New results from the Shaban, Kebrit, Atlantis II, and Discovery Deep. *Marine Geology* 144: 311–330.

Hasan, Faria, Aamer A. Shah, and Abdul Hameed. 2006. Industrial applications of microbial lipases. *Enzyme and Microbial Technology* 39: 235–251.

Horn, Svein J., Gustav Vaaje-Kolstad, Bjørge Westereng, and Vincent G. H. Eijsink. 2012. Novel enzymes for the degradation of cellulose. *Biotechnology for Biofuels* 5: 45.

Joint, Ian, Martin Mühling, and Joel Querellou. 2010. Culturing marine bacteria—An essential prerequisite for biodiscovery. *Microbial Biotechnology* 3: 564–575.

Jongsma, Derk, Anne R. Fortuin, W. Huson et al. 1983. Discovery of an anoxic basin within the Strabo trench, eastern Mediterranean. *Nature* 305: 795–797.

Kennedy, Jonathan, Burkhardt Flemer, Stephen A. Jackson et al. 2010. Marine metagenomics: New tools for the study and exploitation of marine microbial metabolism. *Marine Drugs* 8: 608–628.

Kormas, Konstantinos A., Maria P. Pachiadaki, Hera Karayanni, Edward R. Leadbetter, Joan M. Bernhard, and Virginia P. Edgcomb. 2015. Inter-comparison of the potentially active prokaryotic communities in the halocline sediments of Mediterranean deep-sea hypersaline basins. *Extremophiles* 19: 949–960.

Laatsch, Hartmut. 2010. *Antibase, a Database for Rapid Dereplication and Structure Determination of Microbial Natural Products.* Weinheim, Germany: Wiley-VCH.

La Cono, Violetta, Francesco Smedile, Giovanni Bortoluzzi et al. 2011. Unveiling microbial life in new deep-sea hypersaline Lake Thetis. Part I: Prokaryotes and environmental settings. *Environmental Microbiology* 13: 2250–2268.

Larock, Paul A., Ray D. Lauer, John R. Schwarz, Kathleen K. Watanabe, and Denis A. Wiesenburg. 1979. Microbial biomass and activity distribution in an anoxic hypersaline basin. *Applied and Environmental Microbiology* 37: 466–470.

Laurila, Tea E., Mark D. Hannington, Sven Petersen, and Dieter Garbe-Schönberg. 2014. Trace metal distribution in the Atlantis II Deep (Red Sea) sediments. *Chemical Geology* 386: 80–100.

Lens Piet, N. L., and Johannes G. Kuenen. 2001. The biological sulfur cycle: Novel opportunities for environmental biotechnology. *Water Science & Technology* 44: 57–66.

MEDRIFF Consortium. 1995. Three brine lakes discovered in the seafloor of the eastern Mediterranean. *EOS, Transactions, American Geophysical Union* 76: 313–318.

Miller, Arthur R., Charles D. Densmore, Egon T. Degens et al. 1966. Hot brines and recent iron deposits in deeps of the Red Sea. *Nature* 30: 341–350.

Mohamed, Yasmine M., Mohamed A. Ghazy, Ahmed Sayed, Amged Ouf, Hamza El-Dorry, and Rania Siam. 2013. Isolation and characterization of a heavy metal-resistant, thermophilic esterase from a Red Sea brine pool. *Scientific Reports* 3: 3358.

Ngugi, David K., Jochen Blom, Intikhab Alam et al. 2015. Comparative genomics reveals adaptations of a halotolerant thaumarchaeon in the interfaces of brine pools in the Red Sea. *ISME Journal* 9: 396–411.

Oh, Ki-Hoon, Giang-Son Nguyen, Eun-Young Kim et al. 2012. Characterization of a novel esterase isolated from intertidal flat metagenome and its tertiary alcohols synthesis. *Journal of Molecular Catalysis B: Enzymatic* 80: 67–73.

Owen, Jeremy G., Zachary Charlop-Powers, Alexandra G. Smith et al. 2015. Multiplexed metagenome mining using short DNA sequence tags facilitates targeted discovery of epoxyketone proteasome inhibitors. *Proceedings of the National Academy of Science of the United States of America* 112: 4221–4226.

Pautot, Guy, Pol Guennoc, Alain Coutelle, and Nikos Lyberis. 1984. Discovery of a large brine deep in the northern Red Sea. *Nature* 310: 133–136.

Poli, Annarita, Gianluca Anzelmo, and Barbara Nicolaus. 2010. Bacterial exopolysaccharides from extreme marine habitats: Production, characterization and biological activities. *Marine Drugs* 8: 1779–1802.

Prakash, Om, Yogesh Shouche, Kamlesh Jangid, and Joel E. Kostka. 2013. Microbial cultivation and the role of microbial resource centers in the omics era. *Applied Microbiology Biotechnology* 97: 51–62.

Quyen, Dinh T., Thi T. Dao, and Sy L. T. Nguyen. 2007. A novel esterase from *Ralstonia* sp. M1: Gene cloning, sequencing, high-level expression and characterization. *Protein Expression and Purification* 51: 133–140.

Raddadi, Noura, Ameur Cherif, Daniele Daffonchio, Mohamed Neifar, and Fabio Fava. 2015. Biotechnological applications of extremophiles, extremozymes and extremolytes. *Applied Microbiology Biotechnology* 99: 7907–7913.

Sagar, Sunil, Luke Esau, Tyas Hikmawan et al. 2013a. Cytotoxic and apoptotic evaluations of marine bacteria isolated from brine-seawater interface of the Red Sea. *BMC Complementary & Alternative Medicine* 13: 29.

Sagar, Sunil, Luke Esau, Karie Holtermann et al. 2013b. Induction of apoptosis in cancer cell lines by the Red Sea brine pool bacterial extracts. *BMC Complementary & Alternative Medicine* 13: 344.

Sass, Andrea M., Henrik Sass, Marco J. L. Coolen, Heribert Cypionka, and Jörg Overmann. 2001. Microbial communities in the chemocline of a hypersaline deep-sea basin (Urania Basin, Mediterranean Sea). *Applied and Environmental Microbiology* 67: 5392–5402.

Schardt, Christian. 2015. Hydrothermal fluid migration and brine pool formation in the Red Sea: The Atlantis II Deep. *Mineralium Deposita* 51: 89–111.

Shock, Everett L. 1992. Chemical environments of submarine hydrothermal systems. *Origin of Life and Evolution of Biosphere* 22: 67–107.

Sorokin, Dimitry Y., Ilya V. Kublanov, Sergei N. Gavrilov et al. 2015. Elemental sulfur and acetate can support life of a novel strictly anaerobic haloarchaeon. *ISME Journal* 10: 240–252.

Stoeck, Thorsten, Sabine Filker, Virginia Edgcomb et al. 2014. Living at the limits: Evidence for microbial eukaryotes thriving under pressure in deep anoxic, hypersaline habitats. *Advances in Ecology* 2014: 9.

Swallow, John C., and J. Crease. 1965. Hot salty water at the bottom of the Red Sea. *Nature* 205: 165–166.

Swift, Stephen A., Amy S. Bower, and Raymond W. Schmitt. 2012. Vertical, horizontal, and temporal changes in temperature in Atlantis II and Discovery hot brine pools, Red Sea. *Deep-Sea Research I* 64: 118–128.

Takai, Ken, Kentaro Nakamura, Tomohiro Toki et al. 2008. Cell proliferation at 122 degrees C and isotopically heavy CH_4 production by a hyperthermophilic methanogen under high-pressure cultivation. *Proceedings of the National Academy of Sciences of the United States of America* 105: 10949–10954.

van der Wielen, Paul W. J. J., Henk Bolhuis, Sara Borin et al. 2005. The enigma of prokaryotic life in deep hypersaline anoxic basins. *Science* 307: 121–123.

van der Wielen, Paul W. J. J., and Sander K. Heijs. 2007. Sulfate-reducing prokaryotic communities in two deep hypersaline anoxic basins in the Eastern Mediterranean deep sea. *Environmental Microbiology* 9: 1335–1340.

Vengosh, Avner, Gert J. De Lange, and Abraham Starinsky. 1998. Boron isotope and geochemical evidence for the origin of Urania and Bannock brines at the eastern Mediterranean: Effect of water–rock interactions. *Geochimica et Cosmochimica Acta* 62: 3221–3228.

Vengosh, Avner, and Abraham Starinsky. 1993. Relics of evaporated sea water in deep basins of the eastern Mediterranean. *Marine Geology* 1–2: 15–19.

Wang, Yong, Huiluo Cao, Guishan Zhang et al. 2013. Autotrophic microbe metagenomes and metabolic pathways differentiate adjacent Red Sea brine pools. *Scientific Reports* 3: 1748.

Werner, Johannes, Manuel Ferrer, Gurvan Michel et al. 2014. *Halorhabdus tiamatea*: Proteogenomics and glycosidase activity measurements identify the first cultivated euryarchaeon from a deep-sea anoxic brine lake as potential polysaccharide degrader. *Environmental Microbiology* 16: 2525–2537.

Yakimov, Michail M., Violetta La Cono, Renata Denaro et al. 2007. Primary producing prokaryotic communities of brine, interface and seawater above the halocline of deep anoxic lake L'Atalante, Eastern Mediterranean Sea. *ISME Journal* 1: 743–755.

Yakimov, Michail M., Violetta La Cono, Vladen Z. Slepak et al. 2013. Microbial life in the Lake Medee, the largest deep-sea salt-saturated formation. *Scientific Reports* 3: 3554.

Yakimov, Michail M., Violetta La Cono, Gina La Spada et al. 2015. Microbial community of the deep-sea brine Lake Kryos seawater–brine interface is active below the chaotropicity limit of life as revealed by recovery of mRNA. *Environmental Microbiology* 17(2): 364–382.

Zeng, Xian, Jean-Louise Birrien, Yves Fouquet et al. 2009. *Pyrococcus* CH1, an obligate piezophilic hyperthermophile: Extending the upper pressure-temperature limits for life. *ISME Journal* 3: 873–876.

CHAPTER 3

Extremophile Diversity and Biotechnological Potential from Desert Environments and Saline Systems of Southern Tunisia

Hanene Cherif, Mohamed Neifar, Habib Chouchane, Asma Soussi, Chadlia Hamdi, Amel Guesmi, Imen Fhoula, Afef Najjari, Raoudha Ferjani, Mouna Mahjoubi, Darine El Hidri, Besma Ettoumi, Khaled Elmnasri, Amor Mosbah, Atef Jaouani, Ahmed Slaheddine Masmoudi, Hadda Imen Ouzari, Abdellatif Boudabous, and Ameur Cherif

CONTENTS

3.1 GENERAL OVERVIEW ON TUNISIAN DESERT ENVIRONMENTS

In Tunisia, covering 164,103 km², arid and desert regions form three-quarters of its area (Figure 3.1a). The landscape of the desert features is made of three main types of desert: rocky desert (*Hamada*), sand dune desert (*Ergs*), and salt-lake depressions (Chotts and Sebkhas) (Figure 3.2). The Ergs lies at the eastern end of the Great Eastern Erg. It is characterized by wind-borne sandy soils generating striking sand dunes. Damaging ultraviolet (UV) rays, poor nutrient soil, and day and night temperature fluctuations present the main limiting factors. Saline lakes present 29%

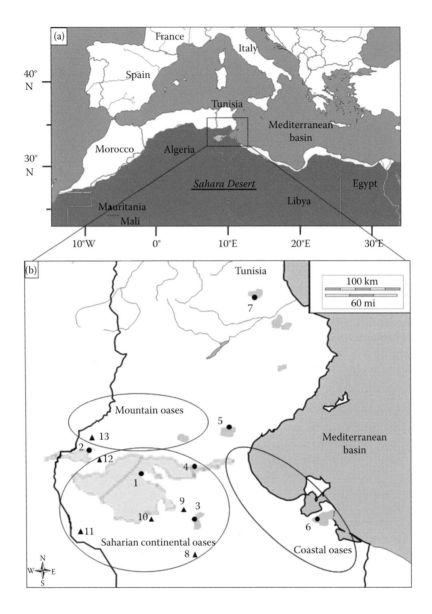

Figure 3.1 (a) General map of the Mediterranean Sea and location of the Sahara Desert in the northern part of the African continent. (b) Distribution of the Tunisian hypersaline systems and oases where the studies were conducted. (1–4) Continental Chotts: (1) Chott Djerid, (2) Chott El Gharsa, (3) Chott Douz, and (4) Chott El Fjej. (5–7) Coastal Sebkhas: (5) Sebkha Ennaouel, (6) Sebkha El Melah, (7) Ras Ellasma. (8–13) Oases: (8) Ksar Ghilane, (9) Douz, (10) El Faouar, (11) Regim Maatoug, (12) Tozeur, and (13) Tamerza/Chbika. Chotts and Sebkha are indicated by dots. Oases are labeled by triangles. The geographic distribution of the mountain, Saharan continental, and coastal oases is indicated by circles.

of the total Tunisian wetland area and account for 54 coastal Sebkhas and 17 continental Chotts (representing 22% and 7% of the total wetland area, respectively). The southwest includes the large areas of Chott Djerid, Gharsa, and Erg (Figure 3.1b). These areas have a limited environmental stability and are under the influence of the rainy season, which leads to periodic salinity dilution, followed by the dry season (Stivaletta et al. 2009). The evaporate deposition processes of arid saline lakes (such as continental Sebkha, Chotts, and playa) have been proposed for

Figure 3.2 Examples of hypersaline environments in southern Tunisia: (a–c) Chott Djerid: Accumulation of salt in artificial saltern and ponds. (b) The red-brown coloration is due to the high concentration of carotenoid pigments. (c) Halite crystals with trapped microorganisms. Scale bar: 0.5 cm. (d) Chott El Gharsa: The white spots indicate the salt deposits. (e, f) Sebkha El Melah: (e) Transect biofilm sample (black arrows) and (f) biofilm submerged with water (black arrow). (g, h) The intertidal zone of Ras Ellamsa. (h) Cyanobacterial MISS.

explaining some of the sulfate-rich deposits of Mars (Gendrin et al. 2005; Murchie et al. 2009) (Figure 3.1b and 3.2). In these harsh environments, and in the middle of the desert oases, the date palm (*Phoenix dactylifera* L.) is installed. The latter is considered a key plant for determining the specific oasis microclimate where the temperature is softened and the air is cooler and humid (Figure 3.3).

Figure 3.3 Examples of desert plants in southern Tunisia. (a) Saharan oasis in Douz. (b) Mountain oasis of Chbika. (c) Coastal oasis of Teboulbou; white arrows indicate the pomegranate plantation between palm trees. (d) Olive tree field in the arid region. (e) *Salicornia* sp. growing at the limit of the Sebkha; the white coloration is due to a salt deposit. (f) Desert truffle *Terfezia boudieri* in association with the host plant *Helianthemum kahiricum*. (From Sbissi, I. et al., *Can. J. Microbiol.*, 57, 599–605, 2011.)

3.1.1 Desert Oases

Oases cover 40,803 ha of Tunisia's land area. According to their geographical locations, Tunisian desert oases are divided into three categories, Saharan continental (Tozeur and Kebili), mountain (Tamaghza, Chebika, and El Faouar), and coastal (Gabes, Jerba, and Medenine) oases, covering 65%, 13%, and 19% of the total oases area, respectively (Figure 3.1) (Ben Salah 2015). Oases are subjected to different aggressive weather conditions, including storms (wind, sirocco, and sandstorms), temperature fluctuations reaching 30°C of variance between day and night, and the sporadic distribution of rainfall, with an average about 90 mm per year. Soil and groundwater salinization present another major limiting factor of the desert environment. This phenomenon is caused mainly by the brine intrusion from Chott, salt water coming up from saline underlying aquifers, and the leaching of agricultural drainage water (Zammouri et al. 2007). Besides, the oasis ecosystem equilibrium is largely influenced by agricultural management practices. The use of traditional methods with three layers of production, including palm trees, fruit trees (olive and pomegranate) (Figure 3.3c), and annual crops (soft wheat, sorghum, and barley), was replaced by modern and industrial systems based on a mono-culture of the "Deglet Nour" date palm variety (Rhouma 1996; Benaoun et al. 2014). Furthermore, within the oases, the arboreal floor presents the dominance of some newly adapted species to oasis abiotic and biotic factors, mainly pomegranate to the detriment of other traditional ones, such as olive tree (Figure 3.3c). These modern oases are characterized by high water consumption originating from deep aquifer and fossil water and high use of chemical inputs (Benaoun et al. 2014).

3.1.2 Chott Ecosystem

Chott, which is Arabic, literally means "beach" and stands for a saline basin with groundwater input. In Tunisia, Chott Djerid is the largest and the most investigated ephemeral lake (Figures 3.1 and 3.2). It is a modern terrestrial and evaporitic environment that consists of salty shallow pools and marshes, covered by a large salt pan during the dry season (Bryant et al. 1994). It is a closed salt playa situated in an arid zone and has an area of approximately 5360 km^2 (Millington et al. 1989), despite the fact that the basin in which it resides is much larger (10,500 km^2) (Gueddari 1984). This Chott is bounded to the north and east by alluvial fans and on the southern margin by eolian sands of the Grand Erg Oriental of the Sahara Spring mounds (Roberts and Mitchell 1987). The strong harsh conditions of the Chott Djerid hypersaline environment, mainly its dryness, the elevated UV irradiation and relevant salt concentration of chlorides and sulfates, making this ecosystem similar to the environment of Mars (Kereszturi et al. 2014). Environmental measurements displayed a water salinity ranging between 29% and 37% and pH values between 7.4 and 7.8 (Stivaletta et al. 2009). Giving its hydrologic allocation, Chott Djerid is an example of salt accumulation for evaporite deposit formation on the desert edge (Bryant et al. 1994; Drake 1994). In addition, theatre-shaped valleys were distinguished in this region of Chott Djerid that are similar to some Martian valleys, which are suspected to have been formed by subsurface sapping (Kereszturi et al. 2014). In evaporite ecosystems, microbial colonizers can either be engaged in specific biochemical processes or just use the evaporite crusts and minerals as a physical dwelling, keeping an enduring record that can be considered a Martian biological signature (Cockell et al. 2002; Martinez-Frias et al. 2006). Chott Djerid has an elongated arm named Chott Fjej that stretches eastward toward the coastal city of Gabes (Figure 3.1) (Stivaletta and Barbieri 2009). It forms the northeastern side of the immense artesian Bas Sahara basin, which consists of complex aquifers that reach the surface in and around the Chott area (Roberts and Mitchell 1987). Chott El Gharsa is the second most important Chott after Chott Djerid, but much smaller (620 km^2) (Figure 3.1) (Stivaletta et al. 2009). In the center of Chott El Gharsa basin, the water table position is relatively low, ranging between –20 and –30 m below sea level (Swezey 2003). Gypsum represents most of the evaporate deposits. On the other hand, dolomite precipitation was found in fossil continental Chott (Barbieri et al. 2006). In this ecosystem, dolomite crystals have a bacterial-related morphology formed by sulfate-reducing bacteria, characterized by the dumbbell shape of their hollow cores (Buczynski and Chafetz 1991), which can serve as distinct biosignatures. These explored microbiological signatures are preserved in the fossil evaporites of this Chott in bedded laminated gypsum deposits and include mineralized microbial morphologies, mainly microfibers, mucilage, and rods (Barbieri et al. 2006).

3.1.3 Sebkha Ecosystems

At the opposite of the Chott, the Sebkhas ecosystems are affected by marine water penetration and tidal-related deposits. Southeastern Tunisian coasts are distinguished by the presence of many microbial mats, preferentially colonizing intertidal zones, making them a modern siliciclastic peritidal environment. Ras Ellamsa is an intertidal zone near Bahira El Bibane, situated between the Zarzis and Tunisian–Libyan border (Figure 3.1). This intertidal region is characterized by an arid climate. The tide is semidiurnal with an amplitude between 80 and 150 cm. Above the arid climate, the continuous rocking movements of tides make this zone an extreme environment (Lakhdar and Soussi 2007). Ras Ellamsa is geologically investigated, wherein microbially induced sedimentary structures (MISSs) are described as "skin elephant structures" formed by tufts and bulges (Gehling 1999; Steiner and Reitner 2001). Sebkha El Melah is an extreme saline environment situated in southeastern Tunisia with an area of approximately 150 km^2 (Jaouani et al. 2014). Within this Sebkha, MISSs are well distinguished (Figure 3.2e–h). These structures are formed by the interaction of microorganisms with the physical agents of erosion, deposition, and transport, or traces of deformation of microbial activities (Noffke et al. 1996; Gerdes et al. 2000).

3.2 DIVERSITY OF DESERT FREE-LIVING MICROBES AND THEIR BIOTECHNOLOGICAL POTENTIAL

In the desert environments, microbial diversity is thought to be influenced by both abiotic factors, such as extreme fluctuations in temperature, elevated UV radiation, low nutrient levels, high soil salinity, and low soil moisture content, and biotic factors, such as plant abundance and species composition (Nagy et al. 2005; Bachar et al. 2012; Andrew et al. 2012). The study of the complex microbial communities living in desert environments and saline systems cannot be carried out in a comprehensive way by using only the traditional microbiological approaches since most of these microorganisms are difficult to isolate and/or are intrinsically uncultivable. The application of new methods in microbial ecology based on the analysis of molecular chronometers like the 16S rRNA gene led to the overcoming of the limits of bacterial cultivability (Dar et al. 2007). With the application of denaturing gradient gel electrophoresis (DGGE), a high diversity of microorganisms (Gram-positive spore formers, Actinobacteria, Cyanobacteria, Archaea, etc.) that are well adapted to thrive in hypersaline environments were recovered from the Tunisian desert (Mapelli et al. 2013; Cherif et al. 2015). Furthermore, the development of DNA microarray technology allowed the quantification of expressed genes of a particular function within complex microbial communities (Dennis et al. 2003; J.S. Kim et al. 2008). Understanding the diversity in such microbial communities can be used to assess the potential effects of extremophiles on soil ecosystem services like plant health (Köberl et al. 2011). Multiple extremophiles are known to promote plant growth in crop saline ecosystems. The principal mechanisms of growth promotion include the production of growth-stimulating phytohormones, the solubilization and mobilization of phosphate, siderophore production, the inhibition of plant ethylene synthesis, and the induction of plant systemic resistance to pathogens (Glick 1995; Ahmad et al. 2008; Díaz and Fernández 2009).

Desert extremophiles, particularly halophilic microorganisms, including Archaea and Bacteria, are important not only because of the fundamentals of their ecological and biochemical biodiversities, but also because of their enormous potential as sources of novel enzymes and other biological materials, with applications in biotechnology and industry (Niehaus et al. 1999; Van Den Burg 2003; Das et al. 2007).

3.2.1 Desert Archaea

Archaea, the third domain of life, are single-celled prokaryotic microorganisms that have unique phenotypic and molecular characteristics, separating them from the two other domains of life, Bacteria and Eukarya (Woese and Fox 1977). Based on 16S rRNA gene classification, Archaea are subdivided into four major phyla: (1) The Euryarchaeota, the most diverse group, colonizing a wide range of ecological niches, include the methanogens, salt-lover cells (the haloarchaea) (Figure 3.4a),

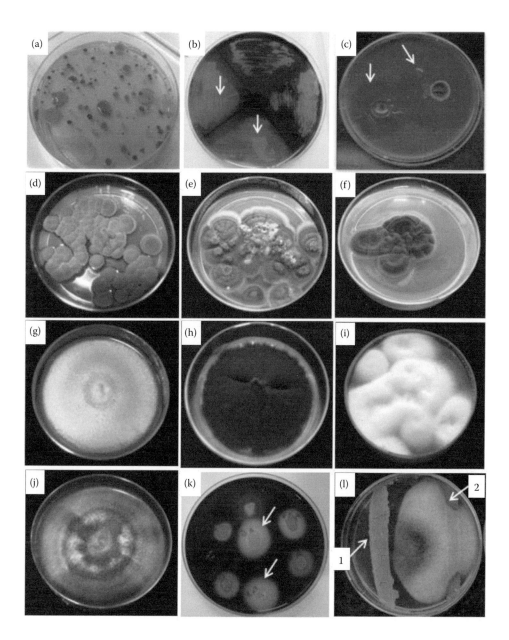

Figure 3.4 Examples of microorganisms isolated from different saline systems and their biotechnological potential. (a–c) Haloarchaea: (a) 10−5 dilution of saturated water from the salt crust of Chott Djerid plated on DSC-97 medium (MgSO$_4$·7H$_2$O, 20 g/L, NaCl, 250 g/L, pH 7.4) after 15 days of incubation at 30°C. Haloarchaea colonies show pink, red, and purple pigmentation due to the bacterioruberin (black and gray). (b) Amylolytic activity of Archaea isolates in the presence of 0.2% soluble starch and 23% NaCl; white arrows indicate the clear zones of starch solubilization. (c) Halocin (Archaea bacteriocin) activities of *Halorubrum* sp. isolates against the same species; white arrows indicate the inhibition zones. (d–j) Fungi: (d) *Cladosporium halotolerans*, (e) *Penicillium chrysogenum*, (f) *Alternaria alternata*, (g) *Aspergillus fumigatus*, (h) *Ulocladium consortiale*, (i) *Engyodontium album*, and (j) *Embellisia phragmospora*. (From Jaouani, A. et al., *BioMed Res. Int.*, 439197, 2014.) (k) Amylolytic activity of spore-forming isolates in the presence of 0.2% soluble starch and 3% NaCl; white arrows indicate the clear zones of starch solubilization. (l) Antifungal activity of *Bacillus* sp. isolates against *Aspergillus brasiliensis*; white arrows indicate the plated active (1) and nonactive (2) isolates.

and other thermophilic microorganisms (Woese et al. 1990). (2) The Crenarchaeota, characterized by their ability to tolerate extremes in temperature and acidity, are isolated from warm springs and soils that contain sulfur and sulfides (Madigan et al. 2005). (3) The Korarchaeota are uncultured microorganisms that have only been identified in high-temperature hydrothermal environments. (4) The Nanoarchaeaota, considered unusual microorganisms because they are known to grow on the surface of another archaea, *Ignicoccus hospitalis* (Huber et al. 2002), contain the smallest known archaeal genome (Waters et al. 2003).

Among these microorganisms, the Haloarchaea (Figure 3.4a), members of the family Halobcteriaceae within the Euryarchaeota phylum, have been largely studied on different hyper-saline ecosystems, including thalassohaline environments (resulting from evaporation of seawater) and athalassohaline environments (that are fed by surface water or rivers) (Satyanarayana et al. 2005), where they constitute, besides halophilic bacteria and halotolerant eukaryotes such as the alga *Dunaliella*, the main dominant microbial population (Oren 1999; Ventosa 2004; DasSarma 2006; Maturrano et al. 2006).

The Haloarchaea is a monophyletic group, including those known as aerobic and obligatory halophilic archaea. They are chimioorganotrophs, and most of them use carbohydrates or amino acids as a carbon source and grow optimally from 3.4 to 4.5 M NaCl, with minimal salinity of 1.5 M for growth (Grant et al. 2001). They are pigmented in pink, red, or purple (Figure 3.4a) due to the presence of bacterioruberin at their membrane, which protects cells from photooxidation (Kamekura 1993; Grant et al. 1998). Often, the ionic composition of these saline environments, which varies considerably depending on climatic, physical, and geological conditions (Madigan et al. 2005), has a direct influence on the distribution and abundance of the Haloarchaea com-munity, as demonstrated based on culture-dependent and -independent approaches. Hence, these microorganisms have been identified in a large range of hypersaline ecosystems, such as the Dead Sea, the Great Salt Lake (Utah), the alkaline brines of Wadi Ntarun (Egypt), the solar saltern of Sfax Sebkhas (Tunisia), the Algerian Sebkhas, and the Greek solar saltern (Baati et al. 2008; Tsiamis et al. 2008; Trigui et al. 2011; Boujelben et al. 2012; Baricz et al. 2014).

The diversity and community structure of members of the halophilic Archaea (Class Halobacteria) were described in four Tunisian hypersaline systems in southern and central Tunisia, using a targeted 16S rRNA gene diversity survey (pyrosequencing). A high diversity of Halobacteria members was revealed among the different samples exhibiting different salt concentrations. By an isolation procedure, 45 Haloarchaea genera were detected. In fact, the salinity was shown to be the most important factor shaping the Halobacteria community structure, as revealed by the strong neg-ative correlation between the salinity and halobacterial diversity (Najjari et al. 2015). In addition, isolation of halobacterial strains showed a clear correlation between the recovered archaeal genera and the origin of isolation (Najjari et al. unpublished), where *Halorubrum* sp. and *Haloarcula* sp. occurred mainly in saline systems, whereas other genera were encountered in sand and bulk soil samples.

To survive in high-salinity concentrations, different mechanisms are adopted by halophilic Archaea, in order to maintain an intracellular osmotic pressure equal to or higher than the extra-cellular environment (Oren 2002). Among them, two mechanisms have been described, includ-ing (1) the "salting in" strategy, where cells accumulate molar concentrations of potassium ions to counter the high extracellular osmotic pressure, and (2) the synthesis and/or uptake of highly soluble compatible solutes, such as "trehalose synthesis or betaine uptake" (Youssef et al. 2014). The trehalose synthesis strategy is studied in the samples harboring different salt concentrations from various hypersaline systems of southern Tunisia (Najjari et al. 2015), and has demonstrated that the relative abundance of genera capable of the biosynthesis of the compatible solutes treha-lose and 2-sulfotrehalose decreased with increasing salinities. Indeed, genera lacking trehalose biosynthetic capabilities are more adapted to the growth and colonization of hypersaline (>25%) environments than trehalose producers (Najjari et al. 2015). These results demonstrate that the recently recognized divergence between trehalose producers and nonproducers is ecologically relevant.

Archaea that live under extremely halophilic conditions are of significant biotechnological importance. They are often considered a dependable source for deriving novel biomolecules. Particular attention has been devoted to peptides (archaeocins and diketopiperazines [DKPs]), biosurfactants (glycolipids), pigments (carotenoids), biopolymers (extracellular polysaccharides [EPSs] and polyhydroxyalkanoates [PHAs]), and haloenzymes (Figure 3.4b).

Archaeocins are antibiotic peptides sourced from archaea being found among haloarchaea (termed halocins) (Figure 3.4c) and more recently from the *Sulfolobus* genus (sulfolobicins). As described by O'Connor and Shand (2002), halocins can be divided into two classes based on their size: the smaller microhalocins, which can be as small as 3.6 kDa, and the larger halocins, of 35 kDa. The antimicrobial activity of these halocins can also vary, with some halocins having a narrow range of activity affecting only close relatives, as opposed to a more broadly active A4 halocin, capable of inhibiting the growth of *Sulfolobus solfataricus*, a representative of another phylum of archaea (Haseltine et al. 2001). A particularly interesting use of archaeocins is in the textile industry, which uses considerable amounts of salt in the tanning process. These conditions allow haloarchaea and halobacteria to grow, which in turn can damage the leather product, and halocins have been used to inhibit the undesired growth (Birbir et al. 2004).

As reported by Tommonaro and coworkers (2012), DKPs are cyclic dipeptides produced by the haloarchaeon *Haloterrigena hispanica*. These DKPs have a variety of useful biological activities with potential significance for industrial and medical purposes, such as antibacterial, antifungal, and antiviral, as well as antitumor, activities. The ability to inhibit quorum sensing systems (QSSs) in bacteria is an interesting activity of DKP, reported by Martins and Carvalho (2007). The inhibition of these QSSs in bacteria is thought to be a potential therapy for a range of pathogens, such as *Pseudomonas aeruginosa* infections of cystic fibrosis patients. There is also potential for QSS blockers to be used more broadly in industries where biofilms can cause biofouling, a problem for a variety of industries, particularly shipping. Biofilms have also been linked to difficulties in implants and catheters, as well as contaminating water pipe systems (Lehtola et al. 2004; Raad et al. 2007; Schultz et al. 2011). DKPs sourced from archaea could potentially be used to block these QSSs in order to prevent biofilm development.

Biosurfactants are produced by a wide range of organisms, including archaea, and they are a mixture of glycolipids, fatty acids, proteins, and sugars. These biomolecules show many advantages to chemically derived surfactants by being ecofriendly, renewable, and active under a range of extreme conditions. Biosurfactants are able to assist bioremediation of oil spills in soil and water samples, as well as a range of other uses in the food, pharmaceutical, and cosmetic industries (Kebbouche-Gana et al. 2009; Sachdev and Cameotra 2013). Kebbouche-Gana et al. (2013) described a biosurfactant produced by *Natrialba* sp. isolated from a solar saltern in Algeria, which constituted sugar protein and lipids, including rhamnolipids. It was suggested that this *Natrialba* species or possibly other halophilic archaea might be an ideal choice for assisting in the bioremediation of oil spills in saline and hypersaline environments, which can often be contaminated through industrial processes.

Carotenoids are naturally occurring pigments that are commonly found in haloarchaea and are responsible for the pigmentation of the organisms. These biomolecules are widely used as food supplements and coloring agents. Some human health potential benefits have been described, such as the prevention of chronic diseases like cancer and osteoporosis (Chandi and Gill 2011; Tanaka et al. 2012). As an example, canthaxanthin is a carotenoid used as a feed additive for chickens, fish, and crustacean farms and used in cosmetics. The halophilic archaeon *Haloferax alexandrinus* produces canthaxanthin at a high enough level that it could be considered for commercial production of canthaxanthin (Asker and Ohta 2002).

Two main types of biopolymers are also produced by such peculiar microorganisms, that is, the EPSs, considered a protection against desiccation and predation, and the endocellular PHAs, which provide an internal reserve of carbon and energy. EPSs are high-molecular-weight carbohydrates produced and released by many different microorganisms, including Archaea. These biopolymers have a number of industrial applications, such as uses in the food industry as gelling or emulsifying agents (Patel and Prajapati 2013), as alternative sources of dietary fibers (Chouchane et al. 2015b), and as bioflocculants in wastewater treatments (Chouchane et al. 2015a). Healthy activities attributed to EPSs, including antioxidant, antimicrobial, anti-inflammatory, antidiabetic, and anticancer activities, are also reported (Liu et al. 2011; Jin et al. 2012; Mahendran et al. 2013). PHAs are water-insoluble polymers used as a means of carbon and energy storage in bacteria and archaea. They have received considerable attention in biotechnology as a potential alternative to petrochemical-based plastics due to their structural properties and biodegradability. Many halophilic Archaea were established as being EPS and PHA producers, such as *Haloferax*, *Haloarcula*, *Halococcus*, *Natronococcus*, and *Halobacterium* (Nicolaus et al. 2010; Legat et al. 2010; Quillaguaman et al. 2010; Poli et al. 2011).

More recently, in a screening program to obtain new archaea exopolysaccharides, producers (Chouchane et al. 2016, unpublished) extracted an exopolysaccharide-based bioflocculant produced by the newly isolated *Halogeometricum borinquense* from the Tunisian desert. Chemical analyses of the bioflocculant showed 30.4% carbohydrate and 50.2% protein. Fourier transform infrared spectroscopy (FTIR) indicated the presence of carboxyl, hydroxyl amino, and amide groups, among others, which likely contribute to the flocculating activity and in dye removal. The bioflocculant was effective in flocculating some soluble anionic dyes in aqueous solution, in particular Reactive Blue 4 and Acid Yellow, with a decolorization efficiency of more than 74%. The decolorization efficiency was dependent on the flocculant dosage and solution pH. *H. borinquense* appears to hold promise as a source of new bioflocculant that could stand as an alternative to inorganic and synthetic organic flocculants.

In conclusion, Archaea display a wide diversity of metabolites representing an untapped resource in the field of natural product discovery and could contribute to different areas of industry and medicine.

3.2.2 Desert Cyanobacteria

Cyanobacteria are the first organisms, having inhabited the earth for 3.5 billion years and having colonized both aquatic and terrestrial ecosystems (Schopf and Packer 1987). They had a distinguished role in our planet evolution, being the principal players of earth oxygenation, which made them among the most important organisms on earth (Kremer et al. 2008). Thanks to their versatility, these photosynthetic bacteria are able to pursue adaptive strategies and survive in extremely harsh conditions (Stivaletta and Barbieri 2009), mainly high rates of salinity, temperature, light, and solar irradiation (Dvornyk and Nevo 2003; Javor 2012; Strunecký et al. 2012; Whitton 2012; Hu et al. 2012; Chu et al. 2015). These ubiquitous microorganisms can endure the most severe ecosystems and live in a diverse range of environments, including arid systems that have been extensively studied worldwide (Lange et al. 1998; Bouvy et al. 1999; Garcia-Pichel and Pringault 2001; Büdel et al. 2004; Warren-Rhodes et al. 2006; Wierzchos et al. 2006; Dong et al. 2007; Nisha et al. 2007; Abed et al. 2008; Lacap et al. 2011; Hu et al. 2012; Whitton 2012; Lin and Wu 2014; Patzelt et al. 2014; Cámara et al. 2015; Kumar and Adhikary 2015; Kutovaya et al. 2015; Powell et al. 2015; Robinson et al. 2015). Unicellular and filamentous Cyanobacteria, including genera *Aphanothece*, *Lyngbya*, *Microcoleus*, *Phormidium*, and *Spirulina*, and synechococcal-like species, were mainly found in the sediment–water interface of salt lakes (Caumette et al. 1994; Oren et al. 1996; Javor 2012). It has been demonstrated that the green and red-orange coloration of Chott Djerid water ponds is caused by the high concentration of these unicellular and filamentous cyanobacteria (Stivaletta et al. 2009) (Figure 3.2b). Generally, unicellular cyanobacteria that develop beneath the interface have an orange-brown color due to their high content of carotenoid pigments, whereas the cyanobacterial mats living in slightly deeper layers have a dark green color (Oren et al. 1996). Besides, filamentous cyanobacteria belonging to the order Oscillatoriales have been observed by transmitted light microscopy of saturated waters and salt crusts of Chott Djerid (Stivaletta et al. 2009). Samples (salt, water, and substrate) collected from Chott Djerid provided an interesting cyanobacterial diversity in extreme saline ecosystems (Soussi et al. unpublished). Based on DGGE analysis, it has been demonstrated that cyanobacteria are moderately abundant in Chott El Fjej compared with the Chott Djerid ecosystem (Soussi et al. unpublished).

The effect of water presence on the diversity of cyanobacterial communities has been evaluated by the analysis of biofilm samples collected from a transect on the edge of Sebkha El Melah going from vegetation to water (Figure 3.2e). Interestingly, based on a culture-independent approach, it has been demonstrated that cyanobacterial diversity increases upon reaching water. However, biofilms collected from the saline system of Sebkha Ennaouel showed cyanobacteria from different orders, namely, *Oscillatoria refringens*, *Synechoccus* sp., *Spirulina* sp., *Nodularia* sp., *Chroococcus* sp., *Cylindrospermum* sp., *Halospirulina tapeticola*, *Prochlorococcus marinus* subsp. *pastoris*, and *Planktorocoides raciborskii* (Soussi et al. unpublished).

Microscopic observation of typical MISS samples collected from the intertidal zone of Ras Ellamsa (Figure 3.2h) displayed a remarkable morphotypic diversity of cyanobacteria dominated by the Oscillatoriales order. Sediment samples present the highest number of morphotypes ($n = 9$), namely *Lyngbya intermedia*, *Spirulina flavovirens*, *Chroococcus turgidus*, *Leptolyngbya perforans*, *Nostoc hormogonium*, *Oscillatoria* sp., *Spirulina subsalsa*, *Oscillatoria refringens*, and *Chroococcus* subg. *Limnococcus*. On the other hand, filamentous cyanobacteria of fresh MISS samples collected from this intertidal zone emitted bright red fluorescence when observed under fluorescence microscopy with an excitation of 633 nm (red fluorescence), indicating the presence of chlorophyll-containing filamentous cyanobacteria in MISS samples (Soussi et al. unpublished).

Zaouiet Lâaouenes, a hydrothermal station situated in southwest Tunisia in the governorate of Kebili, represents another example of a typical extreme environment where warm spring water flows at 43.2°C. Based on light microscopic observations, sediment samples collected from the output of the hydrothermal station cooling tower showed mainly a rare presence of *Calothrix* sp. cyanobacteria compared with the abundance of the genus *Spirulina* (Soussi et al. unpublished).

Interestingly, the culture-independent approach showed common bands in the DGGE profiles of the different studied samples of southern Tunisia extreme environments, although they had different origins (Soussi et al. unpublished). These results are consistent with the unique phylotype detected across three distinct desert sites of the Atacama, the driest and world's oldest continuously arid desert localized in Chile (Yungay in the southernmost site, Salar Llamara in the central zone of the central depression, and Salar Grande close to the western coastal range) (Ríos et al. 2010). These outcomes suggest that some cyanobacterial species may be common for extreme regions of southern Tunisia. Under different extreme conditions, cyanobacteria might evolve specific ecological adaptations to enhance their survivability according to the geoclimatic and environmental circumstances.

The application of cyanobacteria can touch almost all domains, mainly the food industry, agricultural sustainability, medical remedies, and ecological stability (Sharma et al. 2011; Sciuto and Moro 2015; Raja et al. 2016). These bacteria are exploited on the industrial scale as a nutraceutical and functional food, and they are considered an excellent source of vitamins and proteins (Pfeiffer et al. 2011; Ohmori and Ehira 2013). In addition, cyanobacteria have a relevant potential in medical applications. For instance, *Spirulina* can produce high levels of ergothioneine, a very stable antioxidant with unique properties of protecting against oxidative stress. It is a natural amino acid that has been proven under laboratory conditions to be very beneficial for humans and other vertebrates that are not able to produce it and need to absorb it from food (Pfeiffer et al. 2011). Moreover, several cyanobacteria have antiviral potential that can be of clinical interest. For example, *Microcystis* genus showed a notable antiviral activity against influenza A, HIV, and HSV (Zainuddin et al. 2002; Singh et al. 2011). Cyanobacteria can be also applied as antibacterial and antifungal agents. It has been demonstrated that the cyclic peptide calophycin, produced by *Calothrix*, is an antifungal compound active against *Candida albicans* and *Trichophyton mentagrophytes*, as well as an antibacterial agent against several marine bacteria (Moon et al. 1992). In agriculture, cyanobacteria play a crucial role in making nitrogen available for the host plant, hence to be exploited as biofertilizers. It has been reported that *Anabaena fertilissima* and *Anabaena doliolum* alone, or in combination with urea, enhanced greatly the grain yield, biomass, and nutritive value of rice, saving 25% of the chemical nitrogen need of the crop (Dubey and Rai 1995). Cyanobacteria play a significant role in soil structure improvement. They produce mucilage, which binds soil particles by means of EPSs that associate with surrounding sand grains to form considerable resilient matrices, resistant to erosion (De Philippis et al. 1998; Belnap 2001). In addition, after cell death, they increase the humus content of the soil, leading to strong reducing conditions. This endowment is particularly important in arid lands, where cyanobacteria can play a crucial role in desertification control and soil rehabilitation (Isichei 1990). The work of Liu et al. (2008) represents a very nice example of artificial desert crust formation via the application of mass-cultured cyanobacterial strains from the Inner Mongolia Desert in China. On the other hand, in arid lands, cyanobacteria prevent evapotranspiration by enhancing moisture retention, fertilize desert substrate via nitrogen fixation, and block solar radiation by the synthesis of UV-absorbing compounds (Powell et al. 2015).

3.2.3 Spore-Forming Bacteria

The aerobic spore-forming bacteria were originally assigned to the *Bacillus* genus, but molecular and chemical analyses have shown that they form several phylogenetically distinct lineages, and they have been described into many different novel genera (Márquez et al. 2011 and references therein). *Bacillus subtilis* was the earliest studied and described bacteria in the *Bacillus* genus. In 1872, Cohn renamed the organism *Vibrio subtilis*, as described by Ehrenberg in 1835, to *Bacillus subtilis* and studied its ability to produce endospores (Gordon et al. 1973). Later, Koch described, in *Bacillus anthracis*, the cycle of spore formers: vegetative cell to spore and spore to vegetative cell (Gould 2006). After more than 130 years of describing the *Bacillus* genus, endospores have attracted many microbiologists by their resistance to many factors, including heat and radiation, and their ability to survive over long periods. Endospore-forming bacteria are widely distributed in many different environments (hot, cold, arid, etc.). This ubiquity is attributed to their ability to produce spores, their wide-ranging nutritional requirements and growth conditions, and their metabolic diversities. Until now, a massive number of *Bacillus* species have been isolated from different ecosystems around the world, including saline systems (Euzeby 2013).

In southern Tunisia, culture-dependent approaches revealed that arid and saline systems harbor a huge diversity of Gram-positive aerobic spore-forming bacteria. Nine different genera were recovered: *Bacillus*, *Halobacillus*, *Virgibacillus*, *Oceanobacillus*, *Gracilibacillus*, *Peanibacillus*, *Piscibacillus*, *Brevibacillus*, and *Pontibacillus* (Hedi et al. 2009; El Hidri et al. 2013; Guesmi et al. 2013). Compared with similar studies carried out on a salt lake in Algeria (Hacène et al. 2004), a solar saltern (Yeon et al. 2005; Ghozlan et al. 2006), marine environments (Siefert et al. 2000; Ettoumi et al. 2009), and other hypersaline soils (Ventosa et al. 1998) where spore formers were represented by a limited number of genera, arid saline systems in Tunisia revealed a highly diverse population. The work of Guesmi and colleagues (2013) showed that a huge diversity of spore-forming bacilli could be found if pasteurization steps were used in the isolation procedure. As a result, the spore-forming bacteria can grow and are not outcompeted by the fast-growing non-sporulating halophilic bacteria.

The *Bacillus* genus was the most encountered in Tunisian samples, with 19 different species being identified. The two most recovered species were *B. mojavensis* and *B. firmus*. They have been isolated from samples collected from Ksar Ghilan, Sebkha Ennaouel, and Chott Djerid (Guesmi et al. 2013). The type strain of *B. mojavensis* was originally recovered from samples collected from the Mojave Desert of California, with endophytic traits and antifungal capacity (Roberts et al. 1994; Bacon and Hinton 2002, 2011). Other *Bacillus* species with type strains were found to be originally isolated from desert samples like *B. sonorensis* from the Sonoran Desert in the United States (Palmisano et al. 2001), *B. deserti* from the Xinjiang desert in China (Zhang et al. 2011), and *B. coahuilensis* from the Chihuahuan Desert (Cerritos et al. 2008). This finding indicates that a desert environment is a source of new taxa for aerobic spore-forming bacteria that constitute part of the active microflora.

The second most encountered genus in saline systems in southern Tunisia was *Halobacillus*, with five species having been isolated: *H. trueperi*, *H. litoralis*, *H. profundi*, *H. salinus*, and *H. locisalis* (Hedi et al. 2009; El Hidri et al. 2013; Guesmi et al. 2013). Since the discovery of the genus *Halobacillus* by Spring and workers (1996), a number of papers have shown its wide distribution and scientific interest (Burja et al. 1999; Pinar et al. 2001; Yang et al. 2002; Rivadeneyra et al. 2004). These species can tolerate a wide range of salinities, which may be due to their adaptation to environments characterized by fluctuations in salt concentrations. Recently, it was shown that *H. trueperi* plays an active role in the precipitation of carbonates at different salt concentrations and different magnesium/calcium ratios. This can lead to the creation of the supersaturation of carbonates, which induces their precipitation and possibly the formation of bioliths (Rivadeneyra et al. 2004). Evaporate mineral precipitation can constitute microenvironments where microorganisms are protected from exposure to extreme temperatures, UV radiation, and desiccation.

A remarkable difference in the distribution of the aerobic spore-forming bacteria in arid saline systems in Tunisia was found. In Chott Djerid, Sebkha Ennaouel, and Ksar Ghilane, the Bacillales community was dominated by strains affiliated with the *Bacillus* genus. On the other hand, the coastal sites of Sebkha Sahline and Sebkha El Melah were dominated by strict halophilic strains related to the *Halobacillus* genus (Guesmi et al. 2013). The Chott Djerid sample was the most diverse, displaying a mixture of strains affiliated with eight genera; among them, eight species were detected only at this site: *Bacillus aquimaris*, *Bacillus alcalophilus*, *Bacillus beijingensis*, *Bacillus*, *Piscibacillus salipiscarius*, *Paenibacillus illinoisensis*, *Paenibacillus xylanilyticus*, and *Geobacillus orientalis* (El Hidri et al. 2013; Guesmi et al. 2013).

It is interesting to note the uneven geographical distribution of *H. trueperi* strains, which is revealed in the work of Guesmi and colleagues (2013). This species was found to be dominant in coastal sites near the seashore (Sebkhas Sahline and El Melah) and less prevalent in Sebkha Ennaouel, more distant from the Mediterranean coast, suggesting a marine rather than terrestrial origin of this species (Guesmi et al. 2013). The different features of Chotts and Sebkhas, including locations and their distance to the costal sea and the different climatic conditions, explain the uneven distribution of *H. trueperi* and indicate that it is a marine bacterium adapted to high salty ecological niches. This finding is consistent with the previous studies of other microorganisms present in soil (Franklin and Mills 2003; Limmathurotsakul et al. 2010). This uneven distribution of some species can be affected by a nonuniform distribution of organic matter, samples size (Kang and Mills 2006), or biotic and abiotic factors (Wall 2008).

Many species in the Bacillales member are economically very important regarding their ability to produce a wide range of molecules and metabolites with high biotechnological applications (Chaabouni et al. 2012; Raddadi et al. 2015). The spore-forming bacteria isolated from saline systems in southern Tunisia showed an important capability to produce a wide range of enzymes, like protease, amylase, lipase, and DNase (El Hidri et al. 2013) (Figure 3.4k). Some other strains showed potential plant growth-promoting traits and represent promising candidates to enhance the plant growth even in the presence of salt and water stresses (Mapelli et al. 2013; Guesmi et al. unpublished) (Figure 3.4l). Overall, the huge phenotypic and phylogenetic diversity detected in spore-forming bacteria confirms that southern Tunisia represents a valuable source of new lineages and metabolites.

3.2.4 Desert Fungi

The discovery of fungi growing in diverse extreme environments, like desert and saline ecosystems, has broadened the study area of extreme microbiology, which has been traditionally dedicated to the study of prokaryotic microorganisms (Zajc et al. 2012). The best-studied fungal extremophiles are halophilic and halotolerant fungi (Cantrell et al. 2006; Gostinčar et al. 2010, 2011). The dominant extremophilic fungi are meristematic black yeasts, represented by *Hortaea werneckii*, *Phaeotheca triangularis*, *Trimmatostroma salinum*, *Aureobasidium pullulans*, and different species of the genus *Cladosporium*, taxonomically and phylogenetically closely related to black yeasts (Zalar et al. 2007; El-Said and Saleem 2008; Gostinčar et al. 2010, 2011). Different species of *Aspergillus* and *Penicillium* and diverse nonmelanized yeasts appear in hypersaline desert ecosystems less consistently (Jaouani et al. 2014). The most halophilic fungus known to date is *Wallemia ichthyophaga*, as it requires at least 10% NaCl and grows also in solutions saturated with NaCl (Zalar et al. 2005; Zajc et al. 2014). The black yeast-like fungus *H. werneckii* is one of the most halotolerant fungi, with a broad growth optimum from 1 to 3 M NaCl, and it can grow in nearly saturated salt solutions, as well as without sodium chloride (Butinar et al. 2005a, 2005b; Gunde-Cimerman et al. 2000). The halophilic and halotolerant fungi are well adapted to hypersaline environments with low water activities, through several of their traits: plasma membrane composition, enzymes involved in fatty acid modifications, osmolyte composition and accumulation of ions, melanization of the cell wall, differences in the high osmolarity glycerol signaling pathway, and differential gene expression (Petrovic et al. 2002; Turk et al. 2004, 2007; Kogej et al. 2005, 2007; Vaupotic and Plemenitas 2007a, 2007b; Gostinčar et al. 2010). Extremophiles isolated from hypersaline areas are important not only because of the fundamentals of their biochemical adaptations to extreme growth conditions and their structural biodiversities, but also because of their enormous potential as sources of biological materials with applications in bioremediation and biotechnology (Chouchane et al. 2015a, 2015b; Neifar et al. 2015a, 2015b; Raddadi et al. 2015). Indeed, they are widely used in fermentative industries for the production of ethanol, organic acids, polysaccharides, and enzymes (Gorjan and Plemenitaš 2006; Luziatelli et al. 2014). Screening of extremotolerant fungi from different areas of Sebkha El Melah led to the isolation of 21 moderately haloalkali-tolerant fungi, which were able to grow in media with an initial pH of 10 and with 10% salt (Jaouani et al. 2014). The fungi were identified, by microscopic observations and molecular sequencing of the internal transcribed spacers between the 16s and 23s rDNA region (ITS) and specific amplification, as members of seven genera of Ascomycetes: *Cladosporium*, *Penicillium*, *Alternaria*, *Aspergillus*, *Ulocladium*, *Embellisia*, and *Engyodontium* (Figure 3.4d–j). The most salinity-tolerant fungi were found to belong to the two genera *Penicillium and Alternaria*, which were able to grow in the presence of 15% NaCl in liquid medium. Contrary to many reports on hypersaline environments, no species belonging to the genera *Eurotium*, *Thrimmatostroma*, *Emericella*, and *Phaeotheca* have been obtained, probably because of the initial alkaline pH of the Sebkha El Melah salt lake (Jaouani et al. 2014). The 21 isolates produced a broad variety of extracellular hydrolytic and oxidative enzymes. Laccase producers were members of the genus *Cladosporium*, whereas most of the isolates are able to produce hydrolytic enzymes, such as cellulase, amylase, protease, lipase, and keratinase. The results also revealed novel enzymes with interesting pH and salinity profiles that have considerable potential for many industrial applications (Jaouani et al. 2014; Neifar et al. 2015b; Raddedi et al. 2015). Laccase production in the presence of 10% salt by the *Cladosporium* group may be of biotechnological interest, for example, in bioremediation of high salty environments contaminated by colored pollutants (Neifar et al. 2015a). The growing number of genomes available from extremophilic fungi will greatly aid the discovery and identification of novel metabolites that have not been detected by functional screening procedures (Zajc et al. 2013; Gostinčar et al. 2014).

3.3 DESERT PLANT- AND INSECT-ASSOCIATED MICROORGANISMS AND THEIR BIOTECHNOLOGICAL POTENTIAL

In recent years, attention on plant- and insect-associated microbial communities has been increased, with an emphasis on their promoting role and beneficial effects. Special focus has been accorded to extreme environments regarding the astonishing adaptability of fauna and flora to their harsh conditions. An understanding of the plant- and insect-associated microbial diversity perceptions in environmental, agricultural, and industrial contexts is useful to evolve strategies for their better exploitation.

3.3.1 Lactic Acid Bacteria

Lactic acid bacteria (LAB) are among the most important groups of microorganisms used in the preservation and development of many food products. LAB are a group of Gram-positive bacteria characterized by catalase negative, nonmotile, and non-spore-forming cocci or rods, which produce lactic acid during fermentation (Hugenholtz 2008). This group includes many species belonging to the genera *Aerococcus, Carnobacterium, Streptococcus, Enterococcus, Lactobacillus, Lactococcus, Oenococcus, Pediococcus, Tetragenococcus, Vagococcus, Leuconostoc,* and *Weissella* (Wood and Holzapfel 1995). LAB are characterized by the production of other metabolites, such as organic acids, acetoin, carbon dioxide, diacetyl, hydrogen peroxide, low molecular mass compounds with antimicrobial activity, and bacteriocins (Vanderbergh 1993). In addition to the food matrix, LAB are found in the gastrointestinal tract and mucosal cavities of humans and animals (Wood and Holzapfel 1995). They have gained more attention with respect to their health promotion, particularly as probiotics with beneficial effects on human and animal health (Salminen et al. 1998; Parvez et al. 2006; Chaucheyras-Durand and Durand 2009).

The occurrence of LAB from soils and plants, particularly from desert environments, was less reported. *Enterococcus, Lactococcus,* and *Streptococcus* were detected in forage plants (Pahlow et al. 2003). Fhoula et al. (2013) revealed the identification of LAB from rhizosphere samples of olive trees and desert truffles (Figure 3.3d and f) with members of the genera *Enterococcus, Lactobacillus, Pediococcus, Lactococcus, Weissella,* and *Leuconostoc.* In a similar work, the isolation of *Weissella kandleri* was reported from desert springs and desert plants (Holzapfel and van Wyk 1982). Interestingly, a number of genera of LAB have been detected and identified in the gut of the desert ant *Cataglyphis* (Figure 3.5e), including *Enterococcus, Leuconostoc, Pediococcus,* and *Weissella* (Fhoula et al. 2015a, unpublished).

The implication of LAB associated with the gut microbiome of many living organisms, especially for insects, in the survival and adaptation of their hosts, becomes noticeable and well argumented (Douglas 2009; Lenoir et al. 2009; Fhoula et al. 2015a, unpublished). Fhoula et al. (2015b, unpublished) found that some *Weissella halotolerans* strains isolated from arid land living hosts are shown to possess interesting probiotic properties and were generally able to metabolize the oligosaccharides, leading to the maintenance of a balanced gut homeostasis, as reported for probiotic bacteria (Rivière et al. 2014). Several other genera of LAB have been identified in the *Cataglyphis* gut, including members of the families Lactobacillaceae (*Leuconostoc* and *Pediococcus*) and Enterococcaceae (*Enterococcus*) (Fhoula et al. 2015a, unpublished), in accordance with the work of Eilmus and Heil (2009). The presence of LAB within the gut microbiome supports the ants' access to plant-derived sugar-rich material such as floral nectar or hemipteran honeydew (Engel and Moran 2013). According to Husseneder et al. (2007), LAB can be considered as essential for the ecological balance in the ant gut.

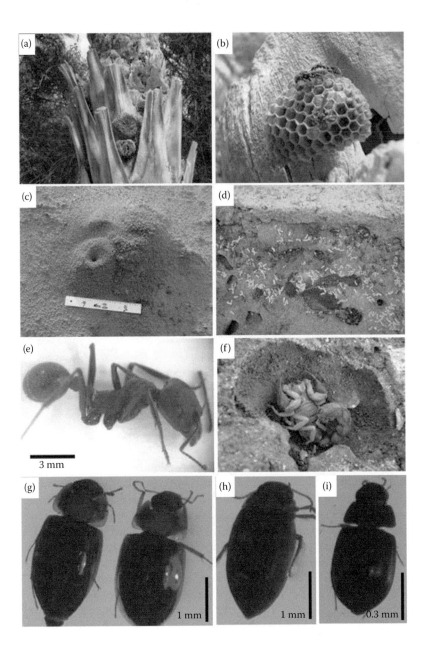

Figure 3.5 Examples of desert insects recovered from desert areas in southern Tunisia. (a, b) Honeybee (*Apis mellifera*) natural hives in a palm tree. (c) Nest of the desert ant *Cataglyphis* sp. located at the limit of Sebkha El Melah. (d) Underground excavation showing *Cataglyphis* galleries and their larvae. (e) Adult form of *Cataglyphis*. Scale bar: 0.3 cm. (f) Scorpion of the species *Androctonus australis* encountered in the area of Sebkha Ennaouel. (g–i) Mediterranean halophilic aquatic Coleoptera from a salt-saturated water–sediment interface (biofilm crust) recovered from Sebkha Ennaouel: (g) *Enochrus nigritus*, (H) *Nebrioporus cerisyi*, and (i) *Paracymus scutellaris*.

Few studies have reported the evaluation of *Weissella kimchii*, *Weissella confuse*, and *Weissella cibaria* strains as potential probiotics (M.J. Kim et al. 2008; Lee et al. 2012). *Weissella* species were isolated from a variety of fermented foods, soil, plants, animal products, human feces, and the gastrointestinal tract of humans and animals (Björkroth et al. 2002; Lee 2005; Lee et al. 2012). Besides, different *Weissella* strains were commonly described for their ability to produce EPSs (M.J. Kim et al. 2008). This ability is one of the distinctive phenotypic features of the genus *Weissella* as prebiotic-producing strains (Collins et al. 1993; Björkroth and Holzapfel 2006; Fhoula et al. 2015b, unpublished). In particular, strains of *Weissella* spp. have received interest due to their ability to produce significant quantities of EPSs (De Bruyne et al. 2008; Björkroth et al. 2009; Maina et al. 2011), fructan and heteropolysaccharides (Di Cagno et al. 2006; Malang et al. 2015), and nondigestible oligosaccharides (Chun et al. 2007; Immerzeel et al. 2014) that may decrease the risk of infections and diarrhea, increase bowel function and metabolism, and persist through the gastrointestinal tract and stimulate the growth of beneficial bacteria (Rastall and Gibson 2015). These overall data make EPS-producing strains of *Weissella* spp., particularly those isolated from extreme environments, very interesting for a broad range of industrial applications (Di Cagno et al. 2006; Wolter et al. 2014; Kajala et al. 2015).

3.3.2 Desert Plant-Associated Bacteria

Despite the harsh environment of the desert, several plants are able to withstand arid conditions by adopting diverse physical and behavioral mechanisms (the presence of few or no leaves, and the development of extremely long roots). In the last few years, the investigation of microorganism–desert host–plant interactions gave new insights into a complex relationship that makes both partners highly competitive and adaptive (Basil et al. 2004). Desert plant-associated microorganisms, mainly rhizospheric and endospheric bacteria, were widely investigated for their potential use as biocontrol agents against soilborne diseases (Sadfi et al. 2001), for the bioremediation of polluted soils (Narasimhan et al. 2003; Mapelli et al. 2012) and for phytoremediation (Moore et al. 2006). Much attention has been paid to their use as biofertilizers to stimulate crop growth in environments that are undergoing climate change and to meet the world's increasing demand for food (Saharan and Nehra 2011). Plant growth-promoting bacteria/rhizobacteria (PGPB/PGPR) exert beneficial effects on the growth of the host plant via direct and indirect mechanisms. They directly promote plant growth by increasing the availability of nutrients through atmospheric nitrogen fixation, inorganic phosphate solubilization, phytohormone production (Graham and Vance 2000; Richardson 2001; Spaepen et al. 2007), and siderophore release, which increases the availability of iron (Neilands 1995). Indirect mechanisms are related to the plant protection against phytopathogens, including antibiotic production and the synthesis of extracellular enzymes that hydrolyze the fungal cellular wall (Figure 3.4l) (Siddiqui 2005).

Numerous plants from extreme natural systems have been explored for their associated bacterial communities, such as date palm tree, *Salicornia*, cactus (Andrew et al. 2012; Mapelli et al. 2013; Cherif et al. 2015; Ferjani et al. 2015) (Figure 3.3). In the oasis ecosystem of southern Tunisia, the rhizosphere of the date palm tree showed a large number of Gammaproteobacteria and Actinobacteria (Ferjani et al. 2015). The diversity of the palm root endophytic bacterial community showed a clear segregation along a north–south gradient in correlation with geoclimatic parameters of the different investigated oases in the Tunisian Sahara. Using DGGE and cultivation-dependent methods, members of the Proteobacteria phylum, mainly Gammaproteobacteria, were identified as a major component, with *Pseudomonas* spp. being the most common (Cherif et al. 2015).

Strains recovered from palm root samples displayed numerous abovementioned PGP traits. The most retrieved activities were auxin synthesis (83%), followed by ammonia production (63%) and biofertilization activities, such as phosphate solubilization (48%) and siderophore release (44%). Furthermore, investigated strains have been shown to tolerate different abiotic stresses, such as drought stress, salinity, and high temperature (Ferjani et al. 2015). The potentiality of two endophytic strains, *Pseudomonas frederickbergensis* and *Pseudomonas brassicacearum*, as plant growth promoters under drought stress was investigated *in vitro* during a 9-month-long growth experiment. Bacterized date palm plantlets showed a significant increase in fresh (root and shoot) and dry biomasses (Cherif et al. 2015).

Salicornia plants (Figure 3.3e), collected from Sebkha and Chott hypersaline ecosystems in Tunisia, namely, Sebkha Ennaouel, Chott El Gharsa, and Chott Djerid, unveiled a large collection of halophilic and halotolerant bacteria with the prevalence of the *Halomonas* genus, typical of hypersaline environments (Mapelli et al. 2013). The established bacterial collection was tested *in vitro* for diverse PGP activities. Ninety-six percent of the strains showed the ability to produce indole-3-acetic acid (IAA), whereas only 2% of the strains belonging to the species *Halomonas taeheungii* and *Halomonas xinjiangensis* displayed ACC (I-Aminocyclopropane-I-Carboxylate)-deaminase activity in the presence of 5% NaCl. Besides, the bacterial collection has been analyzed for its capability to solubilize phosphate, fix nitrogen, and produce ammonia. The phosphate solubilization activity was confirmed in 65% of the total collection, including all the genera except for *Kushneria*. On the other hand, nitrogen fixation capacity has been phenotypically tested by the strain ability to develop in nitrogen-free medium. This test has been confirmed by molecular analysis by polymerase chain reaction (PCR) amplification of the *nifH* gene, codifying for a subunit of the nitrogenase enzyme. Six percent of the analyzed bacterial strains were positive to both tests, proving their potential ability to fix nitrogen. Ammonia production was also a common PGP trait shown by 93% of the isolates. Besides direct PGP activities, 11% of all the taxonomic classes of the collection belonging to the genera *Chromohalobacter*, *Halomonas*, *Kushneria*, *Marinococcus*, *Nesterenkonia*, and *Virgibacillus* displayed *in vitro* protease activity, showing their potential role as biocontrol players.

These outcomes show that the cultivable halophilic and halotolerant bacteria associated with plants inhabiting salty and arid ecosystems have considerable potential in promoting plant growth under the harsh conditions of salinity and drought stresses. Hence, they could be exploited as biofertilizers to sustain crop production in degraded and arid lands (Mapelli et al. 2013).

Several other studies demonstrate the promising attributes of using plant-associated bacteria as biofertilizers. Qin et al. (2015) investigated cultivable endophytic Actinobacteria associated with *Jatropha curcas* L. growing in Panxi dry, hot valley soil in southwest Sichuan, China. Similarly, bacteria (endophytic and epiphytic) associated with *Zygolhyllum dumosum* and *Atriplex halimus* from the Negev Desert exhibited activities typical of PGPB such as cellulose and chitin degradation and release of siderophores. The same study highlighted the potential of two strains, *Sinorhizobium medicae* and *Sinorhizobium meliloti*, isolated from *Trigonella stellate* nodules, to serve as inoculants for providing fixed nitrogen to alfalfa varieties living in arid environments (Kaplan et al. 2013).

3.3.3 Desert Arthropod-Associated Bacteria

The gut microbiome of insects and arthropods in general is a reservoir of microbes that have significant effects on the hosts' ecology and where several host–microbe symbiotic interactions, from strict parasitism to obligate mutualism, can subsist (Rappé and Giovannoni 2003). Gut microbial communities are known to play important roles in nutrition and health. They can contribute to the protection of the host from enemies, the immune stimulation, the increased tolerance to abiotic stresses, and the host sensing and communication (Dale and Moran 2006; Engel and Moran 2013). Several recognized examples of such insect–bacteria symbiosis comprise genera such as *Spiroplasma*, *Cardinium*, and *Wolbachia*, able to manipulate the reproduction of their host (Zindel et al. 2011).

The social insect honeybee (*Apis mellifera*) (Figure 3.5a and b) plays an important ecological role as pollinator of many crops worldwide (Kevan et al. 1999; Batra 1995; Dedej and Delaplane 2003; Jaffé et al. 2010). Through this pollinating activity, *A. mellifera* ensures the diversity and maintenance of plant species (Kevan et al. 1999; Nicholls and Altieri 2013). The pollination service is particularly important and determinant in arid lands where crop productivity yields are very low (Minckley et al. 2013). In these environments, wild populations of desert bees constitute the main pollinators (Figure 3.5a and b). Like many other insects, the gut of *A. mellifera* includes a high diversity of microbes, including beneficial or commensal symbionts and occasional disease-causing entomopathogens. In recent years, honeybees are facing a dramatic decline due to intensive use of pesticides (Pham-Delègue et al. 2002; Bailey et al. 2005; Johnson et al. 2010) and to their exposition to many infective diseases, such as American foulbrood (AFB). The latter is caused by *Paenibacillus larvae*, the primary bacterial pathogen of honeybees (Hamdi et al. 2013); viral diseases; parasites like mites (*Varroa destructor*) (Fries et al. 2003; Conte et al. 2010) and moths (*Galleria melonella*) (Hood et al. 2003; Coskun et al. 2006); and the colony collapse disorder (CCD) (Tentcheva et al. 2004; Cox-Foster et al. 2007).

Beside pathogens, several bacterial species, belonging to Proteobacteria, Firmicutes, and Actinobacteria groups, were detected in honeybee gut (Babendreier et al. 2006; Mohr and Tebbe 2006; Crotti et al. 2010, Hamdi et al. 2011). Among honeybee symbionts, several bacterial species showed a high ability to protect their host as LAB (detailed in Section 3.3.1) (Jeyaprakash et al. 2003; Mohr and Tebbe 2006; Olofsson and Vásquez 2008). Hence, LAB are the most commonly used probiotics, proposed to enhance bee immunity against invading bacteria (Forsgren et al. 2010). They are known to produce antimicrobial substances and others metabolites that could be useful for protecting honeybees from pathogens like *P. larvae*, the causal agent of AFB. Other associated honeybee bacteria belonging to *Bacillus* (Evans and Armstrong 2006; Cherif et al. 2008) and *Brevibacillus* (Alippi and Reynaldi 2006) have also been reported to be able to inhibit the growth of the honeybee pathogen *P. larvae in vitro*. In Tunisia, complex bacterial communities associated with honeybee have been detected, including *Lactobacillus*, *Gluconacetobacter*, Alphaproteobacteria, and spore-forming bacteria. Some isolated strains showed a high ability to stop the growth of *P. larvae in vitro*. These active bacteria were proposed as potential candidates to control the pathogen *P. lavae* on *in vitro*–reared larvae and then at the colony level (in the field) (Hamdi et al. 2015, unpublished).

Beside honeybees, ants (Formicidae) present another highly developed social insect (Schultz 2000). Desert ants belonging to the genus *Cataglyphis* are mainly recognized as one of the most typical in arid environments (Agosti 1994) (Figure 3.5c–e). The *Cataglyphis* ants consist of species that are well adapted to live on desert lands, characterized by high temperatures, salt, stress, and high UV irradiations (Wehner et al. 1983, 1992; Dillier and Wehner 2004). Several *Cataglyphis* species inhabiting the Sahara Desert have become a study model in research for navigation, orientation behavior, and ecology (Wehner 2003; Steck et al. 2009; Stieb et al. 2012). Some *Cataglyphis* species are capable of living in the salt pan areas of Algeria and Tunisia (Dillier and Wehner 2004).

Little is acknowledged regarding the gut bacterial communities of the desert *Cataglyphis* ants in Tunisia. Based on culture-independent and -dependent approaches, from the entire set of gut ant samples, a total of five phyla were identified: Actinobacteria, Firmicutes, Tenericutes, Alphaproteobacteria, and Gammaproteobacteria (Fhoula et al. 2015a, unpublished). Among the genera of Firmicutes, *Paenibacillus* presented an uncommon occurrence in some sampling stations. This last genus was reported for control of white grubs (Petersen 1985). Several members of Tenericutes (Entomoplasmataceae) and Alphaproteobacteria (Acetobacteraceae) have been shown to establish symbiotic association with several organism hosts (Kommanee et al. 2008; Crotti et al. 2010; Funaro et al. 2011). A number of genera of LAB have been identified in the *Cataglyphis* gut (Fhoula et al. 2015a, unpublished). Their presence may contribute to protection against pathogens and improve *Cataglyphis* grazing on plant material remains (Potrikus and Breznak 1981; Bauer et al. 2000; Cox and Gilmore 2007; Fhoula et al. 2015a, unpublished). Furthermore, several of the bacterial species have been proposed to contribute to ant lignin and cellulose degradation, and nitrogen-fixing activities (Bugg et al. 2011; Kalyani et al. 2011). The desert *Cataglyphis*-associated bacteria play important roles, yet to unravel, in ant life in harsh environments.

Another very intriguing, rare, and endemic group of insects is the Mediterranean aquatic beetles. These halotolerant and halophilic species of aquatic Coleoptera, which are typical of saline waters, were mainly investigated for their genetic diversity (Abellán et al. 2007, 2009) and tolerance to salinity and extreme temperatures (Sánchez-Fernández et al. 2010). Three dominant species were recovered from Sebkha Ennaouel: *Enochrus nigritus*, *Paracymus scutellaris*, and *Nebrioporus cerisyi* (Figure 3.5g–i). Using DGGE, species-specific bacterial communities were detected (Elmnasri et al. unpublished). The phylogenetic identification of some selected DGGE bands showed the affiliation of related bacteria to *Thiothrix* sp. within the Gammaproteobacteria phylum. These filamentous sulfur-oxidizing bacteria (*Thiothrix*-like bacteria) were described as epibionts of aquatic invertebrates using fluorescent *in situ* hybridization (FISH) and DGGE (Gillan and Dubilier 2004). Other bacteria, closely related to *Tepidibacter thalassicus* of the Firmicutes phylum, were able to reduce elemental sulfhur and grow organotrophically on a number of protein-aceous substrates and carbohydrates (Slobodkin et al. 2003). In light of the preliminary data, studies are in progress to determine the role and functions of these specific bacteria in the ecology of the aquatic beetles.

The last example of well-adapted desert arthropods is Scorionidae that thrive in harsh conditions thanks to their ability to control their metabolism. Studies were focused mainly on scorpion venom extracts (Schwartz et al. 2007) and the presence and distribution of *Wolbachia* (Baldo et al. 2007). The commensal and symbiotic microbial communities characterizing *Androctonus austra-lis*, the main scorpion species recovered from the arid region of Sidi Bouzid in the south of Tunisia, were studied (Figure 3.5f). Using DGGE and culture-dependent approaches (Elmnasri et al. unpub-lished), the gut, gonads, and venom gland of *A. australis* showed the presence of bacteria affili-ated with Firmicutes, Betaproteobacteria, Gammaproteobcteria, Flavobacteria, Actinobacteria, and Mollicutes phyla with a predominance of LAB. Many of the scorpion-associated strains are endowed with interesting probiotic traits, particularly in relation to drought and salinity. Studies are currently in progress to unravel the biotechnological potential of desert arthropod-associated microorganisms and to understand their role in the ecological fitness of their hosts.

3.4 CONCLUDING REMARKS

The astonishing geographical (topographical) nature of southern Tunisia showing an alternation of oases, salt lakes, and desert Sahara, in a relatively restricted space with diverse biotic and abi-otic factors, received considerable attention to investigate microbial diversity. As discussed in this chapter, a huge microbial diversity within the Archaea and Bacteria domains, including halophilic, halotolerant, PGPB/PGPR, and fungi, was recovered. The overall data provide further evidence of the driven selective effect of pedo-geo-climatic factors in shaping bacterial communities. Thus, the structure of the microbiota associated with diverse microcosms or niches (biofilm, salt crust, oases soil, plant, insect, etc.) provides a valuable tool as a bioindicator for detecting environmental changes. The considerable biotechnological potential, including environmental, agricultural, and industrial applications, displayed by extremophiles and insect- and plant-associated microbiota offers new perceptions on how these microorganisms manage their metabolic and enzymatic path-ways and will contribute to the development of straightforward strategies of microbial resource management for sustainable agriculture in arid environments. In the field of the discovery of new metabolites, Cyanobacteria and Archaea still present promising untapped resources. Furthermore, transcriptomic, proteomic, and genomic approaches are currently providing new avenues to eluci-date the functional interactions between plants and insects and their associated microbiomes.

ACKNOWLEDGMENTS

The authors acknowledge financial support from the European Union (EU Project BIODESERT, FP7-CSA-SA REGPOT-2008-2, grant agreement no. 245746); the South African/Tunisia research partnership program, bilateral agreement 2014 ("An African Desert Soils Microbial Ecology Research

Network" and "Exploration for Novel Xylanases and Ligninases Produced by Actinobacterial Species for Biomass Valorization" research projects); and the Tunisian Ministry of Higher Education and Scientific Research in the ambit of the laboratory projects LR MBA206 and LR11ES31. English editing and corrections were performed by Meriem Attyaoui and Sarra Marouani

REFERENCES

Abed RM, Kohls K, Schoon R et al. Lipid biomarkers, pigments and cyanobacterial diversity of microbial mats across intertidal flats of the arid coast of the Arabian Gulf (Abu Dhabi, UAE). *FEMS Microbiology Ecology* 65 (2008): 449–462.

Abellán P, Gómez-Zurita J, Millán A et al. Conservation genetics in hypersaline inland waters: Mitochondrial diversity and phylogeography of an endangered Iberian beetle (Coleoptera: Hydraenidae). *Conservation Genetics* 8 (2007): 79–88.

Abellán P, Millán A, Ribera I. Parallel habitat-driven differences in the phylogeographical structure of two independent lineages of Mediterranean saline water beetles. *Molecular Ecology* 18 (2009): 3885–3902.

Agosti D. A new inquiline ant (Hymenoptera: Formicidae) in *Cataglyphis* and its phylogenetic relationship. *Journal of Natural History* 28 (1994): 913–919.

Ahmad F, Ahmad I, Khan MS. Screening of free-living rhizospheric bacteria for their multiple plant growth promoting activities. *Microbiological Research* 163 (2008): 173–181.

Alippi AM, Reynaldi FJ. Inhibition of the growth of *Paenibacillus larvae*, the causal agent of American foulbrood of honeybees, by selected strains of aerobic spore-forming bacteria isolated from apiarian sources. *Journal of Invertebrate Pathology* 91 (2006): 141–146.

Andrew DR, Fitak RR, Munguia-Vega A, Racolta A, Martinson VG, Dontsovag K. Abiotic factors shape microbial diversity in Sonoran Desert soils. *Applied and Environmental Microbiology* 78 (2012): 7527–7537.

Asker D, Ohta Y. *Haloferax alexandrinus* sp. nov., an extremely halophilic canthaxanthin-producing archaeon from a solar saltern in Alexandria (Egypt). *International Journal of Systematic and Evolutionary Microbiology* 52 (2002): 729–738.

Baati H, Guermazi S, Amdouni R, Gharsallah N, Sghir A, Ammar E. Prokaryotic diversity of a Tunisian multipond solar saltern. *Extremophiles* 12 (2008): 505–518.

Babendreier D, Joller D, Romeis J, Bigler F, Widmer F. Bacterial community structures in honeybee intestines and their response to two insecticidal proteins. *FEMS Microbiology Ecology* 59 (2006): 600–610.

Bachar A, Ines M, Soares M, Gillor O. The effect of resource islands on abundance and diversity of bacteria in arid soils. *Microbial Ecology* 63 (2012): 694–700.

Bacon CW, Hinton DM. Endophytic and biological control potential of *Bacillus mojavensis* and related species. *Biological Control* 23 (2002): 274–284.

Bacon CW, Hinton DM. *Bacillus mojavensis*: Its endophytic nature, the surfactins and their role in the plant response to infection by *Fusarium verticillioides*. In *Bacteria in Agrobiology: Plant Growth Responses*, ed. Maheswari DK. Berlin: Springer, 2011, pp. 21–39.

Bailey J, Scott-Dupree C, Harris R, Tolman J, Harris B. Contact and oral toxicity to honey bees (*Apis mellifera*) of agents registered for use for sweet corn insect control in Ontario, Canada. *Apidologie* 36 (2005): 623–633.

Baldo L, Prendini L, Corthals A, Werren JH. Wolbachia are present in Southern African scorpions and cluster with supergroup F. *Current Microbiology* 55 (2007): 367–373.

Barbieri R, Stivaletta N, Marinangeli L, Ori GG. Microbial signatures in Sabkha evaporite deposits of Chott el Gharsa (Tunisia) and their astrobiological implications. *Planetary and Space Science* 54 (2006): 726–736.

Baricz A, Coman C, Andrei AS et al. Spatial and temporal distribution of archaeal diversity in meromictic, hypersaline Ocnei Lake (Transylvanian Basin, Romania). *Extremophiles* 18 (2014): 399–413.

Basil A, Strap JL, Knotek-Smith HM, Crawford DL. Studies on the microbial populations of the rhizosphere of big sagebrush (*Artemisia tridentata*). *Journal of Industrial Microbiology and Biotechnology* 31 (2004): 278–288.

Batra SWT. Bees and pollination in our changing environment. *Apidologie* 26 (1995): 361–370.

Bauer S, Tholen A, Overmann J, Brune A. Characterization of abundance and diversity of lactic acid bacteria in the hindgut of wood- and soil-feeding termites by molecular and culture-dependent techniques. *Archives of Microbiology* 173 (2000): 126–137.

Belnap J. Biological soil crusts and wind erosion. In *Biological Soil Crusts: Structure, Function, and Management*, ed. J Belnap, OL Lange. Berlin: Springer, 2001, pp. 339–347.

Benaoun A, Elbakkey M, Ferchichi A. Change of oases farming systems and their effects on vegetable species diversity: Case of oasian agro-systems of Nefzaoua (South of Tunisia). *Scientia Horticulturae* 180 (2014): 167–175.

Ben Salah M. Food value of soft dates cultivated in Tunisian coastal oases. *Journal of Life Sciences* 9 (2015): 234–241.

Birbir M, Eryilmaz S, Ogan A. Prevention of halophilic microbial damage on brine cured hides by extremely halophilic halocin producer strains. *Journal of the Society of Leather Technologists and Chemists* 88 (2004): 99–104.

Björkroth J, Holzapfel WH. Genera *Leuconostoc, Oenococcus* and *Weissella*. In *The Prokaryotes*, ed. M Dworkin, S Falcow, E Rosenberg, KH Schleifer, E Stackebrandt. New York: Springer, 2006, pp. 267–319.

Björkroth JA, Dicks LMTD, Holzapfel WH. Genus III. *Weissella* Collins, Samelis, Metaxopoulos and Wallbanks 1994, 370VP (effective publication: Collins, Samelis, Metaxopoulos, and Wallbanks 1993, 597). In *Bergey's Manual of Systematic Bacteriology*, vol. 3, *The Firmicutes*, ed. P de Vos, GM Garrity, D Jones et al. 2nd ed. New York: Springer, 2009, pp. 643–654.

Björkroth KJ, Schillinger U, Geisen R et al. Taxonomic study of *Weissella confusa* and description of *Weissella cibaria* sp. nov., detected in food and clinical samples. *International Journal of Systematic and Evolutionary Microbiology* 52 (2002): 141–148.

Boujelben I, Gomariz M, Martínez-García M et al. Spatial and seasonal prokaryotic community dynamics in ponds of increasing salinity of Sfax solar saltern in Tunisia. *Antonie Van Leeuwenhoek* 101 (2012): 845–857.

Bouvy M, Molica R, De Oliveira S, Marinho M, Beker B. Dynamics of a toxic cyanobacterial bloom (*Cylindrospermopsis raciborskii*) in a shallow reservoir in the semi-arid region of northeast Brazil. *Aquatic Microbial Ecology* 20 (1999): 285–297.

Bryant RG, Sellwood BW, Millington AC, Drake NA. Marine-like potash evaporite formation on a continental playa: Case study from Chott el Djerid, southern Tunisia. *Sedimentary Geology* 90 (1994): 269–291.

Buczynski C, Chafetz HS. Habit of bacterially induced precipitates of calcium carbonate and the influence of medium vicosity on mineralogy. *Journal of Sedimentary Research* 61 (1991): 226–233.

Büdel B, Weber B, Kühl M, Pfanz H, Sültemeyer D, Wessels D. Reshaping of sandstone surfaces by crypto-endolithic cyanobacteria: Bioalkalization causes chemical weathering in arid landscapes. *Geobiology* 2 (2004): 261–268.

Bugg TD, Ahmad M, Hardiman EM, Rahmanpour R. Pathways for degradation of lignin in bacteria and fungi. *Natural Product Reports* 28 (2011): 1883–1896.

Burja AM, Webster NS, Murphy PT, Hill RT. Microbial symbionts of Great Barrier Reef sponges. In *Proceedings of the 5th International Sponge Symposium: Memoirs of the Queensland Museum*, Brisbane, Australia, 1999, pp. 63–75.

Butinar L, Santos S, Spencer-Martins I, Oren A, Gunde-Cimerman N. Yeast diversity in hypersaline habitats. *FEMS Microbiology Letters* 244 (2005a): 229–234.

Butinar L, Sonjak S, Zalar P, Plemenitas A, Gunde-Cimerman N. Melanized halophilic fungi are eukaryotic members of microbial communities in hypersaline waters of solar salterns. *Botanica Marina* 48 (2005b): 73–79.

Cámara B, Suzuki S, Nealson KH et al. Ignimbrite textural properties as determinants of endolithic colonization patterns from hyper-arid Atacama Desert. *International Microbiology* 17 (2015): 235–247.

Cantrell SA, Casillas-Martinez L, Molina M. Characterization of fungi from hypersaline environments of solar salterns using morphological and molecular techniques. *Mycological Research* 110 (2006): 962–970.

Caumette P, Matheron R, Raymond N, Relexans JC. Microbial mats in the hypersaline ponds of Mediterranean salterns (Salins-de-Giraud, France). *FEMS Microbiology Ecology* 13 (1994): 273–286.

Cerritos R, Vinuesa P, Eguiarte LE et al. *Bacillus coahuilensis* sp. nov., a moderately halophilic species from a desiccation lagoon in the Cuatro Cienegas Valley in Coahuila, Mexico. *International Journal of Systematic and Evolutionary Microbiology* 58 (2008): 919–923.

Chaabouni I, Guesmi A, Cherif A. Secondary metabolites of *Bacillus*: Potentials in biotechnology. In *Bacillus thuringiensis Biotechnology*, ed. E Sansinenea. Amsterdam: Springer, 2012, pp. 347–366.

Chandi GK, Gill BS. Production and characterization of microbial carotenoids as an alternative to synthetic colors: A review. *International Journal of Food Properties* 14 (2011): 503–513.

Chaucheyras-Durand F, Durand H. Probiotics in animal nutrition and health. *Beneficial Microbes* 1 (2009): 3–9.

Cherif A, Rezgui W, Raddadi N, Daffonchio D, Boudabous A. Characterization and partial purification of entomocin 110, a newly identified bacteriocin from *Bacillus thuringiensis* subsp. Entomocidus HD110. *Microbiological Research* 163 (2008): 684–692.

Cherif H, Marasco R, Rolli E et al. Oasis desert farming selects environment-specific date palm root endophytic communities and cultivable bacteria that promote resistance to drought. *Environmental Microbiology Reports* 7 (2015): 668–678.

Chouchane H, Neifar M, Jouani A, Mahjoubi M, Masmoudi AS, Cherif A. Microbial exopolysaccharides as efficient bioflocculants in wastewater treatment. In *Wastewater Treatment: Processes, Management Strategies and Environmental/Health Impacts*, ed. LM Barrett. New York: Nova Publishers, 2015a, pp. 163–189.

Chouchane H, Neifar M, Raddadi N, Fava F, Masmoudi AS, Cherif A. Microbial exopolysaccharides as alternative sources of dietary fibers with interesting functional and healthy properties. In *Dietary Fiber: Production Challenges, Food Sources and Health Benefits*, ed. EC Marvin. New York: Nova Publishers, 2015b, pp. 159–178.

Chu D, Tong J, Song H et al. Early Triassic wrinkle structures on land: Stressed environments and oases for life. *Scientific Reports* 5 (2015).

Chun J, Kim GM, Lee KW et al. Conversion of isoflavone glucosides to aglycones in soymilk by fermentation with lactic acid bacteria. *Journal of Food Science* 72 (2007): M39–M44.

Cockell CS, Lee P, Osinski G, Horneck G, Broady P. Impact-induced microbial endolithic habitats. *Meteoritics and Planetary Science* 37 (2002): 1287–1298.

Collins MD, Samelis J, Metaxopoulos J, Wallbanks S. Taxonomic studies on some leuconostoc-like organisms from fermented sausages: Description of a new genus *Weissella* for the *Leuconostoc paramesenteroides* group of species. *Journal of Applied Bacteriology* 75 (1993): 595–603.

Conte YL, Ellis M, Ritter W. *Varroa* mites and honey bee health: Can *Varroa* explain part of the colony losses? *Apidologie* 41 (2010): 353–363.

Coskun M, Kayis T, Sulanc M, Ozalp P. Effects of different honeycomb and sucrose levels on the development of greater wax moth *Galleria mellonella* larvae. *International Journal of Agriculture and Biology* 8 (2006): 855–858.

Cox CR, Gilmore MS. Native microbial colonization of *Drosophila melanogaster* and its use as a model of *Enterococcus faecalis* pathogenesis. *Infection and Immunity* 75 (2007): 1565–1576.

Cox-Foster DL, Conlan S, Holmes EC et al. A metagenomic survey of microbes in honey bee colony collapse disorder. *Science* 318 (2007): 283–287.

Crotti E, Rizzi A, Chouaia B et al. Acetic acid bacteria, newly emerging symbionts of insects. *Applied and Environmental Microbiology* 76 (2010): 6963–6970.

Dale C, Moran NA. Molecular interactions between bacterial symbionts and their hosts. *Cell* 126 (2006): 453–465.

Dar SA, Yao L, van Dongen U, Kuenen JG, Muyzer G. Analysis of diversity and activity of sulfate-reducing bacterial communities in sulfidogenic bioreactors using 16S rRNA and dsrB genes as molecular markers. *Applied and Environmental Microbiology* 73 (2007): 594–604.

Das A, Yoon SH, Lee SH, Kim JY, Oh DK, Kim SW. An update on microbial carotenoid production: Application of recent metabolic engineering tools. *Applied Microbiology and Biotechnology* 77 (2007): 505–512.

DasSarma, S. Extreme halophiles are models for astrobiology. *Microbe* 1 (2006): 120–126.

De Bruyne K, Camu N, Lefebvre K, De Vuyst L, Vandamme P. *Weissella ghanensis* sp. nov., isolated from a Ghanaian cocoa fermentation. *International Journal of Systematic and Evolutionary Microbiology* 58 (2008): 2721–2725.

Dedej S, Delaplane KS. Honey bee (Hymenoptera: Apidae) pollination of rabbiteye blueberry *Vaccinium ashei* var. "Climax" is pollinator density-dependent. *Journal of Economic Entomology* 96 (2003): 1215–1220.

Dennis P, Edwards EA, Liss SN, Fulthorpe R. Monitoring gene expression in mixed microbial communities by using DNA microarrays. *Applied and Environmental Microbiology* 69 (2003): 769–778.

De Philippis R, Margheri MC, Materassi R, Vincenzini M. Potential of unicellular cyanobacteria from saline environments as exopolysaccharide producers. *Applied and Environmental Microbiology* 64 (1998): 1130–1132.

Díaz Z M, Fernández CMV. Field performance of a liquid formulation of Azospirillum brasilense on dryland wheat productivity. *European Journal of Soil Biology* 45 (2009): 3–11.

Di Cagno R, De Angelis M, Limitone A et al. Glucan and fructan production by sourdough *Weissella cibaria* and *Lactobacillus plantarum*. *Journal of Agricultural and Food Chemistry* 54 (2006): 9873–9881.

Dillier FX, Wehner R. Spatio-temporal patterns of colony distribution in monodomous and polydomous species of North African desert ants, genus *Cataglyphis*. *Insectes Sociaux* 51 (2004): 186–196.

Dong H, Rech JA, Jiang H, Sun H, Buck BJ. Endolithnic cyanobacteria in soil gypsum: Occurrences in Atacama (Chile), Mojave (United States), and Al-Jafr Basin (Jordan) Deserts. *Journal of Geophysical Research: Biogeosciences* 112 (G2) (2007).

Douglas AE. The microbial dimension in insect nutritional ecology. *Functional Ecology* 23 (2009): 38–47.

Drake NA. Playa sedimentology and geomorphology: Mixture modelling applied to Landsat thematic mapper data of Chott el Djerid, Tunisia. In *Sedimentology and Geochemistry of Modern and Ancient Saline Lakes*. Society for Sedimentary Geology Special Publication No. 50. Tulsa, OK: Society for Sedimentary Geology, 1994, pp. 125–131.

Dubey AK, Rai AK. Application of algal biofertilizers (Aulosira fertilissima tenuis and Anabaena doliolum Bhardawaja) for sustained paddy cultivation in Northern India. *Israel Journal of Plant Sciences* 43 (1995): 41–51.

Dvornyk V, Nevo E. Genetic polymorphism of cyanobacteria under permanent natural stress: A lesson from the "Evolution Canyons." *Research in Microbiology* 154 (2003): 79–84.

Eilmus S, Heil M. Bacterial associates of arboreal ants and their putative functions in an obligate ant-plant mutualism. *Applied and Environmental Microbiology* 75 (2009): 4324–4332.

El Hidri D, Guesmi A, Najjari A et al. Cultivation-dependent assessment, diversity, and ecology of haloalkaliphilic bacteria in arid saline systems of southern Tunisia. *BioMed Research International* (2013).

El-Said AHM, Saleem A. Ecological and physiological studies on soil fungi at Western Region, Libya. *Mycobiology* 36 (2008): 1–9.

Engel P, Moran NA. The gut microbiota of insects—Diversity in structure and function. *FEMS Microbiology Reviews* 37 (2013): 699–735.

Ettoumi B, Raddadi N, Borin S, Daffonchio D, Boudabous A, Cherif A. Diversity and phylogeny of culturable spore-forming bacilli isolated from marine sediments. *Journal of Basic Microbiology* 49 (2009): 1–11.

Euzeby JP. List of prokaryotic names with standing in nomenclature—Genus. *Proteus* (2013).

Evans JD, Armstrong TN. Antagonistic interactions between honey bee bacterial symbionts and implications for disease. *BMC Ecology* 6 (2006): 1.

Ferjani R, Marasco R, Rolli E et al. The date palm tree rhizosphere is a niche for plant growth promoting bacteria in the oasis ecosystem. *BioMed Research International* 2015 (2015): 153851.

Fhoula I, Najjari A, Turki Y, Jaballah S, Boudabous A, Ouzari H. Diversity and antimicrobial properties of lactic acid bacteria isolated from rhizosphere of olive trees and desert truffles of Tunisia. *BioMed Research International* (2013): 405708.

Forsgren E, Olofsson TC, Vásquez A, Fries I. Novel lactic acid bacteria inhibiting *Paenibacillus larvae* in honey bee larvae. *Apidologie* 41 (2010): 99–108.

Franklin RB, Mills AL. Multi-scale variation in spatial heterogeneity for microbial community structure in an eastern Virginia agricultural field. *FEMS Microbiology Ecology* 44 (2003): 335–346.

Fries I, Hansen H, Imdorf A, Rosenkranz P. Swarming in honey bees (*Apis mellifera*) and *Varroa destructor* population development in Sweden. *Apidologie* 34 (2003): 389–397.

Funaro CF, Kronauer DJC, Moreau CS, Goldman-Huertas B, Pierce NE, Russell JA. Army ants harbor a host-specific clade of Entomoplasmatales bacteria. *Applied and Environmental Microbiology* 773 (2011): 346–350.

Garcia-Pichel F, Pringault O. Microbiology: Cyanobacteria track water in desert soils. *Nature* 413 (2001): 380–381.

Gehling JG. Microbial mats in terminal Proterozoic siliciclastics; Ediacaran death masks. *Palaios* 14 (1999): 40–57.

Gerdes G, Thomas K, Noffke N. Microbial signatures in peritidal siliciclastic sediments: a catalogue. *Sedimentology* 47 (2000): 279–308.

Gendrin A, Mangold N, Bibring JP et al. Sulfates in Martian layered terrains: The OMEGA/Mars express view. *Science* 307 (2005): 1587–1591.

Ghozlan H, Deif H, Abu Kandil R, Sabry S. Biodiversity of moderately halophilic bacteria in hypersaline habitats in Egypt. *Journal of General and Applied Microbiology* 52 (2006): 63–72.

Gillan DC, Dubilier N. Novel epibiotic *Thiothrix* bacterium on a marine amphipod. *Applied and Environmental Microbiology* 70 (2004): 3772–3775.

Glick BR. The enhancement of plant growth by free-living bacteria. *Canadian Journal of Microbiology* 41 (1995): 109–117.

Gordon RE, Haynes WC, Pang CH-N. The genus *Bacillus*. No. 427. Washington, DC: Agricultural Research Service, U.S. Department of Agriculture, 1973.

Gorjan A, Plemenitaš A. Identification and characterization of ENA ATPases HwENA1 and HwENA2 from the halophilic black yeast *Hortaea werneckii*. *FEMS Microbiology Letters* 265 (2006): 41–50.

Gostinčar C, Grube M, De Hoog S, Zalar P, Gunde-Cimerman N. Extremotolerance in fungi: Evolution on the edge. *FEMS Microbiology Ecology* 71 (2010): 2–11.

Gostinčar C, Lenassi M, Gunde-Cimerman N, Plemenitaš A. Fungal adaptation to extremely high salt concentrations. *Advances in Applied Microbiology* (2011): 71–96.

Gostinčar C, Ohm R, Kogej T et al. Genome sequencing of four *Aureobasidium pullulans* varieties: Biotechnological potential, stress tolerance, and description of new species. *BMC Genomics* 15 (2014): 549.

Gould GW. History of science—Spores. *Journal of Applied Microbiology* 101 (2006): 507–513.

Graham PH, Vance CP. Nitrogen fixation in perspective: An overview of research and extension needs. *Field Crops Research* 65 (2000): 93–106.

Grant WD, Gemmell RT, McGenity TJ. Halobacteria: The evidence for longevity. *Extremophiles* 2 (1998): 279–287.

Grant WD, Kamekura M, McGenity TJ, Ventosa A. Halobacteriaceae. In *Bergey's Manual of Systematic Bacteriology*, ed. DR Boone, RW Castenholz. 2nd ed., vol. I. Berlin: Springer, 2001, pp. 299–301.

Gueddari M. Géochimie et thermodynamique des évaporites continentales: Etude du lac Natron (Tanzanie) et du Chott El Jerid (Tunisie). Strasbourg, France: Université Louis Pasteur de Strasbourg, Institut de Géologie, 1984.

Guesmi A, Ettoumi B, El Hidri D et al. Uneven distribution of *Halobacillus trueperi* species in arid natural saline systems of southern Tunisian Sahara. *Microbial Ecology* 66 (2013): 831–839.

Gunde-Cimerman N, Zalar P, De Hoog S, Plemenitas A. Hypersaline waters in salterns—Natural ecological niches for halophilic black yeasts. *FEMS Microbiology Ecology* 32 (2000): 235–240.

Hacĕne H, Rafa F, Chebhouni N et al. Biodiversity of prokaryotic microflora in El Golea salt lake, Algerian Sahara. Journal of Arid Environments 58 (2004): 273–284.

Hamdi C, Balloi A, Essanaa J et al. Gut microbiome dysbiosis and honeybee health. *Journal of Applied Entomology* 135 (2011): 524–533.

Hamdi C, Essanaa J, Sansonno L et al. Genetic and biochemical diversity of *Paenibacillus larvae* isolated from Tunisian infected honey bee broods. *BioMed Research International* 2013 (2013).

Haseltine C, Hill T, Montalvo-Rodriguez R, Kemper SK, Shand RF, Blum P. Secreted euryarchaeal microhalocins kill hyperthermophilic crenarchaea. *Journal of Bacteriology* 183 (2001): 287–291.

Hedi A, Sadfi N, Fardeau ML et al. Studies on the biodiversity of halophilic microorganisms isolated from El-Djerid salt lake (Tunisia) under aerobic conditions. *International Journal of Microbiology* 2009 (2009): 731786.

Holzapfel WH, van Wyk EP. *Lactobacillus kandleri* sp. nov., a new species of the subgenus *Betabacterium* with glycine in the peptidoglycan. *Zentralblatt für Bakteriologie Mikrobiologie und Hygiene: I. Abt. Originale C: Allgemeine, Angewandte und Ökologische Mikrobiologie* 3 (1982): 495–502.

Hood WM, Horton PM, Mc Creadie JW. Field evaluation of the red imported fire ant (Hymenoptera: Formicidae) for the control of wax moths (Lepidoptera: Pyralidae) in stored honey bee comb. *Journal of Agricultural and Urban Entomology* 20 (2003): 93–103.

Hu C, Gao K, Whitton BA. Semi-arid regions and deserts. In *Ecology of Cyanobacteria II: Their Diversity in Space and Time*, ed. BA Whitton. Amsterdam: Springer, 2012, pp. 345–369.

Huber H, John MJ, Rachel R, Fuchs T, Wimmer VC, Setter KO. A new phylum of Archaea represented by a nanosized hyperthermophilic symbiant. *Nature* 417 (2002): 27–28.

Hugenholtz J. The lactic acid bacterium as a cell factory for food ingredient production. *International Dairy Journal* 18 (2008): 466–475.

Husseneder C, Wise BR, Higashiguchi DT. Bugs in bugs: The microbial diversity of the termite gut. *Proceedings of the Hawaiian Entomological Society* 39 (2007): 143–144.

Immerzeel P, Falck P, Galbe M, Adlercreutz P, Nordberg Karlsson E, Stålbrand H. Extraction of water-soluble xylan from wheat bran and utilization of enzymatically produced xylooligosaccharides by *Lactobacillus*, *Bifidobacterium* and *Weissella* spp. *LWT Food Science and Technology* 56 (2014): 321–327.

Isichei AO. The role of algae and cyanobacteria in arid lands. A review. *Arid Soil Research and Rehabilitation* 4 (1990): 1–17.

Jaffé R, Dietemann V, Allsopp MH et al. Estimating the density of honeybee colonies across their natural range to fill the gap in pollinator decline censuses. *Conservation Biology* 24 (2010): 583–593.

Jaouani A, Neifar M, Prigione V et al. Diversity and enzymatic profiling of halotolerant micromycetes from Sebkha El Melah, a Saharan salt flat in southern Tunisia. *BioMed Research International* 2014 (2014): 439197.

Javor B. *Hypersaline Environments: Microbiology and Biogeochemistry*. Berlin: Springer Science & Business Media, 2012.

Jeyaprakash A, Hoy MA, Allsopp MH. Bacterial diversity in worker adults of *Apis mellifera capensis* and *Apis mellifera scutellata* (Insecta: Hymenoptera) assessed using 16S rRNA sequences. *Journal of Invertebrate Pathology* 84 (2003): 96–103.

Jin M, Lu Z, Huang M, Wang Y. Effects of Se-enriched polysaccharide produced by *Enterobacter cloacae* Z0206 on alloxan-induced diabetic mice. *International Journal of Biological Macromolecules* 50 (2012): 348–352.

Johnson RM, Ellis MD, Mullin CA, Frazier M. Pesticides and honey bee toxicity—USA. *Apidologie* 41 (2010): 312–331.

Kajala I, Shi Q, Nyyssölä A et al. Cloning and characterization of a *Weissella confusa* dextransucrase and its application in high fibre baking. *PLoS One* 10 (2015): 1.

Kalyani D, Phugare S, Shedbalkar U, Jadhav J. Purification and characterization of a bacterial peroxidase from the isolated strain *Pseudomonas* sp. SUK1 and its application for textile dye decolorization. *Annals of Microbiology* 61 (2011): 483–491.

Kamekura M. Lipids of extreme halophiles. In *The Biology of Halophilic Bacteria*, ed. RH Vreeland, LI Hochstein. Boca Raton, Florida: CRC Press, 1993, pp. 135–161.

Kang S, Mills AL. The effect of sample size in studies of soil microbial community structure. *Journal of Microbiological Methods* 66 (2006): 242–250.

Kaplan D, Maymon M, Agapakis CM et al. A survey of the microbial community in the rhizosphere of two dominant shrubs of the Negev Desert highlands, *Zygophyllum dumosum* (*Zygophyllaceae*) and *Atriplex halimus* (*Amaranthaceae*), using cultivation-dependent and cultivation-independent methods. *American Journal of Botany* 100 (2013): 1713–1725.

Kebbouche-Gana S, Gana ML, Ferrioune I et al. Production of biosurfactant on crude date syrup under saline conditions by entrapped cells of *Natrialba* sp. strain E21, an extremely halophilic bacterium isolated from a solar saltern (Ain Salah, Algeria). *Extremophiles* 17 (2013): 981–993.

Kebbouche-Gana S, Gana ML, Khemili S et al. Isolation and characterization of halophilic Archaea able to produce biosurfactants. *Journal of Industrial Microbiology and Biotechnology* 36 (2009): 727–738.

Kereszturi A, Dulai S, Marschall M, Pócs T. The Chott el Jerid Mars analog expedition. In *Lunar and Planetary Science Conference*, Texas, 2014, p. 1357.

Kevan PG. Pollinators as bioindicators of the state of the environment: Species, activity and diversity. *Agriculture, Ecosystems and Environment* 74 (1999): 373–393.

Kim JS, Dungan RS, Crowley D. Microarray analysis of bacterial diversity and distribution in aggregates from a desert agricultural soil. *Biology and Fertility of soils* 44 (2008): 1003–1011.

Kim MJ, Seo HN, Hwang TS, Lee SH, Park DH. Characterization of exopolysaccharide (EPS) produced by *Weissella hellenica* SKkimchi3 isolated from Kimchi. *Journal of Microbiology* 46 (2008): 535–541.

Köberl, M, Müller H, Ramadan EM, Berg G. Desert farming benefits from microbial potential in arid soils and promotes diversity and plant health. *PLoS One* 6 (2011): e24452.

Kogej T, Ramos J, Plemenitas A, Gunde-Cimerman N. The halophilic fungus *Hortaea werneckii* and the halotolerant fungus *Aureobasidium pullulans* maintain low intracellular cation concentrations in hypersaline environments. *Applied and Environmental Microbiology* 71 (2005): 6600–6605.

Kogej T, Stein M, Volkmann M, Gorbushina AA, Galinski EA, Gunde-Cimerman N. Osmotic adaptation of the halophilic fungus *Hortaea werneckii*: Role of osmolytes and melanization. *Microbiology* 153 (2007): 4261–4273.

Kommanee J, Akaracharanya A, Tanasupawat S et al. Identification of *Acetobacter* strains isolated in Thailand based on 16S-23S rRNA gene ITS restriction and 16S rRNA gene sequence analyses. *Annals of Microbiology* 58 (2008): 319–324.

Kremer B, Kazmierczak J, Stal L. Calcium carbonate precipitation in cyanobacterial mats from sandy tidal flats of the North Sea. *Geobiology* 6 (2008): 46–56.

Kumar D, Adhikary SP. Diversity, molecular phylogeny, and metabolic activity of cyanobacteria in biological soil crusts from Santiniketan (India). *Journal of Applied Phycology* 27 (2015): 339–349.

Kutovaya OV, Lebedeva MP, Tkhakakhova AK, Ivanova EA, Andronov EE. Metagenomic characterization of biodiversity in the extremely arid desert soils of Kazakhstan. *Eurasian Soil Science* 48 (2015): 493–500.

Lacap DC, Warren-Rhodes KA, McKay CP, Pointing SB. Cyanobacteria and chloroflexi-dominated hypo-lithic colonization of quartz at the hyper-arid core of the Atacama Desert, Chile. *Extremophiles* 15 (2011): 31–38.

Lakhdar R, Soussi M. Les tapis microbiens du littoral du sud est de la Tunisie: Répartition spatiale et structures sédimentaires associées (MISS). *Revue Méditerranéenne de l'environnement* 2 (2007): 293–311.

Lange OL, Belnap J, Reichenberger H. Photosynthesis of the cyanobacterial soil-crust lichen *Collema tenax* from arid lands in southern Utah, USA: Role of water content on light and temperature responses of CO^2 exchange. *Functional Ecology* 12 (1998): 195–202.

Lee KW, Park JY, Jeong HR, Heo HJ, Han NS, Kim JH. Probiotic properties of *Weissella* strains isolated from human faeces. *Anaerobe* 18 (2012): 96–102.

Lee Y. Characterization of *Weissella kimchii* PL9023 as a potential probiotic for women. *FEMS Microbiology Letters* 250 (2005): 157–162.

Legat A, Gruber C, Zangger K, Wanner G, StanLotter H. Identification of polyhydroxyalkanoates in *Halococcus* and other haloarchaeal species. *Applied Microbiology and Biotechnology* 87 (2010): 1119–1127.

Lehtola MJ, Miettinen IT, Keinanen MM et al. Microbiology, chemistry and biofilm development in a pilot drinking water distribution system with copper and plastic pipes. *Water Research* 38 (2004): 3769–3779.

Lenoir A, Aron S, Cerdà X, Hefetz A. *Cataglyphis* desert ants: A good model for evolutionary biology in Darwin's anniversary year—A review. *Israel Journal of Entomology* 39 (2009): 1–32.

Limmathurotsakul D, Wuthiekanun V, Chantratita N et al. *Burkholderia pseudomallei* is spatially distributed in soil in northeast Thailand. *PLoS Neglected Tropical Diseases* 4 (2010): 694.

Lin CS, Wu JT. Environmental factors affecting the diversity and abundance of soil photomicrobes in arid lands of subtropical Taiwan. *Geomicrobiology Journal* 31 (2014): 350–359.

Liu CF, Tseng KC, Chiang SS et al. Immunomodulatory and antioxidant potential of *Lactobacillus* exopoly-saccharides. *Journal of the Science of Food and Agriculture* 91 (2011): 2284–2291.

Liu Y, Cockell CS, Wang G, Hu C, Chen L, Philippis R. Control of lunar and Martian dust—Experimental insights from artificial and natural cyanobacterial and algal crusts in the desert of Inner Mongolia, China. *Astrobiology* 8 (2008): 75–86.

Luziatelli F, Crognale S, D'Annibale A, Moresi M, Petruccioli M, Ruzzi M. Screening, isolation, and character-ization of glycosyl-hydrolase-producing fungi from desert halophyte plants. *International Microbiology* 17 (2014): 41–48.

Madigan MT, Martinko J, Parker J. *Brock Biology of Microorganisms*. New York: Prentice Hall, 2005.

Mahendran S, Saravanan S, Vijayabaskar P, Ananda Pandian KTK, Shankar T. Antibacterial potential of microbial exopolysaccharide from *Ganoderma lucidum* and *Lysinibacillus fusiformis*. *International Journal of Recent Scientific Research* 4 (2013): 501–505.

Maina NH, Virkki L, Pynnönen H, Maaheimo H, Tenkanen M. Structural analysis of enzyme-resistant iso-maltooligosaccharides reveals the elongation of α-$(1{\rightarrow}3)$-linked branches in *Weissella confusa* dextran. *Biomacromolecules* 12 (2011): 409–418.

Malang SK, Maina NH, Schwab C, Tenkanen M, Lacroix C. Characterization of exopolysaccharide and ropy capsular polysaccharide formation by *Weissella*. *Food Microbiology* 46 (2015): 418–427.

Mapelli F, Marasco R, Balloi A et al. Mineral-microbe interactions: Biotechnological potential of bioweath-ering. *Journal of Biotechnology* 157 (2012): 473–481.

Mapelli F, Marasco R, Rolli E et al. Potential for plant growth promotion of rhizobacteria associated with *Salicornia* growing in Tunisian hypersaline soils. *BioMed Research International* 2013 (2013): 248078.

Márquez MC, Sánchez-Porro C, Ventosa A. Halophilic and haloalkaliphilic, aerobic endospore-forming bacteria in soil. In *Endospore-Forming Soil Bacteria*, ed. NA Logan, P Devos. Berlin: Springer, 2011, pp. 309–339.

Martinez-Frias J, Amaral G, Vázquez L. Astrobiological significance of minerals on Mars surface environment. *Reviews in Environmental Science and Bio/Technology* 5 (2006): 219–231.

Martins MB, Carvalho I. Diketopiperazines: Biological activity and synthesis. *Tetrahedron* 63 (2007): 9923–9932.

Maturrano L, Santos F, Rosselló-Mora R, Antón J. Microbial diversity in Maras salterns, a hypersaline environment in the Peruvian Andes. *Applied and Environmental Microbiology* (2006) 72: 3887–3895.

Millington AC, Drake N, Townshend JR, Quarmby N, Settle JJ, Reading AJ. Monitoring salt playa dynamics using Thematic Mapper data. *IEEE Transactions on Geoscience and Remote Sensing*, 27 (1989): 754–761.

Minckley RL, Roulston TH, William NM. Resource assurance predicts specialist and generalist bee activity in drought. *Proceedings of the Royal Society of London B: Biological Sciences* 280 (2013): 20122703.

Mohr KI, Tebbe CC. Diversity and phylotype consistency of bacteria in the guts of three bee species (*Apoidea*) at an oilseed rape field. *Environmental Microbiology* 8 (2006): 258–272.

Moon SS, Chen JL, Moore RE, Patterson GM. Calophycin, a fungicidal cyclic decapeptide from the terrestrial blue-green alga *Calothrix fusca*. *Journal of Organic Chemistry* 57 (1992): 1097–1103.

Moore FP, Barac T, Borremans B et al. Endophytic bacterial diversity in poplar trees growing on a BTEX-contaminated site: The characterisation of isolates with potential to enhance phytoremediation. *Systematic and Applied Microbiology* 29 (2006): 539–556.

Murchie SL, Mustard JF, Ehlmann BL et al. A synthesis of Martian aqueous mineralogy after 1 Mars year of observations from the Mars Reconnaissance Orbiter. *Journal of Geophysical Research: Planets* 114 (2009): E2.

Nagy ML, Pérez A, Garcia-Pichel F. The prokaryotic diversity of biological soil crusts in the Sonoran Desert (Organ Pipe Cactus National Monument, AZ). *FEMS Microbiology Ecology* 54 (2005): 233–245.

Najjari A, Elshahed MS, Cherif A, Youssef NH. Patterns and determinants of halophilic Archaea (Class Halobacteria) diversity in Tunisian endorheic salt lakes and sebkhet systems. *Applied and Environmental Microbiology* 81 (2015): 4432–4441.

Narasimhan K, Basheer C, Bajic VB, Swarup S. Enhancement of plant-microbe interactions using a rhizosphere metabolomics-driven approach and its application in the removal of polychlorinated biphenyls. *Plant Physiology* 132 (2003): 146–153.

Neifar M, Chouchane H, Jaouani A, Masmoudi AS, Cherif A. Extremozymes as efficient green biocatalysts in bioremediation of industrial wastewaters. In *Wastewater Treatment: Processes, Management Strategies and Environmental/Health Impacts*, ed. LM Barrett. New York: Nova Publishers, 2015a, pp. 191–213.

Neifar M, Maktouf S, Ghorbel RE, Jaouani A, Cherif A. Extremophiles as source of novel bioactive compounds with industrial potential. In *Biotechnology of Bioactive Compounds—Sources and Applications*, ed. VK Gupta, MG Tuohy. Hoboken, NJ: Wiley, 2015b, pp. 245–267.

Neilands JB. Siderophores: Structure and function of microbial iron transport compounds. *Journal of Biological Chemistry* 270 (1995): 26723–26726.

Nicholls CI, Altieri MA. Plant biodiversity enhances bees and other insect pollinators in agroecosystems. A review. *Agronomy for Sustainable Development* 33 (2013): 257–274.

Nicolaus B, Kambourova M, Oner ET. Exopolysaccharides from extremophiles: From fundamentals to biotechnology. *Environmental Technology* 31 (2010): 1145–1158.

Niehaus F, Bertoldo C, Kähler M, Antranikian G. Extremophiles as a source of novel enzymes for industrial application. *Applied Microbiology and Biotechnology* 51 (1999): 711–729.

Nisha R, Kaushik A, Kaushik C. Effect of indigenous cyanobacterial application on structural stability and productivity of an organically poor semi-arid soil. *Geoderma* 138 (2007): 49–56.

Noffke N, Gerdes G, Klenke T, Krumbein WE. Microbially induced sedimentary structures—Examples from modern sediments of siliciclastic tidal flats: *Zentralblatt für Geologie und Paläontologie Teil* 1 (1996): 307–316.

O'Connor E, Shand R. Halocins and sulfolobicins: The emerging story of archaeal protein and peptide antibiotics. *Journal of Industrial Microbiology and Biotechnology* 28 (2002): 23–31.

Ohmori M, Ehira S. Spirulina: An example of cyanobacteria as nutraceuticals. *Cyanobacteria: An Economic Perspective* (2013): 103–118.

Olofsson TC, Vásquez A. Detection and identification of a novel lactic acid bacterial flora within the honey stomach of the honeybee *Apis mellifera*. *Current Microbiology* 57 (2008): 356–363.

Oren A. Microbiology and biogeochemistry of halophilic microorganisms. In *Microbiology and Biogeochemistry of Hypersaline Environments*, ed. A Oren. Boca Raton, FL: CRC Press, 1999, pp. 1–9.

Oren A. Diversity of halophilic microorganisms: Environments, phylogeny, physiology, and applications. *Journal of Industrial Microbiology and Biotechnology* 28 (2002): 55–63.

Oren A, Kuhl M, Karsten U. An endoevaporitic microbial mat within a gypsum crust: Zonation of phototrophs, photopigments, and light penetration. *Oceanographic Literature Review* 6 (1996): 593.

Pahlow G, Muck RE, Driehuis F, Oude Elferink SJWH, Spoelstra SF. Microbiology of ensiling. *Silage Science and Technology* 42 (2003): 31–93.

Palmisano MM, Nakamura LK, Duncan KK, Isotock CA, Cohan FM. *Bacillus sonorensis* sp. nov. a close relative of *Bacillus licheniformis* isolated from soil in the sonorant desert Arizona. *International Journal of Systematic and Evolutionary Microbiology* 51 (2001): 1671–1679.

Parvez S, Malik KA, Ah Kang S, Kim HY. Probiotics and their fermented food products are beneficial for health. *Journal of Applied Microbiology* 100 (2006): 1171–1185.

Patel A, Prajapati JB. Food and health applications of exopolysaccharides produced by lactic acid bacteria. *Advances in Dairy Research* 1 (2013): 107.

Patzelt DJ, Hodač L, Friedl T, Pietrasiak N, Johansen JR. Biodiversity of soil cyanobacteria in the hyper-arid Atacama Desert, Chile. *Journal of Phycology* 50 (2014): 698–710.

Petersen, JJ. Nematode parasites. In *Biological Control of Mosquitoes*, ed. HC Chapman. 1985, pp. 110–122.

Petrovic U, Gunde-Cimerman N, Plemenitas A. Cellular responses to environmental salinity in the halophilic black yeast *Hortaea werneckii*. *Molecular Microbiology* 45 (2002): 665–672.

Pfeiffer C, Bauer T, Surek B, Schömig E, Gründemann D. Cyanobacteria produce high levels of ergothioneine. *Food Chemistry* 129 (2011): 1766–1769.

Pham-Delègue MH, Decourtye A, Kaiser L, Devillers J. Behavioural methods to assess the effects of pesticides on honey bees. *Apidologie* 33 (2002): 425–432.

Pinar G, Ramos C, Rolleke S et al. Detection of indigenous *Halobacillus* populations in damaged ancient wall painting and building materials: Molecular monitoring and cultivation. *Applied and Environmental Microbiology* 67 (2001): 4891–4895.

Poli A, Di Donato P, Abbamondi GR, Nicolaus B. Synthesis, production, and biotechnological applications of exopolysaccharides and polyhydroxyalkanoates by Archaea. *Archaea* 2011 (2011): 693253.

Potrikus CJ, Breznak JA. Gut bacteria recycle uric acid nitrogen in termites: A strategy for nutrient conservation. *Proceedings of the National Academy of Sciences of the United States of America* 78 (1981): 4601–4605.

Powell JT, Chatziefthimiou AD, Banack SA, Cox PA, Metcalf JS. Desert crust microorganisms, their environment, and human health. *Journal of Arid Environments* 112 (2015): 127–133.

Qin S, Miao Q, Feng WW et al. Biodiversity and plant growth promoting traits of culturable endophytic actinobacteria associated with *Jatropha curcas* L. growing in Panxi dry-hot valley soil. *Applied Soil Ecology* 93 (2015): 47–55.

Quillaguaman J, Guzm H, Van-Thuoc D, Hatti-Kaul R. Synthesis and production of polyhydroxyalkanoates by halophiles: Current potential and future prospects. *Applied Microbiology and Biotechnology* 85 (2010): 1687–1696.

Raad I, Hanna H, Jiang Y et al. Comparative activities of daptomycin, linezolid, and tigecycline against catheter-related methicillin-resistant *Staphylococcus* bacteremic isolates embedded in biofilm. *Antimicrobial Agents and Chemotherapy* 51 (2007): 1656–1660.

Raddadi N, Cherif A, Daffonchio D, Neifar M, Fava F. Biotechnological applications of extremophiles, extremozymes and extremolytes. *Applied Microbiology and Biotechnology* 99 (2015): 7907–7913.

Raja R, Hemaiswarya S, Ganesan V, Carvalho IS. Recent developments in therapeutic applications of cyanobacteria. *Critical Reviews in Microbiology* 42 (2016): 394–405.

Rappé MS, Giovannoni SJ. The uncultured microbial majority. *Annual Reviews in Microbiology* 57 (2003): 369–394.

Rastall RA, Gibson GR. Recent developments in prebiotics to selectively impact beneficial microbes and promote intestinal health. *Current Opinion in Biotechnology* 32 (2015): 42–46.

Rhouma, A. Le palmier dattier en Tunisie: Un secteur en pleine expansion. In *Le palmier-dattier dans l'agriculture d'oasis des pays méditerranéens*, ed. M Ferry, D Greinier. CIHEAM/Estacion Phoenix A/28. Zaragoza, Spain: CIHEAM, 1996, pp. 85–104.

Richardson AE. Prospects for using soil microorganisms to improve the acquisition of phosphorus by plants. *Functional Plant Biology* 28 (2001): 897–906.

Ríos Adl, Valea S, Ascaso C et al. Comparative analysis of the microbial communities inhabiting halite evaporites of the Atacama Desert. *International Microbiology* 13 (2010): 79–89.

Rivadeneyra MA, Párraga J, Delgado R, Ramos-Cormenzana A, Delgado G. Biomineralization of carbonates by *Halobacillus trueperi* in solid and liquid media with different salinities. *FEMS Microbiology Ecology* 48 (2004): 39–46.

Rivière A, Moens F, Selak M, Maes D, Weckx S, De Vuyst L. The ability of *Bifidobacteria* to degrade arabinoxylan oligosaccharide constituents and derived oligosaccharides is strain dependent. *Applied and Environmental Microbiology* 80 (2014): 204–217.

Roberts CR, Mitchell CW. Spring mounds in southern Tunisia. *Geological Society, London, Special Publications* 35 (1987): 321–334.

Roberts MS, Nakamura LK, Frederick MC. *Bacillus mojavensis* sp. nov. distinguishable from *Bacillus subtilis* by sexual isolation divergence in DNA sequence and differences in fatty acid composition. *International Journal of Systematic and Evolutionary Microbiology* 44 (1994): 256–264.

Robinson CK, Wierzchos J, Black C et al. Microbial diversity and the presence of algae in halite endolithic communities are correlated to atmospheric moisture in the hyper-arid zone of the Atacama Desert. *Environmental Microbiology* 17 (2015): 299–315.

Sachdev DP, Cameotra SS. Biosurfactants in agriculture. *Applied Microbiology and Biotechnology* 97 (2013): 1005–1016.

Sadfi N, Chérif M, Fliss I, Boudabbous A, Antoun H. Evaluation of bacterial isolates from salty soils and *Bacillus thuringiensis* strains for the biocontrol of *Fusarium* dry rot of potato tubers. *Journal of Plant Pathology* 83 (2001): 101–118.

Saharan B, Nehra V. Plant growth promoting rhizobacteria: A critical review. *Life Sciences and Medicine Research* 21 (2011): 1–30.

Salminen S, von Wright A, Morelli L et al. Demonstration of safety of probiotics—A review. *International Journal of Food Microbiology* 44 (1998): 93–106.

Sánchez-Fernández D, Calosi P, Atfield A et al. Reduced salinities compromise the thermal tolerance of hypersaline specialist diving beetles (Coleoptera, Dytiscidae). *Physiological Entomology* 34 (2010): 265–373.

Satyanarayana T, Raghukumar C, Shivaji S. Extremophilic microbes: Diversity and perspectives. *Current Science* 89 (2005): 78–90.

Sbissi I, Ghodbane-Gtari F, Neffati M, Ouzari H, Boudabous A, Gtari M. Diversity of the desert truffle *Terfezia boudieri* Chatin. in southern Tunisia. *Canadian Journal of Microbiology* 57 (2011): 599–605.

Schopf JW, Packer BM. Early Archean (3.3-billion to 3.5-billion-year-old) microfossils from Warrawoona Group, Australia. *Science* 237 (1987): 70–73.

Schultz MP, Bendick JA, Holm ER, Hertel WM. Economic impact of biofouling on a naval surface ship. *Biofouling* 27 (2011): 87–98.

Schultz TR. In search of ant ancestors. *Proceedings of the National Academy of Sciences of the United States of America* 97 (2000): 14028–14029.

Schwartz EF, Diego-Garcia E, de la Vega RCR, Possani LD. Transcriptome analysis of the venom gland of the Mexican scorpion *Hadrurus gertschi* (Arachnida: Scorpiones). *BMC Genomics* 8 (2007): 119.

Sciuto K, Moro I. Cyanobacteria: The bright and dark sides of a charming group. *Biodiversity and Conservation* 24 (2015): 711–738.

Sharma NK, Tiwari SP, Tripathi K, Rai AK. Sustainability and cyanobacteria (blue-green algae): Facts and challenges. *Journal of Applied Phycology* 23 (2011): 1059–1081.

Siddiqui ZA. PGPR: Prospective biocontrol agents of plant pathogens. In *PGPR: Biocontrol and Biofertilization*, ed. ZA Siddiqui Amsterdam: Springer, 2005, pp. 111–142.

Siefert JL, Larios-Sanz M, Nakamura LK, Slepecky RA. Phylogeny of marine *Bacillus* isolates from the Gulf of Mexico. *Current Microbiology* 41 (2000): 84–88.

Singh RK, Tiwari SP, Rai AK, Mohapatra TM. Cyanobacteria: An emerging source for drug discovery. *Journal of Antibiotics* 64 (2011): 401–412.

Slobodkin AI, Tourova TP, Kostrikina NA et al. *Tepidibacter thalassicus* gen. nov., sp. nov., a novel moderately thermophilic, anaerobic, fermentative bacterium from a deep-sea hydrothermal vent. *International Journal of Systematic and Evolutionary Microbiology* 53 (2003): 1131–1134.

Spaepen S, Vanderleyden J, Remans R. Indole-3-acetic acid in microbial and microorganism-plant signaling. *FEMS Microbiology Reviews* 31 (2007): 425–448.

Spring S, Ludwing W, Maquez MC, Ventosa A, Schleifer KH. *Halobacillus* gen. nov., with description of *Halobacillus litoralis* sp. nov. and *Halobacillus truperi* sp. nov., and transfer of *Sporosarcina halophila* to *Halobacillus halophilus* comb. nov. *International Journal of Systematic and Evolutionary Microbiology* 46 (1996): 492–496.

Steck K, Hansson BS, Knaden M. Smells like home: Desert ants, *Cataglyphis fortis*, use olfactory landmarks to pinpoint the nest. *Frontiers in Zoology* 6 (2009): 5.

Steiner M, Reitner J. Evidence of organic structures in Ediacara-type fossils and associated microbial mats. *Geology* 29 (2001): 1119–1122.

Stieb SM, Hellwig A, Wehner R, Rössler W. Visual experience affects both behavioral and neuronal aspects in the individual life history of the desert ant *Cataglyphis fortis*. *Developmental Neurobiology* 72 (2012): 729–742.

Stivaletta N, Barbieri R. Endolithic microorganisms from spring mound evaporite deposits (southern Tunisia). *Journal of Arid Environments* 73 (2009): 33–39.

Stivaletta N, Barbieri R, Picard C, Bosco M. Astrobiological significance of the sabkha life and environments of southern Tunisia. *Planetary and Space Science* 57 (2009): 597–605.

Strunecký O, Elster J, Komárek J. Molecular clock evidence for survival of Antarctic cyanobacteria (Oscillatoriales, *Phormidium autumnale*) from Paleozoic times. *FEMS Microbiology Ecology* 82 (2012): 482–490.

Swezey CS. The role of climate in the creation and destruction of continental stratigraphic records: An example from the northern margin of the Sahara Desert. In *Climate Controls on Stratigraphy*, ed. CB Cecil, NT Edgar. SEPM Special Publication 77. Tulsa, OK: Society for Sedimentary Geology: 2003, pp. 207–225.

Tanaka T, Shnimizu M, Moriwaki H. Cancer chemoprevention by carotenoids. *Molecules* 17 (2012): 3202–3242.

Tentcheva D, Gauthier L, Zappulla N et al. Prevalence and seasonal variations of six bee viruses in *Apis mellifera* L. and *Varroa destructor* mite populations in France. *Applied and Environmental Microbiology* 70 (2004): 7185–7191.

Tommonaro G, Abbamondi GR, Iodice C et al. Diketopiperazines produced by the halophilic archaeon, Haloterrigena hispanica, activate AHL bioreporters. *Microbial Ecology* 63 (2012): 490–495.

Trigui H, Masmoudi S, Brochier-Armanet C, Maalej S, Dukan S. Characterisation of *Halorubrum sfaxense* sp. nov., a new halophilic archaeon isolated from the solar saltern of Sfax in Tunisia. *International Journal of Microbiology* 2011 (2011): 240191.

Tsiamis G, Katsaveli K, Ntougias S et al. Prokaryotic community profiles at different operational stages of a Greek solar saltern. *Research in Microbiology* 159 (2008): 609–627.

Turk M, Abramovic Z, Plemenitas A, Gunde-Cimerman N. Salt stress and plasma-membrane fluidity in selected extremophilic yeasts and yeast-like fungi. *FEMS Yeast Research* 7 (2007): 550–557.

Turk M, Mejanelle LS, Entjurc M, Grimalt JO, Gunde-Cimerman N, Plemenitas A. Salt-induced changes in lipid composition and membrane fluidity of halophilic yeast-like melanized fungi. *Extremophiles* 8 (2004): 53–61.

Van Den Burg B. Extremophiles as a source for novel enzymes. *Current Opinion in Microbiology* 6 (2003): 213–218.

Vanderbergh PA. Lactic acid bacteria, their metabolic products and interference with microbial growth. *FEMS Microbiology Reviews* 12 (1993): 221–238.

Vaupotic T, Plemenitas A. Osmoadaptation-dependent activity of microsomal HMG-CoA reductase in the extremely halotolerant black yeast *Hortaea werneckii* is regulated by ubiquitination. *FEBS Letters* 581 (2007a): 3391–3395.

Vaupotic T, Plemenitas A. Differential gene expression and Hog1 interaction with osmoresponsive genes in the extremely halotolerant black yeast *Hortaea werneckii*. *BMC Genomics* 8 (2007b): 280–295.

Ventosa A. *Halophilic Microorganisms*. Berlin: Springer Verlag, 2004.

Ventosa A, Marquez CM, Garabito MJ, Arahal DR. Moderately halophilic gram-positive bacterial diversity in hypersaline environments. *Extremophiles* 2 (1998): 297–304.

Wall DH. Biodiversity: Extracting lessons from extreme soils. In *Microbiology of Extreme Soils*. Berlin: Springer, 2008, pp. 71–84.

Warren-Rhodes KA, Rhodes KL, Pointing SB et al. Hypolithic cyanobacteria, dry limit of photosynthesis, and microbial ecology in the hyperarid Atacama Desert. *Microbial Ecology* 52 (2006): 389–398.

Waters E, Hohn MJ, Ahel I et al. The genome of *Nanoarchaeum equitans*: Insights into early archaeal evolution and derived parasitism. *Proceedings of the National Academy of Sciences of the United States of America* 100 (2003): 12984–12988.

Wehner R. Desert ant navigation: How miniature brains solve complex tasks. *Journal of Comparative Physiology* 189 (2003): 579–588.

Wehner R, Harkness RD, Schmid-Hempel P. Foraging strategies in individually searching ants, *Cataglyphis bicolor* (Hymenoptera: Formicidae). Stuttgart: Gustav Fischer Verlag, 1983, pp. 1–79.

Wehner R, Marsh AC, Wehner S. Desert ants on a thermal tightrope. *Nature* 357 (1992): 586–587.

Whitton BA, ed. *Ecology of Cyanobacteria II: Their Diversity in Space and Time*. Berlin: Springer Science and Business Media, 2012.

Wierzchos J, Ascaso C, McKay CP. Endolithic cyanobacteria in halite rocks from the hyperarid core of the Atacama Desert. *Astrobiology* 6 (2006): 415–422.

Woese CR, Fox GE. Phylogenetic structure of the prokaryote domain: The primary kingdoms. *Proceedings of the National Academy of Sciences of the United States of America* 74 (1977): 5088–5090.

Woese CR, Kandler O, Wheelis ML. Towards a natural system of organism. Proposal for the domains Archaea, Bacteria, and Eucarya. *Proceedings of the National Academy of Sciences of the United States of America* 87 (1990): 4576–4579.

Wolter A, Hager AS, Zannini E, Czerny M, Arendt EK. Influence of dextran-producing *Weissella cibaria* on baking properties and sensory profile of gluten-free and wheat breads. *International Journal of Food Microbiology* 172 (2014): 83–91.

Wood BJD, Holzapfel WH. *The Genera of Lactic Acid Bacteria*. London: Chapman & Hall, 1995.

Yang L, Tan RX, Wang Q, Huang WY, Yin YX. Antifungal cyclopeptides from *Halobacillus litoralis* YS3106 of marine origin. *Tetrahedron Letters* 43 (2002): 6545–6548.

Yeon SH, Jeong WJ, Park JS. The diversity of culturable organotrophic bacteria from local solar salterns. *Journal of Microbiology* 43 (2005): 1–10.

Youssef NH, Savage-Ashlock KN, McCully AL et al. Trehalose/2-sulfotrehalose biosynthesis and glycine-betaine uptake are widely spread mechanisms for osmoadaptation in the Halobacteriales. *The ISME Journal* 8 (2014): 636–649.

Zainuddin E, Mundt S, Wegner U, Mentel R. Cyanobacteria a potential source of antiviral substances against influenza virus. *Medical Microbiology and Immunology* 191 (2002): 181–182.

Zajc J, Kogej T, Ramos J, Galinski EA, Gunde-Cimerman N. The osmoadaptation strategy of the most halophilic fungus *Wallemia ichthyophaga*, growing optimally at salinities above 15% NaCl. *Applied and Environmental Microbiology* 80 (2014): 247–256.

Zajc J, Liu Y, Dai W et al. Genome and transcriptome sequencing of the halophilic fungus *Wallemia ichthyophaga*: Haloadaptations present and absent. *BMC Genomics* 14 (2013): 617.

Zajc J, Zalar P, Plemenitaš A, Gunde-Cimerman N. The mycobiota of the salterns. In *Biology of Marine Fungi*, ed. C Raghukumar. Berlin: Springer, 2012, pp. 133–158.

Zalar P, De Hoog GS, Schroers HJ, Crous PW, Groenewald JZ, Gunde-Cimerman N. Phylogeny and ecology of the ubiquitous saprobe *Cladosporium sphaerospermum*, with descriptions of seven new species from hypersaline environments. *Studies in Mycology* 58 (2007): 157–183.

Zalar P, de Hoog GS, Schroers HJ, Frank JM, Gunde-Cimerman N. Taxonomy and phylogeny of the xerophilic genus *Wallemia* (Wallemiomycetes and Wallemiales, cl. et ord. nov.). *Antonie van Leeuwenhoek* 87 (2005): 311–328.

Zammouri M, Siegfried T, Fahem T, Kriaa S, Kinzelbach W. Salinization of groundwater in the Nefzawa oases (Tunisia): Results of a regional-scale hydrogeologic approach. *Hydrogeology Journal* 15 (2007): 1357–1375.

Zhang L, Wu GL, Wang Y, Dai J, Fang CX. *Bacillus deserti* sp. nov., a novel bacterium isolated from the desert of Xinjiang China. *Antonie Van Leevwenhoek* 99 (2011): 221–229.

Zindel R, Gottlieb Y, Aebi A. Arthropod symbiosis, a neglected parameter in pest and disease control programs. *Journal of Applied Ecology* 48 (2011): 864–872.

CHAPTER 4

Culture Studies on a Halophile *Dunaliella salina* from Tropical Solar Salterns, Bay of Bengal, India

Suman Keerthi

CONTENTS

4.1 INTRODUCTION

Hypersaline coastal water bodies, including solar salterns that serve as subplots for incremental evaporation of seawater for the harvest of common salt, are at times colored pink, dark red, or orange in certain seasons. This discoloration is due to the presence of dense communities of halophiles, such as the carotenoid-rich prokaryotes, archaea, and β-carotene-rich unicellular alga *Dunaliella salina* (Elevi Bardavid et al. 2008). These halophilic inhabitants are fascinating because of their ecological, physiological, and biotechnological characteristics.

While diverse prokaryotic and archaea populations exist in salterns, ubiquitous in these is the eukaryotic halophilic carotenogenic *D. salina*. Salterns are the simplest saltwater bodies in which to investigate the relationship between the primary producer *Dunaliella* and heterotrophic microbes (Elevi Bardavid et al. 2008). The general microbial properties of crystallizer brines in salterns worldwide, despite minor variations attributable to differences in the availability of nutrients, are markedly similar (Elevi Bardavid et al. 2008). The most abundant heterotrophs in salterns are *Haloquadratum* and *Salinibacter* species (Elevi Bardavid et al. 2008). These authors discounted the contention of Borowitzka (1981) that *Halococcus* and *Halobacterium* are the dominant heterotrophs in salterns. Further, it has been indicated that *Halobacterium* is present in few numbers in salterns and can be cultured only through special enrichment procedures (Oren and Litchfield 1999).

From the Indian Seas, there have been a few reports on the presence of five or more species, which includes *Dunaliella* based on a morphology and molecular approach, including two promising strains of *D. salina* and *Dunaliella viridis* (Raja et al. 2007; Mishra and Jha 2009; Jayapriyan et al. 2010; Preetha et al. 2012; Keerthi et al. 2015). The present study describes *Dunaliella* strains from Indian strains using a combined morphological, physiological, and molecular approach and their potential utility in biotechnology.

I present here the ecology of the halophile *D. salina* from evaporation salt pans in four regions from a 600 km belt along the Bay of Bengal in the state of Andhra Pradesh, southeast India. Standard methods for collection, culturing, growth media modification, rRNA typing, growth rate determination, pigment analysis, biomass estimates, and lipids were employed in this study. As modifications of the medium affect the growth conditions and biochemical composition of cells, these are briefly mentioned as well. Utilizing local isolates of *Dunaliella*, growth rates and pigment composition in response to stress and their utility in biotechnology are presented.

4.2 MATERIALS AND METHODS

4.2.1 Collection Points

The samples were collected from evaporation salt pans in four regions of southeast India: Bheemunipatnam, Visakhapatnam, Kakinada, and Iskapalli (a hamlet near Nellore) (Figure 4.1). The water in the collection sites was in various shades of green, orange, and pink in the ponds in different subblocks of salterns (Figure 4.2). The diverse colors of the ponds in different subblocks are due to the characteristic microbial community in each of them, suited to the prevailing salt concentration (Elevi Bardavid et al. 2008).

The water salinity in these pans ranges from 90 to 300 ppt, and the temperature from 26°C to 42°C.

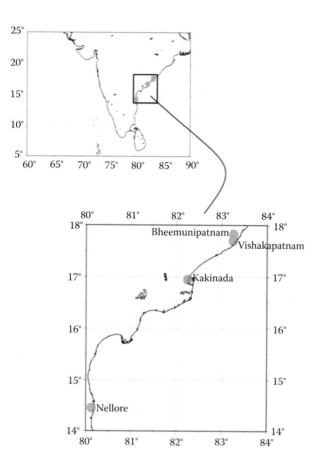

Figure 4.1   Collection points of extremophiles along the southeastern coast of India.

Water samples from hypersaline brine subblocks in the evaporation salt pans with a chocolate brown or brick red or thick green color were collected. The rusty patches on crystallizing salt were also collected on some occasions (Figure 4.2).

The water was green due to presence of a mixture of unicellular green algae *Dunaliella* species; water that was more orange in color was due to the abundance of the unicellular green alga *D. salina*, and later on it was pink when halobacteria were abundant.

About 60 strains of *Dunaliella* were established, of which 16 were carotenogenic. A detailed study of the morphology and identification through molecular typing of ribosomal DNA (rDNA) of the sample of 16 isolates was done (Keerthi et al. 2015). For optimum growth studies, one of the highly carotenogenic isolates was used in Walne's medium (Walne 1970). The isolates differed in their cell morphology (Figure 4.3). Basically, three cell types were recognized and detailed morphological characters were discussed by Keerthi et al. (2015).

4.2.2 rRNA Typing: *Dunaliella* and *Halobacterium*

The 16 isolates of carotenogenic *Dunaliella* were rDNA typed to identify the species. Three regions—the 18S rRNA gene; ITS1 (internal transcribed spacer), 5.8S rRNA; and ITS2, conserved region in ITS2—were studied (Table 4.1) (Keerthi et al. 2015). The primers designed and extensively used for *Dunaliella* spp. were not found to be genus or species specific. Four types of 18S

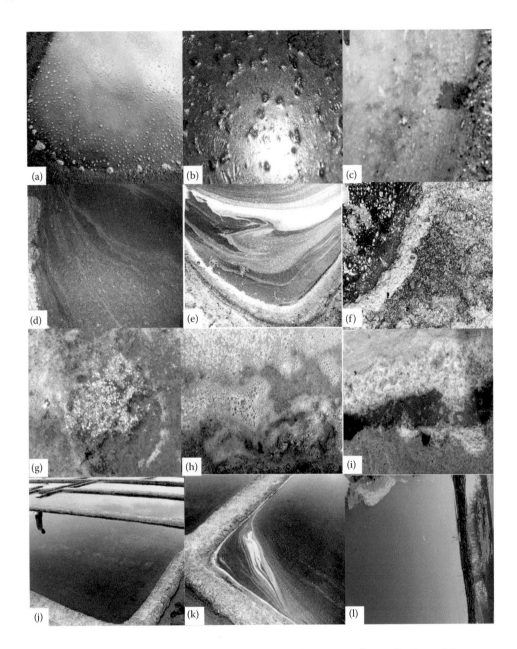

Figure 4.2   Different colored waters (a–l) in the evaporation salt pans at various collection points.

rRNA genes and two types of ITS1 were identified in carotenogenic *Dunaliella* species (Keerthi et al. 2015).

4.2.3 Growth Studies of Carotenogenic *Dunaliella* Isolates

Growth of the 16 isolates was assessed in Walne's medium (Walne 1970) with 12.5% NaCl. Growth was computed as generation time and number of divisions per day (Furnas et al. 2002).

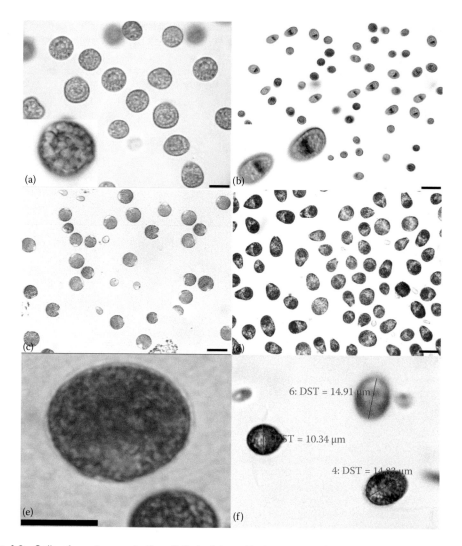

Figure 4.3 Cells of carotenogenic *Dunaliella* isolates with β-carotene when stressed. The bar represents 10 μm. (a–d) 40× stressed cells. (e) Same cells at 100×. (f) Radial measurements.

4.2.4 Media

4.2.4.1 Dunaliella: *Autotrophic*

For autotrophic cultures, Walne's medium and Johnson's medium were used. Walne's medium is a semidefined medium made from seawater (Walne 1970). Walne's medium with 12.5% NaCl was taken as the base medium for improvisation.

D. salina strain I3 (identified through rDNA fingerprint) (Keerthi 2012) isolated from evaporation salt pans was selected for optimization of culture medium. First, the inorganic mineral medium required for optimal culture under photoautotrophic conditions was developed as follows. The optimized medium was fortified with organic carbon to serve as a medium for mixotrophic culture.

The concentration of phosphate used in different media varies only in the range of 0.5 μM–10 mM (Subba Rao 2009). Vitamin mix (thiamine hydrochloride [B 1], biotin [H], and cyanocobalamine

Table 4.1 Details of the Carotenogenic *Dunaliella* Isolates Collected from Salterns in a 600 km belt along the Coast of the Bay of Bengal in South India (cultures were raised in Walne's medium)

Isolate ID	Collection Site and Time	Sample Nature	Cell Type and Motility[a]	G = 24 μ^{-1b}	Carotene Content mg L^{-1}	Carotene Productivity[c]
B12	Bheemunipatnam April 7, 2008, 3:30 p.m.	Reddish brown water with salt crystals	C	62.57	17.15 (±1.08)	Low
B22	Bheemunipatnam June 3, 2008, 3:30 p.m.	Green water with a gelatinous surface layer	C	35.64	12.03 (±0.98)	Low
B24	Bheemunipatnam June 3, 2008, 3:30 p.m.	Green water with a gelatinous surface layer	C	49.46	17.39 (±0.82)	Low
B25	Bheemunipatnam July 26, 2008, 3:30 p.m.	Green water with a gelatinous surface layer	C	38.63	13.23 (±1.45)	Low
B31	Bheemunipatnam September 3, 2008, 3:30 p.m.	Green water with a gelatinous surface layer	C	37.04	23.18 (±0.91)	Low
B32	Bheemunipatnam September 3, 2008, 3:30 p.m.	Green water with a gelatinous surface layer	C	39.64	27.14 (±0.94)	Low
B33	Bheemunipatnam September 3, 2008, 3:30 p.m.	Green water with a gelatinous surface layer	C	35.56	46.49 (±2.01)	Moderate
B34	Bheemunipatnam September 3, 2008, 3:30 p.m.	Brown water with a greasy surface layer	A	38.63	15.71 (±1.18)	Low
I1	Iskapalli May 11, 2008, 10:00 a.m.	Reddish brown water with oily film on the surface	B	35.56	65.76 (±0.99)	High
I2	May 11, 2008, 10:00 a.m.	Brown crystallized salt in greasy water	B	38.63	56.39 (±1.83)	High
I3	May 11, 2008, 10:00 a.m.	Brown crystallized salt in greasy water	B	33.91	66.11 (±1.55)	High
N1	Paravada NTPC March 17, 2008, 3:30 p.m.	Reddish brown water with oily film on the surface	C	51.23	19.69 (±0.62)	Low
N5	Paravada NTPC March 17, 2008, 3:30 p.m.	Reddish brown water with oily film on the surface	C	52.78	23.59 (±0.81)	Low
N7	Paravada NTPC March 17, 2008, 3:30 p.m.	Reddish brown water with oily film on the surface	A	46.75	16.50 (±0.66)	Lowest
K1	Kakinada May 8, 2008, 11:30 a.m.	Dark brown water	A	71.48	64.34 (±1.21)	High
K2	Kakinada May 8, 2008, 11:30 a.m.	Dark brown water with salt crystals	A	75.00	38.19 (±1.17)	Moderate

[a] A: Oblong cells of 12–16 μm length, 4–6 μm width, cells do not change shape frequently, faster cell movement than B type; B: Cell size of 14–18 μm length, 9–12 μm width with flagella as long as the cell and cells often taking a shape with radial symmetry, slow movement; C: Pyriform cells of 13–16 μm length, 6–8 μm width, which quickly change shape into various forms, cells show very fast movement frequently rotating both clockwise and anti-clockwise.

[b] Generation time in Walne's medium with 12.5% NaCl, 24 ± 1°C, 90 μmol photons m^{-2} s^{-1}, light/dark cycle of 12/12 hours.

[c] All isolates are carotene rich; among them, gradation was done as lowest, <25 mg L^{-1}; low, <25 mg L^{-1}; moderate, 25–50 mg L^{-1}; and high, >50 mg L^{-1}.

[B_{12}]) is used in Walne's medium. Borate is used in a much higher concentration (33.6 mg L^{-1}) than in other media used for the culture of *Dunaliella*. Therefore, the following components of the medium were optimized.

- Best nitrogen source and its optimal concentration among sodium nitrate, potassium nitrate, urea, and ammonium chloride.
- Requirement of micronutrient mix, borate, and vitamin mix.
- Optimal sodium chloride concentration. In the optimized mineral medium, different complex nutrients with carbon or nitrogen as a major component were used. Organic carbon sources were sodium acetate, glycerol, and malt extract, and organic nitrogen sources were yeast extract and peptone; different concentrations were tested.

4.2.4.2 Mineral Medium

The inorganic nitrogen sources (sodium nitrate, potassium nitrate, and ammonium chloride) and organic nitrogen source (urea) were tested at four concentrations: 50, 100, 150, and 200 mg L^{-1}. In the medium with an optimal concentration of the best nitrate source, growth was monitored with and without vitamin mix, trace elements, and the major (used in high concentration) trace element borate. Salt concentrations ranging from 10% to 30% (molarity values of ~2.2, 2.6, 3.0, 3.9, 4.8, and 5.6 M) were tested. A medium with components and their optimal concentration in these experiments was thus devised, and the growth of *D. salina* in this medium and in the control (Walne's with 12.5% NaCl) was compared. Based on growth assays, this mineral medium was found to be most suitable for the culture of *Dunaliella*.

4.2.4.3 Experimental Setup

Cultures were maintained at 25°C ± 2°C, 150 µmol photons m^{-2} s^{-1} (8.1 Klux), with a light/dark cycle of 12/12 hours, and were gently mixed twice a day to prevent flocculation. The pH of the culture during the experiments ranged from 7.9 to 8.2.

4.2.4.4 Dunaliella: *Mixotrophic*

Concentration ranges from 2 to 10 g L^{-1} for sodium acetate, 1 to 10 g L^{-1} for malt extract, 1%–5% for glycerol, and 1 to 4 g L^{-1} for yeast extract and peptone were used for enrichment; this medium with complex nutrients was utilized to test whether organic carbon and nitrogen would promote growth in a mixotrophic mode.

4.2.5 Pigment Analysis: Spectrophotometry

4.2.5.1 Chlorophyll a and Carotenoids

Detailed analysis of the pigments was done for the culture of I3 (used in 16S rRNA typing). Pigments were extracted on cells from a 2 mL pellet at 12,000 rpm for 10 minutes. To the pellet, 5 mL ice cold 90% (v/v) acetone was added and vortexed for five minutes and left overnight in a refrigerator at 4°C. The samples were centrifuged for 10 minutes at 12,000 rpm. The supernatant was used for spectrophotometric and high-performance liquid chromatography (HPLC) analysis. An absorption spectrum was obtained with a dual-beam spectrophotometer (Shimadzu UV 1800, Kyoto, Japan) using a 1 cm light path.

The following equations were used to calculate pigment concentrations ($\mu g\ mL^{-1}$ culture):

The equation of Jeffrey and Humphrey (1975) was used to determine the chlorophyll a (Chl a) concentration:

$$Chl\ a = 11.93D_{664} - 1.93D_{647}\ (Vs/Vc)$$

$$Carotenoid = 3.86*D452\ (Vs/Vc)$$

where Vc = volume of culture sample (mL) used and Vs = volume of extract (mL) (Borowitzka and Siva 2007).

4.2.5.2 HPLC Analysis

HPLC analysis was done on cultures of the I3 *Dunaliella* strain in an Agilent 1200 HPLC system (Santa Clara, California) equipped with quaternary pump, auto injector, Peltier column thermostat, temperature-controlled auto sampler, and Chemstation software. Pigments were detected with the diode array detector at 450 nm wavelength with a 20 nm bandwidth. An injector program was optimized to deliver sample extract, and buffer was composed of 28 mM aqueous tetrabutyl ammonium acetate (TBAA) (AR Grade, Fluka) at pH 6.5 and methanol (GC Assay, 99.7% pure, Merck) in a 90:10 ratio. The method proposed by Van Heukelem and Thomas (2001) was adopted. A volume of 400 μL was injected. The oven temperature was set to 60°C, and the solvent flow rate was 1.1 mL min^{-1}. A ZORBAX eclipse XDB-C8, 4.6–150 mm (diameter by length) PN: 963967-906 column was used. Binary gradient elution as described by Chakraborty et al. (2010) was used. Chl a was quantified at 665 nm, and carotenoids and xanthophylls were detected and quantified at 445 nm (Van Heukelem and Thomas 2001).

4.2.6 Mass Culture

The seed inoculum for mass culture was developed both in the optimized mineral medium and in the mixotrophic medium through gradual upscaling from 10 to 250 mL to 2 L in glass flasks, followed by 3 and 5 L carboys in the laboratory. They were further scaled up from 20 L carboys to 80, 500, and 1000 L fiber-reinforced plastic (FRP) tanks in a glass house. At all steps of upscaling, the ratio of inoculum to culture volume was 1:5 (Figure 4.4).

Seed culture developed in mixotrophic mode (up to 150 L) was transferred to (a) mixotrophic medium and (b) optimized mineral medium. As a control, seed culture developed in autotrophic mode was transferred to optimized mineral medium. The cultures reached the end of the exponential phase by the 26th day. At this stage, the salinity of the medium was increased to ~30% (300 ppt) by addition of crude crystal salt to stress the cells and to induce accumulation of β-carotene. Cultures were grown under direct sunlight (76–82 Klux, ~10 hours light duration) for three days, by which time the cultures turned from orange to brown. The mass culture schedule lasted for ~60 days, during which the physical conditions in the glass house were as follows: temperature, 26–32°C day/16–22°C night; light, 33–58 Klux; and duration, ~10 hours.

4.2.7 Harvesting

Cultures were flocculated with ferric chloride solution (0.8–1 g L^{-1}) and left overnight for settling of the biomass (Figure 4.5a–c). The clear supernatant in the culture tank was siphoned off with a 0.5 hp motor. The remaining culture was filtered through a 100-micron pore size filter Nitex screen (Figure 4.5d). The resultant biomass slurry was washed once with 10 L of isotonic ammonium format (0.5 M) to remove salts without causing the cells to burst (Lavens and

Figure 4.4   Serial upscaling of *D. salina* culture from (a) 10 mL to (b) 100 mL to (c) 1.5 L to (d) 20 L to (e) 80 L to (f) 150 L to (g) 1000 L. (h) *D. salina* culture when stressed.

Sorgeloos 1996). The wet biomass (Figure 4.5e and f) was dried in a hot air oven for 12 hours at 50°C (Figure 4.5g–j).

4.2.8 *Dunaliella* sp.: Lipids

Lipid estimation was by fluorescence measurement with a spectrofluorometer (Horiba JobinYvon Fluoromax 4). First, the autofluorescence of the cells was measured at an excitation wavelength of 475 nm and an emission wavelength of 580 nm with a slit width of 2 nm and an

Figure 4.5   Harvesting of mass-cultured *D. salina*. (a, b) Flocculation trial in a 3 L volume. (c) Flocculation in a
1000 L tank. (d) Filtering of biomass. (e) Filtered biomass. (f) Wet biomass. (g) Biomass collected
in petridish. (h) Drying of biomass in an oven: (i) dry powder and (j) final combination of the product
in a storage bottle.

integration time of 0.5 seconds (Priscu et al. 1990). Then, 5 μL of 1% Nile red (NR) (Sigma-
Aldrich) was added to a 5 mL cell suspension and vortexed. The time course of fluorescence
development of the well-mixed cell suspension was recorded at five-minute intervals and used
in lipid quantification as triolein equivalents (Kimura et al. 2004). A calibration curve was
drawn from the fluorescence emission values of triolein (Sigma-Aldrich). The lipid content was
calculated by the formula: μeq triolein mL^{-1} = [FL − (FA + FDW)]/b (Priscu et al. 1990), where
FL = maximum fluorescence of cell suspension in DW after NR addition, FA = autofluorescence
of cell suspension in distilled water (DW) before NR addition, FDW = fluorescence of DW after
NR addition, and b = slope of the standard curve drawn for fluorescence (μg^{-1} triolein mL^{-1}).

4.3 RESULTS

4.3.1 Growth

Two types of cells existed in our cultures: the green cells, identified as *D. salina* and carotenogenic, and the pink, rod-shaped, heterotrophic cells. By 16S rRNA typing, these cells were confirmed as eubacterium (*Halomonas* sp.), and their pigment profile matched that of the archaea *Halobacterium*. Bacterioruberin derivatives were found. Thus, it was concluded that *Halomonas* sp. and *Halobacterium* sp. are among the coinhabitant heterotrophs of *D. salina*.

Pure cultures of *D. salina* established from these salterns showed the typical three colors in salterns, green until 30 days, turning dark orange by 60 days, and pink by 90 days. The 90-day-old cultures had rod-shaped cells, similar to the cells from the pink waters of the ponds and the pink colonies established in the laboratory. The glycerol and proteins released from the degenerating cells of *D. salina* in old laboratory cultures in which the salt concentration increased due to evaporation of the medium led to the gregarious appearance of these heterotrophs.

In Walne's medium, there were three types of cells, designated as A, B, and C. Type A cells were oblong, 12–16 μm in length, and 4–6 μm in width. They did not change shape frequently and had faster movement than type B cells. Category B cells were 14–18 μm in length and 9–12 μm in width, with flagella as long as the cell. They often took shape with radial symmetry and had slow movement. Type C cells were pyriform, 13–16 μm in length, and 6–8 μm in width. They quickly changed shape into various forms and moved very fast, frequently rotating both clock- and counterclockwise. Growth rates ($G = 24\,\mu^{-1}$) varied between these cells; 38.63–75.00 for type A cells, 33.91–38.63 for type B cells, and 35.64–51.23 for type C cells. Ranges of carotenoids (pg cell^{-1}) for categories A, B, and C corresponded to 81.70–187.02, 33.91–38.63, and 35.64–51.23. Carotene productivity (mg L^{-1}) ranged from 16.50 to 64.34 for type A cells, 56.39 to 66.11 for type B cells, and 17.39 to 46.49 for type C cells (Table 4.1).

4.3.2 Mineral Medium

4.3.2.1 NaNO₃, KNO₃, Urea, and NH₄Cl

Growth of *D. salina* was similar in media with 100 mg L^{-1} of both potassium nitrate and urea and corresponded to 1.15 and 1.20×10^6 cells mL^{-1} and specific growth rates of 0.43 and 0.42 (Table 4.2). These are higher than those obtained in media with NaNO₃ and NH₄Cl (Table 4.1). At higher

Table 4.2 Growth of I3 Strain of *D. salina* in Walne's Medium Substituted with Different Nitrogen Sources (mg L⁻¹) at Different Concentrations

Concentration ▶	Max. Cell No. × 10⁶ mL⁻¹				Specific Growth Rate (μ)				G = 24 μ⁻¹ (h)			
	50	100ᵃ	150	200	50	100	150	200	50	100ᵃ	150	200
N₂ Source ▼												
NaNO₃	0.63	0.95	0.90	0.81	0.31	0.42	0.31	0.31	74.93	58.40	75.18	77.47
KNO₃	1.11	1.15	1.14	1.02	0.39	0.43	0.37	0.38	60.02	55.15	64.19	64.48
Urea	1.07	1.20	1.10	1.00	0.38	0.42	0.33	0.32	62.59	57.67	71.69	76.12
NH₄Cl	0.92	0.86	0.86	0.85	0.35	0.30	0.29	0.28	68.28	80.92	82.53	84.52
ANOVA and Tukey HSD column-wise for optimal concentration (100 mg L⁻¹)	df = 3, f = 32.19, p < 0.001; [0.05] = 5.85				df = 3, f = 10.41, p < 0.0015				df = 3, f = 19.75, p < 0.001; [0.05] = 1.23			

concentrations of enrichments (150 and 200 mg L^{-1}) of NaNO$_3$, KNO$_3$, urea, and NH$_4$Cl, growth was low, and so were their specific growth and division. The HSD (honestly significant difference) [0.05] value not available because $K > 2$ and F has a significant ratio. For an equal weight of the compound, the amount of nitrogen available from urea is more than that from potassium nitrate because each molecule of urea has two nitrogen atoms, while potassium nitrate has one nitrogen.

4.3.2.2 Trace Metals, Vitamins, and Boron

Growth was drastically reduced to 0.33 × 10^6 mL in medium without trace elements compared with 2.07 × 10^6 mL in medium with boron, 1.90 × 10^6 mL with vitamin mix, and 1.85 × 10^6 mL in Walne's medium (Table 4.3). The trend with Chl a (pg cell^{-1}) was just the opposite; Chl a (pg cell^{-1}) was 14.62, 3.39, 1.00, and 1.19 corresponding to media without (a) trace metals, (b) boron, (c) vitamin mix and (d) Walne's medium. The specific growth rates were 0.26, 0.33, 0.36, and 0.31 in media a, b, c, and d, respectively. Omission of boron resulted in low carotenoid (3.07) per cell compared with 24.83, 14.22, and 17.54 for media without trace metals, vitamin mix, and Walne's medium, respectively.

In the Table 4.2, HSD [0.05] = the absolute [unsigned] difference between any two sample means required for significance at 0.05 probability; HSD [0.05] values were not obtained when $K > 2$ and F has a significant ratio.

The maximum cell numbers, specific growth rate, and growth in optimized mineral medium (KNO$_3$, 100 mg L^{-1}; KH$_2$PO$_4$, 0.35 mg L^{-1}; TM [trace metal] [trace elements] [stock of Walne's] without H$_3$BO$_3$, 1 mL L^{-1}) were significantly higher ($p < 0.001$); thus, they were 2.01 × 10^6 mL^{-1}, 0.43, and 55.14 compared with 1.86 × 10^6 mL^{-1}, 0.31, and 74.93 in Walne's medium.

4.3.2.3 Salinity (NaCl%)

Among the different NaCl concentrations tested (10%–30%), 12.5% (used in Walne's medium with KNO$_3$, 100 mg L^{-1}; KH$_2$PO$_4$, 0.35 mg L^{-1}; trace element mix [Walne's] without H$_3$BO$_3$, 1 mL L^{-1}). The pattern of growth in media with a salinity range of 10%–30% NaCl was similar. The maximum number of cells attained was the highest in 12.5% NaCl and decreased in 10%, 15%, 20%, 25%, and 30% NaCl (Figure 4.6). Specific growth rates were 0.36, 0.31, 0.26, and 0.35 corresponding to 10%, 15%, 20%, and 12.5% NaCl; at 25% and 30% NaCl, the growth rates were lower and

Table 4.3 Growth of I3 Strain of *D. salina* in Walne's Medium with Elimination of Trace Elements, Boron, and Vitamin Mix

Component Eliminated	Max. Cell No. × 10^6 mL^{-1}	Chl a Max. (pg cell^{-1})	Carotene Max. (pg cell^{-1})	Specific Growth Rate (μ)	G = 24 μ^{-1} (h)
a. TM	0.33 (±0.50)^c	14.62 (±0.52)^a	24.83 (±3.40)^a	0.26 (±0.01)^c	92.18 (±3.47)^a
b. Boron	2.07 (±1.92)^a	3.39 (±0.42)^b	3.07 (±0.20)^d	0.33 (±0.01)^b	58.82 (±0.16)^d
c. Vitamin mix	1.90 (±0.02)^b	1.00 (±0.07)^c	14.22 (±1.11)^c	0.36 (±0.01)^a	66.51 (±0.16)^c
d. Walne's (control)	1.85 (±0.02)^b	1.19 (±0.16)^c	17.54 (±1.54)^b	0.31 (±0.12)^b	77.27 (±0.18)^b
ANOVA and Tukey HSD	$df = 3$, $f = 2007.40$, $p < 0.001$; HSD [0.05] = 0.07	$df = 3$, $f = 4583.85$, $p < 0.001$; HSD [0.05] = 0.38	$df = 3$, $f = 34491.34$, $p < 0.001$; HSD [0.05] = 0.2	$df = 3$, $f = 33.74$, $p < 0.001$; HSD [0.05] = 0.03	$df = 3$, $f = 1530.39$, $p < 0.001$; HSD [0.05] = 1.6

Note: HSD [0.05] = absolute [unsigned] difference between any two sample means required for significance at 0.05 probability. Means with the same superscript letter within a column are similar to the Tukey HSD value being not significant at 0.05 probability.

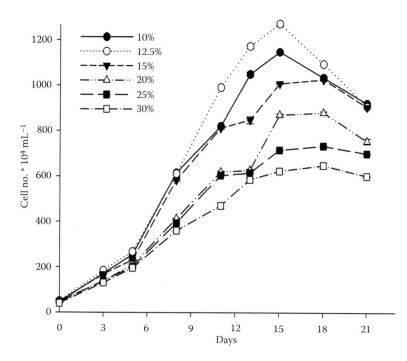

Figure 4.6   Growth of I3 strain of *D. salina* in Walne's medium (12.5%) with different concentrations of NaCl.

corresponded to 0.18 and 0.17. The maximum Chl *a* (pg cell^{-1}) was low, 0.99 and 1.07 at 25% and 30% NaCl, compared with lower salinities.

4.3.2.4 Acetate

Enrichment of cells with 4 and 6 g L^{-1} sodium acetate yielded higher cell numbers (1.96 × 10^6 mL^{-1}) and a high specific growth rate of 0.38 (Table 4.4). At higher levels of enrichment, cell densities and the specific growth rate were low.

Table 4.4   Growth of I3 Strain of *D. salina* in Mixotrophy Mode with Optimized Mineral Medium

Sodium Acetate	Max. Cell No. × 10^6 mL^{-1}	Specific Growth Rate (μ)	$G = 24\ \mu^{-1}$ (h)
2 g L^{-1}	1.67 (±0.71)[b]	0.37 (±0.00)[ab]	62.96 (±0.58)[d]
4 g L^{-1}	1.96 (±0.74)[a]	0.38 (±0.00)[a]	64.09 (±0.58)[cd]
6 g L^{-1}	1.95 (±0.43)[a]	0.37 (±0.01)[ab]	65.28 (±0.71)[cd]
8 g L^{-1}	1.62 (±0.95)[c]	0.35 (±0.01)[bc]	66.49 (±0.76)[c]
10 g L^{-1}	1.66 (±0.95)[b]	0.34 (±0.00)[c]	69.89 (±0.59)[b]
Optimum mineral	1.34 (±0.60)[d]	0.31 (±0.01)[d]	77.19 (±0.60)[a]
ANOVA and Tukey	*df* = 5, *F* = 933.83, *p* < 0.001; HSD [0.05] = 3.53	*df* = 5, *F* = 24.22, *p* < 0.001; HSD [0.05] = 0.02	*df* = 5, *F* = 66.98, *p* < 0.001; HSD [0.05] = 2.34

Note: HSD [0.05] = the absolute [unsigned] difference between any two sample means required for significance at 0.05 probability. Means with the same superscript letter within a column are similar to the Tukey HSD value being not significant at 0.05 probability. Optimized mineral medium substituted with different concentrations of sodium acetate as organic carbon source. Optimized mineral medium: KNO$_3$, 100 mg L^{-1}; KH$_2$PO$_4$, 0.35 mg L^{-1}; trace elements (stock of Walne's) without H$_3$BO$_3$, 1 mL L^{-1}.

4.3.3 Malt

Malt extract was chosen to test in combination with organic nitrogen sources for formulation of medium for mixotrophy. To obtain archaea- and bacteria-free cultures, lower than optimal concentrations of malt extract and organic nitrogen were tried in different proportions. Addition of 4–6 g L^{-1} of malt extract supported growth of more *D. salina* cells (~1.95 × 10^6 mL^{-1}) than at 2, 8, and 10 g L^{-1} (Table 4.5). In general, cell densities were higher in cultures enriched with malt extract than those with sodium acetate. Chl *a* maximum (pg cell^{-1}) was not significantly different at malt extract concentrations of 1–10 g L^{-1}. In these treatments, specific growth rates were in consonance with the cell numbers and ranged between 0.45 and 0.54, and were significantly higher than in Walne's medium. Anthrone test results indicated a gradual decrease in concentration of malt extract in the medium, with a steep decrease observed in samples collected on the sixth day after algal inoculation (results not shown). With a rise in concentration of malt extract, along with *D. salina*, an increased growth of small rod-shaped cells was observed. Cultures had a pink tinge and were identified (from the HPLC pigment profile) as *Halobacterium* spp.—a halophilic archaea.

4.3.3.1 Malt and Peptone Extract

Malt extract was chosen to test in combination with organic nitrogen sources for formulation of medium for mixotrophy. Different proportions of malt (M), peptone (P), and yeast (Y) were used (Table 4.6).

Growth was better than in mineral (Walne's) medium in all media substituted with malt extract and organic nitrogen source (Table 4.6). Maximum growth (4.65 × 10^6 mL^{-1} cells ± 1.45) was observed when medium was substituted with malt extract, peptone, and yeast extract in the ratio of 1:1:3 (Table 4.6). However, the presence of peptone promoted copious bacterial growth. A combination of malt and yeast extract in 1:3 proportions was found to be optimal for bacteria-free growth of *D. salina* (Table 4.6). Malt, peptone, and yeast ratios of 1:1:1 to 1:1:3 yielded higher cell numbers, cellular pigments, specific growth rates, and dry biomass than cultures grown in M:Y media at ratios of 1:1 to 1:3 or those grown in mineral medium (Table 4.6). However, carotenoids in M:Y media at ratios of 1:1 to 1:3 were low compared to the other treatments with malt, yeast, or the mineral medium.

Table 4.5 Growth of I3 Strain of *D. salina* in Mixotrophy Mode with Optimized Mineral Medium

Malt Extract Concentration	Max. Cell No. × 10^6 mL^{-1}	Chl *a* Max. (pg cell^{-1})	Carotenoid Max. (pg cell^{-1})	Specific Growth Rate (μ)	G = 24 μ^{-1} (h)
1 g L^{-1}	2.95 (±1.42)^e	2.63 (±0.01)^b	3.26 (±0.02)^b	0.45 (±0.00)^e	54.58 (±0.17)^b
2 g L^{-1}	3.74 (±2.06)^d	2.66 (±0.01)^b	3.34 (±0.01)^a	0.49 (±0.00)^d	49.34 (±0.14)^c
4 g L^{-1}	4.32 (±1.54)^c	2.64 (±0.01)^b	3.05 (±0.01)^c	0.52 (±0.00)^d	47.41 (±0.17)^d
5 g L^{-1}	4.54 (±1.15)^b	2.26 (±0.02)^c	2.53 (±0.01)^d	0.53 (±0.00)^b	46.39 (±0.11)^e
6 g L^{-1}	4.84 (±1.41)^a	2.23 (±0.02)^c	2.15 (±0.01)^e	0.54 (±0.00)^a	45.52 (±0.14)^f
10 g L^{-1}	4.56 (±1.17)^{a,b}	2.26 (±0.01)^c	2.55 (±0.01)^d	0.54 (±0.00)^a	45.33 (±0.13)^f
Optimum mineral	1.94 (±1.67)^f	2.84 (±0.01)^a	3.25 (±0.02)^b	0.35 (±0.00)^f	70.40 (±0.13)^a
ANOVA and Tukey HSD	df = 6, f = 4847.92, p < 0.001; HSD [0.05] = 5.91	df = 6, f = 258.09, p <0.001; HSD [0.05] = 0.06	df = 6, f = 1077.52, p < 0.001.05] = 0.06	df = 6, f = 996.61, p < 0.001; K > 2 significant F-ratio	df = 6, f = 3999.14, p < 0.001; HSD [0.05] = 0.58

Note: HSD [0.05] = the absolute [unsigned] difference between any two sample means required for significance at 0.05 probability. Means with the same superscript letter within a column are similar to the Tukey HSD value being not significant at 0.05 probability. Optimized mineral medium substituted with different concentrations of sodium acetate as organic carbon source. Optimized mineral medium: KNO$_3$, 100 mg L^{-1}; KH$_2$PO$_4$, 0.35 mg L^{-1}; trace elements (stock of Walne's) without H$_3$BO$_3$, 1 mL L^{-1}.

Table 4.6 Growth of I3 Strain of *D. salina* in Mixotrophy Mode with Optimized Mineral Medium

Medium (g L^{-1})	Max. Cell No. × 10^6 mL^{-1}	Chl *a* Max. (pg cell^{-1})	Carotenoid Max. (pg cell^{-1})	Specific Growth Rate (μ)	G = 24 μ^{-1} (h)	Dry Biomass (g L^{-1})
M:Y 1:1	3.19 (±1.48)^e	2.96 (±0.08)^e	0.33 (±0.01)^b	0.48 (±0.01)^f	49.73 (±0.74)^c	3.05 (±0.05)^f
M:Y 1:2	3.47 (±0.99)^d	3.83 (±0.48)^d	0.19 (±0.01)^d	0.51 (±0.01)^{bc}	47.79 (±0.55)^d	3.47 (±0.06)^e
M:Y 1:3	3.88 (±1.71)^c	3.95 (±0.54)^{cd}	0.18 (±0.03)^d	0.53 (±0.01)^{ab}	45.86 (±0.37)^e	3.76 (±0.54)^d
M:P:Y 1:1:1	3.43 (±0.95)^d	4.22 (±0.03)^{bc}	0.22 (±0.02)^c	0.52 (±0.01)^a	46.38 (±0.19)^b	4.0 (±0.07)^c
M:P:Y 1:1:2	4.15 (±1.12)^b	4.49 (±0.09)^{ab}	0.16 (±0.00)^e	0.55 (±0.01)^e	43.95 (±0.27)^b	4.40 (±0.03)^b
M:P:Y 1:1:3	4.65 (±1.45)^a	4.65 (±0.07)^a	0.15 (±0.00)^e	0.58 (±0.01)^{de}	41.92 (±0.31)^{bc}	6.17 (±0.54)^a
Optimum mineral	1.81 (±1.05)^b	3.03 (±0.10)^e	0.38 (±0.01)^a	0.35 (±0.01)^{bc}	73.16 (±0.49)^a	2.40 (±0.07)^g
ANOVA and Tukey HSD	df = 7, f =3864.52, p < 0.001; HSD [0.05] = 5.55	df = 7, f =84.66, p < 0.001; HSD [0.05] = 0.1	df = 7, f =306.67, p < 0.001; HSD [0.05] = 0.02	df = 7, f =82.93, p < 0.001; HSD [0.05] = 0.03	df = 7, f = 487.82, p < 0.001; HSD [0.05] = 2.18	df = 7, f =351.88, p < 0.001; HSD [0.05] = 0.29

Note: 1 g L^{-1} of stock solutions of M:Y:P added at the rate of mL L^{-1} (e.g., 1:1:1, 1 mL L^{-1} each of optimal mineral medium). HSD [0.05] = the absolute [unsigned] difference between any two sample means required for significance at 0.05 probability. Means with the same superscript letter within a column are similar to the Tukey HSD value being not significant at 0.05 probability. Optimized mineral medium was substituted with malt extract (M) as an organic carbon source and peptone (P) and yeast extract (Y) as a complex organic nitrogen source. Optimized mineral medium: KNO$_3$, 100 mg L^{-1}; KH$_2$PO$_4$, 0.35 mg L^{-1}; trace elements (stock of Walne's) without H$_3$BO$_3$, 1 mL L^{-1}.

4.3.4 Yeast

Enrichment with yeast at 1–4 g L^{-1} did not promote higher growth (Table 4.7); on the contrary, the cell densities decreased from 1.94 × 10^6 mL^{-1} at 1 g L^{-1} enrichment to 1.18 × 10^6 mL^{-1}, and specific growth rates decreased from 0.43 in medium with 1 g L^{-1} to 0.34 in medium with 4 g L^{-1}.

In mineral medium supplemented with peptone at all concentrations tested, bacterial contamination was observed. These bacteria were coiled (unlike the rod-shaped *Halobacterium* spp.) and the cultures became turbid.

Table 4.7 Growth of I3 Strain of *D. salina* in Mixotrophy Mode with Optimized Mineral Medium

Yeast Extract Concentration	Max. Cell No. × 10^6 mL^{-1}	Specific Growth Rate (μ)	G = 24 μ^{-1} (h)
1 g L^{-1}	1.94 (±1.28)^a	0.43 (±0.00)^a	55.69 (±0.17)^e
1.5 g L^{-1}	1.65 (±1.71)^b	0.42 (±0.01)^a	59.59 (±0.11)^d
2 g L^{-1}	1.49 (±1.13)^c	0.39 (±0.00)^b	61.71 (±0.12)^c
2.5 g L^{-1}	1.36 (±1.11)^d	0.37 (±0.00)^c	65.42 (±0.11)^b
3 g L^{-1}	1.24 (±1.30)^e	0.35 (±0.00)^d	69.14 (±0.03)^a
3.5 g L^{-1}	1.22 (±1.06)^e	0.35 (±0.00)^d	69.11 (±0.00)^a
4 g L^{-1}	1.18 (±0.95)^e	0.34 (±0.00)^d	69.25 (±0.10)^a
Optimum mineral	1.66 (±1.80)^b	0.41 (±0.00)^b	59.52 (±0.01)^d
ANOVA and Tukey HSD	df = 7, F = 408.73, p < 0.001; HSD [0.05] = 5.82	df = 7, F =114.21, p < 0.001; HSD [0.05] = K > 2, significant F-ratio	df = 7, F = 286.25, p < 0.001; HSD [0.05] = 0.4

Note: 1 g L^{-1} of stock solutions of M:Y:P added at the rate of mL L^{-1} (e.g., 1:1:1, 1 mL L^{-1} each of optimal mineral medium). HSD [0.05] = the absolute [unsigned] difference between any two sample means required for significance at 0.05 probability. Means with the same superscript letter within a column are similar to the Tukey HSD value being not significant at 0.05 probability. Optimum mineral medium was substituted with different concentrations of yeast extract, a complex organic nitrogen source. Optimized mineral medium: KNO$_3$, 100 mg L^{-1}; KH$_2$PO$_4$, 0.35 mg L^{-1}; trace elements (stock of Walne's) without H$_3$BO$_3$, 1 mL L^{-1}.

Table 4.8   Pigments and Biomass in I3 Strain of _D. salina_ Mass Cultured in Different Modes

Pigments	Mixotrophy[a] pg cell^{-1}	Mixotrophy to Autotrophy pg cell^{-1}	Autotrophy[b] Wet Biomass pg cell^{-1}	Autotrophy[b] Dry Biomass mg g^{-1}
Violaxanthin	0.74	0.21	0.39	0
Zeaxanthin	1.37	0.25	0.18	12
Chl _b_	11.28	2.68	1.92	32
Chl _a_	19.74	5.66	4.63	0
β-Carotene	74.67	40.51	27.14	9.49

[a] Mixotrophy: Malt to yeast extract, 1:3 g L^{-1} in optimized mineral medium.
[b] Autotrophy: Optimized mineral medium: KNO_3, 100 mg L^{-1}; KH_2PO_4, 0.35 mg L^{-1}; trace elements (stock of Walne's) without H_3BO_3, 1 mL L^{-1}.

4.3.5 Mixotrophy and Autotrophy

4.3.5.1 Growth Kinetics and Biomass Yield

In mixotrophic mass culture, growth was much higher with 2.45×10^6 cells mL^{-1}, with specific growth rates of 0.61 and 2.06 dry biomass (g L^{-1}), than in mixo-autotrophic and autotrophic cultures. Cell numbers in the mixo-autotrophic cultures were 1.97×10^6 cells mL^{-1}, compared with 1.68×10^6 cells mL^{-1} in autotrophic cultures (Table 4.8). The specific growth rates for these corresponded to 0.42 and 0.31. The biomass was 1.77 and 1.25 (g L^{-1}) in mixo-autotrophic and autotrophic cultures (Table 4.8). The cell size was uniform in these cultures, and they were stressed two days earlier than the control.

4.3.5.2 Quantitative and Qualitative Analysis of Pigments in Biomass Harvested from Mass Culture

The mass culture I3 strain of _D. salina_, grown for 21 days, was used for pigment analysis using HPLC profiles. Six pigments, Chl _a_ β-carotene, lutein, violaxanthin, zeaxanthin, and neoxanthin, could be identified in the HPLC profiles of the wet biomass (Keerthi et al. 2015). The highest values of all pigments per cell were in mixotrophic cultures and decreased in mixotrophic + autotrophic cultures. In autotrophic cultures, all pigments per cell were the lowest. In dried algal culture biomass, Chl _a_, violoxanthin, and neoxanthin were totally destroyed (Keerthi et al. 2015). Except lutein and neoxanthin, the other pigments were quantified because standards were available. Cells cultured in mixotrophy mode contained the highest concentration of all pigments that were quantified (Table 4.8).

4.4 DISCUSSION

The growth rate and generation time of our 16 isolates (Table 4.9) are comparable to the levels reported in other carotenogenic isolates of _D. salina_ (Gómez and González 2005; Mendoza et al. 2008; Subba Rao 2009). Growth rates depended on the salinity. Specific growth rates were 0.36, 0.31, 0.26, and 0.35, corresponding to 10%, 15%, 20%, and 12.5% NaCl, which are comparable to the literature values for _D. salina_ (Figure 6 in Subba Rao 2009). At 25% and 30% NaCl, the growth rates were lower and corresponded to 0.18 and 0.17. In Walne's medium, the growth rates ranged between 0.31 and 0.41, cell morphology type A, B, and C divisions per day. The highest growth rate, that is, 0.58 divisions per day in our cultures, was in Walne's medium fortified with malt, peptone,

Table 4.9   Summary of Growth of *D. salina* in Different Media with Different Sources of Nitrates

Medium	Modification	Max. Cells 10^6 mL^{-1}	Growth Rate $G = 24\ \mu^{-1}$	Division Day	Chl *a* pg cell^{-1}	Carotene ng cell^{-1}
Walne's	NA		A: 38.63–75.00			81.70–187.02 ng
	NA		B: 33.91–38.63			33.91–38.63 ng
	NA		C: 35.64–51.23			35.64–51.23 ng
Walne's with NaNO$_3$	50 mg L^{-1}	0.63	79.43	0.31		
	100 mg L^{-1}	0.95	58.40	0.42		
	150 mg L^{-1}	0.90	75.18	0.31		
	200 mg L^{-1}	0.81	77.47	0.31		
KNO$_3$	50 mg L^{-1}	1.11	60.02	0.39		
	100 mg L^{-1}	1.15	55.15	0.43		
	150 mg L^{-1}	1.14	64.10	0.37		
	200 mg L^{-1}	1.02	64.48	0.38		
Urea	50 mg L^{-1}	1.07	62.59	0.38		
	100 mg L^{-1}	1.20	57.67	0.42		
	150 mg L^{-1}	1.10	71.69	0.33		
	200 mg L^{-1}	1.00	76.12	0.32		
NH$_4$Cl	50 mg L^{-1}	0.92	68.28	0.35		
	100 mg L^{-1}	0.86	80.92	0.30		
	150 mg L^{-1}	0.86	82.53	0.29		
	200 mg L^{-1}	0.85	84.52	0.28		
Walne's	Complete	1.85	77.27	0.31	1.19	17.54
Without	Trace elements	0.33	92.18	0.26	14.62	24.83
	Boron	2.07	58.82	0.33	3.39	3.07
	Vitamin mix	1.90	66.51	0.36	1.00	14.22
+ Acetate	0	1.34	77.19	0.31		
	2	1.67	62.96	0.37		
	4	1.96	64.09	0.38		
	6	1.95	65.28	0.37		
	8	1.62	66.49	0.35		
	10	1.66	69.89	0.34		
Walne's	None	1.94	70.40	0.35	2.84	3.25
	Malt 1 g L^{-1}	2.95	54.58	0.45	2.63	3.26
	2 g L^{-1}	3.74	49.34	0.49	2.66	3.34
	4 g L^{-1}	4.32	47.41	0.52	2.64	3.05
	5 g L^{-1}	4.54	46.39	0.53	2.26	2.53
	6 g L^{-1}	4.84	45.52	0.54	2.23	2.15
	10 g L^{-1}	4.56	45.33	0.54	2.26	2.55
Walne's	None	1.66	59.52	0.41		
	Yeast 1 g L^{-1}	1.94	55.69	0.43		
	1.5 g L^{-1}	1.65	59.59	0.42		
	2 g L^{-1}	1.49	61.71	0.39		
	2.5 g L^{-1}	1.36	65.42	0.37		
	3 g L^{-1}	1.24	69.14	0.35		
	3.5 g L^{-1}	1.22	69.11	0.35		
	4 g L^{-1}	1.18	69.25	0.34		
Walne's	None	1.81	73.16	0.35	3.03	0.38

(Continued)

Table 4.9 (Continued)   Summary of Growth of *D. salina* in Different Media with Different Sources of Nitrates

Medium	Modification	Max. Cells 10^6 mL^{-1}	Growth Rate $G = 24\ \mu^{-1}$	Division Day	Chl *a* pg cell^{-1}	Carotene ng cell^{-1}
	M:Y 1:1	3.19	49.73	0.48	2.96	0.33
	M:Y 1:2	3.47	47.79	0.51	3.83	0.19
	M:Y 1:3	3.88	45.86	0.53	3.95	0.18
	M:P:Y 1:1:1	3.43	46.38	0.52	4.22	0.22
	M:P:Y 1:1:2	4.15	43.95	0.55	4.49	0.16
	M:P:Y 1:1:3	4.65	41.92	0.58	4.65	0.15

Note: NA, not available.

and yeast (M:P:Y 1:1:3). Growth rates in Walne's medium modified with KNO_3 ranged from 0.31 to 0.42, and 0.32 to 0.42 for urea, and 0.28 to 0.35 for NH_4Cl. Omission of trace elements reduced the division rates to 0.26. Omission of boron and vitamin mix did not seem to have any negative impact on cell division. Cultures supplied with organic nitrogen source, either with malt, yeast, and peptone or in combination, yielded higher division rates. Division rates reported here are comparable to those (0.3–0.5) reported for *D. salina* (Giordano and Bowes 1997). The growth rate and biomass productivity in both the mixotrophic and mixo-autotrophic cultures were much higher than those of the autotrophic cultures. Their generation time is 57 hours, and thus ~45% less than in the photoautotrophic cultures. Also, the cultures reached the end of the exponential phase of growth ~6 days earlier.

In the literature, wide variations exist in units of measurement, and there is a need for their standardization to facilitate comparisons between investigators (Cornet 2010). Often, growth conditions such as the phase of cultures, state of nutrition, temperature and light conditions, and division rates are not given, which makes comparison of the results difficult. Our results show that omission of the vitamin mix did not reduce the division rates (Table 4.10), suggesting that vitamin mix was not essential for growth of *D. salina*, which is in agreement with the results of Borowitzka and Borowitzka (1990). Media without vitamin mix are used for the culture of *D. salina* (Vorst et al. 1994; Subba Rao 2009). It is of interest to note that while Croft et al. (2006) are of the opinion that prokaryotes are autotrophic in vitamin requirements, alternatively, when symbiosis with the vitamin-producing prokaryotes happens, eukaryotic algae like *Porphyridium purpureum*, a red microalgal species, and *Amphinidium operculatum*, a dinoflagellate, are able to meet their vitamin requirements (Croft et al. 2005). The bacterium *Halomonas* sp. was implicated as the symbiotic source of vitamin B_{12} (Croft et al. 2005). It is possible the *Halomonas* sp. and the archaea *Halobacterium* harbored in our *D. salina* cultures are symbiotic and provided the vitamin requirements.

From a biotechnological standpoint, more is important than the high division rates, higher carotene per cell, and higher lipid per cell. The carotene content (~56 mg L^{-1}) in culture seeded with culture grown mixotrophically and autotrophically (~27 mg L^{-1}) is much higher than values (8.1–15.1 mg L^{-1}) reported for commercial strains of *D. salina* (García-González et al. 2003). The carotene content (112 pg cell^{-1}) in mixotrophically cultured *D. salina* was much higher than the reported value of 70 pg cell^{-1} in medium supplemented with acetate and $FeSO_4$ (Mojaat et al. 2008). *D. salina* grown in Walne's medium contained 0.38–17.54 pg cellular carotene, which is comparable to the values summarized in Tables 4.9 and 4.11. Del Campo et al. (2001) recorded the highest cellular carotene (15 pg). Cellular lipid ranged from 0.21 to 3.45 pg for cells grown for 16 days and from 0.04 to 44.85 pg for 21-day cells. In marine *Tetraselmis* sp., the highest lipid per cell recorded was 29.11 pg (Huerlimann et al. 2010).

Omission of the vitamin mix to Walne's medium would save the vitamin mix (B_1, B_{12}, and H) per 1000 L medium about US$45 (estimated from prevailing prices at Sigma-Aldrich). The cost–benefit ratios of the three modes of mass culture tested are computed. The cost of mixo-autotrophic cultures was 8.7% (US$0.6 per 1000 L culture) less than that of autotrophic cultures. There is also a

Table 4.10 Comparison of Lipids and Carotenoids per Cell in Algal Cultures

	Growth (phase)	Medium	Lipids pg cell^{-1}	Lipids L^{-1}	Carotenoids pg cell^{-1}	Carotenoids mg L^{-1}	Reference
D. salina CCAP 19/30	Stationary	f/2 medium	NA	0.50–0.71 lipid harvest rate (mg L^{-1} h^{-1})	0.35–1.77	11.97 (± 3.01)	Mendoza et al. 2008 Chen et al. 2015
D. salina Conc-001	Stationary		NA	NA	9.0 ± 0.3		Pisal and Lele et al. 2005
D. salina Conc-006	Stationary		NA	NA	44.3 ± 0.4		Del Campo et al. 2001
D. salina Conc-007	Stationary		NA	NA	72.1 ± 1.1		Pisal and Lele et al. 2005
D. salina Australian strain	Stationary		NA	NA	13.2 ± 0.1	40.7 mg L^{-1}	Del Campo et al. 2001
Chlorella sp.	Stationary	Mixotrophy inorganic	NA	99 ± 17.2 mg lipid L^{-1} day^{-1}			Moheimani 2013
Scenedesmus sp. *NOVO*	Stationary		63–94.3		0.95–3.58		Durvasula et al. 2015
Nannochloropsis sp.	Stationary	f/2 medium	0.07–0.35	0.45–0.99 mg lipid L^{-1} day^{-1}			Chen et al. 2015
Tetraselmis sp.	Stationary	Mixotrophy inorganic	4.37–29.11	92 ± 13.1 mg lipid L^{-1} day^{-1}			Huerlimann et al. 2010 Moheimani 2013
Haematococcus pluvialis	Stationary	KM1 medium	NA	NA	181 ± 4.0	NA	Cifuentes et al. 2003
D. salina I3	Stationary	M:P:Y 1:1:3	NA	NA	74.67 ± 1.1[®],	112.01 ± 1.8[®],	Present study
D. salina I3	Stationary	Optimum mineral	NA	NA	27.14 ± 0.6[®],	40.51 ± 1.4[®],	Present study

Note: NA, not available; [®], HPLC data.

Table 4.11   Lipid and Carotene in 8 Strains of *D. salina*

S. No.	Strain ID	Lipid, pg cell^{-1}		Carotene, pg cell^{-1}	
		16th day	21st day	16th day	21st day
1	B9	1.45 (±0.02)	44.85 (±3.5)	0.289 (±0.011)	0.325 (±0.011)
2	B25	1.61 (±0.005)	40.30 (±0.94)	0.508 (±0.025)	0.619 (±0.03)
3	B30	1.47 (±0.020)	16.33 (±1.66)	0.493 (±0.014)	0.391 (±0.025)
4	B34	0.22 (±0.003)	2.07 (±0.21)	2.870 (±0.34)	2.356 (±0.39)
5	K1	0.44 (±0.004)	0.07 (±0.001)	5.84 (±0.41)	2.238 (±0.37)
6	N5	3.45 (±0.50)	0.43 (±0.011)	1.56 (±0.36)	4.26 (±0.76)
7	N6	0.21 (±0.002)	0.04 (±0.001)	2.56 (±0.27)	2.44 (±0.19)
8	K10	0.52 (±0.017)	0.22 (±0.005)	2.77 (±0.14)	2.64 (±0.19)

reduction of the production cycle by 11 days, which saves in electricity due to less time needed for generating seed culture. Further, in the mixo-autotrophic cultures, the quality of the biomass with respect to pigments other than β-carotene is also good, with a higher content of chlorophylls and zeaxanthin than autotrophic cultures.

In biotechnological applications of microalgae, Durvasula et al. (2013) stressed the need for "designer microalgae," developing inexpensive media, and the isolation of strains of high-yielding microalgae. Borowitzka (2008) pointed out that while synthesized β-carotene is priced at US$330 per kilogram, carotene from *D. salina* biomass should not exceed US$50–100 per kilogram, and therefore good economic potential exists. This study showed that the extremophile algae isolated from the solar tropical solar salterns of the Bay of Bengal, because of their amenability in mass culture under ambient conditions, have good potential for lipids and carotenoids.

ACKNOWLEDGMENTS

I am grateful to Drs. Subba Rao and Ravi V. Durvasula for inviting me to contribute this chapter, and for their continued encouragement and help. Many thanks to an anonymous reviewer for suggesting improvements to presentation. The experimental part of the work was done at Andhra University, Visakhapatnam, to whom I express my thanks for the facilities.

REFERENCES

Borowitzka, L., and M. Borowitzka. Commercial production of β-carotene by *Dunaliella salina* in open ponds. *Bull. Mar. Sci.* 47 (1990):244–252.

Borowitzka, L.J. The microflora. Adaptation to life in extremely saline lake. *Hydrobiologia* 81 (1) (1981):33–46.

Borowitzka, M.A. The mass culture of *Dunaliella salina*. Technical resource papers, seaweed culture. Fisheries and Aquaculture Department, Rome, 2008. http://www.fao.org/docrep/field/003/AB728E/AB728E06.htm.

Borowitzka, M.A., and L.J. Borowitzka. Algal growth media and sources of cultures. In M.A. Borowitzka and L.J. Borowitzka (eds.), *Dunaliella Micro-Algal Biotechnology*, pp. 27–58. Cambridge University Press, Cambridge, 1988.

Borowitzka, M.A., and C.J. Siva. The taxonomy of the genus *Dunaliella* (Chlorophyta, Dunaliellales) with emphasis on the marine and halophilic species. *J. Appl. Phycol.* 19 (5) (2007):567–590.

Chakraborty, P., Raghunadh Babu, P.V., Tamoghna, A., and Bandyopadhyay, D. Stress and toxicity of biologically important transition metals (Co, Ni, Cu and Zn) on phytoplankton in a tropical freshwater system: An investigation with pigment analysis by HPLC. *Chemosphere* 80 (2010):548–553.

Chen, Y., X. Tang, R.V. Kapoore, C. Xu, and S. Vaidyanathan. Influence of nutrient status on the accumulation of biomass and lipid in *Nannochloropsis salina* and *Dunaliella salina*. *Energy Convers. Manag.* 106 (2015):61–72.

Cifuentes, A.S., M.A. Gonzalez, S. Vargas, M. Hoeneisen, and N. Gonzalez. Optimization of biomass, total carotenoids and astaxanthin production in *Haematococcus pluvialis* Flotow strain Steptoe (Nevada, USA) under laboratory conditions. *Biol. Res.* 36 (3–4) (2003):343–357.

Cornet, J.-F. Calculation of optimal design and ideal productivities of volumetrically lightened photobioreactors using the constructal approach. *Chem. Eng. Sci.* 65 (2) (2010):985–998.

Croft, M.T., A.D. Lawrence, E. Raux-Deery, M.J. Warren, and A.G. Smith. Algae acquire vitamin B_{12} through a symbiotic relationship with bacteria. *Nature* 438 (2005):90–93.

Croft, M.T., M.J. Warren, and A.G. Smith. Algae need their vitamins. *Eukaryot. Cell* 5 (2006):1175–1183.

Del Campo, J.A., H. Rodríguez, J. Moreno, M.Á. Vargas, J. Rivas, and M.G. Guerrero. Lutein production by *Muriellopsis* sp. in an outdoor tubular photobioreactor. *J. Biotechnol.* 85 (3) (2001):289–295.

Durvasula, R., I. Hurwitz, A. Fieck, and D.V. Subba Rao. Culture, growth, pigments and lipid content of *Scenedesmus* species, an extremophile microalga from Soda Dam, New Mexico in wastewater. *Algal Res.* 10 (2015):128–133.

Durvasula, R.V., D.V. Subba Rao, and V.S. Rao. Microalgal biotechnology: Today's (green) gold rush. In F. Bux (ed.), *Biotechnical Applications of Microalgae: Biodiesel and Value-Added Products*, pp. 199–225. London: Taylor & Francis Publishers, 2013.

Elevi Bardavid, R., P. Khristo, and A. Oren. Interrelationships between *Dunaliella* and halophilic prokaryotes in saltern crystallizer ponds. *Extremophiles* 12 (1) (2008):5–14.

Furnas, M. Measuring the growth rates of phytoplankton in natural populations. In D.V. Subba Rao (ed.), *Pelagic Ecology Methodology*, pp. 221–249. Amsterdam: Balkema Publishers, 2002.

García-González, M., J. Moreno, J. Canavate, V. Anguis, A. Prieto, C. Manzano, F.J. Florencio, and M.G. Gurrero. Conditions for open-air outdoor culture of *Dunaliella salina* in southern Spain. *J. Appl. Phycol.* 15 (2003):177–184.

Giordano, M., and G. Bowes. Gas exchange and C allocation in *Dunaliella salina* cells in response to the N source and CO_2 concentration used for growth. *Plant Physiol.* 115 (1997):1049–1056.

Gómez, P.I., and M.A. González. The effect of temperature and irradiance on the growth and carotenogenic capacity of seven strains of *Dunaliella salina* (Chlorophyta) cultivated under laboratory conditions. *Biol. Res.* 38 (2005):151–162.

Huerlimann, R., R. De Nys, and K. Heimann. Growth, lipid content, productivity, and fatty acid composition of tropical microalgae for scale-up production. *Biotechnol. Bioeng.* 107 (2) (2010):245–257.

Jayapriyan, K.R., R. Rajkumar, L. Sheeja, S. Nagaraj, S. Divya, and R. Rengasamy. Discrimination between the morphological and molecular identification in the genus *Dunaliella*. *Int. J. Curr. Res.* 8 (2010):73–78.

Jeffrey, S.W., and G.F. Humphrey. New spectrophotometric equations for determining chlorophylls a, b, c1 and c2 in higher plants, algae and natural phytoplankton. *Biochem. Physiol. Pflanz.* 167(1975):1–194.

Keerthi, S. Bioprospecting of biotechnologically important microalgae form hyper saline and marine regions along south eastern coast of India. PhD dissertation, Andhra University, Visakhapatnam, India, 2012.

Keerthi, S., U.D. Koduru, and N.S. Sarma. A nutrient medium for development of cell dense inoculum in mixotrophic mode to seed mass culture units of *Dunaliella salina*. *Algol. Stud.* 147 (1) (2015):7–28.

Kimura, K., M. Yamaoka, and Y. Kamisaka. Rapid estimation of lipids in oleaginous fungi and yeasts using Nile red fluorescence. *J. Microbiol. Methods* 56 (3) (2004):331–338.

Lavens, P., and P. Sorgeloos. Manual on the production and use of live food for aquaculture. FAO Fisheries Technology Paper 295. Rome: Food and Agriculture Organization, 1996.

Mendoza, H., A. De la Jara, K. Freijanes et al. Characterization of *Dunaliella salina* strains by flow cytometry: A new approach to select carotenoid hyper producing strains. *Electronic J. Biotechnol.* 11 (2008):5–6.

Mishra, A., and B. Jha. Isolation and characterization of extracellular polymeric substances from micro-algae *Dunaliella salina* under salt stress. *Bioresour. Technol.* 100 (13) (2009):3382–3386.

Moheimani, N.R. Inorganic carbon and pH effect on growth and lipid productivity of *Tetraselmis suecica* and *Chlorella* sp (Chlorophyta) grown outdoors in bag photobioreactors. *J. Appl. Phycol.* 25 (2) (2013):387–398.

Mojaat, M., T.J. Pruvos, A. Foucault, and J. Legrand. Effect of organic carbon sources and Fe^{+2} ions on growth and carotene accumulation by *Dunaliella salina*. *Biochem. Eng. J.* 39 (2008):177–184.

Oren, A., and C.D. Litchfield. A procedure for the enrichment and isolation of *Halobacterium*. *FEMS Microbiol. Lett.* 173 (2) (1999):353–358.

Pisal, D.S., and S. Lele. Carotenoid production from microalga, *Dunaliella salina*. *Ind. J. Biotechnol.* 4 (2005):476–483.

Preetha, K., L. John, C.S. Subin, and K.K. Vijayan. Phenotypic and genetic characterization of *Dunaliella* (Chlorophyta) from Indian salinas and their diversity. *Aquat. Biosyst.* 8 (1) (2012):1.

Priscu, J.C., L.R. Priscu, A.C. Palmisano, and C.W. Sullivan. Estimation of neutral lipid levels in Antarctic sea ice microalgae by Nile red fluorescence. *Antarct. Sci.* 2 (02) (1990):149–155.

Raja, R., S.H. Iswarya, D. Balasubramanyam, and R. Rengasamy. PCR-identification of *Dunaliella salina* (Volvocales, Chlorophyta) and its growth characteristics. *Microbiol. Res.* 162 (2) (2007):168–176.

Subba Rao, D.V. Cultivation, growth media, division rates and applications of *Dunaliella* species. In J.E.W. Polle, D.V. Subba Rao, and A. Ben-Amotz (eds.), *The Alga Dunaliella: Biodiversity, Physiology, Genomics and Biotechnology*, pp. 44–89. Enfield, NH: Science Publishers, 2009.

Van Heukelem, L., and C.S. Thomas. Computer-assisted high-performance liquid chromatography method development with applications to the isolation and analysis of phytoplankton pigments. *J. Chromatogr.* 910 (2001):31–49.

Vorst, P., R.L. Baard, L.R. Mur, H.J. Korthals, and H. van den Ende. Effect of growth arrest on carotene accumulation and photosynthesis in *Dunaliella*. *Microbiology* 140 (6) (1994):1411–1417.

Walne, P.R. Studies on the food value of nineteen genera of algae to juvenile bivalves of the genera *Ostrea, Crassostrea, Mercenaria* and *Mytilus*. *Fish. Invest. Ser. 2* 26 (5) (1970).

CHAPTER 5

Heterotrophic Production of Phycocyanin in *Galdieria sulphuraria*

Niels Thomas Eriksen

CONTENTS

5.1 INTRODUCTION

Phycocyanin is a light-harvesting protein in cyanobacteria, red algae, and cryptomonads. It has a deep blue color and a strong, red fluorescence. Phycocyanin belongs to the group of phycobiliproteins that are all multichain holo-proteins composed of an apo-protein with covalently bound phycobilin chromophores (see, e.g., MacColl 1998; Stadnichuk et al. 2015). In phycocyanin and the related phycobiliprotein, allophycocyanin, it is phycocyanobilin chromophores that provide the blue color, as well as the fluorescence (Figure 5.1). Phycobilins are open-chain tetrapyrroles, and other phycobiliproteins may carry different types of phycobilins as chromophores. MacColl (1998) renamed phycobiliproteins with prefixes corresponding to their phycobilin content. All phycocyanins carrying only phycocyanobilin chromophores, the most common phycocyanin type, were named C-phycocyanin.

Phycobiliproteins have been utilized in a number of different ways. The clear blue color of phycocyanin is useful in some types of food and beverages, while a number of health-promoting effects have made phycocyanin increasingly attractive as a nutraceutical (Eriksen 2016). C-Phycocyanin is commercially available from phototrophic cultures of cyanobacteria, predominantly *Arthrospira platensis* (syn. *Spirulina platensis*). One microalga, the extremophilic cyanidiophyte (a subphylum to the red algae) *Galdieria sulphuraria*, has, however, turned out to be a prospective candidate for synthesis of phycocyanin under heterotrophic conditions (Eriksen 2008a). The phycocyanin made by *G. sulphuraria* is also C-phycocyanin. Not only does this alga grow well heterotrophically, but it may also synthesize phycocyanin even in darkness. *G. sulphuraria* cultures have phycocyanin productivity potentials

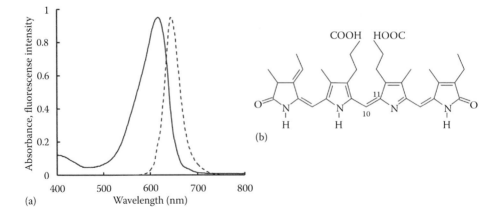

Figure 5.1 (a) Absorbance spectrum (solid line) and fluorescence emission spectrum (dashed line) of phyco-cyanin extracted from heterotrophic *G. sulphuraria*, precipitated by ammonium sulfate and redis-solved at pH 7. (b) Structure of phycocyanobilin. Reducible double bond between carbon atoms C-10 and C-11 indicated. (Redrawn from Eriksen, N.T., *Appl. Microbiol. Biotechnol.*, 80, 1–14, 2008.)

exceeding those of *A. platensis* cultures by one to two orders of magnitude (Graverholt and Eriksen 2007), but commercial production of phycocyanin in *G. sulphuraria* has not yet been established. This chapter focuses on the current status on phycocyanin synthesis in heterotrophic *G. sulphuraria* cultures, the opportunities and challenges of this process compared with phycocyanin synthesis in *A. platensis* cultures, and the potential usefulness of phycocyanin from heterotrophic *G. sulphuraria*.

5.2 PHOTOAUTOTROPHIC PRODUCTION OF PHYCOCYANIN IN *ARTHROSPIRA PLATENSIS*

The main commercial source of phycocyanin is *A. platensis* grown photoautotrophically in outdoor open ponds or raceways in tropical and subtropical regions (Lee 1997; Pulz 2001; Spolaore et al. 2006). *A. platensis* tolerates pH values up to pH 10.5 (Richmond and Grobbelaar 1986). It is therefore one of the few photoautotrophic microorganisms that is able to grow in open cultures without being outcompeted by contaminating organisms, although contaminants do appear in open *A. platensis* cultures (Richmond and Grobbelaar 1986; Richmond et al. 1990). Photoautotrophic cultures depend on externally supplied light, a source of energy that must be supplied at the surface of the culture. Light cannot be homogenously distributed in microalgal cultures, and almost all light is absorbed within a narrow photic zone close to the surface of the culture. In this photic zone, the light intensity is high, and therefore the light utilized for photosynthesis is at a rather low efficiency (see, e.g., Eriksen 2008b, 2013). Below the photic zone, darkness prevails. Biomass concentrations in phototrophic cultures must therefore be low, often close to 1 g L^{-1}, to maximize the extension of the photic zone where photosynthesis is possible.

Low biomass concentrations result in low biomass productivities. In outdoor open ponds and raceways, productivities are most often reported in terms of areal productivity. The productivity of phycocyanin is determined by the biomass concentration, the specific growth rate of the cells in the culture, and the specific phycocyanin content in the biomass. Dry biomass concentrations in cultures of *A. platensis* and *Anabaena* sp., a second cyanobacterium, will be in the order of 0.3–1.6 g L^{-1}, with areal biomass and phycocyanin productivities of 15–25 and 0.82–1.32 g m^{-2} day^{-1}, respectively (Pushparaj et al. 1997; Jiménez et al. 2003; Moreno et al. 2003). These values correspond to volumetric productivities of only 0.05–0.32 g biomass L^{-1} day^{-1} and 3–24 mg phycocyanin L^{-1} day^{-1}, respectively. Such productivities are orders of magnitude lower than what is routinely obtained in cultures of heterotrophic microorganisms where biomass productivities are not limited by externally supplied light (see, e.g., Eriksen 2008b, 2013).

A. platensis can also grow mixotrophically or heterotrophically (Mühling et al. 2005). Heterotrophic cultivation seems, however, to be a nonviable strategy for phycocyanin synthesis due to low specific growth rates and pigment contents in heterotrophic *A. platensis* cells (Marquez et al. 1993; Marquez et al. 1995; Chojnacka and Noworyta 2004). Mixotrophic cultivation may be a way to increase specific growth rate and biomass concentration without an apparent negative effect on the phycocyanin content (Marquez et al. 1993; Marquez et al. 1995; Chen et al. 1996; Chen and Zhang 1997). Mixotrophic cultivation of *A. platensis* seems, however, not to have been employed on a large scale, probably because cultivations have to be carried out at higher costs in enclosed reactors to minimize the introduction of contaminating organisms that would compete with the comparatively slow-growing *A. platensis* for organic carbon and other nutrients and result in decreased productivities and in the formation of products of lower hygienic quality.

5.3 HETEROTROPHIC CULTIVATION OF *GALDIERIA SULPHURARIA*

In heterotrophic cultures, the productivity of phycocyanin will also depend on the concentration of biomass in the culture, the specific growth rate of the biomass, and the product content within the biomass. The greatest advantage of heterotrophic compared with phototrophic cultures is that heterotrophic cultures do not depend on externally supplied light, and their production potentials are therefore much higher than that of light-dependent photoautotrophic cultures. Biomass concentrations in high-cell-density cultures of heterotrophic microorganisms may reach 100–200 g L^{-1} (Riesenberg and Guthke 1999), which is 50–100 times higher than the biomass concentrations normally found in photoautotrophic cultures (see, e.g., Eriksen 2011). Heterotrophic cultures are also easier to scale up with regards to reactor size, mixing, gas transfer, productivity, and axenicity since bioreactors for heterotrophic microorganisms do not need to be primarily designed as solar collectors with large surface-to-volume ratios. It is far from all microalgae that will grow heterotrophically, but quite a few species do thrive well under heterotrophic conditions. Some of these have been grown in heterotrophic cultures with productivities up to 50 g L^{-1} day^{-1} (Eriksen 2011), which is comparable to the productivities of high-cell-density cultures of heterotrophic bacteria and yeast (30–150 g L^{-1} day^{-1}) (Riesenberg and Guthke 1999).

G. sulphuraria is a promising candidate for heterotrophic production of phycocyanin (Eriksen 2008a). *G. sulphuraria* is an extremely acidophilic and also thermophilic organism found naturally in hot, acidic springs in volcanic areas. Its optimal growth conditions are found at temperatures close to 45°C and at pH 1–3. It is able to utilize a great variety of carbohydrates and polyols as carbon sources (Gross and Schnarrenberger 1995).

G. sulphuraria contains phycocyanin and minor amounts of allophycocyanin. Particularly two characteristics make *G. sulphuraria* a suitable host for heterotrophic phycocyanin production. Most importantly, some *G. sulphuraria* strains do retain their photosynthetic apparatus when grown heterotrophically even in darkness (Gross and Schnarrenberger 1995; Marquardt 1998; Sloth et al. 2006). Second, is *G. sulphuraria* surprisingly well suited for heterotrophic cultivation in highly productive high-cell-density cultures where cell dry weight (DW) exceeds 100 g L^{-1} (Graverholt and Eriksen 2007; Schmidt et al. 2005). In its natural habitats, concentrations of sulfate and other solutes are high and *G. sulphuraria* is well adapted to a life in concentrated growth media (Albertano et al. 2000; Gross et al. 2002; Schmidt et al. 2005). Even though the availability of organic substrates in hot, acidic springs is extremely low and *G. sulphuraria* naturally thrives under strong nutrient limitations (Gross et al. 1998), it still tolerates concentrations of glucose or fructose of up to 200 g L^{-1} with no apparent effect on specific growth rate (Schmidt et al. 2005). Even at 400 g L^{-1} glucose or fructose, it still grows. Adaptation to nutrient limitation has probably forced *G. sulphuraria* to economize with its carbon sources, and high growth yields of 0.46–0.54 g DW g^{-1} have been found in batch cultures grown with glucose, fructose, or sucrose as the carbon source (Schmidt et al. 2005). Overflow metabolism, which is a common challenge in the cultivation of cell and microbial cultures, seems not to exist in *G. sulphuraria* even at the highest nutrient concentrations, nor does *G. sulphuraria* seem to secrete inhibitory metabolic by-products. Graverholt and Eriksen (2007) found similar growth yields during exponential growth in glucose-sufficient batch cultures at biomass concentrations up to 25 g DW L^{-1}, in glucose-limited fed-batch cultures at biomass concentrations up to 109 g DW L^{-1}, and in glucose-limited continuous-flow cultures at steady-state biomass concentrations between 24 and 83 g L^{-1}. Also, complex carbon substrates such as molasses containing up to 500 g L^{-1} of sugars are suitable substrates for fed-batch cultivation of this alga (Schmidt et al. 2005). A constant growth yield during nutrient-sufficient and nutrient-limited conditions is a great advantage. Not only will substrates be converted into biomass at maximal efficiency at all conditions, but also the output of the cultures can be accurately predicted, for example, when conditions are modified or cultures are taken to a different scale.

The specific growth rate of *G. sulphuraria* at 42°C is in the order of 1.1–1.2 day^{-1} on glucose, fructose, or glycerol (Schmidt et al. 2005; Sloth et al. 2006). This is lower than found in many industrial microorganisms. If biomass concentrations are high, substantial biomass productivities can still be obtained. In a carbon-limited continuous-flow *G. sulphuraria* culture grown at a dilution rate of 0.6 day^{-1} and a steady-state biomass concentration of 83 g L^{-1}, the biomass productivity was 50 g L^{-1} day^{-1} (Graverholt and Eriksen 2007), something like 50 times above the productivity observed in most cultures of photoautotrophs. *G. sulphuraria* also grows photoautotrophically but at a specific growth rate of only 0.1 day^{-1} (Sloth et al. 2006). Illumination does not affect the specific growth rate of mixotrophic cultures with access also to organic carbon (Sloth et al. 2006). Photosynthesis, however, still contributes to the energy and carbon metabolism of *G. sulphuraria*. In mixotrophic cultures grown with glucose or glycerol as limiting substrates, biomass yields rose to 0.84–1.05 g g^{-1} (0.91–1.2 mol carbon incorporated into biomass per mole of organic carbon metabolized) at the optimal light intensity of approximately 100 mol photons m^{-2} s^{-1}. At higher light intensities, cultures were inhibited by the light.

The low pH values and high concentrations of sulfuric acid in the growth medium will keep most contaminants out of the cultures but may also cause corrosion in bioreactors and auxiliary equipment. The harshness of the growth medium may actually be one of the main technological challenges with

Table 5.1 Characteristics of Phycocyanin and Phycocyanin Production in Photoautotrophic *Arthrospira platensis* and Heterotrophic *Galdieria sulphuraria*

	Arthrospira Platensis	*Galdieria Sulphuraria*
Cultivation	Photoautotrophic	Heterotrophic
Bioreactors	Open ponds or raceways	Closed, stirred tank bioreactors
Technological status	Established technology Relatively simple reactors	Established technology Relatively complex reactors
Suitable location	Outdoor, tropical and subtropical regions	Indoor, climate independent, everywhere
Phycocyanin content in biomass	High (>10%)	Low (1%–2%)
Biomass concentration	Low (~1 g L^{-1})	High (50–100 g L^{-1})
Biomass productivity	Low (~1 g L^{-1} day^{-1})	High (~50 g L^{-1} day^{-1})
Phycocyanin productivity	Low	High
Risk of contamination	Unavoidable	Controllable
Production status	Established	Not established
Phycocyanin properties	Comparable to phycocyanin from *G. sulphuraria*	Comparable to phycocyanin from *A. platensis*
Food safety	High *A. platensis* has a long history as food and feed	Probably high Few, recent studies indicate that *G. sulphuraria* is safe to eat
Nutraceutical potential	Comparable to phycocyanin from *G. sulphuraria* Provided by phycocyanobilin	Comparable to phycocyanin from *A. platensis* Provided by phycocyanobilin

respect to large-scale cultivation of *G. sulphuraria*. Table 5.1 provides a comparison of some of the most important differences and resemblances of production and applications of phycocyanin produced in photoautotrophic cultures of *A. platensis* and in heterotrophic cultures of *G. sulphuraria*.

5.4 PHYCOCYANIN SYNTHESIS IN *GALDIERIA SULPHURARIA*

Phycocyanin synthesis depends on the formation of the apo-protein, as well as the phycocya-nobilin chromophore. Formation of both parts is affected by a number of environmental factors. In cyanobacteria, the C-phycocyanin α- and β-chains are encoded by the *cpcA* and *cpcB* genes on the cpc operon (de Lorimier et al. 1984; Bryant et al. 1985; Belknap and Haselkorn 1987; Liu et al. 2005b). In red algae like *G. sulphuraria*, the *cpcA* and *cpcB* genes are located on the chloroplast genome (Zetsche and Valentin 1993–1994; Ohta et al. 2003). Several putative promoter sequences (Liu et al. 2005b; Guo et al. 2007) and positive response elements, of which one is responding to light (Lu and Zhang 2005), are present in the upstream region of the operon, allowing multiple environmental variables to effect expression of the *cpcA* and *cpcB* genes. The same is the case for phycocyanobilin synthesis and insertion. This chromophore is synthesized from heme and inserted into the C-phycocyanin apo-protein by three enzymatic steps (Cornejo and Beale 1997; Tooley et al. 2001): heme is linearized into biliverdin IXα by heme oxygenase, which is further converted into phycocyanobilin by 3Z-phycocyanobilin–ferredoxin oxidoreductase, and finally inserted into the C-phycocyanin apo-protein by phycocyanobilin lyase. The cellular activities of these enzymes are stimulated by heme precursors (Troxler et al. 1989) and affected by, for example, light intensity and the presence of organic substrates (Rhie and Beale 1994). *G. sulphuraria* possesses an oxygen-dependent and an oxygen-independent heme synthesis pathway, and evidence suggests that oxygen limitation affects heme synthesis negatively, and possibly also the synthesis of phycocyanobilin (Sarian et al. 2016). In all, carbon substrate type and availability, nitrogen availability, light inten-sity, and possibly dissolved oxygen tension are the most important environmental conditions affect-ing the synthesis of C-phycocyanin in *G. sulphuraria* cultures.

In *G. sulphuraria*, *G. partita*, and the related *Cyanidium caldarium*, phycocyanin synthesis is repressed by glucose (Rhie and Beale 1994; Stadnichuk et al. 1998, 2000; Sloth et al. 2006) and heterotrophic grown cells are normally pale. It has, however, been noticed that some *G. sul-phuraria* isolates retain their photosynthetic apparatus when grown heterotrophically (Gross and Schnarrenberger 1995; Marquardt 1998). These isolates have apparently lost their natural ability to downregulate their photosynthetic apparatus. One such isolate, *G. sulphuraria* strain 074G, described as a stable mutant with respect to pigmentation by Gross and Schnarrenberger (1995), has formed the basis for the later studies of phycocyanin production in heterotrophic cultures (Schmidt et al. 2005; Sloth et al. 2006; Graverholt and Eriksen 2007; Sørensen et al. 2013). In glucose-restricted batch cultures of heterotrophic *G. sulphuraria* 074G, the specific phycocyanin content reaches its maximal level at approximately 10 mg g^{-1} shortly after glucose is depleted. This is less than 10% of the phycocyanin content of *A. platensis* (Bhattacharya and Shivaprakash 2005). After years of subcultivation in the laboratory, *G. sulphuraria* 074G cultures are now composed of cells with quite variable pigment content and have been the source for novel isolates of more highly pigmented strains. Maximal phycocyanin contents have reached 30 mg g^{-1} in batch cultures of subisolates derived from strain 074G after glucose depletion (Graverholt and Eriksen 2007; Eriksen 2008a; Sørensen et al. 2013). Systematic screenings for highly pigmented isolates or actual strain improvements have, however, not yet been carried out, so the full potential for phycocyanin accu-mulation in *G. sulphuraria* cells has probably not yet been reached.

Nitrogen sufficiency is also a prerequisite for the synthesis of phycocyanin. Phycobiliproteins may be among the most abundant proteins in many cyanobacteria and algae, but they are not essen-tial to the functioning of the cells. They have therefore also obtained a secondary role as internal nitrogen storage compounds that can be mobilized and selectively degraded in cyanobacteria exposed to nitrogen starvation (Allen and Smith 1969; Yamanaka and Glazer 1980). Ammonium is the preferred nitrogen source in *Galdieria* cultures, and ammonium depletion also results in almost complete depletion of phycobiliproteins in batch cultures of *G. sulphuraria* (Sloth et al. 2006).

Phycocyanin synthesis is also affected by light, as would be expected for a photosynthetic pigment. Low light intensities are normally needed to stimulate phycocyanin synthesis in *Galdieria* (see, e.g., Rhie and Beale 1994; Chaneva et al. 2007), while phycocyanin synthesis is repressed at high light intensities (Rodríguez et al. 1991; Venugopal et al. 2006). Even though heterotrophic cultures of *G. sulphuraria* 074G produce phycocyanin when grown in darkness, phycocyanin synthesis is not necessarily unaffected by light in this strain. In mixotrophic batch cultures grown on glucose or fructose, light does not stimulate phycocyanin synthesis, but when glycerol is the carbon source, phycocyanin synthesis is stimulated at photon flux densities up to 80 μmol m^{-2} s^{-1} (Sloth et al. 2006). Higher light intensities around 200 μmol photons m^{-2} s^{-1} repress phycocyanin synthesis irrespective of the carbon source. When *G. sulphuraria* 074G was grown in glucose- or glycerol-limited mixotrophic continuous-flow cultures where substrate uptake maintained low concentrations of the carbon sources, the specific pigment content was maximal at light intensities of approximately 100 μmol m^{-2} s^{-1} (Sloth et al. 2006). This was also the light intensity where photosynthesis had the largest positive effect on the yield of biomass. Although light intensities at this level seem low when compared with, for example, the intensity of sunlight, it would be very difficult to obtain such high light intensities inside high-cell-density cultures because self-shading will allow light to penetrate only a few millimeters below culture surfaces (see, e.g., Gittelson et al. 1996). Still, Wan et al. (2016) recently described that a light intensity of 250 μmol photons m^{-2} s^{-1} induced phycocyanin synthesis up to 132 mg g^{-1} in cells transferred from heterotrophic to phototrophic conditions over a 7-day period.

5.5 PHYCOCYANIN PRODUCTION IN HETEROTROPHIC *GALDIERIA SULPHURARIA* CULTURES

Schmidt et al. (2005) and Graverholt and Eriksen (2007) investigated growth and phycocyanin production in glucose-limited and ammonium-sufficient, heterotrophic, high-cell-density fed-batch and continuous-flow cultures of *G. sulphuraria* 074G. In these cultures, biomass concentrations reached values above 100 g L^{-1} (Figure 5.2). Although the specific phycocyanin content in heterotrophic *G. sulphuraria* 074G was considerably lower than in *A. platensis*, high biomass productivities up to 50 g L^{-1} day^{-1}, combined with the use of highly pigmented variant strains derived from *G. sulphuraria* 074G, resulted in phycocyanin productivities as high as 0.86 g L^{-1} day^{-1} (Graverholt and Eriksen 2007), more than 10 times above the productivities seen in *A. platensis* cultures.

Numerous protocols have been used to purify phycocyanin from the remaining biomass of cyanobacterial cultures. Most commonly, phycocyanin is extracted from the biomass in aqueous media, precipitated by ammonium sulfate, redissolved, and further purified by different chromatographic principles. Also, ultrafiltration and aqueous two-phase extraction have successfully been employed as operational steps in the preparation of pure phycocyanin solutions. The purity of phycocyanin preparations is conveniently reported by a purity number that describes the ratio between the absorbance from phycocyanobilin at 620 nm and the absorbance from all proteins in the preparation at 280 nm. Phycocyanin preparations with purity numbers greater than 0.7 have been described as food grade by Rito-Palomares et al. (2001), while purity numbers of 3.9 and above 4.0 have been named reactive and analytical grade, respectively. Phycocyanin extraction from *G. sulphuraria* 074G depends on mechanical disruption of the cells (Schmidt et al. 2005), while repeated freeze–thaw cycles may release phycocyanin from other strains (Myounghoon et al. 2014). The same protocols as used for the purification of cyanobacterial phycocyanin can be used to also obtain

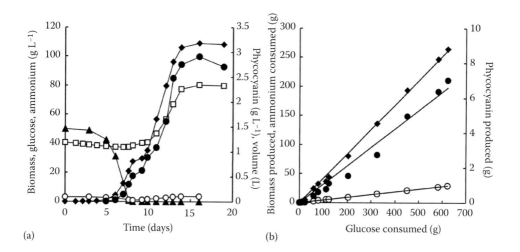

Figure 5.2 (a) Heterotrophic high-cell-density fed-batch culture of highly pigmented variant strain derived from *G. sulphuraria* 074G. Concentrations of biomass DW (◆), glucose (▲), and ammonium (○); volume (□); and concentration of phycocyanin (●). The fed-batch phase from Days 9–14 is observable by increasing culture volume. (b) Amounts of produced biomass (◆), consumed ammonium (○), and produced phycocyanin (●) in relation to the amount of glucose consumed by the culture in panel a. (Redrawn from Graverholt, O.S., and Eriksen, N.T., *Appl. Microbiol. Biotechnol.*, 77, 69–75, 2007.)

analytical-grade preparations of phycocyanin from *G. sulphuraria* aqueous extracts, that is, various combinations of ammonium sulfate precipitation, ultrafiltration, ion exchange chromatography, and aqueous two-phase extraction. A two-step protocol combining $(NH_4)_2SO_4$ fractionation followed by anion exchange chromatography has resulted in phycocyanin preparations from *G. sulphuraria* 074G with purity numbers of 3.5, while purity numbers up to 4.5 and yields of 31%–42% have been obtained when either ultrafiltration or aqueous two-phase extraction was also included as a step in the purification procedure (Sørensen et al. 2013). Myounghoon et al. (2014) has been able to obtain phycocyanin preparations from phototrophic *G. sulphuraria* UTEX 2919 with a purity number of 4.14 employing $(NH_4)_2SO_4$ fractionation as the sole purification method. If needed, phycocyanin can also be separated from allophycocyanin by anion exchange chromatography (Sørensen et al. 2013). The properties of phycocyanin from *G. sulphuraria* are therefore similar to the properties of cyanobacterial phycocyanin with respect to the different purification methodologies, and the same operational steps and procedures that are used for the purification of cyanobacterial phycocyanin work equally well for the purification of phycocyanin from *G. sulphuraria*.

5.6 PROPERTIES AND INTERESTS IN PHYCOCYANIN

The properties that make phycocyanin and other phycobiliproteins useful in many types of applications are linked to the structure of the molecules. Although the blue color of phycocyanin is provided by its phycocyanobilin chromophores (Figure 5.1), the apo-protein part of the molecule serves an equally important role for the color intensity, as well as the fluorescent properties of phycocyanin (Fukui et al. 2004). The protein part also makes phycocyanin water soluble. As mentioned above, C-phycocyanin is the type of phycocyanin present in *A. platensis*, as well as in *G. sulphuraria*. C-Phycocyanin is composed of two subunits; the α-chain, with 162 amino acids and 1 phycocyanobilin attached at cysteine 84, and the β-chain, with 172 amino acids and 2 phycocyanobilins attached at cysteine 84 and cysteine 155 (Troxler et al. 1981; Stec et al. 1999; Adir et al. 2001; Contreras-Martel et al. 2007). The amino acid sequences are highly similar in C-phycocyanin to those in cyanobacteria and red algae (Stec et al. 1999; Adir et al. 2001; Contreras-Martel et al. 2007). The two subunits form αβ monomers that further aggregate into $\alpha_3\beta_3$ trimers and $\alpha_6\beta_6$ hexamers, the functional unit of C-phycocyanin (Stec et al. 1999; Adir et al. 2001, 2002; Padayana et al. 2001; Nield et al. 2003; Contreras-Martel et al. 2007). In cyanobacteria and red algae, phycobiliprotein $\alpha_6\beta_6$ hexamers finally aggregate into structures known as phycobilisomes located on the thylakoid membranes.

Fluorescent chromophores, dyes, and nutraceuticals are the major phycobiliprotein applications. Allophycocyanin and the red-colored phycoerythrin are mostly used as fluorescent dyes (Glazer 1994; Sekar and Chandramohan 2008). Phycoerythrin and allophycocyanin form strong aggregates that provide a large number of fluorescent chromophores in one aggregate (Glazer 1994). Because phycocyanin aggregates are less stable, this phycobiliprotein has not gained similar widespread usage as a fluorescent dye despite its being strongly fluorescent and the most readily available of the phycobiliproteins. The main interests in the utilization of phycocyanin have therefore predominantly been centered on food and health applications. The clear blue color of phycocyanin is attractive in some types of food and beverages, and a number of nutraceutical effects are attributed phycocyanin.

The scientific activities and interest in phycocyanin may be interpreted from the number of scientific papers and the number of citations to these papers recorded by the Web of Science (WOS) (Thomson Reuters 2015), a database recognized as a valid indicator of scientific activities within biological sciences (Carpenter and Narin 1981). WOS had by the end of 2015 registered about 2400 publications on phycocyanin in the core collection of scientific literature. Almost 25% of all publications on phycocyanin also include the genus name of *Arthrospira* or *Spirulina*. Only 10 articles (0.4%) link phycocyanin to *Galdieria* in their topics. Publications on phycocyanin started to be released yearly from 1956, but only from 1995 to 2000 and after 2001 has phycocyanin regularly been linked to food or health and nutraceuticals in the scientific literature (Figure 5.3).

The rates by which the yearly numbers of publications and their citations have increased can be evaluated by fitting a first-order exponential equation to the data points in Figure 5.3:

$$n = e^{k \cdot (t - t_0)} \qquad (5.1)$$

where n is yearly number of publications or citations, t is the time measured in years, t_0 represents the first year the publications on a given topic started to appear on a yearly basis, and k is the rate constant for the yearly growth in numbers of publications or citations. The yearly number of papers on phycocyanin has increased at a rate of 8% per year since 1952, and the number of times these papers are cited has increased by 13% per year in the same period. The increasing publication frequencies, as well as the increasing frequencies by which these publications have been cited, demonstrate rapidly increasing interests in phycocyanin. For comparison, the growth rate of the overall numbers of publications in WOS was only 2.7% per year from 1997 to 2006 (Larsen and von Ins 2010). Only a fraction of the more recent papers on phycocyanin are related to food or health applications, but food or health-related papers on phycocyanin increased at even higher rates of 15% and 22% annually, and they received 39% and 48% more citations each year. These numbers reflect how food and health-related aspects of phycocyanin have become "hot" and fast-growing topics within the last decade.

5.7 PHYCOCYANIN IN FOOD

Phycocyanin from *A. platensis* has been marketed as a food and cosmetics colorant for several years in Japan (Prasanna et al. 2007), but in most parts of the world, phycocyanin is quite a novel food ingredient. Phycocyanin from *Arthrospira* extracts were, however, in 2013 and 2014 approved

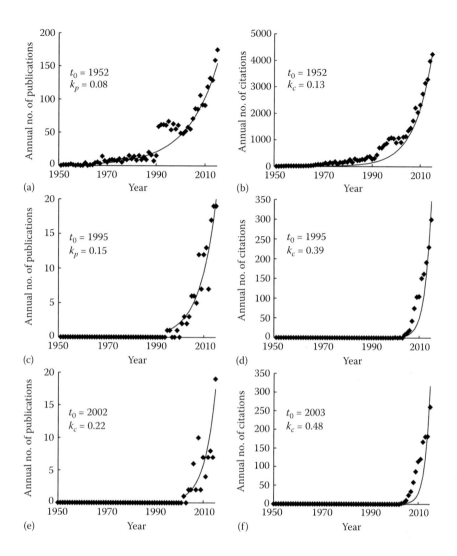

Figure 5.3 Yearly number of publications (a, c, e) or yearly number of citations to these publications (b, d, f) registered by WOS until the end of 2015, where the topic (title, key words, or abstract) includes phycocyanin (a and b), phycocyanin and food (b and c), or phycocyanin and health or nutraceutical (e and f). Curves represent best fits by Equation 5.1. Coefficients t_0, k_p, and k_c represent the first year the publications or citations to the topic started to appear annually, and rate constants for the yearly growth in numbers of publications or citations, respectively.

for use in candy, chewing gum. and other types for confection in the United States (U.S. Food and Drug Administration 2015), and new guidance notes for the use of coloring foodstuffs released by the European Commission in 2013 may provide the opportunity for the use of phycocyanin-rich *Arthrospira* extracts as a so-called food coloring in Europe, although phycocyanin itself is not on the list of approved food additives (European Commission 2015). There are therefore reasons to believe that phycocyanin will gain an increasingly important role as a food ingredient in the coming years.

In purified phycocyanin, the color depends strongly on the intact structure of the protein part of the molecule (Fukui et al. 2004). A few studies have addressed color stability and other properties of phycocyanin from *A. platensis* in foods (see, e.g., Jespersen et al. 2005; Batista et al. 2006; Mishra et al. 2008; de O Moreira et al. 2012; Martelli et al. 2014). Phycocyanin from *A. platensis* is sensitive to light, as well as to temperatures higher than approximately 50°C. Phycocyanin from *G. sulphuraria* is at least as stable and survives temperatures up to approximately 55°C (Myounghoon et al. 2014), but undergoes rapid and irreversible denaturation at 60–60°C (Eisele et al. 2000; Eriksen 2008a). Detailed studies on the use of phycocyanin from *G. sulphuraria* in food and beverages are still missing. Nothing, however, indicates that phycocyanin from *G. sulphuraria* should be a lesser useful coloring agent than phycocyanin from *A. platensis*. The major obstacle against the introduction of *G. sulphuraria* phycocyanin into the food market is probably the lack of approval from the food agencies.

5.8 PHYCOCYANIN IN HEALTH

The largest route for human intake of phycocyanin is in a nonpurified form via *A. platensis* health food products (Spolaore et al. 2006), where phycocyanin is a dominating antioxidant component (Pinero Estrada et al. 2001; Bermejo et al. 2008). Whole cyanobacteria are suggested to stimulate the immune defense system and possess antioxidant, anti-inflammatory, antiviral, anticancer, and cholesterol-lowering effects (see, e.g., Jensen et al. 2001), largely due to their high contents of phycocyanin (Bhattacharya and Shivaprakash 2005), in combination with other biologically active biomass components (Jensen et al. 2001; Sing et al. 2005). *A. platensis* has a long tradition as food and feed. This is not the case for *G. sulphuraria*. A recent study has, however, indicated that *G. sulphuraria* will be a suitable and safe human food component (Graziani et al. 2013). *G. sulphuraria* has also shown nutraceutical effects in rats (Carfagna et al. 2015).

The potential positive effects on human health that have given phycocyanin status as a nutraceutical were recently reviewed by Fernández-Rojas et al. (2014). Phycocyanin is an antioxidant and a radical scavenger (Romay et al. 1998; Bhat and Madyastha 2000; Benedetti et al. 2004; Bermejo et al. 2008; Soni et al. 2008). Positive health effects have in most cases been attributed to these properties. The antioxidant and radical scavenging activities are mainly attributed to the phycocyanobilin groups (Figure 5.1). Phycocyanin is bleached during scavenging of peroxyl radicals (Atanasiu et al. 1998; Bhat and Madyastha 2000; Hirata et al. 2000), antioxidative activities of free phycocyanobilin are comparable to those of phycocyanobilin bound in phycocyanin (Lissi et al. 2000), and antioxidative activity is increased by denaturing or trypsin digestion of phycocyanin (Zhou et al. 2005).

McCarty (2007a) proposed that it is actually a second compound, phycocyanorubin, that is the true antioxidant species *in vivo*. Phycocyanorubin is produced from phycocyanobilin by reduction of the double bond between carbon C-10 and C-11 (Figure 5.1) and is structurally quite similar to bilirubin (see, e.g., Eriksen 2008a), a natural antioxidant in plasma. Bilirubin is synthesized from biliverdin by biliverdin reductase, and it binds to albumin and protects lipids from oxidation, while it itself is reoxidized to biliverdin (Stocker et al. 1987). Phycocyanobilin can also be reduced to phycocyanorubin by biliverdin reductase (Terry et al. 1993), and phycocyanorubin is also a radical scavenger, being reoxidized to phycocyanobilin (Bhat and Madyastha 2000). Elevated plasma bilirubin levels have been linked to lower frequencies of certain types of cancer (McCarty 2007a) and diabetes (Takayanagi et al. 2011).

In addition, bilirubin (Kwak et al. 1991; Fujii et al. 2010) and also phycocyanorubin (McCarty 2007a, 2007b) inhibit formation of superoxide radicals by NADPH oxidase and may therefore play additional protective roles by reducing the generation of reactive oxygen species in the body, a process that has been shown to protect against influenza infections (McCarty et al. 2010; Vlahos et al. 2011). Phycocyanin may also inhibit other enzymes and biochemical pathways and affect gene regulation in mammalian cell lines (Madhyastha et al. 2006; Cherng et al. 2007) and hamsters (Riss et al. 2007), inhibit cell proliferation (Liu et al. 2000), and induce apoptosis in cancerogenic cell lines (Subhashini et al. 2004; Roy et al. 2007).

Ox bile is the current source for biliverdin, but it is not available in sufficient quantities to supply biliverdin to larger human populations. Therefore, McCarty (2007a, 2007b) suggested that phycocyanin can be used as a therapeutic replacement for biliverdin, and that a regular, prophylactic intake of phycocyanobilin or phycocyanin may provide protection against cancer (McCarty 2007a) and other diseases (Farooq et al. 2004). From this perspective, *G. sulphuraria* may appear as a particularly attractive host for phycocyanin synthesis since this alga can be grown equally well in all parts of the world, the high phycocyanin productivity potentials of heterotrophic *G. sulphuraria* cultures will be important if demands for phycocyanin are high, and the closed bioreactors needed for its production are a well-proven technology that been used for a long time in the manufacturing of numerous pharmaceuticals and food ingredients under strong hygienic conditions.

REFERENCES

Adir N, Dobrovetsky Y, Lerner N (2001) Structure of C-phycocyanin from thermophilic cyanobacterium *Synechococcus vulcanus* at 2.5 Å: Structural implications for thermal stability in phycobilisome assembly. *J Mol Biol* 313: 71–81.

Adir N, Vainer R, Lerner N (2002) Refined structure of c-phycocyanin from the cyanobacterium *Synechococcus vulcanus* at 1.6 Å: Insights into the role of solvent molecules in thermal stability and co-factor structure. *Biochim Biophys Acta* 1556: 168–174.

Albertano P, Ciniglia C, Pinto G, Pollio A (2000) The taxonomic position of *Cyanidium*, *Cyanidioschyzon* and *Galdieria*: An update. *Hydrobiologia* 433: 137–143.

Allen MM, Smith AJ (1969) Nitrogen chlorosis in blue-green algae. *Arch Microbiol* 69: 114–120.

Atanasiu R, Stea D, Mateescu MA, Vergely C, Dalloz F, Briot F, Maupoil V, Nadeau R, Rochette L (1998) Direct evidence of caeruloplasmin antioxidant properties. *Mol Cell Biochem* 189: 127–135.

Batista AP, Raymundo A, Sousa I, Empis José (2006) Rheological characterization of coloured oil-in-water food emulsions with lutein and phycocyanin added to the oil and aqueous phases. *Food Hydrocol* 20: 44–52.

Belknap WR, Haselkorn R (1987) Cloning and light regulation of expression of the phycocyanin operon of the cyanobacterium *Anabaena*. *EMBO J* 6: 871–884.

Benedetti S, Benvenutti F, Pagliarani S, Francogli S, Scoglio S, Canestrari F (2004) Antioxidant properties of a novel phycocyanin extract from the blue-green alga *Aphanizomenon flos-aquae*. *Life Sci* 75: 2353–2362.

Bermejo P, Piñero E, Villar ÁM (2008) Iron-chelating ability and antioxidant properties of phycocyanin isolated from a protean extract of *Spirulina platensis*. *Food Chem* 110: 436–445.

Bhat VB, Madyastha KM (2000) C-phycocyanin: A potent peroxyl radical scavenger in vivo and in vitro. *Biochem Biophys Res Commun* 275: 20–25.

Bhattacharya S, Shivaprakash MK (2005) Evaluation of three *Spirulina* species grown under similar conditions for their growth and biochemicals. *J Sci Food Agric* 85: 333–336.

Bryant DA, Dubbs JM, Fields PI, Porter RD, de Lorimier R (1985) Expression of phycobiliprotein genes in *Escherichia coli*. *FEMS Microbiol Lett* 29: 242–249.

Carfagna S, Napolitano G, Barone D, Pinto G, Pollio A, Venditti P (2015) Dietary supplementation with the microalga *Galdieria sulphuraria* (Rhodophyta) reduces prolonged exercise-induced oxidative stress in rat tissues. *Oxid Med Cell Longev* 2015: 732090.

Carpenter MP, Narin F (1981) The adequacy of the Science Citation Index (SCI) as an indicator of international scientific activity. *J Am Soc Inf Sci* 32: 430–439.

Chaneva G, Furnadzhieva S, Minkova K, Lukavsky J (2007) Effect of light and temperature on the cyanobacterium *Arthronema africanum*—A prospective phycobiliprotein-producing strain. *J Appl Phycol* 19: 537–544.

Chen F, Zhang Y (1997) High cell density mixotrophic culture of *Spirulina platensis* on glucose for phycocyanin production using a fed-batch system. *Enzyme Microb Technol* 20: 221–224.

Chen F, Zhang Y, Guo S (1996) Growth and phycocyanin formation of *Spirulina platensis* in photoheterotrophic culture. *Biotechnol Lett* 18: 603–608.

Cherng S-C, Cheng S-N, Tarn A, Chou T-C (2007) Anti-inflammatory activity of c-phycocyanin in lipopolysaccharide-stimulated RAW 264.7 macrophages. *Life Sci* 81: 1431–1435.

Chojnacka K, Noworyta A (2004) Evaluation of *Spirulina* sp. growth in photoautotrophic, heterotrophic and mixotrophic cultures. *Enzyme Microb Technol* 34: 461–465.

Contreras-Martel C, Matamala A, Bruna C, Poo-Caamaño G, Almonacid D, Figueroa M, Martínez-Oyanedel J, Bunster M (2007) The structure at 2 Å resolution of phycocyanin from *Gracilaria chilensis* and the energy transfer network in a PC-PC complex. *Biophys Chem* 125: 388–396.

Cornejo J, Beale SI (1997) Phycobilin biosynthetic reactions in extracts of cyanobacteria. *Photosynth Res* 51: 223–230.

de Lorimier R, Bryant DA, Porter RD, Liu WY, Jay E, Stevens SE (1984) Genes for the α and β subunits of phycocyanin. *Proc Natl Acad Sci USA* 81: 7946–7950.

de O Moreira I, Passos T, Chiapinni C, Silveira GK, Souza JCM, Coca-Vellarde LG, Deliza R, de Lima Araújoa KG (2012) Colour evaluation of a phycobiliprotein-rich extract obtained from *Nostoc* PCC9205 in acidic solutions and yogurt. *J Sci Food Agric* 92: 598–605.

Eisele LE, Bakhru SH, Liu X, Robert MacColl, Edwards ME (2000) Studies on C-phycocyanin from *Cyanidium caldarium*, a eukaryote at the extremes of habitat. *Biochim Biophys Acta* 1456: 99–107.

Eriksen NT (2008a) Production of phycocyanin—A pigment with applications in biology, biotechnology, foods, and medicine. *Appl Microbiol Biotechnol* 80: 1–14.

Eriksen NT (2008b) The technology of microalgal culturing. *Biotechnol Lett* 30: 1525–1536.

Eriksen NT (2011) Heterotrophic microalgae in biotechnology. In Johansen MN (ed), *Microalgae: Biotechnology, Microbiology and Energy*. New York: Nova Science Publishers, pp. 387–412.

Eriksen NT (2013) Growth in photobioreactors. In Razeghifard R (ed), *Natural and Artificial Photosynthesis. Pathways to Clean, Renewable Energy*. Hoboken, NJ: Wiley, pp. 285–305.

Eriksen NT (2016) Research trends in the dominating microalgal pigments, β-carotene, astaxanthin, and phycocyanin used in feed, in foods, and in health applications. *J Nutr Food Sci* 6: 507.

Farooq SM, Asokan D, Kalaiselvi P, Sakthivel R, Varalakshmi P (2004) Prophylactic role of phycocyanin: A study of oxalate mediated renal cell injury. *Chem Biol Interact* 149: 1–7.

Fernández-Rojas B, Hernández-Juárez J, Pedraza-Chaverri J (2014) Nutraceutical properties of phycocyanin. *J Funct Foods* 11: 375–392.

Fujii M, Inoguchi T, Sasaki S, Maeda S, Zheng J, Kobayashi K, Takayanagi R (2010) Bilirubin and biliverdin protect rodents against diabetic nephropathy by downregulating NAD(P)H oxidase. *Kidney Int* 78: 905–919.

Fukui K, Saito T, Noguchi Y, Kodera Y, Matsushima A, Nishimura H, Inada Y (2004) Relationship between color development and protein conformation in the phycocyanin molecule. *Dyes Pigments* 63: 89–94.

Gittelson A, Quiang H, Richmond A (1996) Photic volume in photobioreactors supporting ultrahigh population densities of the photoautotroph *Spirulina platensis*. *Appl Environ Microbiol* 62: 1570–1573.

Glazer AN (1994) Phycobiliproteins—A family of valuable, widely used fluorophores. *J Appl Phycol* 6: 105–112.

Graverholt OS, Eriksen NT (2007) Heterotrophic high cell-density fed-batch and continuous flow cultures of *Galdieria sulphuraria* and production of phycocyanin. *Appl Microbiol Biotechnol* 77: 69–75.

Gross W, Küver J, Tischendorf G, Bouchaala N, Büsch W (1998) Cryptoendolithic growth of the red alga *Galdieria sulphuraria* in volcanic areas. *J Eur Phycol* 33: 25–31.

Gross W, Oesterhelt C, Tishendorf G, Lederer F (2002) Characterization of a non-thermophilic strain of the red algal genus *Galdieria* isolated from Soos (Czech Republic). *Eur J Phycol* 37: 477–482.

Gross W, Schnarrenberger C (1995) Heterotrophic growth of two strains of the acido-thermophilic red alga *Galdieria sulphuraria*. *Plant Cell Physiol* 36: 633–638.

Guo N, Zhang X, Lu Y, Song X (2007) Analysis on the factors affecting start-up intensity in the upstream sequence of phycocyanin β subunit gene from *Arthrospira platensis* by site-directed mutagenesis. *Biotechnol Lett* 29: 459–464.

Graziani G, Schiavo S, Nicolai MA, Buono S, Fogliano V, Pinto G, Pollio A (2013) Microalgae as human food: Chemical and nutritional characteristics of the thermo-acidophilic microalga *Galdieria sulphuraria*. *Food Funct* 144–152.

Hirata T, Tanaka M, Ooike M, Tsunomura T, Sakaguchi M (2000) Antioxidant activities of phycocyanobilin prepared from *Spirulina platensis*. *J Appl Phycol* 12: 435–439.

Jensen GS, Ginsberg DI, Drapeau (2001) Blue-green algae as an immuno-enhancer and biomodulator. *J Am Nutraceutical Assoc* 3: 24–30.

Jespersen L, Strømdahl LD, Olsen K, Skibsted LH (2005) Heat and light stability of three natural blue colorants for use in confectionery and beverages. *Eur Food Res Technol* 220: 261–266.

Jiménez C, Cossío BR, Labella D, Niell FX (2003) The feasibility of industrial production of *Spirulina* (*Arthrospira*) in southern Spain. *Aquaculture* 217: 179–190.

Kwak JY, Takeshige K, Cheung BS, Minakami S (1991) Bilirubin inhibits the activation of superoxide-producing NADPH oxidase in a neutrophil cell-free system. *Biochim Biophys Acta* 1076: 369–373.

Larsen PO, von Ins M (2010) The rate of growth in scientific publication and the decline in coverage provided by Science Citation Index. *Scientometrics* 84: 575–603.

Lee Y-K (1997) Commercial production of microalgae in the Asia-Pacific rim. *J Appl Phycol* 9: 403–411.

Lissi EA, Pizarro M, Romay C (2000) Kinetics of phycocyanine bilin groups destruction by peroxyl radicals. *Free Radical Biol Med* 28: 1051–1055.

Liu J, Zhang X, Sui Z, Zhang X, Mao Y (2005b) Cloning and characterization of c-phycocyanin operon from the cyanobacterium *Arthrospira platensis* FACHB341. *J Appl Phycol* 17: 181–185.

Liu Y, Xu L, Cheng N, Lin L, Zhang C (2000) Inhibitory effect of phycocyanin from *Spirulina platensis* on the growth of human leukemia K562 cells. *J Appl Phycol* 12: 125–130.

Lu Y, Zhang X (2005) The upstream sequence of the phycocyanin β subunit gene from *Arthrospira platensis* regulates expression of *gfp* gene in response to light intensity. *Electronic J Biotechnol* 8: 63–70.

MacColl R (1998) Cyanobacterial phycobilisomes. *J Struct Biol* 124: 311–334.

Madhyastha HK, Radha KS, Sugiki M, Omura S, Maruyama M (2006) C-Phycocyanin transcriptionally regulates uPA mRNA through cAMP mediated PKA pathway in human fibroblast WI-38 cells. *Biochim Biophys Acta* 1760: 1624–1630.

Marquardt J (1998) Effects of carotenoid-depletion on the photosynthetic apparatus of a *Galdieria sulphuraria* (Rhodophyta) strain that retains its photosynthetic apparatus in the dark. *J Plant Physiol* 152: 372–380.

Marquez FJ, Nishio N, Nagai S (1995) Enhancement of biomass and pigment production during growth of *Spirulina platensis* in mixotrophic culture. *J Chem Tech Biotechnol* 62: 159–164.

Marquez FJ, Sasaki K, Kakizono T, Nishio N, Nagai S (1993) Growth characterization of *Spirulina platensis* in mixotrophic and heterotrophic conditions. *J Ferm Bioeng* 76: 408–410.

Martelli G, Folli C, Visai L, Daglia M, Ferrari D (2014) Thermal stability improvement of blue colorant C-phycocyanin from *Spirulina platensis* for food industry applications. *Process Biochem* 49: 154–159.

McCarty MF (2007a) "Iatrogenic Gilbert syndrome"—A strategy for reducing vascular and cancer risk by increasing plasma unconjugated bilirubin. *Med Hypothesis* 69: 974–994.

McCarty MF (2007b) Clinical potential of *Spirulina* as a source of phycocyanobilin. *J Med Food* 10: 566–570.

McCarty MF, Barroso-Aranda J, Contreras F (2010) Practical strategies for targeting NF-kappaB and NADPH oxidase may improve survival during lethal influenza epidemics. *Medical Hypothesis* 74: 18–20.

Mishra SK, Shrivastav A, Mishra S (2008) Effects of preservatives for food grade C-PC from *Spirulina platensis*. *Process Biochem* 43: 339–345.

Moreno J, Vargas MA, Rodríguez H, Rivas J, Guerrero MG (2003) Outdoor cultivation of a nitrogen-fixing marine cyanobacterium, *Anabaena* sp. ATCC 33047. *Biomol Eng* 20: 191–197.

Mühling M, Belay A, Whitton BA (2005) Screening *Arthrospira* (*Spirulina*) stains for heterotrophy. *J Appl Phycol* 17: 129–135.

Myounghoon M, Mishra SK, Kim CW, Suh WI, Park MS, Yang J-W (2014) Isolation and characterization of thermostable phycocyanin from *Galdieria sulphuraria*. *Korean J Chem Eng* 31: 490–495.

Nield J, Rizkallah PJ, Barber J, Chayen NE (2003) The 1.45 Å three-dimensional structure of C-phycocyanin from the thermophilic cyanobacterium *Synechococcus elongatus*. *J Struct Biol* 141: 149–155.

Ohta N, Matsuzaki M, Misumi O, Miyagishima S-Y, Nozaki H, Tanaka K, Shin-i T, Kohara Y, Kuroiwa T (2003) Complete sequence and analysis of the plastid genome of the unicellular red alga *Cyanidioschyzon merolae*. *DNA Res* 10: 67–77.

Padayana AK, Bhat VB, Madyastha KM, Rajashankar KR, Ramakumar S (2001) Crystal structure of a light-harvesting protein C-phycocyanin from *Spirulina platensis*. *Biochem Biophys Res Commun* 282: 893–898.

Pinero Estrada JE, Bermejo Bescós P, Villar del Fresno AM (2001) Antioxidant activity of different fractions of *Spirulina platensis* protean extract. *Il Farmaco* 56: 497–500.

Prasanna R, Sood A, Suresh A, Kaushik BD (2007) Potentials and applications of algal pigments in biology and industry. *Acta Bot Hung* 49: 131–156.

Pulz O (2001) Photobioeractors: Production systems for phototrophic microorganisms. *Appl Microbiol Biotechnol* 57: 287–293.

Pushparaj B, Pelosi E, Tredici MR, Pinzani E, Materassi R (1997) An integrated culture system for outdoor production of microalgae and cyanobacteria. *J Appl Phycol* 9: 113–119.

Rhie G, Beale SI (1994) Regulation of heme oxygenase activity in *Cyanidium caldarium* by light, glucose, and phycobilin precursors. *J Biol Chem* 269: 9620–9626.

Richmond A, Grobbelaar JU (1986) Factors affecting the output rate of *Spirulina platensis* with reference to mass cultivation. *Biomass* 10: 253–264.

Richmond A, Lichtenberger E, Stahl B, Vonshak A (1990) Quantitative assessment of the major limitations on productivity of *Spirulina platensis* in open raceways. *J Appl Phycol* 2: 195–206.

Riesenberg D, Guthke R (1999) High-cell-density cultivation of microorganisms. *Appl Microbiol Biotechnol* 51: 422–430.

Riss J, Décordé K, Sutra T, Delage M, Baccou J-C, Jouy N, Brune J-P, Oréal M, Cristol J-P, Rouanet J-M (2007) Phycobiliprotein C-phycocyanin from *Spirulina platensis* is powerfully responsible for reducing oxidative stress and NADPH oxidase expression induced by an atherogenic diet in hamsters. *J Agric Food Chem* 55: 7962–7967.

Rito-Palomares M, Nunez L, Amador D (2001) Practical application of aqueous two-phase systems for the development of a prototype process for c-phycocyanin recovery from *Spirulina maxima*. *J Chem Technol Biotechnol* 76: 1273–1280.

Rodríguez H, Rivas J, Guerrero MG, Losada M (1991) Enhancement of phycobiliprotein production in nitrogen-fixing cyanobacteria. *J Bacteriol* 20: 263–270.

Romay C, Armesto J, Remirez D, González R, Ledon N, García I (1998) Antioxidant and anti-inflammatory properties of C-phycocyanin from blue-green algae. *Inflammatory Res* 47: 36–41.

Roy KR, Arunasree KM, Reddy NP, Dheeraj B, Reddy GV, Reddanna P (2007) Alteration of mitochondrial membrane potential by *Spirulina platensis* C-phycocyanin induces apoptosis in the doxorubicin-resistant human hepatocellular-carcinoma cell line HepG2. *Biotechnol Appl Biochem* 47: 159–167.

Sarian FD, Rahman DY, Schepers O, van der Maarel MJEC (2016) Effects of oxygen limitation on the biosynthesis of photo pigments in the red microalgae *Galdieria sulphuraria* strain 074G. *PLoS One* 11 (2): 1–10.

Schmidt RA, Wiebe MG, Eriksen NT (2005) Heterotrophic high cell-density fed-batch cultures of the phycocyanin producing red alga *Galdieria sulphuraria*. *Biotechnol Bioeng* 90: 77–84.

Sekar S, Chandramohan M (2008) Phycobiliproteins as a commodity: Trends in applied research, patents and commercialization. *J Appl Phycol* 20: 113–136.

Sing S, Kate BN, Banerjee UC (2005) Bioactive compounds from cyanobacteria and microalgae: An overview. *Crit Rev Biotechnol* 25: 73–95.

Sloth JK, Wiebe MG, Eriksen NT (2006) Accumulation of phycocyanin in heterotrophic and mixotrophic cultures of the acidophilic red alga *Galdieria sulphuraria*. *Enzyme Microb Technol* 38: 168–175.

Soni B, Trivedi U, Madamwar D (2008) A novel method of single step hydrophobic interaction chromatography for the purification of phycocyanin from *Phormidium fragile* and its characterization for antioxidant property. *Bioresour Technol* 99: 188–194.

Sørensen L, Hankte A, Eriksen NT (2013) Purification of the photosynthetic pigment C-phycocyanin from heterotrophic *Galdieria sulphuraria*. *J Sci Food Agric* 93: 2933–2938.

Spolaore P, Joannis-Cassan C, Duran E, Isambert A (2006) Commercial applications of microalgae. *J Biosci Bioeng* 101: 87–96.

Stadnichuk IN, Krasilnikov PM, Zlenko DV (2015) Cyanobacterial phycobilisomes and phycobiliproteins. *Microbiology* 84: 101–111.

Stadnichuk IN, Rakhimberdieva MG, Boichenko VA, Karapetyan NV, Selyakh IO, Bolychevtseva YV (2000) Glucose-induced inhibition of the photosynthetic pigment apparatus in heterotrophically-grown *Galdieria partita*. *Russ J Plant Physiol* 47: 585–592.

Stadnichuk IN, Rakhimberdieva MG, Bolychevtseva YV, Yurina NP, Karapetyan NV, Selyakh IO (1998) Inhibition by glucose of chlorophyll *a* and phycocyanobilin biosynthesis in the unicellular red alga *Galdieria partita* at the stage of coproporphyrinogen III formation. *Plant Sci* 136: 11–23.

Stec B, Troxler RF, Teeter MM (1999) Crystal structure of C-phycocyanin from *Cyanidium caldarium* provides a new perspective in phycobilisome assembly. *Biophys J* 76: 2912–2921.

Stocker R, Glazer AN, Ames BN (1987) Antioxidant activity of albumin-bound bilirubin. *Proc Natl Acad Sci USA* 84: 5918–5922.

Subhashini J, Mahipal VK, Reddy MC, Reddy MM, Rachamallu A, Reddanna P (2004) Molecular mechanisms in C-phycocyanin induced apoptosis in human chronic myeloid leukemia cell line-K562. *Biochem Pharmacol* 68: 453–462.

Takayanagi R, Inoguchi T, Ohnaka K (2011) Clinical and experimental evidence for oxidative stress as an exacerbating factor of diabetes mellitus. *J Clin Biochem Nutr* 48: 72–77.

Terry MJ, Maines MD, Lagarias JC (1993) Inactivation of phytochrome- and phycobiliprotein-chromophore precursors by rat liver biliverdin reductase. *J Biol Chem* 268: 26099–26106.

Troxler RF, Ehrhardt MM, Brown-Mason AS, Offner GD (1981) Primary structure of phycocyanin from the unicellular rhodophyte *Cyanidium caldarium*. II. Complete amino acid sequence of the β subunit. *J Biol Chem* 256: 12176–12184.

Troxler RF, Lin S, Offner GD (1989) Heme regulates expression of phycobiliprotein photogenes in the unicellular rhodophyte, *Cyanidium caldarium*. *J Biol Chem* 264: 20596–20601.

Tooley AJ, Cai YA, Glazer AN (2001) Biosynthesis of a fluorescent cyanobacterial C-phycocyanin holo-α subunit in a heterologous host. *Proc Natl Acad Sci USA* 98: 10560–10565.

Venugopal V, Prasanna R, Sood A, Jaiswal P, Kaushik BD (2006) Stimulation of pigment accumulation in *Anabaena azollae* strains: Effect of light intensity and sugars. *Folia Microbiol* 51: 50–56.

Vlahos R, Stambas J, Bozimovski S, Broughton BRS, Drummond GR, Selemidis S (2011) Inhibition of Nox2 oxidase activity ameliorates influenza A virus-induced lung inflammation. *PLoS Pathogens* 7: 1–12.

Wan M, Wang Z, Zhang Z, Wang J, Li S, Yu A, Li Y (2016) A novel paradigm for the high-efficient production of phycocyanin from *Galdieria sulphuraria*. *Bioresour Technol* 218: 272–278.

Yamanaka G, Glazer AN (1980) Dynamic aspects of phycobilisome structure. Phycobilisome turnover during nitrogen starvation in *Synechococcus* sp. *Arch Microbiol* 124: 39–47.

Zetsche K, Valentin K (1993/94) Structure, coding capacity and gene sequence of the plastid genome from red algae. *Endocytobiosis Cell Res* 10: 107–127.

Zhou Z-P, Liu L-N, Chen X-L, Wang J-X, Chen M, Zhang Y-Z, Zhou B.C (2005) Factors that affect antioxidant activity of C-phycocyanin from *Spirulina platensis*. *J Food Biochem* 29: 313–322.

REFERENCES TO INTERNET HOMEPAGES

European Commission (2013) Guidance notes on the classification of food extracts with colouring properties. http://ec.europa.eu/food/food/fAEF/additives/docs/guidance_en.pdf.

European Commission (2015) Food additives. https://ec.europa.eu/food/safety/food_improvement_agents/additives/database_en.

Thomson Reuters (2015) Web of Science. http://apps.webofknowledge.com.

U.S. Food and Drug Administration (2015) Summary of color additives for use in the United States in foods, drugs, cosmetics, and medical devices. www.fda.gov/forindustry/coloradditives/coloradditiveinventories/ucm115641.htm#table1A.

CHAPTER 6

Biology and Applications of Halophilic and Haloalkaliphilic Actinobacteria

Sangeeta D. Gohel, Amit K. Sharma, Kruti G. Dangar,
Foram J. Thakrar, and Satya P. Singh

CONTENTS

6.1 INTRODUCTION

Actinobacteria constitutes one of the main phyla within the prokaryotes on the basis of its branching position in 16S rRNA gene trees. The ecology of actinobacteria revealed their widespread distribution in varied habitats of extreme nature (Thumar and Singh 2007; Singh et al. 2010, 2012, 2013; Gohel and Singh 2012a, 2013, 2016; Guan et al. 2013a; Bouras et al. 2015; Meklat et al. 2015). They have been shown to occur in different niches, such as the soil of coastal regions (Singh et al. 2013; Suthindhiran et al. 2014), seawater (Tian et al. 2012), marine sediment samples (Acharyabhatta et al. 2013), desert soil (Selama et al. 2014; Saker et al. 2015), brine lakes (Sahay et al. 2012), salt ponds (Chang et al. 2012), salt-fermented seafood (Yun et al. 2011), sandy rhizospheric soil (Hamedi et al. 2010), and marine animals (Chen et al. 2010b).

Halophilic actinomycetes constitute a relatively heterogeneous physiological group of different genera. The identification of these microorganisms is possible by a polyphasic approach that includes molecular, chemotaxonomic, genotypic, and phenotypic chracteristics. Genotypic methods include DNA base ratio (mole percent G + C), DNA-DNA hybridization, rRNA homology, and DNA-based typing methods. The phenotypic methods are based on the conventional phenotypic analysis, numerical analysis, cell wall composition, cellular fatty acids, and whole-cell protein analysis (Vandamme et al. 1996; Yun et al. 2011; Guan et al. 2013b; Bouras et al. 2015; Meklat et al. 2015). Polymerase chain reaction (PCR)–based methods of amplified rDNA restriction analysis (ARDRA), restriction fragment length polymorphism (RFLP), denaturing gradient gel electrophoresis (DGGE), and temperature gradient gel electrophoresis (TGGE) are used for the assessment of the diversity and molecular phylogeney of the actinomycetes (Singh et al. 2013; Gohel et al. 2015).

The members of the phylum Actinobacteria are characterized with a high GC content and valuable secondary metabolites (Ventura et al. 2007). Actinobacterial taxa comprise many bacteria that are major human, animal, or plant pathogens, such as *Mycobacterium*, *Actinomyces*, *Renibacterium*, *Atopobium*, *Gordonia*, *Gardnerella*, *Leifsonia*, and *Clavibacter*, while other actinobacterial taxa are the richest source of natural products, novel compounds, and industrially important enzymes (Berdy 2005; Fiedler et al. 2005; Bull and Stach 2007; Baltz 2007; Singh et al. 2013; Gohel et al. 2015).

The production of antibiotics by Actinobacteria has been identified as an important determining factor in the evolution of both Archaea and Gram-negative bacteria from the Gram-positive bacteria (Gupta 2001; Ludwig and Klenk 2001; Valas and Bourne 2011). However, the enzymatic potential of these microbes has not received considerable attention. Over the last several years, the extracellular alkaline serine proteases of the haloalkaliphilic actinobacteria from the saline habitats have been studied with respect to their occurrence and enzymatic characteristics (Thumar and Singh 2007, 2009; Gohel and Singh 2012a, 2012b, 2013, 2015). Besides proteases, other enzymes, such as amylases, cellulases, and lipases, from the haloalkaliphilic actinobacteria have been identified for their application potential in the food, fermentation, textile, paper, detergent, pharmaceutical, and sugar industries. In this chapter, we describe the occurrence, diversity, adaptation, and phylogeny of the haloalkaliphilic actinobacteria. Further, the industrial applications of these microbes and their biological role in harsh environments are also discussed.

6.2 OCCURRENCE AND ADAPTATION OF THE HALOALKALIPHILIC ACTINOBACTERIA

High metabolic capabilities and unique features of the haloalakliphilic actinobacteria allow them to survive under different physiological niches. Recently, 14 different actinobacteria belonging to three different genera, *Streptomyces*, *Pseudonocardia*, and *Actinoalloteichus*, have been reported from Sambhar Salt Lake, India (Jose and Jebakumar 2013a). More recently, the diversity of the cultivable halophilic actinobacteria in two Algerian arid ecosystems, M'zab and Zibans, of the

Septentrional Sahara has been described (Saker et al. 2015). A total of 69 halophilic strains from 19 soil samples were studied using a polyphasic approach based on the morphological, physiological, chemotaxonomic, and molecular features. The 16S rRNA gene sequences revealed that the strains belonged to six clusters corresponding to six genera: *Actinopolyspora*, *Nocardiopsis*, *Prauserella*, *Saccharomonospora*, *Saccharopolyspora*, and *Streptomonospora*. Besides, four unknown actinobacterial strains have been proposed (Saker et al. 2015).

At higher salt concentrations, the microorganisms encounter the problem of higher osmotic pressure in the surrounding environment, which leads to shrinkage and ultimately the death of the organism. Thus, the cytoplasm should be at least iso-osmotic with its surroundings. The majority of the halophilic microorganisms show potent transport mechanisms, generally based on Na^+/H^+ antiporters, to expel sodium ions from the cytoplasm (Oren 1999). Halophilic microorganisms retrieve two main strategies for their adaptation at higher salt concentrations: high salt in strategy and accumulation of organic osmotic solutes.

6.2.1 High Salt in Strategy

Accumulation of potassium and chloride within the cytoplasm allows cell machinery to adapt very high salt concentrations. In fact, proteins retain their structural conformation and activity at high salt concentrations. Such organisms are unable to survive in lower salt concentrations. Cells need to maintain a positive turgor pressure for growth, which is generally maintained by the intake of potassium ions (Whatmore and Ree 1990). Osmotic pressure in halophilic actinomycetes is relatively high, which allows them to survive in soils under a moisture deficit and high salinity. Only a few actinomycete genera are extremely halophilic in nature (Oren 2002). *Actinopolyspora lacussalsi* sp. nov., an extremely halophilic actinomycete isolated from a hypersaline habitat in Xinjiang Province, northwest China, can grow up to 20% NaCl (Guan et al. 2013b).

6.2.2 Organic Osmotic Solute Accumulation Strategy

The second strategy is to accumulate organic osmotic solutes, a ubiquitous feature among the halophiles. Such organisms exclude salt as much as possible from their cytoplasm. In saline habitats, halophilic actinomycetes accumulate potassium (K^+) ions and one or more low-molecular-mass organic solutes, referred to as organic compatible solutes (Welsh 2000). The higher concentrations of the organic compatible solutes do not restrict the normal enzymatic activities. Thus, these organisms adapt to the broader salt range (Ventosa et al. 1998). In order to balance the osmotic pressure of the growth medium and to maintain cell turgor pressure, the organic compatible solutes are either synthesized or accumulated from the environment (Santos and Costa 2002). Among the compatible solutes, amino acids and polyols, for example, glycine, betaine, ectoine, sucrose, trehalose, and glycerol, are commonly accumulated in halophiles. Some of these molecules, especially glycine, betaine, and ectoines, have considerable biotechnological importance (Ventosa et al. 1998). The osmoprotective effect of glycine betaine, proline, and trehalose on *Streptomyces* sp. mado2 and *nocardiopsis* sp. mado3 is reported (Ameur et al. 2011). Glycine betaine, proline, and trehalose play a critical role in osmotic adaptation at high osmolarity (1 M NaCl).

Halophilic actinomycete *Actinopolyspora halophila* can de novo synthesize the compatible solute glycine betaine, whereas *Nocardiopsis halophila* uses a hydroxy derivative of ectoine and β-glutamate as compatible solutes (DasSarma and Arora 2001). *Streptomyces coelicolor* A3 (2) synthesizes ectoine and 5-hydroxyectoine in the presence of salt or heat stress. The cells produced the highest cellular levels of these compatible solutes under the dual stress of salt and heat. The gene cluster (*ectABCD*) encoding the enzymes for ectoine and 5-hydroxyectoine biosynthesis is present in the genome of many *Streptomyces* species, suggesting that these compatible solutes play an important role as the stress protectants in the genus *Streptomyces* (Bursy et al. 2007). A halotolerant *Rubrobacter xylanophilus* accumulates trehalose and mannosylglycerate (Empadinhas et al. 2007). The humicolous actinomy-cete *Streptomyces coelicolor* adapts to a wide variety of habitats and upon salt stress, and the organ-ism increases the levels of various compatible solutes (Kol et al. 2010). The biosynthetic pathway for the compatible solute ectoinein, a soil actinomycete *Rhodococcus jostii* RHA1, is described (LeBlanc et al. 2008). *Actinopolyspora* sp., isolated from the *Suaeda* (salt marsh) region of the Vellar estuary, southeast coast of India, produces osmolytes at 20% salt (Kundu et al. 2008).

The compatible solutes are excellent stabilizers, useful in various industrial applications. In cosmetics, ectoine is used as a moisturizer and as other skin care products (Montitsche et al. 2000). Recently, the role of ABC transporters in the accumulation of compatible solutes in halophilic *Nocardiopsis xinjiangensis* has been suggested (G. Zhang et al. 2015). It has been demonstrated that the Na^+-dependent transporters and cell motility proteins acted as adaptive proteins that actively counteracted higher salinity stress for adaptation in halophilic *N. xinjiangensis*. Moreover, *A. halophila* (MTCC 263) has been used to treat acid since it produces glycine, betaine, and treha-lose as compatible solutes (Kar et al. 2015). A profile of the haloalkaliphilic actinobacteria adapted at different ranges of salt, pH, and temperature is displayed in Table 6.1.

6.3 CHEMOTAXONOMIC CHARACTERISTICS OF HALOALKALIPHILIC ACTINOBACTERIA

Taxonomy, generally considered a synonym of systematic, is divided into three components: (1) classification, (2) nomenclature, and (3) identification of unknown organisms (Cowan 1968). Haloalkaliphilic actinobacteria represent diverse taxonomic groups. Their taxonomic diversity has been evaluated by a polyphasic approach from different habitats (Govender et al. 2013; Guan et al. 2013b; Bouras et al. 2015; Meklat et al. 2015; Saker et al. 2015).

On the basis of the polyphasic taxonomic approaches, a new actinobacterial strain can be pro-posed. The taxonomic status of a novel halophilic and filamentous actinobacterium strain TRM 40139 from a hypersaline habitat of Xinjiang Province in northwest China was determined using a polypha-sic approach (Guan et al. 2013b). Phylogenetic analysis based on the 16S rRNA gene sequence of the strain demonstrated a well-separated subbranch within the radiation of the genus *Actinopolyspora*, and the organism was related most closely to the type strains of *Actinopolyspora alba* (97.6% simi-larity), *Actinopolyspora xinjiangensis* (97.6%), and *Actinopolyspora erythraea* (97.1%). However, it had a relatively lower mean DNA-DNA relatedness (36.4%, 31.3%, and 26.1%, respectively) with the above strains. Strain TRM 40139T therefore represents a novel species of the genus *Actinopolyspora*, with the proposed name of *Actinopolyspora lacussalsi* sp. nov. (Guan et al. 2013a). Shendure and Lieberman (2012) proposed that next-generation DNA sequencing could be useful in the develop-ment of new sequencing applications. Recently, a novel halophilic actinobacterium strain H8T, iso-lated from a Saharan soil sample of El-Golea, south Algeria, was identified as representative of a new genus *Bounagaea algeriensis* using a polyphasic taxonomic approach (Meklat et al. 2015).

On the basis of the phenotypic, genotypic, and DNA-DNA hybridization data, strain TRM F103T was considered to represent a novel species of the genus *Amycolatopsis*, for which the name *Amycolatopsis salitolerans* sp. nov. was proposed (Guan et al. 2012). Phylogenetic analysis of 16S rRNA gene sequences from strain 104T and reference species of the genus *Kocuria* indicated that strain 104T formed an independent line. On the basis of the G + C content and the predominant fatty acids, strain 104T was most closely related to *Kocuria rhizophila* TA68T (98.9% 16S rRNA gene sequence similarity). The DNA-DNA hybridization value between strain 104T and *K. rhizophila* TA68T was 14.1 ± 3.4%. On the basis of this polyphasic taxonomic analysis, strain 104T appears to represent a novel species in the genus *Kocuria*. The name *Kocuria salsicia* sp. nov. was proposed (Yun et al. 2011). Taxonomic characteristics of some recently reported haloalkaliphilic actinobac-teria are described in Table 6.2.

Table 6.1 Salt, pH, and Temperature Profile of the Haloalkaliphilic Actinobacteria

Sr No.	GeneBank Number	Name of Organism	Isolation Site	Salt (% w/v)	pH	T (°C)	References
			Family: Actinopolysporaceae				
1.	KJ574193	*Actinopolyspora* H259	Zibans, Algeria (salty soil)	7–30	7	30	Saker et al. 2015
2.	KJ409655	*Actinopolyspora* sp. M5A	Saharan soil, Algeria	10–20	5–12	30–40	Selama et al. 2014
3.	EU551237	*Actinopolyspora* sp. VITSDK2	Marine saltern, southern India	6–18	10	15–37	Suthindhiran et al. 2014
4.	HQ918195	*Actinopolyspora algeriensis* H19	Saharan soils, Algeria	7–32	7	25	Meklat et al. 2011
5.	HQ918198	*Actinopolyspora saharensis* H32		15–25	6–7	28–32	Meklat et al. 2013
6.	GQ480940	*Actinopolyspora alba*	Salt lake, Xinjiang, China	10–25	7–8	37	Tang et al. 2011
7.	DQ883811	*Actinopolyspora erythraea*					Tang et al. 2011
			Family: Bogoriellaceae				
8.	FJ717681	*Georgenia halophila*	Salt lake, Xinjiang, China	1–15	6–9	10–45	Tang et al. 2010a
			Family: Glycomycetaceae				
9.	HQ651156	*Glycomyces halotolerans*	Xinjiang, China	0–11	5–11	20–50	Guan et al. 2011a
			Family: Micrococcaceae				
10.	FJ948172	*Nesterenkonia suensis* Sua-BAC020T	Brine—Sua salt pan, Botswana	2.5	9	35–37	Govender et al. 2013
11.	GQ352404	*Kocuria salsicia* 104	Salt-fermented seafood, Korea	2	7–8	30–37	Yun et al. 2011
12.	FJ425902	*Zhihengliuella salsuginis* JSM 071043	Salt mine, Hunan Province, China	5–10	6.5–10.5	10–40	Chen et al. 2010b
			Family: Micromonosporaceae				
13.	AY528866	*Verrucosis poramaris* AB-18-032	Sediment samples, Sea of Japan	7.5	7–9	30–37	Goodfellow et al. 2012
			Family: Nocardiopsaceae				
14.	EU710555	*Nocardiopsis alba* OK-5	Saline soil, Okha site, Gujarat, India	0–15	7–11	28–37	Gohel and Singh 2015
15.	HM560975	*Nocardiopsis xinjiangensis* OM-6	Saline soil, Okha Madhi site, Gujarat, India	0–20	8–11	28–37	Gohel and Singh 2012b

(Continued)

Table 6.1 (Continued) Salt, pH, and Temperature Profile of the Haloalkaliphilic Actinobacteria

Sr No.	GeneBank Number	Name of Organism	Isolation Site	Salt (% w/v)	pH	T (°C)	References
16.	KJ470139	*Nocardiopsis algeriensis B32*	Saharan soil, Algeria	0–5	7–10	25–35	Bouras et al. 2015
17.	KJ574177	*Streptomonospora sp. H238*	Highly saline soils, Zelfana	7–20	7	30	Saker et al. 2015
18.	GU253338	*Salinactinospora qingdaonensis*	Salt pond, Qingdao, China	9–12	7–8	37	Chang et al. 2012
19.	JN006759	*Nocardiopsis coralliicola*	Gorgonian coral, Menellapraelonga	0–18	7–10	20–45	Li et al. 2012
20.	GU997639	*Nocardiopsis flavescens*	Marine sediment, Jiangsu, China	0–10	5.5–11	23–40	Fang et al. 2011
21.	GU112453	*Spinactinospora alkalitolerans*	Marine sediment (17.5 m depth), China	3–8	7–8	37	Chang et al. 2011
22.	EU410477	*Nocardiopsis arvandica*	Sandy soil, Arvand, Khoramshahr, Iran	0–17.5	4–13	15–45	Hamedi et al. 2011
23.	EU410476	*Nocardiopsis sinuspersici*	Sandy rhizospheric soil, Persian Gulf	0–15.0	5–12	15–37	Hamedi et al. 2010
24.	EU583726	*Nocardiopsis litoralis* JSM073097T	Sea anemone tidal flat, South China Sea	1–15	6–10	20–35	Chen et al. 2009
25.	YIM 90022	*Nocardiopsis terrae YIM 90022*	Saline soil of Qaidam Basin, China	1–15	6–10.5	10–45	Chen et al. 2010b
		Family: Nitriliruptoraceae					
26.	EF422408	*Nitriliruptor alkaliphilus ANL-iso2*	Soda lake sediments, Russia	1.2–1.8	8.4–10.6	32	Sorokin et al. 2009
		Family: Promicromonosporaceae					
27.	JX316007	*Myceligenerans salitolerans*	Salt lake, Xinjiang, China	0–16	5–10	15–40	Guan et al. 2013b
28.	EU910872	*Myceligenerans halotolerans*	Xinjiang, China	0–10	6–8.0	10–40	Wang et al. 2011
29.	EU910872	*Myceligenerans halotolerans*	Salt lake, Xinjiang, China	0–10	6–8	10–40	Wang et al. 2011
		Family: Pseudonocardiaceae					
30.	KF981441	*Bounagaea algeriensis H8*	Sauth Algeria (saline soil)	15–25	6–7	28–35	Meklat et al. 2015

(Continued)

Table 6.1 (Continued) Salt, pH, and Temperature Profile of the Haloalkaliphilic Actinobacteria

Sr No.	GeneBank Number	Name of Organism	Isolation Site	Salt (% w/v)	pH	T (°C)	References
31.	KJ574146	Saccharomonospora sp. H233	M'zab region, Algeria (saline soil)	7–20	7	30	Saker et al. 2015
32.	KJ574161	Prauserella sp. H137		7–25	7	30	
33.	CM001439	Saccharomonospora marina XMU15	Ocean sediment, Zhaoan Bay, China	5	7	28–37	Klenk et al. 2012
34.	HQ436534	Amycolatopsis salitolerans	Tarim basin, Xinjiang, China	13	5–8	25–45	Guan et al. 2012
35.	FJ526746	Yuhushiella deserti	Desert, Xinjiang, China	3.5	9	45	Mao et al. 2011
36.	GQ366705	Haloechinothrix alba YIM 93221	Salt lake, Xinjiang, China	9–23	4–8	20–45	Tang et al. 2010b
37.	FJ444996	Prauserella marina MS498	Ocean sediment, South China Sea	10	6–9	15–45	J. Wang et al. 2010
		Family: Streptomycetaceae					
38.	JQ690542	Streptomyces sp. GB3	Saharan soil, Algeria	7	5–12	20–37	Selama et al. 2014
39.	JX284411	Streptomyces rochei strain BTSS 1001	Diviseema, Bay of Bengal (sediments)	3–10	9.5	50	Acharyabhatta et al. 2013
40.	JN645842	Streptomyces viridiviolaceus	Sambhar Lake, India	2–12	8–10	37	Sahay et al. 2012
41.	JN645850	Streptomyces radiopugnans		2.5–11	6–12	37	
42.	HQ585117	Streptomyces glycovorans YIM M10366	Xisha Islands, south China (marine sediments)	3	7	28–37	Xu et al. 2012
43.	HQ585118	Streptomyces xishensis sp. YIM M10378		3	8	28–38	
44.	HQ585121	Streptomyces abyssalis sp. YIM M10400		3	9	28–39	
45.	JN389519	Streptomyces oceani	Sea sediment, China (depth 578 m)	2–12.5	6–8	28	Tian et al. 2012
46.	FR693804	Streptomyces pharmamarensis	Marine sediment, Mediterranean Sea	2–9	7	28	Carro et al. 2012

Table 6.2 Comparison of Morphological and Chemotaxonomic Characteristics of Haloalkaliphilic Actinobacteria

Sr No.	Actinobacteria	Morphology	Cell Wall Peptidoglycan and Whole-Cell Sugar	Polar Lipids	Predominant Menaquinones	Major Fatty Acids in Major Amount	G + C%	References
Genus: *Actinopolyspora*								
1	*Actinopolyspora lacussalsi* TRM 40139^T	Filaments with aerial mycelium, nonmotile	*meso*-DAP Xylose, glucose, ribose, and rabinose	DPG, PG, PC, PI, and two unknown PLs	MK-9(H$_4$) MK-10(H$_4$)	Iso-C$_{16:0}$ and anteiso-C$_{17:0}$	66.4	Guan et al. 2013a
2	*Actinopolyspora alba* YIM 90480^T *Actinopolyspora erythraea* YIM 90600^T	Filamentous and forms long chains of spores on aerial mycelium	*meso*-DAP Galactose, arabinose, and ribose *meso*-DAP Galactose and arabinose	DPG, PC, PL, PGL, and PIM	MK-10(H$_4$) MK-9(H$_4$) MK-9(H$_4$) MK-9(H$_2$)	Iso-C$_{16:0}$ and anteiso-C$_{17:0}$ Anteiso-C$_{15:0}$ and anteiso-C$_{17:0}$	68.3 66.4	Tang et al. 2011
3	*Actinopolyspora xinjiangensis* TRM 40136^T	Irregular branched vegetative mycelium	*meso*-DAP Xylose, glucose, and arabinose	PC, DPG, PG, and two unknown PLs	MK-6	Anteiso-C$_{15:0}$, anteiso-C$_{17:0}$, and iso-C$_{15:0}$	68.9	Guan et al. 2010
Genus: *Blastococcus*								
4	*Blastococcus endophyticus* YIM 68236^T	Nonmotile, non–spore forming	*meso*-DAP Arabinose and galactose	DPG, PC, PE, PI, and PIM	MK-9(H$_4$) MK-(H$_6$) MK-8	Iso-C$_{16:0}$, iso-C$_{15:0}$, and C$_{18:1\omega9C}$	71.6	Zhu et al. 2013
5	*Blastococcus jejuensis* sp. nov. KST3-10^T	Motile, non spore–forming cocci that occur in pairs	*meso*-DA Arabinose and galactose	PC, DPG, PE, PME, and PI	MK-9(H$_4$)	Iso-C$_{16:0}$, C$_{17:1\omega8c}$, and iso-C$_{15:0}$	72.3	Lee 2006
6	*Blastococcus saxobsidens* BC444^T	Single, paired, tetrad, and small aggregates of coccoi; motile or nonmotile	*meso*-DAP	DPG, PG, PD, PI, and PE	MK-9(H$_4$) MK-9	C$_{17:1\omega8c}$, iso-C$_{15:0}$, and C$_{17:0}$	ND	Urzi et al. 2004
Genus: *Jiangella*								
7	*Jiangella mangrovi* sp. nov. 3SM4-07^T	Branching mycelia fragmented into short or elongated rod	LL-2,6-*meso*-DAP Glucose and ribose	DPG, PG, PC, PI, and PIM	MK-9(H$_4$)	Anteiso-C$_{15:0}$, iso-C$_{15:0}$, and iso-C$_{16:0}$	72.3	Suksaard et al. 2015

(Continued)

Table 6.2 (Continued)   Comparison of Morphological and Chemotaxonomic Characteristics of Haloalkaliphilic Actinobacteria

Sr No.	Actinobacteria	Morphology	Cell Wall Peptidoglycan and Whole-Cell Sugar	Polar Lipids	Predominant Menaquinones	Major Fatty Acids in Major Amount	G + C%	References
8	*Jiangella alkaliphila* D8-87[T]	Vegetative mycelia fragmented into short rods	LL-2,6-DAP Glucose as the major sugar	DPG, PG, PI, PC, PIM, and unknown PLs	MK-9(H$_4$)	Anteiso-C$_{15:0}$, iso-C$_{16:0}$, and iso-C$_{14:0}$	71.5	Lee 2008
9	*Jiangella gansuensis* YIM 002[T]	Substrate mycelium fragmented into short rods	LL-2,6-DAP Ribose and glucose as cell wall sugars	PIM, DPG, and PI	MK-9(H$_4$)	Anteiso-C$_{15:0}$, anteiso-C$_{17:0}$, iso-C$_{15:0}$, iso-C$_{16:0}$, and C$_{17:1\omega8c}$	70	Song et al. 2005
			Genus: *Nesterenkonia*					
10	*Nesterenkonia suensis* BAC020[T]	Nonmotile, non-spore forming	Peptidoglycan type is A4α (L-Lys–Gly–D-Asp)	DPG, PG, PI, and unidentified GL	MK-8 MK-9 MK-7	Anteiso-C$_{17:0}$ and anteiso-C$_{15:0}$	64.8	Govender et al. 2013
11	*Nesterenkonia xinjiangensis* YIM 70097[T]	Nonmotile, non–spore forming, diphtheroid, irregular rods	Peptidoglycan type is A4α (L-Lys–Gly–L-Glu) Ribose and galactose as cell wall sugars	DPG, PG, and PC	MK-7 MK-8 MK-9	Anteiso-C$_{15:0}$ and anteiso-C$_{17:0}$	66.7	Li et al. 2004a
	Nesterenkonia halotolerans YIM 70084[T]	Non–spore forming, motile cocci	Peptidoglycan type is A4α (L-Lys–Gly–Asp) Xylose and galactose as cell wall sugars	DPG, PG, PI, and unidentified GL	MK-7 MK-8	Anteiso-C$_{15:0}$, i-C$_{16:0}$, and anteiso-C$_{17:0}$	64.4	
12	*Nesterenkonia lacusekhoensis* IFAM EL-30[T]	Short rods or cocci occasionally branching	Peptidoglycan type is A4α (L-Lys-L-Glu) Cell wall sugars not determined	DPG, PG, PC, and unidentified GL	MK-7 MK-8	Anteiso-C$_{15:0}$, anteiso-C$_{17:0}$, and i-C$_{16:0}$	66.1	Collins et al. 2002

(Continued)

Table 6.2 (Continued) Comparison of Morphological and Chemotaxonomic Characteristics of Haloalkaliphilic Actinobacteria

Sr No.	Actinobacteria	Morphology	Cell Wall Peptidoglycan and Whole-Cell Sugar	Polar Lipids	Predominant Menaquinones	Major Fatty Acids in Major Amount	G + C%	References
			Genus: *Nocardiopsis*					
13	*Nocardiopsis fildesensis* sp. nov. GW9-2^T	Long spore Chains on aerial mycelia Spores: rod Shaped, rough surface, and nonmotile	*meso*-DAP No diagnostic sugars	PG, PC, PME, PE, and unidentified PLs	MK-9(H$_4$)	Iso-C$_{16:0}$, anteiso-C$_{17:0}$, C$_{18:1\omega9c}$, iso-C$_{15:0}$, and C$_{17:0}$	76.8	Xu et al. 2014
14	*Nocardiopsis arvandica* sp. nov. HM7^T	Non-acid-fast, branched substrate mycelium and abundant aerial mycelium	*meso*-DAP No diagnostic sugars	PC, PE, PI, PG, and DPG	MK-10(H$_2$) MK-10(H$_4$) MK-10(H$_0$) MK-9(H$_2$)	Iso-C$_{16:0}$ and anteiso-C$_{17:0}$	71.5	Hamedi et al. 2011
15	*Nocardiopsis* sp. JSM 073097^T	Nonmotile and filamentous	*meso*-DAP No diagnostic sugars	DPG, PC, and PG	MK-10(H$_4$) MK-10(H$_6$) MK-10(H$_8$)	Iso-C$_{15:0}$, iso-C$_{16:0}$, anteiso-C$_{16:0}$, and 10-methyl C$_{18:0}$	70.4	Chen et al. 2009
16	*Nocardiopsis salina* YIM 90010T	Nonmotile and aerial mycelium	*meso*-DAP No diagnostic sugars	PG and PI	MK-10(H$_6$) MK-10(H$_8$) MK-12	Iso-C$_{16:0}$, C$_{18:1\omega9c}$, and 10-methyl C$_{18:0}$	73.1	Li et al. 2004b
17	*Nocardiopsis kunsanensis* HA-9^T	Yellow substrate mycelia and white aerial mycelia, which fragment into elongated smooth-surface nonmotile spores	*meso*-DAP No diagnostic sugars	PC, PG, and DPG	MK-10(H$_8$)	Iso-C$_{15:0}$, anteiso-C$_{16:0}$, and 10-methyl-C$_{18:0}$	71	Chun et al. 2000
			Genus: *Prauserella*					
18	*Prauserella shujinwangii* sp. nov. XJ46^T	Nonmotile, aerial mycelium present	*meso*-DAP Glucose, arabinose, and galactose	DPG, PE, PME, PG, PI, and unknown PLs	MK-9(H$_4$) MK-8(H$_4$) MK-9(H$_6$) MK-10(H$_4$)	Iso-C$_{16:0}$, C$_{17:1\omega6c}$, and C$_{16:0}$	67.5	Liu et al. 2014

(Continued)

Table 6.2 (Continued)   Comparison of Morphological and Chemotaxonomic Characteristics of Haloalkaliphilic Actinobacteria

Sr No.	Actinobacteria	Morphology	Cell Wall Peptidoglycan and Whole-Cell Sugar	Polar Lipids	Predominant Menaquinones	Major Fatty Acids in Major Amount	G + C%	References
19	*Prauserella halophila* sp. nov. YIM 90001^T *Prauserella alba* sp. nov. YIM 90005^T	Branched substrate mycelium is fragmented	*meso*-DAP Galactose, arabinose, and ribose	PI, PG, DPG, PE, MPE	MK-9(H$_4$)	Iso-C$_{16:0}$, *cis*-11C$_{17:1}$, *cis*-9C$_{16:1}$, anteiso-C$_{17:0}$, *cis*-9$_{C17:1}$ Iso-C$_{16:0}$ *cis*-11C17:1, *cis*-9C$_{16:1}$, anteiso-i-C$_{17:0}$, *cis*-9C$_{17:1}$, C$_{16:0}$	65.8 66.7	Li et al. 2003
				Genus: *Verrucosispora*				
20	*Verrucosispora sediminis* MS426^T	Aerobic, nonmotile	*meso*-DAP Glucose and mannose	DPG, PE, PI, PIM, and unknown PLs	MK-9(H$_4$)	C$_{17:0}$, iso-C$_{16:0}$, and iso-C$_{15:0}$	66.8	Dai et al. 2010
21	*Verrucosispora lutea* YIM 013^T	Branching hyphae, substrate mycelia form single or clustered spores	*meso*-DAP Glucose and xylose	PE, DPG, PIM, PI, PL, and unknown PLs	MK-9(H$_4$)	Iso-C$_{15:0}$ and iso-C$_{16:0}$	69.3	Liao et al. 2009
22	*Verrucosispora gifhornensis* HR1-Z^T	Aerobic, spore forming	*meso*-DAP Xylose and mannose	PE, DPG, PIM, PS, and unknown PL	MK-9(H$_6$) MK-10(H$_4$) MK-9(H$_2$)	Iso-C$_{16:0}$, iso-C$_{15:0}$, anteiso-C$_{17:0}$	70	Rheims et al. 1998
				Genus: *Zhihengliuella*				
23	*Zhihengliuella flava* sp. nov. H85-3^T	Nonmotile, non-endospore forming, rod shaped	Peptidoglycan A4α type with interpeptide bridge of (L-Ala–L-Glu) Cell wall sugar: galactose Whole-cell sugars: ribose, galactose, glucose, mannose, and rhamnose	PG, DPG, and unknown GL	MK-9 MK-10 MK-8	Anteiso-C$_{15:0}$, iso-C$_{15:0}$, and anteiso-C$_{17:0}$	70.3	Hamada et al. 2013

(Continued)

Table 6.2 (Continued) Comparison of Morphological and Chemotaxonomic Characteristics of Haloalkaliphilic Actinobacteria

Sr No.	Actinobacteria	Morphology	Cell Wall Peptidoglycan and Whole-Cell Sugar	Polar Lipids	Predominant Menaquinones	Major Fatty Acids in Major Amount	G + C%	References
24	*Zhihengliuellasalsuginis* sp. nov. JSM 071043[T]	Nonmotile, nonsporulating, coccoid	Peptidoglycan A4α type with interpeptide bridge of (L-Lys-L-Ala-L-Glu) Tyvelose and mannose as whole-cell sugars	DPG, PG, PI	MK-9 MK-8	Anteiso-$C_{15:0}$, iso-$C_{16:0}$, and anteiso-$C_{17:0}$	67.8	Chen et al. 2010a
25	*Zhihengliuella halotolerans* YIM 70185[T]	Short rod-shaped cells	Peptidoglycan A4α type with interpeptide bridge of (L-Lys-L-Ala-L-Glu) Glucose and tyvelose as cell wall sugars	PG, DPG, PI, and unknown GL and PL	MK-9	Anteiso-$C_{15:0}$ and iso-$C_{15:0}$	66.5	Zhang et al. 2007

Note: DAP, diaminopimelic acid; DPG, diphosphatidylglycerol; PC, phosphatidylcholine; PE, phosphatidylethanolamine; PG, phosphatidylglycerol; PI, phosphatidylinositol; PIM, phosphatidylinositolmannoside; PL, phospholipid; PME, phosphatidylmethylethanolamine; MPE, methylphosphatidylethanolamine; ND, not determined; Lys, lysine; Ala, alanine; Glu, glutamic acid; Asp, aspartic acid.

6.4 MOLECULAR DIVERSITY OF THE ALOALKALIPHILIC ACTINOBACTERIA

6.4.1 Molecular Aspects and Genomic Characteristics

Genomic characteristics of limited Actinobacteria have been described (Bentley et al. 2004; Klijn et al. 2005; Ventura et al. 2007; Yukawa et al. 2007; Li et al. 2013). In a comprehensive review, Ventura et al. (2007) summarized features of 20 actinobacterial genomes. Recently, Gao and Gupta (2012) described a few molecular signatures of the phylum Actinobacteria based on the complete genome studies. Actinomycetes have a relatively large genome size. The chromosome size varies from species to species, 0.93 Mb in *Tropheryma whipplei* to 12 Mb in *Streptomyces bingcheng-gensis* (Bentley et al. 2003; X.-J. Wang et al. 2010). The 12 Mb genome is twice as large as that of *Escherichia coli*. The genetic properties of the actinomycetes can be reflected in the genome of the model organisms *Streptomyces coelicolor* M145 and *Streptomyces avermitilis* (Bentley et al. 2002; Ikeda et al. 2003). The genomic profiles of the halophilic organisms display some common characteristics, distinct from those of the nonhalophilic organisms, and hence may be considered the specific genomic signatures for salt adaptation (Paul et al. 2008). There are a few reports on complete genome sequences of Actinobacteria that can tolerate moderate salt. *Saccharomonospora marina* grows well with 5% NaCl concentration, while it grows optimally between 0% and 3% NaCl (w/v) (Klenk et al. 2012). *Salinispora tropica* (Udwary et al. 2007) and *Nocardiopsis salina* (Li et al. 2004b) optimally grow at 10% NaCl (w/v). Similarly, *Nocardiopsis alkaliphila* displays growth in 20% (w/v) NaCl with an optimum pH of 9.5–10, failing to grow at pH 7.0 (Hozzein et al. 2004). *Nocardiopsis halophile*, of a saline soil of Iraq, grows up to 20% (w/v) NaCl in both synthetic and complex media (Al-Tai and Ruan 1994). Four genera of actinobacteria, *Streptomyces*, *Rhodococcus*, *Gordonibacter*, and Kineococcus, have linear chromosomes (Chen et al. 2002; McLeod et al. 2006; Schrempf 2006; Sekine et al. 2006; Bagwell et al. 2008; Kirby 2011). These linear chromosomes are characterized by a central replication origin (oriC) and terminal inverted repeats (Choulet et al. 2006; McLeod et al. 2006; Schrempf 2006; Ventura et al. 2007; Bagwell et al. 2008).

The bacterial genome is considered dynamic, where gene gains, gene losses, and lateral gene transfers are common occurrences to shape the gene repertoire (Daubin and Ochman 2004; Lawrence and Hendrickson 2005; Ochman 2005). The main driving force responsible for the genome expansion or reduction is probably a niche adaptation. Most of the actinobacterial species are free living, and from complex and compact populated soil environments. Thus, the size of the genomes is generally large (5–9 Mb) in order to combat environmental changes and species competition (Bentley et al. 2002; Paradkar et al. 2003; McLeod et al. 2006; Oliynyk et al. 2007; Ventura et al. 2007). Interestingly, several actinobacterial species isolated under harsh conditions, such as *Acidothermus cellulolyticus*, *Thermobifida fusca*, *K. rhizophila*, and *Rubrobacter radiotolerans*, have relatively small genomes compared with other genera of Actinobacteria (approximately 2–3.5 Mb) (Lykidis et al. 2007; Takarada et al. 2008; Barabote et al. 2009).

6.4.2 Molecular Signatures and Identification

The phylum Actinobacteria is extremely diverse and is broadly identified exclusively on the basis of the branching patterns of different species in the 16S gene tree. Actinobacteria have unique molecular signatures, that is, a shared derived character: a homologous insertion of about 100 nucleotides between helices 54 and 55 of the 23S rRNA gene (Roller et al. 1992). However, there are no known unique features or characteristics commonly shared by all or most constituent taxa of this phylum. The unique genetic properties shared by different species of Actinobacteria can be further employed to accurately define the species (Gupta and Griffiths 2002; Gupta 2010; Gao and Gupta 2012). Further, a consistent phylogenetic framework for Actinobacteria in conjunction with specific probes for identifying different groups of Actinobacteria can facilitate the discovery of novel actinobacterial species and strains and their secondary metabolites. In this context, molecular markers specific for different major clades of Actinobacteria are of particular importance, since probes based on them can serve as novel and specific tools for the identification and discovery of novel actinobacterial species.

Efficient tools, such as genomic fingerprinting, have recently gained attention for investigating the closely related species. Further, different molecular markers, such as fragment length analysis of the internal transcribed spacer (ITS) region (Papke et al. 2003), fluorescence in situ hybridization (FISH) (Henckel et al. 1999; Friedrich et al. 2001), amplified fragment length polymorphism (AFLP) analysis (Rademaker et al. 2000), ARDRA (Webb and Maas 2002), multilocus enzyme electrophoresis (MLEEC) (Petursdottir et al. 2000; Whitaker et al. 2003), rep-PCR genomic fingerprinting (Fulthorpe et al. 1998), terminal-restriction fragment length polymorphism (T-RFLP) (Lee et al. 2003), and 16S rRNA sequencing-based phylogenetic analysis (Margot et al. 2002), have emerged as effective tools. They differ from genus to strain level in resolution power and reliability.

Ocean represents an ecosystem with many unique forms of actinomycetes. Actinomycetes have been reported and cultivated from various saline and alkaline habitats, such as ocean, salt lakes, soda deserts, marine snow, sponges, and salterns (Mikami et al. 1982; Groth et al. 1997; Duckworth et al. 1998; Jones et al. 1998; Castillo et al. 2005; Xin et al. 2008; Zhang et al. 2008; Abdelmohsen et al. 2014). Culture independent diversity of the obligate marine actinomycete genus *Salinispora* retrived abundant new phylotype that has yet to be cultured. (Mincer et al. 2005). 16S rRNA-RFLP, a rapid and inexpensive method, has been widely used to study the diversity of the microbial communities (Cook and Meyers 2003; Webster et al. 2004; Zhang et al. 2006, 2008). The diversity of actinobacterial communities associated with sponges *Hymeniacidon perleve* and *Sponge* sp. is investigated using a culture-independent nested PCR technique. Similarly, the phylogenetic affiliation of a large number of cultivable actinobacterial isolates associated with five marine sponges is described (Zhang et al. 2008). The diversity among the actinobacterial species was analyzed based on *Hha*I digestion of the 16S rRNA genes, which indicated 11 different band patterns on the RFLP gel. Zhang et al. (2006) studied the molecular and phylogenetic diversity of the cultivable Actinobacteria of the marine sponge *H. perleve* by 16S rRNA gene amplification and RFLP analysis. The diversity of the actinomycetes from solar salterns in Tuticorin, India, was analyzed using rDNA-ARDRA to distinguish various taxonomic groups (Jose et al. 2013).

6.5 PHYLOGENETICS AND BIOINFORMATICS OF HALOALKALIPHILIC ACTINOBACTERIA

6.5.1 16S rRNA Gene Sequence Analysis

The phylum Actinobacteria constitutes one of the earliest lineages within the prokaryotes, representing one of the largest taxonomic units among the 18 major lineages currently recognized within the domain Bacteria (Gupta 2001; Koch 2003; Lake et al. 2007). It includes 5 subclasses, 14 suborders, and approximately 300 genera (Euzéby 2011). The members of this phylum are characterized by high GC content and valuable secondary metabolites (Ventura et al. 2007). Actinobacteria constitutes one of the main phyla within the prokaryotes on the basis of its branching position in 16S rRNA gene trees. Besides the phylogeny of the 16S rRNA gene, some conserved signatory sequences in 23S rRNA and proteins (e.g., cytochrome-c oxidase I, CTP synthetase, and glutamyl-tRNA synthetase) distinguish them from all other bacteria (Gao and Gupta 2005). Interestingly, some subclasses were removed from the phylum Actinobacteria and projected as different phyla with new reclassification. The rationale behind shifting certain genera stems from the fact that with the increasing number of 16S rRNA gene sequences in the public databases, the prediction of the taxonomic ranks becomes exceedingly difficult (Gao and Gupta 2012).

Consequently, this would affect the taxonomic rank expected due to the genera and higher taxa with a single type strain (Zhi et al. 2009).

The phylogenetic and evolutionary relationships among the members of the Actinobacteria depend on the distinction among the species of the same genera of this phylum. The phylum Actinobacteria is delineated from other bacteria solely depending on the basis of its branching position in 16S rRNA gene trees (Zhi et al. 2009; Gao and Gupta 2012). Recently, Zhi et al. (2009) published and reorganized the taxonomy of Actinobacteria and provided a set of 16S rRNA signature nucleotides between the ranks of family and subclass. The phylum Actinobacteria was alienated into four subclasses, which include 219 genera in 50 families (Garrity et al. 2001; Zhi et al. 2009). Euzéby (2012) updated the taxonomy in the List of Prokaryotic Names with Standing in Nomenclature (http://www.bacterio.net/index.html). Accordingly, the phylum Actinobacteria, at the highest level, is now divided into five subclasses. Recently, Ludwig et al. (2012) reported the taxonomy of the phylum Actinobacteria based on 16S rRNA trees, which would further add some new information to the forthcoming *Bergey's Manual of Systematic Bacteriology* (Vol. 5, Actinobacteria Section).

The phylogenetic information and the basis of the update are not posted on the *Bergey's Manual* website. In the revised taxonomy, the taxonomic ranks of subclasses and suborders are eliminated. At the highest level, the phylum Actinobacteria is now divided into six classes: Actinobacteria (16 orders, 43 families, 128 genera), Acidimicrobiia (1 order, 2 families, 5 genera), Coriobacteriia (1 order, 1 family, 13 genera), Nitriliruptoria (2 orders, 2 families, 2 genera), Rubrobacteria (1 order, 1 family, 1 genus), and Thermoleophilia (2 orders, 4 families, 4 genera) (Gao and Gupta 2012; Selvakumar et al. 2015).

In a recently published review on the phylogenetic framework and molecular signatures for the main clades of the phylum Actinobacteria, the genomic characteristics, phylogeny, and molecular signatures of the phylum Actinobacteria are highlighted (Gao and Gupta 2012). A user-friendly database highlighting the diversity, biocatalytic potential, and phylogeny of the salt-tolerant alkaliphilic Actinobacteria has been created (Sharma 2011; Sharma et al. 2012). A more recent review by Shivlata and Satyanarayana (2015) describes the diversity, biology, and potential applications of the thermophilic and alkaliphilic Actinobacteria.

6.5.2 Halophilic Actinobacteria and Phylogenetic Analysis

Halophiles are distributed in all three phylogenetic domains of life, including the Actinomycetales order. Occurrences of actinomycetes in saline habitats and their tolerance to high salt concentrations were described more than 30 years ago (Tresner et al. 1968). Halophiles and nonhalophiles are usually found together in all phylogenetic trees, and many genera and families have representatives with quite distinct salt requirements and tolerance (Oren 2008). Hamedi et al. (2013) reported several halophilic species under the suborders Streptosporangineae, Streptomycineae, Pseudonocardineae, Actinopolysporineae, and Micrococcineae. However, phylogenetic and physiological coherence is observed only in *Nesterenkonia*, *Nocardiopsis*, *Salinispora*, and *Streptomonospora* genera. Four of the five genera in the family Nocardiopsaceae, including *Nocardiopsis*, *Streptomonospora*, *Haloactinospora*, and *Marinactinospora*, contain a large number of halophilic species (Hamedi et al. 2013). Most prominent and highly described halophilic actinomycetes belong to the genus *Nocardiopsis*, comprising 42 published species, with 13 halophilic and 6 halotolerant species. Interestingly, the genus *Streptomonospora* constitutes only five species, able to optimally grow in media supplemented with NaCl. The family Micrococcaceae includes the genera *Nesterenkonia* and *Kocuria* with halophilic species. Subsequent to *Nocardiopsis*, the genus *Nesterenkonia* contains a high number of halophilic species in the phylum Actinobacteria. The family Pseudonocardiaceae comprises six genera (*Actinopolyspora*, *Saccharomonospora*, *Yuhushiella*, *Prauserella*, *Saccharopolyspora*, and *Amycolatopsis*) containing halophilic species.

A recent review (Hamedi et al. 2013) documents the collection and arrangement of around 70 valid halophilic and halotolerant actinomycetes species reported up to 2011.

Recently, the diversity of the cultivable halophilic Actinobacteria in two Algerian arid ecosystems (M'zab and Zibans) of Septentrional Sahara has been described (Saker et al. 2015). A total of 69 halophilic strains were isolated from 19 different soil samples. The 16S rRNA gene sequence-based phylogenetic tree constructed by the neighbor-joining method (Saitou and Nei 1987) with the model of Jukes and Cantor (1969) revealed six clusters corresponding to six genera: *Actinopolyspora*, *Nocardiopsis*, *Prauserella*, *Saccharomonospora*, *Saccharopolyspora*, and *Streptomonospora*. Interestingly, many strains of the Algerian arid soils displayed similarities of <99%–91%, indicating the existence of new species. Further, the diversity of Actinobacteria associated with the marine ascidian *Eudistoma toealensis* has recently been described (Steinert et al. 2015). It revealed that 51 actinobacterial genera constitute 2%–10% of the total bacterial community, the most notable genera being *Salinispora* and *Verrucosispora*, not reported in any environmental sample. Jose and Jebakumar (2012) reported on the phylogenetic diversity of actinomycetes isolated and cultured from a coastal multipond solar saltern in Tuticorin, India. The 16S rDNA sequence analysis suggested the dominance of the genus *Streptomyces*, with the first-time report of the genus *Nonomuraea* from the Indian solar salterns. Besides the *Streptomyces* and *Nonomuraea* genera, *Nocardia*, *Nocardiopsis*, and *Saccharopolyspora* were also reported on the basis of the phylogenetic analysis (Jose and Jebakumar 2012).

The phylogenetic diversity of the cultivable Actinobacteria from five marine sponge species in the intertidal coast of Yello Sea around Dakian city, China, has been described (Zhang et al. 2008). For their isolation, seawater and media were supplemented with salt (20 g/L) and nalidixic acid to inhibit the fast-growing Gram-negative bacteria. The 16S rDNA gene sequencing and RFLP analysis revealed *Streptomyces* as a main representative genus of the phylum Actinobacteria, and the study demonstrated the 16S rDNA gene as a phylogenetic marker. In another instance, the diversity of the actinomycetes of the Challenger Deep sediment (10,898 m) from the Mariana Trench is reported (Pathom-aree et al. 2006). The phylogenetic analysis based on the 16S rRNA gene sequences confirmed the existence of the genera *Dermacoccus*, *Kocuria*, *Micromonospora*, *Streptomyces*, *Tsukamurella*, and *Williamsia*, while members of the genera *Dermacoccus* and *Tsukamurella* were reported for the first time from the marine habitats.

Jenifer et al. (2015) reported haloalkaliphilic *Streptomyces* spp. AJ8 from the Kovalam solar salt, India. AJ8 was active up to pH 10.5, with an optimum pH of 9.5, and displayed growth in the range of 2%–10% NaCl (w/v), with abundant growth in Knight's agar, N agar, and Actinomycetes agar. The phylogenetic studies confirmed 90% sequence similarity with *Streptomyces* sp., including *Streptomyces* sp. 6G16, *Streptomyces* sp. SCAUKO356, *Streptomyces* spp. 337702, and *Streptomyces fragilis* YJ-RT6. G. Zhang et al. (2015) reported an alkaliphilic and halotolerant *Actinobacteria* from the western Pacific Ocean. The organism was Gram positive, aerobic, motile, and non–spore forming and belonged to the genus *Nesterenkonia*, displaying 96.8% 16S rRNA gene sequence similarity with *Nesterenkonia aethiopica* DSM 17733^T. The organism grew over a wide range of temperatures between 4°C and 50°C, and pH 7.0–12.0, while it tolerated up to 12% (w/v) NaCl. The optimal growth, however, was observed at 40°C, pH 9.0, in 1% (w/v) NaCl. Further, an alkaliphilic actinobacterium from a rhizosphere soil of *Reaumuria soongorica* was reported to belong to the genus *Nesterenkonia* (Wang et al. 2014).

Streptomyces alkalithermotolerans sp., a novel alkaliphilic and thermotolerant actinomycete from Lonar soda lake, India, displays optimum pH and salt requirements for growth at 9.5–10.0 and 2%–4% (w/v) NaCl, respectively (Sultanpuram et al. 2015). The phylogenetic analysis revealed that it belonged to the class Actinobacteria, closely related to *Streptomyces sodiiphilus* JCM13581. On a similar note, the hypersaline Sambhar Lake of Rajasthan, India, was explored for its agriculturally and industrially important haloalkaliphilic bacteria (Sahay et al. 2012). The growth of these bacteria was supported by 2%–25% (w/v) NaCl under alkaline pH. Based on the 16S rRNA gene sequences, 93 isolates were reported from 32 groups, representing different taxa belonging to 3 phyla (Firmicutes, Proteobacteria, and Actinobacteria). Nearly half of the isolates exhibited similarity with the phylum Firmicutes, while Proteobacteria and Actinobacteria followed next in dominance. Phylogenetically, the Actinobacteria exhibited 99% similarity with *Streptomyces viridiviolaceus* and *Streptomyces radiopugnans*, which optimally grow at pH 9–10 and in 6%–7% NaCl (w/v). Meklat et al. (2011) studied the taxonomy and antagonistic properties of halophilic actinomycetes from the Saharan soils of Algeria and reported a diversity of 52 halophilic actinomycetes with the help of a polyphasic approach. The study suggested wide distribution of the genera *Actinopolyspora, Nocardiopsis, Saccharomonospora, Streptomonospora,* and *Saccharopolyspora,* with the *Saccharopolyspora* being identified as a new member.

Sorokin et al. (2009) reported *Nitriliruptor alkaliphilus* gen. nov., sp. nov., a deep lineage haloalkaliphilic actinobacterium, obtained from an enrichment culture inoculated with a mixture of soda lake sediments. *N. alkaliphilus* was capable of growing on aliphatic nitriles with moderate salt (0.1–2.0 M NaCl) and pH 9.0–9.5. On the basis of its unique phenotypes and phylogeny, the strain was considered a novel species of a new family Nitriliruptoraceae family nov., and a novel order Nitriliruptorales order nov., of the class Actinobacteria.

6.5.3 Bioinformatics Tools and Phylogenetic Analysis

The following molecular and bioinformatics approaches are used to judge the phylogeny of the actinobacteria (Alam et al. 2010).

6.5.3.1 Assessment of the Phylogeny Based on Single-Gene Analysis

The majority of the phylogenetic reconstructions are currently based on the 16S rRNA gene sequences because lateral gene transfer is very rare and their molecular size ensures that they carry abundant evolutionary information. In general, for phylogenetic analysis, three different rRNA (5S, 16S, and 23S) genes are taken into account for single-gene analysis. All columns containing gaps in alignment are deleted manually, or sometimes the sequences are edited using PHYDIT (Chun 1995). To identify the nearest neighbor, a search is carried out by the National Center for Biotechnology Information (NCBI) (http://www.ncbi.nlm.nih.gov/) using the Basic Local Alignment Search Tool (BLAST). Sequences can also be aligned and separated with the help of Clustal W (Thompson et al. 1994). The gaps are removed and the tree topology is evaluated by bootstrap analysis (100/1000 bootstrap replicates) of each individual sequence. Later, neighbor-joining (Saitou and Nei 1987), Fitch–Margoliash (Fitch and Margoliash 1967), maximum parsimony (Sourdis and Nei 1988), and maximum likelihood algorithms (Felsenstein 1981) are used in phylogenetic tree building. More recently, phylogenetic dendrograms have been constructed using the DNAPARS and DNAML programs in the PHYLIP package (Felsenstein 2005) and MEGA version 6.0 (Tamura et al. 2013).

6.5.3.2 Assessment of the Phylogeny Based on the Whole-Genome Analysis

For the genome-based approach, all coding sequences from the genome annotation of various species available in the NCBI database (http://www.ncbi.nlm.nih.gov) are selected. Homologs are assigned using the best-hit strategy. Later, the conserved proteins, which are present in all species, are used for gene concatenation and gene order–based phylogenetic reconstruction. Genome-based analysis with 45 Actinomycetales species is reported (Alam et al. 2010).

6.5.3.3 Phylogeny Based on the Gene Concatenation, Gene Order, and Gene Content

Phylogeney based on this approach is an intuitive extension of the single-gene approach, due to the increase in the number of informative sites. The gene order–based phylogenetic analysis is quite independent, as no gene sequences are used. It is largely independent of sequence alignment, and hence unaffected by misalignments that usually cause wrong topologies. Moreover, gene content analysis is a comprehensive manner of the phylogenetic inference that relies on the conservation distributed all over the genome. Therefore, the percentages of genes for each pair of species are calculated and the resulting similarity matrix used for phylogenetic reconstruction by the neighbor-joining method.

6.6 SIGNIFICANCE OF HALOPHILIC ACTINOBACTERIA

6.6.1 Extreme Actinobacteria as a Source of New Bioactive Compounds

Extreme habitats constitute major reservoirs of biological diversity, providing rich sources of novel molecules. Numerous microbial species from extreme environments are isolated, including the unique archaea, bacteria, and fungi with the ability to produce various primary and secondary metabolites (Khazal 2013). The immense diversity and its underutilization are the key reasons for the increasing attention of researchers toward discovering novel bioactive compounds from the haloalkaliphilic actinobacteria. The unique characteristics of these organisms to adapt under the harsh conditions of the environment also provide opportunities to explore novel genes and biochemical pathways for the primary and secondary metabolites.

The actinomycetes produce diverse, unique, and sometimes highly complex compounds exhibiting excellent bioactive potency with low toxicity (Berdy 2005; Kurtboke 2012; Kurtboke et al. 2015; Saadoun et al. 2015). Despite innovative tools and in silico approaches in natural products research, the microbial collection is still key for novel drug discovery (Genilloud et al. 2011). Therefore, the search for the largely unexplored actinobacteria from unusual and extreme environments, such as hypersaline marine environments, extreme inland saline zones, volcanic zones, hyperarid zones, and glaciers, has gained impetus in recent years (Swan et al. 2010; Guan et al. 2011b; Phillips et al. 2012). Discovery of novel compounds from the actinobacteria of the unconventional environments provides an attractive area of research (Carr et al. 2012; Subramani and Aalbersberg 2012; Jose and Jebakumar 2014, 2015). The huge capacity of the halophilic and halotolerant actinobacteria for the production of new bioactive secondary metabolites has been recently demonstrated (Hamedi et al. 2013). The search for persuasive actinobacteria from the relatively less explored extreme environments has gained attention for discovering novel secondary metabolites (Peraud et al. 2009; Jose and Jebakumar 2013a, 2013b, 2013c).

Actinobacteria are particularly known for their ability to synthesize various secondary metabolites, possessing antibacterial (Sharma et al. 2005; Grill et al. 2008; Sengupta et al. 2015), antifungal (Cheng et al. 2010a; Xu et al. 2011; Han et al. 2012; Wu et al. 2012), antiprotozoal (Ravikumar et al. 2011), antiperasitic (Kitani et al. 2011), and antitrypanosomal (Inahashi et al. 2011) activities. Various antimicrobial substances from actinobacteria have been isolated and categorized into several major structural classes that include aminoglycosides (Nanjawade et al. 2010), anthracyclines (Kremer and Dalen 2001), glycopeptides, β-lactams (Kollef 2009), macrolides, nucleosides, peptides, polyesters, polyketides, actionomycins and tetracyclines (Berdy 2005; Harvery and Champe 2009), angucyclinone (Sun et al. 2007), polyenes (Ouhdouch et al. 2001; Lemriss et al. 2003), piercidins (Hayakawa et al. 2007a, 2007b), and octaketides (Radzom et al. 2006). A new antibiotic, anthracimycin from actinomycete in saline culture, with the ability to kill the deadly anthrax bacterium *Bacillus anthracis* and methicillin-resistant *Staphylococcus aureus*, has been discovered (Jang et al. 2013). Similarly, quinoline alkaloid from a novel halophilic actinomycete *Nocardiopsis terrae* YIM 90022, which exhibited antimicrobial activities against some plant pathogens, has been reported (Tian et al. 2014). S.Z. Tian et al. (2013) reported an antimicrobial compound, p-terphenyls displaying antifungal, antibacterial, and antioxidant activities, from a halophilic actinomycete *Nocardiopsis gilva* YIM 90087.

However, the number of actinomycetes has dropped due to the reduced interest of some major pharmaceutical companies in the discovery of new antibiotics. Further, pathogenic microorganisms have evolved sophisticated mechanisms to inactivate antibiotics, necessitating the urgent need for the discovery of new antibiotics (Butler et al. 2013). The rapid emergence of drug resistance, especially multidrug resistance among the pathogenic bacteria, underlines the need for new antibiotics (Alanis 2005; Sharma et al. 2011). Biological activity screening and consequent structural characterization suggested the significance of actinobacteria from the hypersaline solar salterns as a source of new antibiotics. A moderately halophilic actinobacteria, *Saccharopolyspora salina* VITSDK4, producing an extracellular bioactive metabolite, which inhibits the proliferation of HeLa cells as well as fungal and bacterial pathogens, has been described (Suthindhiran and Kannabiran 2009). Elucidation of structures of new linear polyketides, actinopolysporins A, B, and C, besides the known antineoplastic antibiotic tubercidin that stabilizes the tumor suppressor Programmed Cell Death Protein 4 (Pdcd4), from the halophilic actinomycete *Actinopolyspora erythraea* YIM 90600, is reported (Zhao et al. 2011).

A significantly large proportion of the antibiotics produced by the actinobacteria come from the genus *Streptomyces*. The *Streptomyces* genes responsible for the secretion of different antibiotics and extracellular enzymes ensure the continued propagation and dispersal of the species. With the advancement in knowledge of the genetic composition of a couple of streptomycetes, aspects of the linked genetic control of sporulation and antibiotic production have been clarified (Bentley et al. 2002; Ikeda et al. 2003; Chater 2006). Further, *Streptomyces* from stressed environments can be novel candidates for the potential molecules for varied application. For instance, exploration of actinobacteria for antidermatophytic drugs is a novel attempt requiring further investigation (Lakshmipathy and Kannabiran 2010). Park et al. (2010) reported *Streptomyces hygroscopicus* producing rapamycin and its analogs, which exhibited antifungal, immunosuppressive, antitumor, neuroprotective, and antiaging activities. Three different species of *Streptomyces*, that is, *S. griseoruber*, *S. calestis*, *S. bikiniensis*, were characterized for their antagonistic properties (Kannan and Vincent 2011). *Streptoverticillium album*, displaying antibacterial activities, was reported as the dominant actinomycetes species in salt pan soil (Gayathri et al. 2011). *Streptomyces* associated with the marine sponge represent a promising source of antibacterial agents against fish and shellfish pathogens (Dharmaraj 2010). *Streptomyces citricolor* isolated from the coastal soil of Parangipetti showed a significant level of antibacterial effect against *Pseudomonas aeruginosa*, *Escherichia coli*, *Klebsiella pneumoniae*, *Salmonella typhi*, and *Vibrio cholera* (Sathiyaseelan and Stella 2011). Antagonistic marine *Streptomyces* sp. LCJ94 isolated from the Bay of Bengal with inhibitory activity against *Vibrio harveyi*, *Vibrio alginolyticus*, and *Vibrio vulnificus*, has potential as a biocontrol agent in controlling the *Vibrio*-caused diseases in shrimps (Mohanraj and Sekar 2013).

6.6.2 Industrially Important Enzymes from the Halophilic Actinobacteria

Extremophilic actinobacteria exist in environments with two or more extreme conditions. The polyextremophiles are adapted to multiple stress conditions, for instance, alkalithermophilic, thermoacidophilic, thermophilic radiotolerant, haloalkaliphilic, and thermos-alkali-tolerant actinobacteria (Gupta et al. 2014). The enzymes of these actinomycetes are active under a combination of stress factors, and hence have potential applications in the food and feed, textile, waste management, medical, and detergent industries. While most of the studies on the actinobacteria have focused on the antibiotic production, relatively limited reports exist on their enzymatic potential.

Alkaline proteases, amylases, cellulases, and lipases are used in detergents. Alkaline proteases in particular are commercially important for the detergent and other industries. In recent years, haloalkaliphilic actinobacteria from saline habitats have been explored for their enzymes, which can withstand harsh application conditions. Earlier, Thumar and Singh (2009) reported an organic solvent-tolerant alkaline protease from salt-tolerant alkaliphilic *Streptomyces clavuligerus* strain Mit-1. More recently, two salt-tolerant alkaliphilic actinomycetes, OM-6 (EU710555.1) and OK-5 (HM560975), were reported to produce thermostable alkaline proteases (Gohel and Singh 2012a, 2013, 2015). The OK-5 protease activity was substantially enhanced at the optimum temperature of 70°C in the presence of NaCl and Ca^{2+}. Even at 90°C, the enzyme activity was enhanced by 500% and 217% of the original activity with 3 M NaCl and 100 mM Ca^{2+}, respectively. The trends were also analyzed on the basis of the kinetics and thermodynamics establishing the ease and stability of the enzyme-catalyzed reactions (Gohel and Singh 2013, 2015).

The organic solvent tolerance adds to the industrial value of the hydrolytic enzymes in light of the significance of the catalysis under nonaqueous conditions (Klibanov 2001; Alexander 2001; Thumar and Singh 2009). Besides proteases, amylases have ample applications in the food, fermentation, textile, paper, detergent, pharmaceutical, and sugar industries. Since many of the commercially available amylases do not withstand industrial reaction conditions, or not meet industrial demand, the search for novel amylases is highly significant. A calcium-independent α-amylase from haloalkaliphilic marine *Streptomyces* strain A3 has recently been reported (Chakraborty et al. 2012). More recently, a thermostable, oxidant, detergent, and surfactant-stable α-amylase of haloalkaliphilic *Nocardiopsis* sp. strain B2 was isolated from marine sediments of the West coast of India (Chakraborty et al. 2015). Acharyabhatta et al. (2013) isolated marine actinomycete BTSS 1001, which produces alkaline amylase with maximum amylolytic activity at pH 9.5 from Diviseema coast, Bay of Bengal.

Cellulases are extensively used for increasing the yield of juice in the food industry, decreasing discoloration and the fuzzing effects of cloth in the textile industry, and strengthening and whitening paper pulp in the paper industry (Sethi et al. 2013). George et al. (2001) reported an alkalothermophilic actinomycete producing carboxymethyl cellulose. In addition to proteases and amylases, alkali-stable extracellular cellulases have been used for the production of improved laundry detergents (Horikoshi 2006). Halo-alkali-stable cellulases can also be used to release sugars from recalcitrant lignocellulose in agricultural waste for the production of bioethanol. Two moderately haloalkaliphilic cellulose-producing *Streptomyces* isolated from saline-alkaline soils of Ararat Plain (Armenia) were identified as *Streptomyces roseosporus* A3 and *S. griseus* A5 (Hakobyan et al. 2013).

Among the other industrially important enzymes are keratinases and dextranase. Keratinases are used in several industrial processes, such as enzymatic dehairing in the leather industry, the production of slow-release nitrogen fertilizers in agriculture, and the synthesis of biodegradable films and coatings in biomedical applications (Anbu et al. 2007; Freeman et al. 2009; Brandelli et al. 2010). An alkaline keratinase from *Streptomyces*, being optimally active at pH 11, was reported as a potential additive in detergent formulations (Tatineni et al. 2008). Another keratinase from *Actinomadura keratinilytica* strain Cpt29 was active and stable at pH 7–11 (Habbeche et al. 2014). During the sugar manufacturing process, interference of dextrans is a cumbersome and inevitable event. Dextranases are produced by a number of fungal and bacterial species, and many such enzymes have been purified and characterized. Recently, an industrially suitable dextranase from a thermoalkaliphilic actinobacteria, *Streptomyces* sp. NK458, was reported (Purushe et al. 2012). However, by and large, the currently available dextranases in the sugar industry lack feasibility, incurring a high cost of production (Jimenez 2005).

6.6.3 Ecological Significance of Actinobacteria

Certain actinobacteria have been reported to degrade a wide range of hydrocarbons, pesticides, and aliphatic and aromatic compounds, together with their significant role in biogeochemical cycles, bioremediation, bioweathering, and plant growth promotion (Sambasiva et al. 2012; Cockell et al. 2013; Palaniyandi et al. 2013; Chen et al. 2015). Palaniyandi et al. (2013) reported a plant-beneficial actinobacteria associated with yam rhizosphere and studied its genetic and functional characteristics. Further, the actinobacteria can also promote symbiosis between plant and nitrogen-fixing bacteria and promote mycorrhiza, eventually promoting plant growth (Schrey and Tarkka 2008; Dimkpa et al. 2009). Moreover, Dimkpa et al. (2008) reported that siderophores produced by *Streptomyces acidiscabies* E13 bind nickel and promote growth of cowpea (*Vigna unguiculata* L.) under the conditions of nickel stress.

Actinobacteria thriving under stress conditions exhibit considerable ecological importance. A number of stress-tolerant actinobacteria have been reported to mineralize the hydrocarbon and other pollutants. An extremely halophilic actinomycete, *Actinopolyspora* sp. DPD1, with the capability to degrade *n*-alkanes and fluorine was reported by Al-Mueini et al. (2007) from an oil production site in Oman. Similarly, Nakano et al. (2011) reported *Dietzia* species as having the ability to degrade organic pollutants and produce wax ester–like compounds as biosurfactants or bioemulsifiers by degrading *n*-alkanes as a sole carbon. Further, a novel thermotolerant polyester-degrading actinomycetes, *Actinomadura miaoliensis* sp., isolated from soil of Miaoli, Taiwan, degrades poly(D-3-hydroxybutyrate), reducing its environmental impacts (Tseng et al. 2007).

Nocardiopsis metallicus sp., a metal-leaching actinomycete from an alkaline slag dump, can grow at the pH range of 7–10 and tolerate up to 10% salt (Schippres et al. 2002). Metal-leaching properties of the actinobacteria are useful in the process of metal extraction from the alkaline sites. Alkaliphilic and alkalitolerant actinobacterial species, *Kocuria rosea* MG2, *Kineococcus radiotolerans*, *Microbacterium radiodurans*, and *Cellulosimicrobium cellulans*, are resistant to lethal radiation (Phillips et al. 2002; Chen et al. 2004; Zhang et al. 2010; Gabani et al. 2012). Overall, significant progress in the discovery of novel bioactive compounds and different enzymes from the haloalkaliphilic actinobacteria highlights their ecological and industrial importance.

6.7 CONCLUSION

This review attempted to provide an account of the actinobacteria thriving under harsh conditions. The occurrence and distribution of actinobacteria in extreme environments signify their ecological role. The chemotaxonomic studies revealed unique cell wall characteristics with respect to cell sugar and polar lipid profile. The 16S rRNA gene-based phylogenetic analysis and the assessment of the molecular diversity by different fingerprinting methods generated abundant evolutionary information. The chapter also demonstrated the significant role of the haloalkaliphilic actinobacteria in providing a new generation of antibiotics and biocatalysts. A search for new natural products and enzymes from these microorganisms would be a valuable addition to the natural product research.

ACKNOWLEDGMENT

The work cited in this chapter from our own research group has been supported by various programs of University Grants Commission (UGC), including the current support under the Centre of Advanced Study (CAS) Programme, Department of Science and Technology—Fund for Improvement of S&T Infrastructure (DST-FIST), DST-Women Scientist Programme, UGC Start-Up Research Project, Ministry of Earth Sciences (MoES) Net Working Project, Department of Science and Technology—Science and Engineering Research Board (DST-SERB) Early Career Research Project and Saurashtra University. The UGC-BSR Basic Scientific Research meritorious fellowships to S.D.G., A.K.S., and F.J.T. are also acknowledged.

REFERENCES

Abdelmohsen, U.R., K. Bayer, and U. Hentschel. Diversity, abundance and natural products of marine sponge-associated actinomycetes. *Nat. Prod. Rep.* 31 no. 3 (2014): 381–399.

Acharyabhatta, A., S.K. Kandula, and R. Terli. Taxonomy and polyphasic characterization of alkaline amylase producing marine actinomycete *Streptomyces rochei* BTSS 1001. Int. *J. Microbiol.* (2013): 1–8.

Alam, M.T., M.E. Merlo, E. Takano, and R. Breitling. Genome-based phylogenetic analysis of *Streptomyces* and its relatives. *Mol. Phylogenet. Evol.* 54 no. 3 (2010): 763–772.

Alanis, A.J. Resistance to antibiotics: Are we in the post-antibiotic era? *Arch. Med. Res.* 36 no. 6 (2005): 697–705.

Alexander, M.K. Improving enzymes by using them in organic solvents. *Nature* 409 (2001): 241–246.

Al-Mueini, R., M. Al-Dalali, I.S. Al-Amri, and H. Patzelt. Hydrocarbon degradation at high salinity by a novel extremely halophilic actinomycete. *Environ. Chem.* 4 (2007): 5–7.

Al-Tai, A.M., and J.-S. Ruan. *Nocardiopsis halophila* sp. nov., a new halophilic actinomycete isolated from soil. *Int. J. Syst. Bacteriol.* 44 no. 3 (1994): 474–478.

Ameur, H., M. Ghoul, and J. Selvin. The osmoprotective effect of some organic solutes on *Streptomyces* sp. MADO2 and *Nocardiopsis* sp. MADO3 growth. *Braz. J. Microbiol.* 42 no. 2 (2011): 543–553.

Anbu, P., S.C. Gopinath, A. Hilda, T. Lakshmipriya, and G. Annadurai. Optimization of extracellular keratinase production by poultry farm isolate *Scopulariopsis brevicaulis*. *Bioresour. Technol.* 98 (2007): 1298–1303,

Bagwell, C.E., S. Bhat, G.M. Hawkins, B.W. Smith, T. Biswas, T.R. Hoover, E. Saunders, C.S. Han, O.V. Tsodikov, and L.J. Shimkets. Survival in nuclear waste, extreme resistance, and potential applications gleaned from the genome sequence of *Kineococcus radiotolerans* SRS30216. *PLoS One* 3 no. 12 (2008): e3878.

Baltz, R.H. Antimicrobials from actinomycetes: Back to the future. *Microbe* 2 (2007): 125–131.

Barabote, R.D., G. Xie, D.H. Leu, P. Normand, A. Necsulea, V. Daubin et al. Complete genome of the cellulolytic thermophile *Acidothermus cellulolyticus* 11B provides insights into its ecophysiological and evolutionary adaptations. *Genome Res.* 19 no. 6 (2009): 1033–1043.

Bentley, S.D., S. Brown, L.D. Murphy, D.E. Harris, M.A. Quail, J. Parkhill et al. SCP1, a 356 023 bp linear plasmid adapted to the ecology and developmental biology of its host, *Streptomyces* coeliciolor A3 (2). *Mol. Microbiol.* 51 no. 6 (2004): 1615–1628.

Bentley, S.D., K.F. Chater, A.-M. Cerdeno-Tarraga, G.L. Challis, N.R. Thomson, K.D. James et al. Complete genome sequence of the model actinomycete *Streptomyces coelicolor* A3(2). *Nature* 417 no. 6885 (2002): 141–147.

Bentley, S.D., M. Maiwald, L.D. Murphy, M.J. Pallen, C.A. Yeats, L.G. Dover et al. Sequencing and analysis of the genome of the Whipple's disease bacterium *Tropheryma whipplei*. *Lancet* 361 no. 9358 (2003): 637–644.

Berdy, J. Bioactive microbial metabolites. *J. Antibiot.* 58 (2005): 1–26.

Bouras, N., A. Meklat, A. Zitouni, F. Mathieu, P. Schumann, C. Spröer, N. Sabaou, and H.-P. Klenk. *Nocardiopsis algeriensis* sp. nov., an alkalitolerant actinomycete isolated from Saharan soil. *Antonie Van Leeuwenhoek* 107 no. 2 (2015): 313–320.

Brandelli, A., D.J. Daroit, and A. Riffel. Biochemical features of microbial keratinases and their production and applications. *Appl. Microbiol. Biotechnol.* 85 (2010): 1735–1750.

Bull, A.T., and J.E.M. Stach. Marine actinobacteria: New opportunities for natural product search and discovery. *Trends Microbiol.* 15 no. 11 (2007): 491–499.

Bursy, J., A.J. Pierik, N. Pica, and E. Bremer. Osmotically induced synthesis of the compatible solute hydroxyectoine is mediated by an evolutionarily conserved ectoine hydroxylase. *J. Biol. Chem.* 282 no. 43 (2007): 31147–31155.

Butler, M.S., M.A. Blaskovich, and M.A. Cooper. Antibiotics in the clinical pipeline in 2013. *J. Antibiot.* 66 (2013): 571–591.

Carr, G., M. Poulsen, J.L. Klassen, Y. Hou, T.P. Wyche, T.S. Bugni et al. Microtermolides A and B from termite-associated *Streptomyces* sp. and structural revision of vinylamycin. *Org. Lett.* 14 (2012): 2822–2825.

Carro, L., P. Zúñiga, F. De la Calle, and M.E. Trujillo. *Streptomyces pharmamarensis* sp. nov. isolated from a marine sediment. *Int. J. Syst. Evol. Microbiol.* 62 no. 5 (2012): 1165–1170.

Castillo, U., S. Myers, L. Browne, G. Strobel, W.M. Hess, J. Hanks, and D. Reay. Scanning electron micros-copy of some endophytic streptomycetes in snakevine—*Kennedia nigricans. Scanning* 27 no. 6 (2005): 305–311.

Chakraborty, R., Y. Li, L. Zhou, and K.G. Golic. Corp Regulates P53 in *Drosophila melanogaster* via a Negative Feedback Loop. *PLoS Genet* 11 no. 7 (2015): e1005400.

Chakraborty, S., G. Raut, A. Khopade, K. Mahadik, and C. Kokare. Study on calcium ion independent a-amylase from haloalkaliphilic marine *Streptomyces* strain A3. *Indian J. Biotechnol.* 11 (2012): 427–437.

Chang, X., W. Liu, and X.-H. Zhang. *Spinactinospora alkalitolerans* gen. nov., sp. nov., an actinomycete iso-lated from marine sediment. *Int. J. Syst. Evol. Microbiol.* 61 no. 12 (2011): 2805–2810.

Chang, X., W. Liu, and X.-H. Zhang. *Salinactinospora qingdaonensis* gen. nov., sp. nov., a halophilic actino-mycete isolated from a salt pond. *Int. J. Syst. Evol. Microbiol.* 62 no. 4 (2012): 954–959.

Chater, K.F. *Streptomyces* inside-out: A new perspective on the bacteria that provide us with antibiotics. *Phil. Trans. R. Soc. B* 361 (2006): 761–768

Chen, C.W., C.-H. Huang, H.-H. Lee, H.-H. Tsai, and R. Kirby. Once the circle has been broken: Dynamics and evolution of *Streptomyces* chromosomes. *Trends Genet.* 18 no. 10 (2002): 522–529.

Chen, M., P. Xu, G. Zeng, C. Yang, D. Huang, and J. Zhang. Bioremediation of soils contaminated with poly-cyclic aromatic hydrocarbons, petroleum, pesticides, chlorophenols and heavy metals by composting: Applications, microbes and future research needs. *Biotechnol. Adv.* 33 (2015): 745–755.

Chen, M.Y., S.H. Wu, G.H. Lin, C.P. Lu, Y.T. Lin, W.C. Chang et al. *Rubrobacter taiwanensis* sp. nov., a novel thermophilic, radiation-resistant species isolated from hot springs. *Int. J. Syst. Evol. Microbiol.* 54 (2004): 1849–1855.

Chen, Y.-G., S.-K. Tang, Y.-Q. Zhang, Z.-X. Liu, Q.-H. Chen, J.-W. He, X.-L. Cui, and W.-J. Li. *Zhihengliuella salsuginis* sp. nov., a moderately halophilic actinobacterium from a subterranean brine. *Extremophiles* 14 no. 4 (2010a): 397–402.

Chen, Y.-G., Y.-X. Wang, Y.-Q. Zhang, S.-K. Tang, Z.-X. Liu, H.-D. Xiao, L.-H. Xu, X.-L. Cui, and W.-J. Li. *Nocardiopsis litoralis* sp. nov., a halophilic marine actinomycete isolated from a sea anemone. *Int. J. Syst. Evol. Microbiol.* 59 no. 11 (2009): 2708–2713.

Chen, Y.-G., Y.-Q. Zhang, S.-K. Tang, Z.-X. Liu, L.-H. Xu, L.-X. Zhang, and W.-J. Li. *Nocardiopsis terrae* sp. nov., a halophilic actinomycete isolated from saline soil. *Antonie Van Leeuwenhoek* 98 no. 1 (2010b): 31–38.

Cheng, J., S.H. Yang, S.A. Palaniyandi, J.S. Han, T.M. Yoon, T.J. Kim et al. Azalomycin F complex is an antifungal substance produced by *Streptomyces malaysiensis* MJM1968 isolated from agricultural soil. *J. Korean Soc. Appl. Biol. Chem.* 53 (2010): 545–552.

Choulet, F., A. Gallois, B. Aigle, S. Mangenot, C. Gerbaud, C. Truong et al. Intraspecific variability of the terminal inverted repeats of the linear chromosome of *Streptomyces ambofaciens. J. Bacteriol.* 188 no. 18 (2006): 6599–6610.

Chun, J. Computer assisted classification and identification of actinomycetes. PhD disseration, University of Newcastle upon Tyne, 1995.

Chun, J., K.S. Bae, E.Y. Moon, S.-O. Jung, H.K. Lee, and S.-J. Kim. *Nocardiopsis kunsanensis* sp. nov., a mod-erately halophilic actinomycete isolated from a saltern. *Int. J. Syst. Evol. Microbiol.* 50 no. 5 (2000): 1909–1913.

Cockell, C.S., L.C. Kelly, and V. Marteinsson. Actinobacteria an ancient phylum active in volcanic rock weathering. *Geomicrobiol. J.* 30 (2013): 706–720.

Collins, M.D., P.A. Lawson, M. Labrenz, B.J. Tindall, N. Weiss, and Peter Hirsch. *Nesterenkonia lacusek-hoensis* sp. nov., isolated from hypersaline Ekho Lake, East Antarctica, and emended description of the genus *Nesterenkonia. Int. J. Syst. Evol. Microbiol.* 52 no. 4 (2002): 1145–1150.

Cook, A.E., and P.R. Meyers. Rapid identification of filamentous actinomycetes to the genus level using genus-specific 16S rRNA gene restriction fragment patterns. *Int. J. Syst. Evol. Microbiol.* 53 no. 6 (2003): 1907–1915.

Cowan, S.T. A dictionary of microbial taxonomic usage. Edinburgh Oliver & Boyd, 1968.

Dai, H.-Q., J. Wang, Y.-H. Xin, G. Pei, S.-K. Tang, B. Ren, A. Ward, J.-S. Ruan, W.-J. Li, and L.-X. Zhang. *Verrucosispora sediminis* sp. nov., a cyclodipeptide-producing actinomycete from deep-sea sediment. *Int. J. Syst. Evol. Microbiol.* 60 no. 8 (2010): 1807–1812.

DasSarma, S., and P. Arora. Halophiles. In *Encyclopedia of Life Science.* London: Nature Publishing Group, 2001, pp. 1–9.

Daubin, V., and H. Ochman. Bacterial genomes as new gene homes: The genealogy of ORFans in *E. coli*. *Genome Res.* 14 no. 6 (2004): 1036–1042.

Dharmaraj, S. Marine *Streptomyces* as a novel source of bioactive substances. *World J. Microbiol. Biotechnol.* 26 no.12 (2010): 2123–39.

Dimkpa, C., A. Svatos, D. Merten, G. Buchel, and E. Kothe. Hydroxamate siderophores produced by *Streptomyces acidiscabies* E13 bind nickel and promote growth in cowpea (*Vigna unguiculata* L.) under nickel stress. *Can. J. Microbiol.* 54 no. 3 (2008): 163–172.

Dimkpa, C.O., D. Merten, A. Svatos, G. Büchel, and E. Kothe. Siderophores mediate reduced and increased uptake of cadmium by *Streptomyces tendae* F4 and sunflower (*Helianthus annuus*), respectively. *J. Appl. Microbiol.* 107, no. 5 (2009): 1687–1696.

Duckworth, A.W., S. Grant, W.D. Grant, B.E. Jones, and D. Meijer. *Dietzia natronolimnaios* sp. nov., a new member of the genus *Dietzia* isolated from an East African soda lake. *Extremophiles* 2 no. 3 (1998): 359–366.

Empadinhas, N., V. Mendes, C. Simoes, M.S. Santos, A. Mingote, P. Lamosa, H. Santos, and M.S. da Costa. Organic solutes in *Rubrobacter xylanophilus*: The first example of di-myo-inositol-phosphate in a thermophile. *Extremophiles* 11 no. 5 (2007): 667–673.

Euzeby, J.P. List of prokaryotic names with standing in nomenclature. 2011. http://www.bacterio.cict.fr/.

Euzeby, J.P. List of bacterial names with standing in nomenclature: A folder available on the Internet. 2012. http://www.bacterio.cict.fr/.

Fang, C., J. Zhang, H. Pang, Y. Li, Y. Xin, and Y. Zhang. *Nocardiopsis flavescens* sp. nov., an actinomycete isolated from marine sediment. *Int. J. Syst. Evol. Microbiol.* 61 no. 11 (2011): 2640–2645.

Felsenstein, J. Evolutionary trees from DNA sequences: A maximum likelihood approach. *J. Mol. Evol.* 17 no. 6 (1981): 368–376.

Felsenstein, J. PHYLIP (phylogeny inference package). Version 3. Distributed by the author. Department of Genome Sciences, University of Washington, Seattle, 2005.

Fiedler, H.P., C. Bruntner, A.T. Bull., A.C. Ward, M. Goodfellow, O. Potterat et al. Marine actinomycetes as a source of novel secondary metabolites. *Antonie Van Leeuwenhoek* 87 no. 1 (2005): 37–42

Fitch, W.M., and E. Margoliash. Construction of phylogenetic trees. *Science* 155 no. 3760 (1967): 279–284.

Freeman, S.R., M.H. Poore, T.F. Middleton, and P.R. Ferket. Alternative methods for disposal of spent laying hens: Evaluation of the efficacy of grinding, mechanical deboning, and of keratinase in the rendering process. *Bioresour. Technol.* 100 (2009): 4515–4520.

Friedrich, A.B., I. Fischer, P. Proksch, J. Hacker, and U. Hentschel. Temporal variation of the microbial community associated with the Mediterranean sponge *Aplysina aerophoba*. *FEMS Microbiol. Ecol.* 38 no. 2–3 (2001): 105–113.

Fulthorpe, R.R., A.N. Rhodes, and J.M. Tiedje. High levels of endemicity of 3-chlorobenzoate-degrading soil bacteria. *Appl. Environ. Microbiol.* 64 no. 5 (1998): 1620–1627.

Gabani, P., E. Copeland, A.K. Chandel, and O.V. Singh. Ultraviolet-radiation-resistant isolates revealed cellulose-degrading species of *Cellulosimicrobium cellulans* (UVP1) and *Bacilluspumilus* (UVP4). *Biotechnol. Appl. Biochem.* 59 (2012): 395–404.

Gao, B., and R.S. Gupta. Conserved indels in protein sequences that are characteristic of the phylum Actinobacteria. *Int. J. Syst. Evol. Microbiol.* 55 no. 6 (2005): 2401–2412.

Gao, B., and R.S. Gupta. Phylogenetic framework and molecular signatures for the main clades of the phylum Actinobacteria. *Microbiol. Molecul. Biol. Rev.* 76 no. 1 (2012): 66–112.

Garrity, G.M., J.G. Holt, W.B. Whitman, J. Keswani, D.R. Boone, Y. Koga et al. Phylum All. Euryarchaeota phy. nov. In *Bergey's Manual of Systematic Bacteriology*, ed. D.R. Boone, R.W. Castenholtz, and G.M. Garrity. New York: Springer, 2001, pp. 211–355.

Gayathri, A., P. Madhanraj, and A. Panneerselvam. Diversity, antibacterial activity and molecular characterization of actinomycetes isolated from salt pan region of Kodiakarai, Nagapattinam DT. *Asian J. Pharm. Technol.* 1 (2011): 79–81.

Genilloud, O., I. Gonzalez, O. Salazar, J. Martín, J.R. Tormo, and F. Vicente. Current approaches to exploit actinomycetes as a source of novel natural products. *J. Ind. Microbiol. Biotechnol.* 38 (2011): 375–389.

George, P.S., A. Ahmad, and M.B. Rao. Studies on carboxymethyl cellulase produced by an alkalothermophilic actinomycete. *Bioresour. Technol.* 77 (2001): 171–175.

Gohel, S.D., A.K. Sharma, K.G. Dangar, F.J. Thakrar, and S.P. Singh. Antimicrobial and biocatalytic potential of haloalkaliphilic actinobacteria. *Halophiles* (2015): 29–55.

Gohel S.D., and S.P. Singh. Purification strategies, characteristics and thermodynamic analysis of a highly thermostable alkaline protease from a salt-tolerant alkaliphilic actinomycete, *Nocardiopsis alba* OK-5. *J. Chromatogr. B* 889 (2012a): 61–68.

Gohel S.D., and S.P. Singh. Cloning and expression of alkaline protease genes from two salt-tolerant alkaliphilic actinomycetes in *E. coli*. *Int. J. Biol. Macromol.* 50 no. 3 (2012b): 664–671.

Gohel, S.D., and S.P. Singh. Characteristics and thermodynamics of a thermostable protease from a salt-tolerant alkaliphilic actinomycete. *Int. J. Biol. Macromol.* 56 (2013): 20–27.

Gohel, S.D., and S.P. Singh. Thermodynamics of a Ca^{2+}-dependent highly thermostable alkaline protease from a haloalkliphilic actinomycete. *Int. J. Biol. Macromol.* 72 (2015): 421–429.

Gohel, S.D., and S.P. Singh. Morphological, cultural and molecular diversity of the salt-tolerant alkaliphilic actinomycetes from saline habitats. In *Microbial Biotechnology: Technological Challenges and Developmental Trends*. ed. B. Bhukya and A. Tangutur. Boca Raton, FL: CRC Press/Taylor & Francis Group, 2016, pp. 338–350.

Goodfellow, M., J.E.M. Stach, R. Brown, A. Naga, V. Bonda, A.L. Jones, J. Mexson, H.-P. Fiedler, T.D. Zucchi, and A.T. Bull. *Verrucosispora maris* sp. nov., a novel deep-sea actinomycete isolated from a marine sediment which produces abyssomicins. *Antonie Van Leeuwenhoek* 101 no. 1 (2012): 185–193.

Govender, L., L. Naidoo, and M.E. Setati. *Nesterenkonia suensis* sp. nov., a haloalkaliphilic actinobacterium isolated from a salt pan. *Int. J. Syst. Evol. Microbiol.* 63 no. 1 (2013): 41–46.

Grill, S., S. Busenbender, M. Pfeiffer, U. Kohler, and M. Mack. The bifunctional lavokinase/flavin adenine dinucleotide synthetase from *Streptomyces davawensis* produces inactive flavin cofactors and is not involved in resistance to the antibiotic roseoflavin. *J. Bacteriol.* 190 (2008): 1546–1545.

Groth, I., P. Schumann, F.A. Rainey, K. Martin, B. Schuetze, and K. Augsten. *Demetria terragena* gen. nov., sp. nov., a new genus of actinomycetes isolated from compost soil. *Int. J. Syst. Bacteriol.* 47 no. 4 (1997): 1129–1133.

Guan, T.W., Y. Liu, K. Zhao, Z.F. Xia, X.P. Zhang, and L.-L. Zhang. *Actinopolyspora xinjiangensis* sp. nov., a novel extremely halophilic actinomycete isolated from a salt lake in Xinjiang, China. *Antonie van Leeuwenhoek* 98 no. 4 (2010): 447–453.

Guan, T.W., B. Wei, Y. Zhang, Z.-F. Xia, Z.-M. Che, X.-G. Chen, and L.-L. Zhang. *Actinopolyspora lacussalsi* sp. nov., an extremely halophilic actinomycete isolated from a salt lake. *Int. J. Syst. Evol. Microbiol.* 63 no. 8 (2013a): 3009–3013.

Guan, T.W., N. Wu, S.-K. Tang, Z.-M. Che, and X.-G. Chen. *Myceligenerans salitolerans* sp. nov., a halotolerant actinomycete isolated from a salt lake in Xinjiang, China. *Extremophiles* 17 no. 1 (2013b): 147–152.

Guan, T.W., N. Wu, Z.F. Xia, J.S. Ruan, and X.P. Zhang. *Saccharopolyspora lacisalsi* sp. nov., a novel halophilic actinomycetes isolated from a salt lake in Xinjiang, China. *Extremophiles* 15 (2011a): 373–378.

Guan, T.W., Z.F. Xia, S.-K. Tang, N. Wu, Z.-J. Chen, Y. Huang, J.-S. Ruan, W.-J. Li, and L.-L. Zhang. *Amycolatopsis salitolerans* sp. nov., a filamentous actinomycete isolated from a hypersaline habitat. *Int. J. Syst. Evol. Microbiol.* 62 no. 1 (2012): 23–27.

Guan, T.W., Z.F. Xia, J. Xiao, N. Wu, Z.-J. Chen, L.-L. Zhang, and X.P. Zhang. *Glycomyces halotolerans* sp. nov., a novel actinomycete isolated from a hypersaline habitat in Xinjiang, China. *Antonie van Leeuwenhoek* 100 no. 1 (2011b): 137–143.

Gupta, G.N., S. Srivastava, S.K. Khare, and V. Prakash. Extremophiles: An overview of microorganism from extreme environment. *Int. J. Agric. Environ. Biotechnol.* 7 (2014): 371–380.

Gupta, R.S. The branching order and phylogenetic placement of species from completed bacterial genomes, based on conserved indels found in various proteins. *Int. Microbiol.* 4 no. 4 (2001): 187–202.

Gupta, R.S., and E. Griffiths. Critical issues in bacterial phylogeny. *Theor. Popul. Biol.* 61 no. 4 (2002): 423–434.

Gupta, R.S. Applications of conserved indels for understanding microbial phylogeny. *Mol. Phylogeny Microorganisms* (2010): 135–150.

Habbeche, A., B. Saoudi, B. Jaouadi, S. Haberra, B. Kerouaz, M. Boudelaa et al. Purification and biochemical characterization of a detergent-stable keratinase from a newly thermophilic actinomycete *Actinomadura keratinilytica* strain Cpt29 isolated from poultry compost. *J. Biosci. Bioeng.* 117 (2014): 413–421.

Hakobyan A., H. Panosyan, and A. Trchounian. Production of cellulase by the haloalkalophilic strains of *Streptomyces* isolated from saline-alkaline soils of Ararat Plain, Armenia. *Nat. Sci. Biotechnol.* 2 no. 21 (2013): 44–46.

Hamada, M., C. Shibata, T. Tamura, and K.-I. Suzuki. *Zhihengliuella flava* sp. nov., an actinobacterium isolated from sea sediment, and emended description of the genus *Zhihengliuella*. *Int. J. Syst. Evol. Microbiol.* 63 no. 12 (2013): 4760–4764.

Hamedi, J., F. Mohammadipanah, G. Pötter, C. Spröer, P. Schumann, M. Göker, and H.-P. Klenk. *Nocardiopsis arvandica* sp. nov., isolated from sandy soil. *Int. J. Syst. Evol. Microbiol.* 61 no. 5 (2011): 1189–1194.

Hamedi, J., F. Mohammadipanah, and A. Ventosa. Systematic and biotechnological aspects of halophilic and halotolerant actinomycetes. *Extremophiles* 17 no. 1 (2013): 1–13.

Hamedi, J., F. Mohammadipanah, M. von Jan, G. Pötter, P. Schumann, C. Spröer, H.-P. Klenk, and R.M. Kroppenstedt. *Nocardiopsis sinuspersici* sp. nov., isolated from sandy rhizospheric soil. *Int. J. Syst. Evol. Microbiol.* 60 no. 10 (2010): 2346–2352.

Han, Z., Y. Xu, O. McConnell, L. Liu, Y. Li, S. Qi et al. Two antimycin A analogues from marine-derived actinomycete *Streptomyces lusitanus*. *Mar. Drugs* 10 (2012): 668–676.

Harvery, R.A., and P.C. Champe. *Lippincott's Illustrated Reviews: Pharmacology*. 4th ed. Philadelphia: Lippincott, Williams and Wilkins, 2009.

Hayakawa, Y., S. Shirasaki, T. Kawasaki, Y. Matsuo, K. Adachi, and Y. Shizuri. Structures of new cytotoxic antibiotics, piericidins C7 and C8. *J. Antibiot.* 60 (2007a): 201–203.

Hayakawa, Y., S. Shirasaki, S. Shiba, T. Kawasaki, Y. Matsuo, K. Adachi et al. Piericidins C7 and C8, new cytotoxic antibiotics produced by a marine *Streptomyces* sp. *J. Antibiot.* 60 (2007b): 196–200.

Henckel, T., M. Friedrich, and R. Conrad. Molecular analyses of the methane-oxidizing microbial community in rice field soil by targeting the genes of the 16S rRNA, particulate methane monooxygenase, and methanol dehydrogenase. *Appl. Environ. Microbiol.* 65 no. 5 (1999): 1980–1990.

Horikoshi, K. Alkaliphilies: Genetic properties and application of enzymes. Berlin: Springer, 2006.

Hozzein, W.N., W.-J. Li, M.I.A. Ali, O. Hammouda, A.S. Mousa, L.-H. Xu, and C.-L. Jiang. *Nocardiopsis alkaliphila* sp. nov., a novel alkaliphilic actinomycete isolated from desert soil in Egypt. *Int. J. Syst. Evol. Microbiol.* 54 no. 1 (2004): 247–252.

Ikeda, H., J. Ishikawa, A. Hanamoto, M. Shinose, H. Kikuchi, T. Shiba, Y. Sakaki, M. Hattori, and S. Ōmura. Complete genome sequence and comparative analysis of the industrial microorganism *Streptomyces avermitilis*. *Nat. Biotechnol.* 21 no. 5 (2003): 526–531.

Inahashi, Y., A. Matsumoto, S. Omura, and Y. Takahashi. Spoxazomicins, A–C, novel antitrypanosomal alkaloids produced by anendophytic actinomycete, *Streptosporangium oxazolinicum* K07-0460T. *J. Antibiot.* 64 (2011): 297–302.

Jang, K.H., S.J. Nam, J.B. Locke, C.A. Kauffman, D.S. Beatty, L.A. Paul et al. Anthracimycin, a potent anthrax antibiotic from a marine-derived actinomycete. *Angew. Chem. Int. Ed. Engl.* 52 (2013): 7822–7824.

Jenifer, J.S.C.A., M.B.S. Donio, M. Michaelbabu, S.G.P. Vincent, and T. Citarasu. Haloalkaliphilic *Streptomyces* spp. AJ8 isolated from solar salt works and its pharmacological potential. *AMB Express* 5 no. 1 (2015): 1–12.

Jimenez, E.R. The dextranase along sugar-making industry. *Biotecnología Aplicada* 22 (2005): 20–27.

Jones, B.E., W.D. Grant, A.W. Duckworth, and G.G. Owenson. Microbial diversity of soda lakes. *Extremophiles* 2, no. 3 (1998): 191–200.

Jose, P.A., and S.R.D. Jebakumar. Phylogenetic diversity of actinomycetes cultured from coastal multipond solar saltern in Tuticorin, India. *Technology* 1 no. 2 (2012): 13.

Jose, P.A., and S.R.D. Jebakumar. Phylogenetic appraisal of antagonistic, slow growing actinomycetes isolated from hyper saline inland solar salterns at Sambhar Salt Lake, India. *Front. Microbiol.* 4 (2013a): 190.

Jose, P.A., and S.R.D. Jebakumar. Diverse actinomycetes from Indian coastal solar salterns—A resource for antimicrobial screening. *J. Pure Appl. Microbiol.* 7 (2013b): 2569–2575.

Jose, A., and S.R.D. Jebakumar. Non-streptomycete actinomycetes nourish the current microbial antibiotic drug discovery. *Front. Microbiol.* 4 (2013c): 240.

Jose, P.A., and S.R.D. Jebakumar. Unexplored hyper saline habitats are sources of novel actinomycetes. *Front. Microbiol.* 5 no. 242 (2014): 1.

Jose, P.A., and S.R.D. Jebakumar. Taxonomic and antimicrobial profiles of rare actinomycetes isolated from an inland solar saltern (India). *Indian J. Geomar. Sci.* 44 no. 3 (2015).

Jose, P.A., K.K. Sivakala, and S.R.D. Jebakumar. Formulation and statistical optimization of culture medium for improved production of antimicrobial compound by *Streptomyces* sp. JAJ06. *Int. J. Microbiol.* 2013 (2013): 526260.

Jukes, T.H., and C.R. Cantor. Evolution of protein molecules. In *Mammalian Protein Metabolism*, ed. H.N. Munro. New York: Academic Press, 1969, pp. 21–132.

Kannan, R.R., and S.G. Vincent. Molecular characterization of antagonistic *Streptomyces* isolated from a mangrove swamp. *Asian J. Biotechnol.* 3 no.3 (2011): 237–243.

Kar, J.R., J.E. Hallsworth, and R.S. Singhal. Fermentative production of glycine betaine and trehalose from acid whey using *Actinopolyspora halophila* (MTCC 263). *Environ. Technol. Innov.* 3 (2015): 68–76.

Khazal, M.J. Isolation and molecular identification of some microbial communities fouling concrete infrastructures. PhD thesis, Babylon University, 2013.

Kirby, R. Chromosome diversity and similarity within the Actinomycetales. *FEMS Microbiol. Lett.* 319, no. 1 (2011): 1–10.

Kitani, S., K.T. Miyamoto, S. Takamatsu, E. Herawati, H. Iguchi, K. Nishitomi et al. Avenolide, a *Streptomyces* hormone controlling antibiotic production in *Streptomyces avermitilis*. *Proc. Natl. Acad. Sci. U.S.A.* 108 (2011): 16410–16415.

Klenk, H.-P., M. Lu, S. Lucas, A. Lapidus, A. Copeland, S. Pitluck et al. Genome sequence of the ocean sediment bacterium *Saccharomonospora marina* type strain (XMU15T). *Stand. Genom. Sci.* 6 no. 2 (2012): 265–275.

Klibanov, A.M. Improving enzymes by using them in organic solvents. *Nature.* 409 (2001): 241–246.

Klijn, A., A. Mercenier, and F. Arigoni. Lessons from the genomes of bifidobacteria. *FEMS Microbiol. Rev.* 29 no. 3 (2005): 491–509.

Koch, A.L. Were gram-positive rods the first bacteria? *Trends Microbiol.* 11 no. 4 (2003): 166–170.

Kol, S., M.E. Merlo, R.A. Scheltema, M. de Vries, R.J. Vonk, N.A. Kikkert, L. Dijkhuizen, R. Breitling, and E. Takano. Metabolomic characterization of the salt stress response in *Streptomyces coelicolor*. *Appl. Environ. Microbiol.* 76 no. 8 (2010): 2574–2581.

Kollef, M.H. New antimicrobial agents for methicillin-resistant *Staphylococcus aureus*. *Crit. Care Resusc.* 11 (2009): 282–286.

Kremer, L.C., E.C. van Dalen, M. Offringa, J. Ottenkamp, and P.A. Voute. Anthracycline induced clinical heart failure in a cohort of 607 children: Long-term follow up study. *J. Clin. Oncol.* 19 (2001): 191–196.

Kundu, S., S. Das, N. Mondai, P.S. Lyla, and S. Ajmal Khan. Evaluation of halophilic actinoinycete *Actinopolyspora* sp. for osmolyte production. *Res. J. Microbiol.* 3 no. 1 (2008): 47–50.

Kurtboke, D.I. Biodiscovery from rare actinomycetes: An eco-taxonomical perspective. *Appl. Microbiol. Biotechnol.* 93 no. 5 (2012): 1843–1852.

Kurtboke, D.I., J.R. French, R.A. Hayes, Quinnand, R.J. Quinn. Eco-taxonomic insights into actinomycete symbionts of termites for discovery of novel bioactive compounds. *Adv. Biochem. Eng. Biotechnol.* 147 (2015): 111–135.

Lake, J.A., C.W. Herbold, M.C. Rivera, J.A. Servin, and R.G. Skophammer. Rooting the tree of life using nonubiquitous genes. *Mol. Biol. Evol.* 24 no. 1 (2007): 130–136.

Lakshmipathy, D., and K. Kannabiran. Antibacterial and antifungal activity of *Streptomyces* sp. vitddk3 isolated from Ennore coast, Tamil Nadu, India. *J. Pharm. Res. Health Care* 2 (2010): 188–196.

Lawrence, J.G., and H. Hendrickson. Genome evolution in bacteria: Order beneath chaos. *Curr. Opin. Microbiol.* 8 no. 5 (2005): 572–578.

LeBlanc, J.C., E.R. Gonçalves, and W.W. Mohn. Global response to desiccation stress in the soil actinomycete *Rhodococcus jostii* RHA1. *Appl. Environ. Microbiol.* 74 no. 9 (2008): 2627–2636.

Lee, E.-Y., H.K. Lee, Y.K. Lee, C.J. Sim, and J.-H. Lee. Diversity of symbiotic archaeal communities in marine sponges from Korea. *Biomol. Eng.* 20 no. 4 (2003): 299–304.

Lee, S.D. *Blastococcus jejuensis* sp. nov., an actinomycete from beach sediment, and emended description of the genus *Blastococcus* Ahrens and Moll 1970. *Int. J. Syst. Evol. Microbiol.* 56 no. 10 (2006): 2391–2396.

Lee, S.D. *Jiangella alkaliphila* sp. nov., an actinobacterium isolated from a cave. *Int. J. Syst. Evol. Microbiol.* 58 no. 5 (2008): 1176–1179.

Lemriss, S., F. Laurent, A. Couble, E. Casoli, J.M. Lancelin, D. Saintpierre-Bonaccio et al. Screening of nonpolyenic antifungal metabolites produced by clinical isolates of actinomycetes. *Can. J. Microbiol.* 49 (2003): 669–674.

Li, J., Q. Chen, S. Zhang, H. Huang, J. Yang, X.-P. Tian, and L.-J. Long. Highly heterogeneous bacterial communities associated with the South China Sea reef corals *Porites lutea*, *Galaxea fascicularis* and *Acropora millepora*. *PloS One* 8 no. 8 (2013): e71301.

Li, J., J. Yang, W.-Y. Zhu, J. He, X.-P. Tian, Q. Xie, S. Zhang, and W.-J. Li. *Nocardiopsis coralliicola* sp. nov., isolated from the gorgonian coral, *Menella praelonga*. *Int. J. Syst. Evol. Microbiol.* 62 no. 7 (2012): 1653–1658.

Li, W.-J., H.-H. Chen, Y.-Q. Zhang, P. Schumann, E. Stackebrandt, L.-H. Xu, and C.-L. Jiang. *Nesterenkonia halotolerans* sp. nov. and *Nesterenkonia xinjiangensis* sp. nov., actinobacteria from saline soils in the west of China. *Int. J. Syst. Evol. Microbiol.* 54 no. 3 (2004a): 837–841.

Li, W.-J., D.-J. Park, S.-K. Tang, D. Wang, J.-C. Lee, L.-H. Xu, C.-J. Kim, and C.-L. Jiang. *Nocardiopsis salina* sp. nov., a novel halophilic actinomycete isolated from saline soil in China. *Int. J. Syst. Evol. Microbiol.* 54 no. 5 (2004b): 1805–1809.

Li, W.-J., P. Xu, S.-K. Tang, L.-H. Xu, R.M. Kroppenstedt, E. Stackebrandt, and C.-L. Jiang. *Prauserella halophila* sp. nov. and *Prauserella alba* sp. nov., moderately halophilic actinomycetes from saline soil. *Int. J. Syst. Evol. Microbiol.* 53 no. 5 (2003): 1545–1549.

Liao, Z.-L., S.-K. Tang, L. Guo, Y.-Q. Zhang, X.-P. Tian, C.-L. Jiang, L.-H. Xu, and W.-J. Li. *Verrucosispora lutea* sp. nov., isolated from a mangrove sediment sample. *Int. J. Syst. Evol. Microbiol.* 59 no. 9 (2009): 2269–2273.

Liu, M., L. Zhang, B. Ren, N. Yang, X. Yu, J. Wang et al. *Prauserella shujinwangii* sp. nov., from a desert environment. *Int. J. Syst. Evol. Microbiol.* 64 no. 11 (2014): 3833–3837.

Ludwig, W., J. Euzéby, P. Schumann, H.-J. Busse, M.E. Trujillo, P. Kämpfer, and W.B. Whitman. Road map of the phylum *Actinobacteria*. In *Bergey's Manual of Systematic Bacteriology*. 2012, pp. 1–28.

Ludwig, W., and H.-P. Klenk. Overview: A phylogenetic backbone and taxonomic framework for procaryotic systematics. In *Bergey's Manual of Systematic Bacteriology*, ed. D.R. Boone, R.W. Castenholtz, and G.M. Garrity. New York: Springer, 2001, pp. 49–65.

Lykidis, A., K. Mavromatis, N. Ivanova, I. Anderson, M. Land, G. DiBartolo et al. Genome sequence and analysis of the soil cellulolytic actinomycete *Thermobifida fusca* YX. *J. Bacteriol.* 189 no. 6 (2007): 2477–2486.

Mao, J., J. Wang, H.-Q. Dai, Z.-D. Zhang, Q.-Y. Tang, B. Ren, N. Yang, M. Goodfellow, L.-X. Zhang, and Z.-H. Liu. *Yuhushiella deserti* gen. nov., sp. nov., a new member of the suborder Pseudonocardineae. *Int. J. Syst. Evol. Microbiol.* 61 no. 3 (2011): 621–630.

Margot, H., C. Acebal, E. Toril, R. Amils, and J. Fernandez Puentes. Consistent association of crenarchaeal Archaea with sponges of the genus *Axinella*. *Mar. Biol.* 140 no. 4 (2002): 739–745.

McLeod, M.P., R.L. Warren, W.W.L. Hsiao, N. Araki, M. Myhre, C. Fernandes et al. The complete genome of *Rhodococcus* sp. RHA1 provides insights into a catabolic powerhouse. *Proc. Natl. Acad. Sci. U.S.A.* 103 no. 42 (2006): 15582–15587.

Meklat, A., N. Bouras, S. Mokrane, A. Zitouni, P. Schumann, C. Spröer, H.-P. Klenk, and N. Sabaou. *Bounagaea algeriensis* gen. nov., sp. nov., an extremely halophilic actinobacterium isolated from a Saharan soil of Algeria. *Antonie van Leeuwenhoek* 108 no. 2 (2015): 473–482.

Meklat, A., N. Bouras, A. Zitouni, F. Mathieu, A. Lebrihi, P. Schumann, C. Spröer, H.-P. Klenk, and N. Sabaou. *Actinopolyspora saharensis* sp. nov., a novel halophilic actinomycete isolated from a Saharan soil of Algeria. *Antonie Van Leeuwenhoek* 103 no. 4 (2013): 771–776.

Meklat, A., N. Sabaou, A. Zitouni, F. Mathieu, and A. Lebrihi. Isolation, taxonomy, and antagonistic properties of halophilic actinomycetes in Saharan soils of Algeria. *Appl. Environ. Microbiol.* 77 no. 18 (2011): 6710–6714.

Mikami, Y., K. Miyashita, and T. Arai. Diaminopimelic acid profiles of alkalophilic and alkaline-resistant strains of actinomycetes. *J. Gen. Microbiol.* 128 no. 8 (1982): 1709–1712.

Mincer, T.J., W. Fenical, and P.R. Jensen. Culture-dependent and culture-independent diversity within the obligate marine actinomycete genus *Salinispora*. *Appl. Environ. Microbiol.* 71 no. 11 (2005): 7019–7028.

Mohanraj, G., and T. Sekar. Isolation and screening of actinomycetes from marine sediments for their potential to produce antimicrobials *Int. J. Life Sc. Bt. Pharm. Res.* 2 no. 3 (2013): 115–126.

Montitsche L., H. Driller, and E. Galinski. Ectoine and ectoine derivatives as moisturizers in cosmetics. U.S. Patent 060071, 2000.

Nakano, M., M. Kihara, S. Iehata, R. Tanaka, H. Maeda, and T. Yoshikawa. Wax ester-like compounds as biosurfactants produced by *Dietziamaris* from *n*-alkaneas a sole carbon source. *J. Basic Microbiol.* 51 (2011): 490–498.

Nanjawade, B.K., S. Chandrashekhara, M.S. Ali, S.G. Prakash, and V.M. Fakirappa. Isolation and morphological characterization of antibiotic producing actinomycetes. *Trop. J. Pharm. Res.* 9 (2010): 231–236.

Ochman, H. Genomes on the shrink. *Proc. Natl. Acad. Sci. U.S.A.* 102 no. 34 (2005): 11959–11960.

Oliynyk, M., M. Samborskyy, J.B. Lester, T. Mironenko, N. Scott, S. Dickens, S.F. Haydock, and P.F. Leadlay. Complete genome sequence of the erythromycin-producing bacterium *Saccharopolyspora erythraea* NRRL23338. *Nat. Biotechnol.* 25 no. 4 (2007): 447–453.

Oren, A. Bioenergetic aspects of halophilism. *Microbiol. Mol. Biol. Rev.* 63.2 (1999): 334–348.

Oren, A. Diversity of halophilic microorganisms: Environments, phylogeny, physiology, and applications. *J. Ind. Microbiol. Biotechnol.* 28, no. 1 (2002): 56–63.

Oren, A. Microbial life at high salt concentrations: Phylogenetic and metabolic diversity. *Saline Syst.* 4 no. 2 (2008): 13.

Ouhdouch, Y., M. Barakate, and C. Finance. Actinomycetes of Moroccan habitats: Isolation and screening for antifungal activities. *Eur. J. Soil Biol.* 37 (2001): 69–74.

Palaniyandi, S.A., S.H. Yang, L. Zhang, and J.W. Suh. Effects of actinobacteria on plant disease suppression and growth promotion. *Appl. Microbiol. Biotechnol.* 97 (2013): 9621–9636.

Papke, R.T., N.B. Ramsing, M.M. Bateson, and D.M. Ward. Geographical isolation in hot spring cyanobacteria. *Environ. Microbiol.* 5 no. 8 (2003): 650–659.

Paradkar, A., A. Trefzer, R. Chakraburtty, and D. Stassi. *Streptomyces* genetics: A genomic perspective. *Crit. Rev. Biotechnol.* 23 no. 1 (2003): 1–27.

Park, S.R., Y.J. Yoo, Y.H. Ban, and Y.J. Yoon. Biosynthesis of rapamycin and its regulation: Past achievements and recent progress. *J. Antibiot. (Tokyo)* 63 no. 8 (2010): 434–441.

Pathom-aree, W., J.E.M. Stach, A.C. Ward, K. Horikoshi, A.T. Bull, and M. Goodfellow. Diversity of actinomycetes isolated from Challenger Deep sediment (10,898 m) from the Mariana Trench. *Extremophiles* 10 no. 3 (2006): 181–189.

Paul, S., S.K. Bag, S. Das, E.T. Harvill, and C. Dutta. Molecular signature of hypersaline adaptation: Insights from genome and proteome composition of halophilic prokaryotes. *Genome Biol.* 9 no. 4 (2008): R70.

Peraud, O., J.S. Biggs, R.W. Hughen, A.R. Light, G.P. Concepcion, G.P. Concepcion et al. Microhabitats within venomous cone snails contain diverse Actinobacteria. *Appl. Environ. Microbiol.* 75 (2009): 6820–6826.

Petursdottir, S.K., G.O. Hreggvidsson, M.S. Da Costa, and J.K. Kristjansson. Genetic diversity analysis of *Rhodothermus* reflects geographical origin of the isolates. *Extremophiles* 4 no. 5 (2000): 267–274.

Phillips, K., F. Zaidan, O.R. Elizondo, and K.L. Lowe. Phenotypic characterization and 16S rDNA identification of culturable non-obligate halophilic bacterial communities from a hyper-saline lake, La Saldel Rey, in extreme south Texas (USA). *Aquat. Biosyst.* 8 (2012): 5.

Phillips, R.W., J. Wiegel, C.J. Berry, C. Fliermans, A.D. Peacock, D.C. White et al. *Kineococcus radiotolerans* sp. nov., a radiation resistant, gram-positive bacterium. *Int. J. Syst. Evol. Microbiol.* 52 (2002): 933–938.

Purushe, J., D.E. Fouts, M. Morrison, B.A. White, R.I. Mackie, P.M. Coutinho et al. Comparative genome analysis of *Prevotella ruminicola and Prevotella bryantii*: Insights into their environmental niche. Microb Ecol. 60 (2010): 721–729.

Rademaker, J.L., B. Hoste, F.J. Louws, K. Kersters, J. Swings, L. Vauterin, P. Vauterin, and F.J. de Bruijn. Comparison of AFLP and rep-PCR genomic fingerprinting with DNA-DNA homology studies: *Xanthomonas* as a model system. *Int. J. Syst. Evol. Microbiol.* 50 no. 2 (2000): 665–677.

Radzom, M., A. Zeeck, N. Antal, and H. Fiedler. Fogacin, a novel cyclic octaketide produced by *Streptomyces* strain Tu6319. *J. Antibiot.* 9 (2006) 315–317.

Ravikumar, S., S.J. Inbaneson, M. Uthiraselvam, S.R. Priya, A. Ramu, and M.B. Banerjee. Diversity of endophytic actinomycetes from Karangkadu mangrove ecosystem and its antibacterial potential against bacterial pathogens. *J. Pharm. Res.* 4 (2011): 294–296.

Rheims, H., P. Schumann, M. Rohde, and E. Stackebrandt. *Verrucosispora gifhornensis* gen. nov., sp. nov., a new member of the actinobacterial family Micromonosporaceae. *Int. J. Syst. Bacteriol.* 48 no. 4 (1998): 1119–1127.

Roller, C., W. Ludwig, and K.H. Schleifer. Gram-positive bacteria with a high DNA G + C content are characterized by a common insertion within their 23S rRNA genes. *J. Gen. Microbiol.* 138 no. 6 (1992): 1167–1175.

Sahay, H., S. Mahfooz, A.K. Singh, S. Singh, R. Kaushik, A.K. Saxena, and D.K. Arora. Exploration and characterization of agriculturally and industrially important haloalkaliphilic bacteria from environmental samples of hypersaline Sambhar Lake, India. *World J. Microbiol. Biotechnol.* 28 no. 11 (2012): 3207–3217.

Saadoun, I., B. AL-Joubori, and R. Al-Khoury. Testing of production of inhibitory bioactive compounds by soil Streptomycetes as preliminary screening Programssds in UAE for anti-cancer and anti-bacterial drugs. *Int. J. Curr. Microbiol. App. Sci* 4 no. 3 (2015): 446–459.

Saitou, N., and M. Nei. The neighbor-joining method: A new method for reconstructing phylogenetic trees. *Mol. Biol. Evol.* 4 no. 4 (1987): 406–425.

Saker, R., A. Meklat, N. Bouras, A. Zitouni, F. Mathieu, C. Spröer, H.-P. Klenk, and N. Sabaou. Diversity and antagonistic properties of culturable halophilic actinobacteria in soils of two arid regions of septentrional Sahara: M'zab and Zibans. *Ann. Microbiol.* 65 no. 4 (2015): 2241–2253.

Sambasiva, R.K.R., N.K. Tripathy, Y. Mahalaxmi, and R.S. Prakasham. Laccase and peroxidase free tyrosinase production by isolated microbial strain. *J. Microbiol. Biotechnol.* 22 no. 2 (2012): 207–214.

Santos, H., and M.S. Da Costa. Compatible solutes of organisms that live in hot saline environments. *Environ. Microbiol.* 4 no. 9 (2002): 501–509.

Sathiyaseelan, K., and Stella, D. Isolation, identification and antimicrobial activity of marine actinomycetes isolated from Parangipettai. *Recent Res. Sci. Technol.* 3 (2011): 74–77.

Schippres, A., K. Bosecker, S. Willscher, C. Sproer, P. Schumann, and R.M. Kroppenstedt. *Nocardiopsis metallicus* sp. nov., a metal-leaching actinomycetes isolated from an alkaline slag dump. *Int. J. Syst. Evol. Microbiol.* 52 (2002): 2291–2295.

Schrempf, H. The family Streptomycetaceae, Part II: Molecular biology. In *The Prokaryotes*, ed. M. Dworkin, S. Falkow, E. Rosenberg, K.H. Schleifer, E. Stackebrandt. New York: Springer, 2006, pp. 605–622.

Schrey, S.D., and M.T. Tarkka. Friends and foes: Streptomycetes as modulators of plant disease and symbiosis. *Antonie van Leeuwenhoek* 94 (2008): 11–19.

Sekine, M., S. Tanikawa, S. Omata, M. Saito, T. Fujisawa, N. Tsukatani et al. Sequence analysis of three plasmids harboured in *Rhodococcus erythropolis* strain PR4. *Environ. Microbiol.* 8 no. 2 (2006): 334–346.

Selama, O., G.C.A. Amos, Z. Djenane, C. Borsetto, R.F. Laidi, D. Porter, F. Nateche, E.M.H. Wellington, and H. Hacène. Screening for genes coding for putative antitumor compounds, antimicrobial and enzymatic activities from haloalkalitolerant and haloalkaliphilic bacteria strains of Algerian Sahara soils. *BioMed Res. Int.* 2014 (2014): 317524.

Selvakumar, M., S.K. Jaganathan, G.B. Nando, and S. Chattopadhyay. Synthesis and characterization of novel polycarbonate based polyurethane/polymer wrapped hydroxyapatite nanocomposites: Mechanical properties, osteoconductivity and biocompatibility. *J. Biomed. Nanotechnol.* 11 no. 2 (2015): 291–305.

Sengupta, S., A. Pramanik, A. Ghosh, and M. Bhattacharyya. Antimicrobial activities of actinomycetes isolated from unexplored regions of Sundarbans mangrove ecosystem. *BMC Microbiol.* 15 (2015): 170.

Sethi, S., A. Datta, B.L. Gupta, and S. Gupta. Optimization of cellulose production from bacteria isolated from soil. *ISRN Biotechnol.* (2013): 985685.

Sharma, A.K. Assessment of molecular diversity, phylogeny and biocatalytic potential of salt tolerant alkaliphilic actinomycetes using bioinformatics approaches. M.Phil. thesis, Saurashtra University Rajkot, 2011.

Sharma, A.K., S.D. Gohel, and S.P. Singh. Actinobase: Database on molecular diversity, phylogeny and biocatalytic potential of salt tolerant alkaliphilic actinomycetes. *Bioinformation* 8 no. 11 (2012): 535–538.

Sharma, R., C.L. Sharma, and B. Kapoor. Antibacterial resistance: Current problems and possible solutions. *Indian J. Med. Sci.* 59 (2005): 120–129.

Shendure, J., and A.E. Lieberman. The expanding scope of DNA sequencing. *Nat Biotechnol.* 30 (2012): 1084–1094.

Shivlata, L., and T. Satyanarayana. Thermophilic and alkaliphilic Actinobacteria: Biology and potential applications. *Front. Microbiol.* 6 (2015).

Singh, S.P., V.H. Raval, M.K. Purohit, J.T. Thumar, S.D. Gohel, and S. Pandey. Haloalkaliphilic bacteria and actinobacteria from the saline habitats: New opportunities for biocatalysis and bioremediation. *Microorganisms Environ. Manag.* (2012), 415–429.

Singh, S.P., J.T. Thumar, S. Gohel, B.A. Kikani, R. Shukla, A. Sharma, and K. Dangar. Actinomycetes from marine habitats and their enzymatic potential. In *Marine Enzymes for Bicatalysis.* ed. D. K. Maheshwari, M. Saraf. Woodhead Publishing Series in Biomedicine. Oxford: Woodhead Publishing, 2013, pp. 191–214.

Singh, S.P., J.T. Thumar, S.D. Gohel, and M.K. Purohit. Molecular diversity and enzymatic potential of salt-tolerant alkaliphilic actinomycetes. In *Current Research, Technology and Education Topics in Applied Microbiology and Microbial Biotechnology*, ed. A. Mendez-Vilas. Spain: Formatex, 2010, pp. 280–286.

Song, L., W.-J. Li, Q.-L. Wang, G.-Z. Chen, Y.-S. Zhang, and L.-H. Xu. *Jiangella gansuensis* gen. nov., sp. nov., a novel actinomycete from a desert soil in north-west China. *Int. J. Syst. Evol. Microbiol.* 55 no. 2 (2005): 881–884.

Sorokin, D.Y., S. van Pelt, T.P. Tourova, and L.I. Evtushenko. *Nitriliruptor alkaliphilus* gen. nov., sp. nov., a deep-lineage haloalkaliphilic actinobacterium from soda lakes capable of growth on aliphatic nitriles, and proposal of Nitriliruptoraceae fam. nov. and Nitriliruptorales ord. nov. *Int. J. Syst. Evol. Microbiol.* 59 no. 2 (2009): 248–253.

Sourdis, J., and M. Nei. Relative efficiencies of the maximum parsimony and distance-matrix methods in obtaining the correct phylogenetic tree. *Mol. Biol. Evol.* 5 no. 3 (1988): 298–311.

Steinert, G., M.W. Taylor, and P.J. Schupp. Diversity of actinobacteria associated with the marine ascidian *Eudistoma toealensis. Mar. Biotechnol.* (2015): 1–9.

Subramani, R., and W. Aalbersberg. Marine actinomycetes: An ongoing source of novel bioactive metabolites. *Microbiol. Res* 167 (2012): 571–580.

Suksaard, P., K. Duangmal, R. Srivibool, Q. Xie, K. Hong, and W. Pathom-aree. *Jiangella mangrovi* sp. nov., isolated from mangrove soil. *Int. J. Syst. Evol. Microbiol.* 65 no. 8 (2015): 2569–2573.

Sultanpuram, V.R., T. Mothe, and F. Mohammed. *Salisediminibacterium haloalkalitolerans* sp. nov., isolated from Lonar soda lake, India, and a proposal for reclassification of *Bacillus locisalis* as *Salisediminibacterium locisalis* comb. nov., and the emended description of the genus *Salisediminibacterium* and of the species *Salisediminibacterium halotolerans. Arch. Microbiol.* 197 no. 4 (2015): 553–560.

Sun, C., Y. Wang, Z. Wang, J. Zhou, W. Jin, X. You et al. Chemomicin A, a new angucyclinone antibiotic produced by *Nocardia mediterranei* sub sp. *kanglensis* 1747-64. *J. Antibiot.* 60 (2007): 211–215.

Suthindhiran, K., M.A. Jayasri, D. Dipali, and A. Prasar. Screening and characterization of protease producing actinomycetes from marine saltern. *J. Basic Microbiol.* 54 no. 10 (2014): 1098–1109.

Suthindhiran, K., and K. Kannabiran. Cytotoxic and antimicrobial potential of actinomycete species *Saccharopolyspora salina* VITSDK4 isolated from the Bay of Bengal Coast of India. *Am. J. Infect. Dis.* 5 no. 2 (2009): 90–98.

Swan, B.K., C.J. Ehrhardt, K.M. Reifel, L.I. Moreno, and D.L. Valentine. Archaeal and bacterial communities respond differently to environmental gradients in anoxic sediments of a California hyper saline lake, the Salton Sea. *Appl. Environ. Microbiol.* 76 (2010): 757–768.

Takarada, H., M. Sekine, H. Kosugi, Y. Matsuo, T. Fujisawa, S. Omata et al. Complete genome sequence of the soil actinomycete *Kocuria rhizophila. J. Bacteriol.* 190 no. 12 (2008): 4139–4146.

Tamura, K., G. Stecher, D. Peterson, A. Filipski, and S. Kumar. MEGA6: Molecular evolutionary genetics analysis version 6.0. *Mol. Biol. Evol.* 30 no. 12 (2013): 2725–2729.

Tang, S.-K., Y. Wang, H.-P. Klenk, R. Shi, K. Lou, Y.-J. Zhang, C. Chen, J.-S. Ruan, and W.-J. Li. *Actinopolyspora alba* sp. nov. and *Actinopolyspora erythraea* sp. nov., isolated from a salt field, and reclassification of *Actinopolyspora iraqiensis* as a heterotypic synonym of *Saccharomonospora halophila. Int. J. Syst. Evol. Microbiol.* 61 no. 7 (2011): 1693–1698.

Tang, S.-K., Y. Wang, J.-C. Lee, K. Lou, D.-J. Park, C.-J. Kim, and W.-J. Li. *Georgenia halophila* sp. nov., a halophilic actinobacterium isolated from a salt lake. *Int. J. Syst. Evol. Microbiol.* 60 no. 6 (2010a): 1317–1421.

Tang, S.-K., Y. Wang, H. Zhang, J.-C. Lee, K. Lou, C.-J. Kim, and W.-J. Li. *Haloechinothrix alba* gen. nov., sp. nov., a halophilic, filamentous actinomycete of the suborder Pseudonocardineae. *Int. J. Syst. Evol. Microbiol.* 60 no. 9 (2010b): 2154–2158.

Tatineni, R., K.K. Doddapaneni, R.C. Potumarthi, R.N. Vellanki, M.T. Kandathil, N. Kolli, and L.N. Mangamoori. Purification and characterization of an alkaline keratinase from *Streptomyces* sp. *Bioresour. Technol.* 99 (2008): 1596–1602.

Thompson, J.D., D.G. Higgins, and T.J. Gibson. CLUSTAL W: Improving the sensitivity of progressive multiple sequence alignment through sequence weighting, position-specific gap penalties and weight matrix choice. *Nucleic Acids Res.* 22, no. 22 (1994): 4673–4680.

Thumar, J.T., and S.P. Singh. Organic solvent tolerance of an alkaline protease from salt- tolerant alkaliphilic *Streptomyces clavuligerus* strain Mit-1. *J. Ind. Microbiol. Biotechnol.* 36 (2009): 211–218.

Thumar, J.T., and S.P. Singh. Two step purification of highly thermo stable alkaline protease from salt tolerant alkaliphilic *Streptomyces clavuligerus* strain Mit-1. *J. Chromatogr. B* 854 (2007): 198–203.

Tian, S., Y. Yang, K. Liu, Z. Xiong, L. Xu, and L. Zhao. Antimicrobial metabolites from a novel halophilic actinomycete *Nocardiopsis terrae* YIM 90022. *Nat. Prod. Res.* 28 (2014): 344–346.

Tian, S.Z., X. Pu, G. Luo, L.X. Zhao, L.H. Xu, W.J. Li et al. Isolation and characterization of new *p*-terphenyls with antifungal, antibacterial, and antioxidant activities from halophilic actinomycete *Nocardiopsis gilva* YIM 90087. *J. Agric. Food Chem.* 61 (2013): 3006–3012.

Tian, X.P., L.J. Long, S.M. Li, J. Zhang, Y. Xu, J. He et al. *Pseudonocardia antitumoralis* sp. nov., a new deoxynyboquinone-producing actinomycetes isolated from a deep-sea sedimental sample in South China Sea. *Int. J. Syst. Evol. Microbiol.* 63 (2013): 893–899.

Tian, X.-P., Y. Xu, J. Zhang, J. Li, Z. Chen, C.-J. Kim, W.-J. Li, C.-S. Zhang, and S. Zhang. *Streptomyces oceani* sp. nov., a new obligate marine actinomycete isolated from a deep-sea sample of seep authigenic carbonate nodule in South China Sea. *Antonie Van Leeuwenhoek* 102 no. 2 (2012): 335–343.

Tresner, H.D., J.A. Hayes, and E.J. Backus. Differential tolerance of streptomycetes to sodium chloride as a taxonomic aid. *Appl. Microbiol.* 16, no. 8 (1968): 1134–1136.

Tseng, M., K. Hoang, M. Yang, S. Yang, and W.S. Chu. Polyester degrading thermophilic actinomycetes isolated from different environment in Taiwan. *Biodegradation* 18 (2007): 579–583.

Udwary, D.W., L. Zeigler, R.N. Asolkar, V. Singan, A. Lapidus, W. Fenical, P.R. Jensen, and B.S. Moore. Genome sequencing reveals complex secondary metabolome in the marine actinomycete *Salinispora tropica. Proc. Natl. Acad. Sci. U.S.A.* 104 no. 25 (2007): 10376–10381.

Urzì, C., P. Salamone, P. Schumann, M. Rohde, and E. Stackebrandt. *Blastococcus saxobsidens* sp. nov., and emended descriptions of the genus *Blastococcus* Ahrens and Moll 1970 and *Blastococcus aggregatus* Ahrens and Moll 1970. *Int. J. Syst. Evol. Microbiol.* 54 no. 1 (2004): 253–259.

Valas, R.E., and P.E. Bourne. The origin of a derived superkingdom: How a gram-positive bacterium crossed the desert to become an archaeon. *Biol Direct* 6 no. 16 (2011): 494–504.

Vandamme, P., B. Pot, E. Falsen, K. Kersters, and L.A. Devriese. Taxonomic study of lancefield streptococcal groups C, G, and L (*Streptococcus dysgalactiae*) and proposal of *S. dysgalactiae* subsp. *equisimilis* subsp. nov. *Int. J. Syst Bacteriol.* 46 no. 3 (1996): 774–781.

Ventosa, A., J.J. Nieto, and A. Oren. Biology of moderately halophilic aerobic bacteria. *Microbiol. Mol. Biol. Rev.* 62 no. 2 (1998): 504–544.

Ventura, M., C. Canchaya, A. Tauch, G. Chandra, G.F. Fitzgerald, K.F. Chater, and D. van Sinderen. Genomics of Actinobacteria: Tracing the evolutionary history of an ancient phylum. *Microbiol. Mol. Biol. Rev.* 71 no. 3 (2007): 495–548.

Wang, H.-F., Y.-G. Zhang, J.-Y. Chen, W.N. Hozzein, L. Li, M.A.M. Wadaan, Y.-M. Zhang, and W.-J. Li. *Nesterenkonia rhizosphaerae* sp. nov., an alkaliphilic actinobacterium isolated from rhizosphere soil in a saline-alkaline desert. *Int. J. Syst. Evol. Microbiol.* 64 no. 12 (2014): 4021–4026.

Wang, J., Y. Li, J. Bian, S.-K. Tang, B. Ren, M. Chen, W.-J. Li, and L.-X. Zhang. *Prauserella marina* sp. nov., isolated from ocean sediment of the South China Sea. *Int. J. Syst. Evol. Microbiol.* 60 no. 4 (2010): 985–989.

Wang, X.-J., Y.-J. Yan, B. Zhang, J. An, J.-J. Wang, J. Tian et al. Genome sequence of the milbemycin-producing bacterium *Streptomyces bingchenggensis. J. Bacteriol.* 192 no. 17 (2010): 4526–4527.

Wang, Y., S.-K. Tang, Z. Li, K. Lou, P.-H. Mao, X. Jin, H.-P. Klenk, L.-X. Zhang, and W.-J. Li. *Myceligenerans halotolerans* sp. nov., an actinomycete isolated from a salt lake, and emended description of the genus *Myceligenerans. Int. J. Syst. Evol. Microbiol.* 61 no. 4 (2011): 974–978.

Webb, V.L., and E.W. Maas. Sequence analysis of 16S rRNA gene of cyanobacteria associated with the marine sponge *Mycale (Carmia) hentscheli. FEMS Microbiol. Lett.* 207 no. 1 (2002): 43–47.

Webster, N.S., L.D. Smith, A.J. Heyward, J.E.M. Watts, R.I. Webb, L.L. Blackall, and A.P. Negri. Metamorphosis of a scleractinian coral in response to microbial biofilms. *Appl. Environ. Microbiol.* 70 no. 2 (2004): 1213–1221.

Welsh, D.T. Ecological significance of compatible solute accumulation by micro-organisms: From single cells to global climate. *FEMS Microbiol. Rev.* 24 no. 3 (2000): 263–290.

Whatmore, A.M., and R.H. Reed. Determination of turgor pressure in *Bacillus subtilis*: A possible role for K+ in turgor regulation. *Microbiology* 136 no. 12 (1990): 2521–2526.

Whitaker, R.J., D.W. Grogan, and J.W. Taylor. Geographic barriers isolate endemic populations of hyperthermophilic archaea. *Science* 301 no. 5635 (2003): 976–978.

Wu, H., S. Qu, C. Lu, H. Zheng, X. Zhou, L. Bai, and Z. Deng. Genomic and transcriptomic insights into the thermo-regulated biosynthesis of validamycin in *Streptomyces hygroscopicus* 5008. *BMC Genomics* 13 (2012): 337.

Xin, Y., J. Huang, M. Deng, and W. Zhang. Culture-independent nested PCR method reveals high diversity of actinobacteria associated with the marine sponges *Hymeniacidon perleve* and *Sponge* sp. *Antonie Van Leeuwenhoek* 94 no. 4 (2008): 533–542.

Xu, L.Y., X.S. Quan, C. Wang, H.F. Sheng, G.X. Zhou, B.R. Lin et al. Antimycins A(19) and A(20), two new antimycins produced by marine actinomycete *Streptomyces antibioticus* H74–18. *J. Antibiot.* 64 (2011): 661–665.

Xu, S., L. Yan, X. Zhang, C. Wang, G. Feng, and J. Li. *Nocardiopsis fildesensis* sp. nov., an actinomycete isolated from soil. *Int. J. Syst. Evol. Microbiol.* 64 no. 1 (2014): 174–179.

Xu, Y., J. He, X.-P. Tian, J. Li, L.-L. Yang, Q. Xie, S.-K. Tang, Y.-G. Chen, S. Zhang, and W.-J. Li. *Streptomyces glycovorans* sp. nov., *Streptomyces xishensis* sp. nov. and *Streptomyces abyssalis* sp. nov., isolated from marine sediments. *Int. J. Syst. Evol. Microbiol.* 62 no. 10 (2012): 2371–2377.

Yukawa, H., C.A. Omumasaba, H. Nonaka, P. Kós, N. Okai, N. Suzuki et al. Comparative analysis of the *Corynebacterium glutamicum* group and complete genome sequence of strain R. *Microbiology* 153 no. 4 (2007): 1042–1058.

Yun, J.-H., S.W. Roh, M.-J. Jung, M.-S. Kim, E.-J. Park, K.-S. Shin, Y.-D. Nam, and J.-W. Bae. *Kocuria salsicia* sp. nov., isolated from salt-fermented seafood. *Int. J. Syst. Evol. Microbiol.* 61 no. 2 (2011): 286–289.

Zhang, G., Y. Zhang, X. Yin, and S. Wang. *Nesterenkonia alkaliphila* sp. nov., an alkaliphilic, halotolerant actinobacteria isolated from the western Pacific Ocean. *Int. J. Syst. Evol. Microbiol.* 65 no. 2 (2015): 516–521.

Zhang, H., Y.K. Lee, W. Zhang, and H.K. Lee. Culturable actinobacteria from the marine sponge *Hymeniacidon perleve*: Isolation and phylogenetic diversity by 16S rRNA gene-RFLP analysis. *Antonie van Leeuwenhoek* 90 no. 2 (2006): 159–169.

Zhang, H., W. Zhang, Y. Jin, M. Jin, and X. Yu. A comparative study on the phylogenetic diversity of culturable actinobacteria isolated from five marine sponge species. *Antonie Van Leeuwenhoek* 93 no. 3 (2008): 241–248.

Zhang, W., H. Zhu, M. Yuan, Q. Yao, R. Tang, M. Lin et al. *Microbacterium radiodurans* sp. nov., a UV radiation-resistant bacterium isolated from soil. *Int. J. Syst. Evol. Microbiol.* 60 (2010): 2665–2670.

Zhang, Y., Y. Li, Y. Zhang, Z. Wang, M. Zhao, N. Su et al. Quantitative proteomics reveals membrane protein-mediated hypersaline sensitivity and adaptation in halophilic *Nocardiopsis xinjiangensis*. *J. Proteome Res.* 4 (2015): 68–85.

Zhang, Y.-Q., P. Schumann, L.-Y. Yu, H.-Y. Liu, Y.-Q. Zhang, L.-H. Xu, E. Stackebrandt, C.-L. Jiang, and W.-J. Li. *Zhihengliuella halotolerans* gen. nov., sp. nov., a novel member of the family Micrococcaceae. *Int. J. Syst. Evol. Microbiol.* 57 no. 5 (2007): 1018–1023.

Zhao, L.X., S.X. Huang, S.K. Tang, C.L. Jiang, Y. Duan, J.A. Beutler et al. Actinopolysporins A–C and tubercidin as a Pdcd4 stabilizer from the halophilic actinomycete *Actinopolyspora erythraea* YIM 90600. *Nat. Prod.* 74 (2011): 1990–1995.

Zhi, X.-Y., W.-J. Li, and E. Stackebrandt. An update of the structure and 16S rRNA gene sequence-based definition of higher ranks of the class Actinobacteria, with the proposal of two new suborders and four new families and emended descriptions of the existing higher taxa. *Int. J. Syst. Evol. Microbiol.* 59 no. 3 (2009): 589–608.

Zhu, W.-Y., J.-L. Zhang, Y.-L. Qin, Z.-J. Xiong, D.-F. Zhang, H.-P. Klenk, L.-X. Zhao, L.-H. Xu, and W.-J. Li. *Blastococcusendophyticus* sp. nov., an actinobacterium isolated from *Camptotheca acuminata*. *Int. J. Syst. Evol. Microbiol.* 63 no. 9 (2013): 3269–3273.

CHAPTER 7

Adaptation Strategies in Halophilic Bacteria

Vikram H. Raval, Hitarth B. Bhatt, and Satya P. Singh

CONTENTS

7.1 INTRODUCTION

Microorganisms are exposed to diverse environmental conditions of habitats. In order to survive, microorganisms sense environmental changes and react with various adaptive mechanisms (Harrison et al. 2013). However, microorganisms are found over a range of environmental conditions because of their ability to survive and proliferate under extreme conditions. These microorganisms are referred to as extremophiles (Cavicchioli et al. 2011). In the last two decades, scientists have shown interest in exploring halophilic microorganisms for their biotechnological potential. Halophilic and halotolerant bacteria are a major class of extremophilic organisms that live in habitats of high ionic strength, such as marine water, salt lakes, brines, saline deserts, salterns, saline soils, and salted foods (Ma et al. 2010; Singh et al. 2010; Raval et al. 2013; Purohit et al. 2014; Bhatt and Singh 2016).

The most widely used definitions for halophiles were formulated by Donn Kushner (1978), who distinguished different categories: extreme halophiles (growing best at 2.5–5.2 M salt), borderline extreme halophiles (growing best at 1.5–4.0 M salt), moderate halophiles (growing best at 0.5–2.5 M salt), and halotolerants, which do not have an absolute requirement of salt for growth but are able to grow well up to high salt concentrations. Similarly, classification of microorganisms in relation to their salt requirement is also suggested (Ventosa et al. 1998; Oren 2002, 2006a). However, there are no sharp boundaries regarding the minimum salt requirement for growth, maximum salt tolerance limit, or optimum concentration of salinity for growth.

Extreme halophilic bacteria exhibit physiology, cell machinery, and molecular systems adapted to a high salinity. They can easily follow the "salt in" strategy by accumulating salt inside the cell. On the contrary, the moderate halophilic bacteria and halotolerant bacteria grow well in lower salt concentrations and the absence of salt, respectively. Moderately halophilic bacteria and halotolerant bacteria frequently face salinity fluctuation in their environment due to rains and drought. To encounter such variations in salinity, they have evolved various physiological strategies (Tsuzuki et al. 2011). A number of physiological changes include an increased uptake of potassium and the accumulation of compatible solutes, and antioxidants in response to high salt concentrations. A low salinity leads to an immediate influx of water, which a cell counteracts by fast efflux of small solutes, thus relieving physical stress. While under high salinity, water efflux is counterbalanced by an increase of compatible solutes, such as proline, glutamate, glycine betaine, ectoine, and trehalose (Krämer 2010). The cellular adaptation and osmotic adjustment to stress are crucial factors in the growth and survival of bacteria under stress conditions. Stress responses are also characterized by measuring the expression level of the specific proteins involved in stress adaptation (Paul 2013).

Adaptation at the molecular level is brought through alterations in protein sequences to sustain them in the extreme conditions of the environment (Tekaia et al. 2002; Brocchieri 2004; Tekaia and Yeramian 2006; Nath et al. 2012; Nath and Subbiah 2014). Analysis of the protein sequences responsible for the stability and functioning of the halophilic proteins can lead to our understanding of adaptation and help in protein engineering (Madigan and Marrs 1997). Studies have indicated the importance of amino acid composition and dipeptide composition, and physicochemical properties of the amino acid can distinguish the halophilic proteins with reference to their nonhalophilic counterparts (Ebrahimie et al. 2011; Purohit and Singh 2011; Smole et al. 2011; Zhang et al. 2012; Zhang and Ge 2013; Zhang and Yi 2013; Raval et al. 2014). Protein sequence properties can be explored to a greater extent with respect to sequence size, molecular mass, isoelectric point, number and frequency of each amino acid, nature of amino acids, hydrophilic and hydrophobic residues, number and frequency of dipeptides, number of α-helices and β-strands, and other secondary protein features. Enzymes produced by the halophilic bacteria are important for their industrial applications due to their stability and functionality under adverse extreme conditions where ordinary enzymes may aggregate and lose activity (Patel et al. 2006; Dodia et al. 2008; Purohit and Singh 2011; Gohel and Singh 2012; Pandey and Singh 2012; Raval et al. 2014, 2015b; DasSarma and DasSarma 2015; L. Yin et al. 2015; Bhatt and Singh 2016).

The halophilic bacteria that follow the "salt out" strategy accumulate high intracellular concentrations of the compatible solutes, which play a significant role in haloadaptation, besides other important biotechnological applications, such as protectants of macromolecules, cells, tissues, and even organs (Da Costa et al. 1998; Welsh 2000). The term *compatible solutes* itself supports the cell architecture without any detrimental effect (Brown 1976; Da Costa et al. 1998). Osmoregulation mechanisms have been reported in extreme halophiles, such as *Halobacterium salinarium* (Coker et al. 2007; Leuko et al. 2009) and *Haloferax volcanii* (Bidle et al. 2008), as well as moderate halophiles, *Halobacillus dabanensis* (Feng et al. 2006; Gu et al. 2008), *Chromobacter salexigens* (Oren et al. 2005), and *Halomonas elongata* (Cánovas et al. 1996, 2000). However, very little is known about the proteins involved in osmoregulation of halotolerant microorganisms. However, there are some studies on the molecular mechanisms adapted by these microorganisms, such as *Escherichia coli* (Lamark et al. 1991), *Bacillus subtilis* (Boch et al. 1996), *Staphylococcus xylosus* (Rosenstein et al. 1999), *Thalassobaculum* (Zhang et al. 2008), *Thalassospira* (López-López et al. 2002), and *Marispirillum* (Lai et al. 2009). Over the years, researchers have come to find halotolerant bacteria fascinating because of their robust nature and application prospects over a wide range of salinity (0%–25%).

In view of the impact of halophiles, this chapter focuses on the strategies adopted by halophilic bacteria and halophilic archaea at the physiological, biochemical, and molecular levels.

7.2 CELL ARCHITECTURE AND ADAPTATIONS

The cellular architecture of the halophilic bacteria in the context of adaptation requires due attention compared with that of their archaeal counterparts. The cellular structure of the halophiles seems to be ill-attended, as only limited reports on the cell organelles and subcellular structures are available (Brown 1960a, 1960b). Biocatalytic features of the halophilic bacteria and archaea are well documented, while the cell wall and other features remain less explored (Oren 2002). Cell envelope, cytosolic and photosynthetic membranes, gas vesicles, capsules, flagella, and endospores are among the cellular structures reported for the halophilic bacteria and their tolerance against various stresses (Oren 2002).

7.2.1 Cell Wall of the Halophilic Bacteria

In 1988, Hart and Vreeland (1988) observed characteristic changes in the cell envelope of *H. elongata* with respect to its hydrophobicity when grown in definite concentrations of NaCl. The hydrophilicity of the cellular surface makes it more selective to water in drought and hypersaline environments. The surface hydrophobicity rises as the cell approaches the stationary phase, which is linked to the structure of the cell wall and cytoplasmic membrane. *Halomonas* possesses leucine, a typical hydrophobic amino acid, in the peptidoglycan, and it adds to the overall hydrophobic nature of the surface (Vreeland et al. 1984). The cell wall of a purple nonsulfur photosynthetic bacterium, *Rhodothalassium salexigens*, has an S-layer with any lipopolysaccharide (LPS) in the cell wall (Tadros et al. 1982). A halophilic actinomycetes, *Actinopolyspora halophila*, possesses a typical peptidoglycan cell wall lacking mycolic acid, resembling chemotype IV actinomycetes (Kates et al. 1987). *Dichotomicrobium thermohalophilum*, isolated from Solar Lake in Sinai, Egypt, produces up to four hyphae from the tetrahedral mother cell, often giving chain or net-like arrangements (Hirsch and Hoffmann 1989). Cells of the halophilic bacteria *Micrococcus halodenitrificans*, *Vibrio costicolus*, and *Pseudomonas salinaria* have a relatively high nitrogen content, indicating that the cell material is predominantly proteinaneous in nature. They contain only small amounts of fat and no detectable amount of carbohydrate other than pentose sugar in nucleic acids. Cell walls are made up of lipoprotein, although certain reports have described the composition as different from that of nonhalophiles, particularly with respect to their nitrogen and carbohydrate contents (Smithies et al. 1955).

7.2.2 Cytosolic and Photosynthetic Membranes and Membrane-Linked Processes

The function and characteristics of the cytosolic membrane are largely regulated by the external solute potential. In halophiles, the external salt concentration regulates the function of the cytoplasmic membrane and integral membrane proteins define the membrane permeability (Russell 1993). Adaptation of halophiles toward the surrounding environment is a response to stress induced by salinity and osmotic shock (Russell and Kogut 1985).

The increasing salinity enhances the amount of the negatively charged polar lipids, such as phosphatidylcholine, cardiolipin, and glycolipids, in place of neutral phosphatidylethanolamine in many halophilic bacteria (Vreeland et al. 1984; Vreeland 1987; Russell 1993). Such changes in lipid composition regulate permeability of the cytosolic membrane for cations and help to maintain the lipid bilayer structure (Ohno et al. 1979). Salt-sensitive mutants contained polar lipids similar to wild-type strains when grown with low salt (Kogut et al. 1992). Temperature along with salt appears to be a determining factor for the lipid profile (Adams and Russell 1992).

Other important aspects that affect the membrane permeability are the presence of straight-chain saturated and monounsaturated fatty acids. Cyclopropane fatty acids are synthesized by deriving methyl moiety from S-adenosylmethionine across the double bond by monounsaturated fatty acids. An increase in unsaturated fatty acids accompanies the increasing membrane fluidity; however, the effect of cyclopropane fatty acids on membrane fluidity is still not clear. Thus, cyclopropane fatty acids juggle between the saturated and unsaturated fatty acids from which they derived. Thus, rising cyclopropane fatty acids leads to a steep reduction in membrane fluidity (Russell 1989, 1993).

It is well known that bacteria and eukaryotes contain phospholipids that have fatty acids linked to glycerol via an ester bond, while archaeal phospholipids with branched isoprene units are linked to glycerol by an ether linkage. The ether linkages render archaeal membranes less permeable to ions, and thus provide natural resistance against high salinity, while bacteria have to evolve definite strategies to avoid excessive water loss. In halophiles, there are two distinct strategies:

1. *High-salt-in strategy*: Halophiles, in order to increase the cytosolic osmolytic activity, accumulate potassium chloride (KCl) up to molar concentrations, hence called high-salt-in strategy (Lanyi 1974). Most enzymes and other proteins of the Halobacteriales get denatured in solutions of less than 1–2 M salt. Many enzymes are more active in the presence of KCl than of NaCl, agreeing well with the findings that K^+ is the intracellularly dominating cation. "Salting-out" salts stabilize, while "salting-in" salts inactivate halophilic enzymes (Oren 2006b). The mesophilic bacteria would undergo desiccation and lose water through osmosis. Halophiles prevent the loss of water by increasing the internal osmolarity of the cell. For accumulation and high influx of K^+ ions in comparison with compatible solute synthesis, the cell spends less ATP (two ATPs for every three molecules of KCl). It is a more radical adaptation for preventing cytosolic water loss and employing selective influx of potassium (K^+) ions into the cytoplasm. To use this method, the entire intracellular machinery—including enzymes, structural proteins, and charged amino acids that allow the retention of water molecules on their surfaces—must be adapted to high salt levels.

2. *Osmolyte synthesis and accumulation (low-salt-in) strategy*: This is the accumulation and synthesis of various organic compounds called compatible solutes. Organic compounds that are osmotically active, low molecular weight, and highly water soluble are either taken from surroundings by specific transport mechanisms or synthesized by the organism as a strategy to prevent protein denaturation by excessive salt (salting-out of proteins) in the vicinity of the cell. Substances like glycine betaine, ectoine, proline, and other amino acid derivatives; N-acetylated diamino acids; and sugars and sugar alcohols are examples of compatible solutes. The strategy is also called *low-salt-in,*

high-organic-solute-in (Cho 2005). In compatible solute-mediated adaptation, only limited adjustment is required of intracellular macromolecules—in fact, the compatible solutes often act as general stress protectants as well as osmoprotectants. Energetically, this is an expensive process. Generally, an autotroph spends ATP in the range of 30–90 molecules for the biosynthesis of a single molecule of the compatible solutes, while in heterotrophs 23–79 ATPs is required for the same reaction.

Both the above strategies are widely adopted by halophiles to survive under salinity and water stress. The organisms accumulating osmolytes tend to tolerate a range of salt concentrations in comparison with those with salt-in strategy (Ventosa et al. 1998). Compatible solutes display a stabilizing effect by preventing the unfolding and denaturation of proteins caused by heating, freezing, and drying (Galinski 1993, 1995). Most compatible solutes are based on amino acids and their derivatives, sugars and sugar alcohols. The majority of the compatible solutes are either uncharged or zwitterionic (Galinski 1985, 1986; Roberts 2005, 2006). The halophilic bacteria are categorized according to their salt and pH tolerance. However, the most widely accepted classification has been proposed by Donn J. Kushner in his editorial articles in *Microbial Life in Extreme Environments* (Kushner 1978, 1985). Based on the literature including the tolerance toward salt and alkalinity, various adaptation strategies, and ecological habitats, the halophiles have been categorized under the following types (Table 7.1).

Certain groups of the photosynthetic microbes, such as cyanobacteria, including a few halophiles, possess thylacoids that harbor elaborated photosynthetic machinery. Variations do exist among the photosynthetic membranes of photosynthetic halophilic anoxygenic bacteria; for example, *Halochromatium*, *Rhodovibrio*, and *Thiohalocapsa* have a vesicular type of photosynthetic membrane, while thylacoid stacks are reported from moderate halophiles *Ectothiorhodospira* and true halophiles *Halorhodospira* (Imhoff and Trüper 1981; Ventosa 1988; Oren et al. 1989; Imhoff and Thiemann 1991).

Table 7.1 Representation of Halophiles, Adaptation Strategies, and Their Ecological Niches

Types of Organisms	Salt Concentrations	Adaptation Strategies	Ecological Habitats
Nonhalophiles	0–0.1 M salt (<0.1 M salt but rarely to 2.5 M)	Osmolyte synthesis	Nonsaline soil, salt waters, slightly saline and alkaline soils in rhizospheric regions
Halotolerant	0.1–0.2 M salt	Osmolyte synthesis	Saline soil, coastal soil and waters
Slightly halophilic	0.2–0.5 M salt	Osmolyte synthesis	Saline soil, coastal soil and waters
Moderate halophiles	0.5–2.5 M salt	Osmolyte synthesis	Saline soil, salterns, marine waters
Borderline extreme halophiles	1.5–4.0 M salt	High salt-in, osmolyte synthesis, acidic amino acids, capsule synthesis, endospore formation, high G + C ratio	Salt lakes, brines, solar salterns, soda lakes
True halophiles	2.5–5.2 M salt	High salt-in, osmolyte synthesis, acidic amino acids, capsule synthesis, endospore formation, high G + C ratio	Dead Sea, salt lakes, brines, solar salterns, soda lakes

7.2.3 Other Structures

7.2.3.1 Gas Vesicles

The halophilic archaea, such as *Haloquadratum walsbyi* (Saponetti et al. 2011), possess gas vesicles, while bacteria rarely contain these structures. Halophilic obligatory anaerobic *Clostridium lortetii* produces gas vesicles attached to endospores at 1–2 M NaCl concentraton in the growth medium (Oren 1983). The cynobacterial halophiles produce gas vesicles but rarely in growing culture, while members of the anaerobic Halobacteriodaceae possess gas vesicles in growing cells.

7.2.3.2 Capsules, S-Layer, EPSs, and Biopolymers

There are certain bacteria that excrete carbohydrate-rich polymeric substances onto the cell wall. These substances are called exopolysaccharides (EPSs) and are observed in the form of either capsules or a slime layer on the cell surface (Sutherland 1990, 2001a, 2001b; Freitas et al. 2011). These polymers are composed of neutral glucans, with additional organic and inorganic substituents imparting charge, that is, uronic acid, pyruvate ketals, and phosphate and sulfate groups. Capsule formation has been reported in halophilic and halotolerant species of *Achromobacter* spp. (Thomson et al. 1972), *Vibrio vulnificus* (Amako et al. 1984), a novel genus *Rhodothermus* (Alfredsson et al. 1988), an obligate halophilic bacterium *Pasteurella piscicida* (Magarinos et al. 1996), and various species of *Halomonas*, such as *H. eurihalina* (Quesada et al. 1993), *H. maura* (Bouchotroch et al. 2001), *H. ventosae* (Martinez-Canovas et al. 2004c), *H. anticariensis* (Martinez-Canovas et al. 2004b), and *Halomonas* spp. NY-011 (Wang et al. 2010). A detailed account of EPSs and biopolymers and their industrial applications is given in Section 7.4.

7.2.3.3 Endospores

Endospores are the structure that helps the cell to survive in various environmental stresses, such as heat, desiccation, and radiation. Endospores have the property to remain dormant for a long time and then become an active vegetative stage under favorable conditions (Cano and Borucki 1995). Spore-forming bacteria have been reported from diverse extreme environments around the world, such as deep sea, high-altitude atmosphere, deserts, and salt lakes (Bae et al. 2005; Shivaji et al. 2006; Hua et al. 2007, 2008). Endospores are composed of several layers of core, cortex, and coat, the majority of which are rich in calcium dipicolinate and modified peptidoglycan.

The halophilic spore-producing bacillus was first reported in 1935 by Hof (solar saltern) and later in 1940 by Elazari-Volcani (Dead Sea) from sources other than cured fish and foods. Bacteria represent halophilic groups: *Clostridium halophilum* (Fendrich et al. 1990), *Desulfotomaculum halophilum* (Tardy-Jacquenod et al. 1998), Halanaerobiaceae (*Natroniella acetigena*) (Zhilina et al. 1996), *Sporohalobacter lortetii* (Oren 1983), *Orenia* spp. (Oren et al. 1987; Zhilina et al. 1999; Moune et al. 2000), and *Halonatronum saccharophilum* (Zhilina et al. 2001). Endospore formation and adaptation toward environmental stress are characteristic features of various genera of halophilic and alkaliphilic bacteria isolated from saline habitats, that is, the genus *Bacillus* and its species (Boyer et al. 1973; Nielsen et al. 1995; Nowlan et al. 2006), *Gracibacillus*, *Halobacillus*, and *Salibacillus* (Spring et al. 1996; Garabito et al. 1997; Arahal et al. 1999; Chaiyanan et al. 1999).

Many endospore-forming novel genera and species, with or without flagellum, have been reported. These bacteria produce oval, round, ellipsoidal spores located terminally, subterminally, and centrally in the cell when grown in medium with at least 10% w/v NaCl. Several novel genera and species reported in the past suggest diverse ecological distribution. For example, *Lentibacillus salicampi* was isolated from a salt field (Yoon et al. 2002), while *Halobacillus locisalis* (Yoon et al.

2004) and *Halobacillus yeomjeoni* (Yoon et al. 2005) were isolated from marine solar salterns in Korea. *Thalassobacillus devorans* is a phenol-degrading bacteria from a saline habitat (Garcia et al. 2005), while *Bacillus seohaeanensis* is reported from solar salterns in Korea (Lee et al. 2006) and *Salsuginibacillus kocurii* from the sediments of Soda Lake in Mongolia (Carrasco et al. 2007).

A number of alkaliphilic and halophilic bacteria have been reported from nonsaline environments. They include *Alkalibacillus silvisoli* from nonsaline soil in Japan (Usami et al. 2007), *Halalkalibacillus halophilus* from nonsaline garden soil (Echigo et al. 2007), *Virgibacillus arcticus* from permafrost of the Canadian high Arctic (Niederberger et al. 2009), *Bacillus neizhouensis* as a symbiont from a marine sea anemone (Chen Yi-G. et al. 2009), and *Bacillus xiaoxiensis* (Chen Yi-G. et al. 2011a) and *Bacillus hunanensis* (Chen Yi-G. et al. 2011b) from nonsaline forest soil. *Bacillus iranensis* (Bagheri et al. 2012), *Alteribacillus bidgolensis* (Didari et al. 2012), and *Ornithinibacillus halophilus* (Bagheri et al. 2013) have been reported from hypersaline lakes in Iran, and *Virgibacillus albus* (Zhang et al. 2012) from Salt Lake in China. Such reports suggest that endospore formation appears to be a favored strategy adopted by the majority of bacteria in saline ecological niches.

7.2.3.4 Flagella

Bacterial motility is due to a proteinaceous structure called flagella. It is also a major means of mobility in halophilic bacteria. Many flagellated bacteria with thick single polar flagella have been reported from saline habitats with NaCl concentrations of 5%–25% w/v. The majority of the reported genera have flagella with unique structures. A single polar flagellum has been reported in *Alkalibacillus silvisoli* from a nonsaline environment in Japan (Usami et al. 2007), *Halalkalibacillus halophilus* from nonsaline garden soil (Echigo et al. 2007), and *Bacillus hunanensis* (Chen Yi-G. et al. 2011a) and *Bacillus xiaoxiensis* (Chen Yi.-G. et al. 2011b) from nonsaline forest soil. A single polar flagellum has also been reported in the novel haloalkaliphilic genus *Idiomarina*, *Idiomarina fontislapidosi* sp. nov. and *Idiomarina ramblicola* sp. nov., isolated from inland hypersaline habitats of Spain (Martinez-Canovas et al. 2004a). Peritrichous flagella have been observed in *Halobacillus locisalis* (Yoon et al. 2005) and *Halobacillus yeomjeoni* (Yoon et al. 2005). On the other hand, few motile bacteria, such as *Bacillus iranensis* (Bagheri et al. 2012) and *Ornithinibacillus halophilus* (Bagheri et al. 2013), are reported without their flagellar position being known.

7.2.3.5 Na+/H+ Antiporter

The Na+/H+ antiporter is a well-known and universal mechanism in bacteria and higher organisms to balance inner cytosolic Na+ levels along with pH. The Na+/H+ antiporter plays a key role in functioning of Na+ or Li+ in exchange over for H+ and maintaining osmotic balance with the surroundings and reduces intoxication (Padan et al. 1989; Nozaki et al. 1996). Other important roles of the antiporter are establishing the electrochemical potential of Na+ across the cytoplasmic membrane, maintaining low Na+ levels (Tsuchiya et al. 1977; Ventosa et al. 1998), and maintaining the smooth functioning of intracellular pH homeostasis under alkaline conditions (Padan and Schuldiner 1994). This mechanism drives many endergonic processes in halophiles, including membrane transport of amino acids and other compounds. It is generally accepted that the negatively charged membrane potential is the driving force for the massive K+ accumulation, in turn maintaining electroneutrality of the cell. Many families of Na+/H+ antiporter genes have been identified in bacteria, that is, nhaC in *Bacillus pseudofirmus* (formerly *B. firmus*) (Ito et al. 1997) and nhaG in *B. subtilis* (Gouda et al. 2001); a single gene is involved in each of the above Na+/H+ antiporters. Another kind of Na+/H+ antiporter contains multiple subunits, such as mrp from *B. subtilis* (Ito et al. 1999) and mnhABCDEFG from *Staphylococcus aureus* (Hiramatsu et al. 1998). Additionally, there are two ATP-dependent chloride pumps that work for inward movement of Cl− in certain halophilic bacteria. One of these chloride pumps is independent of light and has symport movement with Na^+ (Duschl and Wagner 1986), while the other is light driven and requires the involvement of certain photopigments, like bacteriorhodopsin (Schobert and Lanyi 1982; Lanyi 1986).

7.3 MOLECULAR ADAPTATIONS

7.3.1 Amino Acid Composition and Protein Solubility

Compared with other organisms, the halophiles have certain advantages regarding their growth conditions and ease of genetic manipulation. These advantages are applicable to the majority of the moderately halophilic organisms. They have also developed resistance to various toxic substances including heavy metals. Adaptations and the unique physiologies of these organisms are being explored for various biotechnological applications.

The foundation of the adaptation mechanisms was first reported by Christian and Waltho in 1962. While studying the halophilic Archaea of the family Halobacteriaceae, it was observed that accumulation of KCl was very high to balance the excessive concentration of NaCl and other salts in the surrounding medium. The osmotic adaptation by synthesizing osmolytes within the cell and excluding the majority of salts is adopted by aerobic halophilic bacteria (Ventosa et al. 1998). For the first strategy, it is expected that the biosynthetic machinery of the organism should be active at molar concentrations of salts. It has further been observed that halophilic archaea have fewer basic amino acids and low levels of hydrophobic amino acids, except serine (Ser) residues. On the contrary, they have exceptionally high acidic amino acids, and a considerably high content of aspartic acid (Asp) and glutamic acid (Glu) (Lanyi 1974; Dennis and Shimmin 1997; Kennedy et al. 2001). Generally, in halophilic organisms, significantly high levels of valine (Val) and threonine (Thr) residues, along with acidic amino acids, exist. Additionally, a downturn in the amounts of polar uncharged amino acids lysine (Lys), methionine (Met), leucine (Leu), isoleucine (Ile), and cystine (Cys) is also reported in certain halophilic proteomes (Shakhnovich 1994; Eisenberg 1995; Madern et al. 2000; Mevarech et al. 2000; Kennedy et al. 2001; Fukuchi et al. 2003; Paul S. et al. 2008). The halophilic proteins in general are characterized by low hydrophobicity; overrepresentation of the acidic residues, especially aspartate (Asp); underrepresentation of polar uncharged amino acids; a lower proclivity for helix formation; and a higher tendency of the coiled structure (Paul S. et al. 2008). This finding is supported when the sequences of halophilic and nonhalophilic orthologous proteins are compared. It has been observed that the large hydrophobic (Leu and Met) and positively charged (Lys) amino acids with higher helical propensity are significantly underrepresented, whereas Asp residue, with a higher coil-forming propensity, is overrepresented in halophile proteins. There is a significant decrease in Ile and an increase in Val and Thr residues, all of which have higher sheet-forming propensities.

The presence of higher acidic residues has been attributed to hydrate the surface of protein and salt ion network (Eisenberg and Wachtel 1987; Eisenberg et al. 1992; Dym et al. 1995; Eisenberg 1995; Dennis and Shimmin 1997; Paul S. et al. 2008). Additionally, the higher acidic residues contribute towards formation of salt bridges by typically positioning the basic amino acid residues in positions that provide a shielding effect and also give structural firmness to the polypeptide chain, thereby contributing to an overall stabilization of 3D structure of proteins (Dym et al. 1995; Dennis and Shimmin 1997). The native confirmation and stability of any protein is a prime requirement. In order to get insight into the molecular adaptations in saline habitats, protein solvation in the three-dimensional environment rich in salt needs foremost attention. Merely shuffling the amino acid residues in a specific and selective manner may stabilize the protein; however, achieving solubility at high salt concentrations is a difficult task and can be achieved only by swiftly choosing between the acidic and basic amino acids (Frolow et al. 1996; Elcock and McCammon 1998; Jaenicke and Bohm 1998; Jaenicke 2000; Das et al. 2006).

The high intracellular salt requires adaptation of the whole enzymatic machinery of the cell. The cytoplasmic accumulation of molar concentrations of KCl leads to a sharp decrease in the distribution of protein isoelectric points. This allows the formation of an ordered water molecule network and intersubunit salt bridges "locked" in by bound solvent chloride and sodium ions (Richard et al. 2000). Britton and colleagues (1998) discovered that hydrophobicity further decreases by reducing Lys residues on the protein surface, which facilitates such proteins to function at high salt concentrations, while requiring a basal level of salt concentration in the vicinity of protein for its activity and stability.

The proposed mechanism is supported by the fact that electrostatic and hydrophobic interactions play a major role in stability, as well as the refolding of halophilic proteins (Arakawa and Tokunaga 2004). Further, proteins with excess acidic amino acid residues become functional only at high salt concentrations, which have been reported earlier (Oren 1986; Rengpipat et al. 1988; Oren and Gurevich 1993).

The crystal structures of the most extensively studied enzymes, lactate dehydrogenase (LDH) and malate dehydrogenase (MalDH) from halophilic bacteria *Salinibacter ruber* (*Sr*) and haloarchaea *Haloarcula marismorti* (*Hm*), reflect adaptive interaction between proteins in high salt conditions. The *Sr* MalDH show conformational stability that resembles nonhalophilic proteins with a flexible catalytic pocket. This flexibility induced by high KCl concentrations displays two types of nonspecific effects on the structure of the enzyme: (1) weak protein–protein interactions, which are apparently more attractive due to preferential hydration, decreasing protein solubility, and (2) a hydrophobic effect, due to excess salt in the vicinity (Coquelle et al. 2010). Studies suggest that acidic enrichment of the halophilic proteins enhances both weak (repulsive) interparticle interactions and protein–solvent interactions. An acidic surface is therefore required for protein solubility under excessive salt stress. Reducing the number of positively charge residues (Lys in particular) in halophilic proteins improves their solubility by increasing the net negativity on the surface while reducing the hydrophobic surface area of the folded polypeptide (Costenaro et al. 2002; Ebel et al. 2002; Tardieu et al. 2002; Coquelle et al. 2010). Charged amino acid residues (mainly the acidic amino acids) at specific sites in the polypeptide add to the solubility property. Thus, the differences in structural evolution of the proteins lie in the fact that the halophilic proteins have substituted the uncharged and basic amino acids with acidic ones. The nonhalophiles in varying salt concentrations momentarily derive an intermediate charged moiety (similar to acidic amino acid residues) of the protein structures, enhancing the solubility (Coquelle et al. 2010).

7.3.2 Genome Analysis and Stress-Activated Genes

Bacteria with high and low GC content are representatives of extremophilic groups. A high GC content (60% or more) in the genome is a more frequent feature among the extreme halophiles. Genome sequencing of *Halobacterium* sp. NRC-1 (Kennedy et al. 2001) and *Haloquadratum walsbyi* (Bolhuis et al. 2006) as pioneer extremophiles laid the foundation for investigating novel genes, new pathways, and astonishing adaptative strategies. A high GC content helps bacteria to survive UV-induced as well as other possible accumulation of mutations (Kennedy et al. 2001; Soppa 2006). However, it cannot be considered the sole criterion for the extremophiles. It has been observed that *H. walsbyi*, even with low GC content, does not show GC bias in codon selection and still survives in salinity with parallel activity and stability as *Halobacterium* sp. NRC-1, which has a high GC content and strong GC bias in codon usage. In some other organisms, such as *Salinibacter ruber*, the majority of the genes show similarity with their archaeal counterparts, while the codon usage and other genome properties resemble those of the eubacterial domain. It is therefore suggested that since both domains share saline habitats for time immemorial, they share a number of genes betweem them. The notion of a "habitat genome" (or a pool of genes useful for adaptation under a specific set of environmental constraints) is appealing (Rodriguez-Valera 2002; Mongodin et al. 2005). Thus, at the genomic level, the GC bias is not a universal feature for adaptation to high salinity, and other specific features of nucleotide selection may also be involved. The occurrence of dinucleotides in halophilic genomes bears some common characteristics, which are quite distinct from those of nonhalophiles, and hence may be regarded as specific genomic signatures for salt adaptation. The synonymous usage of genetic codes in halophiles exhibits similarity patterns regardless of the long-term evolutionary history of halophiles (Paul M. et al. 2008).

7.3.2.1 Stress-Activated Genes

In nature, the microbial population is confronted with enormous changes in physicochemical parameters, including pH, osmolarity, salinity, oxygen, and temperature. The microbes have developed the ability to adapt in dynamic and extreme environments. The salinity tolerance and osmotic stress have been studied in halophilic, moderately salt-tolerant and nonhalophilic bacteria and archaea (Kempf and Bremer 1998; Wood 1999, Tsuzuki et al. 2011; Schroeter et al. 2013). The common strategy to adapt to high salt concentrations in bacteria is to accumulate organic compatible solutes, which function as osmoprotectants against high salinity without interfering with the normal metabolic processes of the cell (Tsuzuki et al. 2011; Schroeter et al. 2013). The biosynthetic pathways and expression of genes involved are studied in detail under varying concentrations of salt. Salt stress exerts pleiotropic effects on multiple parameters dealing with physiology, membrane composition, cell wall properties (Lopez et al. 1998, 2000, 2006), EPS structural content (Lloret et al. 1995), cell swarming (Steil et al. 2003; Xu et al. 2005), and iron homeostasis (Hoffmann et al. 2002; Argandona et al. 2010; Gancz and Merrell 2011). Synthesis of organic compatible solutes, such as glycine, betaine, ectoine, proline, and trehalose, is activated or upregulated due to salt stress (Sleator and Hill 2002; Hahne et al. 2010).

Genes for general stress responses, such as those involved in the import of compatible solutes and SigB-controlled general stress response, are osmotically upregulated. SigW- and SigM-controlled genes for growth and survival under high salinity are well documented in the various strains of the genus *Bacillus*, that is, alkaliphilic *Bacillus* spp. (Ito et al. 2004), *B. subtilis* (Hahne et al. 2010), *B. subtilis* 168 (Horsburgh and Moir 1999), *B. licheniformis* (Schroeter et al. 2013), *B. halodurans* C125 (Kitada et al. 1994), and *B. pseudofirmus* OF4 (Janto et al. 2011), and *Chromohalobacter salexigens* DSM 3043 (Calderon et al. 2004). Certain responses are sponsored by the chaperone-encoded proteins that become functional only at high salt concentrations.

7.3.2.2 K⁺ Influx and Na⁺ Efflux

The transcriptome analysis of *Bacillus* sp. N16-5 and *B. subtilis* carried out under various salt concentrations suggested that K⁺ uptake transporters (KtrAB and KtrCD) play a critical role against salt stress yet not present in any operon but heavily influenced by salinity. KtrA expression increased with increasing salt and was upregulated at 120 minutes of growth, while KtrC was transcriptionally activated at 30 minutes of growth in *Bacillus* sp. N16-5, unlike *B. subtilis* (Holtmann et al. 2003). Similarly, the Na⁺ exporters Mrp, NhaK, and NhaC are upregulated with increasing salt in *B. subtilis* but not upregulated in *B. licheniformis*, while *Bacillus* sp. N16-5 has homologs of these transporters (Hahne et al. 2010; Schroeter et al. 2013; J. Yin et al. 2015).

7.3.2.3 K⁺/H⁺ and Na⁺/H⁺ Antiporter

The detailed analysis of *Bacillus* sp. N16-5 K⁺/H⁺ and Na⁺/H⁺ antiporters indicates the highest- and lowest-level expressions at different salt concentrations, suggesting their key role in growth under salt stress. In *B. subtilis* a Mrp complex is formed by operon coded proteins that function as a Na⁺/H⁺ antiporter, which in turn is vital for salinity tolerance. Disrupting this complex of any gene from the operon will give bacterial mutants that will be sensitive towards Na⁺ ions and thus salt stress as well (Kajiyama et al. 2007). While under similar salt stress, *Bacillus* sp. N16-5 shows upregulation of mrpABCDEFG operon-coded proteins, while at 15% salt stress other proteins coded by orf0208 and orf 3958 of the mrp operon play a key role in halotolerance (J. Yin et al. 2015).

7.3.2.4 Stress-Activated Genes for Accumulation of Compatible Solutes

Microbes use certain low-molecular-weight compounds, usually organic in nature, to resist stress induced by osmotic shock. As stated earlier, these organic compatible solutes help as osmostress proctectants, generally under extremely saline conditions. Bacteria accumulate high levels of compatible solutes either from the medium or by *de novo* synthesis under various stressful growth conditions (Kempf and Bremer 1998). Study of genes responsible for the synthesis and uptake of organic compatible solutes in *Bacillus* sp. N16-5 indicated overexpression of the genes encoding for glycine betaine transporters OpuD, glycine betaine ABC transporter OpuA, and sodium/glutamate symporter, while in *B. subtilis*, all the genes of the Opu transporters (OpuA to OpuE) involved in the uptake of compatible solute become overexpressed at high salt concentrations (Hahne et al. 2010). However, the expression of the sodium/glutamate symporter GltT was downregulated under salt stress, which may be due to the fact that sodium/solute symporters play an important role in Na$^+$ reentry for completion of the Na$^+$ cycle in alkaliphilic *Bacillus* species (Krulwich et al. 2001). Similarly, the enzymes choline dehydrogenase and betaine aldehyde dehydrogenase, involved in the transformation of choline to glycine betaine, are dramatically increased at high salinity.

The majority of the species of the genus *Bacillus* possess genes for the biosynthesis of ectoine, a well-known osmoprotectant or compatible solute. The biosynthesis of ectoine is mediated by ect-ABC genes that encode diaminobutyric acid acetyltransferase, diaminobutyric acid aminotransferase, and ectoine synthase, respectively (Bursy et al. 2007). The higher expression of ectABC genes under increased osmolality in *Bacillus pasteurii*, studied by Northern blot analysis, is noteworthy. Genes related to compatible solute transport and synthesis show rapid transcriptional activation in response to high salinity (Kuhlmann and Bremer 2002). Thus, for halotolerance and osmoprotection, certain compatible solutes, such as glycine betaine, act with sudden or immediate responses, while ectoine and others play a crucial role in long-term tolerance against osmotic shock and increased salinity.

7.3.2.5 Molecular Chaperones

Molecular chaperones play a crucial role in cellular processes, as well as in stabilizing the nascent polypeptides, *in vivo* protein folding, and maintaining conformation of protein under various stress conditions. Molecular chaperones are widely known stress response factors for prokaryotes. Chaperones have been detected in *Bacillus* sp. N16-5 (J. Yin et al. 2015), *B. subtilis*, and *E. coli* (Hecker et al. 1988). In *E. coli*, the genes encoding for molecular chaperones are overexpressed or upregulated under increased salinity (Diamant et al. 2001). High salt and alkaline pH are considered for protein structural deformation and misfolding.

7.4 STRESS-DRIVEN BIOSYNTHESIS AND COMMERCIAL APPLICATIONS

Under the adverse conditions of salt, halophilic bacteria, more specifically, moderate halophiles, produce certain metabolites to protect the cell from osmotic imbalance. In other words, the negative effect of salt on cells leads to the production of specific metabolites: compatible solutes, polyhydroxyalkanoates (PHAs), polyhydroxybutyrates (PHBs), and carotenoids, which have significant biotechnological applications.

7.4.1 Compatible Solutes

It is a well-established fact that halophilic bacteria that follow the salt-out strategy produce osmotic solutes to maintain the iso-osmotic condition. A compatible solute strategy and osmosolute production were discussed earlier. For extraction of osmotic solutes from microorganisms, a technology of osmotic downshock termed bacterial milking was successfully developed (Tsapis and Kepes 1977; Reed et al. 1986; Fischel and Oren 1993). For instance, a moderate halophile *Halomonas elongate* was applied as a cell factory for ectoine production (Sauer and Galinski 1998). *H. elongata*, with a broad salt tolerance of ~0.1 to ~4 M NaCl, rapidly releases ectoine under hypo-osmotic shock (Vreeland et al. 1980; Cánovas et al. 1997; Pastor et al. 2010). *Halomonas salina* excretes ectoine into the medium even at constant (0.5 M) extracellular osmolarity, without using the bacterial milking method (Zhang et al. 2009).

Hydroxyectoine, with better protection capacity than ectoine, is another compatible solute that has attracted commercial interest (Pastor et al. 2010). The Gram-positive halophilic eubacterium *Marinococcus* M52 can convert ectoine to hydroxyectoine in the stationary phase (Frings et al. 1995). As opposed to the bacterial milking process in ectoine production, *Marinococcus* M52 shows resistance to osmotic downshock; thus, its hydroxyectoine remained intracellular unless organic solvents such as methanol or ethanol were used for extraction (Frings et al. 1995). A method termed thermal permeabilization was described by Schiraldi et al. (2006) to obtain intracellular hydroxyectoine. However, in both the solvent extraction and permeabilization methods, cell recycling does not happen, resulting in lower productivity (Pastor et al. 2010). In Gram-negative strains *H. elongata* and *C. salexigens*, ectoine and hydroxyectoine are produced under high salinity and high temperature (Sauer and Galinski 1998; Vargas et al. 2008).

7.4.2 Polyhydroxyalkanoates and Polyhydroxybutyrates

PHAs are a group of biodegradable and biocompatible polyesters accumulated by many microorganisms. Among these PHAs, poly(3-hydroxybutyrate) and poly(3-hydroxybutyrate-co-3-hydroxyvalerate) (PHBV) are well-studied polymers and have been produced on a large scale (Steinbüchel and Füchtenbusch 1998; Chen 2009). Many halophiles have been shown to synthesize PHAs (Fernandez-Castillo et al. 1986; Legault et al. 2006; Han et al. 2007; Koller et al. 2007). Among the PHA-producing halophiles, the archaeon *Haloferax mediterranei* produced 46 wt% PHA (Lillo and Rodriguez-Valera 1990; Rodriguez-Valera and Lillo 1992). The halobacterium *Halomonas boliviensis* tolerates a wide range of salt concentrations of 0%–25% w/v and pH 6–11, and produces PHBs with a high molecular weight of 1100 kDa under optimized conditions (Quillaguamán et al. 2005, 2006, 2007, 2008). Recently, *Halomonas* sp. TD01, grown optimally at a salt concentration of 5%–6% (w/v) at a pH of 9.0, accumulated above 80% PHBs on glucose salt medium (Tan et al. 2011). In some cases, bacteria produce both PHBs and compatible solute. *Methylarcula marina* and *Methylarcula terricola* synthesize 18% (wt/wt) PHBs and compatible solutes, including ectoine and glutamate, in NaCl concentrations of 6%–10% (w/v) (Doronina et al. 2000). *Halomonas campaniensis* accumulates PHBs and ectoine in 5.8% (w/v) NaCl (Strazzullo et al. 2008).

7.4.3 Carotenoids

Halophilic bacteria and archaea produce carotenoids under salt stress. However, there are only a few instances of microbial production of β-carotene and astaxanthin (Papaioannou et al. 2010). Nevertheless, the extraction of carotenoids is easier and economical from halophilic bacteria, as bacterial cells lyses spontaneously in water (Asker and Ohta 2002). Moreover, although carotenoids have until now been commonly extracted from archaea, bacterioruberin has been found in some bacteria, such as *Rubrobacter radiotolerans* (Saito et al. 1994), *Arthrobacter agilis* (Fong et al. 2001), and *Kocuria rosea* (Chattopadhyay et al. 1997).

Many reports suggest that the salt requirement of the organism largely affects carotenoid production. For instance, *Haloferax mediterranei* ATCC 33500 produced about 20 times more pigments in liquid medium containing 15% total salts than in medium containing 25% total salts (Rodriguez-Valera et al. 1980). In contrast, the archaea *Halobacterium cutirubrum*, currently referred to as *Halobacterium salinarum* ATCC 33170, did not produce carotenoid in the same medium with 15% total salts. Nevertheless, in a medium with 20%, 25%, and 35% total salts, the cells produced more than 1400 µg bacterioruberins g^{-1} cell protein (Kushwaha et al. 1982). In *H. mediterranei* ATCC 33500, a decrease in NaCl concentration led to an increase in pigment content (D'Souza et al. 1997). It has been suggested that *H. mediterranei* could produce bacterioruberin to stabilize the cell membrane and reduce cell lysis as a response to the stress caused by low salt concentrations. *H. volcanii* strain WFD11 (DSM 5716) produces 1.6- to 1.7-fold higher carotenoids in media with low salt content than in media with high salt content (Bidle et al. 2007). Thus, salinity optimization may lead to enhanced carotenoid production at a lower cost.

7.4.4 Commercial Implications

Ectoines are the most widespread compatible solutes, and are commercially available as protectants of proteins, DNA, and mammalian cells (Lippert and Galinski 1992; Kolp et al. 2006; Pastor et al. 2010). RonaCare™ Ectoin, produced by Merck KgaA, Darmstadt, is used as a moisturizer in cosmetics and skin care products. However, the ability of ectoines to stabilize biomolecules may have some specific applications. A German company, Bitop, in collaboration with researchers at the Cologne University Clinic, is exploring the application of solutes in certain cancer therapies.

Betaine has been known for the protection of industrial microbial strains in storage against dehydration and rehydration (Selmer-Olsen et al. 1999). Betaine is also used as a methyl group donor for the microbial production of chemicals with methyl group branches. Further, betaine supplementation in the medium neither significantly increases the production cost nor modifies the fermentation process. Betaine also provides thermal protection to native proteins and benefits the recovery of stress-induced protein aggregates in *E. coli* cells (Diamant et al. 2003). Betaine could also reduce the destabilizing effects of protein under osmotic stress (Sarkar and Pielak 2014). More recently, betaine and its analogs were shown to protect a variety of enzymes, including glucosidase, alkaline phosphatase, lactose dehydrogenase, sulfatase, and horseradish peroxidase (Nakagawa et al. 2015).

PHA has several industrial applications, ranging from bioplastics, biofuels, and fine chemicals to medicine (Chen 2009; Chen and Patel 2011). PHAs display heterogenety in their properties due to more than 150 monomer variations (Steinbüchel and Valentin 1995; Chen and Wu 2005). PHB is rigid and brittle, while PHBV is more flexible, with wider application as medical materials, film products, disposable items, and packaging materials (Philip et al. 2007; Chen 2009). The application of halophiles for PHA and PHB production results in a huge cost reduction in the production and recovery processes for many reasons. Freshwater consumption can be reduced, as seawater itself serves as a salt-containing medium (Tan et al. 2011; Yue et al. 2014). For some halophiles, the cost for PHA recovery can be decreased by cell lysis via hypo-osmotic shock treatment (Quillaguamán et al. 2010). Koller et al. (2007) demonstrated that the cost of PHBV production by *H. mediterranei* could be reduced by 30% above the recombinant PHBV production by *E. coli*. These highlight the importance of halophiles in PHA production.

Carotenoids are extensively used as dyes and functional ingredients in food products, including cosmetics (Hosseini and Shariati 2009). Bacterioruberin acts as a cellular membrane backup since it increases membrane rigidity and decreases water permeability (Lazrak et al. 1988; Fang et al. 2010). It also protects the microorganism from DNA damaging agents, such as ionizing radiation, ultraviolet radiation, and hydrogen peroxide (Oren 2002; Singh and Gabani 2011), probably due to its antioxidant capacity. Extracts of *Halococcus morrhuae* and *Halobacterium salinarum* cells with bacterioruberin and its derivatives have shown high antioxidant capacity (Mandelli et al. 2012). Owing to the properties of the carotenoids, such as antioxidant activity, immunity-boosting activity, and the effect against premature aging, they have wide applications in pharmaceutical and medical fields (Alvarado et al. 2005; Hosseini and Shariati 2009). Carotenoids are also reported to enhance *in vitro* antibody production (Ohyanagi et al. 2009).

7.5 CONCLUSION

In nature, microorganisms develop various adaptation strategies for their sustenance with the changing environmental conditions. Among the extremophilic microbes, the halophilic bacteria and archaea stand out with a multitude of adaptations. Halophilic bacteria have developed multifaceted strategies, such as cell and cell wall, plasma membrane, and cytoplasmic adaptations in terms of ionic strength and concentrations of salts. Halophiles also produce certain essential compounds, such as EPSs and compatible solutes for adaptation, besides being biotechnologically significant.

ACKNOWLEDGMENTS

Our work cited and described in this chapter has been supported by the UGC-CAS Programme, DBT-Multi-Instituional Project, MoES Net Working Project, and Saurashtra University. The UGC-BSR meritorious fellowship to H.B.B. and DST Young Scientist Project to V.H.R. are also acknowledged.

REFERENCES

Adams, Rachel L., and Nicholas J. Russell. Interactive effects of salt concentration and temperature on growth and lipid composition in the moderately halophilic bacterium *Vibrio costicola*. *Canadian Journal of Microbiology* 38, no. 8 (1992): 823–827.

Alfredsson, Gudni A., Jakob K. Kristjansson, Sigridur Hjorleifsdottir, and Karl O. Stetter. *Rhodothermus marinus*, gen. nov., sp. nov., a thermophilic, halophilic bacterium from submarine hot springs in Iceland. *Microbiology* 134, no. 2 (1988): 299–306.

Alvarado, Carmen, Pedro Alvarez, L. Jimenez, and M. De La Fuente. Improvement of leukocyte functions in young prematurely aging mice after a 5-week ingestion of a diet supplemented with biscuits enriched in antioxidants. *Antioxidants & Redox Signaling* 7, no. 9–10 (2005): 1203–1210.

Amako, Kazunobu, Kenji Okada, and Shunji Miake. Evidence for the presence of a capsule in *Vibrio vulnificus*. *Microbiology* 130, no. 10 (1984): 2741–2743.

Arahal, David R., M. Carmen Marquez, Benjamin E. Volcani, Karl H. Schleifer, and Antonio Ventosa. *Bacillus marismortui* sp. nov., a new moderately halophilic species from the Dead Sea. *International Journal of Systematic and Evolutionary Microbiology* 49, no. 2 (1999): 521–530.

Arakawa, Tsutomu, and Masao Tokunaga. Electrostatic and hydrophobic interactions play a major role in the stability and refolding of halophilic proteins. *Protein and Peptide Letters* 11, no. 2 (2004): 125–132.

Argandona, Montserrat, Joaquin J. Nieto, Fernando Iglesias-Guerra, Maria Isabel Calderon, Raul Garcia-Estepa, and Carmen Vargas. Interplay between iron homeostasis and the osmotic stress response in the halophilic bacterium *Chromohalobacter salexigens*. *Applied and Environmental Microbiology* 76, no. 11 (2010): 3575–3589.

Asker, Dalal, and Yoshiyuki Ohta. *Haloferaxalexandrinus* sp. nov., an extremely halophilic canthaxanthin-producing archaeon from a solar saltern in Alexandria (Egypt). *International Journal of Systematic and Evolutionary Microbiology* 52, no. 3 (2002): 729–738.

Bae, Seung Seob, Jung-Hyun Lee, and Sang-Jin Kim. *Bacillus alveayuensis* sp. nov., a thermophilic bacterium isolated from deep-sea sediments of the Ayu Trough. *International Journal of Systematic and Evolutionary Microbiology* 55, no. 3 (2005): 1211–1215.

Bagheri, Maryam, Mohammad Ali Amoozegar, Peter Schumann, Maryam Didari, Malihe Mehrshad, Cathrin Sproer, Cristina Sanchez-Porro, and Antonio Ventosa. *Ornithinibacillus halophilus* sp. nov., a moderately halophilic, gram-stain-positive, endospore-forming bacterium from a hypersaline lake. *International Journal of Systematic and Evolutionary Microbiology* 63, no. 3 (2013): 844–848.

Bagheri, Maryam, Maryam Didari, Mohammad Ali Amoozegar, Peter Schumann, Cristina Sanchez-Porro, Malieh Mehrshad, and Antonio Ventosa. *Bacillus iranensis* sp. nov., a moderate halophile from a hypersaline lake. *International Journal of Systematic and Evolutionary Microbiology* 62, no. 4 (2012): 811–816.

Bhatt, Hitarth B., and Satya P. Singh. Phylogenetic and phenogram based diversity of haloalkaliphilic bacteria from the saline desert. In *Microbial Biotechnology Technological Challenges and Developmental Trends*, ed. Bhima Bhukya, Anjana Devi Tangutur, pp. 373–386. Waretown, NJ: Apple Academic Press, 2016.

Bidle, Kelly A., Thomas E. Hanson, Koko Howell, and Jennifer Nannen. HMG-CoA reductase is regulated by salinity at the level of transcription in *Haloferax volcanii*. *Extremophiles* 11, no. 1 (2007): 49–55.

Bidle, Kelly A., P. Aaron Kirkland, Jennifer L. Nannen, and Julie A. Maupin-Furlow. Proteomic analysis of *Haloferax volcanii* reveals salinity-mediated regulation of the stress response protein PspA. *Microbiology* 154, no. 5 (2008): 1436–1443.

Boch, Jens, Bettina Kempf, Roland Schmid, and Erhard Bremer. Synthesis of the osmoprotectant glycine betaine in *Bacillus subtilis*: Characterization of the gbsAB genes. *Journal of Bacteriology* 178, no. 17 (1996): 5121–5129.

Bolhuis, Henk, Peter Palm, Andy Wende, Michaela Falb, Markus Rampp, Francisco Rodriguez-Valera, Friedhelm Pfeiffer, and Dieter Oesterhelt. The genome of the square archaeon *Haloquadratum walsbyi*: Life at the limits of water activity. *BMC Genomics* 7, no. 1 (2006): 169.

Bouchotroch, Samir, Emilia Quesada, Ana del Moral, Inmaculada Llamas, and Victoria Bejar. *Halomonas maura* sp. nov., a novel moderately halophilic, exopolysaccharide-producing bacterium. *International Journal of Systematic and Evolutionary Microbiology* 51, no. 5 (2001): 1625–1632.

Boyer, E. W., M. B. Ingle, and G. D. Mercer. *Bacillus alcalophilus* subsp. *halodurans* subsp. nov.: An alkaline-amylase-producing, alkalophilic organism. *International Journal of Systematic and Evolutionary Microbiology* 23, no. 3 (1973): 238–242.

Britton, K. Linda, Timothy J. Stillman, Kitty S. P. Yip, Patrick Forterre, Paul C. Engel, and David W. Rice. Insights into the molecular basis of salt tolerance from the study of glutamate dehydrogenase from *Halobacterium salinarum*. *Journal of Biological Chemistry* 273, no. 15 (1998): 9023–9030.

Brocchieri, Luciano. Environmental signatures in proteome properties. *Proceedings of the National Academy of Sciences of the United States of America* 101, no. 22 (2004): 8257–8258.

Brown, A. D. Inhibition by spermine of the action of a bacterial cell-wall lytic enzyme. *Biochimica et Biophysica Acta* 44 (1960a): 178–179.

Brown, A. D. Some properties of a gram-negative heterotrophic marine bacterium. *Microbiology* 23, no. 3 (1960b): 471–485.

Brown, A. D. Microbial water stress. *Bacteriological Reviews* 40, no. 4 (1976): 803.

Bursy, Jan, Antonio J. Pierik, Nathalie Pica, and Erhard Bremer. Osmotically induced synthesis of the compatible solute hydroxyectoine is mediated by an evolutionarily conserved ectoine hydroxylase. *Journal of Biological Chemistry* 282, no. 43 (2007): 31147–31155.

Calderon, M. Isabel, Carmen Vargas, Fernando Rojo, Fernando Iglesias-Guerra, Laszlo N. Csonka, Antonio Ventosa, and Joaquín J. Nieto. Complex regulation of the synthesis of the compatible solute ectoine in the halophilic bacterium *Chromohalobacter salexigens* DSM 3043T. *Microbiology* 150, no. 9 (2004): 3051–3063.

Cano, Raul J., and Monica K. Borucki. Revival and identification of bacterial spores in 25- to 40-million-year-old Dominican amber. *Science* 268, no. 5213 (1995): 1060.

Cánovas, David, Carmen Vargas, Laszlo N. Csonka, Antonio Ventosa, and Joaquin J. Nieto. Osmoprotectants in *Halomonas elongata*: High-affinity betaine transport system and choline-betaine pathway. *Journal of Bacteriology* 178, no. 24 (1996): 7221–7226.

Cánovas, David, Carmen Vargas, Fernando Iglesias-Guerra, Laszlo N. Csonka, David Rhodes, Antonio Ventosa, and Joaquín J. Nieto. Isolation and characterization of salt-sensitive mutants of the moderate halophile *Halomonas elongata* and cloning of the ectoine synthesis genes. *Journal of Biological Chemistry* 272, no. 41 (1997): 25794–25801.

Cánovas, David, Carmen Vargas, Susanne Kneip, Mariá-Jesús Morón, Antonio Ventosa, Erhard Bremer, and Joaquín J. Nieto. Genes for the synthesis of the osmoprotectant glycine betaine from choline in the moderately halophilic bacterium *Halomonas elongata* DSM 3043. *Microbiology* 146, no. 2 (2000): 455–463.

Carrasco, I. J., Melina C. Marquez, Yanfen Xue, Yujie Ma, Donald A. Cowan, B. E. Jones, William D. Grant, and Antonio Ventosa. *Salsuginibacillus kocurii* gen. nov., sp. nov., a moderately halophilic bacterium from soda-lake sediment. *International Journal of Systematic and Evolutionary Microbiology* 57, no. 10 (2007): 2381–2386.

Cavicchioli, Ricardo, Ricardo Amils, Dirk Wagner, and Terry McGenity. Life and applications of extremophiles. *Environmental Microbiology* 13, no. 8 (2011): 1903–1907.

Chaiyanan, Saipin, Sitthipan Chaiyanan, Tim Maugel, Anwarul Huq, Frank T. Robb, and Rita R. Colwell. Polyphasic taxonomy of a novel *Halobacillus, Halobacillus thailandensis* sp. nov. isolated from fish sauce. *Systematic and Applied Microbiology* 22, no. 3 (1999): 360–365.

Chattopadhyay, M. K., M. V. Jagannadham, M. Vairamani, and S. Shivaji. Carotenoid pigments of an Antarctic *Psychrotrophicbacterium micrococcusroseus*: Temperature dependent biosynthesis, structure, and interaction with synthetic membranes. *Biochemical and Biophysical Research Communications* 239, no. 1 (1997): 85–90.

Chen, Guo-Qiang. A microbial polyhydroxyalkanoates (PHA) based bio- and materials industry. *Chemical Society Reviews* 38, no. 8 (2009): 2434–2446.

Chen, Guo-Qiang, and Martin K. Patel. Plastics derived from biological sources: Present and future: A technical and environmental review. *Chemical Reviews* 112, no. 4 (2011): 2082–2099.

Chen, Guo-Qiang, and Qiong Wu. Microbial production and applications of chiral hydroxyalkanoates. *Applied Microbiology and Biotechnology* 67, no. 5 (2005): 592–599.

Chen, Shih-Ya, Mei-Chin Lai, Shu-Jung Lai, and Yu-Chien Lee. Characterization of osmolyte betaine synthesizing sarcosine dimethylglycine N-methyltransferase from *Methanohalophilus portucalensis*. *Archives of Microbiology* 191, no. 10 (2009): 735–743.

Chen, Yi-Guang, Di-Fei Hao, Qi-Hui Chen, Yu-Qin Zhang, Jian-Ben Liu, Jian-Wu He, Shu-Kun Tang, and Wen-Jun Li. *Bacillus hunanensis* sp. nov., a slightly halophilic bacterium isolated from non-saline forest soil. *Antonie Van Leeuwenhoek* 99, no. 3 (2011a): 481–488.

Chen, Yi-Guang, Yu-Qin Zhang, Qi-Hui Chen, Hans-Peter Klenk, Jian-Wu He, Shu-Kun Tang, Xiao-Long Cui, and Wen-Jun Li. *Bacillus xiaoxiensis* sp. nov., a slightly halophilic bacterium isolated from non-saline forest soil. *International Journal of Systematic and Evolutionary Microbiology* 61, no. 9 (2011b): 2095–2100.

Chen, Yi-Guang, Yu-Qin Zhang, Yong-Xia Wang, Zhi-Xiong Liu, Hans-Peter Klenk, Huai-Dong Xiao, Shu-Kun Tang, Xiao-Long Cui, and Wen-Jun Li. *Bacillus neizhouensis* sp. nov., a halophilic marine bacterium isolated from a sea anemone. *International Journal of Systematic and Evolutionary Microbiology* 59, no. 12 (2009): 3035–3039.

Cho, Byung C. Heterotrophic flagellates in hypersaline waters. In *Adaptation to Life at High Salt Concentrations in Archaea, Bacteria, and Eukarya*, ed. Gunde-Cimerman, N., Oren, A., and Plemenitas A. pp. 41–549. Dordrecht: Springer, 2005.

Christian, J. H. B., and Judith A. Waltho. Solute concentrations within cells of halophilic and non-halophilic bacteria. *Biochimica et Biophysica Acta* 65, no. 3 (1962): 506–508.

Coker, James A., Priya DasSarma, Jeffrey Kumar, Jochen A. Müller, and Shiladitya DasSarma. Transcriptional profiling of the model archaeon *Halobacterium* sp. NRC-1: Responses to changes in salinity and temperature. *Saline Systems* 3, no. 1 (2007): 1.

Coquelle, Nicolas, Romain Talon, Douglas H. Juers, Eric Girard, Richard Kahn, and Dominique Madern. Gradual adaptive changes of a protein facing high salt concentrations. *Journal of Molecular Biology* 404, no. 3 (2010): 493–505.

Costenaro, Lionel, Giuseppe Zaccai, and Christine Ebel. Link between protein-solvent and weak protein-protein interactions gives insight into halophilic adaptation. *Biochemistry* 41, no. 44 (2002): 13245–13252.

Da Costa, Milton S., Helena Santos, and Erwin A. Galinski. An overview of the role and diversity of compatible solutes in Bacteria and Archaea. In *Advances in Biochemical Engineering/Biotechnology* 61 (1998): 117–153.

Das, Sabyasachi, Sandip Paul, Sumit K. Bag, and Chitra Dutta. Analysis of *Nanoarchaeum equitans* genome and proteome composition: Indications for hyperthermophilic and parasitic adaptation. *BMC Genomics* 7, no. 1 (2006): 1.

DasSarma, Shiladitya, and Priya DasSarma. Halophiles and their enzymes: Negativity put to good use. *Current Opinion in Microbiology* 25 (2015): 120–126.

Dennis, Patrick P., and Lawrence C. Shimmin. Evolutionary divergence and salinity-mediated selection in halophilic archaea. *Microbiology and Molecular Biology Reviews* 61, no. 1 (1997): 90–104.

Diamant, Sophia, Noa Eliahu, David Rosenthal, and Pierre Goloubinoff. Chemical chaperones regulate molecular chaperones in vitro and in cells under combined salt and heat stresses. *Journal of Biological Chemistry* 276, no. 43 (2001): 39586–39591.

Diamant, Sophia, David Rosenthal, Abdussalam Azem, Noa Eliahu, Anat Peres Ben-Zvi, and Pierre Goloubinoff. Dicarboxylic amino acids and glycine-betaine regulate chaperone-mediated protein-disaggregation under stress. *Molecular Microbiology* 49, no. 2 (2003): 401–410.

Didari, Maryam, Mohammad Ali Amoozegar, Maryam Bagheri, Peter Schumann, Cathrin Sproer, Cristina Sanchez-Porro, and Antonio Ventosa. *Alteribacillus bidgolensis* gen. nov., sp. nov., a moderately halophilic bacterium from a hypersaline lake, and reclassification of *Bacillus persepolensis* as *Alteribacillus persepolensis* comb. nov. *International Journal of Systematic and Evolutionary Microbiology* 62, no. 11 (2012): 2691–2697.

Dodia, Mital, Chirantan Rawal, Hetal Bhimani, Rupal Joshi, Sunil Khare, and Satya Singh. Purification and stability characteristics of an alkaline serine protease from a newly isolated haloalkaliphilic bacterium sp. AH-6. *Journal of Industrial Microbiology and Biotechnology* 35, no. 2 (2008): 121–131.

Doronina, Nina V., Yuri A. Trotsenko, and Tatjana P. Tourova. *Methylarcula marina* gen. nov., sp. nov. and *Methylarculaterricola* sp. nov.: Novel aerobic, moderately halophilic, facultatively methylotrophic bacteria from coastal saline environments. *International Journal of Systematic and Evolutionary Microbiology* 50, no. 5 (2000): 1849–1859.

D'Souza, Sandra E., Wijaya Altekar, and S. F. D'Souza. Adaptive response of *Haloferax mediterranei* to low concentrations of NaCl (<20%) in the growth medium. *Archives of Microbiology* 168, no. 1 (1997): 68–71.

Duschl, Albert, and Gottfried Wagner. Primary and secondary chloride transport in *Halobacterium halobium. Journal of Bacteriology* 168, no. 2 (1986): 548–552.

Dym, Orly, Moshe Mevarech, and Joel L. Sussman. Structural features that stabilize halophilic malate dehydrogenase from an archaebacterium. *Science* 267, no. 5202 (1995): 1344.

Ebel, Christine, Lionel Costenaro, Mihaela Pascu, Pierre Faou, Blandine Kernel, Flavien Proust-De Martin, and Giuseppe Zaccai. Solvent interactions of halophilic malate dehydrogenase. *Biochemistry* 41, no. 44 (2002): 13234–13244.

Ebrahimie, Esmaeil, Mansour Ebrahimi, Narjes Rahpayma Sarvestani, and Mahdi Ebrahimi. Protein attributes contribute to halo-stability, bioinformatics approach. *Saline Systems* 7, no. 1 (2011): 1.

Echigo, Akinobu, Tadamasa Fukushima, Toru Mizuki, Masahiro Kamekura, and Ron Usami. *Halalkalibacillus halophilus* gen. nov., sp. nov., a novel moderately halophilic and alkaliphilic bacterium isolated from a non-saline soil sample in Japan. *International Journal of Systematic and Evolutionary Microbiology* 57, no. 5 (2007): 1081–1085.

Eisenberg, Henryk. Life in unusual environments: Progress in understanding the structure and function of enzymes from extreme halophilic bacteria. *Archives of Biochemistry and Biophysics* 318, no. 1 (1995): 1–5.

Eisenberg, Henryk, Moshe Mevarech, and Giuseppe Zaccai. Biochemical, structural, and molecular genetic aspects of halophilism. *Advances in Protein Chemistry* 43 (1992): 1–62.

Eisenberg, Henryk, and Ellen J. Wachtel. Structural studies of halophilic proteins, ribosomes, and organelles of bacteria adapted to extreme salt concentrations. *Annual Review of Biophysics and Biophysical Chemistry* 16, no. 1 (1987): 69–92.

Elazari-Volcani, B. Algae in the bed of the Dead Sea. *Nature* 145 (1940): 975.

Elcock, Adrian H., and J. Andrew McCammon. Electrostatic contributions to the stability of halophilic proteins. *Journal of Molecular Biology* 280, no. 4 (1998): 731–748.

Fang, Chun-Jen, Kuo-Lung Ku, Min-Hsiung Lee, and Nan-Wei Su. Influence of nutritive factors on C 50 carotenoids production by *Haloferax mediterranei* ATCC 33500 with two-stage cultivation. *Bioresource Technology* 101, no. 16 (2010): 6487–6493.

Fendrich, Claudi, Hans Hippe, and Gerhard Gottschalk. *Clostridium halophilium* sp. nov. and *C. litorale* sp. nov., an obligate halophilic and a marine species degrading betaine in the Stickland reaction. *Archives of Microbiology* 154, no. 2 (1990): 127–132.

Feng, de Q., Bo Zhang, Wei Dong Lu, and Su Sheng Yang. Protein expression analysis of *Halobacillus dabanensis* D-8T subjected to salt shock. *Journal of Microbiology (Seoul, Korea)* 44, no. 4 (2006): 369–374.

Fernandez-Castillo, Rosario, Francisco Rodriguez-Valera, J. Gonzalez-Ramos, and Francisco Ruiz-Berraquero. Accumulation of poly (β-hydroxybutyrate) by halobacteria. *Applied and Environmental Microbiology* 51, no. 1 (1986): 214–216.

Fischel, Uri, and Aharon Oren. Fate of compatible solutes during dilution stress in *Ectothiorhodospira marismortui*. *FEMS Microbiology Letters* 113, no. 1 (1993): 113–118.

Fong, N., Maree Burgess, Kevin Barrow, and D. Glenn. Carotenoid accumulation in the psychrotrophic bacterium *Arthrobacteragilis* in response to thermal and salt stress. *Applied Microbiology and Biotechnology* 56, no. 5–6 (2001): 750–756.

Freitas, Filomena, Vitor D. Alves, and Maria A. M. Reis. Advances in bacterial exopolysaccharides: From production to biotechnological applications. *Trends in Biotechnology* 29, no. 8 (2011): 388–398.

Frings, Eric, Thomas Sauer, and Erwin A. Galinski. Production of hydroxyectoine: High cell-density cultivation and osmotic downshock of *Marinococcus* strain M52. *Journal of Biotechnology* 43, no. 1 (1995): 53–61.

Frolow, Felix, Michal Harel, Joel L. Sussman, Moshe Mevarech, and Menachem Shoham. Insights into protein adaptation to a saturated salt environment from the crystal structure of a halophilic 2Fe-2S ferredoxin. *Nature Structural & Molecular Biology* 3, no. 5 (1996): 452–458.

Fukuchi, Satoshi, Kazuaki Yoshimune, Mamoru Wakayama, Mitsuaki Moriguchi, and Ken Nishikawa. Unique amino acid composition of proteins in halophilic bacteria. *Journal of Molecular Biology* 327, no. 2 (2003): 347–357.

Galinski, Erwin A. Salzadaptation durch kompatible Solute bei halophilen phototrophen Bakterien. PhD thesis, University of Bonn, 1986.

Galinski, Erwin A. Compatible solutes of halophilic eubacteria: Molecular principles, water-solute interaction, stress protection. *Experientia* 49, no. 6–7 (1993): 487–496.

Galinski, Erwin A. Osmoadaptation in bacteria. *Advances in Microbial Physiology* 37 (1995): 273.

Galinski, Erwin A., Heinz-Peter Pfeiffer, and Hans G. Trüper. 1,4,5,6-Tetrahydro-2-methyl-4-pyrimidinecarboxylic acid. *European Journal of Biochemistry* 149, no. 1 (1985): 135–139.

Gancz, Hanan, and D. Scott Merrell. The *Helicobacter pylori* ferric uptake regulator (Fur) is essential for growth under sodium chloride stress. *Journal of Microbiology* 49, no. 2 (2011): 294–298.

Garabito, Maria J., David R. Arahal, Encarnacion Mellado, M. Carmen Marquez, and Antonio Ventosa. *Bacillus salexigens* sp. nov., a new moderately halophilic *Bacillus* species. *International Journal of Systematic and Evolutionary Microbiology* 47, no. 3 (1997): 735–741.

Garcia, María Teresa, Virginia Gallego, Antonio Ventosa, and Encarnacion Mellado. *Thalassobacillus devorans* gen. nov., sp. nov., a moderately halophilic, phenol-degrading, gram-positive bacterium. *International Journal of Systematic and Evolutionary Microbiology* 55, no. 5 (2005): 1789–1795.

Gohel, Sangeeta D., and Satya P. Singh. Purification strategies, characteristics and thermodynamic analysis of a highly thermostable alkaline protease from a salt-tolerant alkaliphilicactinomycete, *Nocardiopsis alba* OK-5. *Journal of Chromatography B* 889 (2012): 61–68.

Gouda, Takiko, Masaynki Kuroda, Toshiaki Hiramatsu, Kaori Nozaki, Teruo Kuroda, Tohru Mizushima, and Tomofusa Tsuchiya. nhaG Na+ H+ antiporter gene of *Bacillus subtilis* ATCC9372, which is missing in the complete genome sequence of strain 168, and properties of the antiporter. *Journal of Biochemistry* 130, no. 5 (2001): 711–717.

Gu, Zhi Jing, Lei Wang, Daniel Le Rudulier, Bo Zhang, and Su Sheng Yang. Characterization of the glycine betaine biosynthetic genes in the moderately halophilic bacterium *Halobacillus dabanensis* D-8T. *Current Microbiology* 57, no. 4 (2008): 306–311.

Hahne, Hannes, Ulrike Mader, Andreas Otto, Florian Bonn, Leif Steil, Erhard Bremer, Michael Hecker, and Dorte Becher. A comprehensive proteomics and transcriptomics analysis of *Bacillus subtilis* salt stress adaptation. *Journal of Bacteriology* 192, no. 3 (2010): 870–882.

Han, Jing, Qiuhe Lu, Ligang Zhou, Jian Zhou, and Hua Xiang. Molecular characterization of the phaECHm genes, required for biosynthesis of poly (3-hydroxybutyrate) in the extremely halophilic archaeon *Haloarcula marismortui*. *Applied and Environmental Microbiology* 73, no. 19 (2007): 6058–6065.

Harrison, Jesse P., Nicolas Gheeraert, Dmitry Tsigelnitskiy, and Charles S. Cockell. The limits for life under multiple extremes. *Trends in Microbiology* 21, no. 4 (2013): 204–212.

Hart, Darrenn J., and Russell H. Vreeland. Changes in the hydrophobic-hydrophilic cell surface character of *Halomonas elongata* in response to NaCl. *Journal of Bacteriology* 170, no. 1 (1988): 132–135.

Hecker, Michael, Christine Heim, Uwe Völker, and Lothar Wölfel. Induction of stress proteins by sodium chloride treatment in *Bacillus subtilis*. *Archives of Microbiology* 150, no. 6 (1988): 564–566.

Hiramatsu, Toshiaki, Kazuyo Kodama, Teruo Kuroda, Tohru Mizushima, and Tomofusa Tsuchiya. A putative multi subunit Na+/H+ antiporter from *Staphylococcus aureus*. *Journal of Bacteriology* 180, no. 24 (1998): 6642–6648.

Hirsch, P., and B. Hoffmann. *Dichotomicrobium thermohalophilum*, gen. nov., spec. nov., budding prosthecate bacteria from the Solar Lake (Sinai) and some related strains. *Systematic and Applied Microbiology* 11, no. 3 (1989): 291–301.

Hof, T. An investigation of the microorganisms commonly present in salted beans. *Recueil des Travaux Botaniques Neerlandais* 32 (1935): 151–173.

Hoffmann, Tamara, Alexandra Schutz, Margot Brosius, Andrea Volker, Uwe Volker, and Erhard Bremer. High-salinity-induced iron limitation in *Bacillus subtilis*. *Journal of Bacteriology* 184, no. 3 (2002): 718–727.

Holtmann, Gudrun, Evert P. Bakker, Nobuyuki Uozumi, and Erhard Bremer. KtrAB and KtrCD: Two K+ uptake systems in *Bacillus subtilis* and their role in adaptation to hypertonicity. *Journal of Bacteriology* 185, no. 4 (2003): 1289–1298.

Horsburgh, Malcolm J., and Anne Moir. σM, an ECF RNA polymerase sigma factor of *Bacillus subtilis* 168, is essential for growth and survival in high concentrations of salt. *Molecular Microbiology* 32, no. 1 (1999): 41–50.

Hosseini Tafreshi, A., and M. Shariati. *Dunaliella* biotechnology: Methods and applications. *Journal of Applied Microbiology* 107, no. 1 (2009): 14–35.

Hua, Ngoc-Phuc, Amel Hamza-Chaffai, Russell H. Vreeland, Hiroko Isoda, and Takeshi Naganuma. *Virgibacillus salarius* sp. nov., a halophilic bacterium isolated from a Saharan salt lake. *International Journal of Systematic and Evolutionary Microbiology* 58, no. 10 (2008): 2409–2414.

Hua, Ngoc-Phuc, Atsuko Kanekiyo, Katsunori Fujikura, Hisato Yasuda, and Takeshi Naganuma. *Halobacillus profundi* sp. nov. and *Halobacillus kuroshimensis* sp. nov., moderately halophilic bacteria isolated from a deep-sea methane cold seep. *International Journal of Systematic and Evolutionary Microbiology* 57, no. 6 (2007): 1243–1249.

Imhoff, Johannes F., and Bernhard Thiemann. Influence of salt concentration and temperature on the fatty acid compositions of *Ectothiorhodospira* and other halophilic phototrophic purple bacteria. *Archives of Microbiology* 156, no. 5 (1991): 370–375.

Imhoff, Johannes F., and Hans G. Trüper. *Ectothiorhodospira abdelmalekii* sp. nov., a new halophilic and alkaliphilic phototrophic bacterium. *Zentralblatt für Bakteriologie Mikrobiologie und Hygiene: I. Abt. Originale C: Allgemeine, Angewandte und Ökologische Mikrobiologie* 2, no. 3 (1981): 228–234.

Ito, Masahiro, Arthur A. Guffanti, Bauke Oudega, and Terry A. Krulwich. mrp, a multigene, multifunctional locus in *Bacillus subtilis* with roles in resistance to cholate and to Na+ and in pH homeostasis. *Journal of Bacteriology* 181, no. 8 (1999): 2394–2402.

Ito, Masahiro, Haoxing Xu, Arthur A. Guffanti, Yi Wei, Lior Zvi, David E. Clapham, and Terry A. Krulwich. The voltage-gated Na+ channel NaVBP has a role in motility, chemotaxis, and pH homeostasis of an alkaliphilic *Bacillus*. *Proceedings of the National Academy of Sciences of the United States of America* 101, no. 29 (2004): 10566–10571.

Jaenicke, Rainer. Stability and stabilization of globular proteins in solution. *Journal of Biotechnology* 79, no. 3 (2000): 193–203.

Jaenicke, Rainer, and Gerald Bohm. The stability of proteins in extreme environments. *Current Opinion in Structural Biology* 8, no. 6 (1998): 738–748.

Janto, Benjamin, Azad Ahmed, Masahiro Ito, Jun Liu, David B. Hicks, Sarah Pagni, Oliver J. Fackelmayer et al. Genome of alkaliphilic *Bacillus pseudofirmus* OF4 reveals adaptations that support the ability to grow in an external pH range from 7.5 to 11.4. *Environmental Microbiology* 13, no. 12 (2011): 3289–3309.

Kajiyama, Yusuke, Masato Otagiri, Junichi Sekiguchi, Saori Kosono, and Toshiaki Kudo. Complex formation by the mrpABCDEFG gene products, which constitute a principal Na+/H+ antiporter in *Bacillus subtilis*. *Journal of Bacteriology* 189, no. 20 (2007): 7511–7514.

Kates, M., Suzanne Porter, and D. J. Kushner. *Actinopolyspora halophila* does not contain mycolic acids. *Canadian Journal of Microbiology* 33, no. 9 (1987): 822–823.

Kempf, Bettina, and Erhard Bremer. Uptake and synthesis of compatible solutes as microbial stress responses to high-osmolality environments. *Archives of Microbiology* 170, no. 5 (1998): 319–330.

Kennedy, Sean P., Wailap Victor Ng, Steven L. Salzberg, Leroy Hood, and Shiladitya DasSarma. Understanding the adaptation of *Halobacterium* species NRC-1 to its extreme environment through computational analysis of its genome sequence. *Genome Research* 11, no. 10 (2001): 1641–1650.

Kitada, Makio, Michizane Hashimoto, Toshiaki Kudo, and Koki Horikoshi. Properties of two different Na+/H+ antiport systems in alkaliphilic *Bacillus* sp. strain C-125. *Journal of Bacteriology* 176, no. 21 (1994): 6464–6469.

Kogut, Margot, Jeremy R. Mason, and Nicholas J. Russell. Isolation of salt-sensitive mutants of the moderately halophilic eubacterium *Vibrio costicola*. *Current Microbiology* 24, no. 6 (1992): 325–328.

Koller, Martin, Paula Hesse, Rodolfo Bona, Christoph Kutschera, Aid Atlić, and Gerhart Braunegg. Potential of various archaea and eubacterial strains as industrial polyhydroxyalkanoate producers from whey. *Macromolecular Bioscience* 7, no. 2 (2007): 218–226.

Kolp, Sonja, Markus Pietsch, Erwin A. Galinski, and Michael Gütschow. Compatible solutes as protectants for zymogens against proteolysis. *Biochimica et Biophysica Acta (BBA)—Proteins and Proteomics* 1764, no. 7 (2006): 1234–1242.

Krämer, Reinhard. Bacterial stimulus perception and signal transduction: Response to osmotic stress. *Chemical Record* 10, no. 4 (2010): 217–229.

Krulwich, Terry A., Masahiro Ito, and Arthur A. Guffanti. The Na$^+$-dependence of alkaliphily in *Bacillus*. *Biochimica et Biophysica Acta (BBA)—Bioenergetics* 1505, no. 1 (2001): 158–168.

Kuhlmann, Anne U., and Erhard Bremer. Osmotically regulated synthesis of the compatible solute ectoine in *Bacillus pasteurii* and related *Bacillus* spp. *Applied and Environmental Microbiology* 68, no. 2 (2002): 772–783.

Kushner, Donn J. Life in high salt and solute concentrations: Halophilic bacteria. In *Microbial Life in Extreme Environments*, ed. Donn J. Kushner, pp. 317–368. London: Academic Press, 1978.

Kushner, D. J. 1985. The Halobacteriaceae. In *The Bacteria. A Treatise on Structure and Function*, vol. VIII. *Archaebacteria*, ed. C. R. Woese and R. S. Wolfe, pp. 171–214. Orlando, FL: Academic Press, Inc.

Kushwaha, S. C., G. Juez-Perez, Francisco Rodriguez-Valera, M. Kates, and D. J. Kushner. Survey of lipids of a new group of extremely halophilic bacteria from salt ponds in Spain. *Canadian Journal of Microbiology* 28, no. 12 (1982): 1365–1372.

Lai, Qiliang, Jun Yuan, Li Gu, and Zongze Shao. *Marispirillum indicum* gen. nov., sp. nov., isolated from a deep-sea environment. *International Journal of Systematic and Evolutionary Microbiology* 59, no. 6 (2009): 1278–1281.

Lamark, T., I. Kaasen, Mark W. Eshoo, P. Falkenberg, J. McDougall, and Arne R. Strøm. DNA sequence and analysis of the bet genes encoding the osmoregulatory choline–glycine betaine pathway of *Escherichia coli*. *Molecular Microbiology* 5, no. 5 (1991): 1049–1064.

Lanyi, Janos K. Salt-dependent properties of proteins from extremely halophilic bacteria. *Bacteriological Reviews* 38, no. 3 (1974): 272.

Lanyi, Janos K. Halorhodopsin: A light-driven chloride ion pump. *Annual Review of Biophysics and Biophysical Chemistry* 15, no. 1 (1986): 11–28.

Lazrak, Tarik, Geneviève Wolff, Anne-Marie Albrecht, Yoichi Nakatani, Guy Ourisson, and Morris Kates. Bacterioruberins reinforce reconstituted *Halobacterium* lipid membranes. *Biochimica et Biophysica Acta (BBA)—Biomembranes* 939, no. 1 (1988): 160–162.

Lee, Jae-Chan, Jee-Min Lim, Dong-Jin Park, Che Ok Jeon, Wen-Jun Li, and Chang-Jin Kim. *Bacillus seohaeanensis* sp. nov., a halotolerant bacterium that contains L-lysine in its cell wall. *International Journal of Systematic and Evolutionary Microbiology* 56, no. 8 (2006): 1893–1898.

Legault, Boris A., Arantxa Lopez-Lopez, Jose C. Alba-Casado, W. Ford Doolittle, Henk Bolhuis, Francisco Rodriguez-Valera, and R. Thane Papke. Environmental genomics of "*Haloquadratum walsbyi*" in a saltern crystallizer indicates a large pool of accessory genes in an otherwise coherent species. *BMC Genomics* 7, no. 1 (2006): 171.

Leuko, Stefan, Mark J. Raftery, Brendan P. Burns, Malcolm R. Walter, and Brett A. Neilan. Global protein-level responses of *Halobacterium salinarum* NRC-1 to prolonged changes in external sodium chloride concentrations. *Journal of Proteome Research* 8, no. 5 (2009): 2218–2225.

Lillo, Jose Garcia, and Francisco Rodriguez-Valera. Effects of culture conditions on poly (β-hydroxybutyric acid) production by *Haloferax mediterranei*. *Applied and Environmental Microbiology* 56, no. 8 (1990): 2517–2521.

Lippert, Karin, and Erwin A. Galinski. Enzyme stabilization by ectoine-type compatible solutes: Protection against heating, freezing and drying. *Applied Microbiology and Biotechnology* 37, no. 1 (1992): 61–65.

Lloret, Javier, Luis Bolanos, M. Mercedes Lucas, Jan M. Peart, Nicholas J. Brewin, Ildefonso Bonilla, and Rafael Rivilla. Ionic stress and osmotic pressure induce different alterations in the lipopolysaccharide of a *Rhizobium meliloti* strain. *Applied and Environmental Microbiology* 61, no. 10 (1995): 3701–3704.

Lopez, Claudia S., Alejandro F. Alice, Horacio Heras, Emilio A. Rivas, and Carmen Sanchez-Rivas. Role of anionic phospholipids in the adaptation of *Bacillus subtilis* to high salinity. *Microbiology* 152, no. 3 (2006): 605–616.

Lopez, Claudia S., Horacio Heras, H. Garda, S. Ruzal, C. Sanchez-Rivas, and E. Rivas. Biochemical and biophysical studies of *Bacillus subtilis* envelopes under hyperosmotic stress. *International Journal of Food Microbiology* 55, no. 1 (2000): 137–142.

Lopez, Claudia S., Horacio Heras, Sandra M. Ruzal, Carmen Sanchez-Rivas, and Emilio A. Rivas. Variations of the envelope composition of *Bacillus subtilis* during growth in hyperosmotic medium. *Current Microbiology* 36, no. 1 (1998): 55–61.

López-López, Arantxa, María J. Pujalte, Susana Benlloch, Manuel Mata-Roig, Ramón Rosselló-Mora, Esperanza Garay, and Francisco Rodríguez-Valera. *Thalassospira lucentensis* gen. nov., sp. nov., a new marine member of the alpha-Proteobacteria. *International Journal of Systematic and Evolutionary Microbiology* 52, no. 4 (2002): 1277–1283.

Ma, Yanhe, Erwin A. Galinski, William D. Grant, Aharon Oren, and Antonio Ventosa. Halophiles 2010: Life in saline environments. *Applied and Environmental Microbiology* 76, no. 21 (2010): 6971–6981.

Madern, Dominique, Christine Ebel, and Giuseppe Zaccai. Halophilic adaptation of enzymes. *Extremophiles* 4, no. 2 (2000): 91–98.

Madigan, M. T., and B. L. Marrs. Extremophiles. *Scientific American* 276, no. 4 (1997): 82–87.

Magarinos, Beatriz, Jesus L. Romalde, Manuel Noya, Juan L. Barja, and Alicia E. Toranzo. Adherence and invasive capacities of the fish pathogen *Pasteurella piscicida*. *FEMS Microbiology Letters* 138, no. 1 (1996): 29–34.

Mandelli, Fernanda, Viviane S. Miranda, Eliseu Rodrigues, and Adriana Z. Mercadante. Identification of carotenoids with high antioxidant capacity produced by extremophile microorganisms. *World Journal of Microbiology and Biotechnology* 28, no. 4 (2012): 1781–1790.

Martinez-Canovas, M. Jose, Victoria Bejar, Fernando Martinez-Checa, Rafael Paez, and Emilia Quesada. *Idiomarina fontislapidosi* sp. nov. and *Idiomarina ramblicola* sp. nov., isolated from inland hypersaline habitats in Spain. *International Journal of Systematic and Evolutionary Microbiology* 54, no. 5 (2004a): 1793–1797.

Martinez-Canovas, M. Jose, Victoria Bejar, Fernando Martinez-Checa, and Emilia Quesada. *Halomonas anticariensis* sp. nov., from Fuente de Piedra, a saline-wetland wildfowl reserves in Malaga, southern Spain. *International Journal of Systematic and Evolutionary Microbiology* 54, no. 4 (2004b): 1329–1332.

Martinez-Canovas, M. Jose, Emilia Quesada, Inmaculada Llamas, and Victoria Bejar. *Halomonas ventosae* sp. nov., a moderately halophilic, denitrifying, exopolysaccharide-producing bacterium. *International Journal of Systematic and Evolutionary Microbiology* 54, no. 3 (2004c): 733–737.

Mevarech, Moshe, Felix Frolow, and Lisa M. Gloss. Halophilic enzymes: Proteins with a grain of salt. *Biophysical Chemistry* 86, no. 2 (2000): 155–164.

Mongodin, Emmanuel F., K. E. Nelson, S. Daugherty, R. T. Deboy, J. Wister, H. Khouri, J. Weidman et al. The genome of *Salinibacter ruber*: Convergence and gene exchange among hyperhalophilic bacteria and archaea. *Proceedings of the National Academy of Sciences of the United States of America* 102, no. 50 (2005): 18147–18152.

Moune, Sophie, Claire Eatock, Robert Matheron, John C. Willison, A. Hirschler, R. Herbert, and P. Caumette. *Orenia salinaria* sp. nov., a fermentative bacterium isolated from anaerobic sediments of Mediterranean salterns. *International Journal of Systematic and Evolutionary Microbiology* 50, no. 2 (2000): 721–729.

Nakagawa, Yuichi, Masahiro Sota, and Kazuya Koumoto. Cryoprotective ability of betaine-type metabolite analogs during freezing denaturation of enzymes. *Biotechnology Letters* 37, no. 8 (2015): 1607–1613.

Nath, Abhigyan, Radha Chaube, and Subbiah Karthikeyan. Discrimination of psychrophilic and mesophilic proteins using random forest algorithm. In *2012 International Conference on Biomedical Engineering and Biotechnology (iCBEB)*, pp. 179–182. Piscataway, NJ: IEEE, 2012.

Nath, Abhigyan, and Karthikeyan Subbiah. Inferring biological basis about psychrophilicity by interpreting the rules generated from the correctly classified input instances by a classifier. *Computational Biology and Chemistry* 53 (2014): 198–203.

Niederberger, Thomas D., Blaire Steven, Sophie Charvet, Beatrice Barbier, and Lyle G. Whyte. *Virgibacillus arcticus* sp. nov., a moderately halophilic, endospore-forming bacterium from permafrost in the Canadian High Arctic. *International Journal of Systematic and Evolutionary Microbiology* 59, no. 9 (2009): 2219–2225.

Nielsen, Preben, Dagmar Fritze, and Fergus G. Priest. Phenetic diversity of alkaliphilic *Bacillus* strains: Proposal for nine new species. *Microbiology* 141, no. 7 (1995): 1745–1761.

Nowlan, Bianca, Mital S. Dodia, Satya P. Singh, and B. K. C. Patel. *Bacillus okhensis* sp. nov., a halotolerant and alkalitolerant bacterium from an Indian saltpan. *International Journal of Systematic and Evolutionary Microbiology* 56, no. 5 (2006): 1073–1077.

Nozaki, Kaori, Kei Inaba, Teruo Kuroda, Masaaki Tsuda, and Tomofusa Tsuchiya. Cloning and sequencing of the gene for Na+/H+ antiporter of *Vibrio parahaemolyticus*. *Biochemical and Biophysical Research Communications* 222, no. 3 (1996): 774–779.

Ohno, Yoshimi, Ikuya Yano, and Masamiki Masui. Effect of NaCl concentration and temperature on the phospholipid and fatty acid compositions of a moderately halophilic bacterium. *Pseudomonas halosaccharolytica*, *J. Biochem. (Tokyo)* 85, (1979): 413.

Ohyanagi, Naho, Miwako Ishido, Fumihito Suzuki, Kayoko Kaneko, Tetsuo Kubota, Nobuyuki Miyasaka, and Toshihiro Nanki. Retinoid ameliorates experimental autoimmune myositis, with modulation of Th cell differentiation and antibody production in vivo. *Arthritis & Rheumatism* 60, no. 10 (2009): 3118–3127.

Oren, Aharon. *Clostridium lortetii* sp. nov., a halophilic obligatory anaerobic bacterium producing endospores with attached gas vacuoles. *Archives of Microbiology* 136, no. 1 (1983): 42–48.

Oren, Aharon. Intracellular salt concentrations of the anaerobic halophilic eubacteria *Haloanaerobium praevalens* and *Halobacteroides halobius*. *Canadian Journal of Microbiology* 32, no. 1 (1986): 4–9.

Oren, Aharon. *Halophilic Microorganisms and Their Environments*. Vol. 5. Berlin: Springer Science & Business Media, 2002.

Oren, Aharon. Life at high salt concentrations. In *The Prokaryotes, A Handbook on the Biology of Bacteria: Ecophysiology and Biochemistry,* ed. Dworkin, M., Falkow, S., Rosenberg, E., Schleifer, K.-H., Stackebrandt, E., 2: pp. 263–282. New York: Springer, 2006a.

Oren, Aharon. The order Haloanaerobiales. In *The Prokaryotes, A Handbook on the Biology of Bacteria: Ecophysiology and Biochemistry,* ed. Dworkin, M., Falkow, S., Rosenberg, E., Schleifer, K.-H., Stackebrandt, E., 4: pp. 809–822. New York: Springer, 2006b.

Oren, Aharon, and Peter Gurevich. Characterization of the dominant halophilic archaea in a bacterial bloom in the Dead Sea. *FEMS Microbiology Ecology* 12, no. 4 (1993): 249–256.

Oren, Aharon, Martin Kessel, and Erko Stackebrandt. *Ectothiorhodospira marismortui* sp. nov., an obligately anaerobic, moderately halophilic purple sulfur bacterium from a hypersaline sulfur spring on the shore of the Dead Sea. *Archives of Microbiology* 151, no. 6 (1989): 524–529.

Oren, Aharon, Frank Larimer, Paul Richardson, Alla Lapidus, and Laszlo N. Csonka. How to be moderately halophilic with broad salt tolerance: Clues from the genome of *Chromohalobacter salexigens*. *Extremophiles* 9, no. 4 (2005): 275–279.

Oren, Aharon, Heike Pohla, and Erko Stackebrandt. Transfer of *Clostridium lortetii* to a new genus *Sporohalobacter* gen. nov. as *Sporohalobacter lortetii* comb. nov., and description of *Sporohalobacter marismortui* sp. nov. *Systematic and Applied Microbiology* 9, no. 3 (1987): 239–246.

Padan, Etana, Noam Maisler, Daniel Taglicht, Rachel Karpel, and Shimon Schuldiner. Deletion of ant in *Escherichia coli* reveals its function in adaptation to high salinity and an alternative Na+/H+ antiporter system(s). *Journal of Biological Chemistry* 264, no. 34 (1989): 20297–20302.

Padan, Etana, and Shimon Schuldiner. Molecular physiology of Na+/H+ antiporters, key transporters in circulation of Na+ and H+ in cells. *Biochimica et Biophysica Acta (BBA)—Bioenergetics* 1185, no. 2 (1994): 129–151.

Pandey, Sandeep, and Satya P. Singh. Organic solvent tolerance of an α-amylase from haloalkaliphilic bacteria as a function of pH, temperature, and salt concentrations. *Applied Biochemistry and Biotechnology* 166, no. 7 (2012): 1747–1757.

Papaioannou, E. H., and M. Liakopoulou-Kyriakides. Substrate contribution on carotenoids production in *Blakeslea trispora* cultivations. *Food and Bioproducts Processing* 88, no. 2 (2010): 305–311.

Pastor, José M., Manuel Salvador, Montserrat Argandoña, Vicente Bernal, Mercedes Reina-Bueno, Laszlo N. Csonka, José L. Iborra, Carmen Vargas, Joaquín J. Nieto, and Manuel Cánovas. Ectoines in cell stress protection: Uses and biotechnological production. *Biotechnology Advances* 28, no. 6 (2010): 782–801.

Patel, Rajesh K., Mital S. Dodia, Rupal H. Joshi, and Satya P. Singh. Purification and characterization of alkaline protease from a newly isolated haloalkaliphilic *Bacillus* sp. *Process Biochemistry* 41, no. 9 (2006): 2002–2009.

Paul, Diby. Osmotic stress adaptations in rhizobacteria. *Journal of Basic Microbiology* 53, no. 2 (2013): 101–110.

Paul, Matthew J., Lucia F. Primavesi, Deveraj Jhurreea, and Yuhua Zhang. Trehalose metabolism and signalling. *Annual Review of Plant Biology* 59 (2008): 417–441.

Paul, Sandip, Sumit K. Bag, Sabyasachi Das, Eric T. Harvill, and Chitra Dutta. Molecular signature of hypersaline adaptation: Insights from genome and proteome composition of halophilic prokaryotes. *Genome Biology* 9, no. 4 (2008): 1.

Philip, Sheryl, Tajalli Keshavarz, and Ipsita Roy. Polyhydroxyalkanoates: Biodegradable polymers with a range of applications. *Journal of Chemical Technology and Biotechnology* 82, no. 3 (2007): 233–247.

Purohit, Megha K., Vikram H. Raval, and Satya P. Singh. Haloalkaliphilic bacteria: Molecular diversity and biotechnological applications. In *Geomicrobiology and Biogeochemistry*, ed. Ajay Singh and Nagina Parmar, pp. 61–79. Berlin: Springer, 2014.

Purohit, Megha K., and Satya P. Singh. Comparative analysis of enzymatic stability and amino acid sequences of thermostable alkaline proteases from two haloalkaliphilic bacteria isolated from coastal region of Gujarat, India. *International Journal of Biological Macromolecules* 49, no. 1 (2011): 103–112.

Quesada, Emilia, V. Bejar, and Conception Calvo. Exopolysaccharide production by *Volcaniella eurihalina*. *Experientia* 49, no. 12 (1993): 1037–1041.

Quillaguaman, Jorge, Osvaldo Delgado, Bo Mattiasson, and Rajni Hatti-Kaul. Poly (β-hydroxybutyrate) production by a moderate halophile, Halomonas boliviensis LC1. *Enzyme and Microbial Technology* 38, no. 1 (2006): 148–154.

Quillaguamán, Jorge, Thuoc Doan-Van, Hector Guzmán, Daniel Guzman, Javier Martín, Akaraonye Everest, and Rajni Hatti-Kaul. Poly (3-hydroxybutyrate) production by *Halomonas boliviensis* in fed-batch culture. *Applied Microbiology and Biotechnology* 78, no. 2 (2008): 227–232.

Quillaguamán, Jorge, Héctor Guzmán, Doan Van-Thuoc, and Rajni Hatti-Kaul. Synthesis and production of polyhydroxyalkanoates by halophiles: Current potential and future prospects. *Applied Microbiology and Biotechnology* 85, no. 6 (2010): 1687–1696.

Quillaguaman, Jorge, Suhaila Hashim, F. Bento, Bo Mattiasson, and Rajni Hatti-Kaul. Poly (β-hydroxybutyrate) production by a moderate halophile, Halomonas boliviensis LC1 using starch hydrolysate as substrate. *Journal of Applied Microbiology* 99, no. 1 (2005): 151–157.

Quillaguamán, Jorge, Marlene Muñoz, Bo Mattiasson, and Rajni Hatti-Kaul. Optimizing conditions for poly (β-hydroxybutyrate) production by *Halomonas boliviensis* LC1 in batch culture with sucrose as carbon source. *Applied Microbiology and Biotechnology* 74, no. 5 (2007): 981–986.

Raval, Vikram H., Megha K. Purohit, and Satya P. Singh. Diversity, population dynamics and biocatalytic potential of cultivable and non-cultivable bacterial communities of the saline ecosystems. In *Marine Enzymes for Biocatalysis: Sources, Biocatalytic Characteristics and Bioprocesses of Marine Enzymes*, ed. A. Trincone, pp. 165–180. Cambridge, UK: Woodhead Publishing, 2013.

Raval, Vikram H., Megha K. Purohit, and Satya P. Singh. Extracellular proteases from halophilic and haloalkaliphilic bacteria: Occurrence and biochemical properties. In *Halophiles*, ed. D.K. Maheshwari and Meenu Saraf, pp. 421–449. Berlin: Springer, 2015a.

Raval, Vikram H., Sumitha Pillai, Chirantan M. Rawal, and Satya P. Singh. Biochemical and structural characterization of a detergent-stable serine alkaline protease from seawater haloalkaliphilic bacteria. *Process Biochemistry* 49, no. 6 (2014): 955–962.

Raval, Vikram H., Chirantan M. Rawal, Sandeep Pandey, Hitarth B. Bhatt, Bharat R. Dahima, and Satya P. Singh. Cloning, heterologous expression and structural characterization of an alkaline serine protease from sea water haloalkaliphilic bacterium. *Annals of Microbiology* 65, no. 1 (2015b): 371–381.

Reed, Robert H., Stephen R. C. Warr, Nigel W. Kerby, and William D. P. Stewart. Osmotic shock-induced release of low molecular weight metabolites from free-living and immobilized cyanobacteria. *Enzyme and Microbial Technology* 8, no. 2 (1986): 101–104.

Rengpipat, Sirirat, S. E. Lowe, and J. G. Zeikus. Effect of extreme salt concentrations on the physiology and biochemistry of *Halobacteroides acetoethylicus*. *Journal of Bacteriology* 170, no. 7 (1988): 3065–3071.

Richard, Stephane B., Dominique Madern, Elsa Garcin, and Giuseppe Zaccai. Halophilic adaptation: Novel solvent protein interactions observed in the 2.9 and 2.6 Å resolution structures of the wild type and a mutant of malate dehydrogenase from *Haloarcula marismortui*. *Biochemistry* 39, no. 5 (2000): 992–1000.

Roberts, Mary F. Organic compatible solutes of halotolerant and halophilic microorganisms. *Saline Systems* 1, no. 1 (2005): 1.

Roberts, Mary F. Characterization of organic compatible solutes of halotolerant and halophilic microorganisms. *Methods in Microbiology* 35 (2006): 615–647.

Rodriguez-Valera, Francisco. Approaches to prokaryotic biodiversity: A population genetics perspective. *Environmental Microbiology* 4, no. 11 (2002): 628–633.

Rodriguez-Valera, Francisco, F. Ruiz-Berraquero, and A. Ramos-Cormenzana. Isolation of extremely halophilic bacteria able to grow in defined inorganic media with single carbon sources. *Microbiology* 119, no. 2 (1980): 535–538.

Rodriguez-Valera, Francisco, and Jos A. G. Lillo. Halobacteria as producers of polyhydroxyalkanoates. *FEMS Microbiology Letters* 103, no. 2–4 (1992): 181–186.

Rosenstein, Ralf, Detlinde Futter-Bryniok, and Friedrich Götz. The choline-converting pathway in *Staphylococcus xylosus* C2A: Genetic and physiological characterization. *Journal of Bacteriology* 181, no. 7 (1999): 2273–2278.

Russell, Nicholas J. Adaptive modifications in membranes of halotolerant and halophilic microorganisms. *Journal of Bioenergetics and Biomembranes* 21, no. 1 (1989): 93–113.

Russell, Nicholas J. Lipids of halophilic and halotolerant microorganisms. In *The Biology of Halophilic Bacteria*, ed. R. H. Vreeland and L. I. Hochstein, pp. 163–210. Boca Raton, FL: CRC Press, 1993.

Russell, Nicholas J., and M. Kogut. Haloadaptation: Salt sensing and cell-envelope changes. *Microbiological Sciences* 2, no. 11 (1985): 345–350.

Saito, T., Hiroaki Terato, and Osamu Yamamoto. Pigments of *Rubrobacter radiotolerans*. *Archives of Microbiology* 162, no. 6 (1994): 414–421.

Saponetti, Matilde Sublimi, Fabrizio Bobba, Grazia Salerno, Alessandro Scarfato, Angela Corcelli, and Annamaria Cucolo. Morphological and structural aspects of the extremely halophilic archaeon *Haloquadratum walsbyi*. *PLoS One* 6, no. 4 (2011): e18653.

Sarkar, Mohona, and Gary J. Pielak. An osmolyte mitigates the destabilizing effect of protein crowding. *Protein Science* 23, no. 9 (2014): 1161–1164.

Sauer, Thomas, and Erwin A. Galinski. Bacterial milking: A novel bioprocess for production of compatible solutes. *Biotechnology and Bioengineering* 57, no. 3 (1998): 306–313.

Schiraldi, Chiara, Carmelina Maresca, Angela Catapano, Erwin A. Galinski, and Mario De Rosa. High-yield cultivation of *Marinococcus* M52 for production and recovery of hydroxyectoine. *Research in Microbiology* 157, no. 7 (2006): 693–699.

Schobert, Brigitte, and Janos K. Lanyi. Halorhodopsin is a light-driven chloride pump. *Journal of Biological Chemistry* 257, no. 17 (1982): 10306–10313.

Schroeter, Rebecca, Tamara Hoffmann, Birgit Voigt, Hanna Meyer, Monika Bleisteiner, Jan Muntel, Britta Jürgen et al. Stress responses of the industrial workhorse *Bacillus licheniformis* to osmotic challenges. *PloS One* 8, no. 11 (2013): e80956.

Selmer-Olsen, E., Terje Sørhaug, Stein-Erik Birkeland, and R. Pehrson. Survival of *Lactobacillus helveticus* entrapped in Ca-alginate in relation to water content, storage and rehydration. *Journal of Industrial Microbiology and Biotechnology* 23, no. 2 (1999): 79–85.

Shakhnovich, Eugene I. Proteins with selected sequences fold into unique native conformation. *Physical Review Letters* 72, no. 24 (1994): 3907.

Shivaji, Sisinthy, Preeti Chaturvedi, Korpole Suresh, G. S. N. Reddy, C. B. S. Dutt, Milton Wainwright, Jayant V. Narlikar, and P. M. Bhargava. *Bacillus aerius* sp. nov., *Bacillus aerophilus* sp. nov., *Bacillus stratosphericus* sp. nov. and *Bacillus altitudinis* sp. nov., isolated from cryogenic tubes used for collecting air samples from high altitudes. *International Journal of Systematic and Evolutionary Microbiology* 56, no. 7 (2006): 1465–1473.

Singh, Om V., and Prashant Gabani. Extremophiles: Radiation resistance microbial reserves and therapeutic implications. *Journal of Applied Microbiology* 110, no. 4 (2011): 851–861.

Singh, Satya, Megha Purohit, Vikram Raval, Sandeep Pandey, Viral Akbari, and Chirantan Rawal. Capturing the potential of haloalkaliphilic bacteria from the saline habitats through culture dependent and metagenomics approaches. In *Current Research, Technology and Education Topics in Applied Microbiology and Microbial Biotechnology*, ed. A. Mendez-Vilas, pp. 81–87. Badajoz, Spain: Formatex, 2010.

Sleator, Roy D., and Colin Hill. Bacterial osmoadaptation: The role of osmolytes in bacterial stress and viru-lence. *FEMS Microbiology Reviews* 26, no. 1 (2002): 49–71.

Smithies, W. R., N. E. Gibbons, and S. T. Bayley. The chemical composition of the cell and cell wall of some halophilic bacteria. *Canadian Journal of Microbiology* 1, no. 8 (1955): 605–613.

Smole, Zlatko, Nela Nikolic, Fran Supek, Tomislav Šmuc, Ivo F. Sbalzarini, and Anita Krisko. Proteome sequence features carry signatures of the environmental niche of prokaryotes. *BMC Evolutionary Biology* 11, no. 1 (2011): 26.

Soppa, Jorg. From genomes to function: Haloarchaea as model organisms. *Microbiology* 152, no. 3 (2006): 585–590.

Spring, S., W. Ludwig, M. C. Marquez, A. Ventosa, and K-H. Schleifer. *Halobacillus* gen. nov., with descrip-tions of *Halobacillus litoralis* sp. nov. and *Halobacillus trueperi* sp. nov., and transfer of *Sporosarcina halophila* to *Halobacillus halophilus* comb. nov. *International Journal of Systematic and Evolutionary Microbiology* 46, no. 2 (1996): 492–496.

Steil, Leif, Tamara Hoffmann, Ina Budde, Uwe Volker, and Erhard Bremer. Genome-wide transcriptional profiling analysis of adaptation of *Bacillus subtilis* to high salinity. *Journal of Bacteriology* 185, no. 21 (2003): 6358–6370.

Steinbüchel, Alexander, and Bernd Füchtenbusch. Bacterial and other biological systems for polyester produc-tion. *Trends in Biotechnology* 16, no. 10 (1998): 419–427.

Steinbüchel, Alexander, and Henry E. Valentin. Diversity of bacterial polyhydroxyalkanoic acids. *FEMS Microbiology Letters* 128, no. 3 (1995): 219–228.

Strazzullo, Giuseppe, Agata Gambacorta, Filomena Monica Vella, Barbara Immirzi, Ida Romano, Valeria Calandrelli, Barbara Nicolaus, and Licia Lama. Chemical-physical characterization of polyhydroxyal-kanoates recovered by means of a simplified method from cultures of *Halomonas campaniensis*. *World Journal of Microbiology and Biotechnology* 24, no. 8 (2008): 1513–1519.

Sutherland, Ian W. *Biotechnology of Microbial Exopolysaccharides*. Vol. 9. Cambridge: Cambridge University Press, 1990.

Sutherland, Ian W. Exopolysaccharides in biofilms, flocs and related structures. *Water Science and Technology* 43, no. 6 (2001a): 77–86.

Sutherland, Ian W. Biofilm exopolysaccharides: A strong and sticky framework. *Microbiology* 147, no. 1 (2001b): 3–9.

Tadros, Monier Habib, Gerhart Drews, and Dierk Evers. Peptidoglycan and protein, the major cell wall con-stituents of the obligate halophilic bacterium *Rhodospirillum salexigens*. *Zeitschrift für Naturforschung C* 37, no. 3–4 (1982): 210–212.

Tan, Dan, Yuan-Sheng Xue, Gulsimay Aibaidula, and Guo-Qiang Chen. Unsterile and continuous production of polyhydroxybutyrate by *Halomonas* TD01. *Bioresource Technology* 102, no. 17 (2011): 8130–8136.

Tardieu, Annette, Françoise Bonnete, Stephanie Finet, and Denis Vivares. Understanding salt or PEG induced attractive interactions to crystallize biological macromolecules. *Acta Crystallographica Section D: Biological Crystallography* 58, no. 10 (2002): 1549–1553.

Tardy-Jacquenod, C., M. Magot, B. K. C. Patel, Robert Matheron, and Pierre Caumette. *Desulfotomaculum halophilum* sp. nov., a halophilic sulfate-reducing bacterium isolated from oil production facilities. *International Journal of Systematic and Evolutionary Microbiology* 48, no. 2 (1998): 333–338.

Tekaia, Fredj, and Edouard Yeramian. Evolution of proteomes: Fundamental signatures and global trends in amino acid compositions. *BMC Genomics* 7, no. 1 (2006): 307.

Tekaia, Fredj, Edouard Yeramian, and Bernard Dujon. Amino acid composition of genomes, lifestyles of organ-isms, and evolutionary trends: A global picture with correspondence analysis. *Gene* 297, no. 1 (2002): 51–60.

Thomson, Jennifer A., D. R. Woods, and R. L. Welton. Collagenolytic activity of aerobic halophiles from hides. *Microbiology* 70, no. 2 (1972): 315–319.

Tsapis, Andreas, and Adam Kepes. Transient breakdown of the permeability barrier of the membrane of *Escherichia coli* upon hypoosmotic shock. *Biochimica et Biophysica Acta (BBA)—Biomembranes* 469, no. 1 (1977): 1–12.

Tsuchiya, Tomofusa, Jane Raven, and T. Hastings Wilson. Co-transport of Na+ and methyl-β-D-thiogalactopyranoside mediated by the melibiose transport system of *Escherichia coli*. *Biochemical and Biophysical Research Communications* 76, no. 1 (1977): 26–31.

Tsuzuki, Minoru, Oleg V. Moskvin, Masayuki Kuribayashi, Kiichi Sato, Susana Retamal, Mitsuru Abo, Jill Zeilstra-Ryalls, and Mark Gomelsky. Salt stress-induced changes in the transcriptome, compatible solutes, and membrane lipids in the facultatively phototrophic bacterium *Rhodobacter sphaeroides*. *Applied and Environmental Microbiology* 77, no. 21 (2011): 7551–7559.

Usami, Ron, Akinobu Echigo, Tadamasa Fukushima, Toru Mizuki, Yasuhiko Yoshida, and Masahiro Kamekura. *Alkalibacillus silvisoli* sp. nov., an alkaliphilic moderate halophile isolated from non-saline forest soil in Japan. *International Journal of Systematic and Evolutionary Microbiology* 57, no. 4 (2007): 770–774.

Vargas, Carmen, Montserrat Argandoña, Mercedes Reina-Bueno, Javier Rodríguez-Moya, Cristina Fernández-Aunión, and Joaquín J. Nieto. Unravelling the adaptation responses to osmotic and temperature stress in *Chromohalobacter salexigens*, a bacterium with broad salinity tolerance. *Saline Systems* 4, no. 1 (2008): 1.

Ventosa, Antonio. Taxonomy of moderately halophilic heterotrophic eubacteria. *Halophilic Bacteria* 1 (1988): 71–84.

Ventosa, Antonio, Joaquín J. Nieto, and Aharon Oren. Biology of moderately halophilic aerobic bacteria. *Microbiology and Molecular Biology Reviews* 62, no. 2 (1998): 504–544.

Vreeland, Russell H. Mechanisms of halotolerance in microorganisms. *CRC Critical Reviews in Microbiology* 14, no. 4 (1987): 311–356.

Vreeland, Russell H., R. Anderson, and R. G. Murray. Cell wall and phospholipid composition and their contribution to the salt tolerance of *Halomonas elongata*. *Journal of Bacteriology* 160, no. 3 (1984): 879–883.

Vreeland, Russell H., C. D. Litchfield, E. L. Martin, and E. Elliot. *Halomonas elongata*, a new genus and species of extremely salt-tolerant bacteria. *International Journal of Systematic and Evolutionary Microbiology* 30, no. 2 (1980): 485–495.

Wang, Kuirong, Zhang, Shujun, LI, Shaohe, Sang, Xiaoxue, and Bai, Linhan. Osmotolerance features and mechanisms of a moderately halophilic bacterium Halomonas sp. NY-011*: Osmotolerance features and mechanisms of a moderately halophilic bacterium halomonas sp. NY-011. *Chinese Journal of Applied Environmental Biology* 16 (2010): 256–260. doi:10.3724/SP.J.1145.2010.00256.

Welsh, David T. Ecological significance of compatible solute accumulation by micro-organisms: From single cells to global climate. *FEMS Microbiology Reviews* 24, no. 3 (2000): 263–290.

Wood, Janet M. Osmosensing by bacteria: Signals and membrane-based sensors. *Microbiology and Molecular Biology Reviews* 63, no. 1 (1999): 230–262.

Xu, Ping, Wen-Jun Li, Shu-Kun Tang, Yu-Qin Zhang, Guo-Zhong Chen, Hua-Hong Chen, Li-Hua Xu, and Cheng-Lin Jiang. Naxibacter alkalitolerans gen. nov., sp. nov., a novel member of the family "Oxalobacteraceae" isolated from China. *International Journal of Systematic and Evolutionary Microbiology* 55, no. 3 (2005): 1149–1153.

Yin, Jin, Jin-Chun Chen, Qiong Wu, and Guo-Qiang Chen. Halophiles, coming stars for industrial biotechnology. *Biotechnology Advances* 33, no. 7 (2015): 1433–1442.

Yin, Liang, Yanfen Xue, and Yanhe Ma. Global microarray analysis of alkaliphilic halotolerant bacterium *Bacillus* sp. N16-5 salt stress adaptation. *PloS One* 10, no. 6 (2015): e0128649.

Yoon, Jung-Hoon, Kook Hee Kang, Tae-Kwang Oh, and Yong-Ha Park. *Halobacillus locisalis* sp. nov., a halophilic bacterium isolated from a marine solar saltern of the Yellow Sea in Korea. *Extremophiles* 8, no. 1 (2004): 23–28.

Yoon, Jung-Hoon, Kook Hee Kang, and Yong-Ha Park. *Lentibacillus salicampi* gen. nov., sp. nov., a moderately halophilic bacterium isolated from a salt field in Korea. *International Journal of Systematic and Evolutionary Microbiology* 52, no. 6 (2002): 2043–2048.

Yoon, Jung-Hoon, So-Jung Kang, Choong-Hwan Lee, Hyun Woo Oh, and Tae-Kwang Oh. *Halobacillus yeomjeoni* sp. nov., isolated from a marine solar saltern in Korea. *International Journal of Systematic and Evolutionary Microbiology* 55, no. 6 (2005): 2413–2417.

Yue, Haitao, Chen Ling, Tao Yang, Xiangbin Chen, Yuling Chen, Haiteng Deng, Qiong Wu, Jinchun Chen, and Guo-Qiang Chen. A seawater-based open and continuous process for polyhydroxyalkanoates production by recombinant *Halomonas campaniensis* LS21 grown in mixed substrates. *Biotechnology for Biofuels* 7, no. 1 (2014): 1.

Zhang, Guangya, and Huihua Ge. Protein hypersaline adaptation: Insight from amino acids with machine learning algorithms. *Protein Journal* 32, no. 4 (2013): 239–245.

Zhang, Guangya, and Lin Yi. Stability of halophilic proteins: From dipeptide attributes to discrimination classifier. *International Journal of Biological Macromolecules* 53 (2013): 1–6.

Zhang, Gwang I., Chung Y. Hwang, and Byung C. Cho. *Thalassobaculum litoreum* gen. nov., sp. nov., a member of the family Rhodospirillaceae isolated from coastal seawater. *International Journal of Systematic and Evolutionary Microbiology* 58, no. 2 (2008): 479–485.

Zhang, Ling-Hua, Ya-Jun Lang, and Shinichi Nagata. Efficient production of ectoine using ectoine-excreting strain. *Extremophiles* 13, no. 4 (2009): 717–724.

Zhang, Yun-Jiao, Yu Zhou, Man Ja, Rong Shi, Wei-Xun Chun-Yu, Ling-Ling Yang, Shu-Kun Tang, and Wen-Jun Li. *Virgibacillus albus* sp. nov., a novel moderately halophilic bacterium isolated from Lop Nur Salt Lake in Xinjiang Province, China. *Antonie van Leeuwenhoek* 102, no. 4 (2012): 553–560.

Zhilina, Tatjana N., E. S. Garnova, T. P. Tourova, N. A. Kostrikina, and G. A. Zavarzin. *Halonatronum saccharophilum* gen. nov. sp. nov.: A new haloalkaliphilic bacterium of the order Haloanaerobiales from Lake Magadi. *Microbiology* 70, no. 1 (2001): 64–72.

Zhilina, Tatjana N., T. P. Tourova, B. B. Kuznetsov, N. A. Kostrikina, and A. M. Lysenko. *Orenia sivashensis* sp. nov., a new moderately halophilic anaerobic bacterium from Lake Sivash lagoons. *Microbiology* 68, no. 4 (1999): 452–459.

Zhilina, Tatjana N., Georgy A. Zavarzin, Ekaterina N. Detkova, and Fred A. Rainey. *Natroniella acetigena* gen. nov. sp. nov., an extremely haloalkaliphilic, homoacetic bacterium: A new member of Haloanaerobiales. *Current Microbiology* 32, no. 6 (1996): 320–326.

CHAPTER 8

Deep-Sea Vent Extremophiles
Cultivation, Physiological Characteristics, and Ecological Significance

Sayaka Mino and Satoshi Nakagawa

CONTENTS

8.1 INTRODUCTION

Deep-sea hydrothermal vents and luxuriant animal communities were first discovered along with the Galapagos Spreading Center in 1977 (Spiess et al. 1980). This finding totally changed traditional views that the deep ocean was an environment too extreme to maintain life due to the darkness, high pressure, coldness, and scarcity of foods. Deep-sea vent ecosystems now represent extremely productive ecosystems primarily supported by microbial chemosynthesis at various temperatures generated by mixing between hydrothermal fluids and ambient seawater. Over the past 35 years, physiologically and phylogenetically diverse bacteria and archaea from deep-sea vents have attracted microbiologists. In particular, (hyper)thermophiles have gained wide interest, not only due to their unique physiological characters suitable for biotechnological applications, but also due to the fact that they provide insights into the origin of life. With developments in sample collection schemes from deep-sea hydrothermal environments (the vent fields are listed in the InterRidge Vents Database, http://vents-data.interridge.org) and in cultivation techniques, an increasing number of (hyper)thermophiles have been isolated in pure cultures. Although culture-dependent studies provide direct knowledge about microbial ecophysiological features, they lead to a biased view on the actual microbial diversity (Amann et al. 1995). Hence, culture-independent molecular eco-logical approaches have been advanced to detect "not-yet-cultured" microbial taxa. The progress in both culture-dependent and -independent techniques has improved our understandings of the microbial ecology in deep-sea hydrothermal systems. In this chapter, we review studies on deep-sea vent thermophiles, especially those on community composition, and ecological features.

8.2 CULTURE-DEPENDENT STUDIES ON THERMOPHILES

8.2.1 Metabolic Diversity and Physical and Chemical Characteristics of Growth of (Hyper)Thermophiles from Deep-Sea Hydrothermal Vents

At deep-sea hydrothermal environments, thermal and chemical gradients are established between hydrothermal vent fluids (up to 400°C) and ambient seawater. These environments include suitable habitats for thermophilic and hyperthermophilic microorganisms with optimal temperatures for growth above 50°C and above 80°C, respectively. Since *Methanocaldococcus jannaschii* (formerly *Methanococcus*) was first isolated from a deep-sea hydrothermal field (Jones et al. 1983; Whitman 2001), a large number of (hyper)thermophilic Archaea and Bacteria have been isolated from many types of deep-sea hydrothermal samples, such as chimneys, sediments, and hydrothermal fluids. In Table 8.1, we list both chemoorganotrophic and chemolithotrophic (hyper)thermophiles isolated from deep-sea hydrothermal environments and characterized by different sets of features.

- *Temperature*: All known deep-sea vent hyperthermophiles belong to the domain Archaea (Table 8.1). Their optimum temperatures for growth range from 80°C to 98°C. Thermophiles listed in Table 8.1 grow in a wide temperature range, from 25°C to 110°C. Using a newly developed high-pressure cultivation technique, the growth ability of the novel hyperthermophilic *Methanopyrus* strain was tested. This strain showed the cellular proliferation under 122°C and extended the upper tempera-ture limit for life on the planet (Takai et al. 2008). In addition to the limit of growth temperature, another methanogenic archaeon isolated from the deep-sea hydrothermal vent fluid set the upper temperature limit of nitrogen fixation at 92°C (Mehta and Baross 2006). These increases in tem-perature limits imply that life could occur in a wide range of environments on the planet.
- *pH*: Although pure hydrothermal fluids are characterized by acidic pH, the microbial habitat in deep-sea vents is neutral to weakly acidic due to the mixing of vent fluids with ambient seawater. Most (hyper)thermophiles from deep-sea hydrothermal environments grow best at near-neutral pH, but some acidophilic thermophiles were isolated. *Lebetimonas acidiphila* was for the first time reported as a culturable thermoacidophilic microorganism from deep-sea hydrothermal environ-ments (Takai et al. 2005). *L. acidiphila* grows from pH 4.2 to 7.0 (optimally at pH 5.2). Subsequently, the obligately acidophilic archaeon *Candidatus* 'Aciduliprofundum boonei,' a member of Deep-Sea Hydrothermal Vent Euryarchaeotic Group 2 (DHVE2), was isolated and showed preference for growth in acidic conditions ranging from pH 3.3 to 5.8 (optimally pH 4.5) (Reysenbach et al. 2006). Recently, a novel moderately thermoacidophilic *Mesoaciditoga lauensis*, a member of the order Thermotogales, was isolated from the hydrothermal chimney emitting acidic fluids (pH 2.8). The bacterium grows from pH 4.1 to 6.0 (optimally pH 5.5–5.7) (Reysenbach et al. 2013). On the other hand, for alkaline conditions, the Lost City field in the Mid-Atlantic Ridge represents alkaline deep-sea hydrothermal vents emitting moderately high-temperature (up to 90°C) and high-pH (up to pH 11.2) hydrothermal fluids. Although no alkaliphiles have been isolated from the field so far, this kind of alkaline environment might be suitable for the origin of life (see the review in Martin et al. 2008).
- *Redox potential, electron donors, and acceptors*: The broad redox gradient formed between reduc-tive fluids and oxidative ambient seawater associates with the metabolic versatility of deep-sea vent thermophiles. Deep-sea vent (hyper)thermophiles, including both chemoorganotrophs and chemo-lithotrophs, utilize diverse combinations of electron donors and acceptors (e.g., organic matter, H_2, H_2S, S_0, $S_2O_3^{2-}$, O_2, NO_2^-, SO_4^{2-}, and Fe(III)) and carbon sources (e.g., organic matter, CH_4, and CO_2)

Table 8.1 List of (Hyper)Thermophiles Isolated from Deep-Sea Hydrothermal Environments

Domain	Phylum	Genus	Species	Temperature Range, °C	Electron Donor	Electron Acceptor	C Source	Reference
Archaea	Crenarchaeota	Aeropyrum	camini	70–97 (85)	Org	O_2	Org	Nakagawa et al., 2004
		Ignicoccus	pacificus	70–98 (90)	H_2	S^0	CO_2	Huber et al., 2000
		Staphylothermus	marinus	65–98 (92)	Org	–	Org	Fiala et al., 1986
		Pyrodictium	abyssi	80–110 (97)	Org	–	Org	Pley et al., 1991
		Pyrolobus	fumarii	90–113 (106)	H_2	NO_2^-, $S_2O_3^{2-}$, O_2	CO_2	Blöchl et al., 1997
	Euryarchaeota	Methanopyrus	kandleri	84–110 (98)	H_2	CO_2	CO_2	Kurr et al., 1991
		Archaeoglobus	profundus	65–90 (82)	H_2, Org	SO_3^{2-}, $S_2O_3^{2-}$, SO_4^{2-}	Org	Burggraf et al., 1990
			veneficus	65–85 (80)	H_2, Org	SO_3^{2-}, $S_2O_3^{2-}$	Org, CO_2	Huber et al., 1997
			infectus	60–75 (70)	H_2, Org	SO_3^{2-}, $S_2O_3^{2-}$	Org	Mori et al., 2008
			sulfaticallidus	60–80 (75)	H_2, Org	SO_3^{2-}, $S_2O_3^{2-}$, SO_4^{2-}	Org, CO_2	Steinsbu et al., 2010
		Geoglobus	acetivorans	50–85 (81)	H_2, Org	Fe^{3+}	Org, CO_2	G. Slobodkina et al., 2009b
			ahangari	65–90 (88)	H_2, Org	Fe^{3+}	Org, CO_2	Kashefi et al., 2002
		Methanocaldococcus	bathoardescens	58–90 (82)	H_2	CO_2	CO_2	Stewart et al., 2015
			fervens	48–92 (85)	H_2	CO_2	CO_2	Jeanthon et al., 1999
			indicus	50–86 (85)	H_2	CO_2	CO_2	L'Haridon et al., 2003
			infernus	55–91 (85)	H_2	CO_2	CO_2	Jeanthon et al., 1998
			vulcanius	49–89 (80)	H_2	CO_2	CO_2	Jeanthon et al., 1999
		Methanotorris	formicicus	55–83 (75)	H_2, Org	CO_2	CO_2	Takai et al., 2004c
		Methanothermococcus	okinawarnsis	40–75 (60–65)	H_2, Org	CO_2	CO_2	Takai et al., 2002
		Palaeococcus	ferrophilus	60–88 (83)	Org	–	Org	Takai et al., 2000
		Pyrococcus	glycovorans	75–104 (95)	Org	–	Org	Barbier et al., 1999
			horikoshii	80–102 (98)	Org	–	Org	Gonzalez et al., 1998
			yayanosii	67–102 (96)	Org	–	Org	Birrien et al., 2011
		Thermococcus	aggregans	60–94 (88)	Org	–	Org	Canganella et al., 1998
			barophilus	48–95 (85)	Org	–	Org	Marteinsson et al., 1999

(Continued)

Table 8.1 (Continued) List of (Hyper)Thermophiles Isolated from Deep-Sea Hydrothermal Environments

Domain	Phylum	Genus	Species	Temperature Range, °C	Electron Donor	Electron Acceptor	C Source	Reference
			barossii	60–92 (82.5)	Org	—	Org	Duffaud et al., 1998
			celericrescens	50–85 (80)	Org	—	Org	Kuwabara et al., 2007
			chitonophagus	60–93 (85)	Org	—	Org	Huber et al., 1995
			cleftensis	55–94 (88)	Org	—	Org	Hensley et al., 2014
			coalescens	57–90 (87)	Org	—	Org	Kuwabara et al., 2005
			fumicolans	73–103 (85)	Org	—	Org	Godfroy et al., 1996
			gammatolerans	55–95 (88)	Org	—	Org	Jolivet et al., 2003
			guaymasensis	56–90 (88)	Org	—	Org	Canganella et al., 1998
			hydrothermalis	55–100 (80–90)	Org	—	Org	Godfroy et al., 1997
			nautili	55–95 (87.5)	Org	—	Org	Gorlas et al., 2014
			paralvinellae	50–91 (82)	Org	—	Org	Hensley et al., 2014
			peptonophilus	60–100 (85)	Org	—	Org	Gonzalez et al., 1995
			prieurii	60–95 (80)	Org	—	Org	Gorlas et al., 2013
			profundus	50–90 (80)	Org	—	Org	Kobayashi et al., 1994
			siculi	50–92 (85)	Org	—	Org	Grote et al., 1999
			thioreducens	55–94 (83–85)	Org	—	Org	Pikuta et al., 2007
Bacteria	Aquificae	Hydrogenivirga	okinawensis	65–85 (70–75)	S^0, $S_2O_3^{2-}$	NO_3^-, O_2	CO_2	Nunoura et al., 2008a
		Balnearium	lithotrophicum	45–80 (70–75)	H_2	S^0	CO_2	Takai et al., 2003b
			crinifex	50–70 (60–65)	H_2	NO_3^-, S^0	CO_2	Alain et al., 2003
			pacificum	55–85 (75)	H_2	NO_3^-, S^0, $S_2O_3^{2-}$	CO_2	L'Haridon et al., 2006b
			atlanticum	50–80 (70–75)	H_2	$S_2O_3^{2-}$	CO_2	L'Haridon et al., 2006b
		Phorcysia	thermohydrogeniphila	65–80 (75)	H_2	NO_3^-, S^0	CO_2	Pérez-Rodriguez et al., 2012
		Thermovibrio	ammonificans	60–80 (75)	H_2	NO_3^-, S^0	CO_2	Vetriani et al., 2004
			guaymasensis	50–88 (75–80)	H_2	NO_3^-, S^0	CO_2	L'Haridon et al., 2006b

(Continued)

Table 8.1 (Continued) List of (Hyper)Thermophiles Isolated from Deep-Sea Hydrothermal Environments

Domain	Phylum	Genus	Species	Temperature Range, °C	Electron Donor	Electron Acceptor	C Source	Reference
		Persephonella	marina	55–80 (73)	H_2, $S_2O_3^{2-}$, S^0	NO_3^-, S^0, O_2	CO_2	Götz et al., 2002
			guaymasensis	55–75 (70)	H_2, $S_2O_3^{2-}$, S^0	NO_3^-, O_2	CO_2	Götz et al., 2002
			hydrogeniphila	50–72.5 (70)	H_2	NO_3^-, O_2	CO_2	Nakagawa et al., 2003
	Bacteroidetes	Rhodothermus	profundi	55–80 (70–75)	Org	O_2	Org	Marteinsson et al., 2010
	Deferribacteres	Deferribacter	autotrophicus	52–75 (60)	H_2, Org	Fe^{3+}, Mn^{4+}, NO_3^-, S^0	Org, CO_2	G. Slobodkina et al., 2009a
			abyssi	45–65 (60)	H_2, Org	Fe^{3+}, NO_3^-, S^0	Org, CO_2	Miroshnichenko et al., 2003d
			desulfuricans	40–70 (60–65)	Org	S^0, NO_3^-, AsO_4^{3-}	Org	Takai et al., 2003a
	Unclassified	Caldithrix	abyssi	40–70 (60)	H_2, Org	NO_3^-	Org	Miroshnichenko et al., 2003a
	Deinococcusthermus	Marinithermus	hydrothermalis	50–72.5 (67.5)	Org	O_2	Org	Sako et al., 2003
		Oceanithermus	profundus	40–68 (60)	H_2, Org	NO_3^-, O_2	Org	Miroshnichenko et al., 2003b
			desulfurans	30–65 (60)	Org	NO_3^-, NO_2^-, S^0, O_2	Org	Mori et al., 2004
		Vulcanithermus	mediatlanticus	37–80 (70)	H_2, Org	NO_3^-, O_2	Org	Miroshnichenko et al., 2003c
	Furmicutes	Tepidibacter	thalassicus	33–60 (50)	Org	–	Org	A. Slobodkin et al., 2003
		Caminicella	sporogens	45–65 (55–60)	Org	–	Org	Alain et al., 2002c
		Caloranaerobacter	tepidiprofundi	22–60 (50)	Org	–	Org	G. Slobodkina et al., 2008
			ferrireducens	40–70 (60)	Org	–	Org	Zeng et al., 2015
			azorensis	45–65 (65)	Org	–	Org	Wery et al., 2001b
			subterraneus	45–65 (65)	Org	–	Org	Sokolova et al., 2001
		Vulcanibacillus	modesticaldus	37–60 (55)	Org	NO_3^-	Org	L'Haridon et al., 2006a

(Continued)

Table 8.1 (Continued) List of (Hyper)Thermophiles Isolated from Deep-Sea Hydrothermal Environments

Domain	Phylum	Genus	Species	Temperature Range, °C	Electron Donor	Electron Acceptor	C Source	Reference
	Proteobacteria	Nautilia	lithotrophica	37–68 (53)	H_2, Org	S^0	Org, CO_2	Miroshnichenko et al., 2002
			nitratireducens	25–65 (55)	H_2, Org	NO_3^-, S^0, $S_2O_3^{2-}$, SeO_4^{2-}	Org, CO_2	Pérez-Rodríguez et al., 2010
			abyssi	33–65 (60)	H_2	S^0	Org, CO_2	Alain et al., 2009
		Hydrogenimonas	thermophilus	35–65 (55)	H_2	NO_3^-, S^0, O_2	CO_2	Takai et al., 2004b
		Nitratiruptor	tergarcus	40–57 (55)	H_2	NO_3^-, S^0, O_2	CO_2	Nakagawa et al., 2005c
		Caminibacter	hydrogeniphilus	50–70 (60)	H_2	NO_3^-, S^0	Org, CO_2	Alain et al., 2002d
			profundus	45–65 (55)	H_2	NO_3^-, S^0, O_2	CO_2	Miroshnichenko et al., 2004
			mediatlanticus	45–70 (55)	H_2	NO_3^-, S^0	CO_2	Voordeckers et al., 2005
		Lebetimonas	acidiphila	30–68 (50)	H_2	S^0	CO_2	Takai et al., 2005
		Geothermobacter	ehrlichii	35–65 (55)	Org	NO_3^-, Fe^{3+}	Org	Kashefi et al., 2003
		Deferrisoma	camini	36–62 (50)	Org	Fe^{3+}, S^0	Org	G. Slobodkina et al., 2012
		Hippea	maritima	55–60 (52–54)	H_2, Org	S^0	Org, CO_2	Miroshnichenko et al., 1999
			alviniae	45–75 (60)	H_2, Org	S^0	Org, CO_2	Flores et al., 2012c
			jasoniae	40–72 (60–65)	H_2, Org	S^0	Org, CO_2	Flores et al., 2012c
		Dissulfuribacter	thermophilus	28–70 (61)	S^0, H_2	H_2O	CO_2	A. Slobodkin et al., 2013
	Spirohaetes	Exilispira	thermophila	37–60 (50)	Org	–	Org	Imachi et al., 2008
	Thermo-desulfobacteria	Thermodesulfobacterium	hydrogeniphilum	50–80 (75)	H_2	SO_4^{2-}	CO_2	Jeanthon et al., 2002
		Thermo-desulfatator	indicus	55–80 (70)	H_2	SO_4^{2-}	CO_2	Moussard et al., 2004

(Continued)

Table 8.1 (Continued) List of (Hyper)Thermophiles Isolated from Deep-Sea Hydrothermal Environments

Domain	Phylum	Genus	Species	Temperature Range, °C	Electron Donor	Electron Acceptor	C Source	Reference
			atlanticus	55–75 (65–70)	H_2	SO_4^{2-}	Org, CO_2	Alain et al., 2010
		Thermo-sulfidibacter	takaii	55–78 (70)	H_2	S^0	CO_2	Nunoura et al. 2008b
		Thermo-sulfurimonas	dismutans	50–92 (74)	S^0, SO_3^{2-}, $S_2O_3^{2-}$	H_2O	CO_2	A. Slobodkin et al., 2012
	Thermotogae	Mesoaciditoga	lauensis	45–65 (57–60)	Org	S^0, $S_2O_3^{2-}$	Org	Reysenbach et al., 2013
		Marinitoga	camini	25–65 (55)	Org	–, S^0, L-cysteine	Org	Wery et al., 2001a
			piezophila	45–70 (65)	Org	S^0, $S_2O_3^{2-}$, L-cysteine	Org	Alain et al., 2002a
			hydrogenitolerans	35–65 (60)	Org	–, S^0, $S_2O_3^{2-}$, L-cysteine	Org	Postec et al., 2005
			okinawensis	30–70 (55–60)	Org	–, S^0, L-cysteine	Org	Nunoura et al., 2007
		Thermosipho	affectus	37–75 (70)	Org	–	Org	Podosokorskaya et al., 2011
			melanesiensis	50–75 (70)	Org	–	Org	Antoine et al., 1997
			japonicus	45–80 (72)	Org	–	Org	Takai and Horikoshi, 2000
			atlanticus	45–80 (65)	Org	–	Org	Urios et al., 2004
	Desulfurobacterium	desulfurobacterium	thermolithotrophum	40–75 (70)				L'Haridon et al., 1998
			crinifex	50–70 (60–65)	H_2	NO_3^-, S^0	CO_2	Alain et al., 2003

(Table 8.1). In deep-sea hydrothermal systems around the globe, dominant culturable chemolithotrophic thermophiles are represented by sulfur oxidizers within the class Epsilonproteobacteria, followed by hydrogen oxidizers within the orders Aquificales and Metanococcales (Takai et al. 2001, 2004a; Nakagawa et al. 2005a, 2005b; Nunoura and Takai 2009; Takai and Nakamura 2011). This is consistent with a thermodynamic study, in which thiotrophic chemolithotrophy was suggested to become predominant in any type of hydrothermal systems and temperature, and hydrogenotrophs are possibly fostered only in high-temperature zones (Takai and Nakamura 2011).

Most deep-sea thermophiles listed in Table 8.1 grow under strictly anaerobic conditions, whereas thermophilic bacterium *Marinithermus hydrothermalis* (Sako et al. 2003) and hyperthermophilic archaeon *Aeropyrum camini* (Nakagawa et al. 2004) are strict aerobes. Indeed, they were isolated from the surface part of the chimney structure, indicating that these microorganisms can grow in the oxidative microhabitats in deep-sea hydrothermal vent environments (Nakagawa et al. 2004).

- *Hydrostatic pressure*: Most deep-sea vent thermophiles do not require in situ pressure for their growth. *Thermococcus barophilus* and *Marinitoga piezophila* are known as piezophiles, which grow well under in situ hydrostatic pressure (40 MPa) conditions (Marteinsson et al. 1999; Alain et al. 2002a), although they are not obligate piezophiles. The first obligate piezo-hyperthermophilic *Pyrococcus* strain was isolated and set the upper pressure–temperature limit (up to 120 MPa and 108°C) for life (Zeng et al. 2009).

8.3 METHODOLOGY OF CULTIVATION OF DEEP-SEA VENT MICROORGANISMS

The anaerobic technique is required for cultivating deep-sea thermophilic anaerobes. Details of anaerobic cultivation techniques are described in Nakagawa and Takai (2006). Anaerobic cultivation is usually performed in bottles or tubes sealed with gas-tight butyl rubber stoppers. Media volume should be less than 20% of the whole bottle or tube volume. Dissolved oxygen in media should be purged with oxygen-free gas species. H_2, H_2/CO_2 (80:20), N_2, or N_2/CO_2 (80:20) is usually supplied as headspace gases. Add the headspace pressure of 200 kPa to avoid air intrusion. For inoculation, plastic syringes with needles are used. For isolation, the dilution-extinction method (Baross 1995) is often applied because most thermophiles do not form colonies on solidified media.

Microbiologists have made significant efforts to cultivate the previously "unculturable" deep-sea microbial organisms not only under laboratory conditions but also under in situ environmental conditions using in situ colonization systems (ISCSs). Novel members of microbial groups (e.g., Epsilon- and Gammaproteobacteria) have been isolated from enrichments of ISCSs deployed at hydrothermal environments (Takai et al. 2003c, 2004d, 2005). In addition to the employment of ISCSs for the isolation of novel microorganisms, various types of in situ equipment, for example, the titanium ring for *Alvinella* colonization (TRAC) (Taylor et al. 1999; Alain et al. 2004), vent caps (Reysenbach et al. 2000; Corre et al. 2001), in situ growth chambers (Higashi et al. 2004; Rassa et al. 2009), thermocouples (Pagé et al. 2008), and the Autonomous in situ Instrumented Colonization System (AISICS) (Callac et al. 2013), have enhanced knowledge about the actual microbial communities' composition and the processes of microbial colonization. Reysenbach et al. (2000) captured the existence of thermoacidophiles in deep-sea hydrothermal environments, and these microorganisms were isolated as reviewed above. Alain et al. (2004) showed the importance of Epsilonproteobacteria members as the pioneers of microbial colonization.

8.4 CULTURE-INDEPENDENT STUDIES FOR MICROBIAL ECOLOGY

8.4.1 16S rRNA–Based Molecular Analysis of Deep-Sea Vent Microbial Community

Despite the fact that a great number of thermophiles have been retrieved in laboratory cultures, the diversity of microorganisms has been inevitably underestimated by the culture-based traditional approach, as it is well established that approximately 99% of microbes on earth are "uncultivated" (Vartoukian et al. 2010). Consequently, studies employing molecular-based methods have significantly increased our understanding of microbial diversity associated with deep-sea hydrothermal environments (Haddad et al. 1995; Moyer et al. 1995; Polz and Cavanaugh 1995; Cary et al. 1997; Takai and and Horikoshi 1999; Reysenbach et al. 2000). The use of approaches concentrated on rRNA molecules, typically 16S rRNA, allows us to handle all microorganisms within both the domains Bacteria and Archaea. For instance, Moyer et al. (1995) investigated the bacterial community in microbial mats and successfully identified the dominant epsilonproteobacterial taxon, which was previously suggested as one of the operation taxonomic units (OTUs) by the restriction fragment length polymorphism (RFLP) approach. Takai and Horikoshi (1999) revealed the archaeal diversity occurring in the deep-sea vents, and contributed to the fundamental knowledge of archaeal phylogenetic organization. In addition to free-living microorganisms, the epibiotic microflora associated with vent polychaete *Alvinella pompejana* was examined (Haddad et al. 1995; Cary et al. 1997). Following the molecular phylogenetic studies at the end of the 1990s, many microbiologists have been challenged to enumerate microbial phylogenetic diversity in worldwide deep-sea hydrothermal systems, such as the East Pacific Rise (Longnecker and Reysenbach 2001; Nercessian et al. 2003; Alain et al. 2004), Guaymas Basin (Teske et al. 2002; Dhillon et al. 2003), Mid-Atlantic Ridge (Reysenbach et al. 2000; Corre et al. 2001), Juan de Fuca Ridge (Alain et al. 2002b; Huber et al. 2003), Central Indian Ridge (Hoek et al. 2003; Takai et al. 2004a), Okinawa Trough (Nakagawa et al. 2005a, 2005b), Izu-Bonin Arc (Higashi et al. 2004), Mariana Arc (Nakagawa et al. 2006), Lau Spreading Center (Flores et al. 2012a), and Manus Basin (Takai et al. 2001). These studies highlighted the existence of enormous uncultured microbial diversity and microbial heterogeneity influenced by the fluid physicochemical properties; at the same time, cosmopolitan taxa that occur around the globe were emerged. Especially sulfur-oxidizing mesophilic bacterial taxa have been regarded as the key player in deep-sea hydrothermal vent ecosystems.

8.4.2 From the Molecular Approaches to Cultivation: Characterizing Genetic Diversity, Metabolic Versatility, and Ecological Role of Cosmopolitan Taxa

Molecular-based approaches have indicated predominant sulfur-oxidizing mesophilic members within the classes Epsilonproteobacteria and Gammaproteobacteria. They have been detected in both symbiotic and free-living microbial communities. Dubilier et al. (2008) reviewed symbionts associated with various vent animals in the deep sea. For example, members within Gamma- and Epsilonproteobacteria are hosted as epibiont and/or endobiont of snails, bivalves, polychaetes, shrimps, and crabs. Relatively wide-range host animals are colonized by Gammaproteobacteria rather than Epsilonproteobacteria (Dubilier et al. 2008). Free-living Gamma- and Epsilonproteobacteria inhibit likely similar niches and have been detected in various types of hydrothermal samples, such as chimneys, vent fluids, and plumes (Huber et al. 2003; Sunamura et al. 2004; Nakagawa et al. 2005a, 2005b). However, their physiological properties and ecological performance have long been open questions because of the difficulty of stable cultivation, especially for Epsilonproteobacteria. The successes of cultivation of chemolithoautotrophic Gammaproteobacteria (Jannasch et al. 1985) and Epsilonproteobacteria (Inagaki et al. 2003) have provided the great progress not only in cultivation techniques but also in understanding their ecological importance. The physiological diversity of 16 mesophilic and thermophilic deep-sea epsilonproteobacterial strains in total has been described, and metabolic versatility is one of the reasons that explains their prosperity in deep-sea hydrothermal systems (Campbell et al. 2006). Yamamoto and Takai (2011) showed the difference between Gamma- and Epsilonproteobacteria in sulfur-oxidizing metabolisms and habitable spaces; Gammaproteobacteria possesses two types of sulfur-oxidizing pathways and inhibits at relatively oxidized environments. On the other hand, Epsilonproteobacteria utilizes sulfur both as an electron donor and as an electron acceptor, indicating the ability to grow under reductive environments close to hydrothermal vent orifices. This view was supported by a recent study employing transcriptomic approaches (Lesniewski et al. 2012; Akerman et al. 2013).

Cultivation of various deep-sea vent microbes has also paved the way for genome-wide studies. One of the remarkable researches that completed genomic sequencing of cosmopolitan deep-sea vent Epsilonproteobacteria provided the evolutionary link that connects deep-sea and human pathogenic Epsilonproteobacteria (Nakagawa et al. 2007). This view was also evidenced by conservation of the luxS/autoinducer-2 quorum-sensing system in Epsilonproteobacteria (Pérez-Rodríguez et al. 2014).

Enhanced culture collections, by increasing the number of phylogenetically closely related strains, also allowed us to apply population genetic approaches to deep-sea vent microbiology. Multilocus sequence analysis (MLSA) (Maiden 2006), basically adapted from population genetics, is a powerful tool for detecting the intraspecific genetic diversity. As strain-level genetic diversity, inferred by MLSA, was shown in a terrestrial archaeon *Sulfolobus* population (Whitaker et al. 2003), a fine scale of genetic and genomic diversity has been characterized in thermophilic DHVE2, known as one of the cosmopolitan but hardly culturable archaea (Flores et al. 2012b); *Persephonella* (a cosmopolitan sulfur and/or hydrogen oxidizer) (Mino et al. 2013); and *Lebetimonas* (an anaerobic hydrogen oxidizer) (Meyer and Huber 2013). Although knowledge of the microbial population genetic structure is limited to a few studies with a few taxa, these studies emphasize the necessity of genetic diversity evaluation at a higher taxonomic resolution than previously applied.

8.5 BEYOND THE CULTIVATION OF DEEP-SEA VENT MICROORGANISMS

Culture-independent methods, especially the high-throughput sequencing (HTS) methods, are progressing rapidly, becoming one of the trends in microbiology. It is worth applying HTS technologies to deep-sea environments because deep-sea hydrothermal vents are regarded as a "rare biosphere" (Sogin et al. 2006) and as an environment enriched with microbial "dark matter" (Galperin 2007). Indeed, omics-based research has shown new insights concerning ecological features; metagenomic analysis applied to deep-sea vent chimney samples has shown the abundance of transposes and chemotaxis genes, suggesting their adaptation to dynamic environments through lateral gene transfer and active movement responding to the environmental cues (Xie et al. 2011). In addition, the ecological significance of Epsilonproteobacteria and Gammaproteobacteria is now evidenced by their relationship with animals (Sanders et al. 2013; Nakagawa et al. 2014) or other heterotrophs (Stokke et al. 2015). These recent studies have shown the indispensability of omics-based analysis; however, the understanding of ecological importance cannot be advanced without cultivation, as the history of deep-sea vent microbiology has been developed by great cultivation efforts. Hence, integrative insights from both omics technologies and traditional methods are effective for illustrating the comprehensive microbial ecology occurring in the deep-sea biosphere. In addition, from the experimental viewpoint, improvement in the devices for sample collection and measuring in situ chemical composition has also contributed to understanding these ecosystems, including the relationships between microbial communities and changing environments.

ACKNOWLEDGMENTS

Sayaka Mino was supported by the research fellowship of the Japanese Society for the Promotion of Science (JSPS).

REFERENCES

Akerman, N., Butterfield, D., and Huber, J. Phylogenetic diversity and functional gene patterns of sulfur-oxidizing subseafloor epsilonproteobacteria in diffuse hydrothermal vent fluids. *Frontiers in Microbiology* 4 (2013): 185.

Alain, K., Marteinsson, V., Miroshnichenko, M. et al. *Marinitoga piezophila* sp. nov., a rod-shaped, thermo-piezophilic bacterium isolated under high hydrostatic pressure from a deep-sea hydrothermal vent. *International Journal of Systematic and Evolutionary Microbiology* 52 (2002a): 1331–1339.

Alain, K., Olagnon, M., Desbruyeres, D. et al. Phylogenetic characterization of the bacterial assemblage associated with mucous secretions of the hydrothermal vent polychaete *Paralvinella palmiformis*. *FEMS Microbiology Ecology* 42 (2002b): 463–476.

Alain, K., Pignet, P., Zbinden, M. et al. *Caminicella sporogenes* gen. nov., sp. nov., a novel thermophilic spore-forming bacterium isolated from an East-Pacific Rise hydrothermal vent. *International Journal of Systematic and Evolutionary Microbiology* 52 (2002c): 1621–1628.

Alain, K., Querellou, J., Lesongeur, F. et al. *Caminibacter hydrogeniphilus* gen. nov., sp. nov., a novel thermophilic, hydrogen-oxidizing bacterium isolated from an East Pacific Rise hydrothermal vent. *International Journal of Systematic Evolutionary Microbiology* 52 (2002d): 1317–1323.

Alain, K., Rolland, S., Crassous, P. et al. *Desulfurobacterium crinifex* sp. nov., a novel thermophilic, pinkish-streamer forming, chemolithoautotrophic bacterium isolated from a Juan de Fuca Ridge hydrothermal vent and amendment of the genus *Desulfurobacterium*. *Extremophiles* 7 (2003): 361–370.

Alain, K., Zbinden, M., Le Bris, N. et al. Early steps in microbial colonization processes at deep-sea hydrothermal vents. *Environmental Microbiology* 6 (2004): 227–241.

Alain, K., Postec, A., Grinsard, E. et al. *Thermodesulfatator atlanticus* sp. nov., a thermophilic, chemolithoautotrophic, sulfate-reducing bacterium isolated from a Mid-Atlantic Ridge hydrothermal vent. *International Journal of Systematic and Evolutionary Microbiology* 60 (2010): 33–38.

Amann, R., Luding, W., and Schleifer, K. Phylogenetic identification and in situ detection of individual microbial cells without cultivation. *Microbiological Reviews* 59 (1995): 143–169.

Antoine, E., Cilia, V., Meunier, J. et al. *Thermosipho melanesiensis* sp. nov., a new thermophilic anaerobic bacterium belonging to the order Thermotogales, isolated from deep-sea hydrothermal vents in the southwestern Pacific Ocean. *International Journal of Systematic Bacteriology* 47 (1997): 1118–1123.

Barbier, G., Godfroy, A., Meunier, J. et al. *Pyrococcus glycovorans* sp. nov., a hyperthermophilic archaeon isolated from the East Pacific Rise. *International Journal of Systematic Bacteriology* 49 (1999): 1829–1837.

Baross, J.A. Isolation, growth and maintenance of hyperthermophiles. In *Archaea: A Laboratory Manual, Thermophiles*, ed. F.T. Robb and A.R. Place, 15–23. New York: Cold Spring Harbor Laboratory, 1995.

Birrien, J., Zeng, X., Jebbar, M. et al. *Pyrococcus yayanosii* sp. nov., an obligate piezophilic hyperthermophilic archaeon isolated from a deep-sea hydrothermal vent. *International Journal of Systematic and Evolutionary Microbiology* 61 (2011): 2827–2831.

Blöchl, E., Rachel, R., Burggraf, S. et al. *Pyrolobus fumarii*, gen. and sp. nov., represents a novel group of archaea, extending the upper temperature limit for life to 113°C. *Extremophiles* 1 (1997): 14–21.

Burggraf, S., Jannasch, H., Nicolaus, B., and Stetter, K. *Archaeoglobus profundus* sp. nov., represents a new species within the sulfur-reducing Archaebacteria. *Systematic Applied Microbiology* 13 (1990): 24–28.

Callac, N., Rommevaux-Jestin, C., Rouxel, O. et al. Microbial colonization of basaltic glasses in hydrothermal organic-rich sediments at Guaymas Basin. *Frontiers in Microbiology* 4 (2013): 250.

Campbell, B., Engel, A., Porter, M., and Takai K. The versatile ε-proteobacteria: Key players in sulphidic habitats. *Nature Reviews Microbiology* 4 (2006): 458–468.

Canganella, F., Jones, W., Gambacorta, A., and Antranikian, G. *Thermococcus guaymasensis* sp. nov. and *Thermococcus aggregans* sp. nov., two novel thermophilic archaea isolated from the Guaymas Basin hydrothermal vent site. *International Journal of Systematic Bacteriology* 48 (1998): 1181–1185.

Cary, S., Cottrell, M., Stein, J., Camacho, F., and Desbruyeres, D. Molecular identification and localization of filamentous symbiotic bacteria associated with the hydrothermal vent annelid *Alvinella pompejana*. *Applied Environmental Microbiology* 63 (1997): 1124–1130.

Corre, E., Reysenbach, A.L., and Prieur, D. ε-Proteobacterial diversity from a deep-sea hydrothermal vent on the Mid-Atlantic Ridge. *FEMS Microbiology Letters* 205 (2001): 329–335.

Dhillon, A., Teske, A., Dillon, J., Stahl, D., and Sogin, M. Molecular characterization of sulfate-reducing bacteria in the Guaymas Basin. *Applied and Environmental Microbiology* 69 (2003): 2765–2772.

Dubilier, N., Bergin, C., and Lott, C. Symbiotic diversity in marine animals: The art of harnessing chemosynthesis. *Nature Reviews Microbiology* 6 (2008): 725–740.

Duffaud, G., d'Hennezel, O., Peek, A., Reysenbach, A.L., and Kelly, R. Isolation and characterization of *Thermococcus barossii*, sp. nov., a hyperthermophilic archaeon isolated from a hydrothermal vent flange formation. *Systematic and Applied Microbiology* 21 (1998): 40–49.

Fiala, G., Stetter, K., Jannasch, H., Langworthy, T., and Madon, J. *Staphylothermus marinus* sp. nov. represents a novel genus of extremely thermophilic submarine heterotrophic archaebacteria growing up to 98°C. *Systematic and Applied Microbiology* 8 (1986): 106–113.

Flores, G., Shakya, M., Meneghin, J. et al. Inter-field variability in the microbial communities of hydrothermal vent deposits from a back-arc basin. *Geobiology* 10 (2012a): 333–346.

Flores, G., Wagner, I.D., Liu, Y., and Reysenbach, A.L. Distribution, abundance, and diversity patterns of the thermoacidophilic "Deep-sea hydrothermal vent euryarchaeota 2." *Frontiers in Microbiology* 3 (2012b): 47.

Flores, G., Hunter, R., Liu, Y. et al. *Hippea jasoniae* sp. nov. and *Hippea alviniae* sp. nov., thermoacidophilic members of the class Deltaproteobacteria isolated from deep-sea hydrothermal vent deposits. *International Journal of Systematic and Evolutionary Microbiology* 62 (2012c): 1252–1258.

Galperin, M. Dark matter in a deep-sea vent and in human mouth. *Environmental Microbiology* 9 (2007): 2385–2391.

Godfroy, A., Meunier, J., Guezennec, J. et al. *Thermococcus fumicolans* sp. nov., a new hyperthermophilic archaeon isolated from a deep-sea hydrothermal vent in the north Fiji basin. *International Journal of Systematic Bacteriology* 46 (1996): 1113–1119.

Godfroy, A., Lesongeur, F., Raguenes, G. et al. *Thermococcus hydrothermalis* sp. nov, a new hyperthermophilic archaeon isolated from a deep-sea hydrothermal vent. *International Journal of Systematic Bacteriology* 47 (1997): 622–626.

Gonzalez, J., Kato, C., and Horikoshi, K. *Thermococcus peptonophilus* sp. nov., a fast-growing, extremely thermophilic archaebacterium isolated from deep-sea hydrothermal vents. *Archives of Microbiology* 164 (1995): 159–164.

Gonzalez, J., Masuchi, Y., Robb, F. et al. *Pyrococcus horikoshii* sp. nov., a hyperthermophilic archaeon isolated from a hydrothermal vent at the Okinawa Trough. *Extremophiles* 2 (1998): 123–130.

Gorlas, A., Alain, K., Bienvenu, N., and Geslin, C. *Thermococcus prieurii* sp. nov., a hyperthermophilic archaeon isolated from a deep-sea hydrothermal vent. *International Journal of Systematic Evolutionary Microbiology* 63 (2013): 2920–2926.

Gorlas, A., Croce, O., Oberto, J. et al. *Thermococcus nautili* sp. nov., a hyperthermophilic archaeon isolated from a hydrothermal deep-sea vent. *International Journal of Systematic and Evolutionary Microbiology* 64 (2014): 1802–1810.

Götz, D., Banta, A., Beveridge, T. et al. *Persephonella marina* gen. nov., sp. nov. and *Persephonella guaymasensis* sp. nov., two novel, thermophilic, hydrogen-oxidizing microaerophiles from deep-sea hydrothermal vents. *International Journal of Systematic Evolutionary Microbiology* 52 (2002): 1349–1359.

Grote, R., Li, L., Tamaoka, J. et al. *Thermococcus siculi* sp. nov., a novel hyperthermophilic archaeon isolated from a deep-sea hydrothermal vent at the Mid-Okinawa Trough. *Extremophiles* 3 (1999): 55–62.

Haddad, A., Camacho, F., Durand, P., and Cary, S.C. Phylogenetic characterization of the epibiotic bacteria associated with the hydrothermal vent polychaete *Alvinella pompejana*. *Applied Environmental Microbiology* 61 (1995): 1679–1687.

Hensley, S., Jung, J., Park, C., and Holden, J. *Thermococcus paralvinellae* sp. nov. and *Thermococcus cleftensis* sp. nov. of hyperthermophilic heterotrophs from deep-sea hydrothermal vents. *International Journal of Systematic and Evolutionary Microbiology* 64 (2014): 3655–3659.

Higashi, Y., Sunamura, M., Kitamura, K. et al. Microbial diversity in hydrothermal surface to subsurface environments of Suiyo Seamount, Izu-Bonin Arc, using a catheter-type in situ growth chamber. *FEMS Microbiology Ecology* 47 (2004): 327–336.

Hoek, J., Banta, A., Hubler, F., and Reysenbach, A.L. Microbial diversity of a sulphide spire located in the Edmond deep-sea hydrothermal vent field on the Central Indian Ridge. *Gobiology* 1 (2003): 119–127.

Huber, H., Jannasch, H., Rachel, R., Fuchs, T., and Stetter, K. *Archaeoglobus veneficus* sp. nov., a novel facultative chemolithoautotrophic hyperthermophilic sulfite reducer, isolated from abyssal black smokers. *Systematic and Applied Microbiology* 20 (1997): 374–380.

Huber, H., Burggraf, S., Mayer, T. et al. *Ignicoccus* gen. nov., a novel genus of hyperthermophilic, chemolithoautotrophic archaea, represented by two new species, *Ignicoccus islandicus* sp. nov. and *Ignicoccus pacificus* sp. nov. *International Journal of Systematic Evolutionary Microbiology* 50 (2000): 2093–2100.

Huber, J., Butterfield, D., and Baross, J. Bacterial diversity in a subseafloor habitat following a deep-sea volcanic eruption. *FEMS Microbiology Ecology* 43 (2003): 393–409.

Huber, R., Stohr, J., Hohenhaus, S. et al. *Thermococcus chitonophagus* sp. nov, a novel, chitin-degrading, hyperthermophillic archaeum from a deep-sea hydrothermal vent environment. *Archives of Microbiology* 164 (1995): 255–264.

Imachi, H., Saka, S., Hirayama, H. et al. *Exilispira thermophila* gen. nov., sp. nov., an anaerobic, thermophilic spirochaete isolated from a deep-sea hydrothermal vent chimney. *International Journal of Systematic and Evolutionary Microbiology* 58 (2008): 2258–2265.

Inagaki, F., Takai, K., Kobayashi, H., Nealson, K.H., and Horikoshi, K. *Sulfurimonas autotrophica* gen. nov., sp. nov., a novel sulfur-oxidizing epsilon-proteobacterium isolated from hydrothermal sediments in the Mid-Okinawa Trough. *International Journal of Systematic Evolutionary Microbiology* 53 (2003): 1801–1805.

Jannasch, H., Wirsen, C., Nelson, D., and Robertson, L. *Thiomicrospira crunogena* sp. nov., a colorless, sulfur-oxidizing bacterium from a deep-sea hydrothermal vent. *International Journal of Systematic Bacteriology* 35 (1985): 422–424.

Jeanthon, C., L'Haridon, S., Reysenbach, A.L. et al. *Methanococcus infernus* sp. nov., a novel hyperthermophilic lithotrophic methanogen isolated from a deep-sea hydrothermal vent. *International Journal of Systematic Bacteriology* 48 (1998): 913–919.

Jeanthon, C., L'Haridon, S., Reysenbach, A.L. et al. *Methanococcus vulcanius* sp. nov., a novel hyperthermophilic methanogen isolated from East Pacific Rise, and identification of *Methanococcus* sp. DSM 4213T as *Methanococcus fervens* sp. nov. *International Journal of Systematic Bacteriology* 49 (1999): 583–589.

Jeanthon, C., L'Haridon, S., Cueff, V. et al. *Thermodesulfobacterium hydrogeniphilum* sp. nov., a thermophilic, chemolithoautotrophic, sulfate-reducing bacterium isolated from a deep-sea hydrothermal vent at Guaymas Basin, and emendation of the genus *Thermodesulfobacterium*. *International Journal of Systematic and Evolutionary Microbiology* 52 (2002): 765–772.

Jolivet, E., L'Haridon, S., Corre, E., Forterre, P., and Prieur, D. *Thermococcus gammatolerans* sp. nov., a hyperthermophilic archaeon from a deep-sea hydrothermal vent that resists ionizing radiation. *International Journal of Systematic and Evolutionary Microbiology* 53 (2003): 847–851.

Jones, W., Leigh, J., Mayer, F., Woese, C., and Wolfe, R. *Methanococcus jannaschii* sp. nov., an extremely thermophilic methanogen from a submarine hydrothermal vent. *Archives of Microbiology* 136 (1983): 254–261.

Kashefi, K., Tor, J., Holmes, D. et al. *Geoglobus ahangari* gen. nov., sp. nov., a novel hyperthermophilic archaeon capable of oxidizing organic acids and growing autotrophically on hydrogen with Fe(III) serving as the sole electron acceptor. *International Journal of Systematic Evolutionary Microbiology* 52 (2002): 719–728.

Kashefi, K., Holmes, D., Baross, J., and Lovley, D. Thermophily in the Geobacteraceae: *Geothermobacter ehrlichii* gen. nov., sp. nov., a novel thermophilic member of the Geobacteraceae from the "Bag City" hydrothermal vent. *Applied and Environmental Microbiology* 69 (2003): 2985–2993.

Kobayashi, T., Kwak, Y., Akiba, T., Kudo, T., and Horikoshi, K. *Thermococcus profundus* sp. nov., a new hyperthermophilic archaeon isolated from a deep-sea hydrothermal vent. *Systematic and Applied Microbiology* 17 (1994): 232–236.

Kurr, M., Huber, R., König, H. et al. *Methanopyrus kandleri*, gen. and sp. nov. represents a novel group of hyperthermophilic methanogens, growing at 110°C. *Archives of Microbiology* 156 (1991): 239–247.

Kuwabara, T., Minaba, M., Iwayama, Y. et al. *Thermococcus coalescens* sp. nov., a cell-fusing hyperthermophilic archaeon from Suiyo Seamount. *International Journal of Systematic and Evolutionary Microbiology* 55 (2005): 2507–2514.

Kuwabara, T., Minaba, M., Ogi, N., and Kamekura, M. *Thermococcus celericrescens* sp. nov., a fast-growing and cell-fusing hyperthermophilic archaeon from a deep-sea hydrothermal vent. *International Journal of Systematic and Evolutionary Microbiology* 57 (2007): 437–443.

L'Haridon, S., Cilia, V., Messner, P. et al. *Desulfurobacterium thermolithotrophum* gen. nov., sp. nov., a novel autotrophic, sulphur-reducing bacterium isolated from a deep-sea hydrothermal vent. *International Journal of Systematic Bacteriology* 48 (1998): 701–711.

L'Haridon, S., Reysenbach, A.L., Banta, A. et al. *Methanocaldococcus indicus* sp. nov., a novel hyperthermophilic methanogen isolated from the Central Indian Ridge. *International Journal of Systematic Evolutionary Microbiology* 53 (2003): 1931–1935.

L'Haridon, S., Miroshnichenko, M., Kostrikina, N. et al. *Vulcanibacillus modesticaldus* gen. nov., sp. nov., a strictly anaerobic, nitrate-reducing bacterium from deep-sea hydrothermal vents. *International Journal of Systematic and Evolutionary Microbiology* 56 (2006a): 1047–1053.

L'Haridon, S., Reysenbach, A.L., Tindall, B. et al. *Desulfurobacterium atlanticum* sp. nov., *Desulfurobacterium pacificum* sp. nov. and *Thermovibrio guaymasensis* sp. nov., three thermophilic members of the Desulfurobacteriaceae fam. nov., a deep branching lineage within the Bacteria. *International Journal of Systematic Evolutionary Microbiology* 56 (2006b): 2843–2852.

Lesniewski, R., Jain, S., Anantharaman, K., Schloss, P., and Dick, G. The metatranscriptome of a deep-sea hydrothermal plume is dominated by water column methanotrophs and lithotrophs. *ISME Journal* 6 (2012): 2257–2268.

Longnecker, K., and Reysenbach, A.L. Expansion of the geographic distribution of a novel lineage of Epsilon-Proteobacteria to a hydrothermal vent site on the southern East Pacific Rise. *Microbiology Ecology* 35 (2001): 287–293.

Maiden, M.C. Multilocus sequence typing of bacteria. *Annual Review of Microbiology* 60 (2006): 561–588.

Marteinsson, V., Birrien, J., Reysenbach, A.L. et al. *Thermococcus barophilus* sp. nov., a new barophilic and hyperthermophilic archaeon isolated under high hydrostatic pressure from a deep-sea hydrothermal vent. *International Journal of Systematic Bacteriology* 49 (1999): 351–359.

Marteinsson, V., Bjornsdottir, S., Bienvenu, N., Kristjansson, J., and Birrien, J. *Rhodothermus profundi* sp. nov., a thermophilic bacterium isolated from a deep-sea hydrothermal vent in the pacific ocean. *International Journal of Systematic and Evolutionary Microbiology* 60 (2010): 2729–2734.

Martin, W., Baross, J., Kelley, D., and Russell, M. Hydrothermal vents and the origin of life. *Nature Reviews Microbiology* 6 (2008): 805–814.

Mehta, M., and Baross, J. Nitrogen fixation at 92 degrees C by a hydrothermal vent archaeon. *Science* 314 (2006): 1783–1786.

Meyer, J.L., and Huber, J.A. Strain-level genomic variation in natural populations of *Lebetimonas* from an erupting deep-sea volcano. *ISME Journal* 8 (2013): 867–880.

Mino, S., Makita, H., Toki, T. et al. Biogeography of *Persephonella* in deep-sea hydrothermal vents of the western Pacific. *Frontiers in Microbiology* 4 (2013): 107.

Miroshnichenko, M., Kostrikina, N., L'Haridon, S. et al. *Nautilia lithotrophica* gen. nov., sp. nov., a thermophilic sulfur-reducing epsilon-proteobacterium isolated from a deep-sea hydrothermal vent. *International Journal of Systematic Evolutionary Microbiology* 52 (2002): 1299–1304.

Miroshnichenko, M., Rainey, F., Rhode, M., and Bonch-Osmolovskaya, E. *Hippea maritima* gen. nov., sp. nov., a new genus of thermophilic, sulfur-reducing bacterium from submarine hot vents. *International Journal of Systematic Bacteriology* 49 (1999): 1033–1038.

Miroshnichenko, M., Kostrikina, N., Chernyh, N. et al. *Caldithrix abyssi* gen. nov., sp. nov., a nitrate-reducing, thermophilic, anaerobic bacterium isolated from a Mid-Atlantic Ridge hydrothermal vent, represents a novel bacterial lineage. *International Journal of Systematic and Evolutionary Microbiology* 53 (2003a): 323–329.

Miroshnichenko, M., L'Haridon, S., Jeanthon, C. et al. *Oceanithermus profundus* gen. nov., sp. nov., a thermophilic, microaerophilic, facultatively chemolithoheterotrophic bacterium from a deep-sea hydrothermal vent. *International Journal of Systematic and Evolutionary Microbiology* 53 (2003b): 747–752.

Miroshnichenko, M., L'Haridon, S., Nercessian, O. et al. *Vulcanithermus mediatlanticus* gen. nov., sp. nov., a novel member of the family Thermaceae from a deep-sea hot vent. *International Journal of Systematic and Evolutionary Microbiology* 53 (2003c): 1143–1148.

Miroshnichenko, M.L., Slobodkin, A.I., Kostrikina, N.A. et al. Deferribacter *abyssi* sp. nov., an anaerobic thermophile from deep-sea hydrothermal vents of the Mid-Atlantic Ridge. *International Journal of Systematic and Evolutionary Microbiology* 53 (2003d): 1637–1641.

Miroshnichenko, M., L'Haridon, S., Schumann, P. et al. *Caminibacter profundus* sp. nov., a novel thermophile of Nautiliales ord. nov. within the class "Epsilonproteobacteria," isolated from a deep-sea hydrothermal vent. *International Journal of Systematic Evolutionary Microbiology* 54 (2004): 41–45.

Mori, K., Kakegawa, T., Higashi, Y. et al. *Oceanithermus desulfurans* sp. nov., a novel thermophilic, sulfur-reducing bacterium isolated from a sulfide chimney in Suiyo Seamount. *International Journal of Systematic and Evolutionary Microbiology* 54 (2004): 1561–1566.

Mori, K., Maruyama, A., Urabe, T., Suzuki, K., and Hanada, S. *Archaeoglobus infectus* sp. nov., a novel thermophilic, chemolithoheterotrophic archaeon isolated from a deep-sea rock collected at Suiyo Seamount, Izu-Bonin Arc, western Pacific Ocean. *International Journal of Systematic Evolutionary Microbiology* 58 (2008): 810–816.

Moussard, H., L'Haridon, S., Tindall, B. et al. *Thermodesulfatator indicus* gen. nov., sp. nov., a novel thermophilic chemolithoautotrophic sulfate-reducing bacterium isolated from the Central Indian Ridge. *International Journal of Systematic and Evolutionary Microbiology* 54 (2004): 227–233.

Moyer, C.L., Dobbs, F.C., and Karl, D.M. Phylogenetic diversity of the bacterial community from a microbial mat at an active, hydrothermal vent system, Loihi Seamount, Hawaii. *Applied Environmental Microbiology* 61 (1995): 1555–1562.

Nakagawa, S., and Takai, K. Methods for the isolation of thermophiles from deep-sea hydrothermal environments. In *Methods in Microbiology*, ed. F.A. Rainey and A. Oren, 55–91. Amsterdam: Elsevier Academic Press, 2006.

Nakagawa, S., Takai, K., Horikoshi, K., and Sako, Y. *Persephonella hydrogeniphila* sp. nov., a novel thermophilic, hydrogen-oxidizing bacterium from a deep-sea hydrothermal vent chimney. *International Journal of Systematic Evolutionary Microbiology* 53 (2003): 863–869.

Nakagawa, S., Takai, K., Horikoshi, K., and Sako, Y. *Aeropyrum camini* sp. nov., a strictly aerobic, hyperthermophilic archaeon from a deep-sea hydrothermal vent chimney. *International Journal of Systematic Evolutionary Microbiology* 54 (2004): 329–335.

Nakagawa, S., Takai, K., Inagaki, F. et al. Variability in microbial community and venting chemistry in a sediment-hosted backarc hydrothermal system: Impacts of subseafloor phase-separation. *FEMS Microbiology Ecology* 54 (2005a): 141–155.

Nakagawa, S., Takai, K., Inagaki, F. et al. Distribution, phylogenetic diversity and physiological characteristics of Epsilon-Proteobacteria in a deep-sea hydrothermal field. *Environmental Microbiology* 7 (2005b): 1619–1632.

Nakagawa, S., Takai, K., Inagaki, F., Horikoshi, K., and Sako, Y. *Nitratiruptor tergarcus* gen. nov., sp. nov. and *Nitratifractor salsuginis* gen. nov., sp. nov., nitrate-reducing chemolithoautotrophs of the Epsilon-Proteobacteria isolated from a deep-sea hydrothermal system in the Mid-Okinawa Trough. *International Journal of Systematic Evolutionary Microbiology* 55 (2005c): 925–933.

Nakagawa, S., Takaki, Y., Shimamura, S. et al. Deep-sea vent ε-proteobacterial genomes provide insights into emergence of pathogens. *Proceedings of the National Academy of Sciences of the United States of America* 104 (2007): 12146–12150.

Nakagawa, S., Shimamura, S., Takaki, Y. et al. Allying with armored snails: The complete genome of gammaproteobacterial endosymbiont. *ISME Journal* 8 (2014): 40–51.

Nakagawa, T., Takai, K., Suzuki, Y. et al. Geomicrobiological exploration and characterization of a novel deep-sea hydrothermal system at the TOTO caldera in the Mariana Volcanic Arc. *Environmental Microbiology* 8 (2006): 37–49.

Nercessian, O., Reysenbach, A.L., Prieur, D., and Jeanthon, C. Archaeal diversity associated with in situ samplers deployed on hydrothermal vents on the East Pacific Rise (13°N). *Environmental Microbiology* 5 (2003): 492–502.

Nunoura, T., Oida, H., Miyazaki, M. et al. *Marinitoga okinawensis* sp. nov., a novel thermophilic and anaerobic heterotroph isolated from a deep-sea hydrothermal field, southern Okinawa Trough. *International Journal of Systematic and Evolutionary Microbiology* 57 (2007): 467–471.

Nunoura, T., Miyazaki, M., Suzuki, Y., Takai, K., and Horikoshi, K. *Hydrogenivirga okinawensis* sp. nov., a thermophilic sulfur-oxidizing chemolithoautotroph isolated from a deep-sea hydrothermal field, southern Okinawa Trough. *International Journal of Systematic and Evolutionary Microbiology* 58 (2008a): 676–681.

Nunoura, T., Oida, H., Miyazaki, M., and Suzuki., Y. *Thermosulfidibacter takaii* gen. nov., sp. nov., a thermophilic, hydrogen-oxidizing, sulfur-reducing chemolithoautotroph isolated from a deep-sea hydrothermal field in the southern Okinawa Trough. *International Journal of Systematic Evolutionary Microbiology* 58 (2008b): 659–665.

Nunoura, T., and Takai, K. Comparison of microbial communities associated with phase-separation-induced hydrothermal fluids at the Yonaguni Knoll IV hydrothermal field, the southern Okinawa Trough. *FEMS Microbiology Ecology* 67 (2009): 351–370.

Pagé, A., Tivey, M., Stakes, D., and Reysenbach, A.L. Temporal and spatial archaeal colonization of hydrothermal vent deposits. *Environmental Microbiology* 10 (2008): 874–884.

Pérez-Rodríguez, I., Ricci, J., Voordeckers, J., Starovoytov, V., and Vetriani, C. *Nautilia nitratireducens* sp. nov., a thermophilic, anaerobic, chemosynthetic, nitrate-ammonifying bacterium isolated from a deepsea hydrothermal vent. *International Journal of Systematic Evolutionary Microbiology* 60 (2010): 1182–1186.

Pérez-Rodríguez, I., Grosche, A., Massenburg, L. et al. *Phorcysia thermohydrogeniphila* gen. nov., sp. nov., a thermophilic, chemolithoautotrophic, nitrate-ammonifying bacterium from a deep-sea hydrothermal vent. *International Journal of Systematic Evolutionary Microbiology* 62 (2012): 2388–2394.

Pérez-Rodríguez, I., Bolognini, M., Ricci, J., Bini, E., and Vetriani, C. From deep-sea volcanoes to human pathogens: A conserved quorum-sensing signal in Epsilonproteobacteria. *ISME Journal* 9 (2014): 1222–1234.

Pikuta, E., Marsic, D., Itoh, T. et al. *Thermococcus thioreducens* sp. nov., a novel hyperthermophilic, obligately sulfur-reducing archaeon from a deep-sea hydrothermal vent. *International Journal of Systematic and Evolutionary Microbiology* 57 (2007): 1612–1618.

Pley, U., Schipka, J., Gambacorta, A. et al. *Pyrodictium abyssi* sp. nov. represents a novel heterotrophic marine archaeal hyperthermophile growing at 110°C. *Systematic and Applied Microbiology* 14 (1991): 245–253.

Podosokorskaya, O., Kublanov, I., Reysenbach, A.L., Kolganova, T., and Bonch-Osmolovskaya, E. *Thermosipho affectus* sp. nov., a thermophilic, anaerobic, cellulolytic bacterium isolated from a Mid-Atlantic Ridge hydrothermal vent. *International Journal of Systematic and Evolutionary Microbiology* 61 (2011): 1160–1164.

Polz, M., and Cavanaugh, C. Dominance of one bacterial phylotype at a Mid-Atlantic Ridge hydrothermal vent site. *Proceedings of the National Academy of Sciences of the United States of America* 92 (1995): 7232–7236.

Postec, A., Le Breton, C., Fardeau, M. et al. *Marinitoga hydro*genitolerans sp. nov., a novel member of the order Thermotogales isolated from a black smoker chimney on the Mid-Atlantic Ridge. *International Journal of Systematic and Evolutionary Microbiology* 55 (2005): 1217–1221.

Rassa, A., Mcallister, S., Safran, S., and Moyer, C. Zeta-proteobacteria dominate the colonization and formation of microbial mats in low-temperature hydrothermal vents at Loihi Seamount, Hawaii. *Geomicrobiology Journal* 26 (2009): 623–638.

Reysenbach, A.L., Longnecker, K., and Kirshtein, J. Novel bacterial and archaeal lineages from an in situ growth chamber deployed at a Mid-Atlantic Ridge hydrothermal vent. *Applied Environmental Microbiology* 66 (2000): 3798–3806.

Reysenbach, A.L., Liu, Y., Banta, A. et al. A ubiquitous thermoacidophilic archaeon from deep-sea hydrothermal vents. *Nature* 442 (2006): 444–447.

Reysenbach, A.L., Liu, Y., Lindgren, A. et al. *Mesoaciditoga la*uensis gen. nov., sp. nov., a moderately thermoacidophilic member of the order Thermotogales from a deep-sea hydrothermal vent. *International Journal of Systematic and Evolutionary Microbiology* 63 (2013): 4724–4729.

Sako, Y., Nakagawa, S., Takai, K., and Horikoshi, K. *Marinithermus hydrothermalis* gen. nov., sp. nov., a strictly aerobic, thermophilic bacterium from a deep-sea hydrothermal vent chimney. *International Journal of Systematic Evolutionary Microbiology* 53 (2003): 59–65.

Sanders, J., Beinart, R., Stewart, F., Delong, E., and Girguis, P. Metatranscriptomics reveal differences in in situ energy and nitrogen metabolism among hydrothermal vent snail symbionts. *ISME Journal* 7 (2013): 1556–1567.

Slobodkina, G., Kolganova, T., Tourova, T. et al. *Clostridium tepidiprofundi* sp. nov., a moderately thermophilic bacterium from a deep-sea hydrothermal vent. *International Journal of Systematic and Evolutionary Microbiology* 58 (2008): 852–855.

Slobodkina, G., Kolganova, T., Chernyh, N. et al. *Deferribacter autotrophicus* sp. nov., an iron(III)-reducing bacterium from a deep-sea hydrothermal vent. *International Journal of Systematic and Evolutionary Microbiology* 59 (2009a): 1508–1512.

Slobodkina, G., Kolganova, T., Querellou, J., Bonch-Osmolovskaya, E., and Slobodkin, A. *Geoglobus acetivorans* sp. nov., an iron(III)-reducing archaeon from a deep-sea hydrothermal vent. *International Journal of Systematic Evolutionary Microbiology* 59 (2009b): 2880–2883.

Slobodkina, G., Reysenbach, A.L., Panteleeva, A. et al. *Deferrisoma camini* gen. nov., sp. nov., a moderately thermophilic, dissimilatory iron(III)-reducing bacterium from a deep-sea hydrothermal vent that forms a distinct phylogenetic branch in the Deltaproteobacteria. *International Journal of Systematic and Evolutionary Microbiology* 62 (2012): 2463–2468.

Slobodkin, A., Tourova, T., Kostrikina, N. et al. *Tepidibacter thalassicus* gen. nov., sp. nov., a novel moderately thermophilic, anaerobic, fermentative bacterium from a deep-sea hydrothermal vent. *International Journal of Systematic and Evolutionary Microbiology* 53 (2003): 1131–1134.

Slobodkin, A., Reysenbach, A.L., Slobodkina, G. et al. *Thermosulfurimonas dismutans* gen. nov., sp. nov., an extremely thermophilic sulfur-disproportionating bacterium from a deep-sea hydrothermal vent. *International Journal of Systematic and Evolutionary Microbiology* 62 (2012): 2565–2571.

Slobodkin, A., Reysenbach, A.L., Slobodkina, G. et al. *Dissulfuribacter thermophilus* gen. nov., sp. nov., a thermophilic, autotrophic, sulfur-disproportionating, deeply branching deltaproteobacterium from a deep-sea hydrothermal vent. *International Journal of Systematic and Evolutionary Microbiology* 63 (2013): 1967–1971.

Sogin, M., Morrison, H., Huber, J. et al. Microbial diversity in the deep sea and the underexplored "rare biosphere." *Proceedings of the National Academy of Sciences of the United States of America* 103 (2006): 12115–12120.

Sokolova, T., Gonzalez, J., Kostrikina, N. et al. *Carboxydobrachium pacificum* gen. nov., sp. nov., a new anaerobic, thermophilic, co-utilizing marine bacterium from Okinawa Trough. *International Journal of Systematic and Evolutionary Microbiology* 51 (2001): 141–149.

Spiess, F., Macdonald, K., Atwater, T. et al. East Pacific Rise: Hot springs and geophysical experiments. *Science* 207 (1980): 1421–1433.

Steinsbu, B., Thorseth, I., Nakagawa, S. et al. *Archaeoglobus sulfaticallidus* sp. nov., a thermophilic and facultatively lithoautotrophic sulfate-reducer isolated from black rust exposed to hot ridge flank crustal fluids. *International Journal of Systematic Evolutionary Microbiology* 60 (2010): 2745–2752.

Stewart, L., Jung, J., Kim, Y. et al. *Methanocaldococcus bathoardescens* sp. nov., a hyperthermophilic methanogen isolated from a volcanically active deep-sea hydrothermal vent. *International Journal of Systematic and Evolutionary Microbiology* 65 (2015): 1280–1283.

Stokke, R., Dahle, H., Roalkvam, I. et al. Functional interactions among filamentous Epsilonproteobacteria and Bacteroidetes in a deep-sea hydrothermal vent biofilm. *Environmental Microbiology* 17 (2015): 4063–4077.

Sunamura, M., Higashi, Y., Miyako, C., Ishibashi, J., and Maruyama, A. Two bacteria phylotypes are predominant in the Suiyo Seamount hydrothermal plume. *Applied Environmental Microbiology* 70 (2004): 1190–1198.

Takai, K., and Horikoshi, K. Genetic diversity of archaea in deep-sea hydrothermal vent environments. *Genetics* 152 (1999): 1285–1297.

Takai, K., and Horikoshi, K. *Thermosipho japonicus* sp. nov., an extremely thermophilic bacterium isolated from a deep-sea hydrothermal vent in Japan. *Extremophiles* 4 (2000): 9–17.

Takai, K., Sugai, A., Itoh, T., and Horikoshi, K. *Palaeococcus ferrophilus* gen. nov., sp. nov., a barophilic, hyperthermophilic archaeon from a deep-sea hydrothermal vent chimney. *International Journal of Systematic Evolutionary Microbiology* 2 (2000): 489–500.

Takai, K., Komatsu, T., Inagaki, F., and Horikoshi, K. Distribution of archaea in a black smoker chimney structure. *Applied Environmental Microbiology* 67 (2001): 3618–3629.

Takai, K., Inoue, A., and Horikoshi, K. *Methanothermococcus okinawensis* sp. nov., a thermophilic, methane-producing archaeon isolated from a western Pacific deep-sea hydrothermal vent system. *International Journal of Systematic Evolutionary Microbiology* 52 (2002): 1089–1095.

Takai, K., Kobayashi, H., Nealson, K.H., and Horikoshi, K. *Deferribacter desulfuricans* sp. nov., a novel sulfur-, nitrate- and arsenate-reducing thermophile isolated from a deep-sea hydrothermal vent. *International Journal of Systematic Evolutionary Microbiology* 53 (2003a): 839–846.

Takai, K., Nakagawa, S., Sako, Y., and Horikoshi, K. *Balnearium lithotrophicum* gen. nov., sp. nov., a novel thermophilic, strictly anaerobic, hydrogen-oxidizing chemolithoautotroph isolated from a black smoker chimney in the Suiyo Seamount hydrothermal system. *International Journal of Systematic Evolutionary Microbiology* 53 (2003b): 1947–1954.

Takai, K., Inagaki, F., Nakagawa, S. et al. Isolation and phylogenetic diversity of members of previously uncultivated Epsilon-Proteobacteria in deep-sea hydrothermal fields. *FEMS Microbiology Letters* 218 (2003c): 167–174.

Takai, K., Gamo, T., Tsunogai, U. et al. Geochemical and microbiological evidence for a hydrogen-based, hyperthermophilic subsurface lithoautotrophic microbial ecosystem (hyperslime) beneath an active deep-sea hydrothermal field. *Extremophiles* 8 (2004a): 269–282.

Takai, K., Nealson, K.H., and Horikoshi, K. *Hydrogenimonas thermophila* gen. nov., sp. nov., a novel thermophilic, hydrogen-oxidizing chemolithoautotroph within the ε-proteobacteria, isolated from a black smoker in a Central Indian Ridge hydrothermal field. *International Journal of Systematic Evolutionary Microbiology* 54 (2004b): 25–32.

Takai, K., Nealson, K.H., and Horikoshi, K. *Methanotorris formicicus* sp. nov., a novel extremely thermophilic, methane-producing archaeon isolated from a black smoker chimney in the Central Indian Ridge. *International Journal of Systematic Evolutionary Microbiology* 54 (2004c): 1095–1100.

Takai, K., Hirayama, H., Nakagawa, T. et al. *Thiomicrospira thermophila* sp. nov., a novel microaerobic, thermotolerant, sulfur-oxidizing chemolithomixotroph isolated from a deep-sea hydrothermal fumarole in the TOTO caldera, Mariana Arc, Western Pacific. *International Journal of Systematic Evolutionary Microbiology* 54 (2004d): 2325–2333.

Takai, K., Hirayama, H., Nakagawa, T. et al. *Lebetimonas acidiphila* gen. nov., sp. nov., a novel thermophilic, acidophilic, hydrogen-oxidizing chemolithoautotroph within the 'epsilonproteobacteria,' isolated from a deep-sea hydrothermal fumarole in the Mariana Arc. *International Journal of Systematic Evolutionary Microbiology* 55 (2005): 183–189.

Takai, K., Nakamura, K., Toki, T. et al. Cell proliferation at 122°C and isotopically heavy CH_4 production by a hyperthermophilic methanogen under high-pressure cultivation. *Proceedings of the National Academy of Sciences of the United States of America* 105 (2008): 10949–10954.

Takai, K., and Nakamura, K. Archaeal diversity and community development in deep-sea hydrothermal vents. *Current Opinion in Microbiology* 14 (2011): 282–291.

Taylor, C., Wirsen, C., and Gaill, F. Rapid microbial production of filamentous sulfur mats at hydrothermal vents. *Applied and Environmental Microbiology* 65 (1999): 2253–2255.

Teske, A., Hinrichs, K.U., Edgcomb, V. et al. Microbial diversity of hydrothermal sediments in the Guaymas Basin: Evidence for anaerobic methanotrophic communities. *Applied Environmental Microbiology* 68 (2002): 1994–2007.

Urios, L., Cueff-Gauchard, V., Pignet, P. et al. *Thermosipho atlanticus* sp. nov., a novel member of the Thermotogales isolated from a Mid-Atlantic Ridge hydrothermal vent. *International Journal of Systematic and Evolutionary Microbiology* 54 (2004): 1953–1957.

Vartoukian, S., Palmer, R., and Wade, W. Strategies for culture of 'unculturable' bacteria. *FEMS Microbiology Letters* 309 (2010): 1–7.

Vetriani, C., Speck, M., Ellor, S., Lutz, R., and Starovoytov, V. *Thermovibrio ammonificans* sp. nov., a thermophilic, chemolithotrophic, nitrate-ammonifying bacterium from deep-sea hydrothermal vents. *International Journal of Systematic Evolutionary Microbiology* 54 (2004): 175–181.

Voordeckers, J., Starovoytov, V., and Vetriani, C. *Caminibacter mediatlanticus* sp. nov., a thermophilic, chemolithoautotrophic, nitrate-ammonifying bacterium isolated from a deep-sea hydrothermal vent on the Mid-Atlantic Ridge. *International Journal of Systematic Evolutionary Microbiology* 55 (2005): 773–779.

Wery, N., Lesongeur, F., Pignet, P. et al. *Marinitoga camini* gen, nov., sp. nov., a rod-shaped bacterium belonging to the order Thermotogales, isolated from a deep-sea hydrothermal vent. *International Journal of Systematic and Evolutionary Microbiology* 51 (2001a): 495–504.

Wery, N., Moricet, J., Cueff, V., et al. *Caloranaerobacter azorensis* gen. nov., sp. nov., an anaerobic thermophilic bacterium isolated from a deep-sea hydrothermal vent. *International Journal of Systematic and Evolutionary Microbiology* 51 (2001b): 1789–1796.

Whitaker, R., Grogan, D., and Taylor, J. Geographic barriers isolate endemic populations of hyperthermophilic archaea. *Science* 301 (2003): 976–978.

Whitman, W. Genus I. *Methanocaldococcus* gen. nov. In *Bergey's Manual of Systematic Bacteriology*, vol. 1, *The Archaea and the Deeply Branching and Phototrophic Bacteria*, ed. D. Boone and R. Castenholz, 243–245. 2nd ed. New York: Springer Verlag, 2001.

Xie, W., Wang, F., Guo, L. et al. Comparative metagenomics of microbial communities inhabiting deep-sea hydrothermal vent chimneys with contrasting chemistries. *ISME Journal* 5 (2011): 414–426.

Yamamoto, M., and Takai, K. Sulfur metabolisms in epsilon- and gamma-proteobacteria in deep-sea hydrothermal fields. *Frontiers in Microbiology* 2 (2011): 192.

Zeng, X., Birrien, J., Fouquet, Y. et al. *Pyrococcus* CH1, an obligate piezophilic hyperthermophile: Extending the upper pressure-temperature limits for life. *ISME Journal* 3 (2009): 873–876.

Zeng, X., Zhang, Z., Li, X. et al. *Caloranaerobacter ferrireducens* sp. nov., an anaerobic, thermophilic, iron (III)-reducing bacterium isolated from deep-sea hydrothermal sulfide deposits. *International Journal of Systematic and Evolutionary Microbiology* 65 (2015): 1714–1718.

CHAPTER 9

Physiological and Biochemical Adaptations of Psychrophiles

Amber Grace Teufel and Rachael Marie Morgan-Kiss

CONTENTS

9.1 INTRODUCTION

Psychrophilic organisms are defined as life-forms that not only grow and reproduce in cold environments, but also require low temperatures for survival. The cold biosphere encompasses many microorganism-dominated habitats, including the deep sea (which comprises almost three-quarters of the earth), alpine and permafrost areas, and the polar regions (Siddiqui and Cavicchioli, 2006). Low-temperature ecosystems relying on light-dependent primary production include the Tibetan Plateau (TP) region, the Antarctic and Arctic oceans, sea ice, alpine environments, ice-covered meromictic lakes, and supraglacial aquatic environments (Figure 9.1) (Morgan-Kiss et al., 2006; Morgan-Kiss and Dolhi, 2011). A wide diversity of microorganisms, encompassing all three kingdoms of life and harboring the capacity for diverse metabolic capabilities, dominate the world's cold biomes (Siddiqui et al., 2013).

Numerous physical and chemical factors limit growth and metabolism at low temperatures. As temperatures decrease, the available enthalphy (thermal energy) decreases within the system,

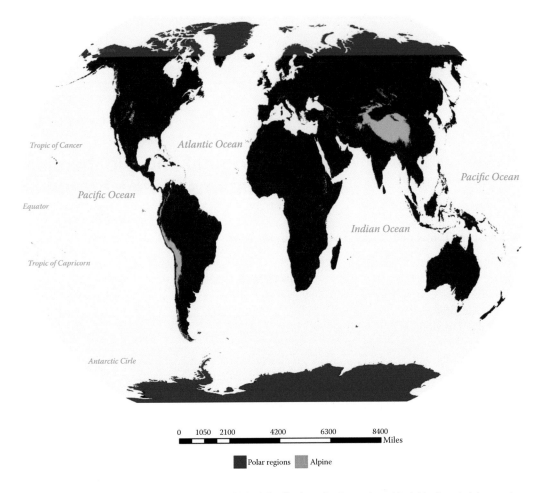

Figure 9.1 Cold-temperature global biomes. Global distribution of polar regions (dark blue) and alpine environments (light blue). The polar regions are separated into the Arctic, at the northern part of the earth within the Arctic Circle, and the Antarctic, which is the southern polar biome. Alpine regions are defined as high elevations of more than 3000 m above sea level, with the largest of these areas being the Tibetan Plateau.

causing increased rigidity of molecules, freezing of water, decreased diffusion, and decreased rates of enzyme-mediated reactions (Cavicchioli, 2006). In addition, reduced enthalpy induces broad-level downstream effects on biological molecules. These effects include the loss of protein and enzyme flexibility, leading to a reduction in catalysis rates and cold denaturation, and increased stability of both DNA and RNA, thus inhibiting replication, transcription, and translation. Cell membranes are also particularly thermally sensitive, as well as their numerous membrane-associated processes, most notably processes associated with energy generation and nutrient acquisition (Cavicchioli, 2006). In addition, cellular survival at low temperatures may require additional adaptations and mechanisms to avoid complications specific to ice formation. As discussed in this chapter, psychrophilic microorganisms integrate a complex spectrum of adaptive strategies to survive these physiological challenges.

This chapter integrates an examination of the diversity and distribution of microorganisms across global low-temperature biomes with a discussion on the spectrum of biochemical and physiological adaptations that these remarkable life-forms have evolved. Special attention is paid to the autotrophic psychrophilic organisms, which, as the primary producers, play a critical role in supporting the higher trophic levels of low-temperature food webs. We also discuss how these adaptations can be potentially exploited for industrial, environmental, or pharmaceutical applications that can benefit many aspects of human life. Finally, we conclude with a discussion about how psychrophiles and their habitats can be used as sentinels for global climate change and why further studies are needed on these unique and extreme organisms.

9.2 GLOBAL DISTRIBUTION OF COLD BIOMES

Permanently cold ecosystems are highly prevalent and exert extreme conditions upon biological processes. As such, low temperatures represent a major environmental factor influencing the global diversity and distribution of microorganisms. The diversity and distribution of cold-adapted life-forms can be influenced directly or indirectly by both abiotic and biotic factors. Direct impacts include low-temperature-associated influences on membrane function and metabolism, while indirect influences are habitat dependent, and include accessibility to organic and inorganic nutrients (e.g., oligotrophic vs. eutrophic environments), water availability (e.g., in ice and snow habitats), and solute concentrations (e.g., in brine channels, and freshwater vs. hypersaline lakes and ponds) (Bakermans, 2012). Many cold ecosystems feature strong variability in their light environment, which limits primary productivity under conditions of low to no light (e.g., subeuphotic zones in polar oceans, subglacial habitats, and 24-hour darkness during polar winters), or induced conditions of photoinhibitory or photooxidative stress at the opposite extreme (e.g., high ultraviolet [UV], high irradiance in ice-free waters and snow, and 24-hour daylight during polar summers). These diverse environmental challenges have selected for specialist life-forms, or "polyextremophiles," which are not only low temperature adapted but also often exhibit additional adaptive strategies to maintain metabolic activity in response to a broad spectrum of abiotic stressors.

9.2.1 Arctic

Polar and alpine climates are characterized by average temperatures below 10°C during all 12 months of the year, and are classified as Group E, according to the Köppen climate classification system (Peel et al., 2007). The Arctic is defined as the area north of the Arctic Circle (66° 33′N) (Figure 9.1) and harbors a variety of microbial habitats, such as sea ice, marine, tundra, permafrost, thaw ponds, and meromictic lakes. Tundra habitats cover more than 20% of the terrestrial landmass on earth and occur in the far north of North America and Eurasia, as well as the TP (Peel et al., 2007). The dominant organisms in this area include methanogenic archaea, which are ubiquitous in anoxic environments within the tundra and permafrost, as these organisms metabolize substrates such as formate, acetate, and methanol and produce methane, a potent greenhouse gas (Garcia et al., 2000). Nutrients are often limiting in permafrost soils, and therefore nitrogen-fixing bacteria, as well as mycorrhizal fungi, play important roles in nutrient availability in the permafrost and tundra soils, respectively. The High Arctic harbors numerous high-latitude meromictic lakes, which are typically classified as oligotrophic or ultraoligotrophic due to the extremely slow weathering processes in polar soils. The combination of ice cover and darkness through the winter months also creates periods of low energy (light) availability combined with permanent chemoclines and permanently anoxic strata. These harsh conditions have selected for bacteria capable of anoxygenic photosynthesis, chemolithotrophic sulfur and hydrogen oxidation, and Fe- or Mn-reducing metabolism (Comeau et al., 2012). Ultimately, the psychrophiles that have adapted to survive the Arctic region retain their own niches to optimize survival and fitness. Although temperatures are similar between the Arctic and the Antarctic, the landscape and biogeochemistry differ such that the overall diversity and biochemical adaptations differ between their representative psychrophiles.

9.2.2 Antarctic

The Antarctic represents the southern polar biome (Figure 9.1) within the EF Köppen climate class (Peel et al., 2007). Antarctica is the windiest, coldest, and driest continent, and is exposed to severe environmental conditions, including low humidity and low liquid water availability, as well as periods of high solar radiation, followed by long periods of complete darkness. Within the continent are numerous extreme habitats, and intensive study has focused on inland ice-covered lakes. These unique habitats represent opportunities to better understand psychrophilic adaptation, as each lake has unique and occasionally extreme biogeochemistry (e.g., hypersalinity, freshwater, and low nitrogen and phosphorus). One of the more well-studied areas in the Antarctic continent, which harbors several ice-covered lakes, is the McMurdo Dry Valleys (MDVs), located in Victoria Land. So named for their characteristic extreme low humidity and lack of precipitation, the extreme climactic conditions are greatly influenced by the presence of katabatic winds, which flow downhill off the Transantarctic Mountains (Nylen et al., 2004). The dry valley landscape includes diverse geological features, such as glaciers, ephemeral streams, soils, permafrost, and ice-covered lakes and ponds. A prominent feature, and of great interest to microbiologists, is the ice-capped lakes, which maintain a year-round liquid water column below a perennial ice cap. The ice coverage on these lakes protects the water column from wind disturbance, thus preventing mixing and allochthonous inputs. In common with the High Arctic lakes, these lakes are oligotrophic, meromictic, and largely unproductive, supporting only microbe-based forms of life (Priscu et al., 1999). This environment exerts numerous environmental stresses for the photosynthetic microbes due to the low temperatures, oligotrophic conditions, and extreme shade environments (Morgan-Kiss and Dolhi, 2011).

Physical and chemical differences within and between the Antarctic lakes support unique assemblages of microorganisms with distinct differences in community function (Dolhi et al., 2015). One such example within the MDVs is Lakes Bonney and Fryxell, which exhibit unique chemical signatures. Although similar in origin, carbon and energy cycling within Lakes Bonney and Fryxell reflects some of the contrasting features of microbial communities catalyzing major biogeochemical cycles within Dry Valley lake food webs. Oligotrophic conditions allow for limited light-driven carbon fixation in oxic epilimnia. Major peaks in primary production occur in the chemoclines, where phytoplankton balance nutrient and light availability. In both lakes, phytoplankton and heterotrophic bacteria are trophically linked through the autochthonous production of dissolved organic carbon, which represents the main input of organic carbon in these closed basin systems. In Lake Fryxell, mixotrophic cryptophytes are abundant phototrophic nanoflagellates that are both key primary producers and active grazers of bacteria. Cryptophytes are in turn predated on the top predators in this truncated food web, the ciliates (Roberts and Laybourn-Parry, 1999). The deep anoxic zone in Lake Fryxell supports dissimilatory sulfate reduction by sulfate-reducing bacteria, which provide a source of energy (sulfides) for sulfur-oxidizing bacteria (Karr et al., 2005; Sattley and Madigan, 2006) and anoxygenic phototrophic bacteria (Karr et al., 2003; Jung et al., 2004). In contrast, bacteria in Lake Bonney favor a dimethyl sulfoniopropionate (DMSP) mineralization pathway, which is a minor pathway in marine systems, but is also prevalent in another hypersaline Antarctic lake located in the Vestfold Hills (Yau et al., 2013). Bacteria-respiring DMSP produce dimethyl sulfide (DMS), which can be utilized as an electron donor by sulfur-oxidizing bacteria, and dimethyl sulfoxide (DMSO), which can be utilized as an alternative electron acceptor. The MDV lakes represent oases for metabolically diverse bacteria, and the adaptations to which these organisms have evolved are models across cold biome habitats.

9.2.3 Alpine

Alpine regions are defined as high-elevation regions, generally more than 3000 m above sea level (Figure 9.1), and are broadly distributed across the globe, including the Rocky and Appalachian Mountains, the Alps, the Himalayas, and the TP. The TP, often referred to as the "Third Pole" region (Kang et al., 2010), possesses many of the extreme features common to both the Arctic and the Antarctic. Alpine regions experience extreme changes in maximal and minimal yearly temperatures, thus leading to frequent freeze–thaw cycles, combined with variable snow cover and exposure to UV radiation and high solar irradiation (Margesin et al., 2004). In addition to low-temperature adaptation, alpine psychrophilic and psychrotolerant organisms must be able to withstand seasonal freeze–thaw cycles, as well as the drawdown of organic nutrients during the winter season (Hoover and Pikuta, 2010). Alpine habitats include snow, permafrost, tundra, glaciers, and high alpine lakes and ponds. In addition, unlike the polar biomes, a diversity of plants grow in alpine meadow regions, including grasses, sedges, cushion plants, mosses, and lichens, which provide an additional habitat, the rhizosphere, for soil microorganisms (Körner, 2003). With increasing altitude, increased numbers of cultivable psychrophilic heterotrophic bacteria are recovered and bacterial and fungal communities shift in response to altitudinal-driven environmental factors (Margesin et al., 2009). For example, Alps soils exhibited an increase in the ratio of gram-negative to gram-positive bacteria at higher altitudes, as well as a lower fraction of *Cytophaga–Flavobacterium–Bacteroides* in alpine versus subalpine soils (Lipson and Schmidt, 2004; Margesin et al. 2009). Within the soils of Mount Everest, ammonia-oxidizing archaea and bacteria have been found, with organism diversity decreasing with altitude (Zhang et al., 2009). In addition to permanent low temperatures, alpine ecosystems represent additional pressures for survival, including oxygen availability and pressure selection.

In general, microorganisms that have adapted to survive within permafrost conditions, soil at or below the freezing point of water, are often aerobic. As microbes in this environment must thrive at constant subzero temperatures, oligotrophic conditions, constant gamma radiation, and periods of darkness, they are often endospore formers, cellulose degraders, sulfate reducers, or nitrifying or denitrifying bacteria. The TP permafrost region contains 10^2–10^5 viable bacteria per gram of dry sample, with 90% of these isolates being gram positive and dominated by *Actinobacteria* (Zhang et al., 2007). Organisms cultivated from this region were found to be adapted to alkaline conditions and produced increased proteases, amylases, and cellulases compared with their mesophilic counterparts (Zhang et al., 2007). Photoautotrophic psychrophiles have also been detected in permafrost, and these communities are often dominated by cyanobacteria. It is thought that cyanobacteria can survive millions of years in a dormant resting state but are readily reversible without losing their photosynthetic capabilities, indicating another unique adaptation found in numerous psychrophiles (Vishnivetskaya et al., 2003; Vishnivetskaya 2009).

The third type of environment often found in alpine regions includes alpine lakes, which are characterized by seasonal ice cover, low temperatures, short growing seasons, high incident solar radiation, and low dissolved organic carbon and major nutrients (Psenner et al., 1999; Rose et al., 2009). Biota residing in alpine aquatic systems are greatly influenced by water transparency because it determines the penetration of photosynthetically active and UV radiation. Alpine lakes are typically ice-covered, dark, and anoxic for seven to nine months each year, followed by a spring turnover, and strong thermal stratification until the autumn turnover, creating seasonal objectives to overcome, as well as permanent environmental conditions. The microbial communities within these lakes often originate from soil or the atmosphere, and seasonal changes in the ice cover extent drive changes in the composition of the microbial communities (Felip et al., 2002).

9.2.4 Sea Ice

Sea ice (i.e., frozen seawater) is a prevalent cold habitat, representing ~7% of the earth's surface, and is associated mainly with polar pack ice at the earth's polar regions. Not only is sea ice a permanent low-temperature habitat (ranging from –1 to –15°C), but also the matrix is highly heterogeneous at the level of salinity, pH, dissolved nutrients and gas, and light penetration. Despite these variable conditions, sea ice represents a biologically dynamic habitat, and sea ice microbial communities have major influences on the productivity of the oceanic food web (Ackley and Sullivan, 1994). Brine inclusions within the sea ice can range in salinity concentration from <10 to >150‰ and harbor diverse microbial assemblages, often dominated by low-light-adapted diatoms, as well as other autotrophic and heterotrophic flagellates, and several metazoa (e.g., amphipods, copepods, and larval stages of ice fish) (Palmisano and Garrison, 1993).

9.2.5 Glacial and Subglacial

The cryosphere on earth represents the solid form of water, such as snow or ice, and includes not only the ice sheets, but also glaciers and snow cover. These permanently frozen environments harbor diverse assemblages of microorganisms distinct from sea ice that are metabolically active and can be recovered by cultivation (Priscu and Christner, 2004; Margesin et al., 2007). The largest proportion of glacier ice occurs in Greenland and Antarctica, which hold 77% of the planet's freshwater (Paterson, 2001). Although microbial cell numbers are generally low (10^2–10^4 cells mL^{-1}), they fluctuate with depth based on insoluble mineral particles, which are considered to be the main microbial carriers (Abyzov et al., 1998).

Supraglacial lakes, bodies of liquid water on top of a glacier, and englacial lakes, those within a glacier, represent short-lived reservoirs for surface meltwater on glaciers (Laybourn-Parry and Wadham, 2014). These lake systems tend to be shallow and highly unproductive and exhibit a broad range of chemical conditions. Englacial lakes are often small areas of water that are remnants of crevasses or meltwater tunnels, whereas supraglacial lakes develop and drain during ablation seasons (Laybourn-Parry and Wadham, 2014). These lakes grow larger over time as they move down the ice valley of the glacier, accumulating new microbiota as they advance. The result is a unique mixed community of psychrophilic microorganisms originating from many different sources. Although these systems are ultra-oligotrophic, numerous organisms have been reported, including a diverse cyanobacterial population, picocyanobacteria, diatoms and desmids, and heterotrophic bacteria, but lack any evidence of benthic mats (Webster-Brown et al., 2010; Laybourn-Parry and Wadham, 2014). Supraglacial systems such as cryoconite holes and supraglacial lakes are found on the surface of glaciers and are exposed to the sun, which provides a habitat for photosynthetic psychrophiles within these isolated and ephemeral food webs. Cryoconite holes are generally eutrophic, providing oases for entire food webs containing bacteria, algae, diatoms, and metazoan in an otherwise highly oligotrophic environment (Miteva, 2008).

Subglacial water is often found as subglacial lakes that exist between the ice sheet and the underlying bedrock (Laybourn-Parry and Wadham, 2014). Biochemical and geochemical analysis has revealed the presence of aerobic heterotrophy, iron oxidation and reduction, sulfur oxidation and reduction, nitrate reduction, and methanogenesis in subglacial communities across the polar regions (Skidmore et al., 2012). Subglacial Lake Whillians is located in West Antarctica under 800 m of ice sheet, and is part of an extensive drainage network with connections to the Southern Ocean. A recent study reported that a significant fraction of 16S rRNA sequences were related to phylotypes of ammonia-oxidizing archaea, indicating that nitrification may be a major chemoautotrophic pathway for organic carbon production in the subglacial aquatic systems (Christner et al., 2014). This project was the first to show that a metabolically active and phylogenetically diverse ecosystem is capable of functioning in this extreme environment of subzero temperatures and the absence of light as an energy source.

9.3 ADAPTATION AND RESPONSE OF CELLULAR PROCESSES IN PSYCHROPHILIC AND PSYCHROTOLERANT MICROORGANISMS

Organisms adapted to cold environments display a broad array of adaptive features at almost all levels of cellular metabolism and function. Psychrophilic enzymes need to remain active at low temperatures, creating a challenge for enzyme reaction rates, which have an exponential loss in kinetic energy with decreasing temperatures. Protein and membrane function, as well as the thermal sensitivity of biochemical processes, is important in determining an organism's success in lower-temperature environments by its changes in reaction rates and the protein's ability to maintain flexibility.

9.3.1 Genome Evolution and Gene Expression

The growth and survival of psychrophilic microorganisms are the result of intrinsic genome-wide changes that accumulate during long-term adaptation to permanently cold temperatures. In general, cold temperatures reduce the enzyme activity, increase DNA and RNA secondary structure stability, and reduce protein folding kinetics. Currently, cold-adapted genomes represent a small fraction of all prokaryotic genomes sequenced (~3%), and few examples of comparative studies exist between mesophilic and psychrophilic organisms. One example of comparative genetic analysis occurred in which the proteome of many psychrophilic bacterial species was compared with that of their mesophilic counterparts (Metpally and Reddy, 2009). This study determined that psychrophilic bacteria maintain higher levels of serine, aspartic acid, threonine, and alanine in the coil regions of proteins, and that a higher proportion of amino acids contribute to protein flexibility, suggesting that amino acid substitution preferences lead to the observed protein adaptations (Metpally and Reddy, 2009). When grown under low temperatures, DNA becomes more negatively supercoiled due to the increased strength of the hydrogen bond interactions (Mizushima et al., 1997). Some nucleoid-associated proteins, like gyrase A, integration host factor, and histone-like nucleoid structuring protein, are critical to relax DNA during cold stress (Gualerzi et al., 2003; Weber and Marahiel, 2003). Several DNA and RNA helicases have also been identified in the polar diatom *Fragilariopsis cylindrus* as important for promoting replication, transcription, and translation at low temperatures by preventing cold denaturation of DNA and RNA templates (Mock and Thomas, 2005).

The expression of cold-shock proteins, which regulate a variety of cellular processes, appears to be a major target for adaptive strategies in psychrophiles (De Maayer et al., 2014). Among these are RNA helicases, which destabilize secondary structures, molecular chaperones, and genes associated with sugar transport and metabolism. Many of these genes are also expressed in mesophiles in response to cold exposure, and thus are also known as cold acclimation proteins (CAPs). High concentrations of reactive oxygen species (ROS) are also prevalent in colder habitats, requiring increased expression of genes encoding antioxidant enzymes, such as catalases and superoxide dismutases (Boutet et al., 2009). Control of both ROS production and detoxification is especially critical for psychrophilic organisms relying on photosynthesis, as the highly unsaturated lipids are susceptible to peroxidation (Morgan-Kiss et al. 2006). Although exact levels differ between organisms, the general rule is to upregulate genes.

Psychrophile DNA sequences often have multiple copies of genes that confer resistance to low temperatures. Transposases catalyze site-specific DNA rearrangements and horizontal gene transfer, and although the potential role of these repeated sequences in regulation remains unknown, it is thought that the transposases may play a role in recombination and genomic evolution (Bakermans et al., 2012). Transposases appear to have played a major role in evolving the genome of the psychrophilic methanogenic archaeaon *Methanooccoides burtonii* (Allen et al., 2009). This Antarctic lake psychrophile also exhibits additional evidence of genomic plasticity, including nucleotide skew and horizontal gene transfer. A metagenomics study from samples obtained from deep, cold ocean waters also reported that transposases were one of the most overrepresented COGs, that is, the clusters of orthologous genes, in the metagenome data (Delong et al., 2006). Conversely, inactivation of transposases has been linked with sensitivity to low temperatures in the deep-sea bacterium *Photobacterium profundum* SS9 (Lauro et al., 2008).

Increased stability of RNA is a fundamental challenge for cold-adapted organisms. The increased strength of hydrogen bonds between nucleic acid bases and the enhanced formation of secondary structures are solutions necessary for critical processes such as genome replication, transcription, and translation (Feller and Gerday, 2003). tRNA nucleotide composition from psychrophilic bacterial species contains 40%–70% more dihydrouridine than in mesophilic *Escherichia coli* strains (Dalluge et al., 1997). This posttranscriptional tRNA modification is a nonplanar base that favors the C-2′-endo sugar conformation, which rearranges the 5′ residue to produce a structure that is more flexible due to the reduced stabilizing effects of base stacking (Dalluge et al., 1997). Small RNA molecules can act as regulatory elements, or riboswitches, which bind to metabolites on the untranslated region (UTR) of mRNA. Riboswitches are thought to act as cellular thermometers that regulate the translation of specific mRNAs in response to a change in temperature (Narberhaus et al., 2006). In colder temperatures, mRNA 5′ UTR forms a temperature-dependent secondary structure to expose the Shine–Dalgarno sequence and AUG start codon, whereas at warmer temperatures this region is hidden in a helical conformation (Giuliodori et al., 2010). Although often unique to psychrophilic organisms, these genetic differences are only the base of the adaptive changes that occur in cold habitats, creating a cascade of events that allow an organism to survive in these extreme conditions.

9.3.2 Enzyme Catalysis, Biochemistry, and Protein Synthesis

In general, all psychrophilic enzymes face the same challenge, that is, to remain active at low temperatures. Temperature creates a significant thermodynamic constraint on enzyme reaction rates due to the exponential loss in kinetic energy with declining temperature. This constraint limits the biochemical and physiological potential of a cellular system. The Arrhenius equation can be applied to approximate the relationship between temperature and a reaction rate:

$$\kappa = \kappa \, (\kappa_B T/h) e^{-\Delta G^*/RT}$$

where κ is the transmission coefficient, κ_B is Boltzmann's constant, h is Planck's constant, ΔG^* is the free energy of activation, and R is the universal gas constant (Arrhenius, 1896). This equation represents the exponential effect that temperature has on the reaction rate of an enzyme (Takacs and Priscu, 1998). Psychrophilic enzymes are generally characterized by higher structural flexibility, lower thermostability, and higher specific activity than their mesophilic homologs (De Maayer et al., 2014). Maintaining functional membranes, evolving cold-adapted enzymes, and adapting structural properties are the basic adaptive strategies of psychrophiles. Cold-evolved enzymes display high catalytic efficiency with low thermal stability, which differentiates them from their thermophilic counterparts.

Physical changes that occur in aqueous environments at low temperatures include a reduced dissociation constant, reduced diffusion, and increased viscosity (Angell, 1982). Reduced diffusion of substrates, combined with the decreased affinity for substrates, creates a formidable challenge to psychrophilic microorganisms that are most often starved for nutrients (Nedwell and Rutter, 1994). Microorganisms need to acquire and sequester these resources more effectively than competing species. Sequestration of substrates is often related to the Monad-type saturation curve, which relates the maximum specific growth rate (μ_{max}) to the concentration of the substrate: $\alpha_A = \mu_{max}/K_s$, where K_s is the substrate affinity constant (Nedwell and Rutter, 1994).

Low temperatures also affect the balance of carbonate and bicarbonate ions in aqueous systems, creating changes in the pH of the environment to which psychrophilic organisms must adapt (Thomas, 1996). This change in pH also has the ability to influence the availability of nutrients, carbon, and trace elements, which can alter the selective advantage of different microorganisms within the system. Last, temperatures below freezing cause the salinity of liquid water to significantly increase due to the salt being excluded based on its different crystalline structure. The brine that results from this exclusion is dense and has a lowered freezing point, often creating pockets of brine that can be formed within ice or permafrost when initial concentrations of solutes are high (Collins et al., 2008). Osmotic stress results from this increase in salinity, with the microorganisms inhibiting the transport of cofactors and substrates into the cell. Thus, psychrophilic microorganisms often have adaptations similar to those seen in a halophile (Ponder et al., 2005).

Lipases are also of high interest in psychrophilic organisms due to their diversity and industrial applications by physiologically catalyzing hydrolysis of insoluble long-chain triacylglycerides to free fatty acids (Tran et al., 2013). These serine hydrolases maintain low thermostability at higher temperatures, and the cold-adapted organisms maintain lipases that have evolved to maintain high biocatalytic activity. This is discussed in further detail in Section 9.4.

In general, psychrophilic proteins maintain catalysis at low temperatures by remaining flexible, while the structure of their thermophilic counterparts is generally more rigid to prevent thermally induced denaturation. Higher protein flexibility around the catalytic site reduces the free energy barrier of the transition state, usually at the expense of substrate affinity (Struvay and Feller, 2012). Destabilization of structures around the active sites or the whole molecule can occur through a number of mechanisms, resulting in increased active site dynamics at low temperatures. Proteins may bury amino acids that are smaller and less hydrophobic in order to maintain core hydrophobicity (Leiros et al., 1999). Alternatively, proteins may display increases in their surface charges and higher proportions of nonpolar amino acids on the surface to maintain surface hydrophillicity (Feller et al., 1997; Bae and Phillips, 2004). In general, psychrophile proteins will also exhibit less hydrogen bonding between amino acid residues (Alvarez et al., 1998), lower numbers of salt bridges (Feller, 2003), decreased arginine content, or an increase in the number of prolines in alpha helices, while decreasing proline in surface loops to help maintain structural flexibility (Feller and Gerday, 2003). This combination of adaptions allows for more accessibility, and thus the accommodation of substrates at a lower energy cost, reducing the activation energy required for the enzyme and substrate to complex. A larger active site also allows for easier release of products, thus decreasing the effects of a rate-limiting step (Struvay and Feller, 2012).

Protein structure and thus function can also be maintained with the help of protein chaperones, which aid in the folding and refolding of denatured proteins. Numerous studies have identified psychrophilic-specific chaperones to assist with the rate-limiting step of protein folding, such as PPIase, DnaJ, Cpn60/10, DnaK, and Clp B, which are important in correcting the protein structure and maintaining enzyme activity in low-temperature conditions (Schmid, 1993; Ferrer, 2004, Goodchild et al., 2005; Strocchi et al., 2006). RNA chaperone proteins can prevent secondary structure formation in RNA transcripts and are very often elevated during growth in cold environments. RNA chaperones in concert with RNA degradosomes act to recycle ribonucleotides to create new RNA molecules (Ting et al., 2010).

Membrane structure and function in response to low temperatures is also a key issue for proper maintenance of critical physiological processes, such as ion transport, energy generation, and cell division. In particular, fluctuations in environmental temperatures perturb the molecular disorder and molecular motion (i.e., membrane fluidity) of cytoplasmic and energy-generating membranes. Low-temperature adaptations, as associated with membrane lipids, prevent membrane rigidification or the so-called "gel phase" and are often associated with the desaturation of membrane lipids by fatty acid desaturases (Nichols et al., 2004). Modulation of the fatty acid unsaturation levels can also be accompanied by other environmentally induced dynamic alterations in the extent of methyl and anteiso-branching, the conversion of cis- to trans-unsaturated fatty acids, the lipid composition, and the lipid/protein ratio (Heipieper et al., 2003; Mullineaux and Kirchhoff, 2009; Los et al., 2013). Most organisms desaturate fatty acids from existing saturated fatty acids, and this phenomenon was first demonstrated in E. coli as "homeoviscous adaptation" (Sinensky, 1974). In later studies, a feedback loop between membrane rigidification and upregulation of fatty acid desaturase genes was discovered (Tasaka et al., 1996). More recently, a broader set of genes and physiological responses has been linked to the perception of changes in membrane properties, in particular processes important for photosynthetic function (Los et al., 2013).

Transcriptomic experiments have also revealed numerous cold-temperature adaptations that maintain the cell membrane, including upregulation of genes important for membrane biogenesis (Gao et al., 2006; Frank et al., 2011). These membrane biogenesis genes include those involved in fatty acid and lipopolysaccharide biosynthesis, peptidoglycan synthesis, outer membrane proteins, and glycosyltransferases. Membrane transport proteins are also reported to be upregulated to counteract the lower diffusion rates in cold temperatures, across the cellular membranes (Cacace et al., 2010). This mechanism enhances the uptake of nutrients, compatible solutes, and peptidoglycan synthesis. Psychrophilic organisms also respond to the cold temperatures by downregulating genes involved in chemotaxis and iron uptake receptors (Durack et al., 2013).

9.3.3 Cryoprotectants

In general, there are two phases of cold shock: moderate and severe. During early-phase cold shock at low temperatures, transcription machinery begins to be affected, followed by translational machinery during midphase cold shock. It is only during late-phase moderate cold shock that heat-shock proteins, metabolism, and transduction genes begin to be affected (Al-Fageeh and Smales, 2006). During severe cold shock, usually occurring at near-freezing conditions, cryoprotectants (substances used to protect an organism from freeze damage, such as antifreeze proteins) become an essential part of an organism's survival. These cryoprotectants include, but are not limited to, trehalose and glycerol.

Trehalose accumulates under various stress conditions and maintains several protective roles in the cell, including membrane and protein stabilization (Aguilera et al., 2007). Trehalose (and its two biosynthesis enzymes, Tps1 and Tps2) has been well documented in cold-adapted fungi as being correlated with freezing resistance (Kandror et al., 2004; Schade et al., 2004). Although the exact mechanism of action of trehalose and its enzymes is unknown, there is evidence suggesting that the enzymes participate in mRNA stabilization and are activated by low-temperature induction of stress response element binding (Moskvina et al., 1999). Much like trehalose, glycerol also protects the cell from damage due to freezing conditions. The glycerol synthesis enzyme GPD1 becomes activated during lower temperatures and accumulates in the cell (Panadero et al., 2006). It is believed that glycerol protects against freezing via osmotic shrinkage resulting from freezing and thawing processes. Although quite different in their mechanisms of actions, both trehalose and glycerol maintain protective roles within psychrophiles in response to decreasing temperatures.

Compared with temperate counterparts, the synthesis of CAPs is continuous at low temperatures. The major CAP in *Pseudoalteromonas haloplanktis* was identified as a chaperone protein that was upregulated 37-fold in 4°C conditions compared with warmer 18°C conditions (Piette et al., 2011). This CAP interacts with the new polypeptides of the ribosome and destabilizes RNA secondary structures to assist in the refolding of cold-denatured proteins (Piette et al., 2011). This chaperone activity may be the rate-limiting step for bacterial growth at low temperatures, as it is involved in the initial protein folding. Late embryogenesis abundant (LEA) proteins are also abundant in polar chlorophyte species such as *Chlorella subellipsoidea* and are thought to aid in H-bond stabilization of enzymes (Honjoh et al., 1995). Cold-shock proteins are only one of the numerous cryoprotectants that organisms have evolved to retain in cold-temperature environments.

Many psychrophiles have adapted to their environments by synthesizing cryoprotective proteins, antinucleating proteins, and ice-binding proteins. These cryoprotectants are capable of changing the freezing point of water within the cell, allowing them to maintain free water at their surface, thus maintaining their structure. One such protectant is glycine betaine, an osmolyte that can prevent the aggregation of proteins, allowing membrane fluidity within the cell (Welsh, 2000).

Psychromonas ingrhamii, for example, can synthesize glycine betaine from choline or take it up from the environment around it (Welsh, 2000; Methé et al., 2005). Cryoprotective proteins are also capable of preventing denaturation by surrounding other proteins and protecting them. These types of proteins have been found in several ice-nucleating bacteria, such as *Pseudomonas fluorescens* (Obata et al., 1998). Antinucleating materials are similar to these proteins, but they allow water to supercool by lowering ice nucleator activity rather than covering proteins. Psychrophiles can discriminate between different absolute and relative low-temperature increments, and as such may either stimulate cryoprotectants for short-term occasions or upregulate cold-acclimated proteins that are synthesized continuously in response to growth at low temperatures.

Another factor threatening psychrophilic organisms is ice recrystallization. This is when larger ice grains grow at the expense of smaller ones, which can then burst compartments of cells or entire cells, leading to cell damage or death (Knight et al., 1995). The combination of ice-binding proteins and antifreeze proteins can inhibit this process and protect the organisms from damage. Ice-binding proteins bind and inhibit the growth of these ice crystals, thus interfering with the recrystallization process. A number of these proteins work by lowering the freezing point of water, thus inhibiting recrystallization within the cell. Although found in many bacteria, such as *Marimonas*, *Rhodococcus*, *Psychrobacter*, and *Sphingomonas*, the bacterial mechanism of action of these proteins has yet to be well studied (Kawahara, 2002). Although antifreeze proteins depress the freezing point, ice-nucleating proteins can induce freezing at temperatures warmer than what would occur naturally or spontaneously. These ice-nucleating proteins help provide access to nutrients leaked, protect from infection, and enhance survival by preventing cell damage due to freezing (Bakermans, 2012).

Exopolysaccharides (EPSs) are believed to be important antinucleating compounds, as they affect ice texture and permeability, prevent recrystallization, and can restrict fluid flow within organisms (Krembs et al., 2002). EPSs are large sugar polymers composed of residues synthesized by cells and secreted extracellularly that are thought to have a role in metal capture and slowing the diffusion of exoenzymes (Qin et al., 2007). It is thought that they are a cryoprotectant by aiding water retention and freeze point depression and/or affecting the concentration of extracellular substrates. One experiment in *Colwellia psychrerythraea* revealed higher rates of protein synthesis at lower temperatures in the presence of EPSs, indicating that they increase metabolic activity by keeping water at the cell surface and/or stabilizing membrane-associated enzymes (Junge et al., 2006). These polymers may also be mediating cell–cell and cell–surface interactions or may be facilitating the exchange of nutrients or genetic material (Junge et al., 2006). Lastly, the genome of the psychrophilic *M. burtonii* encodes for a large number of genes involved in polysaccharide biosynthesis, and high levels of EPSs are produced in this organism when grown at low temperatures relative to high temperatures (Allen et al., 2009; Reid et al., 2006). The mechanisms for cold adaptation discussed thus far have involved organisms throughout the ecosystem and food web. However, the large majority of carbon input into these systems resides within the base of the food web, the photoautotrophs. These important primary producers present their own unique adaptations for evolving to not only survive in cold-temperature habitats but also actively thrive and contribute to these unique food webs.

9.3.4 Photosynthesis and Light Acclimation

Many cold environments are dominated by photosynthetic microorganisms, including cyanobacteria, anoxygenic phototrophic bacteria, many species of algae, and mixotrophic protists (Morgan-Kiss and Dolhi, 2011). Light availability is highly variable in cold environments, and cold-adapted photosynthetic microorganisms possess adaptations to avoid photodamage under excessive light conditions, simultaneously exhibiting the ability for efficient light capture in low light environments. Not only does ice cover reduce downwelling irradiance to phytoplankton communities, but also attenuation within the water column reduces irradiance even further for deeper communities (Vincent, 1981; Lizotte and Priscu, 1992). Photosynthetic organisms grow and reproduce in the presence of these highly variable light environments, indicating that the photosynthetic apparatus of psychrophiles should exhibit the ability to dynamically adjust to maintain maximum efficiency of light energy capture. The term *photostasis* describes the dynamic process of balancing energy capture by photochemical processes with the consumption of stored energy by downstream metabolic reactions.

Oxygenic photosynthetic organisms include photosynthetic eukaryotes, as well as cyanobacteria, that are differentiated from other photosynthetic organisms by their ability to produce oxygen by splitting water in the presence of two membrane-associated photosystems (named photosystem I [PSI] and photosystem II [PSII]). Cyanobacteria represent the only bacterial phylum that harbors oxygenic photosynthetic metabolism. Unlike eukaryotic algae, cyanobacteria have a unique pigment–protein complex, the phycobilisome, which is bound to the cytoplasmic surface of the thylakoids. In addition, the bacterial redox carriers for respiration and electron transport of cyanobacteria are located in thylakoid membranes and share a plastoquinone pool and Cyt b_6f complex (Cooley et al., 2000; Cooley and Vermaas, 2001). Although cyanobacteria are commonly associated with high-temperature environments, cyanobacteria are highly successful photosynthetic organisms in polar and alpine ecosystems. The most luxuriant growth observed in low-temperature habitats is associated with microbial mats. Multiple phycobiliproteins maximize the light-harvesting ability of cyanobacterial mats in low-energy systems, such as littoral mats in Antarctic ice-covered lakes. In addition, mats exposed to the surface resist oxidative damage under high light or UV by the accumulation of high levels of carotenoids. Phycobilisomes are rapidly dynamic in response to the light environment and exhibit the ability to freely move within the thylakoid membranes to redistribute energy between PSII and PSI in a process called energy spillover or state transitions (Mullineaux and Emlyn-Jones, 2005).

The majority of marine phytoplankton biomass comprises unicellular, photosynthetic eukaryotes, including diatoms (Bacillariophyceae) and haptophytes (Prymnesiophycae). Diatoms are very often found in low-temperature environments, such as streams and microbial mats, and are important primary producers in Arctic and Antarctic marine food webs. In particular, large diatom assemblages are associated with sea ice (making up 50%–75% of the protist ice-associated biomass) and are the major source of new biomass during the spring phytoplankton blooms (Horner et al., 1992). Diatoms harbor a light-harvesting complex with a pigment complement of chlorophylls a and c, as well as the carotenoid fucoxanthin (Green and Durnford, 1996). Sea ice diatoms are some of the most low-light-adapted phytoplankton in the world and exhibit photosynthetic activity at the lowest temperatures ever measured. In addition, diatoms utilize a rapid photoprotection system based on a one-step interconversion between two xanthophyll pigments, diadinoxanthin and its deepoxidized form, diatoxanthin, which serve as light-harvesting and energy dissipation functions, respectively. Utilization of this xanthophyll conversion pathway is a critical factor that sea ice communities can exploit to reversibly adjust photosynthetic efficiency as a rapid response to changes in their light environment (Petrou et al., 2011).

Additional contributors to inorganic carbon fixation in low-light, anoxic environments are bacteria capable of anoxygenic photosynthesis. Anoxygenic phototrophic bacteria have only one of two kinds of reaction centers: green bacteria contain type I and purple bacteria contain type II (Madigan, 2005). Although bacteria capable of this process are widely distributed in aquatic ecosystems, the proteins necessary for photosynthesis are only expressed under anaerobic conditions. Oxygen is not produced and alternative electron donors, such as reduced sulfur or iron, are utilized (Madigan, 2005). These types of psychrophilic photosynthetic microorganisms utilize major light-harvesting pigments, like bacteriochlorophyll and carotenoids, but still have electron transport components shared with the respiratory chain, similarly to cyanobacteria (Madigan, 2005).

Photoprotection against high irradiance and UV radiation is important for photosynthetic organisms residing on surfaces. Studies have indicated that carbon fixation can be sensitive to

UVB radiation, and that this sensitivity becomes more pronounced at lower light levels (Kroon 1994). Many polar microalgae are not capable of growing at irradiances above 200 µmol photons m^{-2} s^{-1} (Cota, 1985) and it is thought that this may be a consequence of uncoupling between electron generation and the utilization of electrons for the reduction of carbon dioxide (Davison, 1991). Photoprotection against UV is often provided by photolyases that repair damaged DNA or repair the proteins within PSII, the target of UVB. Production of ROS such as hydrogen peroxide by microorganisms can also result in the damage of these essential proteins. Hydrogen peroxide is photochemically produced under high irradiance levels, resulting in photosynthetic organisms requiring antioxidant enzymes such as catalase, glutathione peroxidase, and glutathione reducatase (Rijstenbil, 2001).

Although some polar algae appear to be restricted to growth under low light conditions and exhibit limited ability to survive high light, ice diatoms have retained the ability to acclimate to high light under subzero temperatures. The polar diatom *Chaetoceros neogracile* responds to a shift in growth irradiance from low to high light by increasing the expression of light harvesting complex (LHCx) proteins and antioxidant proteins. The LHCx proteins bind xanthophyll cycle pigments, which function to dissipate excess energy and prevent photoinhibitory damage to the photosynthetic apparatus (Park et al., 2010). Antioxidant systems play an important role in the protection of photosynthetic organisms from cellular damage induced by oxidative stress. This is particularly important for the protection of photosynthetic membranes, which are typically enhanced with polyunsaturated fatty acids (PUFAs) and are targets for lipid peroxidation. Although it is well known that PUFAs are among the most sensitive biological molecules to ROS, there is also evidence that specific families of PUFAs, such as the long-chain fatty acids docosahexaenoic acid (DHA) and eicosapentaenoic acid (EPA), may have antioxidant properties. In one model, DHA and EPA can shield membranes from the effects of exogenous ROS (Okuyama et al., 2008). In some habitats, for example, the Dry Valley Antarctic lakes, there are environmental factors that facilitate photosynthesis for these cold-adapted organisms, such as the presence of higher nutrient concentrations in lower depths (Priscu et al., 1989a, 1989b). It has also been previously suggested that phytoplankton in these deeper depths support growth at lower irradiances by having higher photosynthetic efficiencies and may be capable of adapting their photosynthetic apparatus to a higher extent than more temperate communities (Vincent, 1981; Priscu et al., 1987).

Models for adaptation of true photosynthetic psychrophiles are limited. One of the few highly studied cold-adapted algae is a photopsychrophile, *Chlamydomonas* sp. UWO241, which was isolated more than 20 years ago from an ice-covered Antarctic lake (Neale and Priscu, 1995). Phytoplankton growing in the water columns of perennially ice-covered lakes in the MDVs grow under an extremely stable regime of low temperatures and low light due to the lack of wind mixing, as well as continuous daylight during the austral summer (Lizotte and Priscu, 1992). The strain UWO241 is adapted to yearly irradiance levels that do not exceed 50 µmol photons m^{-2} s^{-1} and spectral distribution within the blue-green region (440–80 nm) (Lizotte and Priscu, 1992). Long-term physiological studies on this organism have reported a low chlorophyll a/b ratio (reflecting low levels of PSI to PSII) and high rates of PSI-driven cyclic electron transport relative to other mesophilic algae, such as the model species *Chlamydomonas reinhardtii* (Morgan et al., 1998; Morgan-Kiss et al., 2002, 2005).

9.4 BIOTECHNOLOGY AND THE FUTURE

Among all organisms on our planet surviving in extreme environments, psychrophiles are the most abundant in diversity, distribution, and most importantly, biomass. There have been numerous potential applications in biotechnology based on psychrophiles, ranging from industrial to environmental to pharmaceutical. Psychrophilic enzymes can be used for everything from laboratory reagents to medical research, and because these strains can be grown at lower temperatures, researchers can avoid heating fermentation vats, which not only is energy saving but also can prevent aggregation of these enzymes (Lohan and Johnston, 2005). Psychrophilic biomolecules can also be supplied for use as dietary supplements or as applications for antifreeze proteins (Lohan and Johnston, 2005). Another use for psychrophiles includes environmental applications by bioremediation to degrade pollutants with cold-adapted microbes to dispose of wastes or cleanup after environmental spills. Finally, psychrophiles may present interesting avenues for screening for antibiotics or anticancer drugs based on their metabolism in human-driven reactions (Biondi et al., 2008).

There are numerous reasons as to why cold-active enzymes are advantageous for biotechnology, and these reasons can then be exploited for the production of these compounds. Because they remain efficient at lower temperatures, the heating process of most applications can be avoided, creating a cost-effective, energy-saving environment for enzyme production. There are numerous examples of this, such as cooler cycling washing machines or reduced incubation times for lactose hydrolysis or cold pasteurization. Due to a psychrophile's enzymes having higher energy activity, lower concentrations of the catalyst are needed in order to have activity, which then reduces enzyme preparation costs and time (Margesin et al., 2007). Psychrophilic lipases have been explored for many of these industrial applications. In addition to their natural functions, they have potential in catalyzing bioconversion reactions and hydrolysis of organic carbonates without an additional need for cofactors (Rajendran et al., 2009).

The thermolability of psychrophilic proteins has recently been exploited to develop novel temperature-sensitive vaccines within the pharmaceutical industry. Duplantis et al. (2010) engineered live vaccines for tularemia by substituting DNA ligase from a psychrophile, *C. psychrerythraea* into the pathogenic mesophile *Francisella tularensis*. This substitution allowed for the mutant strains to be unable to proliferate at core body temperatures, but can induce protective immunity when injected into a cooler region of the body, such as the tail of a mouse. This provides just one example of how thermolability can be used in such a way as to benefit the human condition.

Currently used psychrophilic microorganisms include the yeast *Candida antarctica*, which produces two lipases, one of which is involved in numerous organosynthetic applications, such as food processing, pharmaceuticals, and cosmetics (Shimada et al., 1999). This lipase is stabilized in its immobilized form and because of its substrate and stereospecificity, it is the preferred lipase for many products (Lohan and Johnston, 2005). This organism is also currently being used for its implications in biodiesel, as this immobilized lipase was found to be the most effective lipase for methanolysis (Shimada et al., 1999). Experiments conducted on this lipase show numerous advantageous of using this cold-adapted enzyme over its mesophilic counterparts, such as lower cost due to continuous reactions within a reactor, decreased risk of explosion since no solvent is required, an enzymatic process that is easier to conduct than any chemical process, and the separation of glycerol and methyl ester layers (Shimada et al., 1999). These are just a few of the applications of a single enzyme isolated from a single organism, indicating endless potential applications for the growth and metabolism of psychrophilic organisms.

Polar proteins are already on the rise, as multiple companies are discovering, processing, and patenting various cold-adapted proteins for scientific and industrial research. For example, New England Biolabs currently has a heat-labile alkaline phosphatase that is sold as Antarctic phosphatase, which was isolated from an Antarctic bacterium found in seawater from the McMurdo Sound (Ipswich, Massachusetts). By isolating this alkaline phosphatase and cloning it into *E. coli*, its properties have been further enhanced by direct evolution for high activity and heat lability (Koutsioulis et al., 2008). Another protein, Antarcticine-NF3, has antifreeze properties commonly used for scar treatment and reepithelialization of wounds (Parente Duena et al., 2006). This glycoprotein was isolated from *Pseudoalteromonas antarctica* and is currently being used in cosmetic regeneration creams as Antarctilyne. These are just a few of the numerous ways in which psychrophilic organisms can be utilized for their molecular biological properties.

Ecologically, psychrophilic organisms are also being researched for their contributions to hydrocarbon bioremediation. These microbial populations mineralize organic pollutants into less harmful, nonhazardous substances that can be integrated into the systems' biogeochemical processes. Many of these studies have focused on the treatment of petroleum hydrocarbons because of their increased risk of accidental spills or release. However, the research has shown that biodegradation activity in contaminated cold environments is provided by numerous hydrocarbon degraders with catabolic pathways for the degradation of various types of hydrocarbons with high mineralization potential (Panicker et al., 2010). Numerous organisms have been researched for their potential, not only in hydrocarbon remediation but also in wastewater treatment. The Antarctic *Arthrobacter psychrolactophilus* has been shown to produce a complete clarification of synthetic wastewater turbid medium, as it was able to hydrolyze proteins, starches, and lipids (Gratia et al., 2009). Both Antarctic and Arctic cyanobacteria, yeasts, bacteria, and fungi have been studied with promising results in the degradation of phenols, which are considered to be the most common toxic pollutants among wastewater samples. Taken together, psychrophiles and their biomolecules are proving to be strong contenders in myriad industrial, environmental, and pharmaceutical applications across a broad spectrum of applications. A large majority of these applications not only are cost-effective but also tend to be more energy efficient and environmentally friendly compared with their mesophilic counterparts.

9.5 CONCLUSION

Herein we have discussed many of the most common biochemical and physical adaptations relevant to psychrophilic microorganisms. It is quickly becoming evident that psychrophilic organisms and their enzymes and products are needed to explore novel biotechnological advances, such as enzymes in detergents, food additives like PUFAs, biosensors for the environment, biotransformations, and environmental bioremediation. New research into these cold-adapted organisms also has the potential to open new applications for their uses in biotechnology processes. Psychrophiles could be used for rapid termination of processes because of their ability to inactivate cold-active enzymes with moderate heat treatment, high yields from thermosensitive components, modulation of stereospecificity, and cost-saving capabilities by eliminating expensive heating and cooling processing steps (Russell and Hamamoto, 1998). Although some of these avenues have been explored, further research is needed to examine the potential applications and benefits from using these organisms.

The average air temperature at the earth's surface has increased drastically over the last century, as much as 0.2°C per decade, but the climate models often predict the polar regions to have an amplified response (Cattle et al., 1995). This becomes especially evident in the polar regions, such that it is projected that there will be significant surface warming over Antarctica over the next 100 years, with an average increase of 0.34°C per decade over land and even more over ice (ACIA, 2005). These changes are projected to increase the number of degree days above freezing in the inland locations, causing drastic changes to both the microhabitat and the biological communities. The largest warming trends across Antarctica have been found in the western and northern parts of the continent, and the Faraday station on the western peninsula has experienced the largest trend of increasing temperatures, with increases of 0.54°C per decade since the 1950s, resulting in fewer extreme cold winters (Turner et al., 2014). Antarctic lakes, for example, are effective sentinels for monitoring these effects of climate change (Morgan-Kiss and Dolhi, 2011). This is due in part to the lakes being closed systems in which inputs and outputs can be defined and measured and in regions that make them more sensitive to climate influences (Wilkins et al., 2013). Polar lakes tend to have simplified food webs with relatively low diversity that rely on key members for ecosystem functioning, and therefore are appropriate targets for climate modeling. Two examples of psychrophilic organisms can be seen in Figures 9.2 and 9.3, which represent a *Brachiomonas* sp. and a dinoflagellate, *Gymnodinium* sp., respectively, both isolated from the Antarctic Lake Bonney.

Although it is predicted that photosynthetic communities in polar or alpine environments will be more responsive to climate warming and episodic events than lower-latitude environments, the complexity within these systems provides numerous challenges to understanding the communities in the future. This prediction is based on the observations that these aquatic environments are often oligotrophic (nutrient limited), psychrophilic (low temperature adapted), and relatively pristine in

Figure 9.2   *Brachiomonas* sp. scanning electron microscope image obtained from a *Brachiomonas* sp. isolated from Lake Bonney within the McMurdo Dry Valleys, Antarctica. Sample obtained from lake water concentrated at 100× using tangential flow filtration. (Image courtesy of Wei Li, Department of Earth Science, Montana State University, Bozeman.)

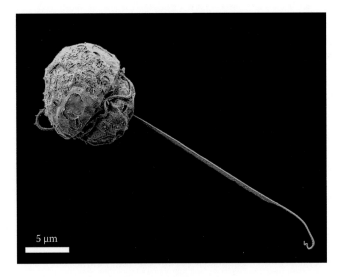

Figure 9.3   Dinoflagellate *Gymnodinium* sp. scanning electron microscope image of a dinoflagellate isolated from Lake Bonney within the McMurdo Dry Valleys, Antarctica. Sample obtained from lake water concentrated at 100× using tangential flow filtration. (Image courtesy of Wei Li, Department of Earth Science, Montana State University, Bozeman.)

comparison with more temperate watersheds (Doran et al., 2008; Morgan-Kiss and Dolhi, 2011). The physiological adaptations of these cold-adapted organisms are directly related to their environment, thus complicating the objective of predicting how future temperatures will affect the acclimation and adaptation of these critical microorganisms. As the earth continues to change, it is important to predict how biogeochemical processes and ecosystem function will be altered in response to future conditions.

Work conducted in the Antarctic Dry Valleys included one such experiment, which hoped to mimic climate change and observe the changes that occurred in the cold-adapted community, in terms of not only community composition but also primary and bacterial production (Teufel et al., 2016). Episodic climatic events occur in the Dry Valley lakes in the form of increased glacial melt, resulting in organic sediment and nutrient input from glacial streams into the closed basin systems. This study examined two lakes varying in their biogeochemical composition to better access how different environments could react to climate change events and determined that although the response varied between the different lakes based on nutrient-dependent responses, overall production increased (Teufel et al., 2016). Furthermore, the authors provided evidence that this increase in production was at the level of the chlorophytes (many of which are characterized psychrophiles or have been observed previously [Christner et al., 2003, Pocock et al., 2004; Jungblut et al., 2005; Morgan-Kiss et al., 2008; Bielewicz et al., 2011; Dolhi et al., 2015]) and that bacterial productivity was generally uncoupled, albeit phytoplankton growth selected for specific bacterial groups without observable changes in heterotrophic production. This research, combined with other advances in the field, suggests that increased global warming could result in diversity shifts, thus altering the availability and composition of autochthonous carbon for heterotrophic production, and provides another example of how scientists can use psychrophilic organisms as models for future climatic event outcomes.

Although psychrophilic microbes exist in numerous habitats and undergo various adaptive strategies, our understanding of what makes an organism psychrophilic is still unknown in a large majority of cold-adapted organisms. Our understanding of these strategies is in its infancy, and thus future investigations are needed regarding cold adaptation and their biotechnological potential to our lives. Although psychrophilic enzymes are often a target for development due to their high efficiency and optimum catalysis at low temperatures, a better understanding of these cold-adapted macromolecules is needed to better design molecules with specific activity at lower temperatures (Doyle et al., 2012). Even as research has increased over the last decade, new technological advances and high-throughput DNA sequencing will continue to provide information about cold adaptation or the mechanisms needed for survival in a changing world.

REFERENCES

Abyzov, Slezov S., N. I. Barkov, N. E. Bobin et al. The ice sheet of central Antarctica as an object of study of past ecological events on the earth. *Biology Bulletin—Russian Academy of Sciences C/C of Izvestiia-Rossiiskoi Akademii Nauk Seriia Biologicheskaia* 25 (1998): 501–6.

Ackley, Steven F., and C. W. Sullivan. Physical controls on the development and characteristics of Antarctic sea ice biological communities—A review and synthesis. *Deep Sea Research Part I: Oceanographic Research Papers* 41, no. 10 (1994): 1583–604.

Aguilera, Jaime, Francisca Randez-Gil, and Jose Antonio Prieto. Cold response in Saccharomyces cerevisiae: New functions for old mechanisms. *FEMS microbiology reviews* 31, no. 3 (2007): 327–41.

Al-Fageeh, Mohamed B., and C. Mark Smales. Control and regulation of the cellular responses to cold shock: The responses in yeast and mammalian systems. *Biochemical Journal* 397, no. 2 (2006): 247–59.

Allen, Michelle A., Federico M. Lauro, Timothy J. Williams et al. The genome sequence of the psychrophilic archaeon, *Methanococcoides burtonii*: The role of genome evolution in cold adaptation. *ISME Journal* 3, no. 9 (2009): 1012–35.

Alvarez, Marco, Johan Ph. Zeelen, Véronique Mainfroid et al. Triose-phosphate isomerase (TIM) of the psychrophilic bacterium *Vibrio marinus* kinetic and structural properties. *Journal of Biological Chemistry* 273, no. 4 (1998): 2199–206.

Angell, Austin. Supercooled water. In *Water: A Comprehensive Treatise*, ed. F. Franks, pp. 1–2. New York: Plenum Press, 1982.

Arctic Council and International Arctic Science Committee. Arctic climate impact assessment. 2005.

Arrhenius, Svante. XXXI. On the influence of carbonic acid in the air upon the temperature of the ground. *London, Edinburgh, and Dublin Philosophical Magazine and Journal of Science* 41, no. 251 (1896): 237–76.

Bae, Euiyoung, and George N. Phillips. Structures and analysis of highly homologous psychrophilic, mesophilic, and thermophilic adenylate kinases. *Journal of Biological Chemistry* 279, no. 27 (2004): 28202–8.

Bakermans, Corien. Psychrophiles: Life in the cold. In *Extremophiles: Microbiology and Biotechnology*, ed. R. Anitoris, pp. 53–56. Hethersett, UK: Horizon Scientific Press, 2012.

Bakermans, Corien, Peter W. Bergholz, Debora F. Rodrigues, Tatiana A. Vishnivetskaya, Hector L. Ayala-del-Río, and James M. Tiedje. Genomic and expression analyses of cold-adapted microorganisms. In *Polar Microbiology: Life in a Deep Freeze*, ed. Robert V. Miller and Lyle G. Whyte, pp. 126–55. Washington, DC: ASM Press, 2012.

Bielewicz, Scott, Elanor Bell, Weidong Kong, Iddo Friedberg, John C. Priscu, and Rachael M. Morgan-Kiss. Protist diversity in a permanently ice-covered Antarctic lake during the polar night transition. *ISME Journal* 5, no. 9 (2011): 1559–64.

Biondi, Natascia, M. R. Tredici, Arnaud Taton, Annick Wilmotte, Dominic A. Hodgson, Daniele Losi, and Flavia Marinelli. Cyanobacteria from benthic mats of Antarctic lakes as a source of new bioactivities. *Journal of Applied Microbiology* 105, no. 1 (2008): 105–15.

Boutet, Isabelle, Didier Jollivet, Bruce Shillito, Dario Moraga, and Arnaud Tanguy. Molecular identification of differentially regulated genes in the hydrothermal-vent species *Bathymodiolus thermophilus* and *Paralvinella pandorae* in response to temperature. *BMC Genomics* 10, no. 1 (2009): 1.

Cacace, Giuseppina, Maria F. Mazzeo, Alida Sorrentino, Valentina Spada, Antonio Malorni, and Rosa A. Siciliano. Proteomics for the elucidation of cold adaptation mechanisms in *Listeria monocytogenes*. *Journal of Proteomics* 73, no. 10 (2010): 2021–30.

Cattle, H., J. Crossley, and D. J. Drewry. Modelling arctic climate change [and discussion]. *Philosophical Transactions of the Royal Society of London A: Mathematical, Physical and Engineering Sciences* 352, no. 1699 (1995): 201–13.

Cavicchioli, Ricardo. Cold-adapted archaea. *Nature Reviews Microbiology* 4, no. 5 (2006): 331–43.

Christner, Brent C., Brian H. Kvitko II, and John N. Reeve. Molecular identification of bacteria and eukarya inhabiting an Antarctic cryoconite hole. *Extremophiles* 7, no. 3 (2003): 177–83.

Christner, Brent C., John C. Priscu, Amanda M. Achberger et al. A microbial ecosystem beneath the West Antarctic ice sheet. *Nature* 512, no. 7514 (2014): 310–13.

Collins, Tony, Frédéric Roulling, Florence Piette, Jean-Claude Marx, Georges Feller, Charles Gerday, and Salvino D'Amico. Fundamentals of cold-adapted enzymes. In *Psychrophiles: From Biodiversity to Biotechnology*, ed. Rosa Margesin, Franz Schinner, and Jean-Claude Marx, pp. 211–227. Berlin: Springer, 2008.

Comeau, André M., Tommy Harding, Pierre E. Galand, Warwick F. Vincent, and Connie Lovejoy. Vertical distribution of microbial communities in a perennially stratified Arctic lake with saline, anoxic bottom waters. *Scientific Reports* 2 (2012): 604.

Cooley, Jason W., Crispin A. Howitt, and Wim F. J. Vermaas. Succinate: Quinol oxidoreductases in the cyanobacterium *Synechocystis* sp. strain PCC 6803: Presence and function in metabolism and electron transport. *Journal of Bacteriology* 182, no. 3 (2000): 714–22.

Cooley, Jason W., and Wim F. J. Vermaas. Succinate dehydrogenase and other respiratory pathways in thylakoid membranes of *Synechocystis* sp. strain PCC 6803: Capacity comparisons and physiological function. *Journal of Bacteriology* 183, no. 14 (2001): 4251–8.

Cota, Glenn F. Photoadaptation of High Arctic ice algae. *Nature* 315 (1985): 219–22.

Dalluge, Joseph J., Tetsuo Hamamoto, Koki Horikoshi, Richard Y. Morita, Karl O. Stetter, and James A. McCloskey. Posttranscriptional modification of tRNA in psychrophilic bacteria. *Journal of Bacteriology* 179, no. 6 (1997): 1918–23.

Davison, Ian R. Environmental effects on algal photosynthesis: Temperature. *Journal of Phycology* 27, no. 1 (1991): 2.

DeLong, Edward F., Christina M. Preston, Tracy Mincer et al. Community genomics among stratified microbial assemblages in the ocean's interior. *Science* 311, no. 5760 (2006): 496–503.

De Maayer, Pieter, Dominique Anderson, Craig Cary, and Don A. Cowan. Some like it cold: Understanding the survival strategies of psychrophiles. *EMBO Reports* 15 (2014): 508–17.

Dolhi, Jenna M., Amber G. Teufel, Weidong Kong, and Rachael M. Morgan-Kiss. Diversity and spatial distribution of autotrophic communities within and between ice-covered Antarctic lakes (McMurdo Dry Valleys). *Limnology and Oceanography* 60, no. 3 (2015): 977–91.

Doran, Peter T., Christopher P. McKay, Andrew G. Fountain et al. Hydrologic response to extreme warm and cold summers in the McMurdo Dry Valleys, East Antarctica. *Antarctic Science* 20, no. 5 (2008): 499–509.

Doyle, Shawn, M. Dieser, E. Broemsen, and B. C. Christner. General characteristics of cold-adapted microorganisms. In *Polar Microbiology: Life in a Deep Freeze*, ed. Robert V. Miller and Lyle G. Whyte, pp. 103–25. Washington, DC: ASM Press, 2012.

Duplantis, Barry N., Milan Osusky, Crystal L. Schmerk, Darrell R. Ross, Catharine M. Bosio, and Francis E. Nano. Essential genes from Arctic bacteria used to construct stable, temperature-sensitive bacterial vaccines. *Proceedings of the National Academy of Sciences of the United States of America* 107, no. 30 (2010): 13456–60.

Durack, Juliana, Tom Ross, and John P. Bowman. Characterisation of the transcriptomes of genetically diverse *Listeria monocytogenes* exposed to hyperosmotic and low temperature conditions reveal global stress-adaptation mechanisms. *PLoS One* 8, no. 9 (2013).

Felip, Marisol, Anton Wille, Birgit Sattler, and Roland Psenner. Microbial communities in the winter cover and the water column of an alpine lake: System connectivity and uncoupling. *Aquatic Microbial Ecology* 29, no. 2 (2002): 123–34.

Feller, Georges. Molecular adaptations to cold in psychrophilic enzymes. *Cellular and Molecular Life Sciences CMLS* 60, no. 4 (2003): 648–62.

Feller, Georges, and Charles Gerday. Psychrophilic enzymes: Hot topics in cold adaptation. *Nature Reviews Microbiology* 1, no. 3 (2003): 200–8.

Feller, Georges, Zoubir Zekhnini, Josette Lamotte-Brasseur, and Charles Gerday. Enzymes from cold-adapted microorganisms—The class C β-lactamase from the Antarctic psychrophile *Psychrobacter immobilis* A5. *European Journal of Biochemistry* 244, no. 1 (1997): 186–91.

Ferrer, Manuel, Heinrich Lünsdorf, Tatyana N. Chernikova et al. Functional consequences of single: Double ring transitions in chaperonins: Life in the cold. *Molecular Microbiology* 53, no. 1 (2004): 167–82.

Frank, Sarah, Frank Schmidt, Jens Klockgether, Colin F. Davenport, Manuela Gesell Salazar, Uwe Völker, and Burkhard Tümmler. Functional genomics of the initial phase of cold adaptation of *Pseudomonas putida* KT2440. *FEMS Microbiology Letters* 318, no. 1 (2011): 47–4.

Gao, Haichun, Zamin K. Yang, Liyou Wu, Dorothea K. Thompson, and Jizhong Zhou. Global transcriptome analysis of the cold shock response of *Shewanella oneidensis* MR-1 and mutational analysis of its classical cold shock proteins. *Journal of Bacteriology* 188, no. 12 (2006): 4560–9.

Garcia, Jean-Louis, Bharat K. C. Patel, and Bernard Ollivier. Taxonomic, phylogenetic, and ecological diversity of methanogenic Archaea. *Anaerobe* 6, no. 4 (2000): 205–26.

Giuliodori, Anna Maria, Fabio Di Pietro, Stefano Marzi et al. The cspA mRNA is a thermosensor that modulates translation of the cold-shock protein CspA. *Molecular Cell* 37, no. 1 (2010): 21–3.

Goodchild, Amber, Mark Raftery, Neil F. W. Saunders, Michael Guilhaus, and Ricardo Cavicchioli. Cold adaptation of the Antarctic archaeon, *Methanococcoides burtonii* assessed by proteomics using ICAT. *Journal of Proteome Research* 4, no. 2 (2005): 473–80.

Gratia, Emmanuelle, Frédéric Weekers, Rosa Margesin, Salvino D'Amico, Philippe Thonart, and Georges Feller. Selection of a cold-adapted bacterium for bioremediation of wastewater at low temperatures. *Extremophiles* 13, no. 5 (2009): 763–8.

Green, Beverley R., and Dion G. Durnford. The chlorophyll-carotenoid proteins of oxygenic photosynthesis. *Annual Review of Plant Biology* 47, no. 1 (1996): 685–714.

Gualerzi, Claudio O., Anna Maria Giuliodori, and Cynthia L. Pon. Transcriptional and post-transcriptional control of cold-shock genes. *Journal of Molecular Biology* 331, no. 3 (2003): 527–39.

Heipieper, Hermann J., Friedhelm Meinhardt, and Ana Segura. The cis–trans isomerase of unsaturated fatty acids in *Pseudomonas* and *Vibrio*: Biochemistry, molecular biology and physiological function of a unique stress adaptive mechanism. *FEMS Microbiology Letters* 229, no. 1 (2003): 1.

Honjoh, Ken-Ichi, Makoto Yoshimoto, Toshio Joh, Taishin Kajiwara, Takahisa Miyamoto, and Shoji Hatano. Isolation and characterization of hardening-induced proteins in *Chlorella vulgaris* C-27: Identification of late embryogenesis abundant proteins. *Plant and Cell Physiology* 36, no. 8 (1995): 1421–30.

Hoover, Richard B., and Elena V. Pikuta. Psychrophilic and psychrotolerant microbial extremophiles in polar environments. In *Polar Microbiology: The Ecology, Biodiversity and Bioremediation Potential of Microorganisms in Extremely Cold Environments*, ed. A. K. Bej, J. Aislabie, and R. M. Atlas, pp. 115–56. Boca Raton, FL: CRC Press, 2010.

Horner, Rita, Stephen F. Ackley, Gerhard S. Dieckmann, Bjorn Gulliksen, Takao Hoshiai, Louis Legendre, Igor A. Melnikov, William S. Reeburgh, Michael Spindler, and Cornelius W. Sullivan. Ecology of sea ice biota. *Polar Biology* 12, no. 3 (1992): 417–27.

Jung, Deborah O., Laurie A. Achenbach, Elizabeth A. Karr, Shinichi Takaichi, and Michael T. Madigan. A gas vesiculate planktonic strain of the purple non-sulfur bacterium *Rhodoferax antarcticus* isolated from Lake Fryxell, Dry Valleys, Antarctica. *Archives of Microbiology* 182, no. 2 (2004): 236–43.

Jungblut, Anne-Dorothee, Ian Hawes, Doug Mountfort, Bettina Hitzfeld, Daniel R. Dietrich, Brendan P. Burns, and Brett A. Neilan. Diversity within cyanobacterial mat communities in variable salinity meltwater ponds of McMurdo Ice Shelf, Antarctica. *Environmental Microbiology* 7, no. 4 (2005): 519–29.

Junge, Karen, Hajo Eicken, Brian D. Swanson, and Jody W. Deming. Bacterial incorporation of leucine into protein down to –20°C with evidence for potential activity in sub-eutectic saline ice formations. *Cryobiology* 52, no. 3 (2006): 417–29.

Kandror, Olga, Nancy Bretschneider, Evgeniy Kreydin, Duccio Cavalieri, and Alfred L. Goldberg. Yeast adapt to near-freezing temperatures by STRE/Msn2,4-dependent induction of trehalose synthesis and certain molecular chaperones. *Molecular Cell* 13, no. 6 (2004): 771–81.

Kang, Shichang, Yanwei Xu, Qinglong You, Wolfgang-Albert Flügel, Nick Pepin, and Tandong Yao. Review of climate and cryospheric change in the Tibetan Plateau. *Environmental Research Letters* 5, no. 1 (2010): 015101.

Karr, Elizabeth A., W. Matthew Sattley, Deborah O. Jung, Michael T. Madigan, and Laurie A. Achenbach. Remarkable diversity of phototrophic purple bacteria in a permanently frozen Antarctic lake. *Applied and Environmental Microbiology* 69, no. 8 (2003): 4910–4.

Karr, Elizabeth A., W. Matthew Sattley, Melissa R. Rice, Deborah O. Jung, Michael T. Madigan, and Laurie A. Achenbach. Diversity and distribution of sulfate-reducing bacteria in permanently frozen Lake Fryxell, McMurdo Dry Valleys, Antarctica. *Applied and Environmental Microbiology* 71, no. 10 (2005): 6353–9.

Kawahara, Hidehisa. The structures and functions of sea ice crystal-controlling proteins from bacteria. *Journal of Bioscience and Bioengineering* 94, (2002): 492–6.

Knight, Charles A., Dingyi Wen, and Richard A. Laursen. Nonequilibrium antifreeze peptides and the recrystallization of ice. *Cryobiology* 32, no. 1 (1995): 23–4.

Körner, Christian. *Alpine Plant Life: Functional Plant Ecology of High Mountain Ecosystems*. Berlin: Springer Science & Business Media, 2003.

Koutsioulis, Dimitris, Ellen Wang, Maria Tzanodaskalaki et al. Directed evolution on the cold adapted properties of TAB5 alkaline phosphatase. *Protein Engineering Design and Selection* 21, no. 5 (2008): 319–27.

Krembs, Christopher, Hajo Eicken, Karen Junge, and Jody Deming. High concentrations of exopolymeric substances in Arctic winter sea ice: Implications for the polar ocean carbon cycle and cryoprotection of diatoms. *Deep Sea Research Part I: Oceanographic Research Papers* 49, no. 12 (2002): 2163–81.

Kroon, Bernd M. A. Variability of photosystem II quantum yield and related processes in *Chlorella pyrenoidosa* (Chlorophyta) acclimated to an oscillating light regime simulating a mixed photic zone. *Journal of Phycology* 30, no. 5 (1994): 841–52.

Lauro, Federico M., Khiem Tran, Alessandro Vezzi, Nicola Vitulo, Giorgio Valle, and Douglas H. Bartlett. Large-scale transposon mutagenesis of *Photobacterium profundum* SS9 reveals new genetic loci important for growth at low temperature and high pressure. *Journal of Bacteriology* 190, no. 5 (2008): 1699–709.

Laybourn-Parry, Johanna, and Jemma Wadham. *Antarctic Lakes*. Oxford: Oxford University Press, 2014.

Leiros, Hanna-Kirsti, Schrøder, Nils Peder Willassen, and Arne O. Smalås. Residue determinants and sequence analysis of cold-adapted trypsins. *Extremophiles* 3, no. 3 (1999): 205–19.

Lipson, David A., and Steven K. Schmidt. Seasonal changes in an alpine soil bacterial community in the Colorado Rocky Mountains. *Applied and Environmental Microbiology* 70, no. 5 (2004): 2867–79.

Lizotte, Michael P., and John C. Priscu. Photosynthesis-irradiance relationship in phytoplankton from the perennially ice-covered lake (Lake Bonney, Antarctica). *Journal of Phycology* 28, no. 2 (1992): 179–85.

Lohan, Dagmar, and Sam Johnston. Bioprospecting in Antarctica. Yokohama, Japan: United Nations University, Institute of Advanced Studies, 2005.

Los, Dmitry A., Kirill S. Mironov, and Suleyman I. Allakhverdiev. Regulatory role of membrane fluidity in gene expression and physiological functions. *Photosynthesis Research* 116, no. 2 (2013): 489–509.

Madigan, Michael T. Anoxygenic phototrophic bacteria from extreme environments. In *Discoveries in Photosynthesis*, ed. Govindjee, J. T. Beatty, H. Gest, and J. F. Allen, pp. 969–83. Vol. 20 of Advances in Photosynthesis. Berlin: Springer, 2005.

Margesin, Rosa, Melanie Jud, Dagmar Tscherko, and Franz Schinner. Microbial communities and activities in alpine and subalpine soils. *FEMS Microbiology Ecology* 67, no. 2 (2009): 208–18.

Margesin, Rosa, G. Neuner, and K. B. Storey. Cold-loving microbes, plants, and animals—Fundamental and applied aspects. *Naturwissenschaften* 94, no. 2 (2007): 77–9.

Margesin, Rosa, Peter Schumann, Cathrin Spröer, and Anne-Monique Gounot. *Arthrobacter psychrophenolicus* sp. nov., isolated from an alpine ice cave. *International Journal of Systematic and Evolutionary Microbiology* 54, no. 6 (2004): 2067–72.

Methé, Barbara A., Karen E. Nelson, Jody W. Deming et al. The psychrophilic lifestyle as revealed by the genome sequence of *Colwellia psychrerythraea* 34H through genomic and proteomic analyses. *Proceedings of the National Academy of Sciences of the United States of America* 102, no. 31 (2005): 10913–8.

Metpally, Raghu Prasad Rao, and Boojala Vijay B. Reddy. Comparative proteome analysis of psychrophilic versus mesophilic bacterial species: Insights into the molecular basis of cold adaptation of proteins. *BMC Genomics* 10, no. 1 (2009): 1.

Miteva, Vanya. Bacteria in snow and glacier ice. In *Psychrophiles: From Biodiversity to Biotechnology*, ed. Rosa Margesin, Franz Schinner, and Jean-Claude Marx, pp. 31–50. Berlin: Springer, 2008.

Mizushima, Tohru, Kazuhiro Kataoka, Yasuyuki Ogata, Ryu-ichi Inoue, and Kazuhisa Sekimizu. Increase in negative supercoiling of plasmid DNA in *Escherichia coli* exposed to cold shock. *Molecular Microbiology* 23, no. 2 (1997): 381–6.

Mock, Thomas, and David N. Thomas. Recent advances in sea-ice microbiology. *Environmental Microbiology* 7, no. 5 (2005): 605–19.

Morgan, Rachael M., Alexander G. Ivanov, John C. Priscu, Denis P. Maxwell, and Norman P. A. Huner. Structure and composition of the photochemical apparatus of the Antarctic green alga, *Chlamydomonas subcaudata*. *Photosynthesis Research* 56, no. 3 (1998): 303–14.

Morgan-Kiss, Rachael M., and Jenna Dolhi. *Temperature Adaptation in a Changing Climate: Nature at Risk*. Wallingford, UK: CAB International, 2011.

Morgan-Kiss, Rachael M., Alexander G. Ivanov, Shannon Modla, Kirk Czymmek, Norman P. A. Hüner, John C. Priscu, John T. Lisle, and Thomas E. Hanson. Identity and physiology of a new psychrophilic eukaryotic green alga, *Chlorella* sp., strain BI, isolated from a transitory pond near Bratina Island, Antarctica. *Extremophiles* 12, no. 5 (2008): 701–11.

Morgan-Kiss, Rachael M., Alexander G. Ivanov, Tessa Pocock, M. Krol, L. Gudynaite-Savitch, and N. P. A. Huner. The Antarctic psychrophile, *Chlamydomonas raudensis* Ettl (UWO241) (Chlorophyceae, Chlorophyta), exhibits a limited capacity to photoacclimate to red light. *Journal of Phycology* 41, no. 4 (2005): 791–800.

Morgan-Kiss, Rachael, Alexander G. Ivanov, John Williams, Mobashsher Khan, and Norman P. A. Huner. Differential thermal effects on the energy distribution between photosystem II and photosystem I in thylakoid membranes of a psychrophilic and a mesophilic alga. *Biochimica et Biophysica Acta (BBA)—Biomembranes* 1561, no. 2 (2002): 251–65.

Morgan-Kiss, Rachael M., John C. Priscu, Tessa Pocock, Loreta Gudynaite-Savitch, and Norman P. A. Huner. Adaptation and acclimation of photosynthetic microorganisms to permanently cold environments. *Microbiology and Molecular Biology Reviews* 70, no. 1 (2006): 222–52.

Moskvina, Eugenia, Esther-Maria Imre, and Helmut Ruis. Stress factors acting at the level of the plasma membrane induce transcription via the stress response element (STRE) of the yeast *Saccharomyces cerevisiae*. *Molecular Microbiology* 32, no. 6 (1999): 1263–72.

Mullineaux, Conrad W., and Daniel Emlyn-Jones. State transitions: An example of acclimation to low-light stress. *Journal of Experimental Botany* 56, no. 411 (2005): 389–93.

Mullineaux, Conrad W., and H. Kirchhoff. Role of lipids in the dynamics of thylakoid membranes. In *Lipids in Photosynthesis Essential and Regulatory Function*, ed. H. Wada and N. Murata, pp. 283–94. Dordrecht: Springer Science, 2009.

Narberhaus, Franz, Torsten Waldminghaus, and Saheli Chowdhury. RNA thermometers. *FEMS Microbiology Reviews* 30, no. 1 (2006): 3–6.

Neale, Patrick J., and John C. Priscu. The photosynthetic apparatus of phytoplankton from a perennially ice-covered Antarctic lake: Acclimation to an extreme shade environment. *Plant and Cell Physiology* 36, no. 2 (1995): 253–63.

Nedwell, David B., and M. Rutter. Influence of temperature on growth rate and competition between two psychrotolerant Antarctic bacteria: Low temperature diminishes affinity for substrate uptake. *Applied and Environmental Microbiology* 60, no. 6 (1994): 1984–92.

Nichols, David S., Matthew R. Miller, Noel W. Davies, Amber Goodchild, Mark Raftery, and Ricardo Cavicchioli. Cold adaptation in the Antarctic archaeon *Methanococcoides burtonii* involves membrane lipid unsaturation. *Journal of Bacteriology* 186, no. 24 (2004): 8508–15.

Nylen, Thomas H., Andrew G. Fountain, and Peter T. Doran. Climatology of katabatic winds in the McMurdo dry valleys, southern Victoria Land, Antarctica. *Journal of Geophysical Research: Atmospheres (1984–2012)* 109, no. D3 (2004).

Obata, Hitoshi, Hitoshi Ishigaki, Hidehisa Kawahara, and Kazuhiro Yamade. Purification and characterization of a novel cold-regulated protein from an ice-nucleating bacterium, *Pseudomonas fluorescens* KUIN-1. *Bioscience, Biotechnology, and Biochemistry* 62, no. 11 (1998): 2091–7.

Okuyama, Hidetoshi, Yoshitake Orikasa, and Takanori Nishida. Significance of antioxidative functions of eicosapentaenoic and docosahexaenoic acids in marine microorganisms. *Applied and Environmental Microbiology* 74, no. 3 (2008): 570–4.

Palmisano, A. C., and D. L. Garrison. Microorganisms in Antarctic sea ice. In *Antarctic Microbiology*, ed. E. I. Friedman, pp. 167–218. New York: Wiley-Liss, 1993.

Panadero, Joaquín, Claudia Pallotti, Sonia Rodríguez-Vargas, Francisca Randez-Gil, and Jose A. Prieto. A downshift in temperature activates the high osmolarity glycerol (HOG) pathway, which determines freeze tolerance in *Saccharomyces cerevisiae*. *Journal of Biological Chemistry* 281, no. 8 (2006): 4638–45.

Panicker, Gitika, Nazia Mojib, Jackie Aislabie, and Asim K. Bej. Detection, expression and quantitation of the biodegradative genes in Antarctic microorganisms using PCR. *Antonie Van Leeuwenhoek* 97, no. 3 (2010): 275–87.

Parente Duena, A., J. Garces Garces, J. Guinea Sanchez et al. Use of a glycoprotein for the treatment and re-epithelialization of wounds. U.S. patent 7,022,668, 2006.

Park, Seunghye, Gyeongseo Jung, Yong-sic Hwang, and EonSeon Jin. Dynamic response of the transcriptome of a psychrophilic diatom, *Chaetoceros neogracile*, to high irradiance. *Planta* 231, no. 2 (2010): 349–60.

Paterson, W. S. B. *The Physics of Glaciers*. London: Butterworth-Heinemann, 2001.

Peel, Murray C., Brian L. Finlayson, and Thomas A. McMahon. Updated world map of the Köppen-Geiger climate classification. *Hydrology and Earth System Sciences Discussions* 4, no. 2 (2007): 439–73.

Petrou, Katherina, R. Hill, M. A. Doblin, and A. McMinn. Photoprotection of sea-ice microalgal communities from the East Antarctic pack ice. *Journal of Phycology* 47 (2011): 77–6.

Piette, Florence, Salvino D'Amico, Gabriel Mazzucchelli, Antoine Danchin, Pierre Leprince, and Georges Feller. Life in the cold: A proteomic study of cold-repressed proteins in the Antarctic bacterium *Pseudoalteromonas haloplanktis* TAC125. *Applied and Environmental Microbiology* 77, no. 11 (2011): 3881–3.

Pocock, Tessa, Marc-André Lachance, Thomas Pröschold, John C. Priscu, Sam Sulgi Kim, and Norman Huner. Identification of a psychrophilic green alga from Lake Bonney Antarctica: *Chlamydomonas raudensis* ETTL. (UWO 241) Chlorophyceae. *Journal of Phycology* 40, no. 6 (2004): 1138–48.

Ponder, Monica A., Sarah J. Gilmour, Peter W. Bergholz et al. Characterization of potential stress responses in ancient Siberian permafrost psychroactive bacteria. *FEMS Microbiology Ecology* 53, no. 1 (2005): 103–15.

Priscu, John C., and Brent C. Christner. Earth's icy biosphere. *Microbial Diversity and Bioprospecting* (2004): 130–45.

Priscu, John C., Anna C. Palmisano, Linda R. Priscu, and Cornelius W. Sullivan. Temperature dependence of inorganic nitrogen uptake and assimilation in Antarctic sea-ice microalgae. *Polar Biology* 9, no. 7 (1989a): 443–6.

Priscu, John C., Linda R. Priscu, Warwick F. Vincent, and Clive Howard-Williams. Photosynthate distribution by microplankton in permanently ice-covered Antarctic desert lakes. *Limnology and Oceanography* 32, no. 1 (1987): 260–70.

Priscu, John C., Warwick F. Vincent, and Clive Howard-Williams. Inorganic nitrogen uptake and regeneration in perennially ice-covered Lakes Fryxell and Vanda, Antarctica. *Journal of Plankton Research* 11, no. 2 (1989b): 335–51.

Priscu, John C., Craig F. Wolf, Cristina D. Takacs, Christian H. Fritsen, Johanna Laybourn-Parry, Emily C. Roberts, Birgit Sattler, and W. Berry Lyons. Carbon transformations in a perennially ice-covered Antarctic Lake. *Bioscience* 49 no. 12 (1999): 997–1008.

Psenner, Roland, Birgit Sattler, Anton Wille et al. Lake ice microbial communities in alpine and Antarctic lakes. In *Cold-Adapted Organisms*, ed. R. Margesin and F. Schinner, pp. 17–31. Berlin: Springer, 1999.

Qin, Guokui, Lizhi Zhu, Xiulan Chen, Peng George Wang, and Yuzhong Zhang. Structural characterization and ecological roles of a novel exopolysaccharide from the deep-sea psychrotolerant bacterium *Pseudoalteromonas* sp. SM9913. *Microbiology* 153, no. 5 (2007): 1566–72.

Rajendran, Aravindan, Anbumathi Palanisamy, and Viruthagiri Thangavelu. Lipase catalyzed ester synthesis for food processing industries. *Brazilian Archives of Biology and Technology* 52, no. 1 (2009): 207–19.

Reid, I. N., W. B. Sparks, S. Lubow et al. Terrestrial models for extraterrestrial life: Methanogens and halophiles at Martian temperatures. *International Journal of Astrobiology* 5, no. 2 (2006): 89–7.

Rijstenbil, Jan. Effects of periodic, low UVA radiation on cell characteristics and oxidative stress in the marine planktonic diatom *Ditylum brightwellii*. *European Journal of Phycology* 36, no. 1 (2001): 1.

Roberts, Emily C., and Johanna Laybourn-Parry. Mixotrophic cryptophytes and their predators in the Dry Valley lakes of Antarctica. *Freshwater Biology* 41, no. 4 (1999): 737–46.

Rose, Kevin C., Craig E. Williamson, Jasmine E. Saros, Ruben Sommaruga, and Janet M. Fischer. Differences in UV transparency and thermal structure between alpine and subalpine lakes: Implications for organisms. *Photochemical & Photobiological Sciences* 8, no. 9 (2009): 1244–56.

Russell, Nicholas J., and Hamamoto T. Psychrophiles. In *Extremophiles: Microbial Life in Extreme Environments*, ed. K. Horikoshi, W. D. Grant, pp. 25–45. Vol. 20. New York: Wiley-Liss, 1998.

Sattley, W. Matthew, and Michael T. Madigan. Isolation, characterization, and ecology of cold-active, chemo-lithotrophic, sulfur-oxidizing bacteria from perennially ice-covered Lake Fryxell, Antarctica. *Applied and Environmental Microbiology* 72, no. 8 (2006): 5562–8.

Schade, Babette, Gregor Jansen, Malcolm Whiteway, Karl D. Entian, and David Y. Thomas. Cold adaptation in budding yeast. *Molecular Biology of the Cell* 15, no. 12 (2004): 5492–502.

Schmid, Franz X. Prolyl isomerase: Enzymatic catalysis of slow protein-folding reactions. *Annual Review of Biophysics and Biomolecular Structure* 22, no. 1 (1993): 123–43.

Shimada, Yuji, Yomi Watanabe, Taichi Samukawa et al. Conversion of vegetable oil to biodiesel using immobilized *Candida antarctica* lipase. *Journal of the American Oil Chemists' Society* 76, no. 7 (1999): 789–93.

Siddiqui, Khawar Sohail, and Ricardo Cavicchioli. Cold-adapted enzymes. *Annual Review of Biochemistry* 75 (2006): 403–33.

Siddiqui, Khawar S., Timothy J. Williams, David Wilkins et al. Psychrophiles. *Annual Review of Earth and Planetary Sciences* 41 (2013): 87–115.

Sinensky, Michael. Homeoviscous adaptation—A homeostatic process that regulates the viscosity of membrane lipids in *Escherichia coli*. *Proceedings of the National Academy of Sciences of the United States of America* 71, no. 2 (1974): 522–5.

Skidmore, Mark, A. Jungblut, M. Urschel, and K. Junge. Cryospheric environments in polar regions (glaciers and ice sheets, sea ice, and ice shelves). In *Polar Microbiology: Life in a Deep Freeze*, ed. Robert V. Miller and Lyle G. Whyte, pp. 218–39. Washington, DC: ASM Press, 2012.

Strocchi, Massimo, Manuel Ferrer, Kenneth N. Timmis, and Peter N. Golyshin. Low temperature-induced systems failure in *Escherichia coli*: Insights from rescue by cold-adapted chaperones. *Proteomics* 6, no. 1 (2006): 193–206.

Struvay, Caroline, and Georges Feller. Optimization to low temperature activity in psychrophilic enzymes. *International Journal of Molecular Sciences* 13, no. 9 (2012): 11643–65.

Takacs, Christian D., and J. C. Priscu. Bacterioplankton dynamics in the McMurdo Dry Valley lakes, Antarctica: Production and biomass loss over four seasons. *Microbial Ecology* 36, no. 3 (1998): 239–50.

Tasaka, Yasushi, Z. Gombos, Y. Nishiyama, P. Mohanty, T. Ohba, K. Ohki, and N. Murata. Targeted mutagenesis of acyl-lipid desaturases in *Synechocystis*: Evidence for the important roles of polyunsaturated membrane lipids in growth, respiration and photosynthesis. *EMBO Journal* 15, no. 23 (1996): 6416.

Teufel, Amber G., Wei Li, Andor J. Kiss, and Rachael M. Morgan-Kiss. Impact of nitrogen and phosphorus on phytoplankton production and bacterial community structure in two stratified Antarctic lakes: A bioassay approach. *Polar Biology* (2016): 1–6.

Thomas, Grant W. Soil pH and soil acidity. In *Methods of Soil Analysis. Part 3: Chemical Methods*, ed. D. L. Sparks, pp. 475–90. Madison, WI: Soil Science Society of America, 1996.

Ting, Lily, Timothy J. Williams, Mark J. Cowley et al. Cold adaptation in the marine bacterium, *Sphingopyxis alaskensis*, assessed using quantitative proteomics. *Environmental Microbiology* 12, no. 10 (2010): 2658–76.

Tran, Dang-Thuan, Ching-Lung Chen, and Jo-Shu Chang. Effect of solvents and oil content on direct transesterification of wet oil-bearing microalgal biomass of *Chlorella vulgaris* ESP-31 for biodiesel synthesis using immobilized lipase as the biocatalyst. *Bioresource Technology* 135 (2013): 213–21.

Turner, John, Nicholas E. Barrand, Thomas J. Bracegirdle et al. Antarctic climate change and the environment: An update. *Polar Record* 50, no. 3 (2014): 237–59.

Vincent, Warwick F. Rapid physiological assays for nutrient demand by the plankton. II. Phosphorus. *Journal of Plankton Research* 3, no. 4 (1981): 699–710.

Vishnivetskaya, Tatiana A. Viable cyanobacteria and green algae from the permafrost darkness. In *Permafrost Soils*, ed. R. Margesin, pp. 73–4. Berlin: Springer, 2009.

Vishnivetskaya, Tatiana A., Sophia Kathariou, and James M. Tiedje. The *Exiguobacterium* genus: Biodiversity and biogeography. *Extremophiles* 13, no. 3 (2003): 541–55.

Weber, Michael H. W., and Mohamed A. Marahiel. Bacterial cold shock responses. *Science Progress* 86, no. 1 (2003): 9–75.

Webster-Brown, J., M. Gall, J. Gibson, S. Wood, and I. Hawes. The biogeochemistry of meltwater habitats in the Darwin Glacier region (80 S), Victoria Land, Antarctica. *Antarctic Science* 22, no. 6 (2010): 646–61.

Welsh, David T. Ecological significance of compatible solute accumulation by micro-organisms: From single cells to global climate. *FEMS Microbiology Reviews* 24, no. 3 (2000): 263–90.

Wilkins, David, Sheree Yau, Timothy J. Williams, Michelle A. Allen, Mark V. Brown, Matthew Z. DeMaere, Federico M. Lauro, and Ricardo Cavicchioli. Key microbial drivers in Antarctic aquatic environments. *FEMS Microbiology Reviews* 37, no. 3 (2013): 303–35.

Yau, Sheree, Federico M. Lauro, Timothy J. Williams, Matthew Z. DeMaere, Mark V. Brown, John Rich, John A. E. Gibson, and Ricardo Cavicchioli. Metagenomic insights into strategies of carbon conservation and unusual sulfur biogeochemistry in a hypersaline Antarctic lake. *ISME Journal* 7, no. 10 (2013): 1944–61.

Zhang, Gaosen, Xiaojun Ma, Fujun Niu, Maoxing Dong, Huyuan Feng, Lizhe An, and Guodong Cheng. Diversity and distribution of alkaliphilic psychrotolerant bacteria in the Qinghai–Tibet Plateau permafrost region. *Extremophiles* 11, no. 3 (2007): 415–24.

Zhang, Li-Mei, Mu Wang, James I. Prosser, Yuan-Ming Zheng, and Ji-Zheng He. Altitude ammonia-oxidizing bacteria and archaea in soils of Mount Everest. *FEMS Microbiology Ecology* 70, no. 2 (2009): 208–17.

CHAPTER 10

Denitrification in Extreme Environments

**Javier Torregrosa-Crespo, Julia María Esclapez Espliego, Vanesa Bautista Saiz,
Carmen Pire, Andrew J. Gates, David J. Richardson, Anna Vegara Luque,
Mónica L. Camacho Carrasco, María José Bonete, and Rosa María Martínez-Espinosa**

CONTENTS

10.1 INTRODUCTION

Denitrification is one of the metabolic pathways belonging to the nitrogen cycle (Figure 10.1). Thanks to this pathway, nitrate is subsequently converted into nitrite, nitric oxide, nitrous oxide, or nitrogen gas. The last three compounds are gases, and they are not readily available for microbial growth; therefore, they are typically released to the atmosphere. Nitrogen gas makes up more than 70% of atmospheric gases; thus, the release of N_2 to the atmosphere is benign. However, nitric oxide and nitrous oxide have important implications in terms of climate change, the chemistry of the atmosphere, human health, and the ecological functioning of natural ecosystems, especially aquatic systems and soils where nitrogen concentrations are increasing, causing eutrophication of lakes or rivers and oceanic dead zones through algal bloom–induced hypoxia (Howarth 2004).

On the other hand, some nitrogen compounds resulting from human activities have a great impact on denitrification, causing several problems at the environmental level (Martínez-Espinosa et al. 2011):

1. NO and N_2O emissions from fertilized soils due to denitrification. These gases are also produced through biomass burning, cattle and feedlots, fossil fuel combustion, and other industrial sources. N_2O, carbon dioxide (CO_2), and methane (CH_4) are the three most important greenhouse gases. Consequently, recent

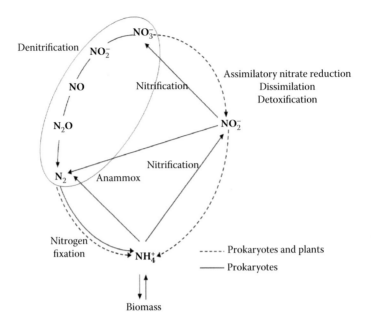

Figure 10.1   Nitrogen cycle. Denitrification pathway is highlighted. (Adapted from Bonete, M.J. et al., *Saline Systems*, 4, 9, 2008.)

strategies to mitigate climate change include the reduction of N_2O emissions. In addition, both N_2O and NO have deleterious effects on the stratosphere, where they are involved in the destruction of atmospheric ozone. Indeed, it has been reported that N_2O is currently the single most important ozone-depleting emission, and it is expected to remain the largest throughout the twenty-first century (Figure 10.2).

2. Excess NO_3^- and NO_2^- derived from fertilizers are leached from soils and enter the groundwater. At concentrations of <5 mM, NO_2^- is toxic for most microorganisms (Shen et al. 2003), and considerably lower concentrations represent a threat to aquatic invertebrates (Alonso and Camargo 2006).

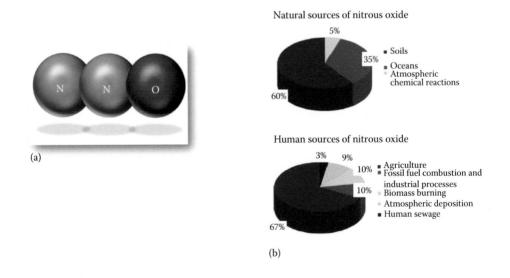

Figure 10.2   (a) Nitrous oxide chemical structure and (b) main sources of nitrous oxide. (From Intergovernmental Panel on Climate Change, Fourth assessment report: Climate change, 2007, https://www.ipcc.ch /publications_and_data/publications_ipcc_fourth_assessment_report_syntthesis_report.htm.)

High levels of nitrate in drinking water are a known risk factor for methemoglobinemia (a potential cause of blue baby syndrome) and colon cancer.

Biological denitrification is an anaerobic respiration reaction in which nitrate $\left(NO_3^-\right)$ is used as a terminal electron acceptor. Most denitrifying microorganisms are aerobic autotrophs or heterotrophs that can switch to anaerobic growth when nitrate is present in the media. Typical assays used to determine whether a new taxon is able to perform denitrification are usually based on the reduction of nitrate under anaerobic conditions. If this reaction takes place, the taxon is characterized as denitrifying. However, this assay is only demonstrating that nitrate reduction could take place anaerobically, but it is not demonstrating if complete or partial denitrification is performed by one specific taxon. Because of this, more effort should be put into the analysis of microbial denitrification capabilities in the near future.

From a biotechnological point of view, denitrification becomes such an important pathway because nitrogen may be completely removed from the system (soil, water, etc.) in gaseous form rather than simply recycled through the system in biomass. Denitrifying microorganisms, and in particular extremophiles able to denitrify, have focused the research community's attention on the design of new strategies for soil and wastewater bioremediation (Nájera-Fernández et al. 2012; Martínez-Espinosa et al. 2015).

In the review presented here, the main features of denitrification in extreme environments are highlighted. The biochemical characteristics of the denitrifying enzymes from extremophilic microorganisms are also described.

10.2 EXTREMOPHILIC MICROBIAL COMMUNITIES IN CHARGE OF DENITRIFICATION

Extreme environments are under conditions that make the life of microorganisms difficult, like high or low values of temperature, pH, salinity, and radiation. These factors affect both the structure of microbial communities (in terms of composition and abundance) and the biochemical and genetic characteristics of denitrification.

10.2.1 Extremely Low pH

The pH range for denitrification is between 6 and 8 (Sánchez-Andrea et al. 2012), so an environment is considered to be acidic when the pH is below 6. Other than natural causes, many of them have been generated because of the use of fertilizers for decades. The microbial biomass is lower in acidic soils than in neutral or alkaline soils (Čuhel et al. 2010), yet in extremely low-pH habitats, there is a broad range of denitrifying bacteria belonging to the genera *Paenibacillus*, *Bacillus*, *Sedimentibacter*, *Lysinibacillus*, *Delftia*, *Alcaligenes*, *Clostidrium*, and *Desulfitobacterium*. An example of this kind of habitat is Tinto River in Huelva, Spain, where the oxidation of metal sulfides results in waters at pH values of around 2.3 (Sánchez-Andrea et al. 2012).

As a general rule, it can be stated that the denitrification rate decreases with low pH values; in contrast, the $N_2O:N_2$ ratio is negatively correlated with soil pH (Čuhel et al. 2010; Liu et al. 2010; Sánchez-Andrea et al. 2012; Huang et al. 2014). However, it is clear that denitrification can occur in soil and water microbial communities at low pH values if they correspond to the optimum pH for optimum growth (Liu et al. 2010). Because of this, acid habitats have an important accumulation of nitrous oxide, and they may contribute to climate change because it is a greenhouse gas that influences the ozone layer stability. In fact, 70% of nitrous oxide emissions to the atmosphere come from acid soils (Mosier 1998).

The accumulation of N_2O at low pH values is associated with the inhibition of the enzyme nitrous oxide reductase, which cannot catalyze the reduction of this gas to N_2 at these values of acidity. The transcription rate of *nosZ* (gene encoding N_2O reductase) compared with that of other denitrification genes like *nirS* is higher at low pH values than at high pH values (Liu et al. 2010), so the inhibition may occur at the posttranscriptional phase. This evidence suggests that there is a bug in the translation or assembly of nitrous oxide reductase, which causes the inhibition of the enzyme. Moreover, the N_2O reductase is the most sensitive enzyme to environmental conditions due to its localization: in the periplasmic space in gram-negative bacteria (Šimek et al. 2002) and between the cell membrane and the S-layer in archaea (some preliminary unpublished results obtained from the haloarchaea *Haloferax mediterranei* support this idea).

10.2.2 Extremely High Salinity

Hypersaline environments have an extensive distribution on earth, like saline soils, alkaline salt lakes, and saline ponds (Oren 2002). Extreme halophiles are generally defined as organisms with optimal growth in media with a concentration of 150–300 g L^{-1} (2.5–5.2 M NaCl) (Andrei et al. 2012).

Halophilic organisms are found in the three domains of life: Archaea, Bacteria, and Eukarya, but at the highest concentrations of salt, the dominant species are archaea, especially the family Halobacteriaceae (Andrei et al. 2012). Despite the fact that low concentrations of NO_3^- are generally found in hypersaline environments (Andrei et al. 2012), some halophilic archaea can grow anaerobically using nitrate as an electron acceptor, forming N_2O or N_2, like *Haloarcula marismortui*, *Haloarcula hispanica*, *Hfx. mediterranei*, *Haloferax volcanii*, or *Halogeometricum borinquense*. In saline and hypersaline environments, the denitrification process has some peculiarities focused on both nitrite reductases: NirK and NirS. No genome from a halophilic denitrifier has been found to date with both types present (Jones and Hallin 2010).

In general terms, there is a negative correlation between salt concentration and NirK richness (Santoro et al. 2006), which indicates some kind of specialization of these enzymes. Studies in coastal ecosystems along salinity gradients conclude that with the increase of NaCl concentration (from 34.5 g L^{-1}), the predominant nitrite reductase in microbial communities is NirS (Santoro et al. 2006; Jones and Hallin 2010). However, in extreme hypersaline environments (from 300 g L^{-1}), the predominant nitrite reductase is NirK, which has been found in archaea species like *Har. marismortui*, *Har. hispanica*, *Hfx. mediterranei*, or *Hfx. volcanii* (Inatomi and Hochstein 1996; Ichiki et al. 2001; Esclapez et al. 2013). This apparent contradiction may be explained by the more cosmopolitan nature of *nirS* sequences compared with *nirK* sequences along salinity gradients. A diversity of genes encoding NirS are found in places with high and low NaCl concentrations, while genes encoding NirK are restricted to low salinity points (Santoro et al. 2006) or really salty environments. Therefore, it is possible that *nirK* communities are more specialized in each habitat and are simply not detected using the currently available primer sets (Jones and Hallin 2010). The supposed advantage of NirK communities over NirS communities at the highest salt concentrations might lie in the different chemical structure of both enzymes. The first has copper centers, which have higher reducing power than the iron sulfur centers of the NirS cytochromes. These differences in cofactors could explain the prevalence of NirK in environments with low oxygen solubility and high reducing power, like hypersaline habitats. Nevertheless, at the time of writing this chapter, there is no scientific evidence to verify this hypothesis.

Finally, with respect to the final stages of denitrification in haloarchaeas, there are not any detailed biochemical studies on nitric oxide reductases and nitrous oxide reductases, but there is evidence that some haloarchaea, such as *Hfx. mediterranei*, are complete denitrifiers because of their ability to reduce nitrous oxide to dinitrogen (Bonete et al. 2008). This evidence opens the doors to the use of these organisms in wastewater treatments with high salt concentrations, transforming large quantities of nitrates and nitrites into dinitrogen (Nájera-Fernández et al. 2012).

10.2.3 Extremely High Temperature

Hyperthermophile organisms grow at temperatures of around 80°C or higher (van Wolferen et al. 2013). They are found in hot terrestrial and marine environments: on land, some of the most common habitats are hot springs that come from volcanic emanations; in the sea, marine environments like hydrothermal systems, abyssal hot sediments, or active seamounts are home to a broad variety of hyperthermophile communities (Stetter 1999). Until today, more than 90 hyperthermophilic species have been discovered (van Wolferen et al. 2013); some of them are bacteria, but the majority belong to the Archaea domain. However, there is little evidence of hyperthermophilic denitrifiers. *Aquifex pyrophilus* is the best-characterized hyperthermophilic dentrifier bacteria, which grow optimally at oxygen concentrations below 5% (v/v) in the gas phase (Amo et al. 2002). Also, other bacteria belonging to the ε-proteobacteria (members of the genera *Sulfurimonas*, *Sulfurovum*, and *Nitratifractor*) and γ-protetobacteria couple the oxidation of reduced sulfur species with NO_3^- reduction (Bourbonnais et al. 2012); examples in Archaea are *Pyrobaculum calidifontis* (Amo et al. 2002) and *Pyrobaculum aerophilum* (Volkl et al. 1993), which are complete denitrifiers, growing because of the reduction of nitrate to dinitrogen as a final product.

The poor knowledge about the denitrification process in hyperthermophilic environments may be due to the difficulty to grow some of these organisms in culture media. However, a complete picture of the functional genes required for denitrification (nitrate, nitrite, nitric oxide, and nitrous oxide reductases) has been detected in hydrothermal vent chimneys (Wang et al. 2009). Moreover, *nirK* seems to override *nirS* (Bourbonnais et al. 2012), which supports the idea that NirK is more resistant in environments with high reducing power. Studies of the genes involved in denitrification will be needed to increase the knowledge about it in these environments.

10.3 EXTREME ENZYMES INVOLVED IN DENITRIFICATION

Denitrification can be considered the modular assemblage of four partly independent respiratory processes: nitrate, nitrite, nitric oxide, and nitrous oxide reduction (Zumft 1997) (Figure 10.3). In the entire denitrification process, nitrate is reduced to N_2 by means of four reaction steps catalyzed by the action of four metalloenzymes: respiratory nitrate reductase, respiratory nitrite reductase, nitric oxide reductase, and nitrous oxide reductase. Physiological, biochemical, and genetic evidence has provided a detailed process for this pathway in the Bacteria domain (Zumft 1997). Nonetheless, the biochemical and genomic data related to the denitrification process in extremophiles are scarce; in fact, at the moment there is not a single archaeon whose denitrification pathway has been described not only at the genetic but also at the enzymatic level. Related to the genetic evidence, during the last years the high number of available genomes of Archaea has allowed scientists to identify denitrification genes by homology search with their bacterial counterparts. However, the few biochemical studies related to denitrification pathways in extremophiles are restricted to the purification and characterization of respiratory nitrate and nitrite reductases from halophilic microorganisms and the hyperthermophilic archaeon *P. aerophilum* (Table 10.1). No methanogenic archaeon has been described as denitrifying up to now. The existence of denitrification in the hyperthermophilic branches indicates an early origin and occurrence of this pathway before the branching of the archaeal and bacterial domains. That is why Archaea and Bacteria exhibit the same denitrification pathway with similar enzymes.

Figure 10.3 Denitrification. Enzymes catalyzing each reaction are indicated under the arrows.

Table 10.1 Summary of the Purified and Characterized Denitrifying Enzymes from Extremophiles

Microorganisms	Domain	Purified and Characterized Enzymes	References
Haloferax mediterranei	Archaea	Respiratory nitrate reductase Respiratory nitrite reductase	Lledó et al. 2004 Martinez-Espinosa et al. 2007 Esclapez et al. 2013
Haloarcula marismortui	Archaea	Respiratory nitrate reductase Respiratory nitrite reductase	Yoshimatsu et al. 2000 Yoshimatsu et al. 2002 Ichiki et al. 2001
Haloarcula denitrificans	Archaea	Respiratory nitrate reductase Respiratory nitrite reductase	Hochstein and Lang 1991 Inatomi and Hochstein 1996
Haloferax volcanii	Archaea	Respiratory nitrate reductase	Bickel-Sandkotter and Ufer 1995
Pyrobaculum aerophilum	Archaea	Respiratory nitrate reductase	Afshar et al. 2001

10.3.1 Respiratory Nitrate Reductases

Denitrifying microorganisms possess nitrate reductase as the terminal enzyme of the nitrate respiration (Zumft 1997). According to the structural and catalytic characteristics, dissimilatory nitrate reductases can be classified into two groups: periplasmic nitrate reductase (Nap) and membrane-bound nitrate reductase (Nar). The Nap enzyme is mainly found in gram-negative bacteria. Its function is related to different processes, depending on the organism in which it is found, for example, the dissipation of excess reducing power for redox balancing, the scavenging of nitrate in nitrate-limited conditions, and aerobic or anaerobic denitrification (Potter et al. 2001; Gavira et al. 2002; Ellington 2003). Generally, Nap enzymes are heterodimers composed of a catalytic subunit (NapA) and a cytochrome *c* (NapB) that receives electrons from NapC, a membrane cytochrome *c* (Richardson et al. 2001). The Nar enzyme is more widely distributed in nitrate-respiring microorganisms and is involved in the generation of metabolic energy using nitrate as a terminal electron acceptor. It is negatively regulated by oxygen, induced by the presence of nitrate and unaffected by ammonium. In general, the Nar complex is a heterotrimer composed of a catalytic subunit (NarG) that binds a bis-molybdpoterin guanine dinucleotide (bis-MGD) cofactor for nitrate reduction, an electron transfer subunit with four iron sulfur centers (NarH), and a di-*b*-heme integral membrane quinol dehydrogenase subunit (NarI). NarG and NarH are membrane-extrinsic domains, while NarI is a hydrophobic membrane protein that attaches the NarGH complex to the membrane (Richardson et al. 2001; Cabello et al. 2004).

At the time of writing, all nitrate reductases purified and characterized from extremophilic microorganisms are membrane-bound Nar enzymes (Table 10.2). In general, enzymatic and physicochemical analysis of these enzymes indicated a marked resemblance to the bacterial NarGH complex, although there was a relevant difference between the archaeal and bacterial enzymes, related to the subcellular localization (Yoshimatsu et al. 2002; Martinez-Espinosa et al. 2007).

In the Archaea domain, the purification of respiratory Nar enzymes has been reported for several denitrifying halophilic microorganisms, including three *Haloferax* species and *Har. marismortui*, and the hyperthermophilic *P. aerophilum*. The *Haloferax denitrificans* membrane-bound Nar was the first extremophilic respiratory nitrate reductase purified and characterized. This enzyme is a heterodimer (Table 10.2) with a K_m for nitrate of 0.2 mM. The enzyme is able to reduce both nitrate and chlorate using methyl viologen (MV) as an electron donor in vitro, and it is inhibited by azide and cyanide. Azide is a competitive inhibitor with respect to nitrate; it may act directly in the molybdenum-containing site of the Nar, probably by metal chelation. On the other hand, cyanide is a noncompetitive inhibitor of nitrate reduction. Curiously, unlike other halophilic enzymes, this nitrate reductase is stable in the absence of salt, and its activity decreases with increasing salt concentrations. Moreover, it was suggested that the enzyme contains molybdenum

Table 10.2 Characteristics of Respiratory Nitrate Reductases from Extremophiles

Microorganism	Structure Features	Optimal Activity Conditions	Substrates	Inhibitors
Haloferax denitrificans	Heterodimer: 116 and 60 kDa	Absence of salt	Nitrate Chlorate	Azide Cyanide
Haloferax volcanii	Heterotrimer: 100, 61, and 31 kDa	Absence of salt Temperature 80°C pH 7.5	Nitrate	Azide Cyanide Thiocyanate
Haloarcula marismortui	Heterodimer: 117 and 47 kDa	2 M NaCl pH 7.0	Nitrate Chlorate	Nondetermined
Haloferax mediterranei	Heterodimer: 112 and 61.5 kDa	Absence of salt pH 7.9 at 40°C pH 8.2 at 60°C	Nitrate Chlorate Perchlorate Bromate	Dithiothreitol Azide Cyanide EDTA
Pyrobaculum aerophilum	Heterotrimer: 130, 52, and 32 kDa	pH 6.5 Temperature 95°C	Nitrate Chlorate	Azide Cyanide

because tungstate represses nitrate reductase synthesis (Hochstein and Lang 1991). *Hfx. volcanii* contains a trimeric respiratory Nar with a K_m for nitrate of 0.36 mM and shows a remarkable grade of thermophilicity, similarly to other halophilic enzymes (Table 10.2). Like the *Hfx. denitrificans* Nar, the enzyme shows optimal activity in the absence of NaCl (Bickel-Sandkötter and Ufer 1995). The *Har. marismortui* Nar was first described as a homotetramer of 63 kDa subunits (Yoshimatsu et al. 2002), but the sequence of the gene, as well as sodium dodecyl sulfate–polyacrylamide gel electrophoresis (SDS-PAGE), in the presence of reducing agent revealed that it is a heterodimer (Table 10.2). The present archaeal enzyme has a K_m for nitrate of 80 μM with 2.0 M NaCl. In relation to salt dependence, the *Har. marismortui* Nar is stable even in the absence of NaCl; however, salt-dependent enhancement of the enzymatic activity was observed. Besides, it was determined by electron paramagnetic resonance (EPR) measurements that the enzyme contains a Mo–molydobterin complex and iron-sulfur centers (Yoshimatsu et al. 2000). Yoshimatsu et al. (2002) proposed the *Har. marismortui* Nar as a new archaeal type of membrane-bound nitrate reductase based on two pieces of evidence: the loss of the NarI membrane-associated protein and the sequence and structure similarity of Nar to dissimilatory selenate reductases from *Thauera selenatis*, although the halophilic enzyme does not reduce selenate. Later, the similarity of Nar to selenite reductase was also supported, since both enzymes have an aspartic residue as a ligand to the molybdenum atom (Jormakka et al. 2004). In *Hfx. mediterranei*, two different nitrate reductases involved in nonassimilatory processes have been described: a dissimilatory nitrate reductase described by Álvarez-Ossorio et al. (1992) and Nar characterized by Lledó et al. (2004). The first one is a salt-requiring enzyme, with an optimal activity at 89°C in 3.2 M NaCl and a K_m for nitrate between 2.5 and 6.7 mM depending on the salt concentration (Álvarez-Ossorio et al. 1992). According to its molecular mass and enzymatic properties, Lledó et al. (2004) proposed that the enzyme purified by Álvarez-Ossorio allows the dissipation of reducing power for redox balancing. The *Hfx. mediterranei* Nar is a heterodimer (Table 10.2), and its K_m for nitrate is 0.82 mM, which is in the range of the values obtained from other nitrate reductases (Zumft 1997). Like other nitrate reductases, cyanide and azide are strong inhibitors of this enzyme. Other compounds, such as dithiothreitol and EDTA, were also tested, but they are not effective inhibitors, since they only partially decrease the activity. The *Hfx. mediterranei* Nar does not exhibit a strong dependence on temperature at the different NaCl concentrations assayed (0–3.8 M NaCl), showing the maximum activity at 70°C for all NaCl concentrations. Therefore, this halophilic enzyme also exhibits a remarkable thermophilicity, although the Nar activity does not show a direct dependence on salt concentration, as described for *Hfx. denitrificans* Nar (Hochstein and Lang 1991) and *Hfx. volcanii* Nar (Bickel-Sandkötter and Ufer 1995). Not all nitrate reductase activities found in halophilic archaea exhibit a similar dependence (Álvarez-Ossorio et al. 1992; Yoshimatsu et al. 2000). Even though most proteins for haloarchaea are stable and active at high salt concentrations, there are some that are either active or stable in the absence of salt. The origin of haloarchaeal enzymes that do not require salt is unclear, but it has been proposed that extreme halophiles could obtain Nar from a eubacterial source (Hochstein and Lang 1991). The absorption spectrum of the *Hfx. mediterranei* Nar shows a broad band of around 400–415 nm, indicating that this enzyme has Fe-S clusters like other Nar purified from denitrifying microorganisms (Lledó et al. 2004). The respiratory nitrate reductase of *P. aerophilum*, a hyperthermophilic microorganism belonging to the Archaea domain, was also purified (Afshar et al. 2001). The hyperthermophilic enzyme is a heterotrimer (Table 10.2) and contains molybdenum, iron, and cytochrome *b* as cofactors, and its K_m for nitrate is 58 µM. Hyperthermophilic microorganisms, such as *P. aerophilum*, are naturally exposed to high concentrations of tungsten, a heavy metal that is abundant in high-temperature environments (Kletzin and Adams 1996). It has been demonstrated that tungstate inactivates molybdoenzymes, for example, the nitrate reductase from *Escherichia coli*, whose function was abolished by the addition of this heavy metal to the medium. Nonetheless, the hyperthermophilic respiratory nitrate reductase remains active in *P. aerophylum* cultured in the presence of high tungstate concentrations (Afshar et al. 1998). Curiously, this nitrate reductase distinguishes itself from the nitrate reductases of mesophilic bacteria and archaea by its very high specific activity (about 7–40 times higher) using reduced benzyl viologen as the electron donor. This fact could be an adaptation of the thermophilic enzyme to counteract the inhibition carried out by the presence of tungsten under physiological growth conditions, since *P. aerophilum* needs to support growth by nitrate respiration even when the concentration of tungsten in the environment is high. As a typical hyperthermophilic enzyme, the *P. aerophilum* nitrate reductase exhibits its maximum activity at or above 95°C. Under this condition, the enzyme could be stabilized by its membrane environment, since detergent extraction results in a fourfold loss of the thermostability of the nitrate reductase activity. From an evolutionary point of view, the enzyme from this hyperthermophilic microorganism is the oldest nitrate reductase purified and characterized. Therefore, the nitrate reductase in the last common ancestor group of microorganisms could be a heterotrimeric enzyme (Afshar et al. 2001).

Classically, it has been considered that NarG and NarH are located in the cytoplasm and associate with NarI at the membrane potential-negative cytoplasmic face of the cytoplasmic membrane, so the nitrate reduction takes place inside of this membrane. This arrangement is conserved in gram-negative bacteria, and indeed, for many years, it was assumed that this orientation would be conserved among prokaryotes in general. However, the presence of a typical twin-arginine signal in *Har. marismortui* and *Hfx. mediterranei* NarG suggests that nitrate reductases from Archaea are translocated across the membrane by the TAT export pathway. Later, the analysis of the N-terminal region of the archaeal nitrate reductases revealed the conservation of a twin-arginine motif (Martinez-Espinosa et al. 2007). The data available suggest that the NarG protein is strongly attached to the membrane fraction and requires detergent solubilization to release it (Yoshimatsu et al. 2000; Afshar et al. 2001; Lledó et al. 2004). To answer the question of whether the subunit is located on the inside or outside of the cytoplasmic membrane, different assays were carried out with intact cells of *Har. marismortui* (Yoshimatsu et al. 2000) and *Hfx. mediterranei* (Martinez-Espinosa et al. 2007). The results obtained revealed that the electron donation to the active site of an enzyme is on the outside, rather than the inside, of the cytoplasmic membrane. These experiments have not yet been reported for the other archaeal Nars with Tat sequences thus far identified. Nonetheless, the available evidence supports the fact that the active site of these archaeal Nar systems is indeed on the outside of the cytoplasmic membrane (Martinez-Espinosa et al. 2007).

Hence, based on subunit composition and subcellular location, it can be suggested that archaeal Nars are a new type of enzyme with the active site facing the outside and attached to the membrane by cytochrome *b* (as proposed for *Hfx. mediterranei*) or stabilized by the lipid environment in the membrane as described for *P. aerophylum*. This system could be an ancient respiratory nitrate reductase, although the nitrite formed may be introduced in the nitrogen assimilation pathway. The outside location of the catalytic site of archaeal NarG has important bioenergetic implications because being energy conserving requires coupling this process to a proton-motive complex, instead of the typical redox-loop mechanism, the NarI subunit described in bacteria. On the other hand, it appears that an active nitrate-uptake system would not be required for respiratory nitrate reduction in archaea, thus increasing the energetic yield of the nitrate reduction process (Bonete et al. 2008).

The last advances related to the knowledge of respiratory Nar have been carried out in *Hfx. mediterranei* (Martínez-Espinosa et al. 2015), where the capacity of the whole cells and pure NarGH to reduce chlorate, perchlorate, bromate, iodate, and selenate was tested. Not only whole *Hfx. mediterranei* but also pure NarGH is able to reduce chlorate, bromate, and perchlorate, but no reduction activity is detected with iodate or selenate. Therefore, the same microorganism is able to reduce nitrate and chlorate thanks to the nitrate reductase under microaerobic or anaerobic conditions. These results are of great interest for wastewater bioremediation purposes since most of the wastewater samples containing nitrate also contain chlorate and other oxyanions. Although the removal process is not really fast (4.8 mM chlorate after 150 hours of incubation), the removed concentration using microorganisms is one of the highest described thus far (Bardiya and Bae 2005; van Ginkel et al. 2005). Moreover, one of the advantages of using *Hfx. mediterranei* cells or its NarGH is that nitrate reduction is not inhibited by the presence of chlorate or perchlorate at high salt concentrations. These results make it possible to create new bioremediation process designs based on the use of haloarchaea, or even to improve the knowledge of biological chlorate reduction in early earth or Martian environments (Martínez-Espinosa et al. 2015).

10.3.2 Respiratory Nitrite Reductases

The nitrite produced by the respiratory nitrate reductase is reduced to nitric oxide by the respiratory nitrite reductases (NiR), a key enzyme used to distinguish between denitrifiers and nitrate reducers. This reaction implies the return of nitrite to the gaseous state, leading to a significant loss of fixed nitrogen from the terrestrial environment. Two types of different enzymes in terms of structure and the prosthetic metal have been reported in denitrifying bacteria: cytochrome cd_1-nitrite reductase (encoded by *nirS*) and Cu-containing dissimilatory nitrite reductase (encoded by *nirK*). The cd_1-nitrite reductase is homodimeric and contains hemes c and d_1 as prosthetic cofactors, whereas the Cu-nitrite reductase is homotrimeric and contains two Cu atoms per subunit molecule. Cu-NiR enzymes can be readily distinguished based on their spectra and their sensitivity to diethyldithiocarbamate (DDC) (Shapleigh and Payne 1985). The two NiR types are functionally and physiologically equivalent, but while the cd_1-nitrite reductase predominates in denitrifying bacteria, the Cu-nitrite reductase is present in a greater variety of physiological groups and bacteria from different habitats (Zumft 1997; Heylen et al. 2006).

The first evidence related to the activity of respiratory nitrite reductase was reported in *Har. marismortui* and *Hfx. denitrificans*. In 1978, the ability of *Har. marismortui* to reduce nitrite to nitric oxide in crude extracts using halophilic ferredoxin as an electron donor was identified (Werber and Mevarech 1978). Later, it was stated that the membranes from *Hfx. denitrificans* reduce nitrite to nitric oxide by a reaction that is inhibited by DDC, which implies that the enzyme is a Cu-NiR (Tomlinson and Hochstein 1988). It was in 1996 when the first extremophilic respiratory nitrite reductase from *Hfx. denitrificans* was purified and characterized (Table 10.3) from soluble and membrane fractions (Inatomi and Hochstein 1996).

The SDS-PAGE analysis of the purified protein resulted in the presence of two peptides of 64 and 51 kDa, and the molecular mass of 127 kDa was determined by gel filtration. The authors suggested that the protein is a dimer and that the lower-weight peptide was a degradation product of the larger subunit, although nowadays it is known that these data are inaccurate. Although the protein shows its maximum activity in the presence of 4 M NaCl (Table 10.3), there is no loss of activity when the enzyme is incubated in the absence of salt. Its absorption spectrum is characterized by

Table 10.3 Characteristics of Respiratory Cu-Nitrite Reductases from Extremophiles

Microorganism	Structure Features	Optimal Activity Conditions	K_m Nitrite	Inhibitors	Absorption Peaks (nm)
Haloferax denitrificans	Homotrimer	4 M NaCl pH 4.8–5.0	4.6 mM	DDC	462, 594, 682
Haloarcula marismortui	Homotrimer	2 M NaCl pH 8.0	ND	DDC EDTA	465, 600
Haloferax mediterranei	Homotrimer	2 M NaCl pH 5.5	4 mM	ND	453, 587

Note: ND, not determined.

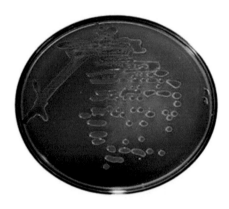

Figure 10.4 *Hfx. volcanii* petri dish.

maxima, located at 462, 594, and 682 nm, which disappeared after the addition of dithionite, concluding that this enzyme belongs to the green Cu-NiR. The assays carried out in the presence of DDC determined that this reagent inhibits the activity of NiR at relatively low concentrations, supporting the fact that this enzyme is a Cu-Nir. Although the membrane-bound Cu-NiR was not totally purified, its characteristics are similar to those of the enzyme purified from the soluble fraction (Inatomi and Hochstein 1996). The respiratory nitrite reductase was also purified from the halophilic archaea *Har. marismortui* (Figure 10.4) (Ichiki et al. 2001). The SDS-PAGE of the purified enzyme gave two protein bands, as in *Hfx. denitrificans*, whose molecular masses are 46 and 42 kDa. N-Terminal amino acid sequences were determined, finding that the sequence of the 46 kDa subunit after the 17th amino acid is identical to the N-terminal sequence of the 42 kDa subunit, except for the 16-amino-acid difference. The absorption spectrum of the purified Cu-NiR shows absorption maxima at 465 and 600 nm, with a small shoulder around 820 nm in the visible region, suggesting that this halophilic enzyme is a blue Cu-NiR. EPR spectroscopy provided evidence that one molecule each of type 1 and type 2 Cu centers is present in a subunit of this enzyme. The Cu-NiR is activated in the presence of high salt concentrations, reaching its maximum at NaCl concentrations higher than 2 M, while being denatured in the absence of salt, as most halophilic enzymes. The physiological electron donor remains unclear, although halocyanin could play this role in some archaea. Analysis of the amino acid sequence of the *Har. marismortui* Cu-NiR suggests that the minimum functional unit of the archaeal enzyme is a trimer constituted by identical subunits. Furthermore, phylogenic analysis indicated that the halophilic enzyme is in a quite close relationship with the enzyme from the gonorrhoeal pathogen *Neisseria gonorrhoeae*. The structural similarities between these two enzymes suggest the lateral transfer of the *nirK* gene between halophilic archaea and the pathogenic proteobacteria (Ichiki et al. 2001). The last studies related to respiratory nitrite reductases in extremophilic microorganisms have been carried out in *Hfx. mediterranei* (Esclapez et al. 2013). The respiratory nitrite reductase from *Hfx. mediterranei* was expressed in the halophilic host *Hfx. volcanii*. The enzymatic activity of the recombinant protein was detected in both cellular fractions (cytoplasmic fraction and membranes) and in the culture media. The enzyme isolated from the cytoplasmic fraction and the culture media were purified and characterized (Figure 10.4). The cytoplasmic NirK is a trimeric protein that shows its maximum activity in the presence of 2 M salt (NaCl or KCl) and at around 70°C. The sequence and structural analysis of this enzyme revealed the presence of four significant regions. The first of them involves the presence of a region similar to the distinctive Tat motif; therefore, it is probable that this region is acting as the Tat motif for the protein to be exported via the Tat system. The second conserved domain shows the presence of two possible cutting targets for proteases, located in positions 27 and 34 from the N-terminal end. The presence of this sequence is associated with the Tat signals, since the mature protein exportation through the cytoplasmic membrane requires the removal of the signal peptide. Finally, seven residues in copper binding were identified sited in a central position inside the chain. These residues may coordinate the type 1 and type 2 copper centers proposed for the Cu-NiR proteins. On the other hand, the ultraviolet-visible (UV-vis) spectrum shows two different maxima absorption at 453 and 587 nm, suggesting that the enzyme belongs to the green Cu-NirK group. In order to elucidate the composition of the native enzyme, an exhaustive study was carried out. A native PAGE of pure enzyme, followed by activity NiR staining, revealed that the intracellular Cu-NiR is composed of at least six different isoforms of the enzyme. The SDS-PAGE of each of the six bands showed that each one exhibits a different combination of two isoforms of 44.3 and 39.8 kDa, the smaller form being the predominant isoform protein in this cellular fraction. Taking into account the two cleavage sites present in the *Hfx. mediterranei* Cu-NiR sequence, it is possible to propose that the expression of recombinant proteins could conclude with the maturation of the initial polypeptide through a cut in one of the two targets present at its N-terminal extreme. Finally, the two possible isoforms could combine to form a pool of active trimers. This maturation mechanism could also explain why it is possible to observe two bands with slightly different masses than those NiR purifications carried out in *Har. marismortui* or *Hfx. denitrificans*. The extracellular pool of recombinant NiR was also purified and characterized. No significant biochemical differences are found between extracellular and intracellular NiR. However, the comparison of the isoform expression pattern of both samples reveals a remarkable difference. In the intracellular fraction, the 39.8 kDa isoform is predominant and the 44.3 kDa isoform appears slightly, while in the extracellular fractions the 44.3 kDa isoform is the predominant or even only one. These data support the fact that the halophilic NiR is involved in a maturation process and in exportation via the Tat system. In order to elucidate the maturation process of the protein and its exportation via the Tat system, the first eight amino acids of the two isoforms that appear in the SDS-PAGE were sequenced. The results show that the 44.3 kDa isoform is obtained as a consequence of the cleavage between the 33rd and 34th residues. Therefore, this isoform may be exported via the Tat system, being cleaved by the twin arginine signal sequence after its translocation to an extracellular medium. The sequence of the small isoform, 39.8 kDa, starts in the 52nd position. No cutting target is predicted in this location, so it seems more likely that this isoform could be obtained as a result of an alternative translation mechanism (Hering et al. 2009) or mRNA processing rather than as a cleavage process. Once the two possible transcripts are translated, a random trimerization occurs between these two possible isoforms. This process originates the pool of possible isoforms found both inside and outside the cell. Finally, the Tat system of *Hfx. volcanii* facilitates the exportation of recombinant Cu-NiR active trimers whenever any of the three contain the signal peptide. In the process of exportation through the membrane, the signal peptides of the large isoform are cleaved. Thus, outside the cell it can find a mixture of the cleaved and signal-avoided NiR, prevailing over the large isoform. In contrast, only the trimers remain inside the cell, exclusively composed of untargeted peptides that not are able to cross the membrane and go outside the cell. This discrimination between targeted and nontargeted peptides looks like a mechanism for regulating the system and final NiR location.

The location of recombinant NiR outside the cell agrees with the results related to the extracellular location of membrane-associated NarGH from *Hfx. mediterranei* detailed above. For this reason, there is increasing evidence that the complete reduction of NO_3^- to N_2 could take place through an extracellular enzymatic complex, which is part of the machinery associated with the outer face of the cytoplasmic membrane, while the rest of the soluble enzymes and metabolites are embedded in the porous S-layer. This atypical respiratory complex orientation offers advantages to these microorganisms in oxygen-poor environments, such as hypersaline ecosystems. With this modification, the presence of NO_3^- transporters is not required and the electron acceptor can be reduced directly in the growth media, increasing the efficiency of the process. Finally, the mobilization of the proteins involved in NO_3^- respiration appears to be regulated by the Tat system so that they are folded and loaded with metallic cofactors inside the cell before being exported out of the cell, where they will take part in their physiological role.

Regarding the cd_1-nitrite reductase, a *nirS* homologous gene has been identified in the genome of the hyperthermophile *P. aerophilum* (Cabello et al. 2004), although the enzyme has not been purified or characterized at the time of this writing. However, polarographic studies carried out with purified membranes revealed that this nitrite reductase uses menaquinol as an electron donor (de Vries and Schroder 2001). These data could suggest the existence of cd_1-nitrite reductase in thermophilic microorganisms, while halophilic microorganisms possess Cu-containing nitrite reductase.

10.3.3 Nitric Oxide Reductases

Nitric oxide is the product of the respiratory nitrite reductase. This compound is toxic to cells, and for that reason, it is immediately reduced to N_2O by nitric oxide reductases (Nor). Several enzymes with Nor activities have been described. In fungal denitrification, Nor enzymes are soluble monomeric proteins belonging to the cytochrome P-450 family (Nakahara et al. 1993). In most denitrifying bacteria, Nor is a heterodimer membrane complex of a cytochrome *c* (encoded by *norC*) and a cytochrome *b* with 12 transmembrane regions (encoded by *norB*). This enzyme is known as cNor. On the other hand, a monomeric Nor with 14 transmembrane regions has been described in other bacteria. This enzyme is called qNor due to its quinol-oxidizing activity. The qNor enzyme is similar to the NorB subunit, although it contains an N-terminal extension, with a quinone-binding site, absent in NorB (Cabello et al. 2004; Bonete et al. 2008).

Despite the fact that there is only one study related to the characterization of Nor in extremophilic microorganisms thus far, gas formation from nitrite has been reported for a number of archaeal microorganisms, for example, *P. aerophilum*, *Hfx. denitrificans*, *Hfx. mediterranei*, *Har. hispanica*, and *Har. marismortui* (Zumft and Kroneck 2007), and for the halophilic bacteria *Halomonas halodenitrificans* (Sakurai et al. 2005). The first studies of Nor in *P. aerophilum* demonstrated the formation of N_2O using menaquinol as an electron donor and the presence of Nor bound to its membrane (de Vries and Schroder 2001). Later, in 2003, the nitric oxide reductase from the hyperthermophilic microorganism was purified and characterized (de Vries et al. 2003). The enzyme is a monomeric protein of 78.8 kDa and contains heme and nonheme iron in a 2:1 ratio. The EPR, resonance Raman, and UV-vis spectroscopy analyses show that one of the hemes has a bis-His-coordinated low-spin configuration, whereas the other heme adopts a high-spin configuration. In comparison with other thermophilic enzymes, the thermostability of the isolated Nor from *P. aerophilum* is very low, while that of the enzyme bound to the membrane is average. It is possible that the removal of the membrane lipids by detergent contributes to the lower thermostability, although it is unclear how this would occur (de Vries et al. 2003).

Regarding the genetic analysis, *nor* genes have been identified in a few genomes of halophilic and hyperthermophilic microorganisms. *Har. hispanica*, *Har. marismortui*, *Hfx. mediterranei*, *Hfx. volcanii*, and *P. aerophilum* contain in their genomes a copy of a *norB* gene, and up to now, there is only one example of a *norZ* gene in the halophile *H. borinquense*, suggesting a possible case of horizontal gene transfer between bacteria and archaea.

10.3.4 Nitrous Oxide Reductases

The reduction of N_2O to N_2 is the last step of denitrification, which is catalyzed by nitrous oxide reductases (Nos). This reaction is of great environmental importance because it closes the N cycle. N_2O is less toxic than NO or nitrite, and the vast majority of microorganisms could manage without converting N_2O to N_2. Nonetheless, there are many bacteria, which contain nitrous oxide reductases encoded by the *nosZ* gene. These bacterial enzymes are located in the periplasm, and they are multicopper homodimers whose electron donor is the cytochrome *c* or pseudoazurin (Zumft 1997). Each monomer contains two copper centers, a di-copper cluster CuA resembling that of cytochrome oxidase, and a CuZ cluster that consists of four Cu atoms ligated by seven His residues (Rasmussen et al. 2000). The putative *nosZ* gene has been identified in the halophilic archaea *Har. marismortui*, *Hfx. mediterranei*, *Har. hispanica*, and *Hfx. denitrificans*. The gene that encodes the nitrous oxide reductase has also been identified in other halophilic archaea, such as *Halopiger xanaduensis*, *H. borinquense*, and *Halorubrum lacusprofundii*. However, in these species, the genes have not bee classified as *nosZ*. It was assumed that *P. aerophilum* has not nitrous oxide reductase, but recently, a thermophilic multicopper oxidase that shows nitrous oxide reductase activity has been purified (Fernandes et al. 2010). This multicopper oxidase is a thermoactive and thermostable metallo-oxidase. It follows a ping-pong mechanism, its sequence contains a putative TAT-dependent signal peptide, and it shows a threefold higher catalytic efficiency when it uses N_2O as an electron acceptor compared to when it uses dioxygen, the typical oxidizing substrate of multicopper oxidases. This fact represents a completely new function among multicopper oxidases, and it could be a novel archaeal nitrous oxide reductase that is probably involved in the final step of the denitrification pathway of *P. aerophilum* (Fernandes et al. 2010).

10.4 FUTURE PROSPECTS

Denitrification is the major biological pathway for N loss from ecosystems, and the gaseous intermediates, nitric oxide and nitrous oxide, have implications in global warming (Prather et al. 2012). Nitrous oxide has become the third most important anthropogenic greenhouse gas (IPCC 2014), and it is today's single most important ozone-depleting emission (Ravishankara et al. 2009). When aiming to mitigate N_2O emissions, an accurate understanding of the biochemical processes responsible for N_2O production is crucial (Richardson et al. 2009).

The potential environmental importance of denitrification has led to numerous measurements of the process in a range of habitats. To know the extent of this process in extreme environments will be essential to understand the contribution of the denitrifying microorganisms to the greenhouse effect. Unfortunately, denitrification is very difficult to measure, mainly in extreme environments or extreme microcosms. So, the existing methods are problematic and the methodology still needs development. Although one review on methods for measuring denitrification is available (Groffman et al. 2006), the development of molecular approaches is necessary. In the pregenomic era, establishing whether a microorganism was a denitrifier entailed testing its ability to grow under O_2-limiting conditions, with nitrate most frequently provided as a terminal oxidant (Payne 1981). Therefore, nearly all denitrifiers characterized were complete denitrifiers that showed robust growth under denitrifying conditions. However, with genome analysis supplanting phenotypic assignment as the principal means of identifying denitrifiers, both complete and partial denitrifiers can be identified (Shapleigh 2013).

The application of molecular methods to study denitrification can lead to understanding how the composition and physiology of the microbial community affects N transformations in the environment. Bacteria, fungi, and archaea are capable of denitrification, and it can be considered to be a community process, as many denitrifying organisms do not produce the complete suite of enzymes and could work together to complete the process (Wallenstein 2006). Understanding the responses of microbial communities to environmental factors and the impact of the community composition on the rate of denitrification is essential to know this process and its impact on gas emissions, and even more so in extreme environments, where the denitrification community has been less studied.

Approaches based on the direct extraction of DNA from the natural environment and polymerase chain reaction (PCR) amplifications can overcome limitations due to archaea and bacteria cultivation and isolation (Demanèche et al. 2009). It must also be taken into account that denitrification is, nearly exclusively, a facultative respiratory pathway and, in some environments, genes for denitrification are often detected where there is no measurable denitrification activity (Groffman 2006). A few studies have attempted to extract mRNA from environmental samples and use reverse transcriptase PCR to measure the active denitrified community (Nogales et al. 2002). This could be a potentially powerful approach.

The most fundamental need for molecular studies of denitrifier communities is an improved database of functional genes. Until now, most of the molecular tools used for studying denitrifier community composition begin by selectively amplifying the target functional genes using PCR. The problem was the degree to which the selective primers target all variants of these genes. The inventory of genes involved in denitrification and the extent of their diversity in extremophilic environments are yet to be explored, and the characterization of whole or partial denitrification pathways with gene sequences becomes necessary (Wallenstein 2006). Previous analysis of genome organization and comparative genomics in bacterial and archaeal genomes indicated complex genetic bases of the process and allowed the identification of new putative denitrifying genes (Philippot 2002). A metagenomic approach has been carried out in order to identify and characterize gene clusters involved in the denitrification process in soil bacteria, and the analysis led to the identification and subsequent characterization of nine denitrification gene clusters (Demanèche et al. 2009). In archaea, there is not an extensive analysis of the gene cluster organization involved in denitrification, but taking into account the previous works (Philipot 2002; Cabello et al. 2004), the variability in archaeal genomes will also be important, and their analysis will shed light on the denitrification processes carried out by extremophiles.

The predictive genomic data must be confirmed by experimental data. Isolation and characterization of the proteins and complexes involved in the denitrification process are compulsory to understand it. As has been mentioned, there are some works focused on archaeal nitrate and nitrite reductases, but much less is known about nitric oxide and nitrous oxide reductases. The characterization of these enzymes and the identification of the electron transport intermediates are necessary to understand the extent of the final step of denitrification in extremophiles and the conditions in which the process is more active.

Another important consideration is the regulation of denitrification. In most denitrifier bacteria, the expression of genes encoding these proteins depends on the presence of nitrogen oxides. In general, Nir and Nor respond to NO stimuli and Nos responds to N_2O. Studies with model organisms have found that reduction of nitrate and nitrite to gaseous products occurs at low O_2 (Zumft 1997; Shapleigh 2013). In particular, the regulation of *nir* and *nor* genes is especially sensitive to O_2. A review of transcriptional regulation in bacteria is available (Shapleigh 2013), but there are no data on transcriptional regulation in archaea.

Knowledge on the regulation of denitrification in archaea would be very important, not only to understand the physiological conditions in which denitrification becomes important to the organisms, but also to improve potential biotechnological applications of the pathway in bioremediation or to understand the contribution of halophilic archaea to N gas emissions (Nájera-Fernández et al. 2012; Martínez-Espinosa et al. 2015).

ACKNOWLEDGMENTS

This work was funded by a research grant from MINECO, Spain (CTM2013-43147-R).

REFERENCES

Afshar, S, C Kim, HG Monbouquette, and I Schroder. 1998. Effect of tungstate on nitrate reduction by the hyperthermophilic archaeon *Pyrobaculum aerophilum*. *Applied and Environmental Microbiology* 64 (8): 3004–3008.

Afshar, S, E Johnson, S de Vries, and I Schroeder. 2001. Properties of a thermostable nitrate reductase from the hyperthermophilic archaeon *Pyrobaculum aerophilum*. *Journal of Bacteriology* 183 (19): 5491–5495.

Alonso, A, and JA Camargo. 2006. Toxicity of nitrite to three species of freshwater invertebrates. *Environmental Toxicology* 21: 90–94.

Álvarez-Ossorio, MC, FJG Muriana, FF de la Rosa, and AV Relimpio. 1992. Purification and characterization of nitrate reductase from the halophile archaebacterium *Haloferax mediterranei*. *Zeitschrift fur Naturforschung* 47C: 670–676.

Amo, T et al. 2002. *Pyrobaculum calidifontis* sp. nov., a novel hyperthermophilic archaeon that grows in atmospheric air. *Archaea (Vancouver, B.C.)* 1 (2): 113–121.

Andrei, AŞ, HL Banciu, and A Oren. 2012. Living with salt: Metabolic and phylogenetic diversity of archaea inhabiting saline ecosystems. *FEMS Microbiology Letters* 330 (1): 1–9.

Bardiya, N, and JH Bae. 2005. Bioremediation potential of a perchlorate-enriched sewage sludge consortium. *Chemosphere* 58 (1): 83–90.

Bickel-Sandkotter, S, and M Ufer. 1995. Properties of a dissimilatory nitrate reductase from the halophilic archaeon *Haloferax volcanii*. *Zeitschrift fur Naturforschung—Section C Journal of Biosciences* 50: 365–372.

Bonete, MJ, RM Martínez-Espinosa, C Pire, B Zafrilla, and DJ Richardson. 2008. Nitrogen metabolism in haloarchaea. *Saline Systems* 4: 9.

Bourbonnais, A et al. 2012. Activity and abundance of denitrifying bacteria in the subsurface biosphere of diffuse hydrothermal vents of the Juan de Fuca Ridge. *Biogeosciences* 9 (11): 4661–4678.

Cabello, P, MD Roldán, and C Moreno-Vivián. 2004. Nitrate reduction and the nitrogen cycle in archaea. *Microbiology* 150: 3527–3546.

Čuhel, J et al. 2010. Insights into the effect of soil pH on N_2O and N_2 emissions and denitrifier community size and activity. *Applied and Environmental Microbiology* 76 (6): 1870–1878.

Demanèche, S, L Philippot, MM David, E Navarro, TM Vogel, and P Simonet. 2009. Characterization of denitrification gene clusters of soil bacteria via a metagenomic approach. *Applied and Environmental Microbiology* 75 (2): 534–537.

de Vries, S, and I Schroder. 2001. Comparison between the nitric oxide reductase family and its aerobic relatives, the cytochrome oxidases. *Biochemical Society Transactions* 30: 662–667.

de Vries, S, MJF Strampraad, S Lu, P Moënne-Loccoz, and I Schröder. 2003. Purification and characterization of the MQH2:NO oxidoreductase from the hyperthermophilic archaeon *Pyrobaculum aerophilum*. *Journal of Biological Chemistry* 278 (38): 35861–35868.

Ellington, MJK. 2003. *Rhodobacter capsulatus* gains a competitive advantage from respiratory nitrate reduction during light-dark transitions. *Microbiology* 149 (4): 941–948.

Esclapez, J, B Zafrilla, RM Martínez-Espinosa, and MJ Bonete. 2013. Cu-NirK from *Haloferax mediterranei* as an example of metalloprotein maturation and exportation via Tat system. *Biochimica et Biophysica Acta—Proteins and Proteomics* 1834 (6): 1003–1009.

Fernandes, AT, JM Damas, S Todorovic, R Huber, MC Baratto, R Pogni, CM Soares, and LO Martins. 2010. The multicopper oxidase from the archaeon *Pyrobaculum Aerophilum* shows nitrous oxide reductase activity. *FEBS Journal* 277: 3176–3189.

Gavira, M, MD Roldan, F Castillo, and C Moreno-Vivian. 2002. Regulation of Nap gene expression and periplasmic nitrate reductase activity in the phototrophic bacterium *Rhodobacter sphaeroides* DSM158. *Journal of Bacteriology* 184 (6): 1693–1702.

Groffman, PM et al. 2006. Methods for measuring denitrification: Diverse approaches to a difficult problem. *Ecological Applications* 16 (6): 2091–2122.

Hering, O, M Brenneis, J Beer, B Suess, and J Soppa. 2009. A novel mechanism for translation initiation operates in haloarchaea. *Molecular Microbiology* 71 (6): 1451–1463.

Heylen, K, D Gevers, B Vanparys, L Wittebolle, J Geets, N Boon, and P de Vos. 2006. The incidence of nirS and nirK and their genetic heterogeneity in cultivated denitrifiers. *Environmental Microbiology* 8 (11): 2012–2021.

Hochstein, LI, and F Lang. 1991. Purification and properties of a dissimilatory nitrate reductase from *Haloferax denitrificans*. *Archives of Biochemistry and Biophysics* 288 (2): 380–385.

Howarth, RW. 2004. Human acceleration of the nitrogen cycle: Drivers, consequences, and steps toward solutions. *Water Science and Technology* 49: 7–13.

Huang, Y et al. 2014. Acidophilic denitrifiers dominate the N2O production in a 100-year-old tea orchard soil. *Environmental Science and Pollution Research* 22 (6): 4173–4182.

Ichiki, H, Y Tanaka, and K Mochizuki. 2001. Analysis of Cu-containing dissimilatory nitrite reductase from a denitrifying halophilic archaeon, *Haloarcula* purification, characterization, and genetic analysis of Cu-containing dissimilatory nitrite reductase from a denitrifying halophilic archaeon. *Journal of Bacteriology* 183 (14): 4149–4156.

Inatomi, K, and LI Hochstein. 1996. The purification and properties of a copper nitrite reductase from *Haloferax denitrificans*. *Current Microbiology* 32 (2): 72–76.

IPCC (Intergovernmental Panel on Climate Change). 2014. http://www.ipcc.ch/.

Jones, CM, and S Hallin. 2010. Ecological and evolutionary factors underlying global and local assembly of denitrifier communities. *ISME Journal* 4 (5): 633–641.

Jormakka, M, D Richardson, B Byrne, and S Iwata. 2004. Architecture of NarGH reveals a structural classification of Mo-bisMGD enzymes. *Structure* 12 (1): 95–104.

Kletzin, A, and MWW Adams. 1996. Tungsten in biological systems. *FEMS Microbiology Reviews* 18 (1): 5–63.

Liu, B et al. 2010. Denitrification gene pools, transcription and kinetics of NO, N2O and N2 production as affected by soil pH. *FEMS Microbiology Ecology* 72 (3): 407–417.

Lledó, B, RM Martínez-Espinosa, FC Marhuenda-Egea, and MJ Bonete. 2004. Respiratory nitrate reductase from haloarchaeon *Haloferax mediterranei*: Biochemical and genetic analysis. *Biochimica et Biophysica Acta—General Subjects* 1674: 50–59.

Martínez-Espinosa RM, JA Cole, DJ Richardson, NJ Watmough. 2011. Enzymology and ecology of the nitrogen cycle. *Biochem Soc Trans* 39 (1): 175–8.

Martinez-Espinosa, RM, EJ Dridge, MJ Bonete, JN Butt, CS Butler, F Sargent, and DJ Richardson. 2007. Look on the positive side! The orientation, identification and bioenergetics of 'archaeal' membrane-bound nitrate reductases. *FEMS Microbiology Letters* 276: 129–139.

Martínez-Espinosa, RM, DJ Richardson, and MJ Bonete. 2015. Characterisation of chlorate reduction in the haloarchaeon *Haloferax mediterranei*. *Biochimica et Biophysica Acta (BBA)—General Subjects* 1850 (4): 587–594. doi: 10.1016/j.bbagen.2014.12.011.

Mosier, AR. 1998. Soil processes and global change. *Biology and Fertility of Soils* 27 (3): 221–229.

Nájera-Fernández, C, B Zafrilla, MJ Bonete, and RM Martínez-Espinosa. 2012. Role of the denitrifying Haloarchaea in the treatment of nitrite-brines. *International Microbiology* 15 (3): 111–119.

Nakahara, K, T Tanimoto, K Hatano, K Usuda, and H Shoun. 1993. Cytochrome P-450 55A1 (P450dNIR) acts as nitric oxide reductase employing NADH as direct electron donor. *Journal of Biological Chemistry* 268: 8350–8355.

Nogales, B, KN Timmis, DB Nedwell, and AM Osborn. 2002. Detection and diversity of expressed denitrification genes in estuarine sediments after reverse transcription-PCR amplification from mRNA. *Applied and Environmental Microbiology* 68 (10): 5017–5025.

Oren, A. 2002. Diversity of halophilic microorganisms: Environments, phylogeny, physiology, and applications. *Journal of Industrial Microbiology & Biotechnology* 28 (1): 56–63.

Payne, WJ. 1981. *Denitrification*. Hoboken, NJ: John Wiley & Sons.

Philippot, L. 2002. Denitrifying genes in bacterial and archaeal genomes. *Biochimica et Biophysica Acta (BBA)—Gene Structure and Expression* 1577 (3): 355–376.

Potter, L, H Angove, DJ Richardson, and JA Cole. 2001. Nitrate reduction in the periplasm of gram negative bacteria. *Advances in Microbial Physiology* 45: 51–112.

Prather, MJ, CD Holmes, and J Hsu. 2012. Reactive greenhouse gas scenarios: Systematic exploration of uncertainties and the role of atmospheric chemistry. *Geophysical Research Letters* 39 (9).

Rasmussen, T, BC Berks, J Sanders-Loehr, DM Dooley, WG Zumft, and AJ Thomson. 2000. The catalytic center in nitrous oxide reductase, CuZ, is a copper-sulfide cluster. *Biochemistry* 39: 12753–12756.

Ravishankara, AR, JS Daniel, and RW Portmann. 2009. Nitrous oxide (N_2O): The dominant ozone-depleting substance emitted in the 21st century. *Science* 326 (5949): 123–125.

Richardson, D, H Felgate, N Watmough, A Thomson, and E Baggs. 2009. Mitigating release of the potent greenhouse gas N(2)O from the nitrogen cycle—Could enzymic regulation hold the key? *Trends in Biotechnology* 27 (7): 388–397.

Richardson, DJ, BC Berks, D Russell, S Spiro, and CJ Taylor. 2001. Functional, biochemical and genetic diversity of prokaryotic nitrate reductases. *Cellular and Molecular Life Sciences: CMLS* 58: 165–178.

Sakurai, T, S Nakashima, K Kataoka, D Seo, and N Sakurai. 2005. Diverse NO reduction by *Halomonas halodenitrificans* nitric oxide reductase. *Biochemical and Biophysical Research Communications* 333: 483–487.

Sánchez-Andrea, I et al. 2012. Screening of anaerobic activities in sediments of an acidic environment: Tinto River. *Extremophiles* 16 (6): 829–839.

Santoro, AE et al. 2006. Denitrifier community composition along a nitrate and salinity gradient in a coastal aquifer denitrifier community composition along a nitrate and salinity gradient in a coastal aquifer. *Applied and Environmental Microbiology* 72 (3): 2102–2109.

Shapleigh, JP. 2013. Denitrifying prokaryotes. In *The Prokaryotes*, ed. E. Rosenberg, E.F. DeLong, S. Lory, E. Stackebrandt, F. Thompson pp. 405–425. Berlin: Springer.

Shapleigh, WJ, and WJ Payne. 1985. Differentiation of cd_1 cytochrome and copper nitrite reductase production in denitrifiers. *FEMS Letter* 26: 275–279.

Shen, QR, W Ran, and ZH Cao. 2003 Mechanisms of nitrite accumulation occurring in soil nitrification. *Chemosphere* 50: 747–753.

Šimek, M, L Jíšová, and DW Hopkins. 2002. What is the so-called optimum pH for denitrification in soil? *Soil Biology and Biochemistry* 34 (9): 1227–1234.

Stetter, KO. 1999. Extremophiles and their adaptation to hot environments. *FEBS Letters* 452 (1–2): 22–25.

Tomlinson, GA, and LI Hochstein. 1988. The enzymes associated with denitrification. In *Abstracts of the Annual Meeting of the American Society for Microbiology*, University of Wisconsin, Madison. *1988*. p. 209.

Van Ginkel, CG, AM van Haperen, and B van der Togt. 2005. Reduction of bromate to bromide coupled to acetate oxidation by anaerobic mixed microbial cultures. *Water Research* 39 (1): 59–64.

van Wolferen, M et al. 2013. How hyperthermophiles adapt to change their lives: DNA exchange in extreme conditions. *Extremophiles* 17 (4): 545–563.

Volkl, P et al. 1993. *Pyrobaculum aerophilum* sp. nov., a novel nitrate-reducing hyperthermophilic archaeum. *Applied and Environmental Microbiology* 59 (9): 2918–2926.

Wallenstein, MD, DD Myrold, M Firestone, and M Voytek. 2006. Environmental controls on denitrifying communities and denitrification rates: Insights from molecular methods. *Ecological Applications* 16 (6): 2143–2152.

Wang, F et al. 2009. GeoChip-based analysis of metabolic diversity of microbial communities at the Juan de Fuca Ridge hydrothermal vent. *Proceedings of the National Academy of Sciences of the United States of America* 106 (12): 4840–4845.

Werber, MM, and M Mevarech. 1978. Induction of a dissimilatory reduction pathway of nitrate in *Halobacterium* of the Dead Sea. *Archives of Biochemistry and Biophysics* 186: 60–65.

Yoshimatsu, K, T Iwasaki, and T Fujiwara. 2002. Sequence and electron paramagnetic resonance analyses of nitrate reductase NarGH from a denitrifying halophilic euryarchaeote *Haloarcula marismortui*. *FEBS Letters* 516: 145–150.

Yoshimatsu, K, T Sakurai, and T Fujiwara. 2000. Purification and characterization of dissimilatory nitrate reductase from a denitrifying halophilic archaeon, *Haloarcula marismortui*. *FEBS Letters* 470 (2): 216–220.

Zumft, WG. 1997. Cell biology and molecular basis of denitrification. *Microbiology and Molecular Biology Reviews: MMBR* 61 (4): 533–616.

Zumft, WG, and PMH Kroneck. 2007. Respiratory transformation of nitrous oxide (N_2O) to dinitrogen by bacteria and archaea. *Advances in Microbial Physiology* 52: 107–227.

CHAPTER 11

Extremophile Enzymes and Biotechnology

Julia María Esclapez Espliego, Vanesa Bautista Saiz, Javier Torregrosa-Crespo,
Anna Vegara Luque, Mónica L. Camacho Carrasco, Carmen Pire,
María José Bonete, and Rosa María Martínez-Espinosa

CONTENTS

11.1 INTRODUCTION

Biotechnology has great significance in many aspects, both industrial and in daily life. The use of enzymes as biocatalysts is well established, and it has been the subject of numerous texts and revisions. The enzymatic properties, such as substrate affinity, solvent tolerance, temperature stability, or selectivity, can be enhanced through genetic engineering to obtain biocatalysts more in line with the needs of each moment (Elleuche et al. 2014). However, the task is tedious and lacks rational concepts. Keeping this in mind, it can be considered that food, power generation, biofuels, or chemical products with a high value in the pharmaceutical industry and medicine, among others, are some of the applications in which extremophilic organisms can play an important role. A new generation of enzymes that can act as biocatalysts, which are stable to different adverse conditions of nature, and which replace the traditional chemical processes, is needed. The problem at the time of this writing is that enzymes used as biocatalysts came from mesophilic organisms, and despite their many advantages, the application of these enzymes was limited because of their poor stability against extreme values of temperature, pH, ionic strength, and so forth (Hough and Danson 1999; Eichler 2001). Extremophiles are microorganisms able to grow and live in environments of extreme temperature (−2°C to 15°C for cold environments, 60°C–110°C in hot environments), ionic strength (2–5 M NaCl), or pH (<4.0 or >9.0). All the organisms mainly belonging to the Bacteria and Archaea domains show their own and specific metabolic pathways adapted to extreme conditions. Because of that, they are considered sources of new enzymes, called extremozymes, with innovative applications and activities. Archaea exist in a wide range of habitats that represent 20% of the biomass on earth (DeLong and Pace 2001). Organisms of the Archaea domain differ from those in the Bacteria and Eukarya domains in structural, genetic, and biochemical properties. Research on the Archaea has greatly increased during recent years, in part initiated by genomic science, as well as by a continuing interest in their proteomics, biochemistry, and metabolism (Bonete and Martínez-Espinosa 2011). Extremozymes are capable of catalyzing their respective reactions in nonaqueous environments, in water–solvent mixtures, at extremely high pressures, at acidic and alkaline pH, at temperatures up to 140°C, or near the freezing point of water (Adams et al. 1995). On the other hand, as a result, the use of traditional protein engineering approaches to produce extremophilic enzymes using mesophilic host cells is not usually successful. Thus, genetic improvements of the extremophilic expression systems are currently the subject of numerous studies. Misfolding and differences in codon usage can interfere with the production of functional extremozyme expression systems commonly used, such as *Escherichia coli* or *Bacillus* sp. Therefore, it is an important prerequisite to establish fine regulated extremophilic hosts, efficient transformations, and suitable expression vectors (Elleuche et al. 2014).

Currently, and more and more with increasing intensity, there are functions that (intend to) apply to archaea-derived materials. For example, the extremely stable lipids of membranes of these organisms represent a novel drug delivery system (Patel and Sprott 1999; Schiraldi et al. 2002; Oren 2010; Zhao et al. 2015). Self-assembling components from archaea, such as the S-layer glycoprotein and bacterioopsin, are of interest for their nanotechnological potential (Oesterhelt et al. 1991; Sleytr et al. 1997). Polysaccharides secreted from haloarchaea could find use in the oil industry (Rodriguez-Valera 1992), while polymers also secreted from haloarchaea have been considered a raw material of biodegradable plastics (Fernández-Castillo et al. 1986). Methanogenic archaea are seen as clean and low-cost energy sources (Reeve et al. 1997).

11.2 PRODUCTION OF ENZYMES FROM EXTREMOPHILES: EXTREMOPHILE BIOMASS PRODUCTION, GENE CLONING, AND OVERPRODUCTION

Extremophiles are adapted to live in different extreme conditions, both physical and geochemical. Therefore, their biochemical machinery has been adapted in the course of evolution to be able to function under conditions at which most biochemical systems would cease to function. Nowadays, the aim of a lot of studies consists of extending the understanding of the molecular basis responsible for extremophilic adaptations. However, this is a hard work because the characterization of extremophilic enzymes is still difficult since most of the extremophile microorganisms have not been isolated from pure cultures (Adrio and Demain 2014).

The production of extremozymes can be done in different ways: through the isolation of microorganisms from pure cultures with subsequent purification of the enzyme from its own host, or by using cloning and expression techniques for the gene of interest. On the other hand, these enzymes can be obtained from environmental DNA by (1) the screening of expression libraries and (2) the cloning of gene fragments to assemble them into functional genes for expressing them in proper hosts (Hough and Danson 1999; Demirjian et al. 2001). In fact, the putative genes discovered have to be verified by means of the expression and characterization of their corresponding protein, although they have been recognized by comparison of sequences. Therefore, information about their substrate specificity, stability, and enantioselectivity can be obtained. Based on the knowledge of the structural bases of stability and activity of extremozymes, the engineering of mesophilic enzymes can be done to modify its stability through site-directed mutagenesis. Actually, to enhance the stability, an evolutionary strategy can also be assumed where random mutations are inserted into the gene of interest and then the mutants containing the expected properties are selected. Otherwise, directed evolution can be performed, which implies sequential steps of random mutagenesis, recombination, and the screening of mutants with the desired features. Additionally, DNA shuffling (the random recombination of fragments from nearly related gene sequences) produces new genes and supplies an option for the directed evolution of enzymes. This last strategy is determined by the proper cloned genes that are available and methods of expression (Zhao et al. 2002; Turner 2003).

The expression of proteins can be performed following two strategies: heterologous expression, when the expression host is an organism different from the one to which the protein belongs, and homologous expression, when the protein is expressed in the same species as the protein belongs. Protein expression processes can be carried out following different protocols, which usually involves gene cloning using different plasmids and host cells.

Due to the problems the occur when the extremophile cultures are grown on a large scale (bioreactors not adapting to extremophilic requirements, low growth rate, etc.), most extremozyme overexpression processes are performed in mesophilic hosts, such as *E. coli*, *Bacillus subtilis*, and yeast (Alquéres et al. 2007). Indeed, the *E. coli*–specific growth rate is 5–10 times higher than the average specific growth observed from extremophilic strains. Additionally, as host, *E. coli* grows effortlessly in the laboratory, producing enough biomass to work on a large scale (Schiraldi and De Rosa 2002). Nevertheless, all extremophilic genes cannot be expressed efficiently in this host, because of the inclusion bodies that are produced during the protein overexpression. Halophilic proteins, for example, contain abundant negatively charged amino acids on their surface. This characteristic can cause problems in the protein expression using mesophilic hosts like *E. coli*, because proteins fold incorrectly and form aggregates in the form of inclusion bodies. This happens in conditions lacking in high ionic strength (Allers 2010). In addition, the inclusion bodies formed must be solubilized with denaturing agents as urea or guanidine hydrochloride. Then, folding processes must be carried out *in vitro* to retrieve the activity of the overproduced enzyme (Pire et al. 2001; Esclapez et al. 2007; Munawar and Engel 2012). That is, overexpressed proteins need to be reactivated and refolded with buffers with high salt concentrations at optimized conditions *in vitro*. When all these steps are optimized, inclusion bodies become an advantage instead of being a disadvantage, because a large amount of protein could be easily purified in a unique purification stage (Pire et al. 2001). Finally, homologous overexpression can also be carried out successfully in extremophiles thanks to the improvements of the molecular biology techniques during the last few years. In halophiles, for example, halophilic protein expression using *Haloferax volcanii* as host has reported positive results because the halophilic enzymes produced are active and completely soluble (Allers 2010; Esclapez et al. 2013).

In order to avoid the problem related to inclusion body production, other bacterial species have been tested as host to overproduce extremophilic enzymes (mainly species of the genera *Bacillus*, *Lactobacillus*, *Pseudomonas*, and *Lactococcus*). In addition, there are eukaryotic expression systems such as *Pichia*, *Kluyveromyces*, *Candida*, and *Hansenula* (van den Burg 2003). Table 11.1 shows some relevant examples of extremozymes produced as recombinant proteins.

It is well known that extremophiles are characterized by low growth and productivity in fermentation processes. This is due to the low production of biomass yield and the low basal levels of enzyme expression. This problem could be solved with fed-batch cultures and systems that eliminate the accumulated toxic metabolites and exchange of nutrients (Schiraldi et al. 2001). In fact, dialysis

Table 11.1 Extremozymes Produced as Recombinant Proteins to Be Used in Biotransformations

Microorganism	Enzyme	Host	Reference
Haloferax mediterranei	Halolysin R4, serine protease	Haloferax volcanii WFD11	Kamekura et al. 1996
Haloferax volcanii	Dihydrolipoamide dehydrogenase	Haloferax volcanii	Jolley et al. 1996
Pyrococcus furiosus	Hyperthermophilic esterase	Escherichia coli	Ikeda and Clark 1998
Archaeoglobus fulgidus	Thermophilic esterase	Escherichia coli	Manco et al. 1998
Pseudomonas sp. B11-1	Psychrophilic lipase	Escherichia coli	Choo et al. 1998
Bacillus stearothermophilus	Thermophilic lipase	Escherichia coli	Kim et al. 2000
Thermococcus aggregans	Hyperthermophilic pullunase	Escherichia coli	Niehaus et al. 2000
Haloferax mediterranei	Halophilic glucose dehydrogenase	Escherichia coli	Pire et al. 2001
Haloferax volcanii	Halophilic isocitrate dehydrogenase	Escherichia coli	Camacho et al. 2002
Thermococcus litoralis	L-Aminoacylase	Escherichia coli	Toogood et al. 2002
Sulfolobus solfataricus	Alpha-glucosidase	Lactococcus lactis	Giuliano et al. 2004
Pyrococcus furiosus	Alcohol dehydrogenase	Escherichia coli	Kube et al. 2006
Haloferax mediterranei	Halophilic glutamate dehydrogenase	Escherichia coli	Díaz et al. 2006
Sulfolobus solfataricus MT4	Maltooligosyl-trehalose synthase	Lactococcus lactis	Cimini et al. 2008
Haloferax volcanii	Halophilic cysteine desulfurase (SufS)	Escherichia coli	Zafrilla et al. 2010
Haloferax mediterranei	Halophilic cyclodextrin glycosyltransferase	Escherichia coli	Bautista et al. 2012
Halobacterium sp. NRC-1	Alcohol dehydrogenase	Haloferax volcanii	Liliensiek et al. 2013
Haloferax mediterranei	Halophilic Cu-nitrite reductase	Haloferax volcanii	Esclapez et al. 2013

membrane fermenters are very useful because (1) they eliminate accumulated toxic metabolites in the fermentation process, and (2) the substrate may be provided directly through the membranes. In such fermenters, the increase of the biomass production is due to the dilution of toxic metabolites, or to the addition of indispensable nutrients for growth provided across the membrane (Krahe et al. 1996). Therefore, different groups of researchers have addressed their research toward physiological studies or to the design of bioreactors and bioprocesses to achieve high productivity and biomass when using extremophiles. Related to the latter, it is important to highlight that the combination of continuous fermentation bioprocesses and the design of different media has achieved good results: the growth of most of the extremophilic strains tested is improved, the production of some specific metabolites increases, and the importance of some distinguished enzymes in unusual metabolic pathways is elucidated. Moreover, research has also centered attention on the study of fed-batch cultures to increase the final concentration of biomass (Schiraldi and De Rosa 2002). These cultures consist of culture media continuously fed with fresh nutrients or any of its components (salts, ions, etc.). If the feeding nutrient is the limiting growth, this technique allows controlling the growth rate of the microorganism. They are especially useful in processes in which cell growth and product formation are sensitive to the concentration of the limiting substrate.

A microfiltration bioreactor can also be used to improve the production of biomass when working with extremophiles. In this case, the system supplies a pressure difference over the membrane, producing a flow that finally maintains the cells with a larger pore size than the membrane and eliminates the entire media consumed. Actually, if all three stages are combined: batch, fed batch, and microfiltration, the growth phase of the microorganisms is extended (Schiraldi and De Rosa 2002).

Table 11.2 Some Examples of Extremophile Cultures in Large-Scale Fermenters

Organism	Cultivation Form	Optimum Temperature (°C)	Reference
Metallosphera sedula	Continuous	74	Rinker et al. 1999
Pyrococcus furiosus	Dialysis	90	Krahe et al. 1996
Sulfolobus solfataricus Gθ	Microfiltration	75	Schiraldi et al. 1999
Thermococcus litoralis	Batch and continuous	85–88	Rinker and Kelly 2000
Sulfolobus solfataricus	Electric water heater	75–80	Worthington et al. 2003
Thermus thermophilus	5 L stirred tank bioreactor	70	Domingueza et al. 2005

Apart from the problems mentioned before about large-scale cultures of extremophiles, it is important to point out that the bioreactors used in those experiments must be constructed of corrosion-resistant materials for growing acidophiles, alkaliphiles, and specially halophiles (Schiraldi et al. 2002). In fact, for halophilic microorganisms, fermenters cannot be constructed of stainless steel, due to the high corrosion potential of the media with high salt concentrations. An example of a corrosion-resistant bioreactor would be one constructed of polyetherether ketone (PEEK), tech glass, and silicium nitrite ceramics for growing halophilic microorganisms (Hezayen et al. 2000). On the other hand, the temperature and pH produce the dissociation of the compounds in the medium. At hyperthermophilic conditions, the rate of diffusion of gases in liquid increases, and some volatile nutrients turn into a deficiency (Krahe et al. 1996). Also, bioreactors can be built to ameliorate the oxygen transfer rate. One example of a bioreactor combining some of these characteristics is the gas-lift reactor used to obtain high cultures of large cells from thermophiles (Holst et al. 1997). Some examples of extremophile large-scale cultures are mentioned in Table 11.2.

11.3 EXTREMOPHILIC ENZYMES VERSUS MESOPHILIC ENZYMES

In many cases, the mesophilic enzymes are not the most appropriate for use in industrial processes, because they do not withstand the harsh conditions to which they are submitted, and consequently, they lose stability and activity. Moreover, it is important to keep in mind that the synthesis of polymeric intermediates, pharmaceuticals, specialty chemicals, and agrochemicals suffers from low selectivity and unwanted subproducts appear. For these reasons, it is very difficult and expensive to develop biotechnological applications (Demirjian et al. 2001). However, the discovery of new extremophilic microorganisms and their enzymes has recently had a major impact on the field of biocatalysis (Bonete et al. 2005; Oren 2010), because these microorganisms produce only biocatalysts operating under conditions in which their mesophilic counterparts could not survive, allowing the development of additional industrial processes.

In the 1970s, Woese and his colleagues concluded that there were three domains in the universe: Bacteria, Archaea, and Eukarya (Woese and Fox 1977; Woese et al. 1990). In recent times, and based on the sequences of the 16s RNAs, the vast microbial diversity has been studied in depth, obtaining many archaeal 16s rRNA genes from environmental studies, increasing the number of new lineages. The Archaea domain, because of its extremophilic nature, has suffered physiological adaptations to be stable in extreme conditions. As a result of these adaptations, great biotechnological potential with many applications in nature has been found in terms of enzymes and other metabolites. In many cases, because of their extremophilic nature, the enzymes are difficult to obtain in large quantities to meet the established biotechnology objectives, but thanks to the techniques of genetic engineering, the overexpression of these genes in mesophilic host cells is carried out, and simultaneously, their properties can be altered to achieve the desired commercial applications.

11.3.1 Factors Affecting the Stability of the Enzymes

The term *extreme conditions* refers to those conditions to which microorganisms have had to adapt to survive, and the biocatalysts obtained from these organisms are usually those used for biotechnological applications. For example, for the degradation of polysaccharides, such as chitin, cellulose, or starch, enzymes that are active and resistant to high temperatures are commonly used. It is known that salt reduces the water activity, and because of that, the enzymes from halophilic microorganisms, for instance, could be the best option for application in nonaqueous media (van den Burg 2003; Bonete and Martínez-Espinosa 2011), as is the case of several pharmaceutical reactions. Regarding thermophiles, different protein structures have been solved, and they have been compared with mesophilic counterparts, with the aim of solving the mechanisms that lead to thermostability. Ten thermophilic and hyperthermophilic serine hydroxymethyltransferases were studied and compared with 53 mesophilic counterparts (Paiardini et al. 2003). Structural alignment and homology modeling have been used to identify the different mechanisms involved in the thermal stability. Psychrophiles are organisms that also have biotechnologically important enzymes, because often the aim is to achieve a reduction in the energy consumption of a process. In this sense, psychrophilic proteases, amylases, or lipases have great commercial potential. The industry of pulp and paper, for example, is also interested in enzymes that degrade polymers that are active at lower temperatures (van den Burg 2003). Regarding alkaliphilic or acidophilic enzymes, their adaptation to the extreme environment depends on their cellular location. If the proteins are intracellular, they are not submitted to any stress due to the cytoplasm of alkaliphilic or acidophilic microorganisms maintaining a neutral pH. However, the extracellular enzymes are exposed to extreme conditions and molecular adaptations are needed (van den Burg 2003). Following this relationship of extreme organisms, we have to mention the piezophiles, whose enzymes stand high pressures and are interesting, for example, for food production, where high pressure is applied to the processing and sterilization of food materials (van den Burg 2003). The most important molecular adaptations of the extremozymes affecting their stability and activity are summarized as follows:

- *Changes in amino acid residues*: When the temperature, pressure, or pH is high, the more labile amino acid residues may suffer covalent modifications, and that involves protein denaturation. If amino acids such as cysteine, aspartic acid, and asparagine are able to maintain their spatial position in a more hydrophobic environment, protected from the outside, it is possible to increase the stability of proteins. This is what happens with extremophile proteins compared with their mesophilic counterparts (Robb and Clark 1999). It has also been found that more stable proteins are achieved, making specific mutations of amino acid residues in the flexible regions of the protein secondary structure.
- *Ionic interactions*: Studies with three-dimensional structures of extremophilic proteins indicate that interactions of ion pairs can play an important role in stabilizing hyperthermophilic proteins in which the hydrophobic effect is minimal. These ionic interactions have a much longer range of action than hydrophobic interactions. Some studies about the hexameric glutamate dehydrogenases from hyperthermophilic organisms such as *Pyrococcus furiosus* and *Thermococcus litoralis* have demonstrated a decrease in the thermostability of the enzyme from *T. litoralis* due to the loss of important ionic pairs between subunits (Vetriani et al. 1998). By altering two residues in the enzyme, this interaction was restored and a significant increase in the thermal stability was observed. In the case of β-glucosidases isolated from hyperthermophilic and mesophilic organisms, it has also been found that the hyperthermophilic protein has significantly more ion pairs on the surface of the protein than their mesophilic counterpart, thereby improving the thermal stability (Bauer et al. 1998).
- *Cooperative association*: The denaturation of the oligomeric proteins under extreme conditions generally begins by subunit dissociation, followed by irreversible denaturation of a monomeric form. There is evidence that the oligomeric structure for some enzymes from extremophiles is often more complex than that of their mesophilic counterpart proteins, because they usually exist as monomers or dimers. This is the case of the iron hydrogenase enzyme isolated from the

hyperthermophilic bacterium *Thermotoga maritima*, which is presented as a homotetramer, while the corresponding enzymes isolated from mesophilic organisms typically consist of one or two subunits (Verhagen et al. 1999).

- *Solvent-exposed surface area*: Another factor influencing the stability of proteins from Archaea is the properties of the surface residues in contact with the solvent. Halophilic enzymes consist of a high negative charged surface, which increases solubility at high salt concentrations, as happens with α-amylase from *Alicyclobacillus acidocaldarius* (Demirjian et al. 2001) or with glucose dehydrogenase from *Haloferax mediterranei* (Britton et al. 2006). The same applies to some extreme acidophilic xylanases that have a high percentage of acidic residues located on the surface compared with other mesophilic xylanases. This property has been used for the implementation of halophilic enzymes in aqueous or organic and nonaqueous media (Klibanov 2001). In this sense, the use of reverse micelles in combination with halophilic enzymes can result in the development of new applications for these enzymes (Marhuenda-Egea and Bonete 2002).

- *Catalytic mechanisms*: All the parameters related to the kinetics and catalytic activity are similar when comparing enzymes from extremophiles and nonextremophilic microorganisms. For instance, recombinant β-glucosidase from hyperthermophilic *P. furiosus* and that from mesophilic *Agrobacterium faecalis* (Bauer et al. 1998) show similar substrate specificities and almost identical pH dependencies with several different substrates, a similar variation of free energy, and similar inhibition constants with various inhibitor types.

11.4 BIOTECHNOLOGICAL USES OF EXTREMOZYMES

11.4.1 Halophilic Enzymes

Hypersaline environments are characterized by extreme conditions of high salinity, low oxygen levels, high and low temperatures, and sometimes high pH values. The enzymes present in halophilic microorganisms are not only halophilic but also thermostable and sometimes alkaliphilic. These properties are desirable for different processing industries. In recent years, halophilic microorganisms have been studied for their potential uses in diverse fields, which can employ either the whole microorganism or their enzymes (Mellado and Ventosa 2003; Oren 2010). The most relevant biotechnological applications of halophilic microorganisms and their enzymes are described below.

11.4.1.1 Hydrolases for Biotechnological Industries

Halophilic enzymes that are being studied for their potential use in biotechnology are primarily hydrolases, such as amylases, cellulases, xylanases, pullulanases, proteases, lipases, DNases, pectinases, and inulinases (Sánchez-Porro et al. 2003a; Cojoc et al. 2009; Rohban et al. 2009; Enache and Kamekura 2010; Litchfield 2011; Makhdoumi Kakhki et al. 2011; Delgado-García et al. 2012; Yin et al. 2015).

The main hydrolytic enzymes employed in industry correspond to moderately halophilic bacteria (Flores et al. 2010; Setati 2010); however, several archaeal enzymes with hydrolytic activity have been reported (Pérez-Pomares et al. 2003; Ventosa et al. 2005; Bautista et al. 2012). Some hydrolase-producing halophilic microorganisms with biotechnological use are summarized in Table 11.3.

The common halophilic amylases are α-amylases and cyclomaltodextrinases; they are mainly used in the detergent and food industries. These enzymes are produced by different archaea and bacteria with a general optimal temperature of 50°C–55°C. They can act in a wide range of pH, temperature, and salinity, and may even be tolerant in the presence of organic solvents (Pérez-Pomares et al. 2003; Fukishima et al. 2005; Setati 2010). An interesting potential use of the archaeal α-amylases is in bread making. Rosell et al. (2001) concluded that most of the bacterial or fungal amylases used in this industry were sensitive to some additives and ingredients, but this did not seem to be the case for those from some haloarchaea strains.

Table 11.3 Halophilic Enzymes in Biotechnological Industries

Enzyme	Organism	Application	Reference
Amylase	*Pseudoalterimonas* sp.	Saccharification of marine microalgae, producing ethanol	Matsumoto et al. 2003
Amylase	*Halococcus* sp.	Starch hydrolysis in industrial processes in saline and organic solvent medium	Fukishima et al. 2005
Amylase	*Streptomyces* sp.	Detergent formulations	Chakraborty et al. 2009
Chitinase	*Halobacterium* sp.	Oligosaccharide synthesis	Hatori et al. 2006
Chitinase	*Virgibacillus* sp.	Bioconversion of chitin from fish, crab, or shrimp; treatment of chitinous waste	Essghaier et al. 2011
Glutaminase	*Micrococcus* sp.	Flavor enhancement in food industries, antileukemic agent	Yoshimune et al. 2010
Inulinase	*Marinimicrobium* sp.	Monosaccharide hydrolysis	Li et al. 2011
Lipase	*Salinivibrio* sp.	Detergent formulations and fatty acid degradations	Amoozegar et al. 2008
Lipase	*Marinobacter* sp.	Hydrolysis of fish oil into free eicosapentaenoic acid	Pérez et al. 2011
Nuclease	*Micrococcus* sp.	Production of the flavoring agent 5'-guanylic acid	Kamekura et al. 1982
Protease	*Halobacterium* sp.	Peptide synthesis	Ryu et al. 1994; Kim and Dordick 1997
Protease	*Natrialba* sp.	Synthesis of tripeptide Ac-Phe-Gly-Phe-NH$_2$	Ruiz et al. 2010
Protease	*Halobacterium* sp.	Fish sauce preparation	Akolkar et al. 2010
Protease	*Geomicrobium* sp.	Peptide synthesis, detergent formulations	Karan et al. 2011
Xylanase	*Bacillus* sp.	Xylan biodegradation in pulp and paper industry	Prakash et al. 2011

There are relatively few halophilic bacteria and archaea that produce xylanases. However, these enzymes showed stability at pH 6.0–11.0, at temperatures above 60°C, and over a wide range of NaCl concentrations (Cojoc et al. 2009; Rohban et al. 2009). They can be used for paper and pulp biobleaching (Mamo et al. 2009; Setati 2010). Recently, Ren et al. (2013) purified a xylanase from *Streptomonospora* sp. YIM 90494 resistant to Ag$^+$ and sodium dodecyl sulfate (SDS).

In recent years, after the characterization of different saline environments, some strains showed cellulase and/or pullulanase activity (Sánchez-Porro et al. 2003b; Cojoc et al. 2009; Rohban et al. 2009). Cellulase activity has been isolated in various strains of the *Bacillus* genus; for example, in *Bacillus* sp. C14 the enzyme is alkaline and thermostable and shows resistance to SDS and chelating solutions (Ashabil and Burhan 2008). Cellulases are useful for the food, laundry, pulp, paper, and textile industries (Yin et al. 2015). A cellulose has recently been characterized in *Halomonas* sp. strain PS47 that is capable of growing in renewable agricultural residues (Shivanand et al. 2013). Thus, this bacterium could be exploited as an economic alternative to existing processes for cellulose production.

Until a few years ago, lipases or esterases from halophilic organisms were not investigated (Jaeger and Eggert 2002). By contrast, to date there has been a considerable amount of research on halophilic bacterial (Amoozegar et al. 2008; Pérez et al. 2011) and archaeal (Bhatnagar et al. 2005; Boutaiba et al. 2006; Muller-Santos et al. 2009; Ozcan et al. 2009; Litchfield 2011) lipases. The lipases or esterases are characterized as thermostable, with activity at a pH from 8.0 to 8.5, NaCl/KCl concentration from 3.0 to 4.0 M, and a temperature of around 50°C–60°C (Ozcan et al. 2009; Schreck and Grunden 2014). These enzymes could be used in the food and neutraceutical industries, in the production of different chemicals, and in microalgae-based biofuel systems (Schreck and Grunden 2014).

Contrary to what happens with other halophilic enzymes, proteases have been studied for years, and they have been isolated from a wide range of microorganisms (Norberg and Hofsten 1969; Izotova et al. 1983; Kamekura et al. 1996; Giménez et al. 2000; Sánchez-Porro et al. 2003b; Vidyasagar et al. 2006; Karbalaei-Heidari et al. 2009). These enzymes often are serine proteases, and they vary widely in optimal salt concentration, pH, or temperature (0–4.0 M NaCl, pH 5.0–10.0, and 40°C–75°C), and some strains are even stable in the presence of organic solvent (Setati 2010). Proteases constitute one of the most important groups of enzymes with industrial interest; in fact, they are the majority of worldwide enzyme sales. They have been widely used in industry for a long time, especially in laundry additives, washing detergent, pharmaceuticals, and the brewing, baking, tanning, and cheese industries (Setati 2010). Recent research is based on the use of proteases in waste management; for example, it has been found that a bacterium is able to grow on chicken feathers and excreted proteases (Gessesse et al. 2003).

Other halophilic hydrolases include chitinases, pectinases, inulinases, and DNases (Hatori et al. 2006; Rohban et al. 2009; Makhdoumi Kakhki et al. 2011). These enzymes, like the hydrolases described above, are thermostable, alkaliphilic, and halophilic, all properties desirable for different industrial processes.

11.4.1.2 Nitrogen Enzymes

Nitrate, nitrite, and ammonium have important implications in agricultural, environmental, and public health. With the use of chemicals such as herbicides, explosives, pesticides, and dyes, residues containing complex mixtures of salts and nitrate or nitrite are generated. The increase of these residues in soils and groundwaters in the last years has focused the research on the physiological and molecular mechanisms responsible for salt tolerance and nitrogen metabolism by microorganisms. Metabolic pathways belonging to the N cycle, such as nitrogen assimilation, nitrification, and denitrification, have been described so far mainly from bacteria. However, knowledge about those pathways from extremophilic microorganisms (mainly those grouped into the Archaea domain) is still scarce, and the potential use of the enzymes involved in such pathways has not really yet been explored. During the last few years, the halophilic archaea *Hfx. mediterranei* has been used as haloarchaea model to look for potential uses of the enzymes supporting denitrification in this species. This microorganism is able to assimilate nitrate and nitrite because of the presence of nitrate and nitrite reductases. It tolerates high nitrate or nitrite concentrations, which makes it a very interesting microorganism for wastewater or brine bioremediation applications (Martínez-Espinosa et al. 2007, 2015; Bonete et al. 2008; Nájera et al. 2012).

11.4.2 Thermophilic Enzymes

Thermophilic enzymes are useful for a large number of commercial applications because of their stability. These enzymes, besides being thermostable, are resistant to denaturation and proteolysis (Kumar and Nussinov 2001). Due to these characteristics, thermophilic enzymes are good candidates for harsh industrial processes (Demirjian et al. 2001). Performing an industrial process at elevated temperatures reduces the risk of contamination by common mesophilic organisms, improves the solubility of organic compounds, and increases reaction rates due to a decrease in viscosity and an increase of the diffusion coefficient of the substrates (Haki and Rakshit 2003). Some examples for thermophilic enzymes and their industrial applications are summarized in Table 11.4.

Table 11.4 Thermophilic Enzymes in Biotechnological Industries

Enzyme	Organism	Application	Reference
Amylase	*Bacillus* sp. WN.11	Starch hydrolysis, brewing, baking, detergents, production of maltose	Mamo and Gessese 1999
	Staphilothermus marinus		Canganella et al. 1994
	Thermus sp.		Shaw et al. 1995
Cellulase	*Bacillus subtilis*	Cellulose hydrolysis, polymer degradation in detergents	Mawadza et al. 2000
	Thermotoga neapoltana		Bok et al. 1998
Chitinase	*Bacillus* strain MH-1	Food, cosmetics, pharmaceuticals, agrochemicals	Kenji et al. 1998
DNA polymerase	*P. furiosus*	Genetic engineering/PCR	Lundberg et al. 1991
	Thermus aquaticus		Jones and Foulkes 1989
Lipase	*Bacillus* sp. RSJ-1	Dairy, oleo chemical, detergent, pulp, pharmaceuticals, cosmetics, and leather industry	Sharma et al. 2002
	Pseudomonas sp.		Rathi et al. 2000
	Pyrococcus horikoshii		Ando et al. 2002
Protease	*Bacillus brevis*	Baking, brewing, detergents, leather industry	Banerjee et al. 1999
	Thermococcus litoralis		Klingberg et al. 1991
	Pyrococcus sp. KODI		Fujiwara et al. 1996
Pullulanase	*Thermococcus litoralis*	Production of glucose syrups	Brown and Kelly 1993
Xylanase	*Bacillus circulans*	Xylan degradation in pulp and paper industry	Ashita et al. 2000

11.4.2.1 Hydrolases for Biotechnological Industries

The starch industry is one of the largest users of thermophilic enzymes, as it requires high temperatures to liquefy starch to make it accessible to hydrolysis. The cost of sugar syrup production is lower if used together with thermostable amylases, pullulanases. and α-glucosidases (Bertoldo and Antranikian 2002; Egorova and Antranikian 2005). Moreover, the amylolytic enzymes can produce other products with commercial interest as linear dextrins that can be employed as texturizers, fat substitutes, prebiotics, or aroma stabilizers (Gupta et al. 2003). Amylases have been isolated from diverse organisms for years. These enzymes can act on a wide range of temperatures (60°C–100°C) and pH values (4.0–7.8). They are useful in the food and cleaning industries for starch hydrolysis, the production of maltose, brewing, baking, and detergents (Haki and Rakshit 2003). Unlike amylases, α-glucosidases act in smaller oligosaccharides with a α-anomeric form. The best-characterized α-glucosidases are those corresponding to species of the genera *Pyrococcus* and *Thermococcus*. The combined use of these enzymes with thermoactive glucose isomerase would produce high-fructose corn syrup in one step at a higher yield, decreasing the industrial cost (Antranikian et al. 2005). The thermostable and thermoactive pullulanases described from *P. furiosus* and *T. litoralis* present a high optimal temperature (90°C–105°C) and a low optimal pH (5.5) and maintain considerable thermostabilty even in the absence of calcium ions (Brown and Kelly 1993). These microorganisms are promising candidates to starch conversion at higher temperatures and lower pH.

Commercial interest in xylanases has increased in recent years because they can improve the yield in the food, paper, and pulp industries. For example, a treatment at high temperature with xylanase disrupts the cell wall in the wood. This process facilitates the removal of lignin and decreases the use of chlorine in different states of bleaching in the pulp industry, thus enhancing the development of lower-pollution processes (Oksanen et al. 2000; Egorova and Antranikian 2005). The xylanases isolated from different organisms are active in a broad range of temperatures between 50°C and 110°C, and at pH 5.0–9.0 (Haki and Rakshit 2003; Egorova and Antranikian 2005).

The enzymes required for the complete hydrolysis of cellulose to glucose include endogluca-nases, exoglucanases, and β-glucosidases (Matsui et al. 2000). Cellulase (endoglucanase) hydro-lyzes cellulose randomly, producing cellobiose, glucose, and oligosacaccharides. This enzyme has been described in bacterial and archaeal organisms, and it is active at temperatures between 65°C and 106°C and pH 5.0–8.0 (Zverlov et al. 1998; Ando et al. 2002). On the other hand, β-glucosidases are able to hydrolyze cellobiose to glucose, which is a very useful reaction in the industry. Several β-glucosidases have been described in *Sulfolobus* and *Pyrococcus* strains (Antranikian et al. 2005; Unsworth et al. 2007). Their optimal temperatures range between 100°C and 120°C and the pH between 4.5 and 6.0. In 2002, two methods were carried out in the industry involving the use of these enzymes. One method was developed to produce novel oligosaccharides from lactose, and in the other, glucose is produced from cellobiose in a reactor with the immobilized enzyme (Schiraldi et al. 2002). In general, cellulolytic enzymes are employed in detergents (color brightening and softening), alcohol production, juices (color extraction), and the pretreatment of wastewater (Haki and Rakshit 2003).

As previously mentioned, proteases (mainly serine proteases optimally working at temperatures between 60°C and 100°C and pH 6.0–12.0) are one of the most used enzymes in the pharmaceuti-cal, food, textile, detergent, and leather industries (Haki and Rakshit 2003).

Lipases from thermophilic organisms perform different bioconversion reactions, so their use in industry is very diverse. The esters produced are used as fuel for diesel engines and in the food, cosmetic, dairy, paper, textile, leather, and pharmaceutical industries. They are able to participate in many different applications because they exhibit a wide range of temperatures and pH, ranging between 50°C and 100°C and 5.0 and 11.0, respectively (Haki and Rakshit 2003).

11.4.2.2 Thermostable DNA Polymerases

The polymerase chain reaction (PCR) was the biggest breakthrough in genetic engineering. To carry out this reaction, heat-resistant and thermostable enzymes are needed for the high tempera-tures reached during the PCR. Although DNA polymerase has been purified from many organ-isms, only few of them are currently used on a large scale in industry (biomedicine, biotechnology, pharmacy, etc.) or in research. DNA polymerases available to be used on a large scale are different in fidelity grade. For example, *Taq* polymerase (*Thermus aquaticus*) presents a 5′-3′ exonucle-ase activity, but 3′-5′ exonuclease activity was not detected; thus, it is a low-fidelity enzyme that is unable to correct misincorporated nucleotides. Unlike *Taq* polymerase, *Pfu* DNA polymerase exhibits 3′-5′ exonuclease proofreading activity; it corrects nucleotide misincorporation errors (Haki and Rakshit 2003).

The research continues, focused on finding new thermophilic enzymes with biotechnologi-cal interest, and many enzymes are still under study: chitinases, alcohol dehydrogenases, nitrile-degrading enzymes, aminoacylases, aldolases, and so forth.

11.4.3 Psychrophilic Enzymes

Most psychrophilic enzymes are cold active and thermolabile, and these characteristics are the focus of its biotechnological interest. The main advantages for the use of cold active enzymes are (1) their high catalytic activity at low temperatures and (2) the lower cost of enzyme preparation. These properties are often improved by genetic engineering, increasing their thermolability or their activity at low temperatures, and even modifying their pH profile or other biochemical properties (Feller 2013). Examples of biotechnological applications of psychrophilic enzymes are provided in Table 11.5.

Psychrophile enzymes are used mainly in the baking, textile, and food industries; in detergents; and in molecular biology. In the food industry, for example, β-galactosidase is used to reduce the

Table 11.5   Applications of Psychrophilic Enzymes in the Industry

Enzyme	Organism	Application	Reference
Polygalacturonase (pectinase)	*Sclerotinia borealis*	Cheese ripening, fruit juice and wine industry	Takasawa et al. 1997
Xylanase	*Cryptococcus adeliae*	Dough fermentation, protoplast formation, wine and juice industry	Petrescu et al. 2000
Alanine racemase	*Pseudomonas fluorescens*	Food storage, antibacterial agent	Yokoigawa et al. 2001
α-Amylase	*Alteromonas haloplanktis*	Detergents, dough fermentation, desizing denim jeans, pulp bleach	Feller et al. 1994
β-Galactosidase	*Carnobacterium piscicola* BA	Dairy industries	Coombs and Brenchley 1999
DNA ligase	*Pseudoalteromonas haloplanktis*	Molecular biology	Georlette et al. 2000
Alkaline phosphatase	*Vibrio* sp. G15-21	Molecular biology	Hauksson et al. 2000

amount of lactose that produces milk intolerance, pectinases reduce juice viscosity, and amylases and xylanases can be used to reduce dough fermentation time. In the textile industry, cellulase treatment is used for biopolishing and stone-washing processes in jeans. The application of enzymes such as amylases, proteases, lipases, or cellulases in detergent formulas is also common. The advantage provided by cold active enzymes is a cold washing that decreases the energy consumption. Finally, in molecular biology the most valuable psychrophilic DNA-modifying enzyme is alkaline phosphatase. This enzyme is used to dephosphorylate DNA vectors before cloning (to prevent self-ligation) and to remove phosphate groups at the 5′ end of DNA strands before end labeling by T4 polynucleotide kinase (Gerday et al. 2000; Cavicchioli et al. 2011).

11.4.4 Acidophilic Enzymes

Acidophilus constitutes a diverse group of organisms, including archaea, bacteria, and eukarya, that grow in acidic conditions (pH less than 4.0). Within acidophile organisms, the most desirable are the thermoacidophiles, as they have enzymes active at low pH and high temperature. These enzymes can be exploited in industrial processes under harsh conditions. Examples of biotechnological applications of acidophilic enzymes are provided in Table 11.6.

As already mentioned, amylases are one of the most important enzymes with a wide range of applications. At present, the α-amylases used in the starch industry are active at pH 6.5 and require calcium for their stability and/or activity. In starch hydrolysis, there are two critical steps that increase the process cost; on the one hand, the pH value must be adjusted between 3.2 and 3.6, and on the other hand, calcium ions must be removed. As a consequence, the whole process would be improved if acid-stable α-amylases were available. The acidic amylases described to date show optimum temperatures and pH between 60°C and 70°C and 3.0 and 11, respectively (Sharma et al.

Table 11.6   Applications of Acidophilic Enzymes in the Industry

Enzyme	Organism	Application	Reference
α-Amylase	*Alicyclobacillus acidocaldarius*	Detergents, bread, and textile industries	Matzke et al. 1997
Glucoamylase	*Thermoplasma acidophilum*	Dextrose and fructose syrups, brewing of low-calorie beer, and in the baking and alcohol industries	Serour and Antranikian 2002
Protease	*Bacillus* sp. NTAP-1	Cheese, pharmaceutical industry	Nakayama et al. 2000

2012). Acid-stable glucoamylases are distributed in all domains with widely varying temperatures and pH values, with the optimum between 70°C and 90°C and 2.0 and 6.0, respectively (Serour and Antranikian 2002).

Other acid-stable hydrolytic enzymes have been described, such as *Sulfolobus solfataricus* xylanase (95°C and pH 3.5), *A. acidocaldarius* cellulase (80°C and pH 4.0), and *Acidobacterium capsulatum* β-glucosidase (55°C and pH 3.0) (Sharma et al. 2012).

Nowadays, the focus of extremophilic enzymes is on preventing contamination of the environment. For instance, there is great interest in the enzymes susceptible for use to produce different biofuels from lignocellulosic biomass. A great advance in this area would be the use of acidic and thermostable cellulases, esterases, and alcohol dehydrogenases, which are currently not available (Galbe and Zacchi 2007).

11.4.5 Other Extremophilic Organisms

Within extremophiles, there are organisms like methanogens whose enzymes are not yet characterized as widely as others'. However, there are other studies in which the entire methanogenic organism is used in some process. For example, methanogenic microorganisms can be used in biogas production or in the decomposition of organic waste by anaerobic fermentation (Zhang et al. 2011; Zhu et al. 2011).

11.4.6 Most Used Enzymes and with More Applications in Biotechnology

The proteolytic enzymes, as already mentioned, are the most widely produced commercially from among all those used in biotechnology. Proteases and peptidases from halophilic organisms have many industrial applications because they are able to stabilize on organic solvent, so that the salt concentration is decreased, and thus the metal corrosion by salt is avoided (Kim and Dordick 1997). Other important enzymes in biotechnology are starch-hydrolyzing enzymes, finding that α-amylases from archaea and mesophile are easy to modify and optimize, and those from *Bacillus* have high thermal stability, such as *B. licheniformis*, *B. amyloliquefaciens*, *B. subtilis*, and *B. stearothermophilus*. In this case, the alignment of the gene sequences coding for α-amylase revealed that the α-amylase gene from *B. subtilis* has 41% sequence similarity and is closely related to the α-amylase gene from *P. furiosus*; similar results were obtained with *Aspergillus oryzae* (Huma et al. 2014). Industrial conversion of starch to obtain glucose units comprises two steps: (1) liquefaction of the raw starch granules, followed by (2) saccharification. The enzymes required for the production of glucose from starch are typically a mesophilic α-amylase and a fungal glucoamylase, with the addition of a pullulanase to break α-1,6 linkages during saccharification. This process starts with a first step of heating essential to facilitate dry starch liquefaction, followed by the saccharification step. Due to the lack of active enzymes at high temperatures and pH in the second stage, cooling and pH adjustment are necessary. This energy and the adjustments at this stage can be optimized by finding more suitable amylolytic enzymes (Elleuche et al. 2014). Furthermore, glycosyl hydrolases are another variety of enzymes that are responsible for the complete or partial degradation of cellulose, agar, agarose, lactose, and amylose (DasSarma et al. 2010). Among these compounds, hemicellulose is the second most abundant renewable polysaccharide found in nature. Its main component is xylan, composed of xylose units with β-1,4 glycosidic bonds. The complete breakdown of xylan requires both xylanase and xylosidase activity, the first useful for the manufacture of coffee, livestock feed, and flour (Elleuche et al. 2014). Belonging to this family are the agarases, which catalyze the hydrolysis of agar and have much application at the laboratory level and in industrial environments for liberating DNA and other molecules embedded in agarose, and extracting bioactive or medicinal compounds from algae and seaweed. They also produce neoagarosaccharides, which inhibit the growth of bacteria, slow starch degradation, and show anticancer and antioxidation activities (Giordano et al. 2006; Elleuche et al. 2014). These agarases have been obtained from several salt-tolerant microbes found in the ocean (*Pseudoalteromonads*, *Pseudomonas*, and *Vibrio*) (DasSarma et al. 2010).

In general, strategies that combine the techniques of molecular biology and bioprocess custom design could greatly improve the performance of archaea bioproducts. This would result in significant savings, and therefore allow industrial applications of these unique biomaterials (Alquéres et al. 2007).

11.5 FUTURE PROSPECTS

A great number of microorganisms containing many important enzymes with applications in industrial processes have been isolated from extremophiles. These microorganisms are good sources for obtaining specialty chemicals, agrochemicals, and pharmaceutical intermediate products, and even for performing organic synthesis, for example. First, several extremophilic enzymes were purified and characterized, but it is the knowledge of the biochemical properties of these unique enzymes that has allowed us to develop creative biotechnological applications. Evolutionary adaptation to extreme conditions has facilitated the selection of a more tolerant metabolism, enzymes, and other products with special features not found in any other prokaryotic. However, the number of biotechnological processes in which these enzymes may have their essential function still remains limited, and knowledge of the physiology, metabolism, enzymology, and genetics of this fascinating group of microorganisms is still scarce. Nevertheless, the growing number of extremophile genomes available today will be helpful for the discovery and identification of useful novel enzymes. This will result in potential enzymes with biotechnological interest, which will be used in faster, more accurate, specific novel biocatalytic processes and according to the environment.

Molecular biology techniques combined with protein engineering will support large-scale extremozyme production, as well as the design of new biocatalysts to perform innovative biotransformations. Furthermore, the production of biocatalysts can be optimized by increasing the biomass production of extremophiles. Alternatively, the gene encoding the biocatalyst can be cloned and expressed in a suitable host. The great challenge of science will be to incorporate new bioprocesses to improve biotechnological applications designated for archaeal enzymes and biomolecules. Enzymes offer a green solution for an industrialized world in the middle of growing environmental concerns. The continued growth of the market for industrial enzymes will depend on technological innovation, the identification and characterization of novel enzymes from natural sources, the modification of these enzymes for optimal performance in selected applications, and the high expression of the enzymes.

REFERENCES

Adams MW, FB Perler, and RM Kelly. 1995. Extremozymes: Expanding the limits of biocatalysis. *Biotechnology (NY)* 13(7):662–668.

Adrio JL and AL Demain. 2014. Microbial Enzymes: Tools for Biotechnological Processes. *Biomolecules* 4(1):117–139.

Akolkar AV, D Durai, and AJ Desai. 2010. *Halobacterium* sp. SP1 (1) as a starter culture for accelerating fish sauce fermentation. *Journal of Applied Microbiology* 109:44–53.

Allers T. 2010. Overexpression and purification of halophilic proteins in *Haloferax volcanii*. *Bioengineered Bugs* 1(4):288–290.

Alquéres SMC, RV Almeida, MM Clementino, RP Vieira, WI Almeida, AM Cardoso, and OB Martins. 2007. Exploring the biotechnological applications in the archaeal domain. *Brazilian Journal of Microbiology* 38(3):398–405.

Amoozegar MA, E Salehghamari, K Khajeh, M Kabiri, and S Naddaf. 2008. Production of an extracellular thermohalophilic lipase from a moderately halophilic bacterium, *Salinivibrio* sp. strain SA-2. *Journal of Basic Microbiology* 48:160–167.

Ando S, H Ishida, Y Kosugi, and K Ishikawa. 2002. Hyperthermostable endoglucanase from *Pyrococcus horikoshi*. *Applied and Environmental Microbiology* 68:430–433.

Antranikian G, C Vorgias, and C Bertoldo. 2005. Extreme environments as a resource for microorganisms and novel biocatalysts. *Advances in Biochemical Engineering/Biotechnology* 96:219–262.

Ashabil A and A Burhan. 2008. A new moderately halo-alkaliphilic, thermostable endoglucanase from moderately halophilic *Bacillus* sp. C14 isolated from Van Soda Lake. *International Journal of Agriculture and Biology* 10:369–374.

Ashita D, J Gupta, and S Khanna. 2000. Enhanced production, purification and characterization of novel cellulase-poor thermostable, alkali tolerant xylanase from *Bacillus circulans* AB 16. *Process Biochemistry* 35:849–856.

Banerjee V, K Saani, W Azmi, and R Soni. 1999. Thermostable alkaline protease from *Bacillus brevis* and its characterization as a laundry additive. *Process Biochemistry* 35:213–219.

Bauer MW, LE Driskill, and RM Kelly. 1998. Glycosyl hydrolases from hyperthermophilic microorganisms. *Current Opinion in Biotechnology* 9(2):141–145.

Bautista V, J Esclapez, F Pérez-Pomares, RM Martínez-Espinosa, M Camacho, and MJ Bonete. 2012. Cyclodextrin glycosyltransferase: A key enzyme in the assimilation of starch by the halophilic archaeon *Haloferax mediterranei*. *Extremophiles* 16(1):147–159.

Bertoldo C and G Antranikian. 2002. Starch-hydrolyzing enzymes from thermophilic archaea and bacteria. *Current Opinion in Chemical Biology* 6(2):151–160.

Bhatnagar T, S Boutaiba, H Hacene, JL Cayol, ML Fardeau, B Ollivier, and JC Baratti. 2005. Lipolytic activity from halobacteria: Screening and hydrolase production. *FEMS Microbiology Letters* 248:133–140.

Bok J, A Dienesh, D Yernool, and D Eveleigh. 1998. Purification, characterization and molecular analysis of thermostable cellulases *CelA* and *CelB* from *Thermotoga neapolitana*. *Applied and Environmental Microbiology* 64:4774–4781.

Bonete MJ, J Ferrer, M Camacho, C Pire, J Esclapez, F Marhuenda-Egea, R Martínez-Espinosa, S Díaz, F Llorca, F Pérez-Pomares, and V Bautista. 2005. Enzymes from halophilic archaea. In *Microorganisms for Industrial Enzymes and Biocontrol*, ed. E Mellado Durán and JL Barredo, 1–24. Kerala, India: Research Signpost.

Bonete MJ, and RM Martínez-Espinosa. 2011. Enzymes from halophilic archaea: Open questions. In *Halophiles and Hypersaline Environments*, ed. A Ventosa, A Oren, and Y Ma, 359–371. Berlin: Springer-Verlag.

Bonete MJ, RM Martínez-Espinosa, C Pire, B Zafrilla, and DJ Richardson. 2008. Nitrogen metabolism in haloarchaea. *Saline Systems* 4:9.

Boutaiba S, T Bhatnagar, H Hacene, DA Mitchell, and JC Baratti. 2006. Preliminary characterization of a lipolytic activity from an extremely halophilic archaeon, *Natronococcus* sp. *Journal of Molecular Catalysis B: Enzymatic* 41:21–26.

Britton KL, PJ Baker, M Fisher, S Ruzheinikov, DJ Gilmour, MJ Bonete, J Ferrer, C Pire, J Esclapez, and DW Rice. 2006. Analysis of protein solvent interactions in glucose dehydrogenase from the extreme halophile *Haloferax mediterranei*. *Proceedings of the National Academy of Sciences of the United States of America* 103:4846–4851.

Brown S and R Kelly. 1993. Characterization of amylolytic enzymes having both α-1,4 and α-1,6 hydrolytic activity from the thermophilic archaea *Pyrococcus furiosus* and *Thermococcus litoralis*. *Applied and Environmental Microbiology* 59:2614–2621.

Camacho M, A Rodríguez-Arnedo, and MJ Bonete. 2002. NADP-dependent isocitrate dehydrogenase from the halophilic archaeon *Haloferx volcanii*: Cloning, sequence determination and overexpression in *Escherichia coli*. *FEMS Microbiology Letters* 209(2):155–160.

Canganella F, C Andrade, and G Antranikian. 1994. Characterization of amylolytic and pullulytic enzymes from thermophilic archaea and from a new *Ferividobacterium* species. *Applied Microbiology and Biotechnology* 42:239–245.

Cavicchioli R, T Charlton, H Ertan, S Mohd Omar, KS Siddiqui, and TJ Williams. 2011. Biotechnological uses of enzymes from psychrophiles. *Microbial Biotechnology* 4(4):449–460.

Chakraborty S, A Khopade, C Kokare, K Mahadik, and B Chopade. 2009. Isolation and characterization of novel α-amylase from marine *Streptomyces* sp. D1. *Journal of Molecular Catalysis B: Enzymatic* 58:17–23.

Choo DW, T Kurihara, T Suzuki, K Soda, and N Esaki. 1998. A cold-adapted lipase of an Alaskan psychrotroph, *Pseudomonas* sp. strain B11-1: Gene cloning and enzyme purification and characterization. *Applied and Environmental Microbiology* 64(2):486–491.

Cimini D, M De Rosa, A Panariello, V Morelli, and C Schiraldi. 2008. Production of a thermophilic maltooligosyl-trehalose synthase in *Lactococcus lactis*. *Journal of Industrial Microbiology and Biotechnology* 35(10):1079–1083.

Cojoc R, S Merciu, G Popescu, L Dumitru, M Kamekura, and M Enache. 2009. Extracellular hydrolytic enzymes of halophilic bacteria isolated from a subterranean rock salt crystal. *Romanian Biotechnological Letters* 14:4658–4664.

Coombs JM and JE Brenchley. 1999. Biochemical and phylogenetic analyses of a cold-active β-galactosidase from the lactic acid bacterium *Carnobacterium piscicola* BA. *Applied and Environmental Microbiology* 65:5443–5450.

Dassarma P, JA Coker, V Huse, and S Dassarma. 2010. Halophiles, industrial applications. In *Encyclopedia of Industrial Biotechnology: Bioprocess, Bioseparation, and Cell Technology*, ed. MC Flickinger, 1–9. Hoboken, NJ: John Wiley & Sons.

Delgado-García M, B Valdivia-Urdiales, C Aguilar-González, J Contreras-Esquivel, and R Rodríguez-Herrera. 2012. Halophilic hydrolases as a new tool for the biotechnological industries. *Journal of the Science of Food and Agriculture* 92:2575–2580.

DeLong EF and NR Pace. 2001. Environmental diversity of bacteria and archaea. *Systematic Biology* 50(4):470–478.

Demirjian D, F Moris-Varas, and C Cassidy. 2001. Enzymes from extremophiles. *Current Opinion in Chemical Biology* 5:144–151.

Díaz S, F Pérez-Pomares, C Pire, J Ferrer, and MJ Bonete. 2006. Gene cloning, heterologous overexpression and optimized refolding of the NAD-glutamate dehydrogenase from *Haloferax mediterranei*. *Extremophiles* 10(2):105–115.

Domingueza A, L Pastrana, MA Longo, ML Rua, and MA Sanroman. 2005. Lipolytic enzyme production by *Thermus thermophilus* HB27 in a stirred tank bioreactor. *Biochemical Engineering Journal* 26:95–99.

Egorova K and G Antranikian. 2005. Industrial relevance of thermophilic Archaea. *Current Opinion in Microbiology* 86(2):649–655.

Eichler J. 2001. Biotechnological uses of archaeal extremozymes. *Biotechnology Advances* 19(4):261–278.

Elleuche S, C Schröder, K Sahm, and G Antranikian. 2014. Extremozymes—Biocatalysts with unique properties from extremophilic microorganisms. *Current Opinion in Microbiology* 29:116–123.

Enache M and M Kamekura. 2010. Hydrolytic enzymes of halophilic microorganisms and their economic values. *Romanian Biotechnology Letters* 47:47–59.

Esclapez J, C Pire, V Bautista, RM Martínez-Espinosa, J Ferrer, and MJ Bonete. 2007. Analysis of acidic surface of *Haloferax mediterranei* glucose dehydrogenase by site-directed mutagenesis. *FEBS Letters* 581(5):837–842.

Esclapez J, B Zafrilla, RM Martínez-Espinosa, and MJ Bonete. 2013. Cu-NirK from *Haloferax mediterranei* as an example of metalloprotein maturation and exportation via Tat system. *Biochimica and Biophysica Acta* 1834(6):1003–1009.

Essghaier B, A Hedi, M Bejji, H Jijakli, A Boudabous, and N Sadfi-Zouaoui. 2011. Characterization of a novel chitinase from a moderately halophilic bacterium, *Virgibacillus marismortui* strain M3-23. *Annals of Microbiology* 62(2):835–841.

Feller G. 2013. Psychrophilic enzymes: From folding to function and biotechnology. *Scientifica* 2013:512840.

Feller G, F Payan, F Theys, M Qian, R Haser, and C Gerday. 1994. Stability and structural analysis of α-amylase from the Antarctic psychrophile *Alteromonas haloplanctis* A23. *European Journal of Biochemistry* 222:441–447.

Fernández-Castillo RF, F Rodriguez-Valera, J González-Ramos, and F Ruiz-Berraquero. 1986. Accumulation of poly(beta-hydroxybutyrate) by halobacteria. *Applied and Environmental Microbiology* 51(1):214–216.

Flores LM, AI Zavaleta, Y Zambrano, L Cervantes, and V Izaguirre. 2010. Halophilic bacteria producers of hydrolases with interest in the biotechnology. *Ciencia e Investigación* 13:42–46.

Fujiwara S, S Okuyama, and T Imanaka. 1996. The world of archaea: Genome analysis, evolution and thermostable enzymes. *Gene* 179:165–170.

Fukishima T, T Mizuki, A Echigo, A Inoue, and R Usami. 2005. Organic solvent tolerance of halophilic α-amylase from a haloarchaeon, *Haloarcula* sp. strain S-1. *Extremophiles* 9:321–331.

Galbe M and G Zacchi. 2007. Pretreatment of lignocellulosic materials for efficient bioethanol production. *Advances in Biochemical Engineering/Biotechnology* 108:41–65.

Georlette D, ZO Jonsson, FV Petegem, JP Chessa, JV Beeumen, U Hubscher, and C Gerday. 2000. A DNA ligase from the psychrophile *Pseudoalteromonas haloplanktis* gives insight into the adaptation of proteins to low temperature. *European Journal of Biochemistry* 267:3502–3512.

Gerday C, M Aittaleb, M Bentahir, JP Chessa, P Claverie, T Collins, S D'Amico et al. 2000. Cold-adapted enzymes: From fundamentals to biotechnology. *Trends in Biotechnology* 18(3):103–107.

Gessesse A, R Hatti-Kaul, BA Gashe, and B Mattiasson. 2003. Novel alkaline proteases from alkaliphilic bacteria grown on chicken feather. *Enzyme and Microbial Technology* 32:519–524.

Giménez MI, CA Studdert, JJ Sánchez, and RE de Castro. 2000. Extracellular protease of *Natrialba magadii*: Purification and biochemical characterization. *Extremophiles* 4:181–188.

Giordano A, G Andreotti, A Tramice, and A Trincone. 2006. Marine glycosyl hydrolases in the hydrolysis and synthesis of oligosaccharides. *Biotechnology Journal* 1(5):511–530.

Giuliano M, C Schiraldi, MR Marotta, J Hugenholtz, and M De Rosa. 2004. Expression of *Sulfolobus solfataricus* alpha-glucosidase in *Lactococcus lactis*. *Applied Microbiology and Biotechnology* 64(6):829–832.

Gupta R, P Gigras, H Mohapatra, V Goswami, and B Chauhan. 2003. Microbial alpha-amylases: A biotechnological perspective. *Process Biochemistry* 38:1599–1616.

Haki GD and SK Rakshit. 2003. Development in industrially important thermostable enzymes: A review. *Bioresource Technology* 89(1):17–34.

Hatori Y, M Sato, K Orishimo, R Yatsunami, K Endo, T Fukui, and S Nakamura. 2006. Characterization of recombinant family 18 chitinase from extremely halophilic archaeon *Halobacterium salinarum* strain NRC-1. *Chitin Chitosan Research* 12:201–210.

Hauksson JB, OS Andresson, and B Asgeirsson. 2000. Heat-labile bacterial alkaline phosphatase from a marine *Vibrio* sp. *Enzyme and Microbial Technology* 27:66–73.

Hezayen FF, BHA Rehm, R Eberhardt, and A Steinbüchel. 2000. Polymer production by two newly isolated extremely halophilic archaea: Application of a novel corrosion-resistant bioreactor. *Applied Microbiology and Biotechnology* 54(3):319–325.

Holst O, A Manelius, M Krahe, H Märlk, N Raven, and R Sharp. 1997. Thermophiles and fermentation technology. *Comparative Biochemistry and Physiology* 118A(3):415–422.

Hough DW and MJ Danson. 1999. Extremozymes. *Current Opinion in Chemical Biology* 3(1):39–46.

Huma T, A Maryam, S Rehman, MT Qamar, T Shaheen, A Haque, and B Shaheen. 2014. Phylogenetic and comparative sequence analysis of thermostable alpha amylases of kingdom archaea, prokaryotes and eukaryotes. *Bioinformation* 10(7):443–448.

Ikeda M and DS Clark. 1998. Molecular cloning of extremely thermostable esterase gene from hyperthermophilic archaeon *Pyrococcus furiosus* in *Escherichia coli*. *Biotechnology and Bioengineering* 57(5):624–629.

Izotova LS, AY Strongin, LN Chekulaeva, VE Sterkin, VI Ostoslavskaya, LA Lyublinskaya, EA Timokhina, and VM Stephanov. 1983. Purification and properties of serine protease from *Halobacterium halobium*. *Journal of Bacteriology* 155:826–830.

Jaeger KE and T Eggert. 2002. Lipases for biotechnology. *Current Opinion in Biotechnology* 13:390–397.

Jolley KA, E Rapaport, DW Hough, MJ Danson, WG Woods, and ML Dyall-Smith. 1996. Dihydrolipoamide dehydrogenase from the halophilic archaeon *Haloferax volcanii*: Homologous overexpression of the cloned gene. *Journal of Bacteriology* 178(11):3044–3048.

Jones M and N Foulkes. 1989. Reverse transcription of mRNA by *Termus aquaticus* DNA polymerase. *Nucleic Acids Research* 17:8387–8388.

Kamekura M, T Hamakawa, and H Onishi. 1982. Application of halophilic nuclease H of *Micrococcus varians* subsp. *halophilus* to commercial production of flavoring agent 5′-GMP. *Applied and Environmental Microbiology* 44:994–995.

Kamekura M, Y Seno, and M Dyall-Smith. 1996. Halolysin R4, a serine proteinase from the halophilic archaeon *Haloferax mediterranei*; gene cloning, expression and structural studies. *Biochimica et Biophysica Acta* 1294:159–167.

Karan R, SP Singh, S Kapoor, and SK Khare. 2011. A novel organic solvent tolerant protease from a newly isolated *Geomicrobium* sp. EMB2 (MTCC 10310): Production optimization by response surface methodology. *New Biotechnology* 2:136–145.

Karbalaei-Heidari HR, MA Amoozegar, and AA Ziaee. 2009. Production, optimization and purification of a novel extracellular protease from the moderately halophilic bacterium. *Journal of Industrial Microbiology and Biotechnology* 36:21–27.

Kenji S, Y Akira, K Hajime, W Mamoru, and M Mitsuaki. 1998. Purification and characterization of three thermostable endochitinases of a noble *Bacillus* strain, MH-1, isolated from chitin containing compost. *Applied and Environmental Microbiology* 64:3397–3402.

Kim J and JS Dordick. 1997. Unusual salt and solvent dependence of a protease from an extreme halophile. *Biotechnology and Bioengineering* 55(3):471–479.

Kim MH, HK Kim, JK Lee, SY Park, and TK Oh. 2000. Thermostable lipase of *Bacillus stearothermophilus*: High-level production, purification, and calcium-dependent thermostability. *Bioscience, Biotechnology and Biochemistry* 64(2):280–286.

Klibanov AM. 2001. Improving enzymes by using them in organic solvents. *Nature* 409(6817):241–246.

Klingberg M, F Hashwa, and G Antrakikian. 1991. Properties of extremely thermostable proteases from anaerobic hyperthermophilic bacteria. *Applied Microbiology and Biotechnology* 34:715–719.

Krahe M, G Antranikian, and H Märkl. 1996. Fermentation of extremophilic microorganisms. *FEMS Microbiology Reviews* 18(2–3):271–285.

Kube J, C Brokamp, R Machielsen, J van der Oost, and H Markl. 2006. Influence of temperature on the production of an archaeal thermoactive alcohol dehydrogenase from *Pyrococcus furiosus* with recombinant *Escherichia coli*. *Extremophiles* 10:221–227.

Kumar S and R Nussinov. 2001. How do thermophilic proteins deal with heat? A review. *Cellular and Molecular Life Sciences* 58:1216–1233.

Li AX, LZ Guo, and WD Lu. 2011. Alkaline inulinase production by a newly isolated bacterium *Marinimicrobium* sp. LS-A18 and inulin hydrolysis by the enzyme. *World Journal of Microbiology and Biotechnology* 28(1):81–89.

Liliensiek AK, J Cassidy, G Gucciardo, C Whitely, and F Paradisi. 2013. Heterologous overexpression, purification and characterisation of an alcohol dehydrogenase (ADH2) from *Halobacterium* sp. NRC-1. *Molecular Biotechnology* 55(2):143–149.

Litchfield CD. 2011. Potential for industrial products from the halophilic Archaea. *Journal of Industrial Microbiology and Biotechnology* 38(10):1635–1647.

Lundberg K, D Shoemaker, M Adams, J Short, J Sorge, and E Marthur. 1991. High-fidelity amplification using a thermostable polymerase isolated from *Pyrococcus furiosus*. *Gene* 108:1–6.

Makhdoumi Kakhki A, MA Amoozegar, and E Mahmodi Khaledi. 2011. Diversity of hydrolytic enzymes in haloarchaeal strains isolated from salt lake. *International Journal of Environmental Science and Technology* 8:705–714.

Mamo G and A Gessese. 1999. A highly thermostable amylase from a newly isolated thermophilic *Bacillus* sp. WN11. *Journal of Applied Microbiology* 86:557–560.

Mamo G, M Thunnissen, R Hatti-Kaul, and B Mattiasson. 2009. An alkaline active xylanase: Insights into mechanisms of high pH catalytic adaptation. *Biochimie* 91:1187–1196.

Manco G, E Adinolfi, FM Pisani, G Ottolina, G Carrea, and M Rossi. 1998. Overexpression and properties of a new thermophilic and thermostable esterase from *Bacillus acidocaldarius* with sequence similarity to hormone-sensitive lipase subfamily. *Biochemical Journal* 332(Pt 1):203–212.

Marhuenda-Egea F and MJ Bonete. 2002. Extreme halophilic enzymes in organic solvents. *Current Opinion in Biotechnology* 13(4):385–389.

Martínez-Espinosa RM, EJ Dridge, MJ Bonete, JN Butt, CS Butler, F Sargent, and DJ Richardson. 2007. Look on the positive side! The orientation, identification and bioenergetics of 'archaeal' membrane-bound nitrate reductases. *FEMS Microbiology Letters* 276(2):129–139.

Martínez-Espinosa RM, DJ Richardson, and MJ Bonete. 2015. Characterisation of chlorate reduction in the haloarchaeon *Haloferax mediterranei*. *Biochimica et Biophysica Acta* 1850(4):587–594.

Matsui I, Y Sakai, E Matsui, H Kikuchi, Y Kawarabayasi, and K Honda. 2000. Novel substrate specificity of a membrane-bound β-glycosidase from the hyperthermophilic archaeon *Pyrococcus horikoshi*. *FEBS Letters* 467:195–200.

Matsumoto M, H Yokouchi, N Suzuki, H Ohata, and T Matsunaga. 2003. Saccharification of marine microalgae using marine bacteria for ethanol production. *Applied Biochemistry and Biotechnology* 108:247–254.

Matzke J, B Schwermann, and EP Baker. 1997. Acidostable and acidophilic proteins: The example of the α-amylase from *Alicyclobacillus acidocaldarius*. *Comparative Biochemistry and Physiology* 118A:475–479.

Mawadza C, R Hatti-Kaul, R Zvauya, and B Mattiasson. 2000. Purification and characterization of cellulases produced by two *Bacillus* strains. *Journal of Biotechnology* 83:177–187.

Mellado E and A Ventosa. 2003. Biotechnological potential of moderately and extremely halophilic microorganisms. In *Microorganisms for Health Care, Food and Enzyme Production*, ed. JL Barredo, 233–256. Kerala, India: Research Signpost.

Muller-Santos M, EM de Souza, F Pedrosa, DA Mitchell, S Longhi, F Carriere, S Canaan, and N Krieger. 2009. First evidence for the salt dependent folding and activity of an esterase from the halophilic archaea *Haloarcula marismortui*. *Biochimica et Biophysica Acta* 1791(8):719–729.

Munawar N and PC Engel. 2012. Overexpression in a non-native halophilic host and biotechnological potential of NAD+-dependent glutamate dehydrogenase from *Halobacterium salinarum* strain NRC-36014. *Extremophiles* 16(3):463–476.

Nájera-Fernández C, B Zafrilla, MJ Bonete, and RM Martínez-Espinosa. 2012. Role of the denitrifying haloarchaea in the treatment of nitrite-brines. *International Microbiology* 15(3):111–119.

Nakayama T, N Tsuruoka, M Akai, and T Nishino. 2000. Thermostable collagenolytic activity of a novel thermophilic isolate, *Bacillus* sp. strain NTAP-1. *Journal of Bioscience and Bioengineering* 89(6):612–614.

Niehaus F, A Peters, T Groudieva, and G Antranikian. 2000. Cloning, expression and biochemical characterisation of a unique thermostable pullulan-hydrolysing enzyme from the hyperthermophilic archaeon *Thermococcus aggregans*. *FEMS Microbiology Letters* 190(2):223–229.

Norberg P and BV Hofsten. 1969. Proteolytic enzymes from extremely halophilic bacteria. *Journal of General Microbiology* 55:251–256.

Oesterhelt D, C Brauchle, and N Hampp. 1991. Bacteriorhodopsin: A biological material for information processing. *Quarterly Reviews of Biophysics* 24(4):425–478.

Oksanen T, J Pere, L Paavilainen, J Buchert, and L Viikari. 2000. Treatment of recycled kraft pulps with *Trichoderma reesei* hemicellulases and cellulases. *Journal of Biotechnology* 78:39–48.

Oren A. 2010. Industrial and environmental applications of halophilic microorganisms. *Environmental Technology* 31:825–834.

Ozcan B, G Ozylmaz, C Cokmus, and M Caliskan. 2009. Characterization of extracellular esterases and lipases activities from five halophilic archaeal strains. *Journal of Industrial Microbiology and Biotechnology* 36:105–110.

Paiardini A, G Gianese, F Bossa, and S Pascarella. 2003. Structural plasticity of thermophilic serine hydroxymethyltransferases. *Proteins* 50(1):122–134.

Patel GB and GD Sprott. 1999. Archaeobacterial ether lipid liposomes (archaeosomes) as novel vaccine and drug delivery systems. *Critical Reviews in Biotechnology* 19(4):317–357.

Pérez D, S Martín, G Fernández-Lorente, M Filice, JM Guisán, A Ventosa, MT García, and E Mellado. 2011. A novel halophilic lipase, LipBL, showing high efficiency in the production of eicosapentaenoic acid (EPA). *PLoS One* 6:e23325.

Pérez-Pomares F, V Bautista, J Ferrer, C Pire, FC Marhuenda-Egea, and MJ Bonete. 2003. Alpha-amylase activity from the halophilic archaeon *Haloferax mediterranei*. *Extremophiles* 7(4):299–306.

Petrescu I, J Lamotte-Brasseur, JP Chessa, P Ntarima, M Claeyssens, B Devreese, G Marino, and C Gerday. 2000. Xylanase from the psychrophilic yeast *Cryptococcus adeliae*. *Extremophiles* 4:137–144.

Pire C, J Esclapez, J Ferrer, and MJ Bonete. 2001. Heterologous expression of glucose dehydrogenase from the halophilic archaeon *Haloferax mediterranei*, an enzyme of the medium chain dehydrogenase family. *FEMS Microbiology Letters* 200(2):221–227.

Prakash P, SK Jayalakshmi, B Prakash, M Rubul, and K Sreeramulu. 2011. Production of alkaliphilic, halotolerent, thermostable cellulase free xylanase by *Bacillus halodurans* PPKS-2 using agro waste: Single step purification and characterization. *World Journal of Microbiology and Biotechnology* 28(1):183–192.

Rathi P, B Sapna, R Sexena, and R Gupta. 2000. A hyperthermostable, alkaline lipase from *Pseudomonas* sp. with the property of thermal activation. *Biotechnology Letters* 22:495–498.

Reeve JN, J Nölling, RM Morgan, and DR Smith. 1997. Methanogenesis: Genes, genomes, and who's on first? *Journal of Bacteriology* 179(19):5975–5986.

Ren W, Z Feng, X Yang, S Tang, H Ming, E Zhou, Y Yin, Y Zhang, and W Li. 2013. Purification and properties of a SDS-resistant xylanase from halophilic *Streptomonospora* sp. YIM 90494. *Cellulose* 20(4):1947–1955.

Rinker KD, CJ Han, and RM Kelly. 1999. Continuous culture as a tool for investigating the growth physiology of heterotrophic hyperthermophiles and extreme thermoacidophiles. *Journal of Applied Microbiology* 85:118S–127S.

Rinker KD and RM Kelly. 2000. Effect of carbon and nitrogen sources on growth dynamics and exopolysaccharide production for the hyperthermophilic archaeon *Thermococcus litoralis* and bacterium *Thermotoga marítima. Biotechnology and Bioengineering* 69(5):537–547.

Robb FT and DS Clark. 1999. Adaptation of proteins from hyperthermophiles to high pressure and high temperature. *Journal of Molecular Microbiology and Biotechnology* 1(1):101–105.

Rodriguez-Valera F. 1992. Biotechnological potential of halobacteria. *Biochemical Society Symposia* 58:135–147.

Rohban R, MA Amoozegar, and A Ventosa. 2009. Screening and isolation of halophilic bacteria producing extracellular hydrolases from Howz Soltan Lake, Iran. *Journal of Industrial Microbiology and Biotechnology* 36:333–340.

Rosell CM, M Haros, C Escrivá, and CB de Barber. 2001. Experimental approach to optimize the use of α-amylases in bread making. *Journal of Agricultural and Food Chemistry* 49:2973–2977.

Ruiz DM, NB Iannuci, O Cascone, and RE De Castro. 2010. Peptide synthesis catalysed by a haloalkaliphilic serine protease from the archaeon *Natrialba magadii* (Nep). *Letters in Applied Microbiology* 51:691–696.

Ryu K, J Kim, and JS Dordick. 1994. Catalytic properties and potential of an extracellular protease from an extreme halophile. *Enzyme and Microbial Technology* 16:266–275.

Sánchez-Porro C, S Martín, E Mellado, and A Ventosa. 2003a. Diversity of moderately halophilic bacteria producing extracellular hydrolytic enzymes. *Journal of Applied Microbiology* 94:295–300.

Sánchez-Porro C, E Mellado, C Bertoldo, G Antranikian, and A Ventosa. 2003b. Screening and characterization of the protease CP1 produced by the moderately halophilic bacterium *Pseudoalteromonas* sp. strain CP76. *Extremophiles* 7:221–228.

Schiraldi C, M Acone, M Giuliano, M Cartenì, and M De Rosa. 2001. Innovative fermentation strategies for the production of extremophilic enzymes. *Extremophiles* 5(3):193–198.

Schiraldi C and M De Rosa. 2002. The production of biocatalysts and biomolecules from extremophiles. *Trends in Biotechnology* 20(12):515–521.

Schiraldi C, M Giuliano, and M De Rosa. 2002. Perspectives on biotechnological applications of archaea. *Archaea* 1(2):75–86.

Schiraldi C, F Marulli, I Di Lernia, A Martino, and M De Rosa. 1999. A microfiltration bioreactor to achieve high cell density in *Sulfolobus solfataricus* fermentation. *Extremophiles* 3:199–204.

Schreck SD and AM Grunden. 2014. Biotechnological applications of halophilic lipases and thioesterases. *Applied and Microbiology Biotechnology* 98(3):1011–1021.

Serour E and G Antranikian. 2002. Novel thermoactive glucoamylase from the thermoacidophilic archaea *Thermoplasma acidophilum, Picrophilustorridus* and *Picrophilus oshimae. Antonie Van Leeuwenhoek* 81(1–4):73–83.

Setati EM. 2010. Diversity and industrial potential of hydrolase-producing halophilic/halotolerant eubacteria. *African Journal of Biotechnology* 9:1555–1560.

Sharma A, Y Kawarabayasi, and T Satyanarayana. 2012. Acidophilic bacteria and archaea: Acid stable biocatalysts and their potential applications. *Extremophiles* 16(1):1–19.

Sharma R, S Soni, R Vohra, L Gupta, and J Gupta. 2002. Purification and characterization of a thermostable alkaline lipase from a new thermophilic *Bacillus* sp. RSJ-1. *Process Biochemistry* 37:1075–1084.

Shaw J, L Pang, S Chen, and I Chen. 1995. Purification and properties of an extracellular α-amylase from *Thermus* sp. *Botanical Bulletin of Academia Sinica* 36:195–200.

Shivanand P, G Mugerava, and A Kumar. 2013. Utilization of renewable agricultural residues for the production of extracellular halostable cellulase from newly isolated *Halomonas* sp. strain PS47. *Annals of Microbiology* 63(4):1257–1263.

Sleytr UB, D Pum, and M Sára. 1997. Advances in S-layer nanotechnology and biomimetics. *Advances in Biophysics* 34:71–79.

Takasawa T, K Sagisaka, K Yagi, K Uchiyama, A Aoki, K Takaoka, and K Yamamato. 1997. Polygalacturonase isolated from the culture of the psychrophilic fungus *Sclerotinia borealis*. *Canadian Journal of Microbiology* 43:417–424.

Toogood HS, EJ Hollingsworth, RC Brown, IN Taylor, SJ Taylor, R McCague, and JA Littlechild. 2002. A thermostable L-aminoacylase from *Thermococcus litoralis*: Cloning, overexpression, characterization and applications in biotransformations. *Extremophiles* 6:111–122.

Turner NJ. 2003. Directed evolution of enzymes for applied biocatalysis. *Trends in Biotechnology* 21(11):474–478.

Unsworth LD, J van der Oost, and S Koutsopoulos. 2007. Hyperthermophilic enzymes: Stability, activity and implementation strategies for high temperature applications. *FEBS Journal* 274(16):4044–4056.

van den Burg B. 2003. Extremophiles as a source for novel enzymes. *Current Opinion in Microbiology* 6(3):213–218.

Ventosa A, C Sánchez-Porro, S Martín, and E Mellado. Halophilic archaea and bacteria as a source of extracellular hydrolytic enzymes. In *Adaptation to Life at High Salt Concentrations in Archaea, Bacteria and Eukarya*, ed. N Gunde-Cimerman, A Oren, and A Plemenitas, 337–350. Dordrecht: Springer, 2005.

Verhagen MF, T O'Rourke, and MW Adams. 1999. The hyperthermophilic bacterium, *Thermotoga maritima*, contains an unusually complex iron-hydrogenase: Amino acid sequence analyses versus biochemical characterization. *Biochimica et Biophysica Acta* 1412(3):212–229.

Vetriani C, DL Maeder, N Tolliday, KSP Yip, TJ Stillman, KL Britton, DW Rice, HH Klump, and FT Robb. 1998. Protein thermostability above 100°C: A key role for ionic interactions. *Proceedings of the National Academy of Sciences of the United States of America* 95(21):12300–12305.

Vidyasagar M, S Prakash, CD Litchfield CD, and K Sreeramulu. 2006. Purification and characterization of a thermostable, haloalkaliphilic extracellular serine protease from the extreme halophilic archaeon *Halogeometricum borinquense* strain TSS101. *Archaea* 2:51–57.

Woese C, O Kandler, and M Wheelis. 1990. Towards a natural system of organisms: Proposal for the domains Archaea, Bacteria and Eucarya. *Proceedings of the National Academy of Sciences of the United States of America* 87(12):4576–4579.

Woese CR and GE Fox. 1977. Phylogenetic structure of the prokaryotic domain: The primary kingdoms. *Proceedings of the National Academy of Sciences of the United States of America* 74(11):5088–5090.

Worthington P, P Blum, F Perez-Pomares, and T Elthon. 2003. Large-scale cultivation of acidophilic hyperthermophiles for recovery of secreted proteins. *Applied and Environmental Microbiology* 69(1):252–257.

Yin J, JC Chen, Q Wu, and GQ Chen. 2015. Halophiles, coming stars for industrial biotechnology. *Biotechnology Advances* 33(7):1433–1442.

Yokoigawa K, Y Okubo, H Kawai, N Esaki, and K Soda. 2001. Structure and function of psychrophilic alanine racemase. *Journal of Molecular Catalysis B: Enzymatic* 12:27–35.

Yoshimune K, Y Shirakihara, M Wakayama, and I Yumoto. 2010. Crystal structure of salt-tolerant glutaminase from *Micrococcus luteus* K-3 in the presence and absence of its product L-glutamate and its activator Tris. *FEBS Journal* 3:738–748.

Zafrilla B, RM Martínez-Espinosa, J Esclapez, F Pérez-Pomares, and MJ Bonete. 2010. SufS protein from *Haloferax volcanii* involved in Fe-S cluster assembly in haloarchaea. *Biochimica et Biophysica Acta* 1804(7):1476–1482.

Zhang Y, EM Zamudio Cañas, Z Zhu, JL Linville, S Chen, and Q He. 2011. Robustness of archaeal populations in anaerobic co-digestion of dairy and poultry wastes. *Bioresource Technology* 102(2):779–785.

Zhao H, K Chockalingam, and Z Chen. 2002. Directed evolution of enzymes and pathways for industrial biocatalysis. *Current Opinion in Biotechnology* 13(2):104–110.

Zhao YX, ZM Rao, YF Xue, P Gong, YZ Ji, and YH Ma. 2015. Poly(3-hydroxybutyrate-co-3-hydroxyvalerate) production by haloarchaeon *Halogranum amylolyticum*. *Applied Microbiology and Biotechnology* 99(18):7639–7649.

Zhu C, J Zhang, Y Tang, X Zhengkai, and R Song. 2011. Diversity of methanogenic archaea in a biogas reactor fed with swine feces as the mono-substrate by mcrA analysis. *Microbiological Research* 166(1):27–35.

Zverlov V, K Riedel, and K Bronnenmeier. 1998. Properties and gene structure of a bifunctional cellulytic enzyme (CelA) from the extreme thermophile *Anaerocellum thermophilum* with separate glycosyl hydrolase family 9 and 48 catalytic domains. *Microbiology* 144:457–465.

CHAPTER 12

Carbonic Anhydrases of Extremophilic Microbes and Their Applicability in Mitigating Global Warming through Carbon Sequestration

Himadri Bose and Tulasi Satyanarayana

CONTENTS

12.1 INTRODUCTION

Extremophile is an anthropocentric term coined about 40 years ago to describe an organism that thrives under extreme conditions hostile for human beings. As the conditions become more demanding, populations of prokaryotes increase, suggesting that the majority of extremophiles are prokaryotes (MacElory 1974). Extremophiles have distinct structural and physiological characteristics that enable them to thrive in extreme habitats. These strategies are often due to the presence of specific biomolecules (DNA, lipids, and enzymes) that serve as novel sources of biotechnological applications. One of the well-known applications is Taq DNA polymerase from *Thermus aquaticus*, which is routinely used in polymerase chain reactions (PCRs) (Canganella and Wiegel 2011). The latter half of the last century witnessed immense interest in research on extremophiles. Apart from intellectual curiosity, their potential applications in industries have been a driving force for the researchers working on extremophiles. One of the most challenging aspects initially was their isolation and maintenance under simulated conditions in the laboratory. Despite encountering several problems, many polyextremophilic microbes were studied by many researchers in the late twentieth century (Mesbah and Wiegel 2012). Techniques of molecular biology and metagenomics techniques have immensely aided in understanding the culturable and nonculturable polyextremophilic microbial diversity and their characterization and applications (Seckbach et al. 2013).

Extremophiles can be divided into two broad categories: obligate extremophiles and facultative extremophiles. The latter tolerate extreme conditions but can grow optimally under normal conditions. The definition of *extremophily* can vary within different domains; for example, presently the upper limit for thermophilic bacteria is 95°C, and for archaea it is about 122°C, and unicellular eukaryotes have a temperature maximum of about 62°C. Each and every domain has witnessed the evolution of polyextremophiles (Horikoshi and Bull 2011). Polyextremophily has been predominantly observed in prokaryotes, but it exists in eukaryotes too. *Cyanidioschyzon*, a red alga, is acidophilic (pH 0.2–3.5) and a moderate thermophile (38°C–57°C). Adaptation of these organisms to the stress conditions is attributed to various factors (Weber et al. 2007). Recent advances in technology have enabled biologists to reach the extreme maxima of earth, sea, and sky for exploring their diversity. Extremophiles have been isolated from a wide array of extreme environments, like deep-sea vents, hot springs, the upper troposphere and stratosphere, and outer space (Wilson and Brimble 2009). There are also reports on the isolation of polyextremophiles from anthropogenic sources, like mines and industries (Onstott et al. 2003; Bhojiya and Joshi 2012). Their characterization and applications have added new dimensions to applied biology. Polyextremophiles are known to produce a variety of useful products (Table 12.1). An enzyme from polyextremophiles, carbonic anhydrase (CA), is useful in carbon capture and storage (CCS). This review focuses on the CAs of polyextremophiles and their utility in CCS.

12.2 CARBON EMISSIONS AND GLOBAL WARMING

The past century has witnessed an alarming rise in the earth's average temperature (~0.8°C) due to global warming. Each of the decades has been considerably warmer than the preceding ones. The period between 1983 and 2012 was the hottest in the last 800 years. The increase in the concentration of greenhouse gases (GHGs) in the atmosphere is one of the reasons for this phenomenon, with CO_2 being the major contributor (Shakun et al. 2012). Anthropogenic emissions of carbon dioxide from the combustion of fossil fuels have been the major culprit, with power plants as the major emitters of CO_2. The concentration of CO_2 has reached to about 398 ppm, compared with 280 ppm during the preindustrial era. Along with CO_2, other gases, like nitrogen and its oxides, and oxides of sulfur have also played a prominent part in global warming. The main source of CO_2 is flue gas from the exhausts of power plants. The composition of flue gas depends on the quality

Table 12.1 Important Products Obtained from Extremophilic Microorganisms

Organism Name	Source	Growth Characteristics	Products Obtained	References
Bacillus halodurans	Paper mill effluent (Pantnagar, India)	pH 9.0 Temperature 45°C	Xylanase	Kumar and Satyanarayana 2011
Bacillus acidicola TSAS1	Soil sample from Rameshwaram Tamilnadu	pH 4.0 Temperature 37°C	α-Amylase	Sharma and Satyanarayana 2010
Geobacillus thermoleovorans NP33	Hot springs from Waimango Volcanic Valley, New Zealand	pH 7.0 Temperature 60°C	Amylopululanases Glucoamylases	Nisha and Satyanarayana 2014
Geobacillus thermoleovorans NP54	Hot springs from Waimango Volcanic Valley, New Zealand	pH 7.0 Temperature 60°C	α-Amylase	Rao and Satyanarayana 2008
Geobacillus thermodinitrificans TSAA1	High-temperature compost from Japan	pH 7.0 Temperature 60°C	Xylanase β-Xylosidase	Anand et al. 2013
Bacillus halodurans	Lonar Lake, Maharashtra	pH 10.0 Temperature 45°C	Carbonic anhydrase	Faridi and Satyanarayana 2016
Paenibacillus tezpurensis	Soil samples of Assam	pH 9.5 Temperature 45°C–50°C	Proteases	Rai et al. 2010
Streptomyces gulbargensis	Soil sample of Gulbarga, Karnataka	pH 8.5–11.0 Temperature 45°C	Amylase	Syed et al. 2009
Rhodothermus marinus	Submarine alkaline hot spring in Iceland	pH 7.5 Temperature 61°C	Xylanase	Dahlberg et al. 1993
Bacillus pumilus	Wood decomposition material, soil	pH 10 Temperature 45°C	Pectinase, carbonic anhydrase, xylanase	Duarte et al. 2000 Prabhu et al. 2009
Nocardia prasina HA-4	Soil and sediment sample of Manipur	pH 7 and 10 Temperature 50°C	Protease	Ningthoujam et al. 2009
Bacillus sp. GUSI	Citrus garden in Fasa	pH 8.0–12.0 Temperature 50°C	Protease	Seifzadeh et al. 2008
Methanosarcina thermophila		pH 6.0–7.0 Temperature 55°C Strict anaerobe	γ-Carbonic anhydrase Carbon monoxide dehydrogenase	Sowers et al. 1984 Alber and Ferry 1994
Methanobacterium thermoautotrophicum	Anaerobic digesters, sewage sludge	pH 6.0–8.5 Temperature 65°C Strict anaerobe	β-Carbonic anhydrase	Smith and Ferry 2000
Sulfurihydrogenibium azorense	Terrestrial hot springs at Furnas, São Miguel Island, Azores, Portugal	pH 6.5 Temperature 68°C	Carbonic anhydrase	Aguiar et al. 2004 De Luca et al. 2012
Thermovibrio ammonificans	Deep-sea hydrothermal vents	Temperature 60°C–80°C	Carbonic anhydrase	Jo et al. 2014

of coal burnt (Solomona et al. 2009). The Industrial Revolution transformed the world in terms of technology and modern facilities. It paved the way toward an ultramodern world. But everything has come with a price. An increase in the earth's surface temperature has led to alarming effects on the global climate. There has been a decrease in cold temperature extremes and an increase in warm temperature extremes. The last 30 years has seen the loss of green land and Antarctic ice masses, which has led to an increase in ocean levels. The rate of the sea level rise in the last millennium has been larger than that of the mid-nineteenth century. The ocean has absorbed about 30% of the anthropogenic CO_2 emissions, leading to its acidification (26% more than the preindustrial era). The oxygen concentration has gradually decreased in coastal waters. The precipitation pattern around the world has been severely affected in the latter half of the past century. This has caused an impact in hydrological systems around the world, in both quality and quantity. Regions with high salt concentrations have become more saline, and regions with high precipitation have become much fresher, thereby disturbing the balance of the water cycle. Warming of the atmosphere and ocean has led to changes in the global carbon cycle. There has been an increase in moisture content in the atmosphere (IPCC 2014). Anthropogenic emissions of GHGs have been increasing at a very fast rate per decade, with a maximum in the last one. About two-thirds of this was contributed by fossil fuels, as they are the driving force needed to channelize the world's energy demands. Cumulative CO_2 emissions from fossil fuel combustion, cement production, flaring, and other uses have increased to about 40% since the 1970s. Climate change due to global warming has also affected the activities, abundance, and interaction patterns of many marine, terrestrial, and freshwater species. There have been an abundance and distribution shifts of many phytoplanktons, marine fishes, and invertebrates toward cooler waters, pole-ward (Houghton 2005). Changes in ecosystems have brought some of the species to the level of extinction. Climate-related hazards also affect poor people's lives directly through an impact on livelihood, reduction in crop yields, or destruction of homes, and indirectly through, for example, increased food prices and food insecurity (Adams et al. 1998). Global warming due to carbon emissions is proving to be a disaster for our survival, which has to be mitigated. There have been efforts by the governments all over the world to mitigate global warming. There is a problem in achieving this. There are more developing countries in the world than developed countries. It is next to impossible to reduce carbon emissions in the atmosphere, because fossil fuels are an integral part of the development agenda of developing countries. Mitigating global warming includes a reduction of GHG emissions and enhancing sinks aimed at reducing the extent of global warming. Scientific consensus on global warming, together with the precautionary principle and fear of nonlinear climate transitions, is leading to increased effort to develop new technologies and carefully manage others (Princiotta 2007). Carbon taxation, carbon trading, and emission trading schemes are some of the policies implemented by governments all over the world, which are aimed at mitigating global warming. CCS is evolving as a major technique in this direction.

12.3 CARBON CAPTURE AND STORAGE

This is a process of capturing carbon dioxide (CO_2) from large point sources, such as power plants using fossil fuels; transporting it to a storage site; and depositing it where it will not leak into the atmosphere, normally an underground geological formation. CCS can also be used to describe the scrubbing of CO_2 from ambient air as a climate engineering technique and storing it as mineral carbonates.

12.3.1 Physical Methods of Carbon Capture

Carbon sequestration by physical means is the most extensively used carbon capture method today. CCS takes up the carbon dioxide from large point sources such as fossil fuel combustion, purifies and concentrates it, and finally stores it deep underground. Three methods of CO_2 capture are currently employed (Haszeldine 2009). Postcombustion capture separates CO_2 from the fossil fuels with the help of chemical solvents. After leaving the power plant, the captured CO_2 is pressurized to about 70 bar, forming a liquid that can be transported to a storage site, where it is injected into rock pores 800 m below the surface. This is called geological carbon sequestration (Orr 2009). The choice of a better storage site can sequester CO_2 for years. Porous and permeable surfaces having nonpotable groundwater or sites having sedimentary rock formations made of such chemicals (e.g., calcium) that can react with CO_2 and convert it into more stable formations are preferred for storage. This technique can also be employed at other sites, such as those for heavy industries, steel plants, refineries, and fertilizer, ethanol fermentation, and cement factories (Benson et al. 2005). The first large-scale CO_2 sequestration project, which started in 1996, was called Sleipner, located in the North Sea, where Norway's StatoilHydro stripped carbon dioxide from natural gas with amine solvents and disposed of this carbon dioxide in a deep saline aquifer. In 2000, a coal-fueled synthetic natural gas plant in Beulah, North Dakota, became the world's first coal-using plant to capture and store carbon dioxide in the Weyburn–Midale Carbon Dioxide Project. CO_2 has been used extensively in enhanced crude oil recovery operations in the United States since 1972. There are more than 10,000 wells that inject CO_2 in the state of Texas alone. CO_2 capture technologies are used in about 120 facilities in the United States, mainly on industrial processes, and the captured CO_2 has a wide range of other uses, including enhanced oil recovery (EOR), food and beverage manufacturing, pulp and paper manufacturing, and metal fabrication. Geological carbon capture is associated with high costs and risk of leakage. It has been estimated that CCS will lead to a rise in power tariff by about 10% in the United States alone. It also requires a large amount of energy, and it can consume 25% of the power plant's output capacity. There are also some major disadvantages associated with postcombustion amine capture method. The equipment will be very large compared with the small size of a coal-fired power plant. Large volumes of solvent are needed, regeneration of solvent by heating may produce carcinogenic and toxic by-products, and columns need to be scrubbed and eliminated. This process also utilizes large volumes of water, which adds to the cost, and there is also the problem of the disposal of leftover residues (Hart and Gnanendran 2009).

12.3.2 Ocean Sequestration

This is another approach currently being employed to capture CO_2. It takes up about 2 billion metric tons of CO_2 each year, that is, the amount of carbon that would double the load in the atmosphere and would increase the concentration in the deep ocean by only 2%. So ocean can be considered an efficient sink for CO_2. Two approaches are used for ocean sequestration: direct injection and ocean fertilization. Direct injection of CO_2 is done either from shore stations or tankers trailing on the ocean floor. Carbon dioxide is denser than seawater at great depths (more than 2500 m below) (Herzog 1998). Phytoplanktons at the surface can fix carbon, which is in turn eaten by sea animals. After their death, their dead remains get deposited on the seafloor. Bacteria present near the seafloor feed on the organic carbon and release CO_2, which dissolves, and the rest of the detritus remains there on the ocean floor. There are some places in the ocean that are rich in nutrients like nitrogen and phosphorus but poor in phytoplankton. Adding a little iron to the mix allows the plankton to use the nutrients and bloom (Traufetter 2009). Natural iron fertilization events (e.g., deposition of iron-rich dust into ocean waters) can enhance carbon sequestration (Monastersky 1995; Jin et al. 2008). Sperm whales act as agents of iron fertilization when they transport iron from the deep ocean to the surface during prey consumption and defecation. Sperm whales have been shown to increase the levels of primary production and carbon export to the deep ocean by depositing iron-rich feces into surface waters of the ocean. The energy for the process is supplied by sunlight (Lavery et al. 2010). Although ocean is a very efficient sink for carbon, several issues must be considered during ocean sequestration. Sequestration efficiency, which is site specific, refers to how long the CO_2 will remain in the ocean before ultimately equilibrating with the atmosphere. Depending on the site, carbon can be stored for as long as a hundred years. Also, direct injection of CO_2 will lead to acidification of deep ocean (Dutreuil et al. 2009). The available data, however, suggest that impacts associated with pH change can be completely avoided if the injection is properly designed to disperse the CO_2 as it dissolves. Direct injection of CO_2 to the ocean will reduce the chance of catastrophic events, such as the shutting down of the ocean's thermohaline circulation (Lovelock and Rapley 2007). Ocean sequestration is also associated with the risk of leakage. Carbon, in the form of CO_2, can be removed from the atmosphere by chemical processes and stored in stable carbonate mineral forms (mineral carbonation). The process involves reacting carbon dioxide with abundantly available metal oxides, either magnesium oxide (MgO) or calcium oxide (CaO), to form stable carbonates. CCS can reduce carbon emissions by about 20%. It is unavoidable if fossil fuels continue to be burnt at more than 10% of the present rate. Efforts are underway to develop processes for directly converting CO_2 in industrial emissions into useful products (Faridi and Satyanarayana 2015; Lim 2015).

12.3.3 Biological Carbon Sequestration

One of the best ways to sequester carbon is by reforestation. It incorporates atmospheric carbon into biomass by photosynthesis (Kindermann et al. 2008). Plants with a C_4 metabolic pathway are much more efficient in sequestering carbon into their biomass. Efforts are also in progress for improving the catalytic efficiency of ribulose 1,5-bisphosphate carboxylase (RuBisCo) enzyme systems so as to increase the rate of CO_2 assimilation (Christer et al. 2010). Trees should be grown perennially for a long time so that the carbon is stored for a long time. As the trees die, their biomass can also be used to produce biochar or phytoliths (Parr and Sullivan 2005). Carbon farms can be developed for this purpose, where soil can be used as carbon sink and the biomass aboveground can be stored in the form of soil organic carbon (SOC) or soil inorganic carbon (SIC). Wetland soil is an important carbon sink; 14.5% of the world's soil carbon is found in wetlands, while only 6% of the world's land is composed of wetlands. Modified agricultural practices may also increase the amount of soil carbon. Use of cover crops such as grasses and weeds and the covering of bare paddocks with hay or dead vegetation are some of the measures that can be employed to increase carbon content in soil (Batjes 1996). Agricultural sequestration practices may have positive effects on soil, air, and water quality; be beneficial to wildlife; and expand food production. On degraded croplands, an increase by 1 ton of the soil carbon pool may increase the crop yield by 20–40 kg ha^{-1} for wheat, 10–20 kg ha^{-1} for maize, and 0.5–1 kg ha^{-1} for cowpeas (Smith 2007). Bioenergy crops are another way by which plants can act as a CO_2 sink. They can also reduce the use of fossil fuels, thereby reducing GHG emissions by fossil fuels. Crops like wheat, maize, and sugarcane are a rich source of bioethanol. Appropriate forest policies could increase the amount of sequestered carbon by up to 100 Gt or up to 2 Gt year^{-1} (Min 2011). Photosynthetic microbes such as algae and cyanobacteria also convert CO_2 into organic carbon, thereby making a significant contribution to the global carbon and oxygen cycle. Algal species such as *Lomentaria* and *Chlorella* grow much better in the presence of excess CO_2 (Kubler et al. 1999). These species can be directly connected to flue gases coming out of power plants, thereby increasing their production. In this way, CO_2 can be incorporated in their biomass. Later, this biomass can be used for different purposes, such as biofuels, food, and enzymes. Cyanobacteria could be harnessed as microbial cell factories with the help of solar energy (Yue and Chen 2005). Methanogens that can thrive in a wide range of temperatures (9°C–110°C and above) can be successfully applied to produce methane commercially (Huber et al. 1989; Zhang et al. 2008). This technology can have a huge socioeconomic impact, as it is cost-effective. Initially, studies by some researchers revealed that CO_2 gets reduced during glycerol fermentation by some genera of heterotrophic bacteria, such as *Propionibacterium*, and further, by combining with pyruvate, it results in the formation of oxaloacetate (Wood et al. 1941).

This pathway can be used for carbon capture using CA, which catalyzes the conversion of CO_2 into bicarbonate, which can be used for amino acid production; that is, carbon can be sequestered in the form of amino acids such as glutamic acid and lysine (Norici et al. 2002). Use of CA as a potential biocatalyst has caught the attention of many researchers, and much work has been done to explore the possibilities of using "nature's own catalyst" for CCS.

12.4 CARBONIC ANHYDRASE

As it is said that "nature has a solution to every problem," we have been endowed with a natural solution to the climate change problem in the form of CA. This is a ubiquitous zinc metalloenzyme (EC No. 4.2.1.1) that catalyzes the interconversion of carbon dioxide to bicarbonates (Smith and Ferry 2000).

$$CO_2 + H_2O \rightleftharpoons H_2CO_3 \rightleftharpoons HCO_3^- + H^+$$

This reaction is slow at physiological pH; therefore, it requires a catalyst (CA) for improving the pace of the reaction. The biocatalyst CA is present in all three domains of life, as this is an essential enzyme for metabolism. It helps in concentrating CO_2 in the living cells. It plays a role in photosynthesis, respiration, ion transport, and the maintenance of acid–base balance. In plants, CA aids in raising the concentration of CO_2 within the chloroplast in order to increase the carboxylation rate of RuBisCO (Smith and Ferry 2000; Smith et al. 2000). It is one of the fast-acting enzymes, with a k_{cat} value in the order of 10^6 s^{-1} and a catalytic efficiency in the order of 10^8 M^{-1} s^{-1}. Six evolutionary distinct classes (α, β, γ, δ, ζ, and ϵ) have been discovered to date, widespread in living beings (Kaur et al. 2012). No specific sequence similarity has been found in these classes (Figure 12.1). Hence, it is a classic case of convergent evolution (Hewett-Emmett and Tashian 1996; Di Fiore et al. 2015). CA was discovered in bovine erythrocytes in 1933 (Meldrum and Roughton 1933; Stadie and O'Brien 1933). This was found to be of α-type. Soon after, many isoforms of α-CA were characterized from many mammalian tissues, such as liver and kidney. They can be subdivided into mitochondrial, cytosolic, and membrane-associated CAs. Until now, about 15 isoforms (CAI–CAXV) of this type have been discovered (Eriksson et al. 1988; Hewett-Emmett and Tashian 1991; Mori et al. 1999). This type of CA has been found in association with some of the primates and aquatic fishes. α-CA was also found in many prokaryotes, plants, algae, and viruses in the latter part of twentieth century (Tripp et al. 2001). The first α-CA characterized from prokaryotic genera was that of *Neisseria gonorrhoeae* (Veitch and Blankenship 1963). The next type (β-CA) was discovered initially in plants. Later, it was also found in the bacterial as well as archaeal genera. This type of CA is the most widely distributed among all the classes (Smith et al. 2000). The γ-CA was discovered in hyperthermophilic archaeon *Methanosarcina* (Alber and Ferry 1994). This CA has been found to be associated mainly with the archaeal domain. Until now, most of the β- and γ-CAs have been characterized from plants and prokaryotic genera. *Synechococcus* harbors multiple CA genes (*ccaA*, *isfA*, and others), which have a close homology with *cynT* of *Escherichia coli* (Kupriyanova et al. 2011). Even the gene encoding γ-CA was also identified and expressed heterologously, but it was not active. δ-, ζ-, and η-CAs have recently been isolated from cyanobacteria, marine diatoms, and some dinoflagellates (Xu et al. 2008). In cyanobacteria and microalgae, CAs are involved in the CO_2 concentrating mechanism (CCM), which helps the cells to photosynthesize in the absence of inorganic carbon (Badger and Price 2003). CAs have also been reported from facultative anaerobes such as *Rhodospirillum rubrum* (Gill et al. 1984). The protein from this bacterium has a molecular mass of 28 kDa. Acetogens such as *Acetobacterium woodii* and *Clostridium thermoaceticum* also display CA activity. Among them, *A. woodii* has the highest CA activity (Smith and Ferry 2000).

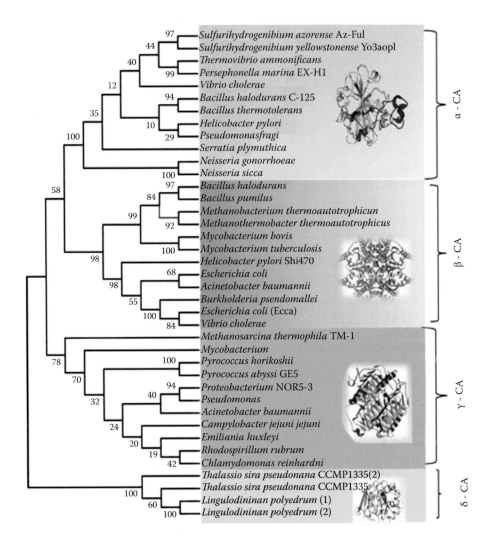

Figure 12.1 Phylogenetic tree showing the comparison of different types of CAs. There is no significant sequence similarity within the different classes of CAs, a classic case of convergent evolution.

12.4.1 Structure and Mechanism of Action

12.4.1.1 α-CA

The first CA that was well characterized was human carbonic anhydrase II (hCAII) from mammals. It is a monomer with a molecular mass of 29 kDa (Andersson et al. 1972; Henderson et al. 1973; Lin et al. 1973, 1974; Hewett-Emmett and Tashian 1991). Its crystal structure revealed that the catalytic center is composed of zinc ion coordinated by three histidine residues (His94, 96, and 119) in a tetrahedral model (Liljas et al. 1972; Eriksson et al. 1988; Huang et al. 1998). The fourth ligand is the water molecule. The secondary structure is dominated by 10 stranded β sheets, which divide the molecule into two equal halves. Another histidine residue, His64, facilitates proton transfer between active site water and solvent water at the mouth of the active site cavity. All the β sheets are in antiparallel orientation; only two pairs are in parallel confirmations. It contains an N-terminal signal peptide for periplasmic localization or extracellular secretion. Hydroxide ion bound to zinc makes a bond with the oxygen atom of the threonine residue present near the active site, which is further attached to glutamic acid. These two amino acid residues are conserved among all α-CAs. The α-CA from *Neisseria gonorrhoeae* was crystallized, and most of its fold and secondary structure closely resemble that of hCAII, apart from a few helical regions of the protein, which has resulted in considerable shortening of three surface loops (Huang et al. 1998). The same deletions were also observed in cyanobacterial α-CAs. It is also a monomer with a molecular weight of 28.6 kDa. Its molecular weight ranges from 29 to 34 kDa. The enzyme is known to play a role in the maintenance of the acid–base balance in mammals. Apart from the CO_2 hydration activity, α-CA also shows esterase and hydrogenase activities. This form of CA is found in plants, green algae, eubacteria, and viruses.

12.4.1.2 β-CA

The β-class enzymes can be divided into seven clades (A–G) based on sequence identity, with the plant enzymes forming two clades representing dicotyledonous and monocotyledonous plants. Enzymes within these clades can differ with respect to structure and their response to inhibitors, signifying different functions and mechanisms of action (Zimmerman and Ferry 2008). The structure of β-class is distinctly different from that of its counterparts. The α-class of enzymes are mainly monomers, while the β-class can be dimers, trimers, tetramers, hexamers, and octamers, with dimer being the fundamental structure (Smith and Ferry 1999, 2000). In prokaryotes, this protein was first discovered in *E. coli* (Guilloton et al. 1992), which has two types of β-CAs. One of them (cynT) is cotranscribed with the cyanase gene under its promoter, which is encoded by a part of the *cyn* operon, whose function allows *E. coli* to utilize cyanate as a nitrogen source. It is an inducible promoter that is activated in the presence of cyanate. It was found localized in cytosol, and the subunit molecular mass is 24 kDa (Sung and Fuchs 1988; Kozliak et al. 1994). The second β-CA gene sequence was detected after genome sequencing of *E. coli*. It was designated as ECCA (Fujita et al. 1994). The crystal structure of ECCA was solved in the early twenty-first century. The active site consists of a Zn^{2+} atom coordinated on four sides by one histidine (His98), two cysteine (Cys42 and Cys101), and one aspartate (Asp44) residue in a tetrahedral arrangement. The aspartate residue makes closest contact to the metal and water molecule away from the active site. Its secondary structure comprises four strands of parallel β sheets flanked with five α-helical segments (Cronk et al. 2001). In photosynthetic bacteria and plant chloroplasts, β-CA is essential for photosynthetic carbon fixation. It functions in three modes: conversion of CO_2 to bicarbonate (to be utilized by RuBisCO in C_4 plants), conversion of bicarbonate to CO_2 (for fixation by phosphoenol pyruvate carboxylase [PEPC]), and help in facilitated diffusion by rapid equilibration between CO_2 and HCO_3^-. It also provides bicarbonate, which is required for the metabolism in plants and many prokaryotes (Monti et al. 2013). This type of CA has also been shown to a play role in the pathogenesis of some bacteria, such as *Salmonella typhimurium* (Valdivia and Falkow 1997). A novel CA was discovered in the carboxysomal shell of chemolithoautotrophic cyanobacterium *Halothiobacillus neapolitanus* (CsoS3). The carboxysomal CA in the shell supplies the active sites of RuBisCO with high concentrations of CO_2 necessary for RuBisCO activity and efficient carbon fixation (So et al. 2004). Initially, it was thought to be a different class of CA, but three-dimensional analysis of the enzyme revealed that it is structurally similar to β-CAs despite amino acid divergence (Sawaya et al. 2006).

12.4.1.3 γ-CA

The first γ-CA to be discovered was Cam CA from thermophilic archaeon *Methanosarcina thermophila* (Alber and Ferry 1994, 1996). Since then, this enzyme has been discovered in several prokaryotic life-forms. Molecular clock analysis suggests that this may be the oldest form of CA (Smith et al. 1999). Only a few γ-class crystal structures have been reported. They are the γ-class archaeal type Cam from *M. thermophila* (MtCam), the CamH homolog from *Pyrococcus horiko-shii* (PhCamH), and the N-terminal domain of the carboxysomal protein CcmM from the thermo-philic β-cyanobacterium *Thermosynechococcus elongatus* BP-1 (TeCcmM209) (Jeyakanthan et al. 2008; Pena et al. 2010). As evidenced by the structural superimposition, all three structures share an overall similar architecture with minor differences. The main differences among three structures are found at the N- and C-termini and in an acidic loop, placed after MtCam strand β10 and contain-ing the proton shuttle residue Glu84 (MtCam numbering). This loop is much shorter in PhCamH (Pena et al. 2010) and TeCcmM209 than in MtCam. These are generally found as homotrimers, with monomers having a distinctive left-handed parallel β-helix fold. The zinc atom is present at its active site, flanked by three histidine residues, His81, 122, and 117 (two from one subunit and the third from the next one). Active site Zn^{2+} is present at the interface between two monomers. Glu62, Asn73, Gln75, and Asn202 also have an active role in catalysis (Iverson et al. 2000). γ-CA is characterized by novel hexapeptide repeats in its β-helix. These repeats are responsible for its novel left-handed confirmation. The Arg59 present 6 Å away from active site Zn^{2+} is important for the association of monomers into the native trimer and is essential for the CO_2 hydration step in catalysis (Ferry 2010). In methanogens, it plays an active role in acetate metabolism by converting the excess carbon dioxide into bicarbonates. It also helps in the conversion of acetate into methane. γ-Class homologs of the CamH subclass are found in mitochondria, where they might have a role in the carbon transport system to increase the efficiency of photosynthetic carbon dioxide fixation (Tripp and Ferry 2000).

12.4.1.4 ζ- and δ-CA

ζ-CA is a Cd(II)-containing enzyme (CDCA1) isolated from the marine diatom *Thalassiosira weissfogii*. It is of cambialistic nature, and the active protein has a molecular mass of about 69 kDa. This CA is majorly confined to diatoms (Lane and Morel 2000; Lane et al. 2005).

Another type of CA, δ-CA, has been discovered in *T. weissfogii*. It is a monomer with a molecu-lar mass of 27 kDa. The gene encoding the enzyme was cloned, and its inhibition studies showed that 3-bromosulfanilamide, acetazolamide, ethoxzolamide, dorzolamide, and brinzolamide are the most effective TweCA inhibitors. Saccharin and hydrochlorothiazide were ineffective inhibitors (Vullo et al. 2014). Its presence has also been recorded in free-living marine dinoflagellate *Lingulodinium polyedrum* (Lapointe et al. 2008). Properties of δ-CA are distinct from those of other known CAs.

12.4.1.5 η-CA

The η-family of CA was recently found in organisms of the genus *Plasmodium*. These are a group of enzymes previously thought to belong to the α-family of CAs; however, it has been demonstrated that η-CAs have unique features, such as their metal ion coordination pattern (Del Prete et al. 2014).

12.4.2 Catalytic Mechanism

The catalytic mechanism has been well understood in α-CAs. Interconversion of CO_2 to bicar-bonate is a two-step process. In the first step, a fourth histidine near the water molecule accepts a proton (H^+), which is released from the water molecule, leaving only hydroxide ion bound to zinc ion. The active site has specific pockets for binding CO_2, which helps in the binding of substrate to the hydroxide ion. The next step is nucleophilic attack on the carbonyl group by the zinc-bound hydroxide, leading to the formation of HCO_3^-. This is the rate-limiting step. The enzyme is then regenerated and the bicarbonate ion is released (Figure 12.2) (Lindskog and Coleman 1973; Steiner et al. 1975; Silverman and Lindskog 1988).

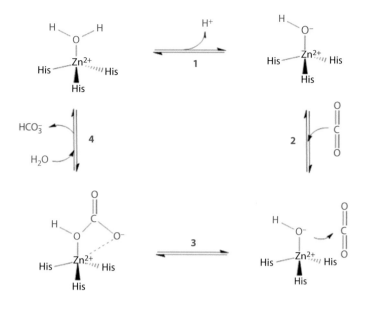

Figure 12.2 Catalytic mechanism of CA.

12.5 BIOMINERALIZATION USING CARBONIC ANHYDRASES

Biomimetic sequestration of CO_2 using the enzyme CA where CO_2 is sequestered into mineral carbonates can be utilized efficiently to capture the carbon emissions. It mimics the mineralization process occurring in nature that is responsible for the presence of a huge amount of limestone on the surface of the earth. This is called silicate weathering. It traps the atmospheric carbon by reacting it with large limestone rocks, such as wollastonite ($CaSiO_3$), serpentine ($Mg_3Si_2O_5(OH)_4$), and olivine (Mg_2SiO_4). It has several advantages over other processes of carbon capture. Silicates and carbonates produced can remain stable for eons. Also, the raw materials in the form of different metal ions are available readily in the form of various minerals. These metals can also be obtained from various metal refining industries where they come out as waste products, thereby providing an effective means to counter these harmful waste products and utilize them in an efficient way (Soong et al. 2006; Mirjafari et al. 2007). It is also extremely environment friendly and more cost-effective than the solvent-based method of carbon capture (Huijgen et al. 2007); also, the mineral carbonates generated can be utilized in other applications (Ciullo 1996). The natural mineralization process is, however, extremely slow, as the conversion of CO_2 to bicarbonate is the rate-limiting step. The CAs can speed up the entire mineralization process. CA catalyzes the hydration of dissolved CO_2 into bicarbonate at a faster rate, as outlined below. The bicarbonate ions thus generated can be converted into mineral carbonates in the presence of divalent metal ions such as Ca^{2+} and Mg^{2+} in alkaline conditions (Favre et al. 2009; Mahinpey et al. 2011).

$$CO_{2(g)} \longleftrightarrow CO_{2(aq)} \tag{12.1}$$

$$CO_{2(aq)} + H_2O \longleftrightarrow HCO_3^- + H^+ \tag{12.2}$$

$$HCO_3^- + H^+ \longleftrightarrow CO_3^{2-} + H_2O \tag{12.3}$$

$$CO_3^{2-} + Ca^{2+}/Mg^{2+} \longrightarrow CaCO_3/MgCO_3 \tag{12.4}$$

Maintenance of the alkaline condition is necessary, since bicarbonate formation and its precipitation occur at alkaline pH because OH⁻ ions are needed for this conversion to occur. This type of sequestration can be more efficient, in case CA can be utilized during geological and ocean sequestration measures deep below the earth's surface. This type of process will require the enzyme to be stable for a long time, and also it should be active at high temperature, as the subsurface temperature is quiet high and increases further as we go down. Immobilization of the enzyme within different matrices (polyutherane foam, alginate, and chitosan beads) is an approach that can be applied to make the enzyme stable (Prabhu et al. 2011). Due to immobilization, the enzyme can be used repeatedly, as it can be recovered to some extent after each reaction. Although an excellent method, immobilization alone does not provide such a high degree of solidarity to the enzyme. This also adds to the cost. Thus, it is always better to look for native CAs from extremophilic organisms (particularly alkaliphiles and thermophiles), which can readily serve the purpose cost-effectively, because the CAs are expected to be stable at alkaline pH and high temperature. These thermostable and alkalistable enzymes coupled with immobilization strategies can be used more efficiently for CCS.

12.6 UTILITY OF CARBONIC ANHYDRASES IN MITIGATING GLOBAL WARMING

CAs have been already utilized as a tool in CO_2 capture technologies. This process requires CA to be immobilized on suitable beads. Utilization of CA for carbon capture requires a series of bioreactors that will have a direct supply of flue gas. It is directly supplied to a bioreactor containing the immobilized enzyme. This enzyme in aqueous condition will break down CO_2 present in flue gas and release HCO^{3-} and H^+. This bicarbonate can be utilized further for various purposes Figure 12.3.

Different techniques have been employed for capturing CO_2, that is, absorption into a liquid, adsorption on a solid, gas phase separation, and membrane systems (Yang et al. 2008), and among these, the chemical absorption technique has efficiently seen the utilization of thermostable CAs (Wang et al. 2011). This process involves the reaction of CO_2 with a chemical solvent to form an intermediate compound, which is then subjected to heat treatment, thereby regenerating the original solvent and releasing a CO_2 stream. The flue gas containing CO_2 is fed from the bottom of the reactor into a packed absorber column at low temperature (30°C–50°C), which gets absorbed into the absorber; after absorption, the CO_2-rich solvent is sent to a stripper operating at high temperatures (120°C–140°C) to separate the absorber and CO_2. After regeneration, the free solvent is pumped back to the absorber so that it can be used again, while the pure CO_2 is compressed for the subsequent transportation and storage (Wang et al. 2011; Yu et al. 2012). The main energy cost associated with this process is the heat required for the desorption and pumping of the absorbing solution around the system (Huang et al. 2001). Alkanolamines are the most extensively used absorbers for CO_2 capture, which includes primary, secondary, and tertiary amines, such as monoethanolamine (MEA), diethanolamine (DEA), and N-methyldiethanolamine (MDEA) (Puxty et al. 2009). MDEA is the one of the most widely used amines. Several studies have demonstrated that the use of tertiary amines in chemical absorption processes is more beneficial with respect to the use of primary and secondary amines, since they have a lower heat of regeneration (Carson et al. 2000) and a higher capture capacity. However, the reaction of CO_2 with tertiary amines is significantly slower than that with primary and secondary amines (Versteeg et al. 1996). Thus, the use of CAs along with MDEA has been proposed to increase the rate of CO_2 absorption (Mirjafari et al. 2007), thereby facilitating the intermolecular transfer of protons (Vinoba et al. 2013). Also, the temperature of flue gas is too high (about 150°C). The initial process requires the cooling of flue gas to about ambient temperatures. Hence, the use of thermostable CAs will not require the cooling of flue gas. Considering the high temperature used in this process, several heat-stable CAs have been employed, which include CAs isolated from *Sulfurihydrogenibium yellowstonense* YO3AOP1 (SspCA) (Daigleand Fradette 2014; Rossi 2014) and from *Caminibacter mediatlanticus* (CmCA), both belonging to the α-class.

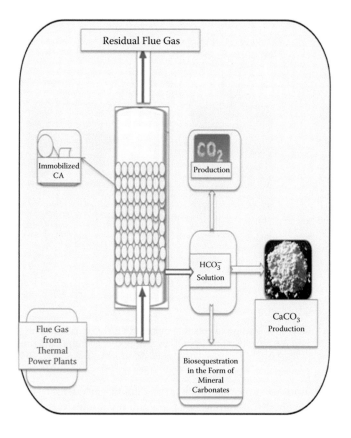

Figure 12.3   Schematic representation of CA-based carbon capture method.

The bicarbonate generated can also be stored in the form of calcium carbonate in underground storage. CAs have also been utilized in bioweathering of silicates by subsequent precipitation of Mg(II) released from the weathering of magnesium silicate rocks.

12.7 BIOREACTORS FOR CARBON CAPTURE

Different reactor systems have been developed for the utilization of CA for CCS. Reactor systems by NASA and Carbozyme Inc. are noteworthy. NASA developed this system to maintain ambient atmosphere in confined inhabited cabins. This system comprises a liquid membrane whose core contains a 330 μm thick layer of enzymatic solution in an appropriate buffer system squeezed between two microporous hydrophobic polypropylene membranes. Two thin metal grids protect the membrane. CO_2 diffuses across the membrane and evaporates out from the other side into a vacuum or carrier gas (Cowan et al. 2003).

Carbozyme Inc. has developed an improved system based on hollow microporous polypropylene microfiber. The enzyme is immobilized directly on the outer face of microfiber, and water vapor under mild vacuum is used as sweeping gas in the released microfibers (Trachtenberg et al. 2007). This technology was found to be very efficient in the case of flue gas containing ~40% CO_2 at a broad range of temperatures (15°C–85°C). This technique utilizes a particular γ-CA isozyme. Some other bioreactors have been also developed, such as a three-phase monolith bioreactor by Iliuta and Larachi (2012), which uses immobilized hCAII enzyme. They demonstrated that gas–liquid

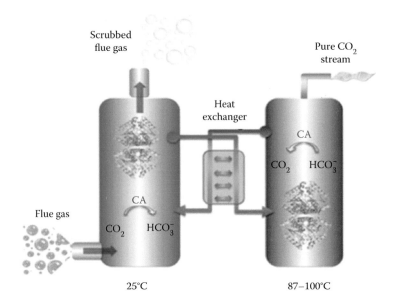

Figure 12.4   Flue gas from a coal-fired power plant is piped into an absorber column (blue) where CO_2 chemisorbs into an amine solvent, catalyzed by CA, and is hydrated to a proton and a bicarbonate ion. (Adopted from Alvizo, O. et al., *Proc. Natl. Acad. Sci. U.S.A.*, 111, 16436–16441, 2014.)

and liquid–solid mass transfer exchange mechanisms could considerably modify CO_2 hydration kinetics. Reactors have been also developed that are equipped with systems that enable them for continuous CO_2 capture. Systems where CA is trapped within membrane pores filled with an ionic liquid that provides low solvent evaporation, thereby maintaining the water level needed for enzyme activity and stability, have been also proposed. A gas-to-liquid membrane CO_2 sequestration system combines a mineralization module with Ca^{2+} for precipitation of bicarbonate to form solid $CaCO_3$.

Another approach was proposed by Alvizo et al. (2014) in which flue gas was transferred to an absorber column where CO_2 chemisorbs into an amine solvent (MDEA), catalyzed by CA, and a proton and a bicarbonate ion are generated. The CO_2-depleted flue gas is released into the atmosphere and the bicarbonate-loaded amine solvent, along with CA, is transferred to a second column, where CO_2 is removed at elevated temperatures (>87°C), resulting in solvent regeneration. The pure CO_2 stream can be compressed and stored in depositories or used in industrial processes. The regenerated solvent is returned to the absorber column to repeat the process. A highly thermostable variant of DvCA (*Desulfovibrio vulgaris* CA) is used in this process (Figure 12.4).

12.8 DEVELOPMENT OF THERMOSTABLE CARBONIC ANHYDRASES BY PROTEIN ENGINEERING

Researchers have attempted to use molecular biology tools to modify proteins to make them tolerant to the extreme environmental conditions prevailing in industrial processes. hCAII has been modified by changing some of its surface amino acid residues to make it more thermostable. Six surface amino acid residues, Leu100, Leu224, Leu240, Tyr7, Asn62, and Asn67, were replaced with His, Ser, Pro, Phe, Leu, and Gln, respectively, by random mutagenesis. These surface residues were chosen because they are far away from the active site and also because leucine confers hydrophobicity to the protein. These mutations on the surface amino acids tend to increase the number of H-bonds and also enhance the hydrophilicity. This strategy increases the thermostability by about

6°C. Also, the proton transfer rate increased by sixfold (Fisher et al. 2012). Per U.S. Patent 7521217 filed by CO_2 Solutions Inc., some other mutations that contributed to increased thermal stability are Ala65Thr, Phe93Leu, Gln136His/Tyr, Lys153Asn, Leu198Met, and Ala247Thr. Study of the crystallographic structure shows that Phe20 and Leu57 are other residues that can be remodeled so that they can also participate in hydrogen bond formation with the adjoining residues. Leu204 and Val135 can be mutated such that they take part in salt bridge formation and stabilize the H-bond further (Daigle and Desrochers 2009).

CO_2 Solutions utilizes CA as an onsite scrubber for separation of carbon from flue gases. This CA, which has been thermally optimized by genetic engineering as described above, is stable at 90°C for 24 hours (CO_2 Solutions Inc. 2012). This technology has increased the carbon adsorption rate 40 times compared with the previous technologies. Another concern is the presence of flue gas contaminants such SOx and NOx, which may affect the stability of the enzyme. Codexis has also developed novel CAs via directed evolution technology that increases the rate of carbon capture 25-fold under industrial conditions (Newman et al. 2010). This strategy has helped in the development of a highly efficient CA enzyme that is highly thermos- and alkalistable. About a 10 million–fold improvement in thermal stability was achieved. It also retains its activity in the presence of a wide variety of flue gas contaminants, such as NOx and SOx. It was also found to be extremely stable when it was subjected to high (4.2 M) concentrations of MDEA. The enzyme was stable at a 4.2 M concentration of MDEA at 50°C for about 14 weeks. This enzyme, initially obtained from *Desulfovibrio vulgaris*, is a β-CA. High-throughput screening has also led to the selection of a highly stable CA variant that can even retain its activity at 107°C (Alvizo et al. 2014). This enzyme has more than 40% residual activity at 82°C at a highly alkaline pH (11.8) in the presence of 5 M MDEA. In another study, CA from *Neisseria gonorrhoeae* (*ng*CA) was engineered in the periplasm of *E. coli*, thereby generating a bacterial whole-cell catalyst. It was highly expressed in soluble form, and even its thermal stability greatly increased. It displayed a distinctly faster CO_2 hydration activity than its cytoplasmic counterpart. This whole-cell catalyst was also found to be much more stable at low pH conditions. So, it might be possible to sequester CO_2 more efficiently using whole-cell enzyme systems even at a pH below the pKa of HCO_3^- or CO_3^{2-}, thereby reducing the expense of maintaining an elevated pH (Jo et al. 2013).

The modified CA thus generated by the aforesaid strategies can aid in efficiently capturing CO_2. This type of mutagenesis approach always adds to the cost of enzyme production. It is always worthwhile to search for enzymes that are stable at multiple extremes. The use of thermoalkalistable CA from polyextremophiles is expected to simplify the process.

12.9 CARBONIC ANHYDRASE FROM EXTREMOPHILIC SOURCES

Microbial enzymes from extremophilic sources have long been the top priority of researchers, and they have always traversed the extreme realms of earth (deep sea, Arctic and Antarctic regions, and geothermal springs) to isolate novel organisms from these sources. Use of thermoalkalistable CAs from extremophilic sources can be an alternative and cost-effective method for CCS. Data mining studies have revealed the presence of all types of CA genes in a wide array of prokaryotic species (both eubacteria and archaea). According to a report, eubacteria and extremophilic archaea harbor genes of all three types of CA (Smith and Ferry 2000). Researchers who have isolated CA from extremophilic sources have used it efficiently in CCS. γ-CA from methanoarcheon *M. thermophila* was one of the first CAs (Cam CA) isolated from archaeal sources. This CA is moderately thermostable and shows activity at 55°C (Alber and Ferry 1994). This methanogen is known to be metabolically diverse, and CA activity is increased by 10-fold when it is grown in the presence of acetate, thus establishing its role in acetate metabolism. This CA was cloned and sequenced in *E. coli*, its open reading frame (ORF) is composed of 234 amino acids, and it codes for 34 kDa protein (Alber and Ferry 1996). The activity of this CA doubled when zinc ion was substituted with cobalt. Also, tis protein was overexpressed and anaerobically purified in *E. coli* (Tripp et al. 2004), and in *Methanosarcina acetivorans* (Macauley et al. 2009), the enzyme contains Fe^{2+} in the active site and is fourfold more active than the zinc metalloenzyme. Around that time, another thermostable β-CA was isolated from *Methanobacterium thermoautotrophicum* (CabCA) (Smith and Ferry 2000). It even retains its CO_2 hydration activity at 75°C. This protein was cloned and heterologously expressed in *E. coli*. Analysis of the amino acid sequence revealed 34% similarity with *E. coli* β-CA. This protein is a tetramer with a molecular weight of 90 kDa. Although it is of the β-type, it shows distinct properties from that of plant β-CA, but its active site configuration is the same as that of β-CAs. A CO_2 hydration activity of 0.8 U mg^{-1} was observed in this enzyme (Smith et al. 2000). These two CAs paved the way for the search of CAs with varied degrees of stabilities. In the last decade, there has been numerous efforts aimed toward the discovery of extremophilic sources of CAs. An extracellular α-CA was discovered in the alkaliphilic photosynthetic cyanobacterium *Microcoleus chthonoplastes* with a molecular mass of 34 kDa. The ECCA enzyme from *E. coli* also shows high activity at alkaline conditions (pH 9.1), and it was found to be inactive at neutral pH (Kupriyanova et al. 2007). Two novel α-CAs have been discovered from thermophilic archaea *Sulfurihydrogenibium azorense* and *Sulfurihydrogenibium yellowstonense* YO3AOP1 (optimum temperature 110°C). These two CAs, namely, SazCA and SspCA, respectively, are highly thermostable. SspCA has a molecular weight of 26 kDa. This enzyme can retain its activity even after incubation at 70°C for three hours. It also retained its activity even after prolonged incubation at 100°C (Capasso et al. 2012; Akdemir et al. 2013). More stability was achieved after immobilization of the enzyme with polyutherane foam (50 hours at 100°C), and a thermoactivity analysis revealed an optimum working temperature of 95°C for this enzyme. The SazCA from *Sulfurihydrogenibium azorense* is known to be the fastest known CA to date, with a k_{cat}/K_m value of 3.5×10^8 M^{-1} s^{-1} (De Luca et al. 2013). It retains its carbon dioxide hydration activity even after incubation at 80°C and 90°C for several hours. The half-life of SazCA at 40°C and 70°C was 53 and 8 days, respectively (Russo et al. 2013). The inhibition constants are in the range of 0.58–0.86 mM for the anions NO_2^-, NO_3^-, and SO_4^{2-}. Both SspCA and SazCA are alkalistable, retaining their activity at pH 9.6 (De Luca et al. 2012; Vullo et al. 2012). These properties make them excellent candidates for biomimetic carbon sequestration. An x-ray crystallographic structure of SspCA was also solved, and factors that are responsible for its high thermostability were studied on the basis of the structure. It was found out that an increased structural compactness, together with an increased number of charged residues on the protein surface and a greater number of ionic networks, seems to be the key factor involved in the higher thermostability of this enzyme with respect to its mesophilic homologs. These efforts have helped us in understanding the structural basis of the thermostability of α-CAs (Di Fiore et al. 2013). The crystallographic structure of SazCA was also solved, and its analysis revealed it to be a dimer. Its structure is similar to that of SspCA with minor differences. The substitution of the SspCA residues Glu2 and Gln207, located on the rim of the cavity, with His2 and His207 in SazCA has been proposed to be responsible of the higher SazCA catalytic activity (De Simone et al. 2015). Efforts are also underway for the construction of a highly efficient super CA by the fusion of SspCA and SazCA (De Luca et al. 2015). Another codon-optimized α-CA gene from *Persephonella marina* EX-H1 (PmCA), a gram-negative bacterium from a deep-sea hydrothermal vent, was cloned and expressed in *E. coli*. The PMCA was cloned with and without the signal peptides (PmCA sp-). PmCA sp- showed five times more activity than the intact gene. Its half-life at 100°C was 88 minutes, and it even retained its activity at variable pH (optimum 7.5). The melting temperature for PMCA was 84.5°C. Kinetic parameters for the CO_2 hydration reaction were determined using stopped-flow spectroscopy at 25°C and pH 7.8, showing that PmCA is less active with respect to the other thermophilic α-CAs, with k_{cat} and k_{cat}/K_m values of 3.2×10^5 s^{-1} and 3.0×10^7 M^{-1} s^{-1}, respectively. Biochemical studies showed that PmCA exists as a dimer (Kanth et al. 2014). *Thermovibrio ammonificans*, another gram-negative thermophilic bacterium from deep-sea ecosystems (Costantino et al. 2004), also harbors an extremely thermostable CA (TaCA). This enzyme was also cloned and heterologously expressed in *E. coli*. The thermoactivities of these CAs (PmCA and TaCA) were determined by esterase assay, and as the temperature increased (up to 95°C), the activities continued to increase. Thus, the activity of TaCA was even elevated after the high-temperature incubation (thermostimulation). The residual activity of TaCA after 100°C incubation was higher than (or comparable to) the activity of the untreated TaCA, but was lower than that of TaCA incubated at 90°C. About 91% and 57% of activity was retained after incubation of TaCA and PmCA at 40°C for 60 days, respectively. At 60°C, TaCA and PmCA retained 62% and 27% of their initial activities, respectively, after 60 days' incubation. At 40°C, the half-lives of TaCA and PmCA were 152 and 75 days, respectively. At 60°C, they were reduced to 77 and 29 days (Jo et al. 2014). TaCA is much more thermostable than SazCA and SspCA.

Crystallographic structure analysis of TaCA revealed a fold very similar to that of the bacterial homologs previously studied (Huang et al. 1998; Di Fiore et al. 2013; De Simone et al. 2015), but having an entirely novel oligomeric pattern. Indeed, TaCA forms a tetramer, comprising two dimers, which are structurally similar to those of SspCA and SazCA (Figure 12.3). The two dimers are joined together by two intermolecular disulfide bridges and by intersubunit ionic interactions (James et al. 2014). This tetrameric state may be a possible reason for the enhanced thermostability of TaCA. This idea is further supported by the observation that other thermostable enzymes present a higher degree of oligomerization with respect to their mesophilic counterparts (Tahirov et al. 1998). It is also worth mentioning that the two conserved cysteine residues, Cys47 and Cys202, which form an intramolecular disulfide bond in SspCA and SazCA, are partially reduced as a result of the insufficiently oxidative expression conditions. It may be possible that in oTaCA, this disulfide bond is fully present and responsible for its extra stability with respect to TaCA (James et al. 2014). Both TaCA and PmCA have been isolated in deep hydrothermal vents of the Pacific Ocean.

There have been several reports of alkalistable CAs. *Dunaliella salina*, a unicellular green alga that thrives in a wide array of natural environments ranging from freshwater to hypersaline lakes, shows the presence of CA localized at its surface. This CA even retains its activity at 4.0 M NaCl and at low salinities (Premkumar et al. 2003). It was cloned and heterologously expressed in *E. coli*. An alkalistable α-CA was characterized from *Serratia* sp., which retains 75% of its activity even at pH 12, but it does not retain its activity at the acidic range. Also, a rapid decline of activity was observed from 40°C to 80°C (Srivastava et al. 2015). An extracellular β-CA has been characterized from alkaliphilic cyanobacterium *Microcoleus chthonoplastes* (CahB1), which is maximally active at pH 9.0 and least active at neutral pH. The gene and nucleotide sequence is similar to that of β-CA from *Synechococcus* sp. PCC 7942 (IcfA) and *Synechocystis* sp. PCC 6803 (CcaA) (Kupriyanova et al. 2011). A putative α-CA localized in the glycocalyx was also characterized from *M. chthonoplastes*. This CA has a molecular mass of about 34 kDa and a pI of 3.5. It showed two peaks of activity at around pH 10.0 and 7.5, with the former being the highest (Kupriyanova et al. 2007). *Helicobacter pylori* encodes a periplasmic α-CA (α-CA-HP) and a cytoplasmic β-CA (β-CA-HP). Both of them are optimally active at acidic pH (3.0) (Mone et al. 2008). The properties and important characteristics of CAs from extremophilic sources are summarized in Table 12.2.

The last three decades have seen the advent and development of genome sequencing. A good number of organisms in all three domains have been already sequenced. The sequencing has given good insight into the basic genomic architecture of each domain. It also aids in the discovery of novel products useful for human welfare. Genome analysis has revealed the presence of CA in all forms of life barring a few proteobacteria (Ueda et al. 2012). Numerous putative CA genes have been identified in various extremophilic life-forms (Ferry 2010). Even some of the prokaryotes have more than one type of CA gene. *Pseudomonas aeruginosa* contains six putative CA genes, three each for β and γ (Smith and Ferry 2000). CAs have their signatures in all three domains. These analyses have demonstrated that CAs are ubiquitous among metabolically

Table 12.2 CAs from Extremophiles and Their Characteristic Features

Enzyme	Organism	Class	k_{cat}/K_m $(M^{-1} s^{-1})$	Characteristic Features	Reference
SspCA	*Sulfurihydrogenibium yellowstonense* YO3AOP1	α	1.1×10^8	Dimer Stable at high temperatures (70°C for 3 hours) Optimum working environment: 95°C, pH 9.6	Capasso et al. 2012 Akdemir et al. 2013
SazCA	*Sulfurihydrogenibium azorense*	α	3.5×10^8 (fastest)	Dimer Half-life at 70°C is 8 days, and 53 days at 50°C Alkalistable (pH 9.6)	De Luca et al. 2013 Russo et al. 2013
TaCA	*Thermovibrio ammonificans*	α	1.6×10^8	Tetramer Highly thermostable Half-life of 77 days at 60°C	Jo et al. 2014
PmCA	*Persephonella marina* EX-H1	α	3.0×10^7	Dimer Half-life at 100°C was 88 minutes	Jo et al. 2014 Kanth et al. 2014
CabCA	*Methanobacterium thermoautotrophicum*	β	5.9×10^6	Tetramer Optimal CO_2 hydration activity at 75°C	Smith and Ferry 2000
MtCam (expressed in *E. coli* and purified aerobically)	*Methanosarcina thermophila*	γ	3.1×10^6	Optimal activity at 55°C Stable at 75°C for 15 minutes Activity doubles on replacing zinc with cobalt	Alber and Ferry 1994
MtCam (expressed in *E. coli* and purified anaerobically)	*Methanosarcina thermophila*	γ	5.4×10^6	NA*	Tripp et al. 2004
MtCam (expressed in *M. acetivorans*)	*Methanosarcina thermophila*	γ	3.9×10^6	Fe^{2+} present at active site enhances its activity	Macauley et al. 2009

* NA: not available.

and phylogenetically diverse prokaryotes from both the Bacteria and the Archaea. There are very few reports of characterization of CAs from extremophiles. The initial results from the study of extremophilic CAs have been startling. The characteristics and properties are distinctly different from those of the CAs from other sources. Inhibition studies of SazCA and SspCA have shown that they are responding to a series of new sulfonamide inhibitors (Alafeefy et al. 2014). Research is underway for exploring distinct properties of these novel enzymes, so that they can be utilized for biosequestration purposes.

The advent of metagenomics has led to the discovery of many new nonculturable life-forms that can withstand multiple extremes. In the last few years, numerous microorganisms have been discovered by metagenomic approaches, and all of them have been a rich source of metabolites useful for human welfare. The search for CAs in these metagenomes may lead to the discovery of more efficient and potent CAs. Metagenomics analysis has revealed the presence of CA-encoding genes in the viral metagenome from the Indian Ocean (Williamson et al. 2012). Jones et al. (2012) also showed the presence of CA and RuBisCo gene clusters in the *Acidithiobacillus* of an extremely acidophilic sulfur-oxidizing biofilm by community genome analysis. Microarray data of the metagenome of acid mine drainage also showed the presence of CA-encoding genes, along with the RuBisCo gene clusters (Guo et al. 2013). At least three copies of genes that code for CA have been reported in the metagenome of the marine ammonium-oxidizing bacterium (Vossenberg et al. 2013). The metagenomes of the serpentinite-hosted Lost City hydrothermal field, Mid-Atlantic Ridge, also showed the presence of CA-encoding genes. These metagenomes were found to be similar to the genome of *Thiomicrospira crunogena* XCL-2, an isolate from a basalt-hosted hydrothermal vent in the Pacific Ocean (Brazelton and Baross 2010). The current available genomic data reveal the presence of many putative CA-encoding genes. The characteristics of the novel CAs are still unknown. Exploring them might reveal an array of new features concerning the enzyme, which will add up to their existing potential. Currently, six classes of CAs with distinct properties have been discovered. Future research may lead to the discovery of CAs with novel properties.

12.10 APPLICABILITY OF THERMOSTABLE CAs IN CARBON CAPTURE AND STORAGE AND THEIR IMMOBILIZATION STRATEGIES

In order to operate bioreactors for sequestering carbon from flue gas, CAs must be thermostable and tolerant to the oxides of nitrogen and sulfur. This necessitates a hunt for CAs that are stable and functional under extreme conditions prevailing in flue gas. CA purified from *M. thermophila* is active at 70°C for 15 minutes. It is a γ-CA with a k_{cat} of 10^5 s^{-1}, and it is optimally active at 55°C (Alber and Ferry 1994). Similarly, highly thermostable CAs from thermophilic archaea are some of the suitable candidates for this purpose. SspCA and SazCA retain activity in the presence of various anionic inhibitors, such as NO_2^-, NO_3^-, and SO_4^{2-}, and high concentrations of carbonates, bicarbonates, and hydrogen sulfides (De Luca et al. 2012; Vullo et al. 2012). PMCA has a high efficiency of CO_2 hydration at higher temperatures. SspCA was characterized as a potential biocatalyst for CO_2 capture processes based on its regenerative absorption ability in alkaline solutions. Its prolonged half-life (53 days at 40°C and 8 days at 70°C) also make it one of the most suitable candidates for CCS. SspCA was therefore tested for its biomimetic carbon sequestration (Russo et al. 2013). Heat-stable CAs from *M. thermophila* and *Caminibacter* sp. have been used in bioreactors for efficient CO_2 removal from flue gases. CAs from *P. marina* and *T. ammonificans* can also be used for sequestration because of their thermostability (Di Fiore et al. 2015).

Immobilization strategies are necessary so that the enzymes can be recycled a number of times. Their stability can be enhanced by various immobilization techniques. Immobilization has long been used as an approach for increasing the stability of mesophilic enzymes. There are several reports of immobilization of mesophilic CAs. These CAs have been proved to be more efficient than the free enzymes for CO_2 capture. CAs from bovine erythrocytes *Pseudomonas fragi* and *Bacillus pumilus* have been shown to display enhanced CO_2 hydration capacity after their immobilization in polyutherane foam (Wanjari et al. 2011). Immobilization of CAs also improved their thermal stabilities; for example, the immobilized CAs retained at least 60% of the initial activity after 90 days at 50°C, compared with about 30% for their free counterparts under the same conditions. The CAs also exhibit a high stability in the presence of inhibitors upon immobilization (Kanbar and Ozdemir 2010; Prabhu et al. 2009, 2011). The $CaCO_3$ precipitation rate doubled in a period of 5 minutes when pure CA from *P. fragi* was immobilized by adsorption on chitosan beads in comparison with the free enzyme (Wanjari et al. 2012; Yadav et al. 2012). Immobilization of CAs on several other matrices, such as ordered mesoporous aluminosilicate, octa(aminophenyl) silsesquioxane-functionalized Fe_3O_4/SiO_2 nanoparticles, and silica nanoparticles, and by single or multiple attachments to polymers deposited on Fe_3O_4 particles was also attempted (Sharma and Bhattacharya 2010; Rayalu et al. 2012). CA immobilized on ordered mesoporous aluminosilicate exhibited a CO_2 sequestration efficiency of 16.14 mg of $CaCO_3$/mg CA compared with that of free enzyme, which sequesters 33.08 mg of $CaCO_3$/mg CA. Immobilized CA even showed enhanced stability, and it retained 67% of its initial activity even after six cycles (Yadav et al. 2011). The kinetics of the immobilized CAs (immobilized on ordered mesoporous aluminosilicate) were studied and compared with that of the free enzyme by Yadav et al. (2010). The K_m, V_{max}, and k_{cat} values of the immobilized enzyme were 0.158 mM, 2.307 μ mole min^{-1} mL^{-1}, and 1.9 s^{-1}, and those for free CA were 0.876 mM, 0.936 μ mole min^{-1} mL^{-1}, and 2.3 s^{-1}, respectively (Yadav et al. 2010). A high CO_2 sequestration and improved stability were achieved when CA was immobilized on core-shell CA-chitosan nanoparticles (SEN-CA) (Rayalu et al. 2012). SspCA was immobilized on a solid matrix made of silica particles (silanized Sipernat®, 22 particles). Enzyme–carrier covalent bonding was adopted as an immobilization technique, and it was observed that the enzyme stability and activity increased upon immobilization. Immobilization of this CA in polyutherane foam also enhanced its stability. The immobilized CA (PU-SspCA) showed exceptional thermostability for a very long duration even at 70°C. The CO_2 absorption capacity of PU-SspCA was verified in a three-phase trickled bed bioreactor, which mimics the postcombustion processes in a thermal power plant. The three-phase reactor was filled with shredded foam with PU-SspCA. The gas mixture (CO_2/N_2) was fed from both sides (i.e., concurrent and countercurrent). Increasing the flow rate of water and decreasing the CO_2 flow rate also greatly improved CO_2 capture in these reactors. SspCA showed good CO_2 capture performance when the PU-SspCA shredded foam was used in the bioreactor (Migliardini et al. 2014). Concerted efforts are needed to develop CA immobilization techniques that allow reuse of the enzyme 100–500 times with sustained activity. Immobilization can also lead to unwanted release of enzyme on the surface of the reactor. In order to overcome this problem, some novel immobilization techniques were designed using γ-CA from *M. thermophila* and *Pyrococcus horikoshii* (Salemme and Weber 2014). The immobilization techniques describe the development of γ-CA nanoassemblies, where single enzyme entities are joined with each other and make numerous linked interactions with the reactor surface. This was achieved by mutating specific enzyme residues to cysteines, in order to create sites for biotinylation, thus ensuring the formation of firm nanostructures by cross-linking biotinylated-γ-CAs with streptavidin tetramers (Salemme and Weber 2014). Further addition of an immobilization sequence at the amino- or carboxy-terminus also allows a controlled and reversible immobilization of the γ-CA to a functionalized surface. The use of immobilized thermostable CAs will thus be useful in the CO_2 capture processes.

12.11 CONCLUSIONS AND FUTURE PERSPECTIVES

Global warming caused due to an enormous increase in carbon dioxide levels in the atmosphere is the main culprit for climate change. It is a tough task to reduce or control CO_2 emissions because of the high energy demands. Developing countries are facing the major wrath of global warming, since their energy demands are constantly increasing. Most of the developing countries are dependent on fossil fuels (coal, petroleum, and natural gas) to meet their energy demands. Harnessing alternative sources of energy is also very expensive. These arguments will not, however, help us in solving the menace of global warming. It is indispensable to develop strategies for mitigating CO_2 in the atmosphere. Biological methods of carbon sequestration are not expected to have any deleterious effects on the environment. Afforestation is one of the easiest methods that can be employed for achieving this goal. Other methods, such as the cultivation of algae, microbes, and biomineralization, are also some of the potential approaches. In this context, the utility of CAs for capturing CO_2 is an efficient method for mitigating global warming. Biomineralization will not only return the carbon to nature in a stable form (carbonates and bicarbonates) but also help in the effective utilization of the captured carbon to produce useful products. Although thermostable CAs engineered from mesophilic analogs such as hCAII and DvCA are alternatives, the development of efficient CAs by genetic engineering will be needed. Thermostable α-CAs from thermophiles (SazCA, SspCA, and TaCA) are some of the potent candidates that can be used in carbon sequestration. These CAs can be employed in developing cost-effective biomimetic carbon capture technologies.

REFERENCES

Adams, M.R., Hurd, H.B., Lenhart, S., Leary, N. Effects of Global Climate Change on Agriculture: An Interpretative Review. *Climate Research* 11 (1998): 19–30.

Aguiar, P., Beveridge, T.J., Reysenbach, A.L. *Sulfurihydrogenibium azorense*, sp. nov., a Thermophilic Hydrogen-Oxidizing Microaerophile from Terrestrial Hot Springs in the Azores. *International Journal of Systematic and Evolutionary Microbiology* 54 (2004): 33–39.

Akdemir, A., Vullo, D., De Luca, V. et al. The Extremo-Alpha-Carbonic Anhydrase (CA) from *Sulfurihydrogenibium azorense*, the Fastest CA Known, Is Highly Activated by Amino Acids and Amines. *Bioorganic & Medicinal Chemistry Letters* 23 (2013): 1087–1090.

Alafeefy, A.M., Abdel-Aziz, H.A., Vullo, D. et al. Inhibition of Carbonic Anhydrases from the Extremophilic Bacteria *Sulfurihydrogenibium yellowstonense* (SspCA) and *S. azorense* (SazCA) with a New Series of Sulfonamides Incorporating Aroylhydrazone-, [1,2,4]Triazolo[3,4-b][1,3,4]Thiadiazinyl- or 2-(Cyanophenylmethylene)-1,3,4-Thiadiazol-3(2H)-yl Moieties. *Bioorganic & Medicinal Chemistry* 22 (2014): 1141–1147.

Alber, B.E., and Ferry, J.G. A Carbonic Anhydrase from the Archaeon *Methanosarcina thermophila*. *Proceedings of the National Academy of Sciences of the United States of America* 91 (1994): 6909–6913.

Alber, B.E., and Ferry, J.G. Characterization of Heterologously Produced Carbonic Anhydrase from *Methanosarcina thermophila*. *Journal of Bacteriology* 178 (1996): 3270–3274.

Alvizo, O., Nguyen, L.J., Savile, C.K. et al. Directed Evolution of an Ultrastable Carbonic Anhydrase for Highly Efficient Carbon Capture from Flue Gas. *Proceedings of the National Academy of Sciences of the United States of America* 111 (2014): 16436–16441.

Anand, A., Kumar, V., and Satyanarayana, T. Characteristics of Thermostable Endoxylanase and β-Xylosidase of the Extremely Thermophilic Bacterium *Geobacillus thermodenitrificans* TSAA1 and Its Applicability in Generating Xylooligosaccharides and Xylose from Agro-Residues. *Extremophiles* 17 (2013): 357–366.

Andersson, B., Nyman, P.O., and Strid, L. Amino Acid Sequence of Human Erythrocyte Carbonic Anhydrase B. *Biochemical and Biophysical Research Communications* 48 (1972): 670–677.

Badger, M.R., and Price, G.D. CO_2 Concentrating Mechanisms in Cyanobacteria: Molecular Components, Their Diversity and Evolution. *Journal of Experimental Botany* 54 (2003): 609–622.

Batjes, N.H. Total Carbon and Nitrogen in the Soils of the World. *European Journal of Soil Science* 47 (1996): 151–163.

Benson, S., Cook, P., Anderson, J. et al. Underground Geological Storage. In Metz, B., Davidson, O., Connick de, H., Loos, M., Meyer, L. (eds), *Special Report on Carbon Dioxide Capture and Storage*, 195–277. Cambridge: Cambridge University Press, 2005.

Bhojiya, A.A., and Joshi, H. Isolation and Characterization of Zinc Tolerant Bacteria from Zawar Mines Udaipur, India. *International Journal of Environmental Engineering and Management* 3 (2012): 239–242.

Brazelton, J.W., and Baross, A.J. Metagenomic Comparison of Two *Thiomicrospira* Lineages Inhabiting Contrasting Deep-Sea Hydrothermal Environments. *PLoS One* 5 (2010): e13530.

Canganella, F., and Wiegel, J. Extremophiles: From Abyssal to Terrestrial Ecosystems and Possibly Beyond. *Naturwissenschaften* 98 (2011): 253–279.

Capasso, C., De Luca, V., Carginale, V. et al. Characterization and Properties of a New Thermoactive and Thermostable Carbonic Anhydrase. *Chemical Engineering Transactions* 27 (2012): 271–276.

Carson, J.K., Marsh, K.N., Mather, A.E. Enthalpy of Solution of Carbon Dioxide in (Water + Monoethanolamine, or Diethanolamine, or *N*-Methyldiethanolamine) and (Water + Monoethanolamine + *N*-Methyldiethanolamine) at T = 298.15 K. *Journal of Chemical Thermodynamics* 32 (2000): 1285–1296.

Christer, J., Wullschleger, D.S., Kalluri, C.U., Tuskan, A.G. Phytosequestration: Carbon Biosequestration by Plants and the Prospects of Genetic Engineering. *BioScience* 60 (2010): 685–696.

Ciullo, P.A. *Industrial Minerals and Their Uses: A Handbook and Formulatory*, 26–28. Westwood, NJ: William Andrew Publishing, 1996.

Costantino, V., Mark, D.S., Susan, V.E., Lutz, A.R., Starovoytov, V. *Thermovibrio ammonificans* sp. nov., a Thermophilic, Chemolithotrophic, Nitrate-Ammonifying Bacterium from Deep-Sea Hydrothermal Vents. *International Journal of Systematic and Evolutionary Microbiology* 54 (2004): 175–181.

CO_2 Solution Inc. 2012. Enzymatic power for carbon capture. *The process*. Details available at http:/www .co2solution.com/en/the-process.

Cowan, R.M., Ge, J.J., Qin, Y.J., McGregor, M.L., Trachtenberg, M.C. CO_2 Capture by Means of an Enzyme-Based Reactor. *Annals of New York Academy of Sciences* 984 (2003): 453–469.

Cronk, J.D., Endrizzi, J.A., Cronk, M.R., O'Neill, J.W., Zhang, K.Y. Crystal Structure of *E. coli* β-Carbonic Anhydrase, an Enzyme with an Unusual pH-Dependent Activity. *Protein Science* 10 (2001): 911–922.

Dahlberg, L., Hoist, O., Kristjansson, K.J. Thermostable Xylanolytic Enzymes from *Rhodothermus marinus* Grown on Xylan. *Applied Microbiology and Biotechnology* 40 (1993): 63–68.

Daigle, R., and Desrochers, M. Carbonic Anhydrase Having Increased Stability under High Temperature Conditions. U.S. Patent 7521217, 2009.

Daigle, R.M.E., and Fradette, S. Techniques for CO_2 Capture Using *Sulfurihydrogenibium* sp. Carbonic Anhydrase. Patent WO 2014066999 A1, May 8, 2014.

De Luca, V., Prete, S.D., Carginale, V. et al. A Failed Tentative to Design a Super Carbonic Anhydrase Having the Biochemical Properties of the Most Thermostable CA (SspCA) and the Fastest (SazCA) Enzymes. *Journal of Enzyme Inhibition and Medicinal Chemistry* 30 (2015): 989–994.

De Luca, V., Vullo, D., Scozzafava, A. et al. Anion Inhibition Studies of an Alpha-Carbonic Anhydrase from the Thermophilic Bacterium *Sulfurihydrogenibium yellowstonense* YO3AOP1. *Bioorganic & Medicinal Chemistry Letters* 22 (2012): 5630–5634.

De Luca, V., Vullo, D., Scozzafava, A. et al. An Alpha-Carbonic Anhydrase from the Thermophilic Bacterium *Sulfurihydrogenibium azorense* is the Fastest Enzyme Known for the CO_2 Hydration Reaction. *Bioorganic & Medicinal Chemistry Letters* 21 (2013): 1465–1469.

Del Prete, S., Vullo, D., Fisher, G.M. et al. Discovery of a New Family of Carbonic Anhydrases in the Malaria Pathogen *Plasmodium falciparum*—The η-Carbonic Anhydrases. *Bioorganic & Medicinal Chemistry Letters* 24 (2014): 4389–4396.

De Simone, G., Monti, S.M., Alterio, V. et al. Crystal Structure of the Most Catalytically Effective Carbonic Anhydrase Enzyme Known, SazCA from the Thermophilic Bacterium *Sulfurihydrogenibium azorense*. *Bioorganic & Medicinal Chemistry Letters* 25 (2015): 2002–2006.

Di Fiore, A., Alterio, V., Monti, S.M., De Simone, G., Ambrosio, K.D. Thermostable Carbonic Anhydrases in Biotechnological Applications. *International Journal of Biological Sciences* 16 (2015): 15456–15480.

Di Fiore, A., Capasso, C., De Luca, V. et al. X-Ray Structure of the First "Extremo-Alpha-Carbonic Anhydrase," a Dimeric Enzyme from the Thermophilic Bacterium *Sulfurihydrogenibium yellowstonense* YO3AOP1. *Acta Crystallographica: Section D, Biological Crystallography* 69 (2013): 1150–1159.

Duarte, T., Cristina, M., Pellegrino, A., Carolina, A., Portugal, P.E., Ponezi, N.A., Franco, T.T. Characterization of Alkaline Xylanases from *Bacillus pumilus*. *Brazilian Journal of Microbiology* 31 (2000): 90–94.

Dutreuil, S., Bopp, L., Tagliabue, A. Impact of Enhanced Vertical Mixing on Marine Biogeochemistry: Lessons for Geo-Engineering and Natural Variability. *Biogeosciences* 6 (2009): 901–912.

Eriksson, A.E., Jones, T.A., Liljas, A. Refined Structure of Human Carbonic Anhydrase II at 2.0 Angstrom Resolution. *Proteins: Structure, Function and Genetics* 4 (1988): 274–282.

Faridi, S., and Satyanarayana, T. Bioconversion of Industrial CO_2 Emissions into Utilizable Products. In Chandra, R. (ed), *Industrial Waste Management*, 111–156. New York: CRC Press, 2015.

Faridi, S., and Satyanarayana, T. Novel Alkalistable α-Carbonic Anhydrase from the Polyextremophilic Bacterium *Bacillus halodurans*: Characteristics and Applicability in Flue Gas CO_2 Sequestration. *Environmental Science and Pollution Research* 23 (2016): 15236–15249.

Favre, N., Christ, M.L., Pierre, A.C. Biocatalytic Capture of CO_2 with Carbonic Anhydrase and Its Transformation to Solid Carbonate. *Journal of Molecular Catalysis B: Enzymatic* 60 (2009): 163–170.

Ferry, J.G. The Gamma Class of Carbonic Anhydrases. *Biochimica et Biophysica Acta* 1804 (2010): 374–381.

Fisher, Z., Boone, D.C., Biswas S.M. et al. Kinetic and Structural Characterization of Thermostabilized Mutants of Human Carbonic Anhydrase II. *Protein Engineering, Design & Selection* 25 (2012): 347–355.

Fujita, N., Mori, H., Yura, T. et al. Systematic Sequencing of the *Escherichia coli* Genome: Analysis of the 2.4–4.1 Min (110,917–193,643 bp) Region. *Nucleic Acids Research* 22 (1994): 1637–1639.

Gill, S.R., Fedorka-C.P.J., Tweten, R.K. et al. Purification and Properties of the Carbonic Anhydrase of *Rhodospirillum rubrum*. *Archives of Microbiology* 138 (1984): 113–118.

Guilloton, M.B., Korte, J.J., Lamblin, A.F., Fuchs, J.A., Anderson, P.M. Carbonic Anhydrase in *Escherichia coli*, a Product of the Cyn Operon. *Journal of Biological Chemistry* 267 (1992): 3731–3734.

Guo, X., Yin, H., Cong, J., Dai, Z., Liang, Y., Liua, X. RubisCO Gene Clusters Found in a Metagenome Microarray from Acid Mine Drainage. *Applied and Environmental Microbiology* 79 (2013): 2019–2026.

Hart, A., and Gnanendran, N. Cryogenic CO_2 Capture in Natural Gas. *Energy Procedia* 1 (2009): 697–706.

Haszeldine, S.R. Carbon Capture and Storage: How Green Can Black Be. *Science* 325 (2009): 1647–1652.

Henderson, L.E., Henriksson, D., Nyman, P.O. Amino Acid Sequence of Human Erythrocyte Carbonic Anhydrase C. *Biochemical and Biophysical Research Communications* 52 (1973): 1388–1394.

Herzog, H.J. Ocean Sequestration of CO_2—An Overview. In *Fourth International Conference on Greenhouse Gas Control Technologies*, Interlaken, Switzerland, August 30–September 2, 1998, 1–5.

Hewett-Emmett, D., and Tashian, R.E. Structure and Evolutionary Origins of the Carbonic Anhydrase Multigene Family. In *The Carbonic Anhydrases: Cellular Physiology and Molecular Genetics*, 15–32. New York: Plenum Press, 1991.

Hewett-Emmett, D., and Tashian, R.E. Functional Diversity, Conservation and Convergence in the Evolution of the α, β and γ Carbonic Anhydrase Gene Families. *Molecular Phylogenetics and Evolution* 5 (1996): 50–77.

Horikoshi, K., and Bull, A.T. Prologue: Definition, Categories, Distribution, Origin and Evolution, Pioneering Studies, and Emerging Fields of Extremophiles. In *Extremophiles Handbook*. Berlin: Springer, 2001.

Houghton, T.J. Global Warming. *Reports on Progress in Physics* 68 (2005): 1343–1403.

Huang, H.P., Shi, Y., Li, W., Chang, S.G. Dual Alkali Approaches for the Capture and Separation of CO_2. *Energy and Fuels* 15 (2001): 263–268.

Huang, S., Xue, Y., Sauer, E.E. et al. Crystal Structure of Carbonic Anhydrase from *Neisseria Gonorrhoeae* and Its Complex with the Inhibitor Acetazolamide. *Journal of Molecular Biology* 283 (1998): 301–310.

Huber, R.K.M, Jannasch, H.W., Stetter, K.O. A Novel Group of Abyssal Methanogenic Archaebacteria (*Methanopyrus*) Growing at 110 C. *Nature* 342 (1989): 833–834.

Huijgen, W.J.J., Comans, R.N.J., Witkamp, G.J. Cost Evaluation of CO_2 Sequestration by Aqueous Mineral Carbonation. *Energy Conversion and Management* 48 (2007): 1923–1935.

Iliuta, I., and Larachi, F. New Scrubber Concept for Catalytic CO_2 Hydration by Immobilized Carbonic Anhydrase II & In-Situ Inhibitor Removal in Three-Phase Monolith Slurry Reactor. *Separation and Purification Technology* 86 (2012): 199–214.

IPCC (Intergovernmental Panel on Climate Change). *Climate Change 2014 Synthesis Report*. Cambridge: Cambridge University Press, 2014.

Iverson, T.M., Alber, B.E., Kisker, C. et al. A Closer Look at the Active Site of γ-Class Carbonic Anhydrases: High-Resolution Crystallographic Studies of the Carbonic Anhydrase from *Methanosarcina thermophila*. *Biochemistry* 39 (2000): 9222–9231.

James, P., Isupov, M.N., Sayer, C. et al. The Structure of a Tetrameric Alpha-Carbonic Anhydrase from *Thermovibrio ammonificans* Reveals a Core Formed around Intermolecular Disulfides That Contribute to Its Thermostability. *Acta Crystallographica: Section D, Biological Crystallography* 70 (2014): 2607–2618.

Jeyakanthan, J., Rangarajan, S., Mridula, P. et al. Observation of a Calcium-Binding Site in the Gamma-Class Carbonic Anhydrase from *Pyrococcus horikoshii*. *Acta Crystallographica: Section D, Biological Crystallography* 64 (2008): 1012–1019.

Jin, X., Gruber, N., Frenzel, H., Doney, S.C., McWilliams, J.C. The Impact on Atmospheric CO_2 of Iron Fertilization Induced Changes in the Ocean's Biological Pump. *Biogeosciences* 5 (2008): 385–406.

Jo, B.H., Kim, I.G., Seo, J.H., Dong, G.K., Hyung, J.C. Engineered *Escherichia coli* with Periplasmic Carbonic Anhydrase as a Biocatalyst for CO_2 Sequestration. *Applied and Environmental Microbiology* 79 (2013): 6697–6705.

Jo, B.H., Seo, J.H., Cha, J.C. Bacterial Extremo-α-Carbonic Znhydrases from Deep-Sea Hydrothermal Vents as Potential Biocatalysts for CO_2 Sequestration. *Journal of Molecular Catalysis B: Enzymatic* 109 (2014): 31–39.

Jones, S.D., Albrecht, L.H., Dawson, S.K. Community Genomic Analysis of an Extremely Acidophilic Sulfur-Oxidizing Biofilm. *ISME Journal* 6 (2012): 158–117.

Kanbar, B., and Ozdemir, E. Thermal Stability of Carbonic Anhydrase Immobilized within Polyurethane Foam. *Biotechnology Progress* 26 (2010): 1474–1480.

Kanth, B.K., Jun, S.Y., Kumari, S. Highly Thermostable Carbonic Anhydrase from *Persephonella marina* EX-H1: Its Expression and Characterization for CO_2 Sequestration Applications. *Process Biochemistry* 49 (2014): 2114–2121.

Kaur, S., Bhattacharya, A., Sharma, A., Tripathi, A.K. Diversity of Microbial Carbonic Anhydrases, Their Physiological Role and Applications. In Satyanarayana, T., Johri, B., Prakash, A. (eds), *Microorganisms in Environmental Management: Microbes and Environment*, 151–173. New York: Springer, 2012.

Kindermann, G., Obersteiner, M., Sohngen, B. et al. Global Cost Estimates of Reducing Carbon Emissions through Avoided Deforestation. *Proceedings of the National Academy of Sciences of the United States of America* 105 (2008): 10302–10307.

Kozliak, E.I., Guilloton, M.B., Gerami, N.M. et al. Expression of Proteins Encoded by the *Escherichia coli cyn* Operon: Carbon Dioxide-Enhanced Degradation of Carbonic Anhydrase. *Journal of Bacteriology* 176 (1994): 5711–5717.

Kubler, J.E., Johnston, A.M., Raven, J.A. The Effects of Reduced and Elevated CO_2 and O_2 on the Seaweed *Lomentaria articulate*. *Plant Cell and Environment* 22 (1999): 1303–1310.

Kumar, V., and Satyanarayana, T. Applicability of Thermo-Alkali-Stable and Cellulase-Free Xylanase from a Novel Thermo-Halo-Alkaliphilic *Bacillus halodurans* in Producing Xylooligosaccharides. *Biotechnology Letters* 33 (2011): 2279–2285.

Kupriyanova, E.V., Sinetova, M.A., Markelova, A.G., Allakhverdiev, S.I., Los, D.A., Pronina N.A. Extracellular β-Class Carbonic Anhydrase of the Alkaliphilic Cyanobacterium *Microcoleus chthonoplastes*. *Journal of Photochemistry and Photobiology B: Biology* 103 (2011): 78–86.

Kupriyanova, E.V., Villarejo, A., Markelova, A. et al. Extracellular Carbonic Anhydrases of the Stromatolite-Forming Cyanobacterium *Microcoleus chthonoplastes*. *Microbiology* 153 (2007): 1149–1156.

Lane, T.W., and Morel, F.M.M. A Biological Function for Cadmium in Marine Diatoms. *Proceedings of the National Academy of Sciences of the United States of America* 97 (2000): 4627–4631.

Lane, T.W., Saito, M.A., George, G.N. et al. Biochemistry: A Cadmium Enzyme from a Marine Diatom. *Nature* 435 (2005): 42.

Lapointe, M., MacKenzie, T.D.B., Morse, D. An External δ-Carbonic Anhydrase in a Free-Living Marine Dinoflagellate May Circumvent Diffusion-Limited Carbon Acquisition. *Plant Physiology* 147 (2008): 1427–1436.

Lavery, T.J., Roudnew, B., Gill, P. et al. Iron Defecation by Sperm Whales Stimulates Carbon Export in the Southern Ocean. *Proceedings of the Royal Society B* 277 (2010): 3527–3531.

Liljas, A., Kannan, K.K., Bergsten, P.C. et al. Crystal Structure of Human Carbonic Anhydrase C. *Nature New Biology* 235 no. 57 (1972): 131–137.

Lim, X. How to Make the Most of Carbon Dioxide. *Nature* 526 (2015): 629–630.

Lin, K.T.D., and Deutsch, H.F. Human Carbonic Anhydrase. XI. The Complete Primary Structure of Carbonic Anhydrase B. *Journal of Biological Chemistry* 248 (1973): 1885–1893.

Lin, K.D., and Deutsch, H.F. Human Carbonic Anhydrase. XII. The Complete Primary Structure of Carbonic Anhydrase C. *Journal of Biological Chemistry* 249 (1974): 2329–2337.

Lindskog, S., and Coleman, E.J. The Catalytic Mechanism of Carbonic Anhydrase. *Proceedings of the National Academy of Sciences of the United States of America* 70 (1973): 2505–2508.

Lovelock, J.E., and Rapley, C.G. Ocean Pipes Could Help the Earth to Cure Itself. *Nature* 449 (2007): 403.

Macauley, S.R., Zimmerman, A.S., Apolinario, E.E. et al., The Archetype Gamma-Class Carbonic Anhydrase (Cam) Contains Iron When Synthesized In Vivo. *Biochemistry* 48 (2009): 817–819.

MacElory, R.D. Some Comments on Evolution of Extremophiles. *Biosystems* 6 (1974): 74–75.

Mahinpey, N., Asghari, K., Mirjafari, P. Bicarbonate Produced from Carbon Capture for Algae Culture. *Chemical Engineering Research and Design* 89 (2011): 1873–1878.

Meldrum, N.U., and Roughton, F.J. Carbonic Anhydrase. Its Preparation and Properties. *Journal of Physiology* 80 (1933): 113.

Mesbah, M.N., and Wiegel, J. Life under Multiple Extreme Conditions: Diversity and Physiology of the Halophilic Alkalithermophiles. *Applied and Environmental Microbiology* 78 (2012): 4074–4081.

Migliardini, F., De Luca, V., Carginale, V. et al. Biomimetic CO_2 Capture Using a Highly Thermostable Bacterial α-Carbonic Anhydrase Immobilized on Polyurethane Foam. *Journal of Enzyme Inhibition and Medicinal Chemistry* 29 (2014): 146–150.

Min, D.H. Carbon Sequestration Potential of Switch Grass as a Bioenergy Crop. East Lansing: Michigan State University, 2011.

Mirjafari, P., Asghari, K., Mahinpey, N. Investigating the Application of Enzyme Carbonic Anhydrase for CO_2 Sequestration Purposes. *Industrial and Engineering Chemistry Research* 46 (2007): 921–926.

Monastersky, R. Iron versus the Greenhouse: Oceanographers Cautiously Explore a Global Warming Therapy. *Science News* 148 (1995): 220.

Mone, S.B., Mendz, G.L., Ball, G.E. et al. Roles of α and β Carbonic Anhydrases of *Helicobacter pylori* in the Urease-Dependent Response to Acidity and in Colonization of the Murine Gastric Mucosa. *Infection and Immunity* 76 (2008): 497–509.

Monti, S.M., De Simone, G., Dathan N.A. et al. Kinetic and Anion Inhibition Studies of a β-Carbonic Anhydrase (FbiCA 1) from the C4 Plant *Flaveria bidentis*. *Bioorganic & Medicinal Chemistry Letters* 23 (2013): 1626–1630.

Mori, K., Ogawa, Y., Ebihara, K. et al. Isolation and Characterization of CA XIV, a Novel Membrane-Bound Carbonic Anhydrase from Mouse Kidney. *Journal of Biological Chemistry* 274 (1999): 15701–15705.

Newman, L., Clark, M.L., Ching, C. et al. Carbonic Anhydrase Polypeptides and Uses Thereof. U.S. Patent WO 10081007, 2010.

Ningthoujam, S.D., Kshetri, P., Sanasam, S., Nimaichand, S. Screening, Identification of Best Producers and Optimization of Extracellular Proteases from Moderately Halophilic Alkalithermotolerant Indigenous Actinomycetes. *World Applied Sciences Journal* 7 (2009): 907–916.

Nisha, M., and Satyanarayana, T. Characterization and Multiple Applications of a Highly Thermostable and Ca^{2+}-Independent Amylopullulanase of the Extreme Thermophile *Geobacillus thermoleovorans*. *Applied Biochemistry and Biotechnology* 174 (2014): 2594–2615.

Norici, A., Dalsass, A., Giordano, M. Role of Phospho-Enolpyruvate Carboxylase in Anaplerosis in the Green Microalga *Dunaliella salina* Cultured under Different Nitrogen Regimes. *Physiologia Plantarum* 116 (2002): 186–191.

Onstott, T.C., Moser, D.P., Pfiffner, S.M. et al. Indigenous and Contaminant Microbes in Ultradeep Mines. *Environmental Microbiology* 5 (2003): 1168–1191.

Orr, J.F.M. CO_2 Capture and Storage: Are we Ready? *Energy and Environmental Science* 2 (2009): 449.

Parr, J.F., and Sullivan, L.A. Soil Carbon Sequestration in Phytoliths. *Soil Biology and Biochemistry* 37 (2005): 117–124.

Pena, K.L., Castel, S.E., de Araujo, C. et al. Structural Basis of the Oxidative Activation of the Carboxysomal Gamma-Carbonic Anhydrase, CcmM. *Proceedings of the National Academy of Sciences of the United States of America* 107 (2010): 2455–2460.

Prabhu, C., Wanjari, S., Gawande, S. et al. Immobilization of Carbonic Anhydrase Enriched Microorganism on Biopolymer Based Materials. *Journal of Molecular Catalysis B: Enzymatic* 60 (2009): 13–21.

Prabhu, C., Wanjari, S., Puri, A. et al. Region Specific Bacterial Carbonic Anhydrase for Biomimetic Sequestration of Carbon Dioxide. *Energy and Fuels* 25 (2011): 1327–1332.

Premkumar, L., Bageshwar U.K., Gokhman, I., Zamir, A., Sussman, L.J. An Unusual Halotolerant Alpha-Type Carbonic Anhydrase from the Alga *Dunaliella salina* Functionally Expressed in *Escherichia coli*. *Protein Expression and Purification* 28 (2003): 151–157.

Princiotta, F. The Role of Power Generation Technology in Mitigating Global Climate Change. In Cen, K., Chi, Y., Wang, F. (eds), *Challenges of Power Engineering and Environment*, 3–13. Berlin: Springer, 2007.

Puxty, G., Rowland, R., Allport, A. et al. Carbon Dioxide Postcombustion Capture: A Novel Screening Study of the Carbon Dioxide Absorption Performance of 76 Amines. *Environmental Science and Technology* 43 (2009): 6427–6433.

Rai, K.S., Roy, K.J., Mukherjee, K.A. Characterisation of a Detergent-Stable Alkaline Protease from a Novel Thermophilic Strain *Paenibacillus tezpurensis* sp. nov. AS-S24-II. *Applied Microbiology and Biotechnology* 85 (2010): 1437–1450.

Rao, J.L.U.M., and Satyanarayana, T. Biophysical and Biochemical Characterization of a Hyperthermostable and Ca^{2+}-Independent α-Amylase of an Extreme Thermophile *Geobacillus thermoleovorans*. *Applied Biochemistry and Biotechnology* 150 (2008): 205–219.

Rayalu, S., Yadav, R., Wanjari, S. et al. Nanobiocatalysts for Carbon Capture, Sequestration and Valorization. *Topics in Catalysis* 55 no. 16 (2012): 1217–1230.

Rossi, M. A New Heat-Stable Carbonic Anhydrase and Uses Thereof. Patent WO 2013064195 A1, May 8, 2014.

Russo, M.E., Oliveri, G., Clemente. C. et al. Kinetic Study of a Novel Thermo-Stable Alpha-Carbonic Anhydrase for Biomimetic CO_2 Capture. *Enzyme and Microbial Technology* 53 (2013): 271–277.

Salemme, F.R., and Weber, P.C. Engineered Carbonic Anhydrase Proteins for CO_2 Scrubbing Applications. U.S. Patent 20140178962 A1, 2014.

Sawaya, M.R., Cannon, G.C., Heinhorst, S. et al. The Structure of Beta-Carbonic Anhydrase from the Carboxysomal Shell Reveals a Distinct Subclass with One Active Site for the Price of Two. *Journal of Biological Chemistry* 281 (2006): 7546–7555.

Seckbach, J., Oren, A., Stan-Lotter, H. *Polyextremophiles: Life under Multiple Forms of Stress.* Dordrecht: Springer, 2013.

Seifzadeh, S., Sajedi, H.R., Sariri, R. Isolation and Characterization of Thermophilic Alkaline Proteases Resistant to Sodium Dodecyl Sulfate and Ethylene Diamine Tetraacetic Acid from *Bacillus* sp. GUS1. *Iranian Journal of Biotechnology* 6 (2008): 214–221.

Shakun, J.D., Clark, P.U., He, F. et al. Global Warming Preceded by Increasing Carbon Dioxide Concentrations during the Last Deglaciation. *Nature* 484 (2012): 49–54.

Sharma, A., and Bhattacharya, A. Enhanced Biomimetic Sequestration of CO_2 into $CaCO_3$ Using Purified Carbonic Anhydrase from Indigenous Bacterial Strains. *Journal of Molecular Catalysis B: Enzymatic* 67 (2010): 122–128.

Sharma, A., and Satyanarayana, T. High Maltose-Forming, Ca^{2+}-Independent and Acid Stable α-Amylase from a Novel Acidophilic Bacterium, *Bacillus acidicola. Biotechnology Letters* 32 (2010): 1503–1507.

Silverman, D.N., and Lindskog, S. The Catalytic Mechanism Of Carbonic Anhydrase: Implications of a rate-Limiting Photolysis of Water. *Accounts of Chemical Research* 21 (1988): 30–36.

Smith, K.S., Cosper, N.J., Stalhandske, C., Scott, R.A., Ferry, J.G. Structural and Kinetic Characterization of an Archaeal Beta Class Carbonic Anhydrase. *Journal of Bacteriology* 182 (2000): 6605–6613.

Smith, K.S., and Ferry, J.G. A Plant Type (β-Class) Carbonic Anhydrase from the Thermophilic Methanoarchaeon *Methanobacterium thermoautotrophicum. Journal of Bacteriology* 181 (1999): 6247–6253.

Smith, K.S., and Ferry J.G. Prokaryotic Carbonic Anhydrases. *FEMS Microbiology Reviews* 24 (2000): 335–366.

Smith, K.S., Jakubzick, C., Whittam, T.S. et al. Carbonic Anhydrase Is an Ancient Enzyme Widespread in Prokaryotes. *Proceedings of the National Academy of Sciences of the United States of America* 96 (1999): 15184–15189.

Smith, P. Soil Organic Carbon Dynamics and Land-Use Change. In Braimoh, A.K., and Vlek, G.L.P. (eds), *Land Use and Soil Resources*, 9–22. Stockholm: Springer, 2007.

So, A.K.C., Espie, G.S., Williams, E.B. et al. A Novel Evolutionary Lineage of Carbonic Anhydrase (ε Class) Is Component of the Carboxysome Shell. *Journal of Bacteriology* 186 (2004): 623–630.

Solomona, S., Plattner, G.K., Knuttic, R. et al. Irreversible Climate Change Due to Carbon Dioxide Emissions. *Proceedings of National Academy of Sciences of the United States of America* 106 (2009): 1704–1709.

Soong, Y., Fauth, D.L., Howard, B.H. CO_2 Sequestration with Brine Solution and Fly Ashes. *Energy Conversion and Management* 47 (2006): 1676–1685.

Sowers, K.R., Nelson, M.J., Ferry, J.G. Growth of Acetotrophic, Methane-Producing Bacteria in a pH Auxostat. *Current Microbiology* 11 (1984): 227–230.

Srivastava, S., Bharti, R.K., Verma, P.K., Thakur, I.S. Cloning and Expression of Gamma Carbonic Anhydrase from *Serratia* sp. ISTD04 for Sequestration of Carbon Dioxide and Formation of Calcite. *Bioresource Technology* 188 (2015): 209–213.

Stadie, W.C., O'Brien, H. The Catalysis of the Hydration of Carbon Dioxide and Dehydration of Carbonic Acid by an Enzyme Isolated from Red Blood Cells. *Journal of Biological Chemistry* 103 (1933): 521–529.

Steiner, H., Jonsson, B.H., Lindskog, S. The Catalytic Mechanism of Carbonic Anhydrase. *European Journal of Biochemistry* 59 (1975): 253–259.

Sung, Y.C., and Fuchs, J.A. Characterization of the *cyn* Operon in *Escherichia coli* K12. *Journal of Biology Chemistry* 263 (1988): 14769–14775.

Syed, G.D., Agasar, D., Pandey, A. Production and Partial Purification of Alpha-Amylase from a Novel Isolate *Streptomyces gulbargensis*. *Journal of Industrial Microbiology and Biotechnology* 36 (2009): 189–194.

Tahirov, T.H., Oki, H., Tsukihara, T. et al. Crystal Structure of Methionine Aminopeptidase from Hyperthermophile, *Pyrococcus furiosus*. *Journal of Molecular Biology* 284 (1998): 101–124.

Trachtenberg, M.C., Smith, D.A., Cowan, R.M. et al. Flue Gas CO_2 Capture by Means of a Biomimetic Facilitated Transport Membrane. Presented at Proceedings of the AIChE Spring Annual Meeting, Houston, TX, 2007.

Traufetter, G. Cold Carbon Sink: Slowing Global Warming with Antarctic Iron. *Spiegel Online*, January 2, 2009.

Tripp, B.C., Bell, C.B., 3rd, Cruz, F., Krebs, C., Ferry, J.G. A Role for Iron in an Ancient Carbonic Anhydrase. *Journal of Biological Chemistry* 279 (2004): 6683–6687.

Tripp, B.C., and Ferry, J.G. A Structure-Function Study of a Proton Transport Pathway in a Novel γ-Class Carbonic Anhydrase from *Methanosarcina thermophila*. *Biochemistry* 39 (2000): 9232–9240.

Tripp, B.C., Smith, K., and Ferry, J.G. Carbonic Anhydrase: New Insights for an Ancient Enzyme. *Journal of Biological Chemistry* 276 (2001): 48615–48618.

Ueda, K., Nishida, H., Beppu, T. Dispensabilities of Carbonic Anhydrase in Proteobacteria. *International Journal of Evolutionary Biology* (2012): 1–5.

Valdivia, R.H., and Falkow, S. Fluorescence-Based Isolation of Bacterial Genes Expressed within Host Cells. *Science* 277 (1997): 2007–2011.

Veitch, F.P., and Blankenship, L.C. Carbonic Anhydrases in Bacteria. *Nature* 197 (1963): 76–77.

Versteeg, G.F., Van Dijck, L.A.J., Van Swaaij, W.P.M. On the Kinetics between CO_2 and Alkanolamines Both in Aqueous and Non-Aqueous Solutions. An Overview. *Chemical Engineering Communications* 144 (1996): 113–158.

Vinoba, M., Bhagiyalakshmi, M., Grace, A.N. et al. Carbonic Anhydrase Promotes the Absorption Rate of CO_2 in Post Combustion Processes. *Journal of Physical Chemistry B* 117 (2013): 5683–5690.

Vossenberg, J. van de, Woebken, D., Maalcke, J. et al. The Metagenome of the Marine Anammox Bacterium '*Candidatus Scalindua profunda*' Illustrates the Versatility of this Globally Important Nitrogen Cycle Bacterium. *Environmental Microbiology* 15 (2013): 1275–1289.

Vullo, D., Del, P.S., Osman, S.M. et al. Sulfonamide Inhibition Studies of the δ-Carbonic Anhydrase from the Diatom *Thalassiosira weissflogii*. *Bioorganic & Medicinal Chemistry Letters* 24 (2014): 275–279.

Vullo, D., De Luca, V., Scozzafava, A. et al. Anion Inhibition Studies of the Fastest Carbonic Anhydrase (CA) Known, the Extremo-CA from the Bacterium *Sulfurihydrogenibium azorense*. *Bioorganic & Medicinal Chemistry Letters* 22 (2012): 7142–7145.

Wang, M., Lawal, A., Stephenson, P., Sidders, J., Ramshaw, C. Post-Combustion CO_2 Capture with Chemical Absorption: A State of the Art Review. *Chemical Engineering Research and Design* 89 (2011): 1609–1624.

Wanjari, S., Prabhu, C., Satyanarayana, T., Vinu, A., Rayalu, S. Immobilization of Carbonic Anhydrase on Mesoporous Aluminosilicate for Carbonation Reaction. *Microporous Mesoporous Materials* 160 (2012): 151–158.

Wanjari, S., Prabhu, C., Yadav, R., Satyanarayana, T., Labhsteswar, N., Rayalu S. Immobilization of Carbonic Anhydrase on Chitosan Beads for Enhanced Carbonation Reaction. *Process Biochemistry* 46 (2011): 1010–1018.

Weber, P.M.A., Horst, J.R., Barbier, G.G., Oesterhe, C. Metabolism and Metabolomics of Eukaryotes Living under Extreme Conditions. *International Review of Cytology* 256 (2007): 1–34.

Williamson, J.S., Allen, L.Z., Lorenzi, A.H. et al. Metagenomic Exploration of Viruses throughout the Indian Ocean. *PLoS One* 7 no. 10. (2012): e42047.

Wilson, Z.E., and Brimble, A.M. Molecules Derived from the Extremes of Life. *Natural Product Reports* 26 (2009): 44–71.

Wood, H.G., Werkman, C.H., Hemingway, A., Nier, A.O. Heavy Carbon as a Tracer in Heterotrophic Carbon Dioxide Assimilation. *Journal of Biological Chemistry* 139 (1941): 367–375.

Xu, Y., Feng, L., Jeffrey, P.D. et al. Structure and Metal Exchange in the Cadmium Carbonic Anhydrase of Marine Diatoms. *Nature* 452 (2008): 56–61.

Yadav, R., Mudliar, S.N., Shekh, A.Y. et al. Immobilization of Carbonic Anhydrase in Alginate and Its Influence on Transformation of CO_2 to Calcite. *Process Biochemistry* 47 (2012): 585–590.

Yadav, R., Satyanarayana, T., Kotwal, S., Rayalu, S. Enhanced Carbonation Reaction Using Chitosan-Based Carbonic Anhydrase Nanoparticles. *Current Science* 100 (2011): 520–524.

Yadav, R., Wanjari, S., Prabhu, C., Rayalu, S. Immobilized Carbonic Anhydrase for the Biomimetic Carbonation Reaction. *Energy and Fuels* 24 (2010): 6198–6207.

Yang, H., Xu, Z., Fan, M. et al. Progress in Carbon Dioxide Separation and Capture: A Review. *Journal of Environmental Science* 20 (2008): 14–27.

Yu, C., Huang, C., Tan, C.S. A Review of CO_2 Capture by Absorption and Adsorption. *Aerosol and Air Quality Research* 12 (2012): 745–769.

Yue, L., and Chen, W. Isolation and Determination of Cultural Characteristics of A New Highly CO_2 Tolerant Fresh Water Microalga. *Energy Conversion and Management* 46 (2005): 1868–1876.

Zhang, G., Jiang, N., Liu, X., Dong, X. Methanogenesis from Methanol at Low Temperatures by a Novel Psychrophilic Methanogen, *Methanolobus psychrophilus* sp. nov., Prevalent in Zoige Wetland of the Tibetan Plateau. *Applied and Environmental Microbiology* 74 (2008): 6114–6120.

Zimmerman, S.A., and Ferry, J.G. The Beta and Gamma Classes of Carbonic Anhydrase. *Current Pharmaceutical Design* 14 (2008): 716–721.

CHAPTER 13

Carotenoid Production in Extremophilic Microalgae and Biotechnological Implications

Sarada Ravi, Vidyashankar Srivatsan, Vikas Singh Chauhan,
Sandeep Narayan Mudliar, Daris P. Simon, and Anila Narayan

CONTENTS

13.1 INTRODUCTION

Extremophilic organisms possess the ability to grow in extreme environments. These could be extremes of pH, temperatures, salinity, pressure, radiation, desiccation, oxygen species or redox potential, metal concentrations, and concentrations of gases, such as CO_2 and other extreme conditions. Polyextremophiles possess the ability to thrive in more than one extreme environment (Rothschild and Mancinelli 2001; Varshney et al. 2015). Microalgae are ubiquitous in their distribution and inhabit various extreme environments. Extremophilic microalgae include halophiles growing in salt pans, for example, *Dunaliella* sp. (Ben-Amotz and Avron 1990); psychrophiles growing in polar regions, for example, *Koliella antarctica* and *Chlamydomonas nivalis* (Remias et al. 2005; Fogliano et al. 2010); alkalophiles growing in alkaline lakes, for example, *Spirulina* sp. and *Scenedesmus* sp. (Durvasula et al. 2015); thermotolerants, for example, *Desmodesmus* sp. and *Coelastrella* sp. F 50 (Hu et al. 2013; Xie et al. 2013); and heavy metal–resistant acidophiles, for example, *Chlamydomonas acidophilus* and *Coccomyxa* sp. (Cuaresma et al. 2011; Garbayo et al. 2012; Vaquero et al. 2014) (Table 13.1). Even the nonextremophilic microalgae face periods of unfavorable growth conditions in their habitats due to seasonal and environmental fluctuations, such

Table 13.1 Extremophilic Microalgae and Their Carotenoids

Algae	Group	Carotenoid	Reference
Salt-tolerant algae			
Dunaliella salina	Green algae	β-Carotene	Ben-Amotz and Avron 1990; Santos et al. 2001
Dunaliella tertiolecta	Green algae	β-Carotene	Pasquet et al. 2011
Chlorella ellipsoidea	Green algae	Violaxanthin	Cha et al. 2008
Tetraselmis suecica	Green algae	Astaxanthin	Ahmed et al. 2014
Isochrysis galbana	Haptophyta	Fucoxanthin, lutein	Kim et al. 2012b; Ahmed et al. 2014
Pavlova salina	Haptophyta	Lutein	Ahmed et al. 2014
Diacronemavlkianum	Haptophyta	Lutein and astaxanthin	Durmaz et al. 2008
Nannochloropsis sp.	Eustigmatophyta	β-Carotene, violaxanthin, and vaucheriaxanthin	Lubián et al. 2000
Synechococcus sp.	Cyanobacteria	β-Carotene, zeaxanthin, cryptoxanthin, and echinenone	Montero et al. 2005
Phaeodactylum tricornutum	Diatom	Fucoxanthin	Kim et al. 2012a
Psychrophilic algae (snow algae)			
Chlamydomonas nivalis	Green algae	Astaxanthin	Remias et al. 2005
Raphidonema sp.	Green algae	Xanthophyll cycle pigments	Leya et al. 2009
Nostoc commune	Cyanobacteria	Myxoxanthophylls and canthaxanthin	Vincent and Quesada 1994
Thermophilic algae			
Desmodesmus sp. F51	Green algae	Lutein	Xie et al. 2013
Acidophilic algae			
Coccomyxa acidophila	Green algae	β-Carotene and lutein	Casal et al. 2011
Coccomyxa onubensis	Green algae	Lutein	Vaquero et al. 2014
Chlamydomonas acidophila	Green algae	Lutein and zeaxanthin	Cuaresma et al. 2011
Cyanidioschyzon merolae	Red algae	β-carotene and zeaxanthin	Cunningham et al. 2007
CO₂-tolerant algae			
Chlorococcum littorale	Green algae	β-Carotene and xanthophylls	Ota et al. 2009

as high light intensity (photooxidation) during summer, low and high nutrient concentrations due to evaporation during summer and dilutions during rains, and diurnal fluctuations in temperatures. The isolation and characterization of extremophilic microalgae are a subject of growing interest, as they are recognized as a potential source of various valuable metabolites. Microalgae, when faced with sub- or supraoptimal conditions, change their metabolic pattern and strategies as an adaptive mechanism, leading to changes in biomass composition and relative content of various metabolites (Hu 2004). The production of secondary metabolites under extreme and stress conditions is one such adaptive strategy for survival (Markou and Nerantzis 2013; Skjånes et al. 2013).

Microalgae synthesize pigments, mainly chlorophylls, carotenoids, and phycobillins, which serve as light energy absorbers in the photosynthetic systems. Carotenoids are important for microalgae, as they also play an important role in photoprotection, $^1O_2$ scavenging, dissipation of excess energy, and membrane structure stabilization (Frank and Cogdell 1996). Carotenoids form pigment–protein complexes by binding with peptides and are located mainly in thylakoid membranes. They may also be located in the cytoplasm or other stromata. Carotenoids can be classified as primary and secondary carotenoids. Primary carotenoids are directly involved in the light absorption process, required for photosynthesis and function within the photosynthetic machinery. Primary carotenoids include neoxanthin, violaxanthin, lutein, zeaxanthin, and β-carotene. While β-carotene is more abundant in the reaction centers of photosystems I and II, other xanthophylls are preferentially distributed in the light-harvesting complexes that transfer excitation energy to the reaction centers (Demmig-Adams and Adams 2002). Secondary carotenoids appear to play a role in stress adaptation, as their synthesis is affected by cultivation conditions and they are accumulated upon exposure of algae to stress conditions like high light intensity, salt stress, or high temperatures (Gouveia et al. 1996; Kobayashi 2003). Secondary carotenoids are a group of compounds, such as canthaxanthin, astaxanthin, and cryptoxanthin, that accumulate as oil droplets in the plastids or cytoplasm. These compounds are esterified with fatty acids and occur along with cytoplasmic triacyl glycerols (TAGs) (Zhekisheva et al. 2002). The culture and environmental stress conditions, mainly the light, temperature, metal ions, and salts, have been shown to enhance the carotenoid content of microalgae (Bhosale 2004). Astaxanthin, β-carotene, and lutein have been identified as the major carotenoids in microalgae. The carotenoids are highly antioxidative in nature and have distinct red, orange, and yellow colors. Carotenoids have found commercial application as food and feed additives, natural food colorant, and nutraceuticals (Armstrong 1994; Ben-Amotz and Fishier 1998). Animals do not synthesize carotenoids and get them through diet. Carotenoids act as precursors for vitamin A and retinols in animals, and as free radical scavengers and antioxidants. Dietary carotenoids have been shown to inhibit the onset of diseases linked to free radicals, for example, arteriosclerosis, cataract, age-related macular degeneration (AMD), multiple sclerosis, and cancer (Peto et al. 1981; Greenberg et al. 1990; Stahl and Sies 1996; Hennekens 1997; Moeller et al. 2000). The commercially available carotenoids are primarily synthetic in nature. However, the demand for natural carotenoids has been increasing mainly owing to the rising negative public perception about the undesirable side effects associated with synthetic food colors. The microalgae could be an important commercial source of natural carotenoids; however, only a few have shown potential for large-scale cultivation for example, *Dunaliella*, *Haematococcus*, *Chlorella vulgaris*, and *Chlorella zofingiensis* (Markou and Nerantzis 2013).

The large-scale outdoor cultivation systems of microalgae face challenges like extremes of climatic conditions (temperatures and sunlight) and acidic, alkaline, or saline waters, depending on the available water source. Also, the integration of algal cultivation systems with industrial effluent and flue gases may expose the microalgae to high concentrations of metals, organic carbon, and CO_2. The extremophilic microalgae therefore have the potential to thrive in outdoor cultivation systems exposed to extreme conditions. The large-scale cultivation of microalgae has been successfully achieved with extremophiles, like the halophilic *Dunaliella* and alkaliphilic *Spirulina*. The unicellular green microalga *Dunaliella* grows at very high NaCl concentrations and is a source of natural β-carotene.

The filamentous microalga *Spirulina* grows at high pH and is widely used as food and feed supplement. The ability of these two microalgae to grow in extreme environments has made it possible to cultivate them in large open outdoor ponds. The high salinity or pH creates an exclusive growth environment for these microalgae, protecting them from other algal contaminants and grazers. However, the extreme or stress culture conditions are known to adversely affect the growth rates of microalgae, leading to decreased productivities and yields. Therefore, the cultivation strategies need to be devised for extremophilic microalgae to provide ideal conditions to maximize their growth and carotenoid content.

13.2 ROLE OF CAROTENOIDS IN MICROALGAE

Carotenoids are essential classes of hydrophobic compounds that play a crucial role in light harvesting and energy transfer during photosynthesis. Carotenoids are generally membrane bound and are mostly associated with the light-harvesting complex (LHC) in the chloroplast, as observed for β-carotene in *Dunaliella* sp. (Ramos et al. 2008; Davidi et al. 2014). However, under certain growth conditions, carotenoids occur in the extra-plastidial regions in lipid globules, as observed in green microalga, *Haematococcus pluvialis* (Boussiba 2000). Carotenoids absorb a broad spectrum of light and initiate the photochemical events and provide stability and functionality to the photosystem (Grossman et al. 2004). They protect the photosystem from photooxidative damage by quenching the excess electrons and free radicals, such as singlet oxygen and triplet chlorophyll (Telfer et al. 2008; Z. Li et al. 2009). In chlorophytes, the xanthophylls, such as violoxanthin and zeaxanthin, dissipate the excess energy as heat; this phenomenon is called as nonphotochemical quenching (NPQ) (Erickson et al. 2015; Goss and Lepetit 2015). The functional role of carotenoids in NPQ and free radical scavenging is well described by Varela et al. (2015).

Dunaliella sp. serves as a model organism for understanding the role of carotenoids in photoprotection, especially against higher light irradiance and ultraviolet (UV) radiation (both UV-A and UV-B) (White and Jahnke 2002). During UV exposure, the photosynthetic apparatus is irreversibly damaged, leading to inhibition of photosynthesis and generation of oxygen free radicals, leading to oxidative stress (Jeffrey and Mitchell 1997). More specifically, under UV radiation, the Mn-containing oxygen-evolving complex (OEC) of photosystem II (PS II) is damaged by loss of Mn, leading to photoinhibition. In addition, UV radiation induces damage of the D1 protein in PS II and the damage of the Rubisco enzyme (Holzinger and Lütz 2006). Carotenoids absorb the excess irradiation and reduce the excited PS II and prevent the loss of Mn (Hakala et al. 2005; Coesel et al. 2008; Kianianmomeni 2014). Shaish et al. (1993) reported that during high light irradiance, the efficiency of the photosynthesis electron transport chain is compromised, leading to the generation of excess electrons and, consequently, oxygen free radicals. Under light-saturating conditions, the ratio of light limited to light-saturated photosynthesis decreases with a concomitant increase in β-carotene. The hypothesis was validated by use of carotenoid pathway inhibitors like norflurazon, which blocked the β-carotene production, leading to the death of cells under light-saturating conditions or UV radiation (Salguero et al. 2005). The presence of conjugated double bonds in β-carotene contributes to the quenching of free radicals. Further, lutein and zeaxanthin have also been reported to protect the photosystem in the case of a few marine macroalgae, such as *Ulva lactuca*. Under high UV irradiation (mainly UV-B), the expression of lutein in primary carotenoids and zeaxanthin in the xanthophyll cycle was observed as an adaptive response during photoprotection (Bischof et al. 2002). The role of carotenoids in stress adaptation in the extra-plastidial region is well elucidated in *H. pluvialis*, a freshwater green microalga. *H. pluvialis* accumulates the ketocarotenoid astaxanthin under unfavorable growth conditions. Astaxanthin synthesis proceeds via canthaxanthin mediated by reactive oxygen species (ROS), whose mechanism is still unknown (Boussiba 2000; Vidhyavathi et al. 2008).

In addition to photoprotection, carotenoids play a major role in the membrane fluidity of the algae under stressful conditions. It was observed that xanthophylls improve the stability of the lipid bilayer under the colder environments usually observed in marine habitats (Dieser et al. 2010). Carotenoids are integrated with membrane proteins and lipids and increase the unsaturation levels, thereby preventing tight packing of lipid molecules and membrane rigidity. It was observed that zeaxanthin was involved in membrane stabilization under heat and salt stress (Ashraf and Harris 2013).

13.3 BIOSYNTHESIS OF CAROTENOIDS

So far, more than 750 carotenoid structures have been isolated and identified from different natural sources (Britton et al. 2004). As such, more than 250 novel carotenoids have originated from marine organisms (Rodrigues et al. 2012), many of which show great potential in commercial applications (Vílchez Lobato et al. 2011). Carotenoids are synthesized as hydrocarbon carotenes (made up of only carbon and hydrogen, e.g., lycopene, α-carotene, and β-carotene) and/or their oxygenated derivatives (e.g., xanthophylls, lutein, α-cryptoxanthin, β-cryptoxanthin, and zeaxanthin) by all photosynthetic organisms, including plants, algae, and cyanobacteria, as well as some nonphotosynthetic bacteria and fungi (Armstrong 1994). The structures of major carotenoids are presented in Figure 13.1. Invariably in all organisms, carotenoids are formed from a five-carbon isopentenyl pyrophosphate (IPP), which is a common precursor irrespective of the pathway by which the carotenoids are biosynthesized, that is, the cytosolic mevalonate (MVA) pathway or plastidial 2-C-methyl-D-erythritol 4-phosphate (MEP) pathway (Schwender et al. 2001). Although several authors have reviewed the biosynthesis of carotenoids in plants (Cunningham and Gantt 1998;

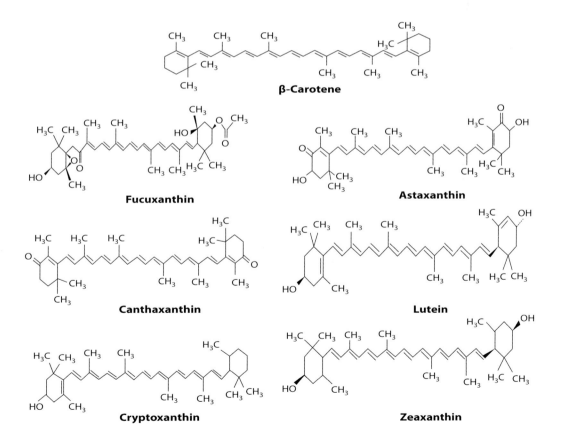

Figure 13.1 Structures of major carotenoids.

Farré et al. 2010), information on the regulation of carotenoid biosynthesis in algae is limited, and it is mainly inferred based on the knowledge acquired from plants (Takaichi 2011). Among algae, the MEP pathway is predominant and the MVA pathway has been reported only for Euglenophyceae (Lichtenthaler 1999; Capa-Robles et al. 2009).

The initial committing step of the carotenoid pathway is the condensation of geranylgeranyl pyrophosphate (GGPP) by the enzyme phytoene synthase (PSY). The gene encoding PSY is characterized in only a few species in microalgae, such as *Chlamydomonas reinhardtii* (McCarthy et al. 2004; Lohr et al. 2005), *H. pluvialis* (Cheng Wei et al. 2006), *C. zofingiensis* (Cordero et al. 2011a), and *Dunaliella* (Yan et al. 2005). As PSY is considered a key enzyme in the carotenoid pathway, several attempts have been made to engineer this gene for regulating the carotenoid production in plants and microbes.

In the second major step, phytoene is converted to lycopene via ζ-carotene. This conversion is facilitated by four condensation reactions by three enzymes: phytoene desaturase (PDS), ζ-carotene desaturase (ZDS), and *cis*-carotene isomerase (CrtISO). The initial condensation is mediated by PDS to convert phytoene to ζ-carotene. The ζ-carotene is further converted to lycopene through condensation and isomerization by ZDS and CrtISO, respectively. The initial condensation enzyme, PDS, has been studied in several microorganisms, including microalgae. The functional characterization of these enzymes has been done in microalgae such as *Chlorella protothecoides* (Li et al. 2012), *C. reinhardtii* (Vila et al. 2008; Liu et al. 2013), *Dunaliella salina* (Zhu et al. 2005, 2007), *C. zofingiensis* (Liu et al. 2010) and *H. pluvialis* (Liang et al. 2006). ZDS is characterized from the microalga *D. salina* (Ye and Jiang 2010).

Cyclization of lycopene is the branching point of the carotenoid pathway. Two related enzymes are responsible for this cyclization and result in the production of β-carotene or α-carotene. β-Carotene is produced by the action of lycopene β-cyclase (LCYB) through γ-carotene, and α-carotene is produced through γ-carotene or δ-carotene by the action of LCYB and lycopene ε-cyclase (LCYE). These enzymes are characterized from different organisms, but the knowledge about the algal lycopene cyclases is limited. The isolation and functional characterization of the LCYB gene is carried out in *C. reinhardtii* (Lohr et al. 2005), *H. pluvialis* (Steinbrenner and Linden 2003), *D. salina* (Ramos et al. 2008), and *C. zofingiensis* (Cordero et al. 2010). The LCYE gene has been characterized by *C. zofingiensis* (Cordero et al. 2012) and *C. protothecoides* (T. Li et al. 2009). The fusion protein of LCYB and LCYE from the green alga *Ostrococcus lucimarinus* was cloned and expressed in *Escherichia coli* and showed the production of β-carotene and α-carotene (Blatt et al. 2015). Mutants of these two enzymes have been used for studying the regulation of NPQ in *C. reinhardtii* (Niyogi et al. 1997).

Xanthophylls are produced from the carotenes by the addition of a hydroxyl group and/or ketone group by hydroxylase and/or ketolase enzymes, respectively. In the initial step, β-carotene and α-carotene are hydroxylated by two structurally distinctive classes of hydroxylases to produce β-xanthophylls and α-xanthophylls. The hydroxylation of α-carotene is mediated by two heme-containing cytochrome P 450 hydroxylases (CYP97A and CYP97C) and nonheme hydroxylases (BCH 1 and BCH 2) to produce lutein. β-Carotene is converted to β-xanthophylls (zeaxanthin) by the action of CYP97A/C or BCH 1/2 (Cui et al. 2013). These enzymes are identified and annotated *in silico* in many algae, but the functional characterization is limited to a few microalgae, such as *C. reinhardtii* (Lohr et al. 2005), *H. pluvialis* (Linden 1999; Cui et al. 2014), *Chlorella kessleri* (Yu et al. 2014), and *D. salina* (Zhang Li et al. 2013).

Ketocarotenoid formation is abundant in plant systems. Ketocarotenoids are formed by the addition of a hydroxyl group and keto group to the carotenes (hydrocarbon carotenoids). The hydroxylation reactions are discussed above. The addition of a keto group is mediated by the enzyme β-carotene ketolase (BKT) in algae. The major ketocarotenoids found in algae are astaxanthin and canthaxanthin. Some microalgae, such as *H. pluvialis* and *Chlorella* sp., are known for their accumulation of astaxanthin. Astaxanthin is produced from zeaxanthin by the action of BKT or through canthaxanthin. BKT can act on β-carotene to form canthaxanthin through echinenone, and further hydroxylation of canthaxanthin results in the formation of astaxanthin. These are the two major routes proposed for the production of astaxanthin in algae. The BKT enzyme is extensively studied in the alga *H. pluvialis*. Three types of this enzyme are reported from this alga (Lemoine and Schoefs 2010). In a recent study, the BKT gene from five microalgae (*Neochloris wimmeri*, *Protosiphon botryoides*, *Scotiellopsis oocystiformis*, *Chlorococcum* sp., and *C. reinhardtii*) was isolated and analyzed for expression in *E. coli* and the plant system (J. Huang et al. 2012).

In a photosynthetic organism, it is important to manage the equilibrium of energy flow in different light regimes. Carotenoids play a major role in this. In plants, algae, and other photosynthetic microbes, it is achieved through the cyclic conversion of xanthophylls (xanthophyll cycles). Two different types of xanthophyll cycles are reported: violaxanthin cycle and diadinoxanthin cycle. The violaxanthin cycle is present in higher plants and some algae (chlorophycea, phaeophyceae, and rhodophyceae). At lower light intensities, the xanthophyll zeaxanthin is converted to violaxanthin through antheraxanthin. This reaction is mediated by the enzyme zeaxanthin epoxidase (ZEP). At higher light intensities, violaxanthin is converted back to zeaxanthin via antheraxanthin. The de-epoxidation of violaxantin is mediated by the enzyme violaxanthin de-epoxidase (VDE) (Goss and Jakob 2010). These enzymes are cloned and characterized from plants, and in algae, ZEP is cloned and characterized from *C. reinhardtii* (Baroli et al. 2003) and *C. zofingiensis* (Couso et al. 2012). The ZEP mutant of *D. salina* constitutively accumulated zeaxanthin in all growth conditions, and other xanthophylls from zeaxanthin were absent in the mutant (Jin et al. 2003). The silencing of VDE in the diatom *Phaeodactylum tricornutum* reduced the production of diatoxanthin and NPQ. Apart from some group of algae that are showing the diadinoxanthin cycle, violaxanthin is converted to neoxanthin by neoxanthin synthase (NXS). Even though neoxanthin is reported from green algae, the isolation and characterization of NXS is limited to some plant species, such as potato (Al-Babili et al. 2000).

The diadinoxanthin cycle is present in some groups of algae (diatoms, haptophytes, and phaeophytes). The operation of the diadinoxanthin cycle has been studied in the diatom *P. tricornutum* (Latowski et al. 2012). This cycle contains only one step of epoxidation and de-epoxidation. The de-epoxidation of diadinoxanthin forms diatoxanthin with the enzyme diadinoxanthin de-epoxidase (DDE) and diatoxanthin converted back to diadinoxanthin by the enzyme diatoxanthin epoxidase (DTE) (Goss and Jakob 2010). The exact mechanism and function of this cycle is still unclear, and the enzymes responsible for this conversion are yet to be analyzed at the molecular level. It is suggested that the reactions in diadinoxanthin cycles are faster than those in the violaxanthin cycle (Lohr and Wilhelm 1999).

Fucoxanthin is a major carotenoid of seaweeds and diatoms. This pigment has wide application in therapeutics. The exact biosynthetic pathway leading to the production of this pigment is yet to be investigated. Two hypotheses have been proposed for the production of fucoxanthin in algae: (1) the neoxanthin hypothesis, where the fucoxanthin is produced from neoxanthin, and (2) the diadinoxanthin hypothesis, where diadinoxanthin is produced from violaxanthin and then branched to form fucoxanthin and diatoxanthin. However, the enzymes and the intermediates for these proposed pathways are still to be identified (Mikami and Hosokawa 2013).

The biosynthetic pathway of carotenoids in microalgae is illustrated in Figure 13.2.

13.4 CAROTENOIDS AND THEIR APPLICATIONS

13.4.1 β-Carotene

β-Carotene ($C_{40}H_{55}$) is carotenoids of great commercial value with wide applications, such as a food coloring agent, a pro-vitamin A (retinol) in food and animal feed, an additive to cosmetics and

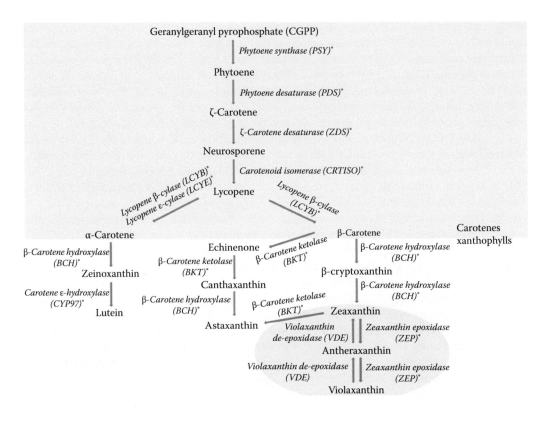

Figure 13.2 Biosynthetic pathway of carotenoids in microalgae.

multivitamin preparations, and a health food product with antioxidant benefits (Edge et al. 1997). It was observed that β-carotene-rich *Dunaliella* supplementation improved egg yolk coloration and was suggested as a source of vitamin A in animal feed (Gómez and González 2004). Kuroiwa et al. (2006) reported on the safety of β-carotene supplementation in rats and suggested its use as a food supplement. The nutraceutical potential of algae-derived β-carotene has been well established and reported by several authors (Dufossé et al. 2005; Sarada et al. 2005; Raja et al. 2007). The interest in the use of β-carotene as an antioxidant comes from its excellent ability to scavenge the ROS. Several epidemiological studies have shown the cancer-preventing ability of β-carotene attributed to its antioxidant potential (Poppel and Goldbohm 1995; Chidambara Murthy et al. 2005). The antioxidant property of β-carotene has also been shown to help in fighting coronary heart disorders, premature aging, and arthritis (Törnwall et al. 2004). The human body converts β-carotene to vitamin A, which is utilized for eye functioning, while the excess β-carotene is deposited or stored in the liver, where it provides hepatoprotection. β-Carotene supplementation improves eye health by preventing eye disorders such as night blindness and cataract (Agarwal and Rao 2000). Vanitha et al. (2007) reported the hepatoprotective activity of β-carotene obtained from *Dunaliella bardawil* (*salina*) against CCl₄-induced toxicity. CCl₄ induces liver necrosis and also affects renal and kidney function. β-Carotene protects the liver by reducing the lipid peroxidation levels (as indicated by a reduction in thiobarbituric acid-reactive substances [TBARSs]) and by preventing DNA damage in hepatocytes. Further supplementation of β-carotene increased the levels of *in vivo* antioxidant enzymes, such as catalase, superoxide dismutase (SOD), and peroxidase (Chidambara Murthy et al. 2005; Ranga Rao et al. 2013). β-Carotene has also been shown to possess immune-modulating function, such as inhibition of the proliferation of human lymphocytes (Moriguchi et al.

1985), and prevention of the decrease in antigen expression in human peripheral blood mononuclear cells (Gruner et al. 1986).

13.4.2 Astaxanthin

Astaxanthin is a powerful antioxidant, with high therapeutic efficacy in animal and human models (Snodderly 1995). Astaxanthin commonly finds application as fish and crustacean feed for its skin coloration. In addition, astaxanthin is considered a vitamin for salmon, as it is essential for the proper development and survival of juveniles under stressful environments. Further, it was observed that natural microalgal astaxanthin shows superior bioefficacy over the synthetic form in fish pigmentation. Astaxanthin has also been promoted as a high-value nutraceutical and a health supplement. The nutritional and therapeutic applications of algae-derived astaxanthin have been thoroughly reviewed by several authors (Dufossé et al. 2005; Sarada et al. 2005; Ambati et al. 2014). The antioxidant potential of astaxanthin is superior to that of other carotenoids, such as lutein, β-carotene, lycopene, and α-tocopherol (Naguib 2000). Further, astaxanthin-rich *Haematococcus* biomass showed inhibitory activity against *Helicobacter pylori*–induced ulcers in mice (Wang et al. 2000).

Microalgae are the largest natural source of astaxanthin, with two green algal species, that is, freshwater *H. pluvialis*, accumulating up to 2%–3% (Rao et al. 2007), and salt-tolerant *C. zofingiensis*, accumulating up to 1% of dry weight (Liu et al. 2014) under autotrophic conditions. However, these species have the ability to also grow in heterotrophic conditions (Han et al. 2013). Psychrophilic *Chlamydomonas nivalis* has also been reported to contain high amounts of astaxanthin and its fatty acid ester derivatives (Remias et al. 2005). As mentioned in earlier sections, unfavorable growth conditions, such as very high light intensities, nutrient deprivation, or high salinity, induce astaxanthin accumulation in these algae.

13.4.3 Xanthophylls (Lutein and Zeaxanthin)

Primary xanthophylls, lutein and its natural isomer zeaxanthin, are the major xanthophylls observed in microalgae. They play an important role in the protection of photosynthetic machinery from photodamage. In humans, lutein and zeaxanthin are deposited in the macular region of the retina and are commonly called macular pigments. Several studies have shown that lutein and zeaxanthin may provide significant protection against the potential damage caused by light on the retina. Lutein and zeaxanthin filter high-energy wavelengths of visible light (mainly blue light) and UV rays and act as antioxidants. They restrict the formation of ROS and subsequent free radicals (Roberts et al. 2009). Further, they reduce the effects of light scatter and chromatic aberration on visual performance. The main mechanism behind the bioactivity of lutein and zeaxanthin is due to their free radical scavenging activity and antilipid peroxidation activity. Ranga Rao et al. (2013) reported antilipid peroxidation activity of lutein obtained from *B. braunii* biomass in a rat model. Lutein and zeaxanthin are anchored to macular xanthophyll binding proteins (XBPs) and play a significant role in preventing AMD (Mozaffarieh et al. 2003; Ma et al. 2012). Lutein prevents the development of AMD by suppressing the inflammation pathway of NF-κB associated with the development of choroidal neovascularization (CNV), a critical pathogenesis in AMD (Izumi-Nagai et al. 2007). Recently, the European Food Safety Authority (EFSA) has suggested a daily lutein intake of 1 mg kg^{-1} body weight (EFSA 2010). The increasing importance of lutein could be gauged from the significant growth in the global lutein market, which is estimated to reach US$309 million in 2018 (Xie et al. 2013). Zeaxanthin has been identified as a therapeutic candidate for cancer and atherosclerosis treatment (Sajilata et al. 2008b; Nishino et al. 2009; Kadian and Garg 2012). In addition to nutritional applications, lutein and zeaxanthin find application as important feed additives for aquaculture and poultry and as natural colorants in food, pharmaceuticals, and cosmetics (Cordero et al. 2011b). They have been used as color additives in animal feed and mainly poultry feed with the classification codes of E 161b and E 161h, respectively. They are mainly concentrated in the egg yolk and give a yellowish to orange hue (Breithaupt 2007). Currently, the main source of commercial lutein is marigold petals, but they have a lower lutein content (can be as low as 0.03%), (Fernández-Sevilla et al. 2010; Yen et al. 2010; Piccaglia et al. 1998). Microalgae could serve as a potential alternative source of lutein, as several microalgal species have been reported for their high lutein content and productivities, for example, *Muriellopsis* sp. (Del Campo et al. 2000), *C. zofingiensis* (Campo et al. 2004), *Chlorella sorokiniana* (Cordero et al. 2011b), and *Scenedesmus obliquus* (Chan et al. 2013). Microalgae thus hold the potential to become an economically viable source of lutein.

13.4.4 Fucoxanthin

Fucoxanthin is another important xanthophyll occurring predominantly in brown algae (Pheophyceae) and heterokont species like *Isochrysis galbana* (Kim et al. 2012b). Fucoxanthin possesses a unique structure, which includes an unusual allenic bond and 5,6-monoepoxide (Maeda et al. 2008), responsible for its high antioxidant activity. Fucoxanthin is metabolized into fucoxanthinol, amarouciaxanthin, and halocynthiaxanthin after absorption into the human body (Sangeetha et al. 2009). Fucoxanthin has been attributed to several bioactive properties, such as anticancer, hypolipidemic, antiobesity, and antidiabetic, owing to its free radical scavenging potential (Hosokawa et al. 2004; Maeda et al. 2005; Miyashita et al. 2011; Peng et al. 2011). It was observed that fucoxanthin intake leads to the oxidation of fatty acids and heat production in the mitochondria of white adipose tissue, leading to the breakdown of fat and consequently obesity (Maeda et al. 2008). In addition, it was observed that fucoxanthin intake also significantly reduces blood glucose and plasma insulin. Apart from increasing thermogenesis, fucoxanthin increases the absorption of docosahexaenoic acid (DHA) in liver. Chung et al. (2013) reported the inhibition of metastasis in melanoma cells by fucoxanthin through the inhibition of tumor invasion by suppressing a metastatic factor MMP-9, and migration and cell adhesion by suppressing surface glycoprotein factor CD44 and CXC chemokine receptor-4 (CXCC-4). The fucoxanthin-fed groups showed a reduction in tumor nodules.

13.4.5 Other Minor Carotenoids

13.4.5.1 Canthaxanthin

Canthaxanthin ($C_{40}H_{52}O$) is a xanthophyll formed from β-carotene by the addition of the keto group by the enzyme BKT. It gives an orange to pink coloration and finds major application as a natural colorant in fish and poultry feed. It is absorbed on the skin and flesh of salmonids (Brizio et al. 2013) and is also used for the coloration of egg yolk in poultry (Esfahani-Mashhour et al. 2009). In addition to its use in animal feed, canthaxanthin has been demonstrated with high antioxidant activity (Paolo et al. 2000).

13.4.5.2 Cryptoxanthin

Cryptoxanthin ($C_{40}H_{56}O$) is a xanthophyll formed from β-carotene by the addition of one hydroxyl group. Cryptoxanthin is considered a pro-vitamin, as it is converted to vitamin A in the animal body. Cryptoxanthin has been reported to possess antioxidant activity and is associated with bone homeostasis (Lian et al. 2006; Yamaguchi 2012). This pigment has been reported to have a stimulatory effect on bone formation and prevents bone loss in menopausal women.

13.5 SOME EXTREMOPHILIC MICROALGAE AND THEIR CAROTENOIDS

Extremophilic microalgae accumulate various carotenoids, and Table 13.1 presents a summarized account of the distribution of carotenoids in them.

Some extremophilic microalgae and their carotenoids are discussed below.

13.5.1 Halophilic *Dunaliella salina*

A halophilic biflagellate green microalga, *D. salina*, has been recognized as an efficient producer of β-carotene (Ben-Amotz 1999). *Dunaliella* lacks a rigid cell wall and is able to grow at extremely high salinities ranging from 0.5 to 5.0 M NaCl (Karni and Avron 1988). *Dunaliella* as a source of β-carotene has been reviewed in detail by several authors (Del Campo et al. 2007; Raja et al. 2007; Hosseini Tafreshi and Shariati 2009). The alga accumulates β-carotene in oil globules in the interthylakoid space under stress conditions, that is, high light intensity, high salinity, and nitrogen limitation. The alga has been shown to accumulate up to 12% β-carotene on a dry weight basis (Ben-Amotz 1999). The β-carotene of *Dunaliella* is composed of mainly two stereoisomers, all-*trans* and 9-*cis*, which are present in approximately equal amounts. This composition of *Dunaliella* β-carotene gives it superior bioavailability, antioxidant properties, and physiological effects compared with synthetic β-carotene, which has an all-*trans* stereogeometry (Ben-Amotz 1999). The outdoor commercial cultivation of *Dunaliella* is carried out in open extensive unmixed or intensive open raceway ponds, where mixing is provided with paddle wheels. The intensive cultivation generally follows a two-stage process. The biomass is cultivated under optimal growth conditions in the first stage and then transferred to the second stage to induce carotenogenesis by stress conditions like nitrogen deficiency and increasing the availability of light by dilution of the culture (Ben-Amotz 1995). A β-carotene concentration of 0.1–1.0 mg L^{-1} and *Dunaliella* productivity of 0.05–0.1 g m^{-2} day^{-1} with maximal β-carotene productivity of 10 mg m^{-2} day^{-1} have been reported for extensive culture systems. In comparison, intensive culture systems achieve a carotene concentration of 10–20 mg L^{-1} and *Dunaliella* productivity of 5–10 g m^{-2} day^{-1} with maximal β-carotene productivity of 400–450 mg m^{-2} day^{-1} (Ben-Amotz 1993, 1995; Tafreshi and Shariati 2009). Del Campo et al. (2007) reported an intensive cultivation under the semicontinuous mode where the yields of β-carotene were higher than those in the conventional two-stage process. Commercial cultivation facilities of *Dunaliella* are in operation in Israel, India, China, and the United States in the regions receiving high solar irradiance, having a warm climate with the availability of hypersaline waters (Ben-Amotz 1999; Del Campo et al. 2007). The global production of *Dunaliella* biomass is estimated as 1200 MT $year^{-1}$ (Pulz and Gross 2004).

The closed photobioreactors, although not yet scaled up to the commercial level, offer the advantages of achieving higher biomass concentration, better control of culture parameters, and protection against grazers and other contaminants. Therefore, these cultivation systems are being studied in detail for *Dunaliella*. Optimization of conditions for outdoor cultivation of *Dunaliella* in a 55 L closed tubular photobioreactor to achieve productivity values of more than 2 g dry biomass and 100 mg β-carotene m^{-2} day^{-1} has been reported (García-González et al. 2005). Further optimization of cultivation conditions in the semicontinuous mode led to a productivity of 5 g dry weight m^{-2} day^{-1}, with biomass having an average β-carotene level of 14% (Del Campo et al. 2007). They compared the performance of open ponds and outdoor closed tubular photobioreactors operated under batch, semicontinuous, and two-stage modes. The study found that the two-stage mode of operation improved the performance of both the open and closed systems, and the tubular photobioreactor's performance was better than the open pond's.

13.5.2 Thermotolerant *Desmodesmus* sp.

Thermotolerant strains of microalgae could offer a cost advantage, as they could be grown in tropical and subtropical regions. These regions are suitable for algal cultivation, as they offer the inherent advantage of year-round stable ambient temperatures and longer sunlight durations (Pan et al. 2011; Franz et al. 2012). In a recent study, Xie et al. (2013) have reported the effects of the culture medium, nitrate concentration, and light intensity on the cell growth and lutein content of four indigenous thermotolerant strains of *Desmodesmus* sp., isolated from freshwaters of southern Taiwan. The strains have been reported to have the ability to survive and grow at 45°C for 24 hours, with the highest growth rate at 35°C. One of the strains had the ability to grow at 46°C and at a high light intensity of 2600 µmol m^{-2} s^{-1} (C.-C. Huang et al. 2012; Xie et al. 2013). Cultivation studies were conducted in a 1 L glass photobioreactor at 35°C, pH of 7.5, and agitation rate of 300 rpm. The cultures were grown under a light intensity of approximately 150–750 µmol m^{-2} s^{-1} and were fed with 2.5% CO_2 as the sole source of inorganic carbon. The photobioreactor was also operated in fed-batch mode where a concentrated nitrate solution was pulse-fed into the cultures when the nitrogen source was depleted to maintain desired nitrate concentrations. The highest lutein content of 3.87 mg g^{-1} was observed in strain F32, and the highest lutein productivity was observed as 1.54 ± 0.03 mg L^{-1} day^{-1} in strain F51. A nitrogen-sufficient condition led to the accumulation of lutein, and high light intensity led to enhanced cell growth with a decrease in lutein content. The fed-batch cultivation with pulse feeding of 2.2 mM nitrate further enhanced the lutein productivity to 3.56 ± 0.10 mg L^{-1} day^{-1} and the content to 5.05 ± 0.20 mg g^{-1} in strain F51. The high lutein content achieved in the study in the fed-batch cultivation mode indicates that the *Desmodesmus* sp. thermotolerant strain 51 has good potential for microalgal-derived lutein. The fed-batch cultivation strategy has been further discussed in detail by Xie et al. (2013), mainly to overcome the retardation of cell growth over a prolonged cultivation due to light limitation and nutrient deficiency. The fed-batch mode consisting of pulse feeding of concentrated fresh medium and a stepwise increase in light intensity was best among all the fed-batch modes and resulted in a CO_2 fixation rate and lutein productivity of 1582.4 mg L^{-1} day^{-1} and 3.91 mg L^{-1} day^{-1}, respectively. The repeated operations led to further improvement in CO_2 fixation rate and lutein productivity.

13.5.3 Thermotolerant *Coelastrella* sp. F 50

The isolation and characterization of a thermotolerant microalga *Coelastrella* sp. F50 has been reported recently (Hu et al. 2013). The microalga can tolerate temperatures over 50°C for more than 8 hours and under stress accumulates reddish pigment–containing carotenoids. The culture changed its color from bright green in the midlog phase to yellowish green in the early stationary phase and then to reddish in the stationary phase after one month of incubation under normal conditions. A culture regime of exposing the culture to high light (400 µmol m^{-2} s^{-1}) and increasing the NaCl concentration in the medium to 1.5% reduced the incubation time for color change to 10–12 days. The pigments were identified as astaxanthin, lutein, canthaxanthin, and β-carotene. The content (percent dry biomass) of astaxanthin and β-carotene is reported as 1.8 and 1.4%, respectively. The alga is also shown to accumulate a significant quantity of lipids. The thermotolerant nature of the alga makes it a potential candidate for the production of natural pigments under large-scale cultivation.

13.5.4 Acidophilic *Chlamydomonas acidophila*

The acidophilic microalga *Chlamydomonas acidophila* has been reported to accumulate lutein and β-carotene. The effect of abiotic stress on the production of lutein and β-carotene by *C. acidophila* isolated from highly acidic (pH 2.5) waters of Tinto River (Huelva, Spain) has been discussed by Garbayo et al. (2012). Abiotic stresses like light intensity and nutrition are known to cause oxidative stress, which induces carotenogenesis in extremophiles (Bhosale 2004). *C. acidophila* was cultured under light intensities of 160, 240, and 1000 µmol m^{-2} s^{-1}. While the high light intensity inhibited the growth, a 20-day incubation of cultures under 240 µmol m^{-2} s^{-1} led to a total carotenoid content of 57.5 ± 1.6 mg L^{-1}. The lutein was the major carotenoid, with a concentration as high as 20.2 mg L^{-1}, which was equivalent to 37.7% of total carotenoids. β-Carotene, violaxanthin, and zeaxanthin were other major carotenoids, comprising 55.6% of the total carotenoids. The study also shows that supplementation of photosynthetically active radiation light of 160 µE m^{-2} s^{-1} with 10 µE m^{-2} s^{-1} of UV-A light leads to a significant enhancement of intracellular carotenoids. The cultures incubated at 40°C showed a higher growth and carotenoid production, and induction of carotenogenesis by Cu^{2+} was also observed.

C. acidophila has also been studied for carotenoid production under supplementation with different carbon sources (Cuaresma et al. 2011). The carbon sources used in the study were CO_2, glucose, glycerol, starch, urea, and glycine. CO_2 and urea were able to support algal growth as the sole carbon source (~20 g dry biomass m^{-2} day^{-1}). Among the organic carbon used in the study, the alga showed an efficient growth (~14 g dry biomass m^{-2} day^{-1}) only with glucose under mixotrophic conditions. Aside from being carbon sources, glycerol and acetate can also be natural precursors for terpenoid biosynthesis (Droop 1974; Garcí et al. 2000; Ceron Garcia et al. 2006). The mixotrophic cultures using glycerol and acetate showed the highest carotenoid content; however, cultures in acetate-containing medium could not be sustained, as they showed no growth. In cultures grown with CO_2, glucose, urea, starch, or glycerol, lutein and β-carotene were the major carotenoids, with about 9–10 and 1.5–4 mg g^{-1}, respectively. β-Carotene was the major carotenoid in cultures grown with glycerol (12 mg g^{-1}), and zeaxanthin was the major carotenoid in glycine-grown cultures (7.5 mg g^{-1}). The total volumetric productivity (mg L^{-1}) of carotenoids was higher in cultures grown with CO_2, urea, and glucose due to the higher biomass concentration in these cultures. Therefore, a two-step cultivation process may be effective where biomass is obtained with either CO_2, glucose, or urea supplementation in the first step, and then in the second step glycerol or glycine is used as a carbon source for β-carotene or zeaxanthin accumulation.

13.5.5 Acidophilic *Coccomyxa* sp.

The acidophilic microalga *Coccomyxa onubensis*, isolated from acidic waters of Tinto River, has been studied for light-mediated enhancement of lutein content (Vaquero et al. 2014). The ability of *C. onubensis* to grow at very low pH gives it an advantage over competing microorganisms, and hence it may be suited for outdoor cultivation. In the study reported by Vaquero et al. (2014), the cultures were cultivated under constant illumination at light intensities of 50, 140, and 400 µmol m^{-2} s^{-1} at 27°C with constant bubbling of 5% (v/v) CO_2 mixed air. The pH of the medium was 2.5. The growth rate of the alga increased with increasing light intensity, and the lutein content increased at low light intensity. At 400 µmol m^{-2} s^{-1}, the lutein content was above 4.5 mg g^{-1} and lutein productivities were 1.22 mg L^{-1} day^{-1}. β-Carotene values ranged between 0.6 and 0.75 mg g^{-1}, with higher values in cultures incubated under low light. The study also suggests a slight dependence of lutein accumulation on biomass concentration, with relatively dense cultures of 0.7 and 1 g L^{-1} showing maximum biomass and lutein productivities.

An isolate of the acidophilic *Coccomyxa* sp. has also been reported to have iron dependence for growth with 0.5 mM Fe^{2+} as the optimal concentration (Garbayo et al. 2012). The same concentration of iron also led to an enhanced carotenoid content in the alga, about 20% higher than in control cultures. The alga accumulated lutein up to 7.2 mg g^{-1} of dry biomass, which is one of the highest contents among algae. The isolate also showed an increase in lipid content at elevated iron concentration. Dissolved metal ions, for example, iron, exert oxidative stress (Nishikawa et al. 2003, 2006), and alga, may be synthesizing the unsaturated fatty acids and carotenoids to face oxidative stress caused by iron (Granado-Lorencio et al. 2009).

13.5.6 Psychrophilic *Koliella antarctica*

The psychrophilic unicellular microalga *Koliella antarctica*, an alga isolated from Ross Sea, Antarctica, has been reported to be rich in carotenoids, especially astaxanthin and lutein (Fogliano et al. 2010). The algal cultures were grown in water jacketed tubes at 10°C and 15°C and showed a higher specific growth rate at 15°C. The cultures grown at both temperatures showed similar concentrations of lutein and astaxanthin (in the ester form). The dry biomass obtained in the late exponential phase showed a higher carotenoid content than the biomass obtained during the early exponential phase. The biomass in the late exponential phase showed a lutein content of 1.41 ± 0.12 g 100 g^{-1} dry biomass and an astaxanthin content of 1.40 ± 0.14 g 100 g^{-1} dry biomass. *K. antarctica* can be cultivated in outdoor conditions at low temperatures, as the alga's ability to grow at lower temperatures may facilitate the maintenance of a monoalgal culture. The alga is also identified as a potential source of omega-3 polyunsaturated fatty acids (PUFAs), especially for eicosapentaenoic acid (EPA) and DHA.

13.6 EXTRACTION METHODS FOR CAROTENOIDS

The extraction of carotenoids from algae is a challenge, as the majority of them contain a thick cell wall. Therefore, to achieve efficient extraction, it is important to pretreat the algal cells and digest the thick cell wall components. Several techniques have been used for the pretreatment of algal cells, such as ultrasound, bead milling, autoclave treatment, abrasion with alumina, acid hydrolysis, and alkali treatment. Alkali pretreatment is used for the saponification of carotenoids that are esterified with fatty acids (Mäki-Arvela et al. 2014). The extraction of carotenoids from the algal cells is considered solid–liquid extraction (SLE), where the soluble constituents are extracted from a solid or semisolid matrix using appropriate solvents. The traditional methods of extraction, for example, soxhlet extraction, maceration, hydrodistillation, infusion, and decoction, are time-consuming and need a high amount of solvent. The new methods, including microwave-assisted extraction (MAE), ultrasound-assisted extraction (UAE), pressurized solvent extraction (PSE), supercritical fluid extraction (SFE), pulse electric field (PEF)–assisted extraction, enzyme-assisted extraction (EAE), and extractions with switchable solvents and ionic liquids (ILs), have been found to be more efficient than traditional methods in one or the other property (Kaufmann and Christen 2002; Grosso et al. 2015).

β-carotene, the major pigment of extremophilic microalga *Dunaliella*, is extracted efficiently with different methods, such as SFE, PSE, UAE, MAE, and saponification, followed by conventional extraction (Mäki-Arvela et al. 2014). From *Dunaliella*, the highest recovery rate was observed with hexane–acetone–ethanol in the ratio 2:1:1 at room temperature, followed by saponification of the extract with KOH (Hu et al. 2008). The highest recovery (79%) of fucoxanthin was achieved from *Odontella aurita* using ethanol extraction at 45°C (Xia et al. 2013). PSE has been shown to reduce the extraction time of fucoxanthin compared with UAE and MAE. PSE is also found to be effective in the extraction of other carotenoids, such as lutein and zeaxanthin from *Chlorella* (Cha et al. 2010; Koo et al. 2011). SFE has been reported to be effective in the extraction of astaxanthin from its major source, *H. pluvialis* (Nobre et al. 2006).

13.7 ANALYSIS OF CAROTENOIDS

The analysis of carotenoids begins with the separation of carotenoid compounds from a total carotenoid extract. Different chromatographic techniques, such as paper chromatography, thin-layer chromatography (TLC), and liquid chromatography (LC), are used for the separation of carotenoids.

Among the chromatographic methods, high-performance liquid chromatography (HPLC) is widely used. The stationary phase (column) of HPLC is important in the separation of compounds, and different types of columns, such as C-18, C-30, and C-34, are used. Different solvents are pumped through the column in combination. This can be run isocratically or as gradient. Methanol, ethyl acetate, tetrahydrofuran, t-butyl methylether, chloroform, dichloromethane, acetone, acetonitrile, and water are some examples of solvents used. Rivera and Canela-Garayoa (2012) reviewed the different HPLC conditions, programs, and solvent systems for the separation of different carotenoids. The C-30 stationary phase is reported to give better separation of carotenoids. It allows the separation of even stereoisomers with the appropriate solvent system.

Coelution of different compounds is the major difficulty in HPLC separation. The major coeluting carotenoids are violaxanthin and neoxanthin, lutein and zeaxanthin, astaxanthin and antheraxanthin, and all-*trans*-isomers of β-carotene and lycopene. Application of low temperature and chemical treatment is reported to be effective to resolve this problem (Breitenbach et al. 2001; Rodriguez-Amaya 2001; Sajilata et al. 2008a). An improved version of HPLC, ultra-high-performance liquid chromatography (UHPLC), has recently been reported (Swartz 2005). In this method, column packing material is smaller than HPLC and the solvent is pumped at a higher pressure, allowing efficient separation. The high-strength silica (HSS) C18 and T3 and ethylene bridged hybrid (BEH) C18 columns are used as the stationary phase in UHPLC. In a recent study (Chauveau-Duriot et al. 2010), a comparison of separation of the same extract in HPLC and UHPLC showed 12 and 23 peaks, respectively. UHPLC is also efficient in separating the abovementioned coeluting compounds (Chauveau-Duriot et al. 2010).

Further confirmation of separated carotenoid compounds can be achieved by mass spectrometry (MS) analysis. The exact molecular mass of compounds can be deduced through MS analysis. LC-MS, which is HPLC coupled with MS, is the efficient method for confirmation of carotenoids. Different ionization conditions can be used to analyze various carotenoids. Atmospheric pressure chemical ionization (APCI) in the positive mode is a widely used ionization technique for carotenoids. Carotenoids with different polarities (xanthophylls, carotenes, and carotenoid esters) are ionized and separated by this approach (Weller and Breithaupt 2003; Ornelas-Paz et al. 2007). The *cis* and *trans* isomers of carotenoids can be distinguished by using tandem mass spectrometry (MS/MS), which is the advanced version of MS (Fang et al. 2003).

There are different methods for the qualitative analysis of compounds, including carotenoids. Raman spectroscopy (Lademann et al. 2011), matrix-assisted laser desorption/ionization with time-of-flight mass spectrometry (MALDI/TOF-MS) (Fraser et al. 2007), atmospheric solids analysis probe mass spectrometry (ASAP-MS) (Fussell et al. 2010), infrared (IR) spectroscopy (Berardo 2004), and nuclear magnetic resonance (NMR) spectroscopy are some of the methods used in the qualitative analysis of carotenoids (Strohschein et al. 1997; McEwen et al. 2005; Schulz et al. 2005).

13.8 CULTIVATION STRATEGIES FOR EXTREMOPHILIC MICROALGAE FOR CAROTENOID PRODUCTION

Carotenoids are produced under stressful conditions. These conditions could be either environmental (such as light) or nutritional (such as nitrogen starvation). Various strategies have been reported in order to maximize the carotenoid production from a variety of extremophilic microalgal strains. The cultivation strategies focus on different combinations of nutrient and light stress, along with photobioreactor configuration and mode of operation, to achieve high carotenoid yields. The photobioreactor systems that can be used for the cultivation of extremophilic microalgae, along with their merits and demerits, are briefly discussed below.

13.9 PHOTOBIOREACTOR SYSTEMS

The economic production of microalgal biomass is desirable. Practically, convenient large-scale production of algal biomass can be achieved by two feasible methods: (1) conventional open pond systems and (2) closed photobioreactors (Carvalho et al. 2006). The large-scale cultivation of microalgae is usually carried out in outdoor open raceway pond systems that rely on natural light for illumination. These open raceway ponds are in use for about four decades and are well established for the production of microalgae, for example, *Spirulina*, *Dunaliella*, and *Chlorella*. The setting up, operating, and maintenance cost of open ponds is lower than that of closed photobioreactors (Chisti 2006). Despite the advantages, in general, the raceway ponds are faced with problems of lower productivity and biomass yields and contamination due to diurnal variations in climatic conditions. The extremophile algae, namely, thermophilic, halophilic, and acidophilic, can be potentially cultivated in open raceway ponds with a lower probability of contamination in the long run. However, the effect of extremophile cultivation conditions, such as higher temperature, salinity, and pH, on the solubility of various macro- and micronutrients, as well as the solubility of desirable gases, such as CO_2, and the desorption of oxygen from the algal culture, needs to be rigorously investigated.

Closed photobioreactors are the closed-culture system and are preferred over open ponds, as they can be established and maintained either indoors or outdoors. These bioreactors allow the cultivation of single microalgal species for a prolonged duration under controlled conditions (Carvalho et al. 2006). Several types of photobioreactors exist for the cultivation of algal biomass. These include widely used tubular photobioreactors, plate reactors, bubble column reactors, and the less commonly used semihollow spheres (Olivieri et al. 2014). Mixing in reactors is important to prevent sedimentation of cells and the even distribution of gases. The closed photobioreactors are advantageous in the way that they facilitate suitable environmental conditions, such as CO_2 concentration, nutritional balance due to proper mixing of growth medium, pH, and temperature and light intensity. Maximum productivity can be achieved by optimizing these culture conditions. The major drawback with these reactor systems is that they have more capital and operating cost. The closed photobioreactors, which are under continuous development for their application to microalgal cultivation, face many limitations during scale-up (Ugwu et al. 2008). The major challenges faced in the scale-up of these closed photobioreactors are the complexity of designs and flow parameters (Bitog et al. 2011).

Large-scale cultivation of microalgal biomass is still a challenge, as they consume large amounts of energy—mainly because of the small cell size and relatively low biomass concentration of microalgal cultures. Research efforts are therefore warranted for cultivation under higher cell densities, which poses engineering challenges concerning cell accessibility to light and gas. To meet the challenge of achieving enhanced production efficiency by open and closed photobioreactors, an extensive system modeling and optimization of process parameters is required. An extensive system modeling and optimization of various input and output parameters, like temperature, light, flow rate, pond depth, nutrient availability, geometry, and reactor design, would lead to an improvement in the cultivation efficiency of existing open and closed photobioreactors. Therefore, there exists a scope for the improvement of production efficiency of the raceway ponds and closed photobioreactors through analysis and optimization of various process and design parameters (James and Boriah 2010). The mathematical model development will help in the realistic prediction and optimization of the response of photobioreactors to a host of variables and contribute toward the development of a high-performance photobioreactor for algae cultivation.

Also, like cells of higher plants and animals, microalgae are damaged by intense hydrodynamic shear fields that occur in high-velocity flow in pipes, pumps, and mixing tanks (Sobczuk et al. 2006). Shear sensitivity can pose a significant problem, as the intensity of turbulence needed in photobioreactors' optimal light–dark cycle is difficult to achieve without damaging algal cells. Some algae, such as *Dunaliella*, are more sensitive to shear damage than others due to the absence of a cell wall. The identification and quantification of the effect of hydrodynamic stress on two different microalgae strains, *Dunaliella tertiolecta* and *D. salina*, cultivated in bench-scale bubble columns has been reported (Barbosa et al. 2004). The cell death rate constant was increased with increasing gas entrance velocity at the sparger. *D. salina* was slightly more sensitive than *D. tertiolecta*. The critical gas entrance velocities reported were ~50 and 30 m s^{-1} for *D. tertiolecta* and *D. salina*, respectively. The effects of gas flow rate, culture height, and nozzle diameter on the death rate constant were also reported. It was reported that bubble rising and bubble bursting were not responsible for cell death, and bubble formation at the sparger was found to be the main event leading to cell death. Also, small-diameter nozzles were more detrimental to cells. Robust methods need to be developed to reduce the damage associated with the turbulence of limited intensity. Intensities of shear stress are not easily determined in bioreactors, but improved methods for doing so are emerging (Sánchez Pérez et al. 2006). Some algae will preferentially grow attached to the internal wall of the photobioreactor tube, thus preventing light penetration into the tube and reducing bioreactor productivity. Robust methods for controlling wall growth are needed. Bioprocess intensification approaches that have proved successful in improving the economics of various biotechnology-based processes have barely been assessed for use with photobioreactors. High-cell-density microalgal cultures can be achieved with a proper reactor design and process optimization. For this reason, it is important to build a scientific knowledge base in the field, with emphasis on the most critical scale-up and operational parameters: light and mass transfer, shear, and mixing rates (Molina Grima et al. 1999), along with linking their effect on the kinetics of algae growth and carotenoid accumulation.

13.10 FUTURE PROSPECTS

Microalgae are a natural resource of carotenoids, as these pigments are inherent components of their photosynthetic machinery. Among the microalgae, extremophiles hold higher potential, as they have adapted to thrive in extreme stress environments, which are also needed for enhanced accumulation of carotenoids. The carotenoid content is enhanced under extreme stress environments, as these pigments also play an important role in photooxidative protection. Microalgae provide sustainable resources, as they can be cultivated in a semicontinuous mode, have the ability to utilize saline and brackish waters, are cultivable on nonarable lands in tropical and subtropical regions, and can also utilize industrial effluent and flue gases. Therefore, further research studies are required to establish extremophilic microalgae as a viable commercial source of carotenoids. Some of the major areas of research interest are identified as follows.

- A vast biodiversity of microalgae in extreme environments remain largely unexplored. Therefore, bioprospecting of extremophilic microalgae is required to identify and assess the prospective carotenoid producers and novel carotenoids.
- The adaptive mechanisms of extremophilic microalgae are not fully understood, and hence the understanding of underlying adaptive mechanisms is necessary to tap their potential.
- The molecular biology tools for microalgae in general and for extremophilic microalgae in particular are needed for metabolic engineering approaches to enhance desired characteristics, for example, growth and carotenoid content.
- High-performance cultivation systems are needed for large-scale production of microalgal biomass. Except *Dunaliella*, the large-scale commercial cultivation of extremophilic microalgae for carotenoid production has yet to be achieved.
- The lower cell densities and small cell size of microalgae pose major limitations in downstream processing, especially harvesting and dewatering. Effective and economically viable processes are required.
- Safety standards of extremophilic microalgal biomass must be established for potential food and nutraceutical applications.
- Food products incorporating microalgal biomass or extract are to be developed to impart the nutritional benefits of carotenoids to target populations.

ACKNOWLEDGMENTS

The authors thank the Council of Scientific and Industrial Research (CSIR), Department of Science and Technology, Department of Biotechnology, government of India, and various industries for their support. The authors thank the director of CSIR-CFTRI for his constant encouragement.

REFERENCES

Agarwal, Sanjiv, and Akkinappally Venketeshwer Rao. 2000. Tomato Lycopene and Its Role in Human Health and Chronic Diseases. *CMAJ: Canadian Medical Association Journal* 163 (6): 739–744.

Ahmed, Faruq, Kent Fanning, Michael Netzel, Warwick Turner, Yan Li, and Peer M. Schenk. 2014. Profiling of Carotenoids and Antioxidant Capacity of Microalgae from Subtropical Coastal and Brackish Waters. *Food Chemistry* 165: 300–306.

Al-Babili, Salim, Philippe Hugueney, Michael Schledz, Ralf Welsch, Hanns Frohnmeyer, Oliver Laule, and Peter Beyer. 2000. Identification of a Novel Gene Coding for Neoxanthin Synthase from *Solanum tuberosum*. *FEBS Letters* 485 (2–3): 168–172.

Ambati, Ranga Rao, Phang Siew Moi, Sarada Ravi, and Ravishankar Gokare Aswathanarayana. 2014. Astaxanthin: Sources, Extraction, Stability, Biological Activities and Its Commercial Applications— A Review. *Marine Drugs* 12 (1): 128–152.

Armstrong, Gregory A. 1994. Eubacteria Show Their True Colors: Genetics of Carotenoid Pigment Biosynthesis from Microbes to Plants. *Journal of Bacteriology* 176 (16): 4795–4802.

Ashraf, Muhammad, and Phil J.C. Harris. 2013. Photosynthesis under Stressful Environments: An Overview. *Photosynthetica* 51 (2): 163–190.

Barbosa, Maria, Hadiyanto, and René Wijffels. 2004. Overcoming Shear Stress of Microalgae Cultures in Sparged Photobioreactors. *Biotechnology and Bioengineering* 85 (1): 78–85.

Baroli, Irene, An D. Do, Tomoko Yamane, and Krishna K. Niyogi. 2003. Zeaxanthin Accumulation in the Absence of a Functional Xanthophyll Cycle Protects *Chlamydomonas reinhardtii* from Photooxidative Stress. *Plant Cell* 15 (4): 992–1008.

Ben-Amotz, Ami. 1993. Production of β-Carotene and Vitamins by the Halotolerant Alga *Dunaliella*. In *Pharmaceutical and Bioactive Natural Products*, ed. David H. Attaway and Oskar R. Zaborsky, 411–417. New York: Springer US.

Ben-Amotz, Ami. 1995. New Mode of *Dunaliella* Biotechnology: Two-Phase Growth for β-Carotene Production. *Journal of Applied Phycology* 7 (1): 65–68.

Ben-Amotz, Ami. 1999. *Dunaliella* β-Carotene. In *Enigmatic Microorganisms and Life in Extreme Environments*, ed. Joseph Seckbach, 399–410. Cellular Origin and Life in Extreme Habitats 1. Dordrecht: Springer.

Ben-Amotz, Ami, and Mordhay Avron. 1990. The Biotechnology of Cultivating the Halotolerant alga *Dunaliella*. *Trends in Biotechnology* 8: 121–126.

Ben-Amotz, Ami, and Rachel Fishier. 1998. Analysis of Carotenoids with Emphasis on 9-Cis β-Carotene in Vegetables and Fruits Commonly Consumed in Israel. *Food Chemistry* 62 (4): 515–520.

Berardo, N., O.V. Brenna, A. Amato, P. Valoti, V. Pisacane, and M. Motto. 2004. Carotenoids Concentration among Maize Genotypes Measured by Near Infrared Reflectance Spectroscopy (NIRS). *Innovative Food Science & Emerging Technologies* 5 (3): 393–398.

Bhosale, Prakash. 2004. Environmental and Cultural Stimulants in the Production of Carotenoids from Microorganisms. *Applied Microbiology and Biotechnology* 63 (4): 351–361.

Bischof, Kai, Gudrun Kräbs, Christian Wiencke, and Dieter Hanelt. 2002. Solar Ultraviolet Radiation Affects the Activity of Ribulose-1,5-Bisphosphate Carboxylase-Oxygenase and the Composition of Photosynthetic and Xanthophyll Cycle Pigments in the Intertidal Green Alga *Ulva lactuca* L. *Planta* 215 (3): 502–509.

Bitog, Jessie Pascual, I.B. Lee, C.G. Lee, Kuk-Hwan Kim, H.S. Hwang, Se-Woon Hong, Il Hwan Seo, K.S. Kwon, and Ehab Mostafa. 2011. Application of Computational Fluid Dynamics for Modeling and Designing Photobioreactors for Microalgae Production: A Review. *Computers and Electronics in Agriculture* 76 (2): 131–147.

Blatt, Andreas, Matthias E. Bauch, Yvonne Pörschke, and Martin Lohr. 2015. A Lycopene β-Cyclase/Lycopene ε-Cyclase/Light-Harvesting Complex-Fusion Protein from the Green Alga *Ostreococcus lucimarinus* Can Be Modified to Produce α-Carotene and β-Carotene at Different Ratios. *Plant Journal: For Cell and Molecular Biology* 82 (4): 582–595.

Boussiba, Sammy. 2000. Carotenogenesis in the Green Alga *Haematococcus pluvialis*: Cellular Physiology and Stress Response. *Physiologia Plantarum* 108 (2): 111–117.

Breitenbach, Jurgen, Gisela Braun, Sabine Steiger, and Gerhard Sandmann. 2001. Chromatographic Performance on a C30-Bonded Stationary Phase of Monohydroxycarotenoids with Variable Chain Length or Degree of Desaturation and of Lycopene Isomers Synthesized by Various Carotene Desaturases. *Journal of Chromatography A* 936 (1–2): 59–69.

Breithaupt, D.E. 2007. Modern Application of Xanthophylls in Animal Feeding—A Review. *Trends in Food Science & Technology* 18 (10): 501–506.

Britton, George, Synnøve Liaaen-Jensen, and Hanspeter Pfander, eds. 2004. *Carotenoids*. Basel: Birkhäuser.

Brizio, Paola, Alessandro Benedetto, Marzia Righetti, Marino Prearo, Laura Gasco, Stefania Squadrone, and Maria Cesarina Abete. 2013. Astaxanthin and Canthaxanthin (Xanthophyll) as Supplements in Rainbow Trout Diet: In Vivo Assessment of Residual Levels and Contributions to Human Health. *Journal of Agricultural and Food Chemistry* 61 (46): 10954–10959.

Campo, J.A. Del, H. Rodríguez, J. Moreno, M.Á. Vargas, J. Rivas, and M.G. Guerrero. 2004. Accumulation of Astaxanthin and Lutein in *Chlorella zofingiensis* (Chlorophyta). *Applied Microbiology and Biotechnology* 64 (6): 848–854.

Capa-Robles, Willian, J. Paniagua-Michel, and Jorge Olmos Soto. 2009. The Biosynthesis and Accumulation of β-Carotene in *Dunaliella salina* Proceed via the Glyceraldehyde 3-Phosphate/Pyruvate Pathway. *Natural Product Research* 23 (11): 1021–1028.

Carvalho, Ana P., Luís A. Meireles, and F. Xavier Malcata. 2006. Microalgal Reactors: A Review of Enclosed System Designs and Performances. *Biotechnology Progress* 22 (6): 1490–1506.

Casal, Carlos, Maria Cuaresma, Jose Maria Vega, and Carlos Vilchez. 2011. Enhanced Productivity of a Lutein-Enriched Novel Acidophile Microalga Grown on Urea. *Marine Drugs* 9 (1): 29–42.

Ceron Garcia, M.C., F. García Camacho, A. Sánchez Mirón, J.M. Fernández Sevilla, Y. Chisti, and E. Molina Grima. 2006. Mixotrophic Production of Marine Microalga *Phaeodactylum tricornutum* on Various Carbon Sources. *Journal of Microbiology and Biotechnology* 16 (5): 689.

Cha, Kwang Hyun, Hee Ju Lee, Song Yi Koo, Dae-Geun Song, Dong-Un Lee, and Cheol-Ho Pan. 2010. Optimization of Pressurized Liquid Extraction of Carotenoids and Chlorophylls from *Chlorella vulgaris*. *Journal of Agricultural and Food Chemistry* 58 (2): 793–797.

Chan, Ming-Chang, Shih-Hsin Ho, Duu-Jong Lee, Chun-Yen Chen, Chieh-Chen Huang, and Jo-Shu Chang. 2013. Characterization, Extraction and Purification of Lutein Produced by an Indigenous Microalga *Scenedesmus obliquus* CNW-N. *Biochemical Engineering Journal, Biorefineries, Biomaterials, and Bio-Based Functional Chemicals* 78: 24–31.

Chauveau-Duriot, Beatrice, Michel Doreau, Pierre Nozière, and Benoit Graulet. 2010. Simultaneous Quantification of Carotenoids, Retinol, and Tocopherols in Forages, Bovine Plasma, and Milk: Validation of a Novel UPLC Method. *Analytical and Bioanalytical Chemistry* 397 (2): 777–790.

Cheng Wei, Liang, Zhao Fang Qing, Shamg Lin Qin, T.A.N. Cong Ping, W.E.I. Wei, and Meng Chun Xiao. 2006. Molecular Cloning and Characterization of Phytoene Synthase Gene from a Unicellular Green Alga *Haematococcus pluvialis*. *Progress in Biochemistry and Biophysics* 33 (9): 854–860.

Chidambara Murthy, K.N., A. Vanitha, J. Rajesha, M. Mahadeva Swamy, P.R. Sowmya, and Gokare A. Ravishankar. 2005. In Vivo Antioxidant Activity of Carotenoids from *Dunaliella salina*—A Green Microalga. *Life Sciences* 76 (12): 1381–1390.

Chisti, Yusuf. 2006. Microalgae as Sustainable Cell Factories. *Environmental Engineering & Management Journal (EEMJ)* 5 (3): 261–274.

Chung, Tae-Wook, Hee-Jung Choi, Ji-Yeon Lee, Han-Sol Jeong, Cheorl-Ho Kim, Myungsoo Joo, Jun-Yong Choi et al. 2013. Marine Algal Fucoxanthin Inhibits the Metastatic Potential of Cancer Cells. *Biochemical and Biophysical Research Communications* 439 (4): 580–585.

Coesel, Sacha Nicole, Alexandra Cordeiro Baumgartner, Licia Marlene Teles, Ana Alexandra Ramos, Nuno Miguel Henriques, Leonor Cancela, and João Carlos Serafim Varela. 2008. Nutrient Limitation Is the Main Regulatory Factor for Carotenoid Accumulation and for Psy and Pds Steady State Transcript Levels in *Dunaliella salina* (Chlorophyta) Exposed to High Light and Salt Stress. *Marine Biotechnology* 10 (5): 602–611.

Cordero, Baldo F., Inmaculada Couso, Rosa León, Herminia Rodríguez, and M. Ángeles Vargas. 2011a. Enhancement of Carotenoids Biosynthesis in *Chlamydomonas reinhardtii* by Nuclear Transformation Using a Phytoene Synthase Gene Isolated from *Chlorella zofingiensis*. *Applied Microbiology and Biotechnology* 91 (2): 341–351.

Cordero, Baldo F., Inmaculada Couso, Rosa Leon, Herminia Rodriguez, and Maria Angeles Vargas. 2012. Isolation and Characterization of a Lycopene ε-Cyclase Gene of *Chlorella* (Chromochloris) *zofingiensis*. Regulation of the Carotenogenic Pathway by Nitrogen and Light. *Marine Drugs* 10 (9): 2069–2088.

Cordero, Baldo F., Irina Obraztsova, Inmaculada Couso, Rosa Leon, Maria Angeles Vargas, and Herminia Rodriguez. 2011b. Enhancement of Lutein Production in *Chlorella sorokiniana* (Chorophyta) by Improvement of Culture Conditions and Random Mutagenesis. *Marine Drugs* 9 (9): 1607–1624.

Cordero, Baldo F., Irina Obraztsova, Lucía Martín, Inmaculada Couso, Rosa León, María Ángeles Vargas, and Herminia Rodríguez. 2010. Isolation and Characterization of a Lycopene β-Cyclase Gene from the Astaxanthin-Producing Green Alga *Chlorella zofingiensis* (Chlorophyta). *Journal of Phycology* 46 (6): 1229–1238.

Couso, Inmaculada, Baldo F. Cordero, María Ángeles Vargas, and Herminia Rodríguez. 2012. Efficient Heterologous Transformation of *Chlamydomonas reinhardtii* npq2 Mutant with the Zeaxanthin Epoxidase Gene Isolated and Characterized from *Chlorella zofingiensis*. *Marine Drugs* 10 (9): 1955–1976.

Cuaresma, María, Carlos Casal, Eduardo Forján, and Carlos Vílchez. 2011. Productivity and Selective Accumulation of Carotenoids of the Novel Extremophile Microalga *Chlamydomonas acidophila* Grown with Different Carbon Sources in Batch Systems. *Journal of Industrial Microbiology & Biotechnology* 38 (1): 167–177.

Cui, Hongli, Xiaona Yu, Yan Wang, Yulin Cui, Xueqin Li, Zhaopu Liu, and Song Qin. 2013. Evolutionary Origins, Molecular Cloning and Expression of Carotenoid Hydroxylases in Eukaryotic Photosynthetic Algae. *BMC Genomics* 14 (1): 457.

Cui, Hongli, Xiaona Yu, Yan Wang, Yulin Cui, Xueqin Li, Zhaopu Liu, and Song Qin. 2014. Gene Cloning and Expression Profile of a Novel Carotenoid Hydroxylase (CYP97C) from the Green Alga *Haematococcus pluvialis*. *Journal of Applied Phycology* 26 (1): 91–103.

Cunningham, Francis X., and Elisabeth Gantt. 1998. Genes and enzymes of carotenoid biosynthesis in plants. *Annual Review of Plant Physiology and Plant Molecular Biology* 49 (1): 557–583.

Cunningham, Francis X., Hansel Lee, and Elisabeth Gantt. 2007. Carotenoid Biosynthesis in the Primitive Red Alga *Cyanidioschyzon merolae*. *Eukaryotic Cell* 6 (3): 533–545.

Davidi, Lital, Eyal Shimoni, Inna Khozin-Goldberg, Ada Zamir, and Uri Pick. 2014. Origin of β-Carotene-Rich Plastoglobuli in *Dunaliella bardawil*. *Plant Physiology* 164 (4): 2139–2156.

Del Campo, José A., Mercedes García-González, and Miguel G. Guerrero. 2007. Outdoor Cultivation of Microalgae for Carotenoid Production: Current State and Perspectives. *Applied Microbiology and Biotechnology* 74 (6): 1163–1174.

Del Campo, José A., José Moreno, Herminia Rodríguez, M. Angeles Vargas, Joaquín Rivas, and Miguel G. Guerrero. 2000. Carotenoid Content of Chlorophycean Microalgae: Factors Determining Lutein Accumulation in *Muriellopsis* Sp. (Chlorophyta). *Journal of Biotechnology* 76 (1): 51–59.

Demmig-Adams, Barbara, and William W. Adams. 2002. Antioxidants in Photosynthesis and Human Nutrition. *Science* 298 (5601): 2149–2153.

Dieser, Markus, Mark Greenwood, and Christine M. Foreman. 2010. Carotenoid Pigmentation in Antarctic Heterotrophic Bacteria as a Strategy to Withstand Environmental Stresses. *Arctic, Antarctic, and Alpine Research* 42 (4): 396–405.

Droop, M.R. 1974. The Nutrient Status of Algal Cells in Continuous Culture. *Journal of the Marine Biological Association of the United Kingdom* 54 (4): 825–855.

Dufossé, Laurent, Patrick Galaup, Anina Yaron, Shoshana Malis Arad, Philippe Blanc, Kotamballi N. Chidambara Murthy, and Gokare A. Ravishankar. 2005. Microorganisms and Microalgae as Sources of Pigments for Food Use: A Scientific Oddity or an Industrial Reality? *Trends in Food Science & Technology* 16 (9): 389–406.

Durmaz, Y., M. Donato, M. Monteiro, L. Gouveia, M.L. Nunes, T. Gama Pereira, Ş Gökpınar, and N.M. Bandarra. 2008. Effect of Temperature on α-Tocopherol, Fatty Acid Profile, and Pigments of *Diacronema vlkianum* (Haptophyceae). *Aquaculture International* 17 (4): 391–399.

Durvasula, Ravi, Ivy Hurwitz, Annabeth Fieck, and D.V. Subba Rao. 2015. Culture, Growth, Pigments and Lipid Content of *Scenedesmus* Species, an Extremophile Microalga from Soda Dam, New Mexico in Wastewater. *Algal Research* 10: 128–133.

Edge, R., D.J. McGarvey, and T.G. Truscott. 1997. The Carotenoids as Anti-Oxidants—A Review. *Journal of Photochemistry and Photobiology: B, Biology* 41 (3): 189–200.

EFSA Panel on Food Additives and Nutrient Sources added to Food (ANS). 2010. Scientific Opinion on the re-evaluation of lutein (E 161b) as a food additive on request of the European Commission. *EFSA Journal* 8 (7): 1678 [57 pp.].

Erickson, Erika, Setsuko Wakao, and Krishna K. Niyogi. 2015. Light Stress and Photoprotection in *Chlamydomonas reinhardtii*. *Plant Journal: For Cell and Molecular Biology* 82 (3): 449–465.

Esfahani-Mashhour, M., H. Moravej, H. Mehrabani-Yeganeh, and S.H. Razavi. 2009. Evaluation of Coloring Potential of *Dietzia natronolimnaea* Biomass as Source of Canthaxanthin for Egg Yolk Pigmentation. *Asian-Australasian Journal of Animal Sciences* 22 (2): 254.

Fang, Liqiong, Natasa Pajkovic, Yan Wang, Chungang Gu, and Richard B. van Breemen. 2003. Quantitative Analysis of Lycopene Isomers in Human Plasma Using High-Performance Liquid Chromatography-Tandem Mass Spectrometry. *Analytical Chemistry* 75 (4): 812–817.

Farré, Gemma, Georgina Sanahuja, Shaista Naqvi, C Baia, Teresa Capell, Changfu Zhu, and Paul Christou. 2010. Travel advice on the road to carotenoids in plants. Plant Science 179(1): 28–48.

Fernández-Sevilla, José M., F.G. Acién Fernández, and E. Molina Grima. 2010. Biotechnological Production of Lutein and Its Applications. Applied Microbiology and Biotechnology 86 (1): 27–40.

Fogliano, Vincenzo, Carlo Andreoli, Anna Martello, Marianna Caiazzo, Ornella Lobosco, Fabio Formisano, Pier Antimo Carlino et al. 2010. Functional Ingredients Produced by Culture of *Koliella antarctica*. *Aquaculture* 299 (1–4): 115–120.

Frank, Harry A., and Richard J. Cogdell. 1996. Carotenoids in Photosynthesis. *Photochemistry and Photobiology* 63 (3): 257–264.

Franz, Anette, Florian Lehr, Clemens Posten, and Georg Schaub. 2012. Modeling Microalgae Cultivation Productivities in Different Geographic Locations—Estimation Method for Idealized Photobioreactors. *Biotechnology Journal* 7 (4): 546–557.

Fraser, Paul D., Eugenia M.A. Enfissi, Michael Goodfellow, Tadashi Eguchi, and Peter M. Bramley. 2007. Metabolite Profiling of Plant Carotenoids Using the Matrix-Assisted Laser Desorption Ionization Time-of-Flight Mass Spectrometry. *Plant Journal: For Cell and Molecular Biology* 49 (3): 552–564.

Fussell, Richard J., Danny Chan, and Matthew Sharman. 2010. An Assessment of Atmospheric-Pressure Solids—Analysis Probes for the Detection of Chemicals in Food. *TrAC Trends in Analytical Chemistry* 29 (11): 1326–1335.

Garbayo, Inés, Rafael Torronteras, Eduardo Forján, María Cuaresma, Carlos Casal, Benito Mogedas, María C. Ruiz-Domínguez, et al. 2012. Identification and Physiological Aspects of a Novel Carotenoid-Enriched, Metal-Resistant Microalga Isolated from an Acidic River in Huelva (Spain). *Journal of Phycology* 48 (3): 607–614.

Garcí, M.C. Cerón, J.M. Fernández Sevilla, F.G. Acién Fernández, E. Molina Grima, and F. García Camacho. 2000. Mixotrophic Growth of *Phaeodactylum tricornutum* on Glycerol: Growth Rate and Fatty Acid Profile. *Journal of Applied Phycology* 12 (3–5): 239–248.

García-González, Mercedes, José Moreno, J. Carlos Manzano, F. Javier Florencio, and Miguel G. Guerrero. 2005. Production of *Dunaliella salina* Biomass Rich in 9-Cis-Beta-Carotene and Lutein in a Closed Tubular Photobioreactor. *Journal of Biotechnology* 115 (1): 81–90.

Gómez, Patricia I., and Mariela A. González. 2004. Genetic Variation among Seven Strains of *Dunaliella salina* (Chlorophyta) with Industrial Potential, Based on RAPD Banding Patterns and on Nuclear ITS rDNA Sequences. *Aquaculture* 233 (1–4): 149–162.

Goss, Reimund, and Torsten Jakob. 2010. Regulation and Function of Xanthophyll Cycle-Dependent Photoprotection in Algae. *Photosynthesis Research* 106 (1–2): 103–122.

Goss, Reimund, and Bernard Lepetit. 2015. Biodiversity of NPQ. *Journal of Plant Physiology* 172: 13–32.

Gouveia, Luisa, Emdio Gomes, and Jos Empis. 1996. Potential Use of a Microalga (*Chlorella vulgaris*) in the Pigmentation of Rainbow Trout (*Oncorhynchus mykiss*) Muscle. *Zeitschrift Für Lebensmittel-Untersuchung Und -Forschung* 202 (1): 75–79.

Granado-Lorencio, F., C. Herrero-Barbudo, G. Acién-Fernández, E. Molina-Grima, J.M. Fernández-Sevilla, B. Pérez-Sacristán, and I. Blanco-Navarro. 2009. In Vitro Bioaccessibility of Lutein and Zeaxanthin from the Microalgae *Scenedesmus almeriensis*. *Food Chemistry* 114 (2): 747–752.

Greenberg, E. Robert, John A. Baron, Thérèse A. Stukel, Marguerite M. Stevens, Jack S. Mandel, Steven K. Spencer, Peter M. Elias et al. 1990. A Clinical Trial of Beta Carotene to Prevent Basal-Cell and Squamous-Cell Cancers of the Skin. *New England Journal of Medicine* 323 (12): 789–795.

Grossman, Arthur R., Martin Lohr, and Chung Soon Im. 2004. *Chlamydomonas reinhardtii* in the Landscape of Pigments. *Annual Review of Genetics* 38 (1): 119–173.

Grosso, Clara, Patrícia Valentão, Federico Ferreres, and Paula B. Andrade. 2015. Alternative and Efficient Extraction Methods for Marine-Derived Compounds. *Marine Drugs* 13 (5): 3182–3230.

Gruner, Stefan, Hans-Dieter Volk, Peter Falck And, and Rudiger Von Baehr. 1986. The Influence of Phagocytic Stimuli on the Expression of HLA-DR Antigens; Role of Reactive Oxygen Intermediates. *European Journal of Immunology* 16 (2): 212–215.

Hakala, Marja, Ilona Tuominen, Mika Keränen, Taina Tyystjärvi, and Esa Tyystjärvi. 2005. Evidence for the Role of the Oxygen-Evolving Manganese Complex in Photoinhibition of Photosystem II. *Biochimica et Biophysica Acta (BBA)—Bioenergetics* 1706 (1–2): 68–80.

Han, Danxiang, Yantao Li, and Qiang Hu. 2013. Astaxanthin in Microalgae: Pathways, Functions and Biotechnological Implications. *Algae* 28 (2): 131–147.

Hennekens, Charles H. 1997. β-Carotene Supplementation and Cancer Prevention. *Nutrition* 13 (7): 697–699.

Holzinger, Andreas, and Cornelius Lütz. 2006. Algae and UV Irradiation: Effects on Ultrastructure and Related Metabolic Functions. *Micron* 37 (3): 190–207.

Hosokawa, Masashi, Masahiro Kudo, Hayato Maeda, Hiroyuki Kohno, Takuji Tanaka, and Kazuo Miyashita. 2004. Fucoxanthin Induces Apoptosis and Enhances the Antiproliferative Effect of the PPARγ Ligand, Troglitazone, on Colon Cancer Cells. *Biochimica et Biophysica Acta (BBA)—General Subjects* 1675 (1–3): 113–119.

Hosseini Tafreshi, A., and M. Shariati. 2009. *Dunaliella* Biotechnology: Methods and Applications. *Journal of Applied Microbiology* 107 (1): 14–35.

Hu, Chao-Chin, Jau-Tien Lin, Fung-Jou Lu, Fen-Pi Chou, and Deng-Jye Yang. 2008. Determination of Carotenoids in *Dunaliella salina* Cultivated in Taiwan and Antioxidant Capacity of the Algal Carotenoid Extract. *Food Chemistry* 109 (2): 439–446.

Hu, Che-Wei, Lu-Te Chuang, Po-Chien Yu, and Ching-Nen Nathan Chen. 2013. Pigment Production by a New Thermotolerant Microalga *Coelastrella* Sp. F50. *Food Chemistry* 138 (4): 2071–2078.

Hu, Qiang. 2004. Environmental Effects on Cell Composition. In *Handbook of Microalgal Culture*, ed. Amos Richmond, 83–94. Oxford: Blackwell Publishing.

Huang, Chai-Cheng, Jia-Jang Hung, Shao-Hung Peng, and Ching-Nen Nathan Chen. 2012. Cultivation of a Thermo-Tolerant Microalga in an Outdoor Photobioreactor: Influences of CO_2 and Nitrogen Sources on the Accelerated Growth. *Bioresource Technology* 112: 228–233.

Huang, Junchao, Yujuan Zhong, Gerhard Sandmann, Jin Liu, and Feng Chen. 2012. Cloning and Selection of Carotenoid Ketolase Genes for the Engineering of High-Yield Astaxanthin in Plants. *Planta* 236 (2): 691–699.

Izumi-Nagai, Kanako, Norihiro Nagai, Kazuhiro Ohgami, Shingo Satofuka, Yoko Ozawa, Kazuo Tsubota, Kazuo Umezawa, Shigeaki Ohno, Yuichi Oike, and Susumu Ishida. 2007. Macular Pigment Lutein Is Antiinflammatory in Preventing Choroidal Neovascularization. *Arteriosclerosis, Thrombosis, and Vascular Biology* 27 (12): 2555–2562.

James, Scott C., and Varun Boriah. 2010. Modeling Algae Growth in an Open-Channel Raceway. *Journal of Computational Biology* 17 (7): 895–906.

Jeffrey, Wade H., and David L. Mitchell. 1997. Mechanisms of UV-Induced DNA Damage and Response in Marine Microorganisms. *Photochemistry and Photobiology* 65 (2): 260–263.

Jin, EonSeon, Brian Feth, and Anastasios Melis. 2003. A Mutant of the Green Alga *Dunaliella salina* Constitutively Accumulates Zeaxanthin under All Growth Conditions. *Biotechnology and Bioengineering* 81 (1): 115–124.

Kadian, Sumita S., and Munish Garg. 2012. Pharmacological Effects of Carotenoids: A Review. *International Journal of Pharmaceutical Sciences and Research* 3: 42–48.

Karni, Leah, and Mordhay Avron. 1988. Ion Content of the Halotolerant Alga *Dunaliella salina*. *Plant & Cell Physiology* 29 (8): 1311–1314.

Kaufmann, Béatrice, and Philippe Christen. 2002. Recent Extraction Techniques for Natural Products: Microwave-Assisted Extraction and Pressurised Solvent Extraction. *Phytochemical Analysis: PCA* 13 (2): 105–113.

Kianianmomeni, Arash. 2014. Cell-Type Specific Light-Mediated Transcript Regulation in the Multicellular Alga *Volvox carteri*. *BMC Genomics* 15: 764.

Kim, Sang Min, Yu-Jin Jung, Oh-Nam Kwon, Kwang Hyun Cha, Byung-Hun Um, Donghwa Chung, and Cheol-Ho Pan. 2012a. A Potential Commercial Source of Fucoxanthin Extracted from the Microalga *Phaeodactylum tricornutum*. *Applied Biochemistry and Biotechnology* 166 (7): 1843–1855.

Kim, Sang Min, Suk-Woo Kang, Oh-Nam Kwon, Donghwa Chung, and Cheol-Ho Pan. 2012b. Fucoxanthin as a Major Carotenoid in *Isochrysis* Aff. *galbana*: Characterization of Extraction for Commercial Application. *Journal of the Korean Society for Applied Biological Chemistry* 55 (4): 477–483.

Kobayashi, Makio. 2003. Astaxanthin Biosynthesis Enhanced by Reactive Oxygen Species in the Green alga *Haematococcus pluvialis*. *Biotechnology and Bioprocess Engineering* 8 (6): 322.

Koo, Song Yi, Kwang Hyun Cha, Dae-Geun Song, Donghwa Chung, and Cheol-Ho Pan. 2011. Optimization of Pressurized Liquid Extraction of Zeaxanthin from *Chlorella ellipsoidea*. *Journal of Applied Phycology* 24 (4): 725–730.

Kuroiwa, Y., A. Nishikawa, T. Imazawa, Y. Kitamura, K. Kanki, Y. Ishii, T. Umemura, and M. Hirose. 2006. A Subchronic Toxicity Study of *Dunaliella* Carotene in F344 Rats. *Food and Chemical Toxicology* 44 (1): 138–145.

Lademann, Juergen, Martina C. Meinke, Wolfram Sterry, and Maxim E. Darvin. 2011. Carotenoids in Human Skin. *Experimental Dermatology* 20 (5): 377–382.

Latowski, Dariusz, Reimund Goss, Monika Bojko, and Kazimierz Strzalka. 2012. Violaxanthin and Diadinoxanthin De-Epoxidation in Various Model Lipid Systems. *Acta Biochimica Polonica* 59 (1): 101.

Lemoine, Yves, and Benoît Schoefs. 2010. Secondary Ketocarotenoid Astaxanthin Biosynthesis in Algae: A Multifunctional Response to Stress. *Photosynthesis Research* 106 (1–2): 155–177.

Leya, Thomas, Andreas Rahn, Cornelius Lütz, and Daniel Remias. 2009. Response of Arctic Snow and Permafrost Algae to High Light and Nitrogen Stress by Changes in Pigment Composition and Applied Aspects for Biotechnology. *FEMS Microbiology Ecology* 67 (3): 432–443.

Li, Meiya, Zhibing Gan, Yan Cui, Chunlei Shi, and Xianming Shi. 2012. Structure and Function Characterization of the Phytoene Desaturase Related to the Lutein Biosynthesis in *Chlorella protothecoides* CS-41. *Molecular Biology Reports* 40 (4): 3351–3361.

Li, Ting, Chunlei Shi, Zhibing Gan, and Xianming Shi. 2009. Cloning and analysis of the gene encoding lycopene epsilon cyclase in *Chlorella protothecoides* CS-41. *Wei Sheng Wu Xue Bao = Acta Microbiologica Sinica* 49 (9): 1180–1189.

Li, Zhirong, Setsuko Wakao, Beat B. Fischer, and Krishna K. Niyogi. 2009. Sensing and Responding to Excess Light. *Annual Review of Plant Biology* 60 (1): 239–260.

Lian, Fuzhi, Kang-Quan Hu, Robert M. Russell, and Xiang-Dong Wang. 2006. β-Cryptoxanthin Suppresses the Growth of Immortalized Human Bronchial Epithelial Cells and Non-Small-Cell Lung Cancer Cells and Up-Regulates Retinoic Acid Receptor β Expression. *International Journal of Cancer* 119 (9): 2084–2089.

Liang, Cheng W., Fang Q. Zhao, Chun X. Meng, Cong P. Tan, and Song Qin. 2006. Molecular Cloning, Characterization and Evolutionary Analysis of Phytoene Desaturase (PDS) Gene from *Haematococcus pluvialis*. *World Journal of Microbiology and Biotechnology* 22 (1): 59–64.

Lichtenthaler, Hartmut K. 1999. The 1-Deoxy-D-Xylulose-5-Phosphate Pathway of Isoprenoid Biosynthesis in Plants. *Annual Review of Plant Physiology and Plant Molecular Biology* 50 (1): 47–65.

Linden, Hartmut. 1999. Carotenoid Hydroxylase from *Haematococcus pluvialis*: cDNA Sequence, Regulation and Functional Complementation. *Biochimica et Biophysica Acta* 1446 (3): 203–212.

Liu, Jin, Henri Gerken, Junchao Huang, and Feng Chen. 2013. Engineering of an Endogenous Phytoene Desaturase Gene as a Dominant Selectable Marker for *Chlamydomonas reinhardtii* Transformation and Enhanced Biosynthesis of Carotenoids. *Process Biochemistry* 48 (5–6): 788–795.

Liu, Jin, Zheng Sun, Henri Gerken, Zheng Liu, Yue Jiang, and Feng Chen. 2014. *Chlorella zofingiensis* as an Alternative Microalgal Producer of Astaxanthin: Biology and Industrial Potential. *Marine Drugs* 12 (6): 3487–3515.

Liu, Jin, Yujuan Zhong, Zheng Sun, Junchao Huang, Gerhard Sandmann, and Feng Chen. 2010. One Amino Acid Substitution in Phytoene Desaturase Makes *Chlorella zofingiensis* Resistant to Norflurazon and Enhances the Biosynthesis of Astaxanthin. *Planta* 232 (1): 61–67.

Lohr, Martin, Chung-Soon Im, and Arthur R. Grossman. 2005. Genome-Based Examination of Chlorophyll and Carotenoid Biosynthesis in *Chlamydomonas reinhardtii*. *Plant Physiology* 138 (1): 490–515.

Lohr, Martin, and Christian Wilhelm. 1999. Algae Displaying the Diadinoxanthin Cycle Also Possess the Violaxanthin Cycle. *Proceedings of the National Academy of Sciences of the United States of America* 96 (15): 8784–8789.

Lubián, Luis M., Olimpio Montero, Ignacio Moreno-Garrido, I. Emma Huertas, Cristina Sobrino, Manuel González-del Valle, and Griselda Parés. 2000. *Nannochloropsis* (Eustigmatophyceae) as Source of Commercially Valuable Pigments. *Journal of Applied Phycology* 12 (3–5): 249–255.

Ma, Le, Shao-Fang Yan, Yang-Mu Huang, Xin-Rong Lu, Fang Qian, Hong-Lei Pang, Xian-Rong Xu et al. 2012. Effect of Lutein and Zeaxanthin on Macular Pigment and Visual Function in Patients with Early Age-Related Macular Degeneration. *Ophthalmology* 119 (11): 2290–2297.

Maeda, Hayato, Masashi Hosokawa, Tokutake Sashima, Katsura Funayama, and Kazuo Miyashita. 2005. Fucoxanthin from Edible Seaweed, *Undaria pinnatifida*, Shows Antiobesity Effect through UCP1 Expression in White Adipose Tissues. *Biochemical and Biophysical Research Communications* 332 (2): 392–397.

Maeda, Hayato, Takayuki Tsukui, Tokutake Sashima, Masashi Hosokawa, and Kazuo Miyashita. 2008. Seaweed Carotenoid, Fucoxanthin, as a Multi-Functional Nutrient. *Asia Pacific Journal of Clinical Nutrition* 17 (Suppl 1): 196–199.

Mäki-Arvela, Päivi, Imane Hachemi, and Dmitry Yu. Murzin. 2014. Comparative Study of the Extraction Methods for Recovery of Carotenoids from Algae: Extraction Kinetics and Effect of Different Extraction Parameters. *Journal of Chemical Technology & Biotechnology* 89 (11): 1607–1626.

Markou, Giorgos, and Elias Nerantzis. 2013. Microalgae for High-Value Compounds and Biofuels Production: A Review with Focus on Cultivation under Stress Conditions. *Biotechnology Advances* 31 (8): 1532–1542.

McCarthy, Sarah S., Marilyn C. Kobayashi, and Krishna K. Niyogi. 2004. White Mutants of *Chlamydomonas reinhardtii* Are Defective in Phytoene Synthase. *Genetics* 168 (3): 1249–1257.

McEwen, Charles N., Richard G. McKay, and Barbara S. Larsen. 2005. Analysis of Solids, Liquids, and Biological Tissues Using Solids Probe Introduction at Atmospheric Pressure on Commercial LC/MS Instruments. *Analytical Chemistry* 77 (23): 7826–7831.

Mikami, Koji, and Masashi Hosokawa. 2013. Biosynthetic Pathway and Health Benefits of Fucoxanthin, an Algae-Specific Xanthophyll in Brown Seaweeds. *International Journal of Molecular Sciences* 14 (7): 13763–13781.

Miyashita, Kazuo, Sho Nishikawa, Fumiaki Beppu, Takayuki Tsukui, Masayuki Abe, and Masashi Hosokawa. 2011. The Allenic Carotenoid Fucoxanthin, a Novel Marine Nutraceutical from Brown Seaweeds. *Journal of the Science of Food and Agriculture* 91 (7): 1166–1174.

Moeller, Suzen M., Paul F. Jacques, and Jeffrey B. Blumberg. 2000. The Potential Role of Dietary Xanthophylls in Cataract and Age-Related Macular Degeneration. *Journal of the American College of Nutrition* 19 (Suppl 5): 522S–527S.

Molina Grima, E., F.G. Acién Fernández, F. García Camacho, and Yusuf Chisti. 1999. Photobioreactors: Light Regime, Mass Transfer, and Scaleup. *Journal of Biotechnology* 70 (1–3): 231–247.

Montero, Olimpio, Maria Dolores Macías-Sánchez, Carmen M. Lama, Luis M. Lubián, Casimiro Mantell, Miguel Rodríguez, and Enrique M. de la Ossa. 2005. Supercritical CO2 Extraction of Beta-Carotene from a Marine Strain of the Cyanobacterium *Synechococcus* Species. *Journal of Agricultural and Food Chemistry* 53 (25): 9701–9707.

Moriguchi, S., J.C. Jacksont, and R.R. Watson. 1985. Effects of Retinoids on Human Lymphocyte Functions in Vitro. *Human & Experimental Toxicology* 4 (4): 365–378.

Mozaffarieh, Maneli, Stefan Sacu, and Andreas Wedrich. 2003. The Role of the Carotenoids, Lutein and Zeaxanthin, in Protecting against Age-Related Macular Degeneration: A Review Based on Controversial Evidence. *Nutrition Journal* 2: 20.

Naguib, Yousry M.A. 2000. Antioxidant Activities of Astaxanthin and Related Carotenoids. *Journal of Agricultural and Food Chemistry* 48 (4): 1150–1154.

Nishikawa, Kahoko, Haruko Machida, Yoko Yamakoshi, Ryo Ohtomo, Katsuharu Saito, Masanori Saito, and Noriko Tominaga. 2006. Polyphosphate Metabolism in an Acidophilic Alga *Chlamydomonas acidophila* KT-1 (Chlorophyta) under Phosphate Stress. *Plant Science* 170 (2): 307–313.

Nishikawa, Kahoko, Yoko Yamakoshi, Isao Uemura, and Noriko Tominaga. 2003. Ultrastructural Changes in *Chlamydomonas acidophila* (Chlorophyta) Induced by Heavy Metals and Polyphosphate Metabolism. *FEMS Microbiology Ecology* 44 (2): 253–259.

Nishino, Hoyoku, Michiaki Murakoshi, Harukuni Tokuda, and Yoshiko Satomi. 2009. Cancer Prevention by Carotenoids. *Archives of Biochemistry and Biophysics* 483 (2): 165–168.

Niyogi, Krishna K., Olle Björkman, and Arthur R. Grossman. 1997. The Roles of Specific Xanthophylls in Photoprotection. *Proceedings of the National Academy of Sciences of the United States of America* 94 (25): 14162–14167.

Nobre, Beatriz, Filipa Marcelo, Renata Passos, Luis Beirão, António Palavra, Luísa Gouveia, and Rui Mendes. 2006. Supercritical Carbon Dioxide Extraction of Astaxanthin and Other Carotenoids from the Microalga *Haematococcus pluvialis*. *European Food Research and Technology* 223 (6): 787–790.

Olivieri, Guiseppe, Salatino Piero, and Marzocchella Antonio. 2014. Advances in photobioreactors for intensive microalgal production: Configurations, operating strategies and applications. *Journal of Chemical Technology & Biotechnology* 89: 178–195.

Ornelas-Paz, J. de Jesus, Elhadi M. Yahia, and Alfonso Gardea-Bejar. 2007. Identification and Quantification of Xanthophyll Esters, Carotenes, and Tocopherols in the Fruit of Seven Mexican Mango Cultivars by Liquid Chromatography-Atmospheric Pressure Chemical Ionization-Time-of-Flight Mass Spectrometry [LC-(APcI(+))-MS]. *Journal of Agricultural and Food Chemistry* 55 (16): 6628–6635.

Ota, Masaki, Hiromoto Watanabe, Yoshitaka Kato, Watanabe, Yoshiyuki Sato, Richard Lee Smith, and Hiroshi Inomata. 2009. Carotenoid Production from *Chlorococcum littorale* in Photoautotrophic Cultures with Downstream Supercritical Fluid Processing. *Journal of Separation Science* 32 (13): 2327–2335.

Pan, Yi-Ying, Suz-Ting Wang, Lu-Te Chuang, Yen-Wei Chang, and Ching-Nen Nathan Chen. 2011. Isolation of Thermo-Tolerant and High Lipid Content Green Microalgae: Oil Accumulation Is Predominantly Controlled by Photosystem Efficiency during Stress Treatments in *Desmodesmus*. *Bioresource Technology* 102 (22): 10510–10517.

Paolo, Palozza, Gabriella Calviello, Mario Emilia De Leo, Simona Serini, and Gianna Maria Bartoli. 2000. Canthaxanthin Supplementation Alters Antioxidant Enzymes and Iron Concentration in Liver of Balb/c Mice. *Journal of Nutrition* 130 (5): 1303–1308.

Pasquet, Virginie, Jean-René Chérouvrier, Firas Farhat, Valérie Thiéry, Jean-Marie Piot, Jean-Baptiste Bérard, Raymond Kaas et al. 2011. Study on the Microalgal Pigments Extraction Process: Performance of Microwave Assisted Extraction. *Process Biochemistry* 46 (1): 59–67.

Peng, Juan, Jian-Ping Yuan, Chou-Fei Wu, and Jiang-Hai Wang. 2011. Fucoxanthin, a Marine Carotenoid Present in Brown Seaweeds and Diatoms: Metabolism and Bioactivities Relevant to Human Health. *Marine Drugs* 9 (10): 1806–1828.

Peto, R., R. Doll, J.D. Buckley, and M.B. Sporn. 1981. Can Dietary Beta-Carotene Materially Reduce Human Cancer Rates? *Nature* 290 (5803): 201–208.

Piccaglia, Roberta, Mauro Marotti, and Silvia Grandi. 1998. Lutein and Lutein Ester Content in Different Types of Tagetes patula and Tagetes erecta. Industrial Crops and Products 8 (1): 45–51.

Poppel, G. van, and R.A. Goldbohm. 1995. Epidemiologic Evidence for Beta-Carotene and Cancer Prevention. *American Journal of Clinical Nutrition* 62 (6): 1393S–1402S.

Pulz, Otto, and Wolfgang Gross. 2004. Valuable Products from Biotechnology of Microalgae. *Applied Microbiology and Biotechnology* 65 (6): 635–648.

Raja, R., S. Hemaiswarya, and R. Rengasamy. 2007. Exploitation of *Dunaliella* for β-Carotene Production. *Applied Microbiology and Biotechnology* 74 (3): 517–523.

Ramos, Ana, Sacha Coesel, Ana Marques, Marta Rodrigues, Alexandra Baumgartner, João Noronha, Amélia Rauter, Bertram Brenig, and João Varela. 2008. Isolation and Characterization of a Stress-Inducible *Dunaliella salina* Lcy-β Gene Encoding a Functional Lycopene β-Cyclase. *Applied Microbiology and Biotechnology* 79 (5): 819–828.

Ranga Rao, Ambati, Vallikannan Baskaran, Ravi Sarada, and Gokare Aswathanarayana Ravishankar. 2013. In Vivo Bioavailability and Antioxidant Activity of Carotenoids from Microalgal Biomass—A Repeated Dose Study. *Food Research International* 54 (1): 711–717.

Rao, Ambati Ranga, Ravi Sarada, and Gokare Aswathanarayana Ravishankar. 2007. Stabilization of Astaxanthin in Edible Oils and Its Use as an Antioxidant. *Journal of the Science of Food and Agriculture* 87 (6): 957–965.

Ravishankar, Gokare Aswathanarayana, Ravi Sarada, B. Sandesh Kamath, and K.K. Namitha. 2006. Food Applications of Algae. In *Food Science and Technology*, 2nd Edition, eds. Kalidas Shetty, Gopinadhan Paliyath, Anthony Pometto, Robert E. Levin, 491–524. Food Science and Technology. Boca Raton, FL: CRC Press, Taylor & Francis Group.

Remias, Daniel, Ursula Lütz-Meindl, and Cornelius Lütz. 2005. Photosynthesis, Pigments and Ultrastructure of the Alpine Snow Alga *Chlamydomonas nivalis*. *European Journal of Phycology* 40 (3): 259–268.

Rivera, S.M., and R. Canela-Garayoa. 2012. Analytical Tools for the Analysis of Carotenoids in Diverse Materials. *Journal of Chromatography A* 1224: 1–10.

Roberts, Richard L., Justin Green, and Brandon Lewis. 2009. Lutein and Zeaxanthin in Eye and Skin Health. *Clinics in Dermatology* 27 (2): 195–201.

Rodrigues, Eliseu, Lilian R.B. Mariutti, and Adriana Z. Mercadante. 2012. Scavenging Capacity of Marine Carotenoids against Reactive Oxygen and Nitrogen Species in a Membrane-Mimicking System. *Marine Drugs* 10 (8): 1784–1798.

Rodriguez-Amaya, Delia B. 2001. *A Guide to Carotenoid Analysis in Foods*. Washington, DC: ILSI Press.

Rothschild, Lynn J., and Rocco L. Mancinelli. 2001. Life in Extreme Environments. *Nature* 409 (6823): 1092–1101.

Sajilata, M.G., R.S. Singhal, and M.Y. Kamat. 2008a. Fractionation of Lipids and Purification of γ-Linolenic Acid (GLA) from *Spirulina platensis*. *Food Chemistry* 109 (3): 580–586.

Sajilata, M.G., R.S. Singhal, and M.Y. Kamat. 2008b. The Carotenoid Pigment Zeaxanthin—A Review. *Comprehensive Reviews in Food Science and Food Safety* 7 (1): 29–49.

Salguero, Alonso, Rosa León, Annalisa Mariotti, Benito de la Morena, José M. Vega, and Carlos Vílchez. 2005. UV-A Mediated Induction of Carotenoid Accumulation in *Dunaliella bardawil* with Retention of Cell Viability. *Applied Microbiology and Biotechnology* 66 (5): 506–511.

Sánchez Pérez, J.A., E.M. Rodríguez Porcel, J.L. Casas López, J.M. Fernández Sevilla, and Y. Chisti. 2006. Shear Rate in Stirred Tank and Bubble Column Bioreactors. *Chemical Engineering Journal* 124 (1–3): 1–5.

Sangeetha, Ravi Kumar, Narayan Bhaskar, Sounder Divakar, and Vallikannan Baskaran. 2009. Bioavailability and Metabolism of Fucoxanthin in Rats: Structural Characterization of Metabolites by LC-MS (APCI). *Molecular and Cellular Biochemistry* 333 (1–2): 299.

Santos, Carla A., Ana M. Vieira, Helena L. Fernandes, Jose A. Empis, and Júlio M. Novais. 2001. Optimisation of the Biological Treatment of Hypersaline Wastewater from *Dunaliella salina* Carotenogenesis. *Journal of Chemical Technology & Biotechnology* 76 (11): 1147–1153.

Schulz, H., M. Baranska, and R. Baranski. 2005. Potential of NIR-FT-Raman Spectroscopy in Natural Carotenoid Analysis. *Biopolymers* 77 (4): 212–221.

Schwender, Jörg, Claudia Gemünden, and Hartmut K. Lichtenthaler. 2001. Chlorophyta Exclusively Use the 1-Deoxyxylulose 5-Phosphate/2-C-Methylerythritol 4-Phosphate Pathway for the Biosynthesis of Isoprenoids. *Planta* 212 (3): 416–423.

Shaish, Aviv, Mordhay Avron, Uri Pick, and Ami Ben-Amotz. 1993. Are Active Oxygen Species Involved in Induction of β-Carotene in *Dunaliella bardawil*? *Planta* 190 (3): 363–368.

Skjånes, Kari, Céline Rebours, and Peter Lindblad. 2013. Potential for Green Microalgae to Produce Hydrogen, Pharmaceuticals and Other High Value Products in a Combined Process. *Critical Reviews in Biotechnology* 33 (2): 172–215.

Snodderly, D.M. 1995. Evidence for Protection against Age-Related Macular Degeneration by Carotenoids and Antioxidant Vitamins. *American Journal of Clinical Nutrition* 62 (6): 1448S–1461S.

Sobczuk, T. Mazzuca, F. García Camacho, E. Molina Grima, and Yusuf Chisti. 2006. Effects of Agitation on the Microalgae *Phaeodactylum tricornutum* and *Porphyridium cruentum*. *Bioprocess and Biosystems Engineering* 28 (4): 243.

Stahl, Wilhelm, and Helmut Sies. 1996. Lycopene: A Biologically Important Carotenoid for Humans? *Archives of Biochemistry and Biophysics* 336 (1): 1–9.

Steinbrenner, Jens, and Hartmut Linden. 2003. Light Induction of Carotenoid Biosynthesis Genes in the Green Alga *Haematococcus pluvialis*: Regulation by Photosynthetic Redox Control. *Plant Molecular Biology* 52 (2): 343–356.

Strohschein, Sabine, Matthias Pursch, Heidrun Händel, and K. Albert. 1997. Structure Elucidation of β-Carotene Isomers by HPLC-NMR Coupling Using a C30 Bonded Phase. *Fresenius' Journal of Analytical Chemistry* 357 (5): 498–502.

Swartz, Michael E. 2005. UPLC™: An Introduction and Review. *Journal of Liquid Chromatography & Related Technologies* 28 (7–8): 1253–1263.

Takaichi, Shinichi. 2011. Carotenoids in Algae: Distributions, Biosyntheses and Functions. *Marine Drugs* 9 (6): 1101–1118.

Telfer, Alison, Andrew Pascal, and Andrew Gall. 2008. Carotenoids in Photosynthesis. In *Carotenoids*, Vol. 4, *Natural Functions*, ed. George Britton, H.C. Synnøve Liaaen-Jensen, and Hanspete Pfander, 265–308. Basel: Birkhauser Verlag.

Törnwall, Markareetta E., Jarmo Virtamo, Pasi A. Korhonen, Mikko J. Virtanen, Philip R. Taylor, Demetrius Albanes, and Jussi K. Huttunen. 2004. Effect of α-Tocopherol and β-Carotene Supplementation on Coronary Heart Disease during the 6-Year Post-Trial Follow-up in the ATBC Study. *European Heart Journal* 25 (13): 1171–1178.

Ugwu, C.U., H. Aoyagi, and H. Uchiyama. 2008. Photobioreactors for Mass Cultivation of Algae. *Bioresource Technology* 99 (10): 4021–4028.

Vanitha, A., K.N. Chidambara Murthy, Vinod Kumar, G. Sakthivelu, Jyothi M. Veigas, P. Saibaba, and Gokare A. Ravishankar. 2007. Effect of the Carotenoid-Producing Alga, *Dunaliella bardawil*, on CCl4-Induced Toxicity in Rats. *International Journal of Toxicology* 26 (2): 159–167.

Vaquero, Isabel, Benito Mogedas, M. Carmen Ruiz-Domínguez, José M. Vega, and Carlos Vílchez. 2014. Light-Mediated Lutein Enrichment of an Acid Environment Microalga. *Algal Research* 6 (Pt A): 70–77.

Varela, João C., Hugo Pereira, Marta Vila, and Rosa León. 2015. Production of Carotenoids by Microalgae: Achievements and Challenges. *Photosynthesis Research* 125 (3): 423–436.

Varshney, Prachi, Paulina Mikulic, Avigad Vonshak, John Beardall, and Pramod P. Wangikar. 2015. Extremophilic Micro-Algae and Their Potential Contribution in Biotechnology. *Bioresource Technology* 184: 363–372.

Vidhyavathi, Raman, Lakshmanan Venkatachalam, Ravi Sarada, and Gokare Aswathanarayana Ravishankar. 2008. Regulation of Carotenoid Biosynthetic Genes Expression and Carotenoid Accumulation in the Green Alga *Haematococcus pluvialis* under Nutrient Stress Conditions. *Journal of Experimental Botany* 59 (6): 1409–1418.

Vila, M., I. Couso, and R. León. 2008. Carotenoid Content in Mutants of the Chlorophyte *Chlamydomonas reinhardtii* with Low Expression Levels of Phytoene Desaturase. *Process Biochemistry* 43 (10): 1147–1152.

Vílchez Lobato, Carlos, Lozano Eduardo Forján, Franco María Cuaresma, Francisco Bédmar, Inés Garbayo Nores, and Vega. M. Jose. 2011. Marine Carotenoids: Biological Functions and Commercial Applications. *Marine Drugs* 9 (3): 319–333.

Vincent, Warwick F., and Antonio Quesada. 1994. Ultraviolet Radiation Effects on Cyanobacteria: Implications for Antarctic Microbial Ecosystems. In *Ultraviolet Radiation in Antarctica: Measurements and Biological Effects*, ed. C. Susan Weiler and Polly A. Penhale, 111–124. Washington, DC: American Geophysical Union.

Wang, Xin, Willén Roger, and Wadström Torkel. 2000. Astaxanthin-Rich Algal Meal and Vitamin C Inhibit *Helicobacter pylori* Infection in BALB/cA Mice. *Antimicrobial Agents and Chemotherapy* 44 (9): 2452–2457.

Weller, Philipp, and Dietmar E. Breithaupt. 2003. Identification and Quantification of Zeaxanthin Esters in Plants Using Liquid Chromatography-Mass Spectrometry. *Journal of Agricultural and Food Chemistry* 51 (24): 7044–7049.

White, Andrea L., and Leland S. Jahnke. 2002. Contrasting Effects of UV-A and UV-B on Photosynthesis and Photoprotection of Beta-Carotene in Two *Dunaliella* spp. *Plant & Cell Physiology* 43 (8): 877–884.

Xia, Song, Ke Wang, Linglin Wan, Aifen Li, Qiang Hu, and Chengwu Zhang. 2013. Production, Characterization, and Antioxidant Activity of Fucoxanthin from the Marine Diatom *Odontella aurita*. *Marine Drugs* 11 (7): 2667–2681.

Xie, Youping, Shih-Hsin Ho, Ching-Nen Nathan Chen, Chun-Yen Chen, I.-Son Ng, Ke-Ju Jing, Jo-Shu Chang, and Yinghua Lu. 2013. Phototrophic Cultivation of a Thermo-Tolerant *Desmodesmus* Sp. for Lutein Production: Effects of Nitrate Concentration, Light Intensity and Fed-Batch Operation. *Bioresource Technology* 144: 435–444.

Yamaguchi, Masayoshi. 2012. Role of Carotenoid β-Cryptoxanthin in Bone Homeostasis. *Journal of Biomedical Science* 19: 36.

Yan, Yuan, Yue-Hui Zhu, Jian-Guo Jiang, and Dong-Lin Song. 2005. Cloning and Sequence Analysis of the Phytoene Synthase Gene from a Unicellular Chlorophyte, *Dunaliella salina*. *Journal of Agricultural and Food Chemistry* 53 (5): 1466–1469.

Ye, Zhi-Wei, and Jian-Guo Jiang. 2010. Analysis of an Essential Carotenogenic Enzyme: ζ-Carotene Desaturase from Unicellular Alga *Dunaliella salina*. *Journal of Agricultural and Food Chemistry* 58 (21): 11477–11482.

Yen, Hong-Wei, Cheng-Hsiung Sun, and Te-Wei Ma. 2010. The Comparison of Lutein Production by *Scenesdesmus* Sp. in the Autotrophic and the Mixotrophic Cultivation. *Applied Biochemistry and Biotechnology* 164 (3): 353–361.

Yu, Xiaona, Hongli Cui, Yulin Cui, Yan Wang, Xueqin Li, Zhaopu Liu, and Song Qin. 2014. Gene Cloning, Sequence Analysis, and Expression Profiles of a Novel β-Ring Carotenoid Hydroxylase Gene from the Photoheterotrophic Green Alga *Chlorella kessleri*. *Molecular Biology Reports* 41 (11): 7103–7113.

Zhang Li, Gong YiFu, Liu XiaoDan, and Zhang WenQing. 2013. Cloning and Expression Analysis of β-Carotene Hydroxylase Gene (chyb) from *Dunaliella salina*. *Journal of Agricultural Biotechnology* 21 (8): 920–930.

Zhekisheva, M., S. Boussiba, I. Khozin-Goldberg, A. Zarka, and Z. Cohen. 2002. Accumulation of Triacylglycerols in *Haematococcus pluvialis* Is Correlated with That of Astaxanthin Esters. *Journal of Phycology* 38: 40–41.

Zhu, Yue-Hui, Jian-Guo Jiang, and Xin-Wen Chen. 2007. cDNA for Phytoene Desaturase in *Dunaliella salina* and Its Expressed Protein as Indicators of Phylogenetic Position of the β-Carotene Biosynthetic Pathway. *Journal of the Science of Food and Agriculture* 87 (9): 1772–1777.

Zhu, Yue-Hui, Jian-Guo Jiang, Yuan Yan, and Xing-Wen Chen. 2005. Isolation and Characterization of Phytoene Desaturase cDNA Involved in the Beta-Carotene Biosynthetic Pathway in *Dunaliella salina*. *Journal of Agricultural and Food Chemistry* 53 (14): 5593–5597.

CHAPTER 14

Secondary Metabolites from Microalgal Extremophiles and Their "Extreme-Loving" Neighbors

Antje Labes

CONTENTS

14.1 INTRODUCTION

Research on extremophiles in the last decades has more and more pushed the "border of life" by discovering organisms able to live in boiling water, in acidic lakes with a pH around zero, and at hypersaline conditions that would dry any other cell to death. Extremophilic microbes and macrobes were found to live under extreme pressures in the depth of the ocean or algae blooming directly in snow (Canganella and Wiegel 2011). These discoveries are fascinating in themselves, but extremophilic microorganisms additionally have been revealed to be research objects delivering revolutionary new concepts for biology. Examples are the revised trees of life comprising the archaea as a completely new domain (Forterre 2015), evidence for horizontal gene transfer as a major driving force in evolution (Daubin and Szöllősi 2016), and some answers to the question of the origin of life (Forterre 2015). Especially, research on extremophiles has helped us to understand how adaptations to the environment are realized on the cell level in general (Cava et al. 2009). It also delivered new approaches and methods for biotechnology, such as the use of thermophilic enzymes for molecular biology (Chen 2014), cold-adapted enzymes for laundry, and antifreeze proteins for cosmetics and the food industry (Schiraldi and De Rosa 2002).

It was early recognized that—besides specific inventions of the extremophiles—many cellular adaptation mechanisms are independent of the grade and type of "extremeness" of the environmental condition. Even more, for any taxonomic group, an extreme habitat condition can be defined as being the "survival border" for most of the organisms. Extremeness from the perspective of the organism always needs an adaptation, no matter how extreme the environmental condition is. This explains the very broad ranges of conditions—which are considered as extreme and which are not necessarily hyperextreme conditions. For instance, adaptation to temperature can be living at 42°C, which is extreme for *Escherichia coli*, but this temperature is far from the hyperextreme condition of the highest temperature optimum observed for an organism, at around 121°C. We have to keep this in mind when discussing properties of extremophiles.

The microbial biodiversity of extreme environments is very specific, with microbes being present, of which we have only seen the tip of the iceberg. Most of the extreme microbial biodiversity has not yet been discovered and characterized—either taxonomically or chemically—which opens room for discovery programs tapping the potential of extremophiles in a more strategic manner.

In the microalga field, extremophiles can be found in hot and cold, saline, deep or pressurized, and acidic or alkaline environments, and a number of studies have described the taxonomy, ecologic adaptations, and growth behavior. New strains are constantly isolated from unique extreme habitats, that is, hot springs, alkaliphilic and acidic vents, and hypersaline lakes. The flexibility of microalgal genomes probably allowed life to adapt to a wide spectrum of extreme environments (Pikuta et al. 2007). Even more, analysis of the phylogenetic relationship of the reported extremophiles suggests that certain protist groups are more tolerant to extremophilic conditions than other taxa (Varshney et al. 2015).

Cyanobacteria (blue-green algae, belonging to the Bacteria domain) and eukaryotic microalgae in general are well recognized as sources of chemically diverse, bioactive compounds—especially toxins in algal blooms—and for their applicability in biotechnological processes (Priyadarshani and Rath 2012). In recent years, the great need for new natural products, mostly being derived from secondary metabolism, led to a revival of natural product research. Natural products are the starting points for drug development. They can also be applied in cosmetics, help to replace toxic chemicals, or be applied as plant-protecting agents and in many other application fields. For example, more than 60% of the currently used anti-infective drugs are derived from natural resources (Newman and Cragg 2012), with an ever-growing number of new structural classes being identified from the vast biodiversity of our planet. Both cyanobacteria and eukaryotic microalgae contributed to this pool of new compounds. Cyanobacteria are a rich and biotechnologically widely unexplored source of structurally diverse bioactive metabolites with potential pharmaceutical and biotechnological applications (Rastogi and Sinha 2009). The best-known example is the dolastatins, microtubule-destabilizing agents, isolated from the genus *Symploca* (Simmons et al. 2005). The cyclic depsipeptide largazole from another *Symploca* sp. is a recent example of a novel chemical scaffold (Hong and Luesch 2012), exemplary for the untapped chemical space of these microbes. A wide variety of chemical structures have been reported from cyanobacteria with a wide variety of biological activities, including antibacterial metabolites (Prasanna et al. 2008). In the case of eukaryotic microalgae, more than 120 species of marine eukaryotic microalgae are known to produce bioactive metabolites with potent toxic and/or other allelochemical properties against other organisms. Again, the known bioactive compounds comprise a wide diversity of metabolite classes, ranging from linear and polycyclic ethers to tetrahydropurine alkaloids (e.g., saxitoxin) and secondary amino acids (e.g., domoic acid) (Krock et al. 2008; Wijffels et al. 2013).

However, none of these examples were derived from extremophilic microalgae. They were scarcely studied with respect to secondary metabolites. As extremophiles are nature's ultimate extreme survivors, they may have evolved new and unique survival mechanisms to serve as a valuable resource for novel metabolites and their biotechnological utilization. The untapped biodiversity in extreme habitats offers the opportunity to identify attractive new bioactive molecules in extremophilic microbes, especially those with new modes of action. Just statistically, extreme environments are scarcely studied, increasing the probability of finding new taxa and thereby also new enzymes, metabolites, and so forth. Realizing that habitat conditions are important triggers of secondary metabolite production, an intensified search for secondary metabolites in organisms living under extreme conditions was started in recent years. The assumption is that extremophilic microalgae accumulate structurally unique bioactive natural products not found in nonextreme organisms to adapt and survive in the extreme environment characterized by very special conditions that differ from those found in other habitats. Hence, the extreme conditions may be answered by small molecules enabling survival. However, if only a very small part of the genetic information has to be adapted to be able to thrive under extreme conditions, the novelty might be only minor. The increasing number of studies on secondary metabolites from extremophilic microalgae, but also on genomes of extremophiles, does not draw a clear picture yet. Especially, comparative genomics is still in its infancy with respect to extremophilic natural products. However, the number of compounds from extremophilic microbes, including microalgae, described in recent years is slowly increasing, but the way from discovery to production and even development of lead compounds is far from being advanced. Some steps of the discovery pipeline need serious attempts for improvement (as outlined in Section 14.8), including studies on the ecophysiology and functional diversity of extreme microbes; others will continue to be time-consuming (legal procedures for, e.g., drug development), risky (product development), or economically nonfeasible.

With respect to the nature of the extreme habitat, differences in secondary metabolite production can be observed and explained by the different lifestyles in adaptation to the condition. Hence, some extreme habitats seem to be much more promising resources for the discovery of secondary metabolites than others. In this chapter, the capability of extremophilic microalgae of producing secondary metabolites is reported and compared with that of other extremophilic microbes.

14.2 HYPERSALINE ENVIRONMENTS

Microbiologists very early recognized that saline and hypersaline environments are well colonized by microbes and some macrobes. Adaptation to the saline conditions was early understood as the production of osmolytes coping with the defense of the cell against osmotic effects. Most of these osmolytes were found to be primary metabolites, such as sugars and amino acids, or to be specifically accumulated ions (Schimel et al. 2007). However, especially phenolic compounds were described as products from secondary biosynthetic pathways. Specific studies on the bioactive secondary metabolites of halophytic plants led to the identification of bioactive compounds known from nonsaline plants (triterpenes, flavonoids, and sterols). Consequently, several salt marsh plants have traditionally been used for medical, nutritional, and even artisanal purposes (Ksouri et al. 2012). In contrast to the studied macroscopic plant world, only few data on saline microalgae are available. However, these compound classes also are well described from microalgae and could be exploited from these in a biotechnological approach.

Especially in the marine environment, hypersaline conditions can occur, affecting microalgae, for example, in sea ice, where saline-containing channels form a specific habitat, but also in hypersaline brines and basins. A biodiversity study on microbial eukaryotes of the anoxic hypersaline deep-sea basin Thetis in the eastern Mediterranean revealed an unexpected diversity (Stock et al. 2012). The halite-saturated brine of this polyextreme basin revealed one of the highest salt concentrations ever reported for such an environment (salinity of 348‰). Cell counts indicated more than 1000 protists per liter of anoxic brine, including fungi, the most diverse taxonomic group of eukaryotes in the brine. This makes deep-sea brines sources of unknown fungal diversity and hot spots for the discovery of novel metabolic pathways and for secondary metabolites. Unexpectedly, the second most diverse phylotypes were ciliates and stramenopiles, the latter being considered algal-related protists. They are well known from nonextreme habitats as prolific producers of omega fatty acids and a variety of bioactive compounds (Gupta et al. 2011).

In arid regions, hypersaline conditions frequently occur in shallow waters. Two microalgae, the diatom *Nitzschia frustule* and the green alga *Chlamydomonas plethora*, isolated from hypersaline waters off Kuwait, yielded growth rates and high levels of technologically interesting metabolites (mainly amino acids, but the study did not focus on secondary metabolite content [Subba Rao et al. 2005]); such strains could serve as the perfect biotechnological host for secondary metabolites produced under saline conditions. *Dunaliella salina* is a major source for β-carotene (Hosseini Tafreshi and Shariati 2009).

To mention the very famous hypersaline neighbors, hypersaline archaea are rather studied for their primary metabolites and compatible solutes, but also serve as a potent source for carotenoids (Rodrigo-Banos et al. 2015). The bacteriorhodopsin from *Halobacterium salinarium* has found its way into a wide range of applications, from dyes to artificial retinas (Charlesworth and Burns 2015). In addition, antimicrobial peptides ("halocins") and diketopiperazines (cyclic dipeptides) seem to be very common in halophilic archaea, exhibiting a broad spectrum of antimicrobial activities (DasSama et al. 2009).

14.3 DEEP SEA

Such deep-sea habitats are extreme with respect to pressure, too. Recently, the culturability and secondary metabolite diversity of extreme microbes were reviewed with respect to the deep sea and deep-sea vent microbes (Pettit 2011). In general, it can be stated that microbes from extreme environments do not necessarily require extreme culture conditions. Perhaps the most extreme environments known, deep-sea hydrothermal vent sites, support an incredible array of archaea, bacteria, and fungi, many of which have now been cultured. Microbes cultured from extreme environments have not disappointed in the natural product arena; diverse bioactive secondary metabolites have been isolated from cultured deep-sea microbes. The recent isolation of new chroman derivatives and siderophores (compounds facilitating iron uptake) from deep-sea hydrothermal vent bacteria can be seen as an example. Both have possible application in the medical field. However, the deep-sea conditions do not allow phototrophic growth of microalgae. Only heterotrophic species will be active under the deep-sea conditions; therefore, not many researchers considered deep-sea microalgal species as producers (Andrianasolo et al. 2009).

14.4 PH EXTREMES: FROM ACID TO SODA

pH extremes serve as a trigger for secondary metabolites; especially low pH seems to be a trigger (Stierle and Stierle 2014). For extremophilic microbes of the Berkeley Pit Lake, a variety of new and interesting secondary metabolites have been described. It is of particular interest that these acidophilic microbes produce small-molecule inhibitors of pathways associated with low pH and high Eh. These same small molecules also inhibit molecular pathways induced by reactive oxygen species (ROS) and inflammation in mammalian cells.

Recent molecular studies on the genomics of the thermoacidophilic red alga *Galdieria sulphuraria* revealed the presence of secondary metabolite biosynthetic pathways. *G. sulphuraria* species are unicellular red microalgae that occur worldwide in hot acidic waters, volcanic calderas, and human-made acidic environments, such as acidic mine drainage. *G. sulphuraria* has a unique position, as it is able to grow photoautotrophically, mixotrophically, and heterotrophically. It is resistant to not only acid (pH 0) and heat (56°C), but also high salt (1.5 M NaCl), toxic metals, and many other abiotic stressors. This unusual combination of features, such as thermophily, acidophily, resistance to a wide array of abiotic stressors, and an extraordinary metabolic plasticity, make *G. sulphuraria* both a highly interesting model organism and an interesting (host) producer for secondary metabolites

(Barbier et al. 2005). *G. sulphuraria* also has the potential to meet the requirement for biodiesel production. Combining secondary metabolite and biodiesel production into one process line would be quite advantageous (Pulz and Gross 2004).

14.5 HOT, HOTTER, HOTTEST: THERMOPHILIC AND HYPERTHERMOPHILIC CONDITIONS

It was established that 73°C–74°C is the maximum temperature enabling development of cyanobacteria (Seckbach 2007). Only a few thermophiles have been exploited commercially, such as the unicellular biflagellate *Dunaliella* sp., cultivated on a mass scale for the production of high-value carotenoids (Hosseini Tafreshi and Shariati 2009). A new strain of the thermophilic alga *Scenedesmus* sp. isolated from Jemez Springs, New Mexico, yielded high levels of lipids and carotenoids (Durvasula et al. 2015).

For the extreme temperature range, which was defined by Stetter in the 1980s to be above 80°C, no secondary metabolites were described until now. Most of these organisms capable of living as hyperthermophiles are archaea (Stetter 1999). A detailed genetic search revealed the absence of typical gene clusters for secondary metabolites. This absence is probably due to the energetic situation within these habitats, which are often driven by chemotrophy. Secondary metabolite synthesis might be too costly in terms of energy demand, especially when living on minimum energetic conditions and in thermodynamic unstable ranges. At temperatures above 80°C, not only molecules like ATP face a stability problem, but also many structural classes of natural products.

14.6 RATHER COLD: PSYCHROPHILIC

Plants, including those from Arctic and alpine areas, protect themselves from herbivores by producing an array of secondary metabolites. In the sea surrounding Antarctica, there are many chemical interactions between invertebrates, algaem and their predators, and in Arctic ecosystems, there are indications that secondary metabolites also play a role in the chemical interactions between species. However, these potential interactions have not been studied in any detail in microbiological systems to reveal their chemical nature (Frisvad 2007). However, secondary metabolites seem to play a crucial role in protection against harmful ultraviolet (UV) and excessive visible (VIS) radiation, especially under psychrophilic conditions, which keeps the general metabolic rates at a low level. *Mesotaenium berggrenii* is one of the few autotrophs that thrive on bare glacier surfaces in alpine and polar regions. Survival of this alga at exposed sites seems to depend on high amounts of a brownish vacuolar pigment with a tannin nature, purpurogallin carboxylic acid-6-O-b-D-glucopyranoside. Attributes and abundances of the purpurogallins found in *M. berggrenii* strongly suggest that they are of principal ecophysiological relevance, like analogous protective pigments of other extremophilic microorganisms (Remias et al. 2012). Similar functionality was assumed for mycosporines and mycosporine-like amino acids (MAAs). MAAs are low-molecular-weight water-soluble molecules absorbing UV radiation. They are accumulated by a wide range of microorganisms, prokaryotic (cyanobacteria) as well as eukaryotic (microalgae, yeasts, and fungi), especially in UV-rich and cold habitats (but also in a variety of marine macroalgae, corals, and other marine life-forms). The role that MAAs play as sunscreen compounds to protect against damage by harmful levels of UV radiation is well established. Evidence was gained that MAAs may have additional functions: they may serve as antioxidant molecules scavenging toxic oxygen radicals, they can be accumulated as compatible solutes following salt stress, and their formation is induced by desiccation or by thermal stress (Oren and Gunde-Cimerman 2007).

14.7 EXAMPLES REALIZED IN BIOTECHNOLOGY

Microalgae are attractive candidates for biotechnology, because bioactive strains are amenable to isolation and culture and scale-up for mass culture is feasible for optimal yield. Algae can be produced via photoautotrophic production that is called autotrophic photosynthesis. Some algae strains are capable of combining autotrophic photosynthesis and heterotrophic assimilation of organic compounds in a mixotrophic process. Currently, the photoautotrophic system is the only one that is technically and economically feasible for large-scale production of algae biomass. It is based on open-pond and closed-photobioreactor technologies. Open-pond production systems have been used since the 1950s to cultivate algae (Borowitzka 1999). Microalgae production using closed-photobioreactor technology is designed to overcome some of the major problems associated with the open-pond production systems. Photobioreactors permit culture of a single microalgae species for prolonged periods of time, with minimal risk of contamination (Mirón et al. 2002), being necessary especially for high-value products and/or high-cost fermentations. Only a few extremophiles are utilized in biotechnology for the exploitation of commercially important bioactive compounds. As mentioned before, examples are *Dunaliella*, a unicellular biflagellate, and the cyanobacteria *Synecocystis* and *Synechococcus*. *Dunaliella* is cultivated on a mass scale for the production of high-value carotenoids and zeaxanthin (Hosseini Tafreshi and Shariati 2009). The heterotrophic marine microalga *Cryptothecodinium cohnii* accumulates lipids that are 30%–50% decosahexaaenoic acid (DHA 22:6 [De Swaaf 2003]).

A true biotechnologically realized process for the production of secondary metabolites with extremophilic microalgae was not realized until today, despite available basic data, such as media optimization and tank reactor adaptation to the needs of microalgae (Greque de Morais et al. 2015).

14.8 CHALLENGES AND FUTURE DEVELOPMENTS

Some of the extremophiles synthesize biotic compounds that can remain most stable under exacting bioprocess engineering, and therefore will find utility in biotechnology. As illustrated, some of the extremophiles have tolerance mechanisms for thermostable proteins, salt tolerance proteins that have biotechnological utility in the production of thermophilic enzymes, detergent formulation, and mineralization of high-salinity organic waste (Table 14.1).

For instance, both cyanobacteria and eukaryotic microalgae produce a variety of metabolites that may have allelochemical functions in the natural environment, for example, in the prevention of biofilm formation during fouling processes. Such activities are extremely relevant for the development of anti-infectives. However, until now, this potential was not explored in detail (Dobretsov et al. 2013). In terms of the biorefinery process, it would be highly recommendable to gain high-value products, as natural products are, in combination with other products (Dong et al. 2016). This will enable sustainable, economically feasible, and environmentally friendly processes.

While extremophilic microalgae are beginning to be explored, much needs to be done in terms of the physiology, molecular biology, metabolic engineering, and outdoor cultivation trials before their true potential is realized (Varshney et al. 2015).

The supply issue seriously restricts the development of natural products into new medicines. Biotechnological production of the new molecules will enable a sustainable and reliable supply of the compounds. With respect to extremophilic microalgae, especially culture conditions for new isolates need to be developed and optimized for the biotechnological production, which is not trivial with respect to the extreme conditions. Despite their unique biochemical and physiological characteristics, many extremophilic microalgae are not investigated because of the difficulties involved in bringing them into culture. In addition, biotechnological fermentation processes need to be adapted, as the extreme conditions are a big threat to the technological setup, for example, by corrosion

Table 14.1 Examples of Secondary Metabolites from Extremophile Microbes and Their Potential Field of Application

Extreme Condition	Compound/Compound Group	Producer	Biological Activity/Application Field
Deep sea, high temperature	Chroman derivatives	*Thermovibrio ammonificans*	Medicine: Antidiabetic, hypolipidemic agents
Deep sea, high temperature	Siderophores	For example, *Thermovibrio ammonificans*	Medicine: Antimicrobials
High temperature	Carotenoids	*Scenedesmus* sp.	Medicine, food supplements, cosmoceuticals
Hypersaline	Bacteriorhodopsin	Halophilic archaea, such as *Halobacterium salinarium*	Medicine, analytical devices, dyes, etc.
Hypersaline	Phenolic compounds	Eukaryotic microalgae	Medicine, food supplements, compatible solutes
Hypersaline	Triterpenes, flavonoids, sterols	*Nitzschia frustula*	Medicine, food supplements, compatible solutes
Hypersaline	Triterpenes, flavonoids, sterols	*Chlamydomonas plethora*	Medicine, food supplements
Hypersaline	β-Carotene, zeaxanthin	*Dunaliella salina*	Food supplement
Hypersaline	Halocines (antimicrobial peptides)	Halophilic archaea	Medicine and veterinary medicine: Antibiotic
Hypersaline	Diketopiperazine	Halophilic archaea	Medicine and veterinary medicine: Antibiotic
Low pH	Small molecules of undefined nature	Acidophilic microbes	Medicine: Anti-inflammatory
High UV, low temperature	Purpurogallin carboxylic acid-6-O-b-D-glucopyranoside	*Mesotaenium berggrenii*	UV protection

(Schiraldi and De Rosa 2002). This holds especially true as the majority of the secondary metabolites from extremophiles today were identified in hypersaline conditions.

In addition to classical culture-based process development, massive sequencing of genomes and transcriptomes would provide not only an insight into the mechanism that would allow the extremophile algae to survive and prosper under stressful—extreme—conditions, such as demonstrated for *Exiguobacterium pavilionensis* (Andrianasolo et al. 2009), but also deliver clues for genetic manipulation for biotechnological applications. A straightforward combination of metabolic profiling (chemical fingerprint methods for the description of the secondary metabolites produced under a given, e.g., environmental situation) and modern "omics-techniques" (comprehensive approaches for the description of the different biological levels of a cell) will lead to a deep understanding of the metabolic potential and the regulation of the cellular level. Application of newly gained knowledge will lead to new discoveries and to advanced biotechnological processing.

In the reviewed field of extremophilic microalgae and their secondary metabolites, extended discovery programs and the respective transfer of early research results into standardized manufacturing processes seem to be the next big steps. Compared with the theoretical potential, extremophiles were neglected by the natural product community. However, combing efforts and applying modern combinations of methods, including metabolomics, can speed up the discovery and development of new natural products from extremophilic microalga. Technologists will need to consult the early-stage researchers and vice versa, to push this development, as much specialized knowledge is necessary with respect to the biology of the extremophilic microbes, but also in terms of natural product discovery and chemistry offering the challenging field structural elucidation techniques, which requires experienced scientists.

REFERENCES

Andrianasolo, E.H., L. Haramaty, R. Rosario-Passapera et al. 2009. Ammonificins A and B, hydroxyethyl-amine chroman derivatives from a cultured marine hydrothermal vent bacterium, *Thermovibrio ammonificans*. *J Nat Prod* 72 (6):1216–1219.

Barbier, G.G., M. Zimmermann, and A.P.M. Weber. 2005. Genomics of the thermo-acidophilic red alga *Galdieria sulphuraria*. In *Astrobiology and Planetary Missions* edited by R.B. Hoover, G.V. Levin, A.Y. Rozanov, and G.R. Gladstone, Proc. of SPIE Vol. 5906, 590609, (2005), 0277-786X/05/$15. doi: 10.1117/12.614532.

Borowitzka, M.A. 1999. Commercial production of microalgae: Ponds, tanks, tubes and fermenters. *J Biotechnol* 70:313–321.

Canganella, F., and J. Wiegel. 2011. Extremophiles: From abyssal to terrestrial ecosystems and possibly beyond. *Naturwissenschaften* 98 (4):253–279.

Cava, F., A. Hidalgo, and J. Berenguer. 2009. *Thermus thermophilus* as biological model. *Extremophiles* 13 (2):213–231.

Charlesworth, J.C., and B.P. Burns. 2015. Untapped resources: Biotechnological potential of peptides and secondary metabolites in archae. *Archaea* 2015:282035.

Chen, C.-Y. 2014. DNA polymerases drive DNA dequencing-by-synthesis technologies: Both past and present. *Front Microbiol* 5:305.

DasSama, P., J.A. Cocker, V. Huse et al. 2009. Halophiles, industrial applications. In *Encyclopedia of Industrial Biotechnology*. Hoboken, NJ: Wiley. P 1–43. doi:10.1002/9780470054581.eib43

Daubin, V., and G.J. Szöllősi. 2016. Horizontal gene transfer and the history of life. *Cold Spring Harb Perspect Biol* 8 (4).

De Swaaf, M.E. 2003. Analysis of docosahexaenoic acid biosynthesis in *Crypthecodinium cohnii* by c13 labeling and desaturase inhibitor experiments. *J Biotechnol* 103:21–29.

Dobretsov, S., R.M. Abed, and M. Teplitski. 2013. Mini-review: Inhibition of biofouling by marine microorganisms. *Biofouling* 29 (4):423–441.

Dong, T., E.P. Knoshaug, R. Davis, L.M.L. Laurens, S. Van Wychen, P.T. Pienkos, and N. Nagle. 2016. Combined algal processing: A novel integrated biorefinery process to produce algal biofuels and bioproducts. *Algal Res* 19:316–323.

Durvasula, R.V., S.R. Durvasula, A. Fieck, and I. Hurwitz. 2015. Cultured extremophilic algae species native to New Mexico. U.S. Patent 20150299646.

Forterre, P. 2015. The universal tree of life: An update. *Front Microbiol* 6:717.

Frisvad, J.C. 2007. Cold-adapted fungi as a source for valuable metabolites. In: Margesin R., Schinner F., Marx JC., Gerday C. (eds), *Psychrophiles: From Biodiversity to Biotechnology*. Berlin, Heidelberg: Springer 381–387.

Greque de Morais, M., B. da Silva Vaz, E. Greque de Morais, and J.A. Vieira Costa. 2015. Biologically active metabolites synthesized by microalgae. *BioMed Res Int* 2015:835761.

Gupta, A., J.C. Barrow, and M. Puri. 2011. Omega-3 biotechnology: Thraustochytrids as a novel source of omega-3 oils. *Biotechnol Adv* 30 (6):1733–1745.

Hong, J., and H. Luesch. 2012. Largazole: From discovery to broad-spectrum therapy. *Nat Prod Rep* 29 (4):449–456.

Hosseini Tafreshi, A., and M. Shariati. 2009. *Dunaliella* biotechnology: Methods and applications. *J Appl Microbiol* 107 (1):14–35.

Krock, B., U. Tillmann, A.I. Selwood, and A.D. Cembella. 2008. Unambiguous identification of pectenotoxin-1 and distribution of pectenotoxins in plankton from the North Sea. *Toxicon* 52 (8):927–935.

Ksouri, R., W.M. Ksouri, I. Jallali, A. Debez, C. Magné, I. Hiroko, and C. Abdelly. 2012. Medicinal halophytes: Potent source of health promoting biomolecules with medical, nutraceutical and food applications. *Crit Rev Biotechnol* 32 (4):289–326.

Mirón, A.S., M.C.C. Garcia, F.C. Camacho, E.M. Grima, and Y. Chisti. 2002. Growth and biochemical characterization of microalgal biomass produced in bubble column and airlift photobioreactors: Studies in fed-batch culture. *Enz Microb Technol* 31:1015–1023.

Newman, D.J., and G.M. Cragg. 2012. Natural products as sources of new drugs over the 30 years from 1981 to 2010. *J Nat Prod* 75 (3):311–335.

Oren, A., and N. Gunde-Cimerman. 2007. Mycosporines and mycosporine-like amino acids: UV protectants or multipurpose secondary metabolites? *FEMS Microb Lett* 269:1–10.

Pettit, R.K. 2011. Culturability and secondary metabolite diversity of extreme microbes: Expanding contribution of deep sea and deep-sea vent microbes to natural product discovery. *Mar Biotechnol* 13 (1):1–11.

Pikuta, E.V., R.B. Hoover, and J. Tang. 2007. Microbial extremophiles at the limits of life. *Crit Rev Microbiol* 33 (3):183–209.

Prasanna, R., L. Nain, R. Tripathi, V. Gupta, V. Chaudhary, S. Middha, M. Joshi, R. Ancha, and B.D. Kaushik. 2008. Evaluation of fungicidal activity of extracellular filtrates of cyanobacteria—Possible role of hydrolytic enzymes. *J Basic Microbiol* 48 (3):186–194.

Priyadarshani, I., and B. Rath. 2012. Commercial and industrial applications of micro algae—A review. *J Algal Biomass Utln* 3 (4):89–100.

Pulz, O., and W. Gross. 2004. Valuable products from biotechnology of microalgae. *Appl Microbiol Biotechnol* 65 (6):635–648.

Rastogi, R.P., and R.P. Sinha. 2009. Biotechnological and industrial significance of cyanobacterial secondary metabolites. *Biotechnol Adv* 27 (4):521–539.

Remias, D., S. Schwaiger, S. Aigner, T. Leya, H. Stuppner, and C. Lutz. 2012. Characterization of an UV- and VIS-absorbing, purpurogallin-derived secondary pigment new to algae and highly abundant in *Mesotaenium berggrenii* (Zygnematophyceae, Chlorophyta), an extremophyte living on glaciers. *FEMS Microbiol Ecol* 79 (3):638–648.

Rodrigo-Banos, M., I. Garbayo, C. Vilchez, M.J. Bonete, and R.M. Martinez-Espinosa. 2015. Carotenoids from haloarchaea and their potential in biotechnology. *Mar Drugs* 13 (9):5508–5532.

Schimel, J., T.C. Balser, and M. Wallenstein. 2007. Microbial stress response physiology and its implications for ecosystem function. *Ecology* 88 (6):1386–1394.

Schiraldi, C., and M. De Rosa. 2002. The production of biocatalysts and biomolecules from extremophiles. *Trends Biotechnol* 20 (12):515–521.

Seckbach, J. 2007. *Algae and Cyanobacteria in Extreme Environments*. Berlin: Springer.

Simmons, T.L., E. Andrianasolo, K. McPhail, P. Flatt, and W.H. Gerwick. 2005. Marine natural products as anticancer drugs. *Mol Cancer Ther* 4 (2):333–342.

Stetter, K.O. 1999. Extremophiles and their adaptation to hot environments. *FEBS Lett* 452:22–25.

Stierle, A.A., and D.B. Stierle. 2014. Bioactive secondary metabolites from acid mine waste extremophiles. *Nat Prod Commun* 9 (7):1037–1044.

Stock, A., H.W. Breiner, M. Pachiadaki, V. Edgcomb, S. Filker, V. La Cono, M.M. Yakimov, and T. Stoeck. 2012. Microbial eukaryote life in the new hypersaline deep-sea basin Thetis. *Extremophiles* 16 (1):21–34.

Subba Rao, D.V., Y. Pan, and F. Al-Yamani. 2005. Growth and photosynthetic rates of *Chlamydomonas plethora* and *Nitzschia frustula* cultures isolated from Kuwait Bay, Arabian Gulf, and their potential as live algal food for tropical mariculture. *Mar Ecol* 26 (1):63–71.

Varshney, P., P. Mikulic, A. Vonshak, J. Beardall, and P.P. Wangikar. 2015. Extremophilic micro-algae and their potential contribution in biotechnology. *Bioresour Technol* 184:363–372.

Wijffels, R.H., O. Kruse, and K.J. Hellingwerf. 2013. Potential of industrial biotechnology with cyanobacteria and eukaryotic microalgae. *Curr Opin Biotechnol* 24:405–413.

CHAPTER 15

Medicinal Utility of Extremophiles

Jane A. Irwin

CONTENTS

15.1 INTRODUCTION

The term *extremophile* was first proposed more than 40 years ago (MacElroy, 1974), and it describes organisms that are adapted to various environmental niches that would be regarded as "extreme," for example, extremes of heat, cold, pH, salinity, hydrostatic pressure, dessication, and ionizing radiation. The concept of extremophily is relative: conditions that are suitable for terrestrial animals and plants would be lethal for many extremophilic microorganisms, and vice versa. Therefore, extreme conditions can only be defined in relation to one set of environmental conditions and are not general, but are usually understood to be relative to conditions appropriate for mammalian survival, that is, moderate temperature, pH values close to neutrality, and low levels of ionizing radiation, salinity, and hydrostatic pressure. Some extremophiles are multiextremophilic, as they are adapted to multiple categories of extreme conditions, for example, thermoalkaliphiles, described by Bowers et al. (2009), or halophilic alkalithermiphiles (Mesbah and Wiegel, 2008). Some microbes do not grow in anaerobic or aerobic environments, and others tolerate extremely high concentrations of heavy metals or toxic gases (Pikuta et al., 2007), so these also qualify as extremophiles. Many deep-sea microbes are also barophilic, or piezophilic, as well as psychrophilic or psychrotrophic, as they are adapted to both high pressures and low temperatures. Such conditions are found deep in the earth's oceans, and bacteria that are tolerant of pressures of >100 MPa have been isolated at depths of 10,500 m and below (Horikoshi, 1998). Other species isolated from hydrothermal vents, for example, *Pyrococcus horikoshii* and *Thermococcus profundus*, are both piezophilic and thermophilic (Horikoshi, 1998).

The diversity of adaptations required for survival under such extreme conditions means that extremophiles produce proteins and many other molecules with specific characteristics, for example, thermostable proteins that are not denatured even at temperatures approaching or exceeding 100°C. As part of their survival "tool kit," some also produce secondary metabolites or exopolysaccharides that are of considerable interest to biotechnology companies due to their potential or established application in diagnostics, food manufacturing, and industrial processes, and as therapeutics for humans and animals. For example, radiation-resistant extremophiles produce various products in response to ionizing radiation that are finding application in medicine, particularly as products for skin protection (Gabani and Singh, 2013). Halophilic microorganisms produce ectoines and other compatible solutes, like glycine betaine, which are used in medicines and cosmetics (Lentzen and Schwarz, 2006). Psychrophilic organisms, which grow preferentially at low temperature, are a source of cold-adapted enzymes and secondary metabolites, and piezophiles display increased levels of polyunsaturated fatty acids, which help to maintain membrane fluidity.

These properties have encouraged research into the capacity of these microorganisms to produce products with medical utility (Babu et al., 2015). Not all organisms with extremophilic properties are bacteria or archaea. Some marine eukaryotes adapted to low temperature, or microalgae like *Dunianella* sp. (Varshney et al., 2015), could also be described as extremophilic, and are exploited as sources of medicinal products.

15.2 DO EXTREMOPHILIC PATHOGENS EXIST?

The vast majority of extremophiles are found in environments inhospitable to mammals, implying that they are not pathogenic to humans or other mammals. However, since many animals live in cold environments, some cold-adapted bacteria can cause disease in poikilothermic animals, such as cold water fish, particularly farmed fish in North Atlantic waters (reviewed by Irwin, 2010). It was reported as long ago as 1887 that fish harbored microbes that grew well at 0°C (Hoyoux et al., 2004). Psychrophilic and psychrotolerant microorganisms also spoil chilled food, thereby causing human disease. Morita (1975) defined psychrophiles as having an optimal growth temperature of ≤15°C, a maximal growth temperature of approximately 20°C, and a minimal growth temperature at or below 0°C, so these organisms readily grow at the temperature of refrigerated food. For example, a range of *Clostridium* species is responsible for "blown-pack" spoilage of vacuum-packed beef (Moschonas et al., 2011), and *Clostridium botulinum* grows from 3°C–48°C, bringing it into the psychrotolerant range (http://food.unl.edu/safety/botulinum). *C. botulinum* produces a type E neurotoxin when grown at 3°C (Peck, 2010), so it can still cause disease when it grows at low temperatures. *Listeria monocytogenes*, which causes listeriosis, is a halotolerant psychrotroph, surviving in up to 20%–30% (w/v) NaCl and growing at temperatures as low as 1°C (http://www.foodsafetywatch.org/e-books/). In some cases, an environmental psychrotroph can shift to being a potential pathogen. *Pseudomonas fluorescens* has been isolated as a causative agent of nosocomial infections, and it has been shown that it can adhere to cultures of human A549 pulmonary cells and form a biofilm at 37°C (Donnarumma et al., 2010). As psychrotrophic bacteria such as *P. fluorescens* are common in raw milk, this has human health implications, as well as food safety considerations (De Oliveira et al., 2015).

Strictly speaking, thermophilic microorganisms are not generally regarded as being pathogenic. However, there are some reports that some thermophilic species may have implications for human, and indeed animal, health. Some *Campylobacter* species are described as "thermophilic," but as their optimum growth temperature is below 50°C, they are not true thermophiles, at least not in the context of extremophilic microorganisms. There are reports of thermophilic bacteria and fungi causing human disease. "Farmer's lung" is a type of hypersensitivity pneumonitis that can be caused by thermophilic actinomycetes, including *Saccharopolyspora rectivirgula*, *Thermoactinomyces vulgaris*, *Thermoactinomyces viridis*, and *Thermoactinomyces sacchari*. These organisms grow at elevated temperatures, such as those found in moldy hay or grain, in which temperatures rise due to microbial metabolism. The temperature optimum of *T. vulgaris* is close to 55°C (Kirillova et al., 1974), enabling it to flourish in those conditions. The disease develops after exposure to the dust and gives rise to influenza-like symptoms, which lead ultimately to granulomatous and fibrous lesions in the lung (Murphy et al., 1995). Thermophilic actinomycetes also cause "bird fancier's lung," and a similar disorder can occur in mushroom workers (Van den Bogart et al., 1993; Hayes and Barrett, 2015). The thermophile *Mycobacterium xenopi* caused an outbreak of pulmonary mycobacteriosis in a hospital in Israel, confirmed by its isolation from the hospital's hot water system (Lavy et al., 1992).

There are fewer examples of halophilic organisms displaying pathogenic properties in humans, although such examples have been known for more than 50 years. Okudaira et al. (1962) reported four cases of death due to food poisoning due to a halophilic microorganism (*Pseudomonas enteritis* Takikawa). *Vibrio parahaemolyticus* is another example of a halophile that causes food poisoning in humans, requiring at least 0.5% (w/v) NaCl to grow (Beuchat, 1974). A recent survey of subgingival microbes in human periodontal disease revealed that the halophile *Halomonas hamiltonii* was a predominant component of the microflora in healthy subjects (Park et al., 2015), suggesting that some types of extremophiles can be components of the indigenous microflora.

15.3 EXTREMOPHILES IN DIAGNOSTIC REAGENTS

One of the first applications of an extremophile product was as the source of thermostable DNA polymerases for use in the polymerase chain reaction (PCR). The thermophile *Thermus aquaticus* was first discovered by Thomas Brock in hot springs in Yellowstone National Park in the United States, and its DNA polymerase, now known to molecular biologists as Taq polymerase, was found to be ideal for PCR, as its thermostability (up to 80°C) (Chien et al., 1976) enables it to withstand the cycles of heating and cooling required for PCR. Since the 1980s, it has been used for many diagnostic applications, and it has been joined by other thermophilic polymerases used for DNA amplification, for example, the *Pyrococcus*-like high-fidelity Phusion DNA polymerase sold by New England Biolabs. The history of development of DNA polymerases from thermophilic and hyperthermophilic microorganisms as diagnostic reagents has been reviewed by Ishino and Ishino (2014), and they discuss the capacity of some of these enzymes to incorporate modified nucleotides for improving PCR performance or for the development of new genetic technologies. DNA polymerases are applied in sequencing whole genomes, or parts of genomes, enabling researchers to discover genetic mutations that underlie a range of human diseases (reviewed by Chen [2014] with regard to their use in next-generation sequencing [NGS] technologies). Another function for extremophiles is as model organisms in the development of *de novo* nucleic acid target discovery methodologies: for example, Zhang and Sun (2014) devised a method for identifying uniquely conserved regions as candidate diagnostic targets, and they used some extremophilic species (*Sulfolobus islandicus* and *Thermus thermophilus*) to assist in the development and validation of this technology. Heat-labile uracil DNA glycosylase derived from the marine psychrophile BMTU 3346UNG, with a half-life of two minutes at 40°C (Jaeger et al., 2000), is commercially available from Roche, New England

Biolabs, and Sigma-Aldrich, for preventing PCR carryover and contamination control. ArcticZymes (www.arcticzymes.com) also sells the same enzyme derived from Atlantic cod for essentially the same purpose.

Biosensors also represent an area in which enzymes of extremophile origin have utility. For example, glucose sensors used in blood glucose monitoring use glucose oxidase, but the enzyme from sources such as *Aspergillus niger* is unstable over the long term with respect to activity and coenzyme utilization (de Champdoré et al., 2007). Improvements in stability could be exploited to construct implantable sensors or sensors that measure glucose in extracted interstitial fluid, all using fluorescence-based technology. D'Auria et al. (2000) used the glucose oxidase from *Thermoplasma acidophilum*, a thermophilic archaeon, as a glucose sensor, as well as glucokinase from *Bacillus stearothermophilus* (D'Auria et al., 2002). These enzymes showed excellent thermostability, and both studies illustrated the capacity of these enzymes, in a coenzyme-depleted state, to act as glucose sensors in which fluorescence was used to monitor substrate binding. de Champdoré et al. (2007) also reported that thermostable enzymes were used as reagents for measurements of blood sodium and potassium ions. Pyruvate kinase (PK) from *Bacillus acidocaldarius* alters its steady-state emission spectrum upon binding of Na^+, and cation-dependent intensities were used to measure cation-binding constants for PK.

Other extremophile bioproducts have potential diagnostic uses. Raveendran et al. (2014b) used mauran, a sulfated polysaccharide derived from the moderately halophilic bacterium *Halomonas maura*, to stabilize $ZnS:Mn^{2+}$ quantum dots. These nanocrystals can be used as fluorescent markers for imaging and clinical diagnostics, but their cytotoxicity has to date precluded widespread application. Polysaccharide conjugation decreased cytotoxicity against two cell lines, and the quantum dots were as effective for imaging as unmodified ones.

15.4 DRUGS AND DRUG CANDIDATES FROM EXTREMOPHILES

Prokaryotes are the source of many secondary metabolites with biological activity against human and animal disease. Other valuable secondary metabolites have been derived from various eukaryotic species, including fungi, invertebrates, and plants. Secondary metabolites are compounds that are not required for growth or reproduction, but function in defense against predators or for eukaryotes and microbial, fungal, and viral pathogens. Many drugs developed over the last 60–70 years have been derived from microbial sources, ranging from antimicrobials to antitumor, anti-inflammatory, antiangiogenic, and antihypertensive agents. According to Giddings and Newman (2013), microbial secondary metabolites comprise the main source of drugs used for cancer treatment, as well as provide molecular scaffolds that can be modified by chemists. By the end of 2012, 189 small molecule antitumor agents were approved by the U.S. Food and Drug Administration (FDA), or the equivalent in other countries, and approximately one-sixth of these were produced directly by microbes or derived from microbial secondary metabolites. Many other drug candidates are in preclinical and clinical development, and clinical trials are ongoing in different countries (see Petit and Biard [2013] and Newman and Cragg [2014] for recent reviews of clinical trials of marine-derived cancer and cancer pain control agents). The development of high-throughput screening and combinatorial chemistry techniques has meant that less natural product screening is now carried out by pharmaceutical companies.

One issue in natural product discovery is the large volume of the source required to extract the secondary metabolite. For example, marine sponges are difficult to collect in sufficient quantities for metabolite isolation, requiring scuba diving or collection at depth using submersibles with robotic arms, and their associated microbiota is largely unculturable (Molinski et al., 2009). Considerable effort is required to devise a chemical or enzyme-based synthesis to facilitate scale-up of production. Other source organisms may be culturable, albeit with difficulty, but may produce different compounds depending on culture conditions. These difficulties may be overcome in the future by the application of genetic technology to natural product drug discovery. The structural differences between metabolites from terrestrial microbes and those derived from marine organisms, extremophilic or otherwise, arise from differences in biosynthetic pathways, and the presence in marine or extremophilic organisms of genes encoding enzymes with novel properties. Salomon et al. (2004) have reviewed this general approach, with emphasis on a number of gene clusters commonly involved in secondary metabolite synthesis, for example, polyketide assembly systems (Staunton and Weissman, 2001) and nonribosomal peptide synthase clusters (Schwartzer et al., 2003). Over the first decade of the twenty-first century, more gene clusters have been identified, with the advent of whole-genome sequencing (Lane and Moore, 2011), and this should lead to identification of pathways to synthesis of useful compounds.

15.5 DRUGS AND DRUG CANDIDATES FROM MARINE SOURCES

Within both marine and terrestrial environments, extremophiles exist within particular niches, and in some cases, organisms may be multiextremophilic. This creates difficulties in defining precisely which drugs are extremophile derived. In particular, marine environments vary greatly in temperature worldwide, and even with the seasons. As a result, not all bioproducts derived from marine sources can be described as coming from extremophiles, as many marine environments, for example, those in tropical or temperate shallow waters, support organisms that do not quite fit the category of extremophile. Psychrophiles inhabit cold water habitats, and polar seas fit this definition, as does the temperate and tropical deep sea, which below the thermocline has a constant temperature close to 4°C (Lalli and Parsons, 1997). The high pressures in deep-sea habitats also mean that many microorganisms and other creatures found at these depths have piezophilic adaptations. Lebar et al. (2007) in their review of cold water marine natural products defined cold water habitats as those in which organisms experience ice at some stage in their annual cycle. Variations in sea temperatures add complexity to defining whether a microorganism (and its associated metabolites) can be described as extremophilic, unless the organism's optimum growth temperature has been determined. Cold regions have in the past been overlooked as sources of bioactive metabolites, not least because of their inhospitable temperatures and assumptions that they contain lower chemical diversity, but work on Antarctic marine microbial communities shows that they are rich and largely uncharacterized (Michaud et al., 2004). Marine organisms may also be halophilic, and those living close to undersea volcanic vents may be thermophilic or hyperthermophilic (Prieur et al., 1995).

Deep-sea organisms include many types of marine fauna, as well as bacteria and fungi. One recent review (Skropeta and Wei, 2014) described the properties and structures of 188 new marine natural products from 2008 to 2013. Seventy-five percent of these had bioactivity, and almost half of these had low micromolar cytotoxicity toward human cancer cell lines. However, caution should be exercised when considering the bioactivity of marine secondary metabolites. The literature abounds in references to compounds that have bioactivity toward cell lines, but many of these have not been tested *in vivo*, or if they have, this has been limited to rodent models. The assumption that all these agents are active *in vivo* and have potential applicability in cancer treatment, or treatment of other disorders, may thus be mistaken. Furthermore, cytotoxic and anticancer agents are not synonymous terms, and agents must display selectivity in terms of their ability to differentiate between normal and cancer cells (Gomes et al., 2015). Some of these compounds, along with drug candidate compounds, are listed in Table 15.1. Their structures are shown in Figure 15.1.

A comprehensive review of drugs and drug candidates derived from marine extremophilic organisms has been published recently by Giddings and Newman (2015a), who provide a review of known biologically active secondary metabolites produced by marine extremophiles, along with the structures of the compounds themselves. The number of clinically approved agents, however, is far lower.

Table 15.1 Drugs in Clinical Use, and Some Compounds with Potential Therapeutic Application, from Marine Organisms

Compound	Source and Type of Organism	Bioactivity	Reference
Trabectidin	*Aplidium* sp. (tunicate)	Anticancer (in clinical use)	D'Incalci and Galmarini, 2010
Halichondrin B	*Halichondria okadai* (desmosponge)	Anticancer (eribulin mesylate, clinical use)	Swami et al., 2015
Dolastatin 10 (part of brentuximab vedotin, in clinical use)	*Dolabella auricularia* (gastropod mollusk)	Anticancer, antimicrobial	Pettit et al., 1998; Giddings and Newman, 2013
Bryostatin 1	*Bugula neritina* (colonial bryozoan)	Anticancer	Bullen et al., 2014; Kollár et al., 2014
Dehydrodidemnin B	*Aplidia albicans* (tunicate)	Anticancer	Lee et al., 2012
Salinosporamide-A	*Salinospora tropica* (actinomycete bacteria)	Anticancer	Jensen P.R. et al., 2015; Russo et al., 2015
Variolin B	*Kirkpatrickia variolosa* (sponge)	Cytotoxic kinase inhibitor	Anderson et al., 2002; Remuiñan et al., 2003
Marinopyrrole A and B	*Streptomyces* CNQ-418 (actinomycete bacteria)	Antimicrobial	Hughes et al., 2008
Spiromastixones	*Spiromastix* sp. (fungus)	Antibacterial	Niu et al., 2014; Wang et al., 2015
Emerixanthones A–D	*Emericella* sp. SCSIO 05240 (actinomycete bacteria)	Antibacterial, antifungal	Fredimoses et al., 2014
Cladosin C	*Cladosporium sphaerospermum* (fungus)	Antiviral	Wu et al., 2014
p-Hydroxyphenopyrrozin	*Chromocleista* sp.(fungus)	Antifungal	Park et al., 2006

Many agents have been tested against cancer cell lines, but only a handful may make it to the clinic. According to Gomes et al. (2015), of marine-derived anticancer agents, only a small number are clinically approved by the FDA: the nucleoside cytarabine from sponges, used to treat leukemia; eribulin mesylate (Halaven®, based on halichondrin B, and used to treat metastatic breast cancer); and brentuximab vedotin. Trabectedin, also known as Yondelis®, is used to treat soft tissue sarcoma and relapsed platinum-sensitive ovarian cancer. It is approved in Europe and is undergoing clinical trials for various other cancers (https://www.clinicaltrialsregister.eu/ctr-search/search?query=trabectedin). Eribulin mesylate is a macrocyclic ketone analogue of halichondrin B, which has undergone phase III studies in breast cancer, soft tissue sarcoma, and non-small-cell lung cancer and is undergoing phase I and II studies for other cancers (Swami et al., 2015). Yondelis (also called trabectedin, ET-743) is a novel marine-derived alkaloid anticancer compound from the ascidian *Ecteinascidia turbinata*. It has undergone phase II and III development for various cancers and is European Union approved for treatment of soft tissue sarcoma (Gerwick and Fenner, 2013; Mayer, 2015).

The marine-derived part of brentuximab vedotin is a derivative of dolastatin 10, a linear pentapeptide microtubule assembly inhibitor initially isolated from *Dolabella auricularia*, the "sea hare," which is a gastropod mollusk. The drug is an anti-CD30 antibody-conjugated drug approved by the FDA for treatment of lymphomas, and has been in clinical trials for other cancers (Giddings and Newman, 2013). However, its true source is not the invertebrate, but a cyanobacterial species that it hosts, *Lyngbya*. Various compounds from this genus are under investigation as anticancer agents (Swain et al., 2015).

The literature on natural products from marine sources, including marine extremophiles, is vast and is the subject of regular reviews. Venugopal (2009) reviewed marine pharmaceuticals,

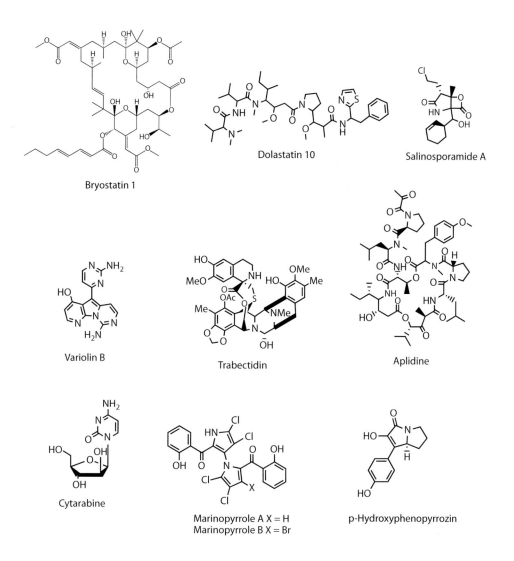

Figure 15.1 Molecular structures of some compounds isolated from marine organisms (see Table 15.1 and the text).

with particular emphasis on nonmicrobial sources. One interesting example is bryostatin, produced by the bryozoan *Bugula neritina*, which is a protein kinase C inhibitor with antineoplastic activity (Kollár et al., 2014). This compound also affects memory formation (Sun and Alkon et al., 2014) and induces transcription of HIV mRNA in infected T cells, facilitating T-cell-mediated killing (Bullen et al., 2014). It is in phase I clinical trials against a range of solid tumors (Mayer, 2015). However, *Bugula* spp. appear to grow optimally in warmer waters, and a study based in British and Irish coastal waters showed that their growth was disrupted when winters were cold and summer temperatures were below 20°C (Ryland et al., 2011) so that it is questionable as to whether this can be considered extremophilic. Dehydrodidemnin B, or aplidine, was first isolated from *Aplidium albicans* (Lee et al., 2012). This drug is a cytotoxic depsipeptide, produced by a Spanish company, PharmaMar (as is trabectidin), and has undergone phase II clinical trials for the treatment of non-Hodgkin lymphoma and phase II trials with dexamethasone for treating multiple myeloma (Giddings and Newman, 2013). The source organism is a tunicate of the genus *Aplidium*, which can live in cold Antarctic water at depths below 200 m (Nuñez-Pons et al., 2012).

Some marine natural products can reverse multidrug resistance in cancer, for example, agosterol A, ecteinascidin 743, sipholane triterpenoids, bryostatin 1, and cyanobacterial-derived welwitin-dolinones (Abraham et al., 2012). A recent systematic review (Mayer et al., 2013) of the marine pharmacology literature between 2009 and 2011 presents the structures and pharmacological action of 162 compounds, 102 of which have antibacterial, antifungal, antiprotozoal, antituberculosis, and antiviral properties. Many of these, however, cannot be defined as having come from extremophilic organisms. Variolins are a class of alkaloid compounds isolated from an Antarctic sponge, and these act as kinase inhibitors (Bharate et al., 2013). In particular, variolin B, from the Antarctic sponge *Kirkpatrickia variolosa* (Perry et al., 1994), has been under development as an antitumor agent (Anderson et al., 2002; Remuiñan et al., 2003).

Deep-sea fungi, which live in cold water at high pressures, produce hundreds of bioactive metabolites, including compounds with anticancer, antimicrobial, antifungal, antiviral, and antiprotozoal activity (Wang et al., 2015). These communities have been investigated using both culture-dependent and sequencing approaches and include cytotoxic polyketides, steroid and indole derivatives, and sesquiterpinoids from Antarctic waters, as well as antibacterial prenylxanthones, depsidone-based analogues, an antiviral polyketide, and an antifungal compound, *p*-hydroxyphenopyrrozin. None of the compounds described by Wang et al. (2015) appear to be at the stage of animal studies or clinical trials.

Marine actinomycetes are recognized as being a reservoir of diverse bioactive compounds. Salinosporamide-A, also known as marizomib, is a salinosporamide, a compound with a functionalized γ-lactam-β-lactone bicyclic core. Such compounds are produced by the genus *Salinispora*, a marine actinomycete genus that requires seawater for growth. One species, *Salinispora tropica*, is found in marine sediments around the world, at depths of up to 1100 m (Mincer et al., 2005). They have also been identified using culture-independent methods at a depth of 5669 m (Prieto-Davó et al., 2013). *Salinospora* spp. produce many different compounds, including the salinosporamides, reviewed by Jensen P.R. et al. (2015). Salinosporamide-A acts by binding to the β subunit of the 20S proteasome, which degrades misfolded proteins in cells. This compound inhibits three different enzymatic activities in the proteasome, and also activates caspases-8 and -9, thereby helping to induce apoptosis and providing a novel neoplastic activity (Russo et al., 2015). This compound is in multiple phase I trials, most of which are for relapsed or refractory multiple myeloma (Russo et al., 2015). Marizomib, produced by Triphase Accelerator Corporation, has received orphan drug designation from both the FDA and the European Union for use in multiple myeloma (http://triphaseco .com/pipeline/).

Marinopyrroles (A and B) are alkaloids that were isolated from a *Streptomyces* strain (strain CNQ-418) from the seafloor off the coast of La Jolla, California. These bispyrroles show antimicrobial activity against methicillin-resistant *Staphylococcus aureus* (Hughes et al., 2008).

15.6 TERRESTRIAL EXTREMOPHILE DRUG CANDIDATES

Giddings and Newman (2015b) have reviewed compounds produced by various terrestrial extremophiles, and described how most extreme environments harbor microorganisms that produce, or have the potential to produce, medicinally interesting compounds. So far, none of the compounds that they describe are as yet available for clinical use. Table 15.2 shows some of these compounds, and the structures are shown in Figure 15.2.

Among the best known are a category of compounds based on berkelic acid, produced by acidophilic bacteria inhabiting Berkeley Pit Lake, Butte, Montana, reviewed recently by Stierle and Stierle (2014). This lake contains metal sulfate–rich, highly acidic water containing mine waste. This environment is too toxic for most organisms to withstand, but harbors a valuable reservoir of acidophiles. The microorganisms found there largely consist of *Penicillium rubrum* and associated

Table 15.2   Compounds with Potential Therapeutic Application from Terrestrial Extremophiles

Compound	Source	Bioactivity	Reference
Acidophiles			
Berkeleydiones, triones	*Penicillium* spp. (acidophiles from mine waste)	Anti-inflammatory, cytotoxic	Stierle et al., 2004, 2006
Berkeleyones A–C	As above	Cytotoxic	Stierle et al., 2011
Berkazaphilones	As above	Cytotoxic	Stierle et al., 2012
Berkeleyacetals	As above	Cytotoxic	Stierle et al., 2007
Purpuriquinones	*Penicillium* spp.	Antiviral	Wang et al., 2011
Purpuresters		Antiviral	
Psychrophiles			
Psychrophilins	*Penicillium* sp. (psychrophilic)	Antifungal, cytotoxic	Dalsgaard et al., 2004
Chetracins	*Oidiodendron truncatum*	Cytotoxic	Li et al., 2012
Violacein, flexorubin	*Janthinobacterium* sp. Ant5–2	Antimycobacterial	Mojib et al., 2010
	Flavobacterium Ant342	Antimycobacterial	
Griseusins	*Nocardopsis* spp. YIM80133, DSM1664	Cytotoxic	Ding et al., 2012
Halophiles			
Halocin H7	Haloarchaea	Treatment of ischemia	Lequerica et al., 2006
Haloduracin	*Bacillus halodurans* C-125	Antimicrobial	Lawton et al., 2007
Acidothermophiles			
Sulfolobicins	*Sulfolobus* spp. (Iceland)	Unclear	Ellen et al., 2011
Xerophiles			
ZA01, ZA02	Actinomycetes (desert)	Antifungal	Zitouni et al., 2005
Terrecyclic acid A derivatives	*Aspergillus terreus*	Cytotoxic	Wijeratne et al., 2003

Penicillium spp., with an optimal growth pH of 2.7, living in the lake sediment or at the surface. The compounds produced include berkeleydiones and triones, which inhibit matrix metalloprotease-3 (MMP-3) and caspase-1, as well as berkelic acid itself (Stierle et al., 2004, 2006); berkeleyones A–C, which inhibit caspase-1 (Stierle et al., 2011); berkazaphilones B and C; and the vermistatin analogue penisimplicissin, which also inhibits caspase-1 (Stierle et al., 2012). The latter compounds showed activity against leukemia cell lines. Berkeleyacetals (Stierle et al., 2007; Etoh et al., 2013) were also shown to have cytotoxic effects, and inhibited interleukin (IL)-1-receptor-associated kinase-4. Berkeleyacetal C inhibited MMP-3 and caspase-1 with IC_{50} values in the micromolar range, and showed cytotoxic activity in the NCI-DTP60 cancer cell line assay (Giddings and Newman, 2015b). It appears that this one environment may produce an enormous range of compounds with cytotoxic and anti-inflammatory potential. Other acidophile products include the purpuriquinones and purpuresters, with activity against influenza virus (Wang et al., 2011).

Some psychrotrophic *Penicillum* species produce psychrophilins, with antifungal and cytotoxic activity, and cycloaspeptide, with antifungal and antiplasmodial activity (Dalsgaard et al., 2004). Other cold-adapted microbes, for example, the soil microbe *Oidiodendron truncatum* GW3-13, produce chetracins, a range of cytotoxic compounds (Li et al., 2012). A lake in eastern Antarctica provided *Janthinobacterium* sp Ant5-2 and *Flavobacterium* Ant342, which produced violacein and flexorubin, respectively. These compounds were tested against various species of *Mycobacteria* and were found to have antimycobacterial activity, suggesting that they might be useful as lead compounds for the synthesis of antimycobacterials, essential for the treatment of tuberculosis (Mojib et al., 2010).

Alkaline environments can also harbor producers of novel compounds. He et al. (2007) examined soil from a strongly alkaline environment in Yunnan, China, with an ambient pH of about

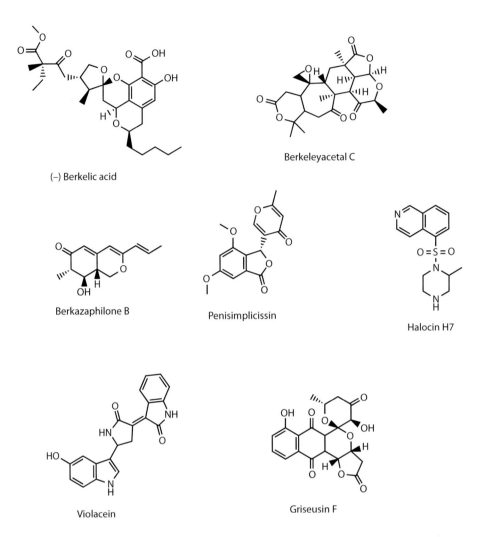

Figure 15.2 Molecular structures of several bioactive compounds isolated from terrestrial extremophiles (see Table 15.2 and the text).

9–10. The actinomycetes *Nocardiopsis* spp. YIM80133 and DSM1664 were isolated and found to produce a range of griseusins. In particular, 4′-dehydro-deacetylgriseusin A was a selective inhibitor of colony formation of tumor cells in semisolid medium, and had a lower IC$_{50}$ for a range of cancer cell lines than the other griseusin derivatives. Griseusins F and G produced by strain YIM DT266 were both cytotoxic and antimicrobial, with minimum inhibitory concentration (MIC) values of 0.8–1.65 µg mL^{-1} against *S. aureus* ATCC 29213, *Bacillus subtilis*, and *Micrococcus luteus* (Ding et al., 2012). Variecolorquinones A and B, produced by *Aspergillus varicolor* B-17, had some cytotoxic activity toward various cancer cell lines (Wang et al., 2007).

Some microbes produce various types of antimicrobial peptides and proteins, which enable them to compete with other microbes, and bacteriocins and archaeocins are produced by bacteria and archaea, respectively (reviewed by Besse et al., 2015). Some halophilic archaea produce halocins. Some of these antimicrobial peptides may have other applications, for example, inhibiting the Na$^+$/H$^+$ antiporter, which is dysfunctional in some diseases (Loo et al., 2012). One of these peptides, halocin H7, prevented ischemic and reperfusion damage in dog heart, suggesting that this and similar compounds might be useful in preventing ischemic damage (Lequerica et al., 2006). Haloduracin was produced by the alkaliphilic soil bacterium *Bacillus halodurans* C-125 (Lawton et al., 2007) and was active against a range of Gram-negative and Gram-positive bacteria, including *Lactobacillus* and *Listeria* spp. Sulfolobicins, isolated from acidophilic and thermophilic *Sulfolobus* strains in Iceland, also have antimicrobial action, but appear to work via preventing bacterial growth, rather than by lysing the cells (Ellen et al., 2011).

Finally, xerophiles—microorganisms that survive in dry conditions—can also produce bioactive compounds. Examples include antifungal compounds ZA01 and ZA02, isolated from actinomycete species from the Sahara Desert (Zitouni et al., 2005), and xerophiles from the Sonora Desert in the United States, which produce metabolites with cytotoxic action against some cancer cell lines (Wijeratne et al., 2003).

15.7 DRUG DELIVERY: NANOPARTICLES AND ARCHAEOSOMES

15.7.1 Nanoparticles

Nanoparticles for medical applications have been defined by the European Technology Platform on Nanomedicine (ETPN) as particles with a size, in at least one direction, of 1–1000 nm (Nanomedicine: Nanotechnology for Health, ftp://ftp.cordis.europa.eu/pub/nanotechnology/docs /nanomedicine_bat_en.pdf). Some of these structures are in clinical trials as agents for drug delivery, gene delivery, *in vivo* imaging, and *in vitro* diagnostics (Ryan and Brayden, 2014). However, few of these structures are in clinical use as of yet (Juliano, 2013). Among these structures are liposomes, nanoshells, metallic nanoparticles, proteins, cyclodextrins, polymeric micelles, and nucleic acid–based nanoparticles (Davis et al., 2008).

Extremophiles show potential as a source of "starter materials" for generating nanoparticles, and nanoparticles are also produced within the cells. Bacterial production of nanoparticles has been reviewed by Zhang et al. (2015). They point out that nanoparticles may be produced from bacteria and in fermentation media, but there are significant challenges with respect to the specific conditions under which they are made. These extremophile products fall into two main categories at present, exopolysaccharides and archaeosomes, made from lipids from halophilic and thermophilic archaea.

Active microbial polysulfated exopolysaccharides are under investigation as nanodrug carriers to encapsulate anticancer agents. Mauran, a polysaccharide isolated from the moderate halophile *Halomonas maura*, was used as the basis for nanoparticles, blended with chitosan, that could scavenge reactive oxygen species *in vitro* and *ex vivo*. They were tagged with fluorescent labels and shown to be taken up by cells. They were also used to encapsulate 5-fluorouracil and were shown to be cytotoxic toward MCF-7 breast cancer cells (Raveendran et al., 2013), and a later study showed that they were cytotoxic toward three cancer cell lines to varying extents (Raveendran et al., 2015). Cancer cells were treated with mauran-coated gold nanoparticles and exposed to infrared radiation from a femtosecond pulse laser at 800 nm. This caused hyperthermia and led to cell death, suggesting that these could be applied to thermal ablation of tumors (Raveendran et al., 2014a).

The halophilic archaeon *Halococcus salifodinae* BK3 has been used to make spherical intracellular silver nanoparticles (Srivastava et al., 2014) and tellurium nanoparticles, catalyzed by the intracellular enzyme tellurite reductase found in this bacterial species (Srivastava et al., 2015). These nanoparticles displayed antibacterial activity against a range of Gram-negative and Gram-positive bacterial species, but were more effective against Gram-negative species, in particular *Escherichia coli* NCIM 2345 and *Pseudomonas aeruginosa* MTCC 2581.

The thermophile *Geobacillus* sp. strain ID17 was used to make gold nanoparticles (Correa-Llantén et al., 2013). Another thermophile, *Geobacillus stearothermophilus*, was used to generate biogenic gold nanoparticles, which were highly stable due to the presence of capping proteins secreted by the bacterium. These were used to augment PCR efficiency, increasing yield and decreasing the reaction time (Girilal et al., 2013). This technology may have application in PCR-based medical diagnostics. *Thermoplasma acidophilum* proteasome nanoparticles, labeled with a peptide and a near-infrared fluorescence dye, were engineered for *in vivo* tumor detection. These labeled tumors in a mouse model after intravenous injection and had low renal clearance (Ahn et al., 2014).

Shivaji et al. (2011) reported the synthesis of antibacterial silver nanoparticles using five species of psychrophilic Antarctic bacteria. Cell-free culture supernatant from *Pseudomonas antarctica* and *Arthrobacter kerguelensis* gave rise to silver nanoparticles, which were bactericidal against three Gram-negative and three Gram-positive species of bacteria. *Deinococcus radiodurans*, a radioresistant bacterial species, was found to accumulate silver nanoparticles in culture media. Interestingly, these were multifunctional, inasmuch as they could inhibit the growth of some Gram-negative bacterial species (Gram-positive species less so), but also inhibited biofilm formation by multi-drug-resistant *S. aureus* and *P. aeruginosa*. They also altered cell morphology in MCF-7 breast cancer cells (Kulkarni et al., 2015).

15.7.2 Archaeosomes

Archaeosomes are liposomes made with ether lipids that are found only in Archaea. These can comprise mixtures of lipids from archaeobacteria and nonarchaeobacterial prokaryotes. They consist of archaeol (diether) and/or caldarchaeol (tetraether) core structures. Branched and saturated phytanyl chains of 20–40 carbons are attached through ether bonds to the sn-2,3 carbon atoms of the glycerol backbone (Kaur et al., 2016). These bilayer structures are, unlike conventional liposomes, stable at low pH and high temperatures and pressures, and resist oxidation. They are also more readily internalized by phagocytic cells, and are resistant to autoclaving (Kaur et al., 2016). These properties make them ideal for the encapsulation of drugs, gene delivery systems, novel antigen delivery systems, and immunoadjuvants, among other applications. There are three major sources of archaea: the methanogens, only some of which are extremophilic; the halophiles, for example, *Halobacterium* sp. and *Natronobacterium* sp. (which are thermoalkaliphilic); and the thermoacidophiles, for example, *Thermoplasma acidophilum* (Benvegnu et al., 2009).

Some archaeobacterial lipids produce remarkably stable liposomes, for example, those made from the polar lipid E fraction of *Sulfolobus acidocaldarius*. These structures are bipolar tetraether molecules with up to four cyclopentane rings in each of the two dibiphytanyl chains. They are stable above pH 4, and retain their vesicular integrity after multiple cycles of autoclaving (Brown et al., 2009). Archaeosomes derived from this archaeon's lipid have also been used as carriers for the oral delivery of peptides. They are stable in simulated gastrointestinal tract fluids *in vitro*, and in an *in vivo* study in rats they facilitated a slow transit of peptide through the GI tract. However, when loaded with insulin, their hypoglycemic effect was small (Li et al., 2010). A later study by the same group showed that these structures elicited increased immune responses against a model antigen, ovalbumin, and also facilitated antigen-specific CD8(+) T-cell responses (Li et al., 2011). Liposomes derived from *S. acidocaldarius* could potently permeate skin, as shown by a Wistar rat skin model (Moghimpour et al., 2013). *Sulfolobus islandicus*, a hyperthermophilic archaeon, produces lipids that were used to make liposomes that withstood treatment with bile salts, as illustrated by their ability to retain a marker trapped within (Jensen S.M. et al., 2015).

15.7.3 Archaeosomes from Halophiles

Archaeal lipids from *Halobacterium salinarium* enriched with soy phosphatidylcholine were used as vehicles for the trans-dermal delivery of betamethasone dipropionate (González-Paredes et al., 2010). The vesicles were effective carriers for this model drug, allowing the drug to penetrate and accumulate in the skin. It was suggested that this was due to the enhanced fluidity of the archaeosomal bilayer, and this strategy may hold promise in enhancing anti-inflammatory drug efficacy in the treatment of skin disorders. Archaeosomes from *Halobacterium halobium* and *Halococcus* sp. differentially modulated dendritic cell activation, inducing IL-12 secretion (Sprott et al., 2003). *Halobacterium tebenquichense*, isolated from an Argentinian salt flat, was used to make archaeosomes (Higa et al., 2013) and was tested with respect to cytotoxicity, intracellular transit, and adjuvant activity for delivery of a bovine serum albumin (BSA) dose to mice. This appeared to be an effective adjuvant delivery system to promote humoral immunity toward BSA, and they also suggested that it might do so with respect to cell-mediated immunity. Archaeosomes made from polar lipids of the same archaeal species were used to immunize mice with antigens from *Trypanosoma cruzi*, the causative agent of Chagas disease. The animals developed higher levels of circulating antibodies with the parasite and were protected against challenge with a normally lethal strain in comparison with animals immunized with just the antigen alone. Halophilic microorganisms have also been used to transfer DNA into cells, and may have potential use as novel systems for gene delivery. A strain of *Haloarcula hispanica* was used as a source for archaeosome formulations to transfer plasmid DNA encoding green fluorescent protein or β-galactosidase into HEK293 cells. The transfection was optimized by adding other molecules, for example, LiCl, $CaCl_2$, and DOTAP, to the lipid formulation (Attar et al., 2015).

15.8 RADIATION-RESISTANT EXTREMOPHILES: NATURE'S SUNSCREEN

One group of extremophiles with medical and biotechnological applications is radioresistant microorganisms. These organisms have molecular defenses against ionizing and nonionizing radiation that would otherwise be damaging or lethal. One effect of radiation is oxidative stress, leading to mutations in DNA, as well as damage to proteins and lipid membranes. The best characterized of these is the bacterium *Deinococcus radiodurans*. This bacterial species, listed by the *Guinness Book of Records* as the "World's Toughest Bacterium," has a strain R1 that can survive chronic radiation exposure of 60 Gy h^{-1}, which exceeds that found in most spent decontamination waste from nuclear reactors, and acute exposure of 10,000–20,000 Gy (Gogada et al., 2015). In comparison, 200 Gy will kill most bacteria (Munteanu et al., 2015). This property makes it an ideal candidate for remediation of radioactive waste. Some strains can survive high levels of ultraviolet (UV) radiation, for example, those occurring at high altitudes, of up to 1000 Jm^{-2} (Yang et al., 2008). The key to its survival lies in a remarkably robust DNA repair mechanism, reviewed by Munteanu et al. (2015). Each cell has at least two copies of its genome, allowing recombination, and repairs its radiation-damaged genome with the aid of an extended synthesis-dependent strand annealing mechanism. The *D. radiodurans* RecA protein, which is found in all bacteria, functions not only in genome repair, but also in genome restoration after bacterial desiccation (Mattimore and Battista, 1996).

Radiation-resistant extremophiles are found in many environments: mountain ranges and open fields, exposed to high UV flux, and environments contaminated with radioactive material. Along with DNA repair mechanisms, these organisms, which comprise bacteria and cyanobacteria, produce a range of primary and secondary metabolites that are under investigation for possible use as drugs and protectants against radiation-induced damage. The applications of these have been reviewed by Singh and Gabani (2011) and Gabani and Singh (2013). The best known of these metabolites are scytonemin and MAAs.

15.9 SCYTONEMIN

This compound, shown in Figure 15.3, is made only by certain extremophilic cyanobacteria, including *Anabaena*, *Calothrix*, *Chroococcus*, *Lyngbya*, *Scytonema*, and *Nostoc* spp. (Rastogi et al., 2015).

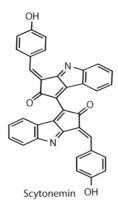

Figure 15.3 Molecular structure of scytonemin.

It is small, yellow-brown to dark red in color, and lipid soluble, and it comprises part of the microorganisms' extracellular sheath, protecting them against UV radiation. It absorbs near-UV and blue radiation, but not red and green wavelengths of light required for photosynthesis (Gao and Garcia-Pichel, 2011). It has also been suggested that scytonemin may be a biosignature for life on Mars, although this is speculative (Mishra et al., 2015).

In terms of structure, it is a dimer comprised of two polycyclic chromophores formed by condensation of cyclopentanone and indole rings, synthesized from tryptophan and *p*-hydroxyphenylpyruvate. It may be present in an oxidized (green) or reduced form, which is red (Rastogi et al., 2015), and according to Garcia-Pichel and Castenholz (1991), it can prevent up to 90% of incident UV from the sun from entering the cell. This UV screening capability makes scytonemin a potentially valuable compound as a component of sunscreens. It can prevent UV-induced formation of reactive oxygen species and cyclobutane thymine dimers (Rastogi et al., 2014), and has been patented by Su (2012) as a component of a beauty product. It also has dual kinase inhibitory activity, with possible implications for its use as an anti-inflammatory and antiproliferative agent (Stevenson et al., 2002). It inhibits polo-like kinase-1 by nonspecifically outcompeting ATP at micromolar concentrations (McInnes et al., 2005) and is active against this enzyme in multiple myeloma and renal cancer cells, thereby arresting the cell cycle (Zhang et al., 2013a, 2013b). The inhibition of this kinase by scytonemin induces apoptosis in osteosarcoma cells (Duan et al., 2010). These properties suggest that this molecule, or derivatives, may have future applications for treating cancer or inflammatory disorders.

15.10 MYCOSPORINE-LIKE AMINO ACIDS

These compounds absorb UV radiation in the 310–365 nm range (Singh et al., 2010), and are found, like scytonemin, in cyanobacteria, but also in some eukaryotic algae. There are various MAAs known, including palythine, palythene, asterina, palythinol, shinorine, and porphyra (Figure 15.4), and their general mode of action involves prevention of pyrimidine dimers in DNA (Singh and Gabani, 2011). The structure of MAAs is based on an aminocyclohexenone, or aminocyclohexene imine to which amino acids are bound (Sinha and Häder et al., 2008). They dissipate energy as heat without producing reactive oxygen species (Conde et al., 2004). Their ability to absorb a wide range of UV wavelengths makes them suitable for inclusion in sunscreens, and their use in preventing skin cancers, for example, melanoma, has been proposed (de la Coba et al., 2009). For example, porphyra-334 from *Porphyra vietnamensis* has a broad-spectrum sunscreen effect (Bhatia et al., 2010),

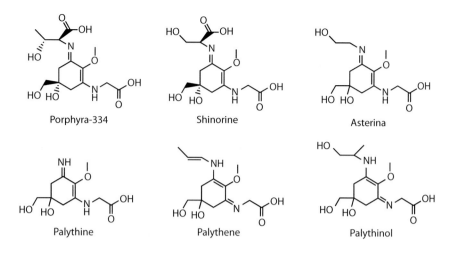

Figure 15.4 Molecular structures of the mycosporine-like amino acids.

and a product called Helioguard® 365, consisting of porphyra-334 from the red alga *Porphyra umbilicalis* encapsulated in liposomes, is commercially available from Mibelle AG Biochemistry. A French company, Gelyma, has also produced a skin-protective UV block called Helionori®, produced from extracts of the same species (known as *nori* in Japan). MAAs from other algal species, for example, *Chlamydomonas hedleyi*, have potential antiskin aging activity, as shown by their anti-inflammatory action in inhibiting production of COX-1 mRNA (Suh et al., 2014). MAAs and related compounds are largely drawn from organisms from marine rather than terrestrial environments, and efforts are ongoing to elucidate the pathways by which they are produced and also to develop heterologous expression systems to facilitate production of different UV-protective compounds that protect against a wider range of wavelengths (Colabella et al., 2014). Compounds with photoprotective activity are not limited to radiation-resistant algae: Martins et al. (2013) have found photoprotective compounds in thermophilic and psychrotolerant bacteria isolated from Mid-Atlantic Ridge deep-sea hydrothermal vents near the Azores Islands. Bacterioruberin, found in *Rubrobacter radiotoleransis*, aids in DNA repair and may also have applications in protection against UVR, along with sphaerophorin and pannarin (Singh and Gabani, 2011).

15.11 COMPATIBLE SOLUTES FROM HALOPHILES: APPLICATIONS OF ECTOINE

The majority of halophilic and halotolerant microorganisms accumulate low-molecular-weight intracellular solutes to provide osmotic balance with the extracellular hypersaline environment. The best known of these include glycine betaine, sucrose and trehalose, various polyols, and a variety of amino acids and their derivatives. One of the most widely found in nature of these compounds is ectoine (1,4,5,6-tetrahydro-2-methyl-4-pyrimidine carboxylic acid). The accumulation of these compounds makes halophiles among the most versatile extremophiles for biotechnological exploitation (reviewed by Oren, 2010). Haloalkaliphiles also produce compatible solutes, for example, glycine betaine, but they do so less efficiently than halophiles in habitats of near-neutral pH, as they have to expend more energy to ensure pH homeostasis. For this reason, there are fewer examples of compatible solute production from these organisms (Zhao et al., 2014). Some hyperthermophiles also synthesize compatible solutes, for example, trehalose and mannosylglycerate in *Thermus thermophilus* or mannosylglycerate in *T. thermophilus* and *Rhodothermus marinus* (Empadinhas and da Costa, 2006).

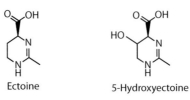

Figure 15.5 Molecular structures of ectoine and 5-hydroxyectoine, two bioactive compounds derived from halophiles.

Ectoine was first discovered in an extreme halophile, *Halorhodospira halochloris*, a phototroph that grows in up to 5 M NaCl (Galinski et al., 1985). Since then, many bacterial species have been found to produce ectoine and a derivative, hydroxyectoine (shown in Figure 15.5), first found in *Streptomyces parvulus* (Inbar and Lapidot, 1988). These microorganisms include various *Actinobacteria* spp.; Firmicutes, including *Bacillus*, *Marinococcus*, and *Halobacillus* spp.; and Proteobacteria, including *Rhodovibrio*, *Halomonas*, and *Vibrio* spp. (Pastor et al., 2010). As demand for ectoine grew for pharmaceutical and other applications, production processes were developed and scaled up to meet demand. The technique of "bacterial milking" was developed to extract metabolites on a large scale from microorganisms. Sauer and Galinski (1998) established this bioprocess using *Halomonas elongata* as an ectoine-producing source. In bacterial milking, the microbes are grown in a high-salinity medium, and then transferred to a medium of lower salinity, which makes them release the ectoine and other compatible solutes to maintain osmotic equilibrium. This process could be repeated with multiple cycles of high- and low-salinity medium, thereby generating high yields of ectoine. This was patented by the German biotechnology firm bitop AG (Galinski et al., 1994) using *H. elongata* ATCC 33173 as the ectoine source, and since then, ectoine has been used in many pharmaceutical and other applications. The production and applications of ectoine have been reviewed comprehensively by Pastor et al. (2010). Ectoine has been proposed to act as a protein stabilizer, with potential (albeit unproven) applications in preventing and inhibiting protein aggregation in neurogenerative diseases, for example, Alzheimer's disease (Kanapathipillai et al., 2005); decreasing nuclear inclusions and thereby preventing apoptosis (Furusho et al., 2005); stable storage of retroviral vectors used in gene therapy (Cruz et al., 2006); and action as a moisturizer, protecting skin against dehydration (Graf et al., 2008). The mechanism by which ectoine exerts an effect was described as "preferential exclusion." In this mechanism, when ectoine is present along with proteins or lipids, membrane surfaces expel ectoine. When this occurs, the membrane surface becomes more hydrated, and the lipid layers become more fluid (Harishchandra et al., 2010). In the tear fluid lipid layer of the eye, ectoine increases the fluidity of the meibomian lipid films, and it was proposed that this property could be used to devise new treatments for dry eye syndrome (Dwivedi et al., 2014).

In recent years, many new medical uses for ectoine have been devised based on these properties. These are in the main immunomodulatory (e.g., Unfried et al., 2014), and particular interest exists in using these to treat allergic disease, particularly those involving mucous membranes in the nose, eyes, and airways. Werkhäuser et al. (2014) published an interesting study where they show that nasal spray and eye drops with ectoine reduced symptoms in patients with allergic rhinitis. Marini et al. (2014) showed that an ectoine-containing cream, in comparison with a nonsteroidal anti-inflammatory cream, was equally efficacious and was well tolerated by patients with atopic dermatitis. Krutmann (2014), with bitop AG, has filed a patent (US08822477) for use of ectoine and hydroxyectoine in products for the treatment of neurodermatitis. Ectoine and 5-α-hydroxyectoine can help to ameliorate inflammatory bowel disease (IBD), at least in a rat model (Abdel-Aziz et al., 2015). It was proposed that this is due to stabilization of the intestinal barrier, which could be helpful in IBD, once it is shown to be safe and effective.

15.12 BIOPROSPECTING AND GENOME MINING OF EXTREMOPHILES

Only a small proportion (<1%) of the earth's microbiome is culturable (Amann, 2000). In genomic technologies, this precluded the sequencing of the genomes of many bacterial and archaeal species, but recent developments in metagenomics (De Maayer et al., 2014) and single-cell genomics (Blainey, 2013) have facilitated access to many genomes from microbial species. This has been further augmented by the availability of other "omic" technologies, for example, proteomics, transcriptomics, and metabolomics, providing a better understanding of how microbial communities function (Cowan et al., 2015). Analysis of community metagenomes can provide data on systems that are not readily accessible, and show that the metagenomes in particular biological niches may have particular types of genes that are essential for functioning in these environments (e.g., sulfur cycling genes in some deep-sea hydrothermal vent communities Cao et al., 2014). Such techniques can also reveal the presence of novel classes of microbes. One of these, derived from a hypersaline environment, is the Nanohaloarchaea, which have unusual properties, such as a novel combination of protein amino acids that might have application as compatible solutes (Narasingarao et al., 2011). Halophilic, acidophilic, and psychrophilic communities from these habitats are beginning to be catalogued, but deep subterranean habitats are poorly understood, due to their inaccessibility to sampling (Cowan et al., 2015).

Recent developments in whole-genome sequencing have led to a rapid expansion of information about extremophilic microorganisms and facilitated the identification of gene clusters. These clusters are frequently implicated in the synthesis of secondary metabolites, some of which have potential medical application. Lane and Moore (2011) have reviewed progress in the field of marine natural product gene discovery from the beginning of the twenty-first century, and discuss the progression from discovery of individual gene clusters to entire sequences. In prokaryotes, the discovery of genes encoding natural products is facilitated by gene clustering into regions of DNA, whereas genes in eukaryotes that are necessary for synthesizing natural products are often scattered throughout genomes. More than 90% of the genes or gene clusters that carry out secondary metabolite biosynthesis are not obvious (Ishikawa, 2008), which makes it harder to identify and exploit them. Among the first clusters to be identified were the sequences encoding enzymes involved in biosynthesis of the enterocins and wailupemycins from the marine actinomycete *Streptomyces maritimus* (Piel et al., 2000). The barbamide gene cluster, derived from cyanobacterium *Lyngbya majuscula* strain 19L, has features of a mixed type II polyketide synthase and nonribosomal peptide synthase biosynthetic structure. These modular-type systems assemble carboxylate and amino acid–based metabolites into polyketides and nonribosomal peptides, which can add different components to chains of natural products (Walsh, 2004). The increased availability of cyanobacterial genome data in public databases now enables them to be mined for multiple clusters, and improvements in bioinformatic techniques allow the identification and comparison of new gene clusters from related species. Micallef et al. (2015) studied 11 publicly available genomes from Subsection V cyanobacteria, along with draft genomes for *Hapalosiphon welwitschii* IC-52-3 and *Westiella intracata* UH strain HT-29-1 for their capacity to produce secondary metabolites. This comparison revealed a putative gene cluster for hapalosin, which reverses P-glycoprotein multiple drug resistance, as well as various NRPS/PKS gene clusters; cyanobactin, microviridin, and bacteriocin clusters; and clusters encoding MAAs, scytonemin, hydrocarbon, and terpenes.

While these sequences encoding the biosynthetic enzymes are becoming available, the regulation of the pathways is less well understood (Lane and Moore, 2011). Some genes may be global regulators that act on multiple pathways, and others may be local (Martín and Liras, 2010). Rapid genome sequencing technologies may make cluster sequencing obsolete, and approaches such as that used by Micallef et al. (2015) may become the norm. NGS technologies will also be aided by improvements in annotation and functional analysis. Alam et al. (2013) have devised a data warehouse system, Integrated Data Warehouse of Microbial Genomes (INDIGO), which serves as a repository for genomes of pure cultures and uncultured single cells of Red Sea bacteria and archaea. In particular, information from *Salinisphaera shabanensis*, *Haloplasma contractile*, and *Halorhabdus tiamatea*, halophiles isolated from deep-sea anoxic brine lakes of the Red Sea, is available (http://www.cbrc.kaust.edu.sa/indigo). Grötzinger et al. (2014) described improvements in annotation for the database using a profile and pattern-matching algorithm, and identified 11 proteins with biotechnological applications. Such data warehouses for organisms from other environments may revolutionize discovery of pathways encoding secondary metabolites and also individual enzymes.

An alternative approach is that of Mohimani et al. (2014). They developed RiPPquest, a tandem mass spectrometry database search tool for identifying microbial ribosomally synthesized and post-translationally modified peptides (RiPPs) and used it for lanthipeptide discovery from *Streptomyces* sp., connecting the RiPPs to genomic data about gene clusters. Glycosylated natural product discovery requires different approaches, and Kersten et al. (2013) described an approach using glycogenomics. They characterized O-linked and N-linked glycans using mass spectrometry and matched them to the corresponding glycosylation genes in secondary metabolic pathways using a mass spectrometry–glycogenetic code. The biosynthetic genes of the glycosylated natural product genotype, without the glycans, were used to classify the natural product to determine the structure. This technique was demonstrated by the characterization of several bioactive glycosylated molecules and their gene clusters, including cinerubin B, an anticancer compound from *Streptomyces* sp. SPB74, and an antibiotic, arenimycin B, from *Salinispora arenicola* CNB-527, which was active against multi-drug-resistant *S. aureus*. *Salinospora tropica* CNB-440 underwent genomic analysis, and it was shown to have 17 diverse pathways, only 4 of which were connected to their known products (Udwary et al., 2007).

15.13 CONCLUSIONS

Extremophiles from many sources produce various metabolites and other molecules (e.g., membrane lipids) that are potentially very valuable to both human and veterinary medicine. Some of these are well established (e.g., bacterial secondary metabolites or their derivatives approved for use as drugs). The scientific literature abounds in extremophile-derived compounds that may have use in the future, some of which are discussed here, but their use has not been established and further experimentation *in vitro* and *in vivo* will be needed to determine their efficacy and safety. For that reason, extremophiles may promise more than they can actually deliver in medical terms, at least at present. The cytotoxic or immunomodulatory effect of a bioproduct on a cell line, for example, may be different in a rodent model, not to mention human clinical trials. Care should be taken in interpreting the literature, with respect to both the effect and the nature of the metabolite-producing organism itself. Not all marine microbes are, technically, extremophiles, at least with respect to their temperature optima and salinity or pressure requirements. This makes the literature on marine-derived compounds somewhat difficult to interpret in terms of whether source organisms are extremophilic.

However, use of some extremophile products in skin care, for example, MAAs and ectoines, is well established, and new applications are constantly under investigation. Furthermore, there is little doubt as to whether extreme halophiles or radiation-resistant prokaryotes are extremophilic. It is likely that other applications will be developed over the coming decades, for example, in drug delivery (archaeosomes) and protein stabilization, and that genome mining will yield more information as to how these organisms produce secondary metabolites and extremolytes, allowing scale-up of industrial production in host organisms. Another possible use for extremophile products is in hostile environments, for example, as protection against UV or other types of radiation in extraterrestrial environments. If life is discovered elsewhere in our solar system, a study of these living organisms may reveal the presence of even more robust molecular defenses that could be exploited for healthcare and other purposes in our own world. It is likely that we have only scratched the surface of knowledge of microbial diversity. Further exploration and bioprospecting are crucial to enable humanity to reveal the full potential of these remarkable entities.

REFERENCES

Abdel-Aziz, H., Wadie, W., Scherner, O. et al. 2015. Bacteria-derived compatible solutes ectoine and 5-alpha-hydroxyectoine act as intestinal barrier stabilizers to ameliorate experimental inflammatory bowel disease. *J Nat Prod* 78: 1309–15.

Abraham, I., El Sayed, K., Chen, Z.S. et al. 2012. Current status on marine products with reversal effect on cancer multidrug resistance. *Mar Drugs* 10: 2312–21.

Ahn, K.Y., Ko, H.K., Lee, B.T. et al. 2014. Engineered protein nanoparticles for in vivo tumor detection. *Biomaterials* 35: 6422–9.

Alam, I., Antunes, A., Kamau, A.A. et al. 2013. INDIGO—INtegrated Data Warehouse of MIcrobial GenOmes with examples from the Res Sea extremophiles. *PLOS One* 8 (12): e82210.

Amann, R. 2000. Who is out there? Microbial aspects of biodiversity. *Syst Appl Microbiol* 23: 1–8.

Anderson, R.J., Manzanares, I., Morris, J.C. et al. 2002. Variolin derivatives as anti-cancer agents. WO200204447 A1, Pharma-Mar S.A., Spain. http://www.google.com/patents/WO2002004447A1?cl=en.

Attar, A., Ogan, A., Yucel, S. et al. 2015. The potential of archaeosomes as carriers of pDNA into mammalian cells. *Artif Cells Nanomed Biotechnol* 44: 710–6.

Babu, P., Chandel, A.K., and Singh, O.V. 2015. *Extremophiles and Their Applications in Medical Processes*. Heidelberg: Springer International Publishing.

Benvegnu, T., Lemiègre, L., Dalençon, S. et al. 2009. Applications of extremophilic archaeal lipids in the field of nanocarriers for oral/topical drug delivery. *Curr Biotechnol* 2: 294–303.

Besse, A., Peduzzi, J., Rebuffat, S. et al. 2015. Antimicrobial peptides and proteins in the face of extremes: Lessons from archaeocins. *Biochimie* 118: 344–55.

Beuchat, L.R. 1974. Combined effects of water activity, solute, and temperature on the growth of *Vibrio parahaemolyticus*. *Appl Microbiol* 27: 1075–80.

Bharate, S.B., Sawant, S.D., Singh, P.P. et al. 2013. Kinase inhibitors of marine origin. *Chem Rev* 113: 6761–815.

Blainey, P.C. 2013. The future is now: Single-cell genomics of bacteria and archaea. *FEMS Microbiol Rev* 37: 407–27

Bowers, K.J., Mesbah, N.M., and Wiegel, J. 2009. Biodiversity of poly-extremophilic bacteria: Does combining the extremes of high salt, alkaline pH and elevated temperature approach a physico-chemical boundary for life? *Saline Systems* 5: 9.

Brown, D.A., Venegas, B., Cooke, P.H. et al. 2009. Bipolar tetraether archaeosomes exhibit unusual stability against autoclaving as studied by dynamic light scattering and electron microscopy. *Chem Phys Lipids* 159: 95–103.

Bullen, C.K., Laird, G.M., Durand, C.M. et al. 2014. New ex vivo approaches distinguish effective and ineffective single agents for reversing HIV-1 latency in vivo. *Nat Med* 20: 425–9.

Bhatia, S., Sharma, K., Namdeo, A.G. et al. 2010. Broad-spectrum sun-protective action of Porphyra-334 derived from *Porphyra vietnamensis*. *Pharmacognosy Res* 2: 45–9.

Cao, H., Wang, Y., Lee, O.O. et al. 2014. Microbial sulfur cycle in two hydrothermal chimneys on the southwest Indian ridge. *MBio* 5: e00980-00913.

Chen, C.-Y. 2014. DNA polymerases drive DNA-sequencing-by-synthesis technologies: Both past and present. *Front Microbiol* 5: 305.

Chien, A., Edgar, D.B., and Trela, J.M. 1976. Deoxyribonucleic acid polymerase from the extreme thermophile *Thermus aquaticus*. *J Bacteriol* 127: 1550–7.

Colabella, F., Moline, M., and Libkind, D. 2014. UV sunscreens of microbial origin: Mycosporines and mycosporine-like amino acids. *Recent Pat Biotechnol* 8: 179–93.

Conde, F.R., Churio, M.S., and Previtali, C.M. 2004. The deactivation pathways of the excited-states of the mycosporine-like amino acids shinorine and porphyra-334 in aqueous solution. *Photochem Photobiol Sci* 3: 960–7.

Correa-Llantén, D.N., Muñoz-Ibacache, S.A., Castro, M.E. et al. 2013. Gold particles synthesized by *Geobacillus* sp. strain ID17, a thermophilic bacterium isolated from Deception Island, Antarctica. *Microb Cell Fact* 12: 75

Cowan, D.A., Ramond, J.-B., Makhalanyane, T.P. et al. 2015. Metagenomics of extreme environments. *Curr Opin Microbiol* 25: 97–102.

Cruz, P.E., Silva, A.C., Roldao, A. et al. 2006. Screening of novel excipients for improving the stability of retroviral and adenoviral vectors. *Biotechnol Prog* 22: 568–76.

Dalsgaard, P.W., Larsen, T.O., Frydenvang, K. et al. 2004. Psychrophilin A and cycloaspeptide D, novel cyclic peptides from the psychrotolerant fungus *Penicillum rubeum*. *J Nat Prod* 67: 878–81.

D'Auria, S., Di Cesare, N., Gryczynski, Z. et al. 2000. A thermophilic apoglucose dehydrogenase as nonconsuming glucose sensor. *Biochem Biophys Res Commun* 274: 727–31.

D'Auria, S., Di Cesare, N., Staiano, M. et al. 2002. A novel fluorescence competitive assay for glucose determinations by using a thermostable glucokinase from the thermophilic microorganism *Bacillus stearothermophilus*. *Anal Biochem* 303: 138–44.

Davis, M.E., Chen, Z.G., and Shin, D.M. 2008. Nanoparticle therapeutics: An emerging modality for cancer. *Nat Rev Drug Discov* 7: 771–82.

de Champdoré, M. Staiano, M., Rossi, M. et al. 2007. Proteins from extremophiles as stable tools for advanced biotechnological applications of high social interest. *Interface* 4: 183–91.

de la Coba, F., Aguilera, J., de Galvez M.V. et al. 2009. Prevention of the ultraviolet effects on clinical and histopathological changes, as well as the heat shock protein-70 expression in mouse skin by topical applications of algal UV-absorbing compounds. *J Derm Sci* 55: 161–9.

De Maayer, P., Valverde, A., and Cowan, D.A. 2014. The current state of metagenomic analysis. In *Genome Analysis: Current Procedures and Applications*, ed. M.S. Poptsova, 183–220. Norfolk, UK: Caister Academic Press.

De Oliveira, G.B., Favarin, L., Luchese, R.H. et al. 2015. Psychrotrophic bacteria in milk: How much do we really know? *Braz J Microbiol* 46: 313–21.

D'Incalci, M., and Galmarini, C.M. 2010. A review of trabectidin (ET-743): A unique mechanism of action. *Mol Cancer Ther* 9: 2157–63.

Ding, Z.-G., Zhao, J.-Y., Li, M.-G. et al. 2012. Griseusins F and G, spiro-napthoquinones from a tin mine tailings-derived *Nocardiopsis* species. *J Nat Prod* 75: 1994–8.

Donnarumma, G., Buommino, E., Fusco, A. et al. 2010. Effect of temperature on the shift of *Pseudomonas fluorescens* from an environmental microorganism to a potential human pathogen. *Int J Immunopathol Pharmacol* 23: 227–34.

Duan, Z., Ji, D., Weinstein, E.J. et al. 2010. Lentiviral shRNA screen of human kinases identifies PLK1 as a potential therapeutic target for osteosarcoma. *Cancer Lett* 293: 220–9.

Dwivedi, M., Backers, H., Harishchandra, R.K. et al. 2014. Biophysical investigations of the structure and function of the tear fluid lipid layer and the effect of ectoine. Part A: Natural meibomian lipid films. *Biochim Biophys Acta* 1838: 2708–15.

Ellen, A.F., Rohulya, O.V., Fusetti, F. et al. 2011. The sulfolobicin genes of *Sulfolobus acidocaldarius* encode novel antimicrobial proteins. *J Bacteriol* 193: 4380–7.

Empadinhas, N., and da Costa, M.S. 2006. Diversity and biosynthesis of compatible solutes in hyper/thermophiles. *Int Microbiol* 9: 199–206.

Etoh, T., Kim, Y.P., Tanaka, H. et al. 2013. Anti-inflammatory effect of berkeleyacetal C through the inhibition of interleukin-1 receptor-associated kinase-4 activity. *Eur J Pharmacol* 698: 435–43.

Fredimoses, M., Zhou, X., Lin, X. et al. 2014. New prenylxanthones from the deep-sea derived fungus *Emericella* sp. SCSIO 05240. *Mar Drugs* 12: 3190–202.

Furusho, K., Yoshizawa, T., and Shoji, S. 2005. Ectoine alters subcellular localisation of inclusions and reduces apoptotic cell death induced by the truncated Machado-Joseph disease gene product with an expanded polyglutamine stretch. *Neurobiol Disorders* 20: 170–8.

Gabani, P., and Singh, O.V. 2013. Radiation-resistant extremophiles and their potential in biotechnology and therapeutics. *Appl Microbiol Biotechnol* 97: 993–1004.

Galinski, E.A., Pfeiffer, H.P., and Trüper, H.G. 1985. 1,4,5,6-Tetrahydro-2-methyl-pyrimidinecarboxylic acid—A novel cyclo amino-acid from halophilic phototrophic bacteria of the genus *Ectothiorhodospora*. *Eur J Biochem* 149: 135–9.

Galinski, E.A., Trueper, H., Sauer, T. et al. 1994. Bitop Ges Biotechnische Optimierung BMH assignee. Production of natural substances, e.g., new hydroxy:ectoin derivative—By in vitro release from cells, esp. microorganisms, following application of stress. Patent DE424580-A.

Gao, Q., and Garcia-Pichel, F. 2011. Microbial ultraviolet sunscreens. *Nat Rev Microbiol* 9: 791–802.

Garcia-Pichel, F., and Castenholz, W.W. 1991. Characterization and biological implications of scytonemin, a cyanobacterial sheath pigment. *J Phycol* 27: 395–409.

Gerwick, G.H., and Fenner, A.M. 2013. Drug discovery from marine microbes. *Microb Ecol* 65: 800–6.

Giddings, L.-A., and Newman, D.J. 2013. Microbial natural products: Molecular blueprints for antitumor drugs. *J Ind Microbiol Biotechnol* 40: 1181–210.

Giddings, L.-A., and Newman, D.J. 2015a. *Bioactive Compounds from Marine Extremophiles*. Cham, Switzerland: Springer.

Giddings, L.-A., and Newman, D.J. 2015b. *Bioactive Compounds from Terrestrial Extremophiles*. Cham, Switzerland: Springer.

Girilal, M., Mohammed Fayaz, M., Mohan Balaji, P. et al. 2013. Augmentation of PCR efficiency using highly thermostable gold nanoparticles synthesized from a thermophilic bacterium, *Geobacillus stearothermophilus*. *Colloids Surf B Biointerfaces* 106: 165–9.

Gogada, R., Singh, S.S., Lunavat, S.K. et al. 2015. Engineered *Deinococcus radiodurans* R1 with NiCoT genes for bioremoval of trace cobalt from spent decontamination solutions of nuclear power reactors. *Appl Microbiol Biotechnol* 99: 9203–13.

Gomes, N.G.M., Lefranc, F., Kijjoa, A. et al. 2015. Can some marine-derived fungal metabolites become actual anti-cancer agents? *Mar Drugs* 13: 3960–91.

González-Paredes, A., Manconi, M., Caddeo C. et al. 2010. Archaeosomes as carriers for topical delivery of betamethasone dipropionate: In vitro skin permeation study. *J Liposome Res* 4: 269–76.

Graf, R., Anzali, S., Bunger, J. et al. 2008. The multifunctional role of ectoine as a natural cell protectant. *Clin Dermatol* 26: 326–33.

Grötzinger, S.W., Alam, I., Ba Alawi, W. et al. 2014. Mining a database of single amplified genomes from Red Sea brine pool extremophiles—Improving reliability of gene function prediction using a profile and pattern matching algorithm (PPMA). *Front Microbiol* 5: 134.

Harishchandra, R.K., Wulff, S., Lentzen, G. et al. 2010. The effect of compatible solute ectoines on the structural organization of lipid monolayer and bilayer membranes. *Biophys Chem* 150: 37–46.

Hayes, J., and Barrett, M. 2015. Bird fancier's lung in mushroom workers. *Ir Med J* 108: 119–20.

He, J., Roemer, E., Lange, C. et al. 2007. Structure, derivatization and antitumor activity of new griseusins from *Nocardiopsis* sp. *J Med Chem* 50: 5168–75.

Higa, L.H., Corral, R.S., Morilla, M.J. et al. 2013. Archaeosomes display immunoadjuvant potential for a vaccine against Chagas disease. *Hum Vaccin Immunother* 9: 409–12.

Horikoshi, K. 1998. Barophiles: Deep-sea microorganisms adapted to an extreme environment. *Curr Opin Microbiol* 1: 291–5.

Hoyoux, A., Blaise, V., Collins, T. et al. 2004. Extreme catalysts from low-temperature environments. *J. Biosci. Bioeng.* 98: 317–30.

Hughes, C.C., Kauffman, C.A., Jensen, P.R. et al. 2008. The marinopyrroles, antibiotics of an unprecedented structure class from a marine *Streptomyces* sp. *Org Lett* 10: 629–31.

Inbar, L. and Lapidot, A. 1988. The structure and biosynthesis of new tetrahydropyrimidine derivatives in actinomycin D producer *Streptomyces parvulus*. Use of 13C- and 15N-labeled L-glutamate and 13C and 15N NMR spectroscopy. *J Biol Chem* 263: 16014–22.

Irwin, J.A. 2010. Extremophiles and their application to veterinary medicine. *Environ Technol* 31: 857–69.

Ishikawa, J. 2008. Genome analysis system for actinomycetes: Development and application. *Actinomycetologica* 22: 46–9.

Ishino, S., and Ishino, Y. 2014. DNA polymerases as useful reagents for biotechnology—The history of developmental research in the field. *Front Microbiol* 5: 465.

Jaeger, S., Schmuck, R., and Sobek, H. 2000. Molecular cloning, sequencing and expression of the heat-labile uracil-DNA glycosylase from a marine psychrophilic bacterium, strain BMTU3346. *Extremophiles* 4: 115–22.

Jensen, P.R., Moore, B.S., and Fenical, W. 2015. The marine actinomycete genus *Salinispora*: A model organism for secondary metabolite discovery. *Nat Prod Rep* 32: 738–51.

Jensen, S.M., Christensen, C.J., Petersen, J.M. 2015. Liposomes containing lipids from *Sulfolobus islandicus* withstand intestinal bile salts: An approach for oral drug delivery? *Int J Pharmaceutics* 493: 63–9.

Juliano, R. 2013. Nanomedicine: Is the wave trending? *Nat Rev Drug Discov* 12: 171–2.

Kanapathipillai, M., Lentzen, G., Sierks, M. et al. 2005. Ectoine and hydroxyectoine inhibit aggregation and neurotoxicity of Alzheimer's beta-amyloid. *FEBS Lett* 579: 4775–80.

Kaur, G., Garg, T., Rath, G. et al. 2016. Archaeosomes: An excellent carrier for drug and cell delivery. *Drug Deliv* 23: 2497–512.

Kersten, R.D., Ziemert, N., Gonzalez, D.J. et al. 2013. Glycogenomics as a mass-spectrometry-guided genome-mining method for microbial glycosylated molecules. *Proc Natl Acad Sci USA* 110: E4407–16.

Kirillova, I.P., Agre, N.S., and Kalakoutskii, L.V. 1974. Spore initiation and minimum temperature for growth of *Thermoactinomyces vulgaris*. *J Basic Microbiol* 14: 69–72.

Kollár, P., Rajchard, J., Balounová, Z. et al. 2014. Marine natural products: Bryostatins in preclinical and clinical studies. *Pharm Biol* 52: 237–42.

Krutmann, J. 2014. Use of osmolytes obtained from extremophilic bacteria for medicine for the external treatment of neurodermatitis. Patent US08822477.

Kulkarni, R.R., Shaiwale, N.S., Deobagkar, D.N. et al. 2015. Synthesis and extracellular accumulation of silver nanoparticles by employing radiation-resistant *Deinococcus radiodurans*, their characterization and determination of bioactivity. *Int J Nanomed* 10: 963–74.

Lalli, C., and Parsons, T. 1997. *Biological Oceanography: An Introduction*. Oxford: Butterworth-Heinemann.

Lane, A.L., and Moore, B.S. 2011. A sea of biosynthesis: Marine natural products meet the molecular age. *Nat Prod Rep* 28: 411–28.

Lavy, A., Rusu, R., and Mates, A. 1992. *Mycobacterium xenopi*, a potential human pathogen. *Isr J Med Sci* 28: 772–5.

Lawton, E.M., Cotter, P.D., Hill, C. et al. 2007. Identification of a novel two-peptide lantibiotic, haloduracin, produced by the alkaliphile *Bacillus halodurans* C-125. *FEMS Microbiol Lett* 267: 64–71.

Lebar, M.D., Heimbegner, J.L., and Baker, B.J. 2007. Cold-water marine natural products. *Nat Prod Rep* 24: 774–97.

Lee, J., Currano, J.N., Carroll, P.J. et al. 2012. Didemnins, tamandarins and related natural products. *Nat Prod Rep* 29: 404–24.

Lentzen, G., and Schwarz, T. 2006. Extremolytes: Natural compounds from extremophiles for versatile applications. *Appl Microbiol Biochem* 72: 623–34.

Lequerica, J.L., O'Connor, J.E., Such, L. et al. 2006. A halocin acting on Na+/H+ exchanger of *Haloarchaea* as a new type of inhibitor in NHE of mammals. *J Physiol Biochem* 62: 253–62.

Li, L., Li, D., Luan, Y. et al. 2012. Cytotoxic metabolites from the Antarctic psychrophilic fungus *Oidiodendron truncatum*. *J Nat Prod* 75: 920–7.

Li Z., Chen, J., Sun W. et al. 2010. Investigation of archaeosomes as carriers for oral delivery of peptides. *Biochem Biophys Res Commun* 394: 412–7.

Li, Z., Zhang, L., Sun, W. et al. 2011. Archaeosomes with encapsulated antigens for oral vaccine delivery. *Vaccine* 29: 5260–6.

Loo, S.Y., Chang, M.K., Chua, C.S. et al. 2012. NHE-1: A promising target for novel anti-cancer therapeutics. *Curr Pharm Des* 18: 1372–82.

MacElroy, R.D. 1974. Some comments on the evolution of extremophiles. *Biosystems* 6: 74–5.

Marini, A., Reinelt, K., Krutmann, J. et al. 2014. Ectoine-containing cream in the treatment of mild to moderate atopic dermatitis: A randomised, comparator-controlled intra-individual double-blind, multi-centre trial. *Skin Pharm Physiol* 27: 57–65.

Martín, J.F., and Liras, P. 2010. Engineering of regulatory cascades and networks controlling antibiotic synthesis in *Streptomyces*. *Curr Opin Microbiol* 13: 263–73.

Martins, A., Tenreiro, T., Andrade, G. et al. 2013. Photoprotective bioactivity present in a unique bacteria collection from Portuguese deep sea hydrothermal vents. *Mar Drugs* 11: 1506–23.

Mattimore, V., and Battista, R. 1996. Radioresistance of *Deinococcus radiodurans*: Functions necessary to survive ionizing radiation are also necessary to survive prolonged desiccation. *J Bacteriol* 178: 633–7.

Mayer, A.M.S. 2015. Marine pharmaceuticals: The clinical pipeline. http://marinepharmacology.midwestern.edu/clinPipeline.htm (accessed November 17, 2015).

Mayer, A.M.S., Rodríguez, A.D., Taglialatela-Scafati, O. et al. 2013. Marine pharmacology in 2009–2011: Marine compounds with antibacterial, antidiabetic, antifungal, anti-inflammatory, antiprotozoal, anti-tuberculosis and antiviral activities; affecting the immune and nervous systems, and other miscellaneous mechanisms of action. *Mar Drugs* 11: 2510–73.

McInnes, C., Mezna, M., and Fischer, P.M. 2005. Progress in the discovery of polo-like kinase inhibitors. *Curr Topics Med Chem* 5: 181–97.

Mesbah, N.M., and Weigel, J. 2008. Life under multiple extreme conditions: Diversity and physiology of the halophilic alkalithermophiles. *Appl Environ Microbiol* 78: 4074–82.

Micallef, M.L., D'Agostino, P.M., Sharma, D. et al. 2015. Genome mining for natural product biosynthetic gene clusters in the Subsection V cyanobacteria. *BMC Genomics* 16: 669.

Michaud, L., di Cello, F., Brilli M. et al. 2004. Biodiversity of cultivable psychrotrophic marine bacteria isolated from Terra Nova Bay (Ross Sea, Antarctica). *FEMS Microbiol Lett* 230: 63–71.

Mincer, T.J., Fenical, W., and Jensen, P.R. 2005. Culture-dependent and culture-independent diversity within the obligate marine actinomycete genus *Salinispora*. *Appl Environ Microbiol* 71: 7019–28.

Mishra, A., Tandon, R., Kesarwani, S. et al. 2015. Emerging applications of cyanobacterial ultraviolet protecting compound scytonemin. *J Appl Phycol* 27: 1045–51.

Moghimpour, E., Kargar, M., Ramezani, Z. et al. 2013. The potent *in vitro* skin permeation of archaeosome made from lipids extracted of *Sulfolobus acidocaldarius*. *Archaea* 2013: 782012.

Mohimani, H., Kersten, R.D., Liu, W.T. et al. 2014. Automated genome mining of ribosomal peptide natural products. *ACS Chem Biol* 9: 1545–51.

Mojib, N., Philpott, R., Huang, J.P. et al. 2010. Antimycobacterial activity in vitro of pigments isolated from Antarctic bacteria. *Antonie van Leeuwenhoek* 98: 531–40.

Molinski, T.F., Dalisay, D.S., Lievens, S.L. et al. 2009. Drug development from marine natural products. *Nat Rev Drug Discov* 8: 69–85.

Morita, R.Y. 1975. Psychrophilic bacteria. *Bacteriol Rev* 39: 144–67.

Moschonas, G., Bolton, D.J., McDowell, D.A. et al. 2011. Diversity of culturable psychrophilic and psychrotrophic anaerobic bacteria isolated from beef abbatoirs and their associated environments. *Appl Environ Microbiol* 77: 4280–4.

Munteanu, A.-C., Uivarosi, V., and Andries, A. 2015. Recent progress in understanding the molecular mechanisms of radioresistance in *Deinococcus* bacteria. *Extremophiles* 19: 707–19.

Murphy, D.M.F., Morgan, W., Keith, C. et al. 1995. Hypersensitivity pneumonitis. In *Occupational Lung Diseases*, ed. W. Morgan, C. Keith, and A. Seaton, 525–67. Philadelphia: W.B. Saunders.

Narasingarao, P., Podell, S., Ugalde, J.A. et al. 2011. *De novo* metagenomic assembly reveals abundant novel major lineage of Archaea in hypersaline microbial communities. *ISME J* 6: 81–93.

Newman, D.J., and Cragg, G.M. 2014. Marine-sourced anti-cancer and cancer pain control agents in clinical and late preclinical development. *Mar Drugs* 12: 255–78.

Niu, S., Liu, D., Hu, X. et al. 2014. Spiromastixones A–O, antibacterial chlorodepsidones from a deep-sea-derived *Spiromastix* sp. fungus. *J Nat Prod* 77: 1021–30.

Nuñez-Pons L., Carbone M., Vásquez J. et al. 2012. Natural products from Antarctic colonial ascidians of the genera *Aplidium* and *Synoicum*: variability and defensive role. *Mar Drugs* 10: 1741–64.

Okudaira, M., Kawamura, H., Ueno, M. et al. 1962. Food poisoning caused by pathogenic halophilic bacteria (*Pseudomonas enteritis* TAKIKAWA). *Pathol Int* 12: 299–304.

Oren, A. 2010. Industrial and environmental applications of halophilic microorganisms. *Environ Technol* 31: 825–34.

Park, O.J., Yi, H., Jeon, J.H. et al. 2015. Pyrosequencing analysis of subgingival microbiota in distinct periodontal conditions. *J Dent Res* 94: 921–7.

Park, Y.C., Gunasekera, S.P., Lopez, J.V. et al. 2006. Metabolites from the marine-derived fungus *Chromocleista* sp. isolated from a deep-water sediment collected in the Gulf of Mexico. *J Nat Prod* 69: 580–4.

Pastor, J.M., Salvador, M., Argandoña, M. et al. 2010. Ectoines in cell stress protection: Uses and biotechnological production. *Biotechnol Adv* 28: 782–801.

Peck, M.W. 2010. *Clostridium botulinum*. In *Pathogens and Toxins in Foods: Challenges and Interventions*, ed. V.K. Juneja and J.N. Sofos, 31–48. Washington, DC: American Society for Microbiology.

Perry, N.B., Ettouati, L., Litaudon, M. et al. 1994. Alkaloids from the Antarctic sponge *Kirkpatrickia variolosa*: Part 1, variolin B, a new antitumor and antiviral compound. *Tetrahedron* 50: 3987–92.

Pettit, R.K., Pettit, G.R., and Hazen, K.C. 1998. Specific activities of dolastatin 10 and peptide derivatives against *Cryptococcus neoformans*. *Antimicrob Agents Chemother* 42: 2961–5.

Petit, K., and Biard, J.F. 2013. Marine natural products and related compounds as anticancer agents: An overview of their clinical status. *Anticancer Agents Med Chem* 13: 603–31.

Piel, J., Hertweck, C., Shipley, P.R. et al. 2000. Cloning, sequencing and analysis of the enterocin biosynthesis gene cluster from the marine isolate '*Streptomyces maritimus*': Evidence for the derailment of an aromatic polyketide synthase. *Chem Biol* 7: 943–55.

Pikuta, E.V., Hoover, R.B., and Tang, J. 2007. Microbial extremophiles at the limits of life. *Crit Rev Microbiol* 33: 183–209.

Prieto-Davó, A., Villareal-Gómez, L.J., Forschner-Dancause, S. et al. 2013. Targted search for actinomytes from nearshore and deep-sea marine sediments. *FEMS Microbiol Ecol* 84: 510–8.

Prieur, D., Erauso, G., and Jeanthon, C. 1995. Hyperthermophilic life at deep-sea hydrothermal vents. *Planet Space Sci* 43: 115–22.

Rastogi, R.P., Sonani, R.R., and Madamwar, D. 2014. The high-energy radiation protectant extracellular sheath pigment scytonemin and its reduced counterpart in the cyanobacterium *Scytonema* sp. R77DM. *Bioresour Technol* 171: 396–400.

Rastogi, R.P., Sonani, R.R., and Madamwar, D. 2015. Cyanobacterial sunscreen scytonemin: Role in photoprotection and biomedical research. *Appl Biochem Biotechnol* 176: 1551–63.

Raveendran, S., Chauhan, N., Palaninathan, V. et al. 2014a. Extremophilic polysaccharide for biosynthesis and passivation of gold nanoparticles and photothermal ablation of cancer cells. *Part Part Syst Charact* 32: 54–64.

Raveendran, S., Girija, A.R., Balasubram, S. et al. 2014b. Green approach for augmenting biocompatibility to quantum dots by extremophilic polysaccharide conjugation and nontoxic bioimaging. *ACS Sustain Chem Eng* 2: 1551–8.

Raveendran, S., Palaninathan, V., Nagaoka, Y. et al. 2015. Extremophilic polysaccharide nanoparticles for cancer nanotherapy and evaluation of antioxidant properties. *Int J Biol Macromol* 76: 310–19.

Raveendran, S., Poulose, A.C., Yoshida, Y. et al. 2013. Bacterial exopolysaccharide based nanoparticles for sustained drug delivery, cancer chemotherapy and bioimaging. *Carbohyd Polym* 91: 22–32.

Remuiñan, M., Gonzalez, J., Del Pozo, C. et al. 2003. Variolin derivatives and their use as antitumour agents. WO2003006457, PharmaMar S.A., Spain. https://patentscope.wipo.int/search/en/detail.jsf?docId=WO2003006457.

Russo, P., Del Bufalo, A., and Fini, M. 2015. Deep sea as a source of novel anti-cancer drugs: Update on discovery and preclinical/clinical evaluation in a systems medicine perspective. *EXCLI J* 14: 228–36.

Ryan, S.M., and Brayden, D.J. 2014. Progress in the delivery of nanoparticle constructs: Towards clinical translation. *Curr Opin Pharmacol* 18: 120–8.

Ryland, J.S., Bishop, J.D.D., De Blauwe, H. et al. 2011. Alien species of *Bugula* (Bryozoa) along the Atlantic coasts of Europe. *Aquatic Invasions* 6: 17–31.

Salomon, C.E., Magarvey, N.A., and Sherman, D.H. 2004. Merging the potential of microbial genetics with biological and chemical diversity: An even brighter future for marine natural product discovery. *Nat Prod Rep* 21: 105–21.

Sauer, T., and Galinski, EA. 1998. Bacterial milking: A novel process for production of compatible solutes. *Biotechnol Bioeng* 57: 306–13.

Schwartzer, D., Finking, R., and Marahiel, M.A. 2003. Nonribosomal peptides: From genes to products. *Nat Prod Rep* 20: 275–87.

Shivaji, S., Madhu, S., and Singh, S. 2011. Extracellular synthesis of antibacterial silver nanoparticles using psychrophilic bacteria. *Process Biochem* 46: 800–7.

Singh, O.V., and Gabani, P. 2011. Extremophiles: Radiation resistance microbial reserves and therapeutic implications. *J Appl Microbiol* 110: 851–61.

Singh, S.P., Klisch, M., Sinha, R.P. et al. 2010. Genome mining of mycosporine-like amino acid (MAA) synthesising and non-synthesizing cyanobacteria: A bioinformatics study. *Genomics* 95: 120–8.

Sinha, R.P., and Häder, D.P. 2008. UV—Protectants in cyanobacteria. *Plant Sci* 174: 278–89.

Skropeta, D., and Wei, L. 2014. Recent advances in deep-sea natural products. *Nat Prod Rep* 31: 999–1025.

Sprott, G.D., Sad, S., Fleming, L.P. et al. 2003. Archaeosomes varying in lipid composition differ in receptor-mediated endocytosis and differentially adjuvant immune responses to entrapped antigen. *Archaea* 1: 151–64.

Srivastava, P., Bragança, J.M., and Kowshik, M. 2014. In vivo synthesis of selenium nanoparticles by *Halococcus salifodinae* BK18 and their anti-proliferative properties against HeLa cell line. *Biotechnol Prog* 30: 1480–7.

Srivastava, P., Nikhil, V., Bragança, J.M. et al. 2015. Anti-bacterial TeNPs biosynthesized by haloarchaeon *Halococcus salifodinae* BK3. *Extremophiles* 19: 875–84.

Staunton, J., and Weissman, K.J. 2001. Polyketide synthesis: A millennium review. *Nat Prod Rep* 18: 380–416.

Stevenson, C.S., Capper, E.A., Roshak, A.K. et al. 2002. The identification and characterization of the marine natural product scytonemin as a novel antiproliferative pharmacophore. *J Pharm Exp Ther* 303: 858–66.

Stierle, A.A., and Stierle, D.B. 2014. Bioactive secondary metabolites from acid mine waste extremophiles. *Nat Prod Commun* 7: 1037–44.

Stierle, A.A., Stierle, D.B., Hobbs, J.D. et al. 2004. Berkeley dione and berkeleytrione, new bioactive metabolites from an acid mine organism. *Org Lett* 6: 1049–52.

Stierle, A.A., Stierle, D.B., and Kelly, K. 2006. Berkelic acid, a novel spiroketal with selective anticancer activity from an acid mine waste fungus extremophile. *J Org Chem* 71: 5357–60.

Stierle, A.A., Stierle, D.B., Patacini, B. et al. 2007. The berkeleyacetals, three meroterpenes from a deep water acid mine waste *Penicillium*. *J Nat Prod* 70: 1820–3.

Stierle, A.A., Stierle, D.B., Patacini, B. et al. 2011. Berkeleyones and related meroterpenes from a deep water acid mine waste fungus that inhibit the production of interleukin 1-β from induced inflammasomes. *J Nat Prod* 71: 856–60.

Stierle, D.B., Stierle, A.A., Girtsman, T. et al. 2012. Caspase-1 and -3 inhibiting drimane sesquiterpinoids from the extremophilic fungus *Penicillium solitum*. *J Nat Prod* 75: 344–50.

Su, Z. 2012. Beauty product containing desert algae radiation-proof ingredient and natural medical whitening ingredient and preparation method thereof. Patent CN102764206 (A). http://worldwide.espacenet.com/publicationDetails/biblio?FT=D&date=20121107&DB=worldwide.espacenet.com&locale=en_EP&CC=CN&NR=102764206A&KC=A&ND=4 (accessed November 4, 2015).

Suh, S.-S., Hwang, J., Park, M., Seo, H.H., Kim, H.-S., Lee, J.H., Moh, S.H., and Lee, T.-K. 2014. Anti-inflammation activities of mycosporine-like amino acids (MAAs) in response to UV radiation suggest potential skin anti-aging activity. *Mar Drugs* 12: 5174–87.

Sun, M.K., and Alkon, D.L. 2014. The 'memory kinases': Roles of PKC isoforms in signal processing and memory formation. *Prog Mol Biol Transl Sci* 122: 31–59.

Swain, S.S., Padhy, R.N., and Singh, P.K. 2015. Anticancer compounds from cyanobacterium *Lyngbya* species: A review. *Antonie van Leeuwenhoek* 108: 223–65.

Swami, U., Shah, U., and Goel, S. 2015. Eribulin in cancer treatment. *Mar Drugs* 13: 5016–58.

Udwary, D.W., Zeigler, L., Asolkar, R.N. et al. 2007. Genome sequencing reveals complex secondary metabolome in the marine actinomycete *Salinospora tropica*. *Proc Natl Acad Sci USA* 104: 10376–81.

Unfried, K., Kroker, M., Autengruber, A. et al. 2014. The compatible solute ectoine reduces the exacerbating effect of environmental model particles on the immune response of the airways. *J Allergy* 2014: 708458.

Van den Bogart, H.G., Van den Ende, G., Van Loon, P.C. et al. 1993. Mushroom worker's lung: Serologic reactions to thermophilic actinomycetes present in the air of compost tunnels. *Mycopathologia* 122: 21–8.

Varshney, P., Miculic, P., Vonshak, A. et al. 2015. Extremophilic micro-algae and their potential contribution in biotechnology. *Bioresour Technol* 184: 363–72.

Venugopal, V. 2009. Drugs and pharmaceuticals from marine sources. In *Functional and Bioactive Nutraceutical Compounds from the Ocean*, ed. V. Venugoapal 371–404. Boca Raton, FL: CRC Press.

Walsh, C.T. 2004. Polyketide and nonribosomal peptide antibiotics: Modularity and versatility. *Science* 303: 1805–10.

Wang, H., Wang, Y., Wang W. et al. 2011. Anti-influenza virus polyketides from the acid-tolerant fungus *Penicillium purporogenum* JS03-21. *J Nat Prod* 74: 2014–8.

Wang, Y.-T., Xue, Y.-R., and Liu, C.-H. 2015. A brief review of bioactive metabolites derived from deep-sea fungi. *Mar Drugs* 13: 4594–616.

Wang, W., Zhu, T., Tao, H. et al. 2007. Two new cytotoxic quinone type compounds from the halotolerant fungus *Aspergillus varicolor*. *J Antibiot* (Tokyo) 60: 603–7.

Werkhäuser, N., Bilstein, A., and Sonnemann, U. 2014. Treatment of allergic rhinitis with ectoine containing nasal spray in comparison with azelastine containing nasal spray and eye drops or with cromoglycic acid containing nasal spray. *J Allergy* 2014: 176597.

Wijeratne, E.M.K., Turbyville, T.J., Zhang Z. et al. 2003. Cytotoxic constituents of *Aspergillus terreus* from the rhizosphere of *Opuntia versicolor* of the Sonoran Desert. *J Nat Prod* 66: 1567–73.

Wu, G., Sun, X., Yu, G. et al. 2014. Cladosins A–E, hybrid polyketides from a deep-sea-derived fungus, *Cladosporium sphaerospermum*. *J Nat Prod* 77: 270–5.

Yang, Y., Itahasih, S., Yokobori, S., and Yamagishi, A. 2008. UV resistant bacteria isolated from upper troposphere and lower stratosphere. *Biol Sci Space* 22: 18–25

Zhang, G., Zhang, Z., and Liu, Z. 2013a. Polo-like kinase is overexpressed in renal cancer and participates in the proliferation and invasion of renal cancer cells. *Tumour Biol* 34: 1887–94.

Zhang, G., Zhang, Z., and Liu, Z. 2013b. Scytonemin inhibits cell proliferation and arrests cell cycle through downregulating Plk1 activity in multiple myeloma cells. *Tumour Biol* 34: 2241–7.

Zhang, J., Wanner, J., and Singh, O.V. 2015. Extremophiles and synthesis of nanoparticles: Current and future perspectives. In *Bio-Nanoparticles: Biosynthesis and Sustainable Biotechnological Implications*, ed. O.V. Singh, 101–22. Hoboken, NJ: Wiley-Blackwell.

Zhang, Y., and Sun, Y. 2014. A method for de novo nucleic acid diagnostic target discovery. *Bioinformatics* 30: 3174–80.

Zhao, B., Yan, Y., Chen, S. 2014. How could haloalkaliphilic microorganisms contribute to biotechnology? *Can J Microbiol* 60: 717–27.

Zitouni, A., Boudjella, H., Lamari, L. et al. 2005. *Nocardiopsis* and *Saccharothrix* genera in Saharan soils in Algeria: Isolation, biological activities and partial characterisation of antibiotics. *Res Microbiol* 156: 984–93.

CHAPTER 16

Extremophile Case Studies
Genomic Organization and Optimized Growth

Ravi Durvasula and D. V. Subba Rao

CONTENTS

16.1 INTRODUCTION

16.1.1 Genomic Insights from *Picochlorum* SENEW3: Parsimony with Efficient Regulation

In a recent study, Foflonker et al. (2016) explored the genomic structure and regulation of *Picochlorum* sp. strain SENEW 3 (*Picochlorum* SE3), a broadly halotolerant green alga that was isolated from mesophilic brackish water lagoons in San Diego County, California. This alga can survive in settings that range from freshwater to salt concentrations that are three times those of seawater. Additionally, *Picochlorum* species have been isolated in broad ranges of light intensity, from 80 to 2000 µmol m^{-2} s^{-1}, and in temperatures from 16°C to 33°C. Despite this "polyextremophile" capacity, the genome of *Picochlorum* SE3 is one of the smallest reported from a eukaryotic organism, at a paltry 13.5 Mbp comprised of 7363 genes. The limited genetic machinery of this organism is organized and regulated with maximum efficiency, to permit tolerance of extreme changes in temperature, light, and salinity.

Other members of the genus *Picochlorum* have been investigated as candidates for biotechnology applications. *P. oklahomensis* and *P. atomus*, for example, may be highly amenable to industrial applications, based on lipid production, carotenoid levels, and bioremediation properties. The responsiveness of *Picochlorum* species to nitrogen depletion, with increases in lipid production, augurs well for biotechnology and algal biofuel efforts. For these reasons, a better understanding of regulatory aspects of the *Picochlorum* SENEW3 genome, with respect to environmental changes, may offer valuable insights that translate to improved culture conditions and yield of bioproducts.

Using two ambient salt concentrations, 1.5 M and 10 mM NaCl, to represent varying conditions during a rainfall event in nature, the investigators studied differential expression profiles. A total of 3681 genes (more than 50% of the gene inventory of this organism) exhibited differential expression during salt flux. This set of differentially expressed genes included 12 of 24 genes obtained by *Picochlorum* SE3 via horizontal gene transfer (HGT) from bacterial sources. This suggests that the alga obtained foreign genetic material that facilitates halotolerance. The authors also noted marked differences in numbers of differentially expressed genes during the initial phase of salt stress, defined as one hour, compared with the later phase, defined as five hours. At one hour, a total of 3256 genes exhibited differential regulation, compared with 1629 genes at five hours. The plasticity of genomic responses in response to salt stress suggests that *Picochlorum* could be cultivated in conditions that are limiting to competing or contaminating organisms, such as open ponds. Furthermore, the prospects of reducing ambient stressors, such as salt concentrations, to increase yields of valuable by-products—through a redirection of cellular machinery—are quite attractive for commercial operations.

Studies of gene colocalization, in gene clusters, provide further insights into the organization of the *Picochlorum* SE3 genome. Colocalized genes that were differentially expressed and shared

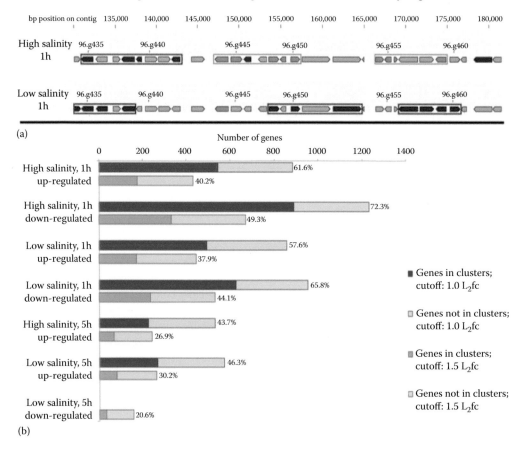

Figure 16.1 (a) Gene clusters and colocalized genes are represented in the boxes. Genes that are upregulated are in blue boxes; downregulated genes are in red boxes. The gray color indicates genes that are not differentially regulated. The relative location of the genes within a contig and the gene number are noted above the boxes. (b) Clustering of genes (as percentage total) that are up- and downregulated at varying salinity and time points. (From Foflonker, F. et al. *Algal Res.*, 16, 465–472, 2016.)

an expression pattern comprised 42%–72% of genes in the coexpressed data sets. Furthermore, colocalization of genes that were either up- or downregulated at high or low salinities was observed more frequently at one hour of salt flux versus five hours (Foflonker et al. 2016) (Figure 16.1a and b).

Upregulation of genes in response to hypersalinity shock was most pronounced in genes that encoded proteins of the photorespiration process, which may play a role in stress protection. Upregulation of glycolate dehydrogenase with concomitant downregulation of glycolate oxidase and catalase suggests the switch to stress-induced photorespiration at the mitochondrion. Production of carbonic anhydrase, another stress response elicited by reduced CO_2 concentrations during hypersaline shock, was significantly upregulated in *Picochlorum* SE3, along with increased production of tetrahydrofolate (THF) and enzymes required for ammonia fixation, notably glutamine synthetase and ferredoxin-dependent glutamate synthase.

The dynamic nature of gene regulation in this organism as a response to rapid changes in salt concentration, coupled with the clustering of key genes, some of which have been obtained through HGT, represents a fascinating feature of extremophiles, namely, the ability of these organisms to efficiently package and regulate compact genomes for survival under harsh environmental conditions.

16.2 CODON USAGE BIAS IN THE GENUS *SULFOLOBUS*: GENETIC ADAPTATIONS IN HYPERTHERMOACIDOPHILIC EXTREMOPHILES

Organisms exhibit widely divergent patterns of codon usage during translational processes. Patterns of codon usage play a key role in recombinant strategies and are therefore of great use in biotechnology. However, the evolution of extremophiles and adaptations to environmental stressors may alter patterns of codon usage. The study of codon bias may shed light on important evolutionary and regulatory elements of extremophiles. Patterns of codon usage bias may be associated with increased translational rate and accuracy. For polyextremophile organisms such as those of the genus *Sulfolobus*, a study of these mechanisms may shed light on survival strategies under the harshest conditions on earth.

Sulfolobus acidocaldarius strain DSM639 is a hyperthermoacidophile described from terrestrial solfataras in 1972. It grows at temperatures between 75°C and 80°C and pH ranges from 2 to 3. This organism has served as a prototype for the study of hyperthermophiles, and its mechanisms of chromatin folding, genome organization, and DNA repair, along with extremozymes that work at very high temperatures, have been the focus of considerable investigation. Nayak (2013) recently described codon and amino acid usage patterns from several species of the genus *Sulfolobus* to better understand genetic mechanisms that could aid in industrial applications, cloning, and heterologous protein expression systems. Using multivariate statistical methods and nonparametric testing, Nayak investigated patterns of codon bias across the *Solfolobus* genus, including nine other species and subspecies, using GenBank sequence data. Four factors were found to affect codon usage variation: (1) expression level of genes, (2) compositional mutational bias, (3) translational selection, and (4) gene length. Of these, gene expression levels and mutational bias played predominant roles. Interestingly, for high-expression genes, the optimal codon structure was not always the same as translationally optimal codon patterns. As shown in Figure 16.2, translationally preferred GC-rich codons were found for genes at higher expression levels and the right side of the Nc curve where NC refers to the effective number of codons of a gene (Wright 1990). Since the GC-rich codons fall below the expected curve, elements of bias toward codon usage remain that, according to the author, merit further studies. Overall, this study offers in-depth analysis of codon usage patterns that appear to correlate with adaptations to polyextremophile environments in the genus *Solfolobus* and provide valuable tools to design heterologous and optimized expression systems that may harness the tremendous genetic potential of these organisms (Nayak 2013).

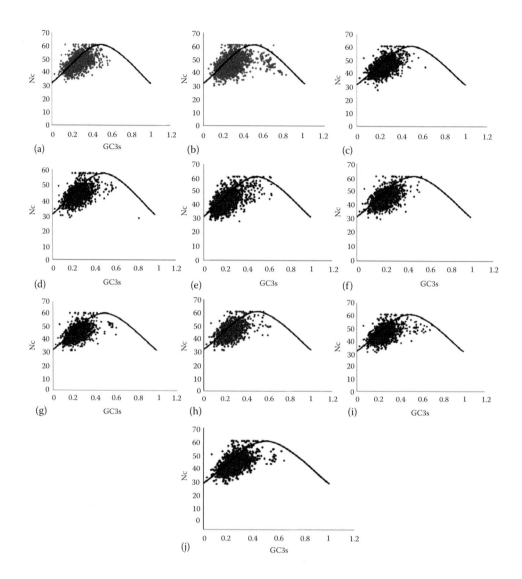

Figure 16.2 Plots of Nc vs. GC3s for *S. acidocaldarius* DSM639 (a), *S. solfatericus* P2 (b), *S. solfolobus tokodaii* (c), *S. islandicus* LD 8.5 (d), *S. islandicus* LS 2.15 (e), *S. islandicus* M 14.25 (f), *S. islandicus* M 16.27 (g), *S. islandicus* M 16.4 (h), *S. islandicus* Y.G. 57.14 (i), and S. *islandicus* Y.N. 15.51 (j). Nc, number of codons used; GC3s, frequency of codons ending in G + C excluding Met, Trp, and stop codons. The curved lines represent expected codon usage patterns based on random usage. The dark squares represent the Nc/GC3s ratio for each gene. Codon usage bias is demonstrated by the very large proportion of squares that fall below the expected curve. (From Nayak, K., *Gene*, 512, 163–173, 2013.)

16.3 TRANSGENIC APPROACHES IN EXTREMOPHILES: AN EXAMPLE USING THE HYPERTHERMOPHILE *PIROCOCCUS FURIOSUS*

In a very elegant series of experiments, Keller et al. (2013) applied recombinant DNA approaches to express molecules in the hyperthermophilic archeon *Pirococcus furiosus*, under its ideal temperature conditions. *P. furiosus* is an obligate heterotroph that grows optimally at 100°C. It can ferment sugars to produce hydrogen, carbon dioxide, and acetate, but cannot directly process carbon dioxide. In this study, five genes of the carbon fixation cycle of another archaeon, *Metallosphaera sedula*, which grows autotrophically at 73°C, were expressed heterologously in *P. furiosus*. The five genes encode three separate enzymes and were delivered to *P. furiosus* as a synthetic operon under the control of P_{slp}, a *P. furiosus* S-layer promoter.

The engineered *P. furiosus* followed a new pathway, using hydrogen as a reducing agent of sugars to produce acetyl-CoA and 3-hydroxypropionic acid, a mass-produced, globally important bulk chemical. There are two very important outcomes of this study. First, heterologous expression of genes from one thermophile in another permits the use of carbon dioxide as a substrate to produce valuable products for chemical synthesis. Second, by inserting genetic material from a relative thermophile, which grows optimally at 73°C, the synthetic processes of *Pirococcus* can be driven at a temperature that is 30° below its optimal temperature. Survival of extremophiles at conditions that represent maximal environmental stress occurs at great metabolic cost, with considerable cellular machinery involved. By reducing levels of ambient stress, there is a theoretical advantage to cells, and pathways related to the production of by-products or heterologous proteins may be increased, with many benefits for industrial processes. Thus, overall metabolic requirements for *Pirococcus* are reduced at lower temperatures and cellular expenditure on metabolic processes will be reduced, allowing for optimal production of heterologous compounds (Keller et al. 2013) (Figures 16.3 and 16.4).

Recombinant DNA approaches that involve extremophile organisms offer tremendous potential for various biotechnology industries. Heterologous expression of proteins, nutraceuticals, biomass, or other bioproducts at extreme temperatures, pH, or salt concentrations that preclude the growth of competing or contaminating organisms can greatly facilitate industrial processes in bioreactors. As demonstrated in the Keller study, extremophiles expend cellular resources to survive under harsh conditions. By reducing environmental stressors—for example, lowering ambient temperature for engineered *Pirococcus*—larger components of cellular energy can be applied to the production of recombinant molecules. Indeed, reducing environmental stress through artificial culture conditions may greatly enhance the production of useful bioproducts by extremophiles. The following section describes a new thermophilic green alga isolated from a hot spring in New Mexico, which offers considerable potential for biotechnological processes.

16.4 *SCENEDESMUS* SPECIES *NOVO*: A THERMOPHILE WITH UNUSUAL POTENTIAL FOR BIOTECHNOLOGY

In 2015, Durvasula et al. described a novel green alga of the genus *Scenedesmus* with several physiologic attributes that are attractive for biotechnology. The setting for this discovery was the village of Jemez Springs, New Mexico, a region with numerous hot springs that originate from a geothermal reservoir beneath the Valles Caldera. Water flow at Soda Dam hot springs, the site where the novel *Scenedesmus* was isolated, is 19 L s^{-1} with a temperature of 50°C. The chemical composition of the water is characterized by sodium and chloride concentrations of 990 and 1500 mg L^{-1}, respectively. The pH is 7.1 with a bicarbonate concentration of 1578 mg L^{-1}. Pure cultures of the *Scenedesmus* were established on agar slants, enriched with f/10, f/50, or TAP (Albert et al. 2002). Following PCR amplification of total cellular DNA, the gene encoding 18S rDNA of the unknown alga was sequenced using primers fwd_seq 5′-ccagtagtcatatgcttgtctcaaa-3′ and rev_seq 5′-taccttgttacgacttctcc-3′. A 1.6 kb contig was generated and compared with other known sequences in the GenBank nucleotide database. BLAST analysis of the unknown Jemez alga showed that it was more distant from *Scenedesmus abundans* and *Scenedesmus communis* but most closely related to G24. The organism was tentatively named *Scenedesmus* species *novo* (Durvasula et al. 2015) (Figure 16.5 and Table 16.1).

Several experiments were conducted with the new *Scenedesmus* to determine its suitability for commercial applications. Given the poor nutrient content of the Soda Dam hot water springs, the

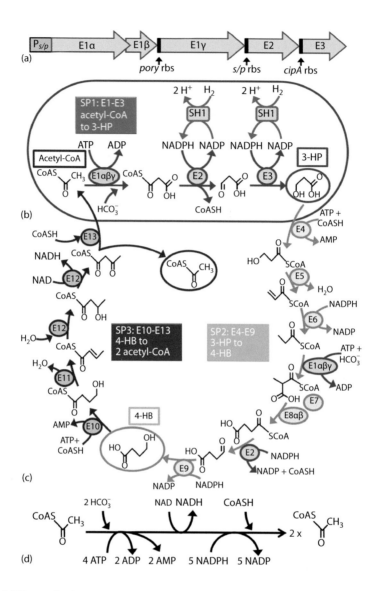

Figure 16.3 (a) The synthetic operon constructed to express the *M. sedula* genes encoding E1 ($\alpha\beta\gamma$), E2, and E3 in *P. furiosus* under the control of Pslp. This includes *P. furiosus* RBSs from highly expressed genes encoding the pyruvate ferredoxin oxidoreductase subunit γ (porγ, PF0971), the S-layer protein (slp, PF1399), and the cold-induced protein A (cipA, PF0190). (b) The first three enzymes of the *M. sedula* 3-HP/4-HB cycle produce the key intermediate 3-HP. E1 is acetyl/propionyl-CoA carboxylase ($\alpha\beta\gamma$, encoded by Msed_0147, Msed_0148, and Msed_1375), E2 is malonyl/succinyl-CoA reductase (Msed_0709), and E3 is malonate semialdehyde reductase (Msed_1993). NADPH is generated by *P. furiosus* soluble hydrogenase 1 (SH1), which reduces NADP with hydrogen gas. (c) The first three enzymes (E1–E3) are shown in the context of the complete 3-HP/4-HB cycle for carbon dioxide fixation by *M. sedula*, showing the three subpathways SP1 (blue), SP2 (green), and SP3 (red). (d) The horizontal scheme shows the amount of energy (ATP), reductant (NADPH), oxidant (NAD), and coenzyme A (CoASH) required to generate 1 mol acetyl-CoA from 2 mol carbon dioxide. (From Keller, M. et al., *Proc. Natl. Acad. Sci. U.S.A.*, 110 (15), 5840–5845, 2013.)

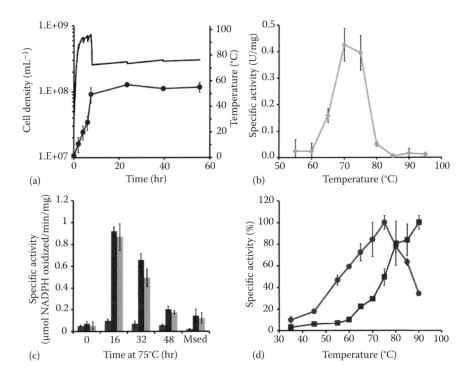

(a)

(b)

(c)

(d)

Figure 16.4 Temperature-dependent production of the SP1 pathway enzymes in *P. furiosus* strain PF506. (a) Growth of triplicate cultures at 98°C (red circles) and temperature (black line) for the temperature shift from 98°C to 75°C are shown. (b) Specific activity (μmol NADPH oxidized min^{-1} mg^{-1}) of the coupled activity of E2 + E3 in cell-free extracts from cultures grown at 95°C to a high cell density of 1 Å to 108 cells mL^{-1} and then incubated for 18 hours at the indicated temperature. (c) Activities of E1, E2 + E3, and E1 + E2 + E3 after the temperature shift to 75°C for the indicated period. The activities of a cell-free extract of autotrophically grown *M. sedula* cells are also shown (labeled Msed) (Keller et al. 2013). The specific activities are an E1 + E2 + E3–coupled assay with acetyl-CoA and bicarbonate (blue), an E2 + E3–coupled assay with malonyl-CoA (red), and E2 with succinyl-CoA (green) as substrates. (d) Temperature dependence of the coupled activity of E2 + E3 (blue circles) in the cell-free extracts after induction at 72°C for 16 hours. The activity of *P. furiosus* glutamate dehydrogenase in the same cell-free extracts is also shown (red squares). (From Keller, M. et al., *Proc. Natl. Acad. Sci. U.S.A.*, 110 (15), 5840–5845, 2013.)

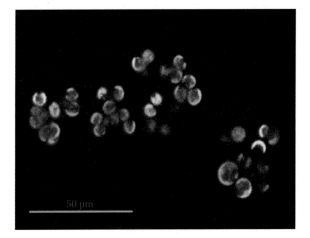

Figure 16.5 Confocal micrograph of *Scenedesmus* species *novo*. (From Durvasula, R. et al., *Algal Res.*, 10, 128–133, 2015.)

Table 16.1 Growth and Production of Chlorophyll a and Carotenoids by *Scenedesmus* Species *novo* under Different Growth Conditions

Growth	Medium	10^6 Cells mL^{-1}	µg Chlorophyll a mL^{-1}	µg Carotenoids mL^{-1}	pg Chlorophyll a 10^6 cells	pg Carotenoids 10^6 cells	Division Day^{-1}
Indoor	TAP	27.42 (7)	49.11 (7)	24.93 (7)	1.79	0.91	0.54
	BG11	0.96 (7)	2.61 (7)	1.30 (7)	2.71	1.35	0.23
	ST	10.21 (9)	14.9 (2)	7.43 (2)	2.14 (2)	1.07(2)	0.24
	NST	5.39 (11)	7.94 (4)	3.65 (4)	1.62 (9)	0.97 (7)	0.12
	WWS	10.18 (4)	12.08 (2)	7.64 (17)	3.11 (2)	1.58 (2)	0.06
	WWNS	4.33 (17)	10.87 (11)	5.55 (17)	2.78 (11)	1.44 (9)	0.14
Outdoor	ST	10.41 (4)	6.92 (4)	3.94 (4)	1.12 (2)	0.55 (2)	0.24
	NST	5.65 (14)	5.93 (4)	3.65 (4)	1.23 (7)	0.76 (7)	0.10
	WWS	8.81 (14)	5.82 (14)	4.49 (14)	0.82 (2)	0.56 (11)	0.19
	WWNS	5.08 (14)	5.41 (4)	3.01 (14)	1.32 (4)	0.73 (4)	0.14

Source: Durvasula, R. et al., *Algal Res.*, 10, 128–133, 2015.
Note: TAP, BG 11, growth media; ST, sterilized water with TAP enrichment; NST, nonsterilized water with TAP enrichment; WWS, sterilized municipal wastewater; WWNS, nonsterilized municipal wastewater. This organism grew at satisfactory rates under harsh outdoor conditions in municipal wastewater. However, growth rates and carotenoid production were significantly increased under controlled indoor conditions.

organism was grown in municipal wastewater from the city of Albuquerque, with and without sterilization or TAP nutrient supplementation. To test growth under harsh environmental conditions, closed cultures of the *Scenedesmus* were maintained outdoors for 14 days during the peak of summer, with ambient temperatures of 40°C and solar radiation of 6524–7360 µmol m^{-2} s^{-1}. Indoor cultures were maintained for 14 days at a temperature of 24°C under continuous artificial fluorescent light at 132–148 µmol m^{-2} s^{-1}. For each experimental condition, growth rates and the production of chlorophyll a, total carotenoids, and lipids were measured.

The *Scenedesmus* yielded very high levels of lipids, which were among the highest ever reported. Lipids (µg mL^{-1}) from indoor cultures that were grown in TAP were 21.29 on day 2, increased to 213.69 by day 5, and peaked at 242 on day 7. This was followed by a decrement to 222.8 on day 9. When cultured in ST, lipids (µg mL^{-1}) were low (5.01) on day 2, increased to 28.54 by day 5, and gradually decreased to 17.68 by day 9; clearly, this was a reduced output. On a dry cell weight basis (pg lipids per pg cell dry weight), lipids varied between 0.57 and 0.85 in cells raised in TAP and from 0.15 to 0.33 in ST medium. The value of 0.85 pg of lipid per pg of dry cell weight is one of the best yields reported in the literature and suggests that under controlled indoor conditions with TAP medium, *Scenedesmus* species *novo* can accumulate very large amounts of lipid. This is significantly greater than the lipid accumulations that result under harsh outdoor conditions, suggesting again that extremophiles, when grown in more favorable environments, may yield very high levels of commercially valuable products.

The ability of *Scenedesmus* species *novo* to grow in nutrient-poor municipal wastewater and under harsh conditions of heat and intense solar radiation characteristic of New Mexico summers, while producing reasonable biomass, lipids, and carotenoids, suggests its suitability for commercial culture. However, the idealized growth conditions offered indoors, with reduced ambient heat and continuous artificial light, resulted in significantly increased cell mass, carotenoids, and lipids. Indeed, cellular lipid production under tightly controlled indoor settings resulted in yields of 242 µg mL^{-1} by day 7, one of the highest values ever reported. These results suggest that extremophiles may offer exceptional opportunities for commercial bioproducts. Furthermore, as noted by Keller et al. (2013) with their studies of *P. furiosus*, the transfer of extremophiles to less harsh growth conditions may result in optimized cellular functioning and production. The shift of cellular machinery from gene transcription directed at survival to processes that are of commercial interest, namely, lipid accumulation, biomass production, and carotenoid synthesis, merits additional studies in organisms such as *Scenedesmus* species *novo*. An optimized culture strategy that permits robust growth with the production of valuable by-products would propel algal biotechnology application with these extremophiles.

16.5 *EUGLENA MUTABILIS*: AN ACIDOPHILE THAT DISPLAYS CIRCADIAN RHYTHMS OF GENETIC ADAPTATION TO SEVERE UV STRESS

Euglena mutabilis is one of the most abundant extreme acidophilic protists, inhabiting acidic volcanic lakes, geothermal water collections, and drainage regions of mines that are often very acidic. It is a photosynthetic organism and plays a key role in nutrient cycling and primary production in harsh environments characterized by high concentrations of heavy metals and depleted nitrogen.

In a recent publication, Puente-Sanchez et al. (2016) applied a metatranscriptomic approach to characterize responses of *E. mutabilis* biofilms found in the hyperacidic environment of Rio Tinto, Spain. This 92 km river in the southwest part of Spain has an unusually acidic composition with very high concentrations of heavy metals. *E. mutabilis* is a predominant organism in this harsh environment and accounts for a significant part of primary production. The regulation of genes associated with photosynthetic pathways and adaptations to environmental stressors, including ultraviolet (UV) radiation, could provide valuable insights for applications in biotechnology.

Metatranscriptomic libraries were constructed from biofilm-associated *E. mutabilis* collected at noon, 8 p.m., and 5 a.m. of the following day from a region exposed to full UV solar radiation. During a typical daily cycle, maximum air photosynthetic active radiation (PaR) irradiance occurred between 2 p.m. and 3 p.m., reaching 2000 μmol m^{-2} s^{-1}. Although the atmospheric temperature ranged from 17°C in the morning to 41°C at midday, the water temperature stayed fairly constant at 20°C.

After analysis, 995 transcripts, from a total of 2884 differentially expressed transcripts, demonstrated a fourfold increase in regulation. Transcripts related to (1) processes of photosynthesis, such as light-harvesting complexes and photosystem II (PSII) stabilization; (2) responses to biotic and abiotic stresses, such as temperature response and protein structure; and (3) responses to solar radiation stress, such as nonphotochemical quenching and blue-light photoreceptors, were preferentially overexpressed at noon. Transcripts that involved DNA replication, DNA polymerase activity, and transcription factors were preferentially overexpressed at 8 p.m. Transcripts involved in protein translation and biosynthesis, RNA binding, translation, elongation, and cell division were preferentially upregulated in the NIGHT library. The results of this metagenomic study are summarized in Figure 16.6.

The study of *Euglena* biofilms from an environment characterized by hyperacidity and severe midday UV stress reveals a very significant adaptation in an extremophile: the ability to regulate cellular processes in a circadian fashion. A synchronized approach to gene transcription has evolved in response to unusually high levels of UV radiation during the early afternoon period. Such radiation can be damaging to many photosynthetic organisms, and indeed, levels of photosynthesis were adversely affected during the 12 p.m. to 3 p.m. period. However, the stress response of *E. mutabilis* involved a carefully orchestrated upregulation of transcription processes involved in photosynthesis, temperature response, protein structural responses, and responses to the stresses of solar radiation. Cellular function switched to processes involved in DNA replication and packaging by the evening, when radiation stress was greatly reduced and cellular replication, presumably, would increase. It comes as no surprise, then, that nighttime cellular functions were largely devoted to cellular elongation, division, and protein synthesis.

A circadian coordination of transcription suggests highly regulated and finely tuned cellular machinery. Ultimately, a greater understanding of the regulatory elements within the genome of this

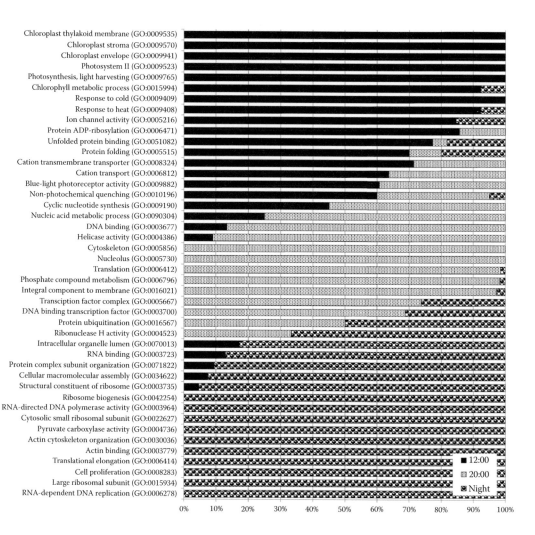

Figure 16.6 Differential GO term distribution. Relative proportions of gene ontologies for the three metatranscriptomic libraries. The graph represents 100% of the GO annotations for each library. The x-axis contains the proportion of genes in each category; the y-axis shows the GO categories. GO IDs are presented in parentheses. (From Puente-Sanchez, F. et al., *Protist*, 167, 67–81, 2016.)

organism and many other extremophiles will yield even more valuable insights into the biology of these remarkable organisms and opportunities for biotechnology.

In this chapter, five distinct extremophiles have been reviewed to reveal three very important themes. First, the genomic organization and regulation of these organisms is precise and tightly linked to survival under the harshest of environmental conditions. Colocalization of important genes, codon usage bias, differential expression of stress response genes, and important diurnal processes all point to evolved mechanisms for the tolerance of extreme environments. Second, the ability to transfer some of these organisms to more favorable conditions, such as reduced temperatures or attenuated solar radiation, may offer tremendous advantages to biotechnology processes by allowing cells to devote fewer resources to stress response, with enhanced production of commercially valuable by-products. Finally, prospects for genetic manipulation of extremophiles and transfer of traits and regulatory elements among organisms could yield huge dividends for a biotechnology industry.

REFERENCES

Alberts B, Johnson A, Lewis J et al. 2002. Molecular Biology of the Cell. 4th edition. New York: Garland Science; Isolating, Cloning, and Sequencing DNA. Available from: https://www.ncbi.nlm.nih.gov/books/NBK26837/

Durvasula R, Hurwitz I, Fieck A, and Subba Rao DV. 2015. Culture, growth, pigments and lipid content of *Scenedesmus* species, an extremophile microalga from Soda Dam, New Mexico in waste water. *Algal Research* 10: 128–133.

Foflonker F, Ananyev G, Qiu H, Morrison A, Palenik B, Dismukes G, and Bhattacharya D. 2016. The unexpected extremophile: Tolerance to fluctuating salinity in the green alga *Picochlorum*. *Algal Research* 16: 465–472.

Keller M, Schut G, Lipscomb G, Menon A, Iwuchukwu I, Leuko T, Thorgersen M, Nixon W, Hawkins A, Kelly R, and Adams M. 2013. Exploiting microbial hyperthermophilicity to produce an industrial chemical, using hydrogen and carbon dioxide. *Proceedings of the National Academy of Sciences of the United States of America* 110 (15): 5840–5845.

Nayak K. 2013. Comparative genome sequence analysis of *Solfolobus acidocaldarius* and 9 other isolates of its genus for factors influencing codon and amino acid usage. *Gene* 512: 163–173.

Puente-Sanchez F, Olsson S, Gomez-Rodriguez M, Souza-Egipsy V, Altamirano-Jeschke M, Amils R, Parro V, and Aguilera A. 2016. Solar radiation stress in natural acidophilic biofilms of *Euglena mutabilis* revealed by metatranscriptomics and PAM fluorometry. *Protist* 167: 67–81.

Wright F. 1990. The 'effective number of codons' used in a gene. *Gene*. 87 (1): 23–29.

Appendix: Culturing Extremophiles: Progress to Date

D. V. Subba Rao, Ravi Durvasula, and Adinarayana Kunamneni

A.1 INTRODUCTION

The chapters presented earlier in this volume discussed microbiota living at extremes of temperature, salinity, pressure, and chemical environments, and the production of exozymes and several bioactive compounds. For bioprospecting active compounds with potential commercial utility, a steady supply of biomass, preferably through culturing, would be necessary. The resiliency of the extremophiles in the face of harsh habitats is phenomenal, allowing them to have survived and reproduced in much the same form for nearly 4 billion years. The famous quotation "Everything is everywhere: but, the environment selects" (Baas Becking 1934) is valid in the case of halophiles, which are distributed worldwide. This also suggests their amenability for culturing. Most extremophiles require specific growth conditions that are either unknown or difficult to achieve in a laboratory. However, the key for successful culturing is simulation of environmental conditions suitable for an extremophile.

A great diversity in the occurrence, structure, and functioning of extremophile groups exists. For example, Wilkins et al. (2013) described the occurrence of groups of cyanobacteria, green sulfur bacteria, green nonsulfur bacteria, actinobacteria, cytophaga (CFNB group), methanogenic Archaea, Haloarchaea, diatoms, phytoflagellates and flagellates, and ciliates from the Antarctic lakes and mats. Depending on the group, their key functions vary and include photosynthesis, anaerobic anoxygenic phototrophy, anoxygenic photoptrophy and heterotrophy, aerobic anoxygenic phototrophy, dissimilation of sulfur or sulfate reduction, aerobic heterotrophy, and hydrogenotrophic, acetoclastic, and methylotrophic methanogenesis. The occurrence of 4000 species in the subglacial Whillans Lake, 800 m beneath the Antarctic ice sheet (Christner et al. 2014; Fox 2014), suggests the existence of several psychrophilic extremophile species. These microbial ecosystems can play an important role in the Southern Ocean geochemical and biological system (Christner et al. 2014). In the Cuatro Cienegas basin in the Chihuahuan desert, Mexico, a system of springs, streams, and pools exists. These ecosystems support greater than 70 endemic species and abundant living stromatolites and other microbial communities, representing a desert oasis of high biodiversity. There were 250 different phylotypes among the 350 cultivated strains (Souza et al. 2006).

A conservative estimate of the number of marine microalgae is 72,500 species (Guiry 2012). Although culturing marine microalgae dates back to 1893 (Miquel 1893), to date, only a few hundred photosynthetic strains have been cultured, usually in medium F and its derivatives (Guillard and Ryther 1962), but only about 30 have been studied in detail. For a long time, it was presumed that life would not exist under extreme physical and chemical barriers, but only in recent years have studies begun on extremophiles.

Because the extremophiles are taxonomically and physiologically diverse, there is no single culture medium comparable to medium F, used with silica for diatoms and without silica for other microalgae. Consequently, no concerted effort was made to culture extremophiles. A few culture studies showed that as a survival mechanism, the halophyte *Dunaliella* accumulates β-carotene as an osmoregulant against high radiation; this organism has been used in mass cultivation for the production of astaxanthin (Fan et al. 1998). Culture studies of extremophiles lagged behind mainly because of a lack of data on their special requirements, which facilitate their survival and growth under extreme conditions.

For successful establishment of cultures, it would be necessary to know modes of nutrition, that is, autotrophic, heterotrophic, or mixotrophic. Equally important is the nature of stressors experienced by extremophiles, because the composition and induction of bioactive compounds are related to stress conditions, as in *Lyngbya* sp. (Rosly et al. 2013). The prokaryotic actinomycetes, which are economically and biotechnologically valuable, form a stable persistent population; although they are metabolically active in their natural environment of marine sediments, they are hard to cultivate (Lam 2006). Several marine extremophiles known to produce bioactive compounds remain uncultivable (Giddings and Newman 2015).

We present methodology utilized for culturing extremophiles. This is not a complete review of culturing extremophiles, but brings together the much diffused and scattered publications on culturing a few selected extremophiles.

A.2 CULTURE COLLECTION CENTERS

The Culture Collection of Cryophilic Algae (CCCryo), Germany, maintains a comprehensive collection of autotrophic cryophilic microalgae. Their website, http://cccryo.fraunhofer.de/web/infos /scope, maintained by Dr. Thomas Leya, provides a wealth of information. This collection maintains 459 strains of 143 species from 81 genera: Ascomycota (2 genera), Klebsomidiophyceae (2), Eubacteria (4), Cyanophycea (10), Dinophyceae (1), Bangiophyceae (2), Euglenophyceae (2), Bacillariophyceae (6), Xanthophyceae (6), Trebouxiophyceae (124), Chlorophyceae (240), Bangiophyceae (2), Glaucophycreae (1), Ascomycota (2), Eubacteria (4), Zygnematophyceae (33), Bryopsidae (mosses) (7), and Ulvophyceae (11). These algae constitute 31% psychrophiles, 21% mesotrophs, 10% cryotrophs, 1% thermophiles, and 37% of undetermined category. Isolates are found in Antarctic, Arctic, and tropical waters and are suitably maintained at temperatures in the range of 2°C–32°C in 16:8 light to dark phases of light. Several media are used, designated as 3N BBM pH 5.5 Ag, Desmid SVCK pH 6.0 Lq., WEES pH 5.5, TAP pH 6.0, BG11 pH 8.5, FW diatom, C medium, Galdieria pH 1.8 and MiEg 1:1 pH 6.7. Most of these psychrophilic algae are clonal, and axenic and details on their 18S rRNA gene sequences are provided.

Leibniz-Institut DSMZ-Deutsche Sammlung von Mikroorganismen und Zellkulturen DSMZ (https://www.dsmz.de), founded in 1969, is one of the most comprehensive bioresource culture collection centers and has approximately 31,000 cultures representing 2,000 genera and 10,000 species grown in 1647 media. These microorganisms include Archaea and bacteria, some of which are hyperthermophiles and extremophiles.

The National Center for Marine Algae and Microbiota Bigelow Laboratory for Ocean Sciences (https:// ncma.bigelow.org) maintains 2647 strains in culture, of which 233 are listed as cold water species, with 118 isolated from the Arctic and the Antarctic, and maintained at –2°C. The World Federation for Culture Collections (WDCM) serves as a networking vehicle for the resource center (http://www.wfcc.info/index .php/collections/display/) for various microbes. There are 589 culture collections in 68 countries registered in the WDCM; most are bacteria, filamentous fungi, and yeasts, and some extremophiles may be included.

A wealth of information is given in Rainey and Oren's *Methods in Microbiology*. For collection, handling, culturing, preparation of media, and equipment, readers are referred to the excellent chapters by Burns and Dyall-Smith (2006), Nakagawa and Takai (2006), Godfroy et al. (2006), Park et al. (2006), Russell and Cowan (2006), Grant (2006), Mesbah and Wiegel (2006), and González-Toril et al. (2006).

A.3 AUTOTROPHS

For autotrophs, conventional microalgal culture techniques, that is, serial dilution and repetitive plating on semisolid agar slants or liquid medium, are suitable (Andersen 2005; Subba Rao 2009;

Table A.1 Comparison of Ranges of Nutrient Concentrations in Marine and Freshwater Media Used for Culturing Autotrophs

		Marine (24 media)	Freshwater (26 media)
Macronutrients	PO_4	0.5 μM–10 mM	0.73 μM–2 mM
	N_2	1 μm–9.9 mM	0.1 μm–17.6 μM
	Si	10 μM–0.7 mM	12.5 μM–30 mM
Trace metals	Fe	1 μM–1.8 mM	0.72 μM–17.9 μM
	Cu	0.24 nM–0.063 mM	0.4 nM–6.29 μM
	Co	0.063 nM–0.04 mM	8.1 nM–0.68 μM
	Zn	0.3 nM–3.48 mM	0.8 nM–30.7 μM
	Mn	0.21 nM–1.4 mM	0.18 μM–7.28 μM
Vitamins	B_{12}	3.69 pM–7.4 nM	0.738 pM–1.84 nM
	Thiamine	0.3 nM–3 mM	73.8 pM–0.148 μM
	Biotin	3.27 nM–0.2 μM	0.41 nM–10.2 nM

Durvasula et al. 2013). A variety of media are available (28 for *Dunaliella* [Subba Rao 2009] and 55 for microalgae [Andersen 2005]). It is always best to enrich and use the same water from which isolations are made. The media are based on artificial seawater or natural seawater enriched with known quantities of nutrients and a soil extract of unknown composition. For culturing halophiles, salinity is adjusted by the addition of sodium chloride or sea salt. In both marine and freshwater culture media (Table A.1), a wide range of macronutrients, trace metals, and vitamins are used. Whether vitamins are required or not, B_{12}, thiamine, and biotin are used routinely in *Dunaliella* cultures. If algae can grow in media without vitamins, the cost of the medium will be reduced.

The halophile diatom *Nitzschia frustula* (plate) and the green alga *Chlamydomonas plethora* isolated from the semiarid hypersaline waters off Kuwait were cultured in F medium. They yielded high division rates and carbon assimilation ratios approaching their theoretical maxima, and contained high levels of leucine, lysine, glutamic acid, and arginine (Subba Rao et al. 2005). High photosynthetic efficiencies were reported in sea ice microalgae (Kottmeir and Sullivan 1988). Photosynthetically, the microbial cyanobacterial mats isolated from the polar waters of King George Island, Antarctica, were efficient at high temperatures (ca. 20°C). While not an adaptation to low temperatures in these cyanobacterial mats, protection against ultraviolet (UV) photoinhibition may be more important (Tang et al. 1997a). Tang et al. (1997b) isolated strains of E18 and *Phormidium* sp., a chlorococcalean assemblage denoted as Vc that grows well in outdoor wastewater systems; the former grows well at low temperature, and the latter prefers temperature greater than 15°C, up to 25°C. He suggested using them suitably with the seasonal temperatures for remedial purposes. The freshwater thermophilic green alga *Scenedesmus* species Novo isolated from Soda Dam, Jemez Springs, New Mexico, grows well in wastewater enriched with TAP and BG11, and yielded high levels of lipids (63–94.3 pg cell^{-1} and 0.95–3.58 pg carotenoids cell^{-1}) (Durvasula et al. 2015) under the harsh ambient temperature of 40°C and light (6524–7360 light μmol photons m^{-2} s^{-1}).

A.4 HALOPHILES

Twenty-two strains of *Dunaliella* species, mostly from the salterns on the east coast of India, were isolated in Walne's medium (Keerthi et al. 2015). These strains yielded a maximum of 44.85 pg lipid cell^{-1} and 74.7 pg β-carotene cell^{-1} (Keerthi et al. 2015). Walne's media with 2.2, 2.6, 3.0, 3.9, 4.8, and 5.6 M NaCl were taken as the base media for improvisation. The range of phosphate was 0.035 g L^{-1}. $NaNO_3$, KNO_3, urea, and NH_4Cl served as the nitrogen source at concentrations of 50, 100, 150, and 200 mg L^{-1}.

To study the mechanisms underlying the stress response in *Dunaliella salina*, Cuaresma et al. (2011) used a solid medium that supported the viability of cells up to 63% in SGM medium that has (L^{-1}) 0.25 M NaCl, 1 M glycerol, 10 mM $NaHCO_3$, 5 mM $MgSO_4$, 0.3 mM $CaCl_2$, 0.2 mM KH_2PO_4, 0.2 mM H_3BO_3, 7 µM $MnCl_2$, 6 µM Na_2 EDTA, 1.5 µM $FeCl_3$, 0.8 µM $ZnCl_2$, 20 nM, $CoCl_2$, and 0.2 nM $CuCl_2$.

The Dead Sea has high levels of magnesium, and accordingly, Oren (1983) enriched the medium while culturing *Halorubrum sodomense*. For extremophile marine cyanobacteria and microbiota, acclimation times in the laboratory environment are important in bringing new cultures into cultivation. Usage of agar plates has resulted in obtaining unialgal cultures and axenic clones (Gerwick et al. 1991) screened for anticancer-type bioactive compounds.

The thermoacidophilic red alga *Galdieria sulphuraria* (plate) isolated from the hot sulfur springs at 0–4 pH and temperatures up to 56°C can survive autotrophically and heterotrophically as well (Weber et al. 2004).

A.5 ACIDOPHILES

The acidophile *Chlamydomonas acidophila* is a green microalga from the acidic Tinto River in Huelva, Spain. It grows in the river between 1.7 and 3.1 pH and was brought into an axenic culture on basal agar medium at pH 2.5 (Cuaresma et al. 2011). The agar medium contained 300 mL 7% agar and 700 mL modified K9 medium (Silverman and Lundgren 1959). Cultures were grown at 200 µE m^{-2} s^{-1} at 25°C. Glucose, glycerol, starch, glycine, and urea were used as carbon sources (Cuaresma et al. 2011).

Using urea ($CO(NH_2)_2$), a small-molecular-weight polar and relatively lipid-insoluble compound, Casal et al. (2010) cultivated the acidophile *Coccomyxa acidophila* from the Tinto River mining area in Huela, Spain, and enhanced the productivity of lutein. Their mixotrophic cultures accumulated up to 3.55 mg g^{-1}. Nutrients seem to affect the growth and cellular stored products of extremophiles as in other microbiota. Stringent phosphate limitation in the extremophile *C. acidophila* (pH 2–3.5) resulted in higher total fatty acid levels and a lower percentage of polyunsaturated fatty acids (Spijkerman and Wacker 2011). *C. acidophila* cultures grown autotrophically on urea as a carbon source yielded high biomass levels (approximately 20 g dry biomass m^2 d^{-1}) compared with approximately 14 g dry biomass m^2 d^{-1} grown mixotrophically utilizing glucose as a carbon source (Cuaresma et al. 2011). However, mixotrophic growth of *C. acidophila* on glucose resulted in better accumulation of carotene and lutein (10 g kg^{-1} dry weight), the highest recorded for a microalga (Cuaresma et al. 2011).

A.6 THERMOPHILES

Cultivation of hyperthermophiles is defined by a temperature optimum for growth of around or above 80°C. Most representatives of this group belong to the Archaea, whereas only a few are found among the bacteria. Meeks and Castenholz (1971) cultured a thermophile, *Synechococcus lividis*, isolated from the drain ways at Hunter's Hot Springs, Oregon, in medium D at 70°C. This thermophile grows at temperatures from 54°C to 72°C, and the optimal growth was between 63°C and 67°C. Growth of this thermophile seems to determine both the upper and lower limits of growth. The generation time was 11 hours, but it was suppressed in the supraoptimal range from 68°C to 72°C.

A.7 MIXOTROPHS

In mixotrophic cultures, both CO_2 and organic carbon are simultaneously assimilated, and both respiration and photosynthesis are carried out simultaneously. Kaplan et al. (1986), Lee (2004), and Bumbak et al. (2011), utilizing heterotrophic cultures of *Chlorella*, *Crypthecodinium*, and *Galdieria*, respectively, obtained cell densities of more than 100 g L^{-1} cell dry weight and discussed the heterotrophic high-cell-density cultivation.

D. salina could be grown mixotrophically in De Walne's culture medium enriched with 2–10 g L^{-1} for sodium acetate, 1–10 g L^{-1} for malt extract, 1–5% for glycerol, and 1–4 g L^{-1} yeast as organic carbon sources. For organic nitrogen sources (50, 100, 150, and 200 mg L^{-1}), yeast extract and peptone at different concentrations were tested (Keerthi et al. 2015). For bioactive compounds, mixotrophic cultures were upscaled gradually from 10–250 mL to 3, 5, and finally 20 L. *D. salina* seems to grow in modified AS100 (MAS 100) medium and TAP medium containing 0.1 M NaCl (Anila et al. 2016).

D. salina was grown on a solid medium (Yang et al. 2000) modified by the addition of 0.1% phytagel (Sigma) as a gelling agent The basic medium developed by Chitlaru and Pick (1989) consisted of 0.25 M NaCl, 1 M glycerol, 10 mM NaHCO$_3$, 5 mM MgSO$_4$, 0.3 mM CaCl$_2$, 0.2 mM KH$_2$PO$_4$, 0.2 mM H$_3$BO$_3$, 7 μM MnCl$_2$, 6 μM Na$_2$ EDTA, 1.5 μM FeCl$_3$, 0.8 μM ZnCl$_2$, 20 nM CoCl$_2$, and 0.2 nM CuCl$_2$.

A.8 HETEROTROPHIC CULTURES

Heterotrophic haloarchaeal bacteria are found in salterns and survive at 42°C–58°C and are categorized as thermohalophiles. Examples are *Halobacterium salinarum*, *Haloarcula hispanica*, and *Haloferax volcanii*. They cause red discoloration due to their carotenoid pigments. Heterotrophic cultures have several major limitations: (1) a limited number of microalgal species that can grow heterotrophically, (2) increased energy expenses and costs by adding an organic substrate, (3) contamination and competition with other microorganism, (4) inhibition of growth by excess organic substrate, and (5) inability to produce light-induced metabolites (Chen 1996).

For heterotrophic cultures, autotrophic culture media enriched with an organic carbon source are used (Tsavalos and Day 1994). Cultures are established using thoroughly cleaned glassware. The media is usually based on artificial seawater enriched with nitrogen [peptone (Oxoid) and yeast extract] and carbon (lactate, succinate, glucose, glycerol, and galactose) sources, several salts, trace elements (see Table A.3), and vitamins. Serial 10-fold dilutions with subsequent isolation of single colonies are done; samples are plated on agar plates (see Dyall-Smith 2009).

There are 150 species of halophilic bacteria belonging to 28 families and 70 genera; a good number of haloalkaliphilic actinobacteria have the potential for antimicrobial and biomolecules that are pharmaceutically active (see Maheswari and Saraf 2015). These biomolecules, besides serving as biopolymers and biofertilizers, are also used in the bioremediation of effluents from tanneries, textiles, and petrochemical industries, and thus have great biotechnological potential. Delgado-García et al. (2015), while discussing biotechnological applications of halophilic microorganisms, presented details on the isolation and screening of halophilic bacteria. The media include (1) halophilic medium, (2) HALO medium, (3) Marine agar medium, (4) nutritive agar medium modified for NaCl as a salt source, (5) specific culture medium for *Halobacterium trueperi* and *Halobacterium karajensis*, and (6) culture medium for extreme halophilic *Halobacterium salinarum*.

For physiological characterization, strains of the halophilic bacterium *Halomonas elongata*, which co-occurs with *D. salina*, were grown aerobically at 30°C on MM63 medium (Larsen et al. 1987) with glucose as a carbon source and in various NaCl concentrations. Ono et al. (1999) used Luria-Bertani agar medium consisting of 1.0% tryptone, 0.5% yeast extract, and 1.0% NaCl. The final concentration of NaCl was 15% (wt/v). A mineral-glucose medium, M63, consisting of 100 mM KH$_2$PO$_4$, 75 mM KOH, 15 mM (NH$_4$)$_2$SO$_4$, 1 mM MgSO$_4$, 3.9 mM FeSO$_4$, and 22 mM glucose, pH 7.2, was modified by the addition of NaCl at a concentration of 3% (M632-3 medium) or 15% (M63S-15 medium). Under nutrient-limiting conditions, in *D. salina* living under high light and salt stress, higher salinity increased carotenogenesis substantially (Coesel et al. 2008); nutrient availability seems to control carotenogenesis and messenger RNA levels.

From the salterns in Goa, India, Ballav et al. (2015) isolated halophilic Actinomycetales in starch casein, R2A, and inorganic salt starch agar at four different salinities (35, 50, 75, and 100 psu), which produced antibacterial metabolites.

Grammann et al. (2002) grew *Halomonas elongata* strains aerobically at 30°C on MM63 medium with various NaCl concentrations supplemented with glucose as a carbon source. They discussed the existence of two mechanisms that halophilic prokaryotes have developed to lower the potential of cytoplasmic water and the cell cytoplasm achieving an osmotic strength similar to that of the surrounding medium.

A.9 HYPERTHERMOPHILES

Organisms that have an optimum growth above 80°C and grow up to 110°C are grouped as hyperthermophiles. These are Archaea, strictly anaerobic or obligately aerobic, and are chemolithotrophic. Culture conditions depend on the environment from which the strains were established; the temperature, light, and salt content were adjusted. Seawater or distilled water with sodium chloride were used for enrichments.

1. The culture media used are Brock's basal salt medium, MG basal medium, SME medium, MGM medium, mineral salt medium (MSM), 2216-2 medium, TB medium, T4-14 medium, half SME medium, and SME modified medium (Table A.2).
2. The recipes are quite complicated, and a variety of inorganic salts are used (Table A.2).
3. Most used 13–20 amino acids (Table A.2).
4. Other constituents included glucose, soluble starch, yeast extract, piperzine, sulfur, resazurin, agar, glycogen, acetate, pyruvate, dextrin, and maltose, and their concentrations varied (Table A.2).
5. Levels of trace elements and vitamins also varied significantly (Table A.3).

Raven et al. (1992), Park et al. (2006), and Godfroy et al. (2006) described in great detail the sophisticated experimental apparatus used for culturing hyperthermophiles. Two strains, *Pyrobolus fumarii* DSM 11204 (anaerobic) and *Sulfobolus solfataricus* DSM 16161 (aerobic), are studied in cultures (see DSMZ website: http://www.dsmz.de/microorganisms/medium/pdf/DSMZ_Medium792 .pdf). *P. fumarii*, isolated from the hydrothermal vent on the sea floor at the Mid-Atlantic Ridge, and *Sulfobolus*, isolated from a volcanic hot spring in Italy (Zillig et al. 1980), belong to the group Thermococcales and are difficult to grow. *P. fumarii* grows under anaerobic conditions, while *S. solfataricus* grows under aerobic conditions and uses substrates like peptone or yeast extract for organotrophic growth. The new isolate, *P. fumarii*, was able to form colonies on plates (at 102°C) and grow at a pressure of 25,000 kPa (250 bar) (Blochl et al. 1997).

Other hyperthermophilics studied (Lee et al. 2011) include *Thermococcus*, *Pyrococcus*, and *Paleococcus*, which are anaerobic and utilize peptides such as carbon and energy sources. Growth is usually better in the presence of sulfur. They are grown in gas-lift bioreactors. Specialized bioreactors that can operate up to 880 atm and 200°C, and pressurization from 7.8 to 500 atm were used for cultivating the hyperthermophile *Methanocaldococcus jannaschii* (formerly *Methanococcus jannaschii*) (Boonyaratanakornkit et al. 2006).

From Bakreshwqar hot springs, West Bengal, India, esterase-producing thermophilic bacteria were cultured (Ghati et al. 2013). Anaerobically collected water samples were serially diluted and plated on tributyrin agar medium, and plates were incubated for 72 hours at 55°C. Distinct colonies were further subcultured and screened for esterase production.

Table A.2 Compilation on the Enrichment of Nutrients, Trace Elements, and Vitamins Used in the Culture Media for Hyperthermophiles

Strain	Nature	Taxa	Enrichment		Reference
DSM 3638	Hyperthermophilic	*Pyrococcus furiosus*—anaerobic			
			Modified minimal medium	L^{-1}	Raven and Sharp 1997
			Maltose	5.4 g	
			Peptone	5 g	
			Yeast extract	1 g	
			Iron chloride tetrahydrate	10 g	
			L-Alanine	30 mg	
			L-Arginine HCl	50 mg	
			L-Asparagine monohydrate	40 mg	
			L-Aspartic acid	20 mg	
			L-Glutamine	20 mg	
			L-Glutamic acid K+	40 mg	
			Glycine	80 mg	
			L-Histidine, L-isoleucine	40 mg	
			L-Methionine	30 mg	
			L-Phenylalanine	30 mg	
			L-Proline	50 mg	
			L-Serine	30 mg	
			L-Threonine	40 mg	
			L-Tryptophan	30 mg	
			L-Tyrosine	40 mg	
			L-Valine	20 mg	
			L-Cysteine; replaced by sulfide—$9H_2O$ elemental sulfur	0.5 g	
			Vitamins	L^{-1}	
			Pyridoxine hydrochloride	200 mg	
			Nicotinic acids	100 mg	
			Riboflavin	100 mg	
			Thiamine hydrochloride	100 mg	
			Lipoic acid	100 mg	
			Biotin	40 mg	
			Folic acid	40 mg	
			Cyanocobalamine	2 mg	
			DL-Calcium pantothenate	100 mg	
Hot spring	Thermophilic	*Sulfolobus*	**Brock's basal salt medium** (Sigma-Aldrich)		Sakai and Kurosawa 2016

(Continued)

Table A.2 (Continued)   Compilation on the Enrichment of Nutrients, Trace Elements, and Vitamins Used in the Culture Media for Hyperthermophiles

Strain	Nature	Taxa	Enrichment		Reference
			Yeast extract	0.01%, 0.1%, 1%	
Hot springs	Acidophilic/ acidtolerant thermophile	*Thermo anaerobacter—* anaerobic	**MG basal medium**	mg L^{-1}	Prokofeva et al. 2005
			$MgCl_2 \cdot 6H_2O$	330	
		Thermo abnerobacterium	$CaCl_2 \cdot 2H_2O$	330	
			KCl	325–330	
			$MgSO_4 \cdot 7H_2O$	3,450.00	
			$CaCl_2 \cdot 2H_2O$	150.00	
			NH_4Cl	250.00	
			K_2HPO_4	150–330	
			Trace elements solution	1 mL mg 500 mL^{-1}	
			$FeSO_4(NH_4)_2SO_4 \cdot 6H_2O$	392	
			$NiSO_4(NH_4)_2SO_4 \cdot 6H_2O$	197.50	
			$CoCl_2 \cdot 6H_2O$	119.00	
			$ZnSO_4 \cdot 7H_2O$	71.80	
			$MnCl_2 \cdot 4H_2O$	49.50	
			Na_2SeO_4	47.30	
			$Na_2WO_4 \cdot 2H_2O$	16.50	
			$Na_2MoO_4 \cdot 2H_2O$	12.00	
			H_3BO_3	3.00	
			$CuCl_2 \cdot 2H_2O$	0.90	
			Vitamin solution (Wolin et al. 1963)	1 mL mg L^{-1}	
			Pyridoxine hydrochloride	10	
			Nicotinic acids	5.00	
			Riboflavin	5.00	
			Thiamine hydrochloride	5.00	
			Thioctic acid	5.00	
			Biotin	2.00	
			Folic acid	2.00	
			p-Aminobenzoic acid	5.00	
			Pantothenic acid	5.00	
			Yeast extract	g L^{-1}	
			Starch	1.5	
			Glucose	2.0	
			Sucrose	2.0	
			Maltose	1.5	

(Continued)

Table A.2 (Continued) Compilation on the Enrichment of Nutrients, Trace Elements, and Vitamins Used in the Culture Media for Hyperthermophiles

Strain	Nature	Taxa	Enrichment		Reference
			Peptone	1.5	
			Acetate	2.0	
			Elemental sulfur	10.0	
			Thiosulfate or nitrate as sodium salt	2.0	
Deep sea	Thermophile	*Pyrococcus*	**SME medium** (Sharp and Raven 1997)	L^{-1}	Postec et al. 2005
Hydrothermal		*Marinitoga* and *Bacillus*	Seawater + Na$_2$S	0.5 mg	
			Yeast extract	0.5 g	
			Peptone	1.0 g	
			Casamino acids	0.5 g	
			Glucose	0.4 g	
			Dextrin	0.2 g	
			D + galactose	0.2 g	
			Dextran	0.1 g	
			Glycogen	0.2 g	
			Pyruvate	0.2 g	
			Acetate	0.1 g	
			Colloidal sulfur	3.0 g	
Andean Lakes	Radiation	88 bacterial isolates	**MGM**	L^{-1}	Ordonez et al. 2009
			MGM, 12%–25% salt water	400–833 mL	
			Pure water, 12%–25%	567–134 mL	
			Peptone (oxoid)	5 g	
			Yeast extract	1 g	
			Yeast extract	1–2 g	
			NaCl	240 g	
			Agar	12 g	
			Tryptone	1 g	
			MgCl$_2$·6H$_2$O	30 g	
			MgSO$_4$·7H$_2$O	35 g	
			KCl	7 g	
			CaCl$_2$	1–5 mL	
			Tris HCl	1–5 mL	
East Pacific rise	Thermophiles,	Epsilonprotobacteria			
	Anaerobic	Aquificales like phylotypes			
		Thermodesulfo-bacteriales	**Artificial seawater**	mM	Houghton et al. 2007
			NaCl	499	
			KCl	6.70	
			MgCl$_2$	12.40	

(Continued)

Table A.2 (Continued)   Compilation on the Enrichment of Nutrients, Trace Elements, and Vitamins Used in the Culture Media for Hyperthermophiles

Strain	Nature	Taxa	Enrichment		Reference
			$CaCl_2$	2.63	
			NH_4Cl	3.70	
			$MgSO_4$	58.00	
			K_2HPO_4	0.99	
			$NaNO_3$	20.00	
			$MgSO_4$	58.00	
			$Na_2S_2O_3$	12.70	
DSM 3638	Thermophile, anaerobic	*Pyrococcus furiosus*	Artificial seawater	L^{-1}	Brown and Kelly 1989
			Yeast extract	0.00 g	
			Tryptone	0.01 g	
			NaCl	47.8 g	
			Na_2SO_4	8.0 g	
			KCl	1.4 g	
			$NaHCO_3$	0.4 g	
			KBr	0.2 g	
			H_3BO_3	0.06 g	
			$MgCl_2 \cdot 6H_2O$	21.6 g	
			$CaCl_2 \cdot 2H_2O$	3.0 g	
			$SrCl_2 \cdot 6H_2O$	0.05 g	
			NH_4Cl	12.5 g	
			K_2HPO_4	7.0 g	
			CH_3CO_2Na	50.0 g	
			Na_2S	0.5 g	
Deep-sea vent	Thermophile, anaerobi	*Pyrococcus abyssi* ST549	**SME medium** (Sharp 1997)	L^{-1}	Godfroy et al. 2000
			Peptone	1–2 g	
			Maltose or cellobiose	5 g	
			Yeast replaced with brain heart infusion	9 g	
			20 classical amino acids	0.1 g each	
			Sulfur	10 g	
Deep-sea vent	Thermophile	*Thermococcus fumicolans*	**2216-2 medium**	g L^{-1}	Godfroy et al. 1996
			Peptone 1	2.0	
			Yeast extract	0.50	
			Sea salt	30.00	
			Piperazine (PIPES)	6.05	
			Sulfur	10.00	
			Resazurin	0.001	
		SME medium	BHI-S medium		
		2216-2 medium	Brain heart infusion	9.00	

(Continued)

Table A.2 (Continued) Compilation on the Enrichment of Nutrients, Trace Elements, and Vitamins Used in the Culture Media for Hyperthermophiles

Strain	Nature	Taxa	Enrichment		Reference
			NaCl	23.00	
			Piperazine (PIPES)	6.05	
			Sulfur	10.00	
			20 classical amino acids	0.2 mM	
Volcanic pool	Thermophile	*Thermotoga neopolitana*	**TB medium**	L^{-1}	Childers et al. 1992
			NaCl	20 g	
			KCl	0.335 g	
			$MgCl_2 \cdot 2H_2O$	2.75 g	
			$MgSO_4 \cdot 7H_2O$	3.45 g	
			NH_4Cl	0.25 g	
			$CaCl_2 \cdot 2H_2O$	0.14 g	
			K_2HPO_4	0.14 g	
			Glucose	5.0 g	
			Yeast extract	0.5 g	
			Piperazine (PIPES)	6.0 g	
			Cysteine hydrochloride	0.5 g	
			Resazurin	0.001 g	
			Nitriloacetic acid	1.5 g	
			$MnCl_2 \cdot 4H_2O$	0.56 g	
			$(NH_4)_6Mo_7O_{24} \cdot 4H_2O$	0.37 g	
			$CoCl \cdot 6H_2O$	0.34 g	
			$ZnSO_4 \cdot 7H_2O$	0.32 g	
			$NiSO_4 \cdot 6H_2O$	0.21 g	
			$NaSeO_4$	0.2 g	
			$CaCl_2 \cdot H_2O$	0.1 g	
			$FeSO_4 \cdot 7H_2O$	0.1 g	
			$AlK(SO_4)_2 \cdot 12H_2O$	0.033 g	
			H_3BO_3	0.01 g	
			$CuSO_4$	0.00 g	
			Vitamins	mg L^{-1}	
			Pyridoxine hydrochloride	10.0	
			Calcium pantothenate	5.00	
			Nicotinic acids	5.00	
			p-Aminobenzoic acid	5.00	
			Riboflavin	5.00	
			Thiamine hydrochloride	5.00	
			Lipoic acid	5.00	
			Biotin	2.00	
			Folic acid	2.00	
			Cyanocobalamine	0.10	

(*Continued*)

Table A.2 (Continued)   Compilation on the Enrichment of Nutrients, Trace Elements, and Vitamins Used in the Culture Media for Hyperthermophiles

Strain	Nature	Taxa	Enrichment		Reference
			TB medium	L^{-1}	
			Soluble starch	5 g	
			Glucose	5 g	
			Yeast extract	0.5 g	
			Piperazine (PIPES), 1.5 Na salt	6 g	
			NaCl	20 g	
			KCl	2 g	
			$MgSO_4 \cdot 7H_2O$	500 mg	
			NH_4Cl	250 mg	
			$CaCl_2 \cdot 2H_2O$	50 mg	
			K_2HPO_4	50 mg	
			Trace metals	L^{-1}	
			$FeSO_4 \cdot 7H_2O$	7 mg	
			$Na_2WO_4 \cdot 2H_2O$	0.3 mg	
			Biotin	20 µg	
			Cysteine hydrochloride	500 mg	
			Resazurin	1 mg	
Crimean reservoirs	Thermophilic	*Acetohalobium arabaticum*		$g \, L^{-1}$	Kevbrin and Zavarzin 1991
			NaCl in distilled water	150	
			KH_2PO_4	0.30	
			KCl	2.70	
			NH_4Cl	0.50	
			$CaCl_2 \cdot 2H_2O$	0.07	
			Magnesium DL-lactate	5.00	
			Yeast extract	0.10	
			$NaHCO_3$	4.00	
			Resazurin	0.00	
			Vitamin solution	15.00	
			Trace elements solution	1 mL	
				mg 500 mL^{-1}	
			$FeSO_4 (NH_4)_2SO_4 \cdot 6H_2O$	392	
			$NiSO_4(NH_4)_2SO_4 \cdot 6H_2O$	197.50	
			$CoCl_2 \cdot 6H_2O$	119.00	
			$ZnSO_4 \cdot 7H_2O$	71.80	
			$MnCl_2 \cdot 4H_2O$	49.50	
			Na_2SeO_4	47.30	
			$Na_2WO_4 \cdot 2H_2O$	16.50	
			$Na_2MoO_4 \cdot 2H_2O$	12.00	
			H_3BO_3	3.00	
			$CuCl_2 \cdot 2H_2O$	0.90	
			Concentrated HCl	0.5 mL	

(Continued)

Table A.2 (Continued)   Compilation on the Enrichment of Nutrients, Trace Elements, and Vitamins Used in the Culture Media for Hyperthermophiles

Strain	Nature	Taxa	Enrichment		Reference
Acidic hot springs		*Sulfurisphaera ohwakunsis*	**T4-14 medium, modified Brock's basal salt mixture** (Brock et al. 1972)	mg L^{-1}	Kurosawa et al. 1998
Japan thermophilic			$(NH_4)_2SO_4$	1,300	
			KH_2PO_4	280.00	
			$MgSO_4 \cdot 7H_2O$	250.00	
			$CaCl_2 \cdot 2H_2O$	70.00	
			$FeCl_3 \cdot 6H_2O$	2.00	
			$MnCl_2 \cdot 4H_2O$	1.80	
			$Na_2B_4O_7 \cdot 10H_2O$	4.50	
			$ZnSO_4 \cdot 7H_2O$	0.22	
			$CuCl_2 \cdot 2H_2O$	0.05	
			$NaMoO_4 \cdot 2H_2O$	0.03	
			$VOSO_4 \cdot 2H_2O$	0.03	
			$CoSO_4$	0.01	
			Yeast extract	1.00	
Mid-Atlantic thermal vent		*Pyrodictium occulatum* and *Pyrodictium abyssi*	**Half SME medium** (Stetter 1982; Pley et al. 1991)	mg mL^{-1}	Blochl et al. 1997
Thermophilic			$NaNO_3$	1,000.00	
			NaCl	13,860.00	
			$MgSO_4 \cdot 7H_2O$	3,530.00	
			$MgCl_2 \cdot 6H_2O$	2,750.00	
			$CaCl_2 \cdot 2H_2O$	751.00	
			KH_2PO_4	500.00	
			KCl	325.00	
			NaBr	50.00	
			H_3BO_3	15.00	
			$SrCl_2 \cdot 6H_2O$	7.50	
			$MnSO_4 \cdot 2H_2O$	5.00	
			$(NH_4)_2Ni(SO_4)_2$	2.00	
			$CoCl_2 \cdot 6H_2O$	1.00	
			$FeSO_4 \cdot 7H_2O$	2.00	
			$ZnSO_4$	1.00	
			Na_2SeO_4	0.10	
			$CuSO_4 \cdot 5H_2O$	0.10	
			KI	0.03	
			Resazurin	1.00	
			Water	1,000 mL	
	Hyperthermophilic, anaerobic	*Pyrococcus furiosus*	**SME Modified** (Stetter et al. 1983)	g L^{-1}	Raven et al. 1992
			Peptone	5.00	
			Yeast extract	1.00	
			NaCl	28.00	

(Continued)

Table A.2 (Continued) Compilation on the Enrichment of Nutrients, Trace Elements, and Vitamins Used in the Culture Media for Hyperthermophiles

Strain	Nature	Taxa	Enrichment		Reference
			1-Cysteine hydrochloride		
			Magnesium salt solution	10 mL	
			Resazurin	1 mg mL^{-1}	
			$MgSO_4$ and H_2O	180.00	
			$MgCl_2 \cdot 6H_2O$	140.00	
			Solution A	1 mL L^{-1}	
			Trisodium citrate	4 g L^{-1}	
			$MnSO_4 \cdot 4H_2O$	9.00	
			$ZnSO_4 \cdot 7H_2O$	2.50	
			$NiCl_2 \cdot 6H_2O$	2.50	
			$ALK(SO_4)_2 \cdot 12H_2O$	0.30	
			$CoCl_2 \cdot 6H_2O$	0.30	
			$CuSO_4 \cdot 5H_2O$	0.15	
			Solution B	1 mL^{-1}	
			$CaCl_2 \cdot 6H_2O$	56 g	
			NaBr	25.00 g	
			KCl	16.00 g	
			KI	10.00 g	
			$SrCl_2 \cdot 6H_2O$	4.00 g	
			Solution C	1 mL^{-1}	
			K_2HPO_4	50 g	
			H_3BO_3	7.50 g	
			$Na_2WO_4 \cdot 2H_2O$	3.30 g	
			$Na_2MoO_4 \cdot 2H_2O$	0.15 g	
			Na_2SeO_3	0.01 g	
			Vitamin solution	mg L^{-1}	
			Pyridoxine hydrochloride	200	
			Nicotinic acids	100	
			Riboflavin	100	
			Thiamine hydrochloride	100	
			Lipoic acid	100	
			Biotin	40	
			Folic acid	40	
			Cyanocobalamine	2.0	
			DL-Calcium pantothenate	100	
Uzon, Kamchatka, Russia	Thermo and organotrophic	*Thermanaerovi-briovelox*		g L^{-1}	Zavarzina et al. 2000

(Continued)

Table A.2 (Continued) Compilation on the Enrichment of Nutrients, Trace Elements, and Vitamins Used in the Culture Media for Hyperthermophiles

Strain	Nature	Taxa	Enrichment		Reference
			NH_4Cl	0.33	
			KH_2PO_4	0.33	
			$MgCl_2 \cdot {}^\wedge H_2O$	0.33	
			$CaCl_2 \cdot 6H_2O$	0.33	
			KCl	0.33	
			Yeast extract	0.10	
			$Na_2S \cdot 9H_2O$	0.50	
			$NaHCO_3$	0.70	
			Na_2SO_4	2.00	
			Resazurin	0.001	
			Sodium lactate 50% solution	10 mL	
			Trace element solution	1 mL	
				mg 500 mL^{-1}	
			$FeSO_4(NH_4)_2SO_4 \cdot 6H_2O$	392	
			$NiSO_4(NH_4)_2SO_4 \cdot 6H_2O$	197.50	
			$CoCl_2 \cdot 6H_2O$	119.00	
			$ZnSO_4 \cdot 7H_2O$	71.80	
			$MnCl_2 \cdot 4H_2O$	49.50	
			$Na_2SeO_4^-$	47.30	
			$Na_2WO_4 \cdot 2H_2O$	16.50	
			$Na_2MoO_4 \cdot 2H_2O$	12.00	
			H_3BO_3	3.00	
			$CuCl_2 \cdot 2H_2O$	0.90	
			Vitamin solution (Wolin et al. 1963)	1 mL	
				mg L^{-1}	
			Pyridoxine hydrochloride	10	
			Nicotinic acids	5.00	
			Riboflavin	5.00	
			Thiamine hydrochloride	5.00	
			Thioctic acid	5.00	
			Biotin	2.00	
			Folic acid	2.00	
			p-Aminobenzoic acid	5.00	
			Pantothenic acid	5.00	

Table A.3 Comparison of Levels of Trace Elements and Vitamins Used in Media for Culturing Extremophiles

	Medium	Concentration, mg L^{-1}	Reference	Medium	Concentration, mg L^{-1}	Reference
Trace Elements	TB		Childers et al. 1992	Half SME		Blochl et al. 1997
$MnCl_2 \cdot 4H_2O$		560				
$(NH_4)_6MoO_{24} \cdot 4H_2O$		370				
$CoCl \cdot 6H_2O$		340			1	
$ZnSO_4 \cdot 7H_2O$		320			1	
$NiSO_4 \cdot 6H_2O$		210				
$NaSeO_4$		200			0.1	
$CaCl \cdot 2H_2O$		100				
$FeSO_4 \cdot 7H_2O$		100			2	
$AlK(SO_4)_2 \cdot 12H_2O$		33				
H_3BO_3		10			15	
$CuSO_4$		4			0.1	
$SrCl_2 \cdot 6H_2O$					7.5	
$MnSO_4 \cdot 2H_2O$					5	
$(NH_4)_2Ni(SO_4)_2$					2	
KI					0.03	
Vitamins						
Pyridoxine hydrochloride		200	Wolin et al. 1963		10	
Nicotinic acids		100			5	
Riboflavin		100			5	
Thiamine hydrochloride		100			5	
Lipoic acid		100				
Biotin		40			5	
Folic acid		40			2	
Cyanocobalamine		2				
DL-Calcium pantothenate		100				
Thioctic acid					5	
p-Aminobenzoic acid					5	
Pantothenic acid					5	

Vitamin levels also varied similarly. For example, basal medium (Kurosawa et al. 1998) has the following vitamins:

	mg L^{-1}
D-Biotin	1
Choline chloride	1
Folic acid	1
myo-Inositol	2
Niacinamide	1
D-Pantothenic acid (hemicalcium)	1
Pyridoxal•HCl	1
Riboflavin	0.01
Thiamine•HCl	1

Bergey's Manual of Systematic Bacteriology (Krieg and Holt 1984) is a good source of culture methods for the Archaea; for the deeply branching and phototrophic bacteria, details are given by Boone and Castenholz (2005). Dyall-Smith (2009) presented a detailed manual on the preparation of modified versions of MGM medium, together with the protocols for halobacterial genetics. Souza et al. (2006) cultivated bacteria and Archaea on Marine agar and H medium with different salt concentrations (0–250 g L^{-1}). Archaeal media and Luria broth were used to cultivate organisms from water samples. The compositions of the media used can be found at www.atcc .org. Colonies with different morphologies were selected from each medium, and axenic cultures were obtained.

The difficulties associated with culturing marine bacteria (Joint et al. 2010) are equally valid for culturing extremophiles. These include (1) a lack of appreciation of the environmental conditions under which the extremophiles live, (2) the domination of fast-growing weed species at the expense of the rest, (3) the nonsuitability of the chemical culture media, (4) allelopathy between species, (5) the deleterious effects of high concentrations of the nutrient enrichments used, and (6) virus infection.

Cultures of the facultatively anaerobic prokaryote halobacterium *Haloferax volcanii*, grown in MGM medium at 37°C, had a generation time of three to four hours, while *Halobacterium* sp. has a generation time of 8–12 h. Seawater (25%) enriched with 0.5% yeast extract supported the growth of most halobacteria; the alkaliphilic halobacteria (*Natronobacterium* and *Natronococcus* spp.) needed very low Mg^{2+}, *Halorubrum sodomense* apparently needs starch and clay minerals, and certain haloarchaea prefer specific carbon sources.

Optimal growth for most halophilic alkalithermophiles is at Na^+ concentrations greater than 2 M, pH > 8.5, and temperatures greater than or equal to 50°C. The halophilic alkalithermophiles of the genera *Natranaerobius*, *Natronovirga*, and *Halonatronum* are obligately anaerobic chemoorganotrophs, living by the fermentation of sugars such as glucose, fructose, sucrose, maltose, starch, glycogen, and *N*-acetyl-D-glucosamine (Zhilina et al. 2001; Mesbah and Wiegel 2012).

Anaerobic halophilic alkalithermophiles have an obligate requirement for yeast extract and tryptone or casamino acids and are unable to synthesize *de novo* some amino acids and essential vitamins and/or cofactors.

Bioreactor systems are used to cultivate halotolerant bacteria and are cultured using plates on Modified Nutrient Agar, Starch Casein Agar, and Ken Knight's Agar prepared with 50% natural seawater and supplemented with an additional 7 g 100 mL^{-1} NaCl (Rohban et al. 2009). Extremophile lake populations from the Andean lakes are cultured in MGM medium (Ordonez et al. 2009).

A.10 PIEZOPHILES

In a series of publications (Mitsuzawa et al. 2005, 2006; Deguchi et al. 2007; Tsudome et al. 2009), the authors discussed highly specialized techniques and apparatus used for culturing piezophiles. They used porous, solid, nonfibrous cellulose plates to culture deep-sea acidophile, alkaliphile, and thermophile bacteria that live in hydrothermal vents and under hypergravity. The photopsychrophile *Chlorella* sp. Strain BI, isolated from Antarctica, is unique in retaining the ability for dynamic short-term adjustment of light energy distribution between photosystem II and photosystem I and can grow as a heterotroph in the dark (Morgan-Kiss et al. 2008).

A.11 SUMMARY AND FUTURE PROSPECTS

A great taxonomic diversity exists among extremophile microbiota, but only a few species are studied in culture. Culturing autotrophic microalgae has been in progress for more than a century, and their culturing conditions, such as nutrients, light, and temperature, are well defined. The culture media show a wide latitude of enrichment; some microalgae utilize various organic sources, including glucose, glycerol, starch, glycine, and urea, as carbon sources. Two green algae, *Chlorella luteoviridis* and *Parachlorella hussii*, because of their tolerance to oxidative stress, utilize exogenous organic carbon sources for mixotrophic growth and grow well in raw wastewater and yield high biomass and lipids (Varshney et al. 2015). A few autotrophic extremophiles are physiologically vigorous and produce carotenoids, such as *Dunaliella* and *Scenedesmus* (Durvasula et al. 2015), as an adaptive mechanism to protect against stress, such as UV photoinhibition.

In contrast, there are several categories of heterotrophic extremophiles, and their stressors vary with their habitats, which suggests their adaptability. For example, members of the genus *Exiguobacterium* were isolated from environments such as hot springs, hydrothermal vents, permafrost, ice, processing plants, moraine, soil, marine water, freshwater, biofilm environments, brine shrimp, and a stromatolite from an Andean lake (White et al. 2013). From a sediment from the Bay of Bengal collected from a depth of 2100 m, bacteria related to *Pseudoalteromonas* sp., *Ruegeria* sp., *Exiguobacterium* sp., and *Actinebacter* sp. were isolated and grown in MSM at 27°C ± 2°C (Kumar et al. 2014). MSM contained 0.1 g NaCl, 0.1 g KCl, 1.0 g KH_2PO_4·1.5 g K_2HPO_4, 0.75 g $(NH_4)_2SO_4$, 0.3 g $MgSO_4$·$7H_2O$, 0.1 g peptone and 1 mL trace salts per liter. The trace salts were 20 mg $CaCl_2$, 30 mg $FeCl_3$, 0.5 mg $CuSO_4$, 0.5 mg $MnSO_4$·$4H_2O$, and 10 mg $ZnSO_4$·$7H_2O$ per liter, and the pH was adjusted to 7.2 ± 0.2. It should be noted that (1) the composition of this medium may not reflect that of the *in situ* natural seawater; (2) the incubations were at 27°C ± 2°C, much higher than the *in situ* ones; and (3) the bacteria grew in the laboratory without any hydrostatic pressure. These growth conditions suggest the adaptability of these deep-sea bacteria to conditions altogether different from those they would experience in nature. Another example is the polyextremotolerant green alga *Picochlorum* SE3 (Foflonker et al. 2016), discussed earlier (in our chapter 16) on extremophile case studies. This shows that extremophiles have a plasticity and have been able to adapt physiologically and biochemically to changes in their environment and survive under extreme and harsh conditions (van Wolferen et al. 2013; Singh et al. 2014). As a survival mechanism, they produce extremolytes, which helps them to maintain their homeostasis. Culturing the heterotrophic extremophiles is complicated, if not challenging. A few halophiles, thermophiles, and hyperthermophiles have been cultured, and their culture media included glucose, soluble starch, yeast extract, piperzine, sulfur, resazurin, agar, glycogen, acetate, pyruvate, dextrin, and maltose. These enrichments, with respect to their wide latitude of variations, resemble those of the autotrophs. Optimal conditions of growth in extremophiles based on nutritional requirements need be established and a standard basic medium developed. Further, investigations of their growth and biochemical and physiological responses to suitable gradient stressors (salinity, pH, temperature, and nutrients), similar to those tested on autotrophs, would be necessary. Extremophiles produce a variety of unique bioactive compounds, which could be used in biotechnology. A good database on extremophile growth and the production of bioactive compounds is needed to recommend any economic and technical evaluation similar to that for microalgae (Stephens et al. 2010) in microalgal biofuels.

REFERENCES

Andersen, R.A. 2005. *Algal Culturing Techniques.* New York: Academic Press.

Anila, N., Simon, D.P., Chandrashekar, A. et al. 2016. Metabolic engineering of *Dunaliella salina* for production of ketocarotenoids. *Photosynth Res* 127(3): 321–333.

Baas Becking, L.G.M. 1934. Geobiologie of inleiding tot de milieukunde. The Hague: W.P. van Stockum and Zoon.

Ballav, S., Kerkar, S., Thomas, S. et al. 2015. Halophilic and halotolerant actinomycetes from a marine saltern of Goa, India producing anti-bacterial metabolites. *J Biosci Bioeng* 119(3): 323–330.

Blochl, E., Rachel, R., Burggraf, S. et al. 1997. *Pyrolobus fumarii*, gen. and sp. nov., represents a novel group of archaea, extending the upper temperature limit for life to 113 degrees C. *Extremophiles* 1(1): 14–21.

Boone, D.R., and Castenholz, R.W. (eds.). 2005. *Bergey's Manual of Systematic Bacteriology*. 2nd ed., vol. 1. New York: Springer-Verlag.

Boonyaratanakornkit, B., Cordova, J., Park, C.B. et al. 2006. Pressure affects transcription profiles of *Methanocaldococcus jannaschii* despite the absence of barophilic growth under gas-transfer limitation. *Environ Microbiol* 8(11): 2031–2035.

Brock, T.D., Brock, K.M., Belly, R.T. et al. 1972. *Sulfolobus*: A new genus of sulfur-oxidizing bacteria living at low pH and high temperature. *Arch Mikrobiol* 84(1): 54–68.

Brown, S.H., and Kelly, R.M. 1989. Cultivation techniques for hyperthermophilic archaebacteria: Continuous culture of *Pyrococcus furiosus* at temperatures near 100 degrees C. *Appl Environ Microbiol* 55(8): 2086–2088.

Bumbak, F., Cook, S., Zachleder, V. et al. 2011. Best practices in heterotrophic high-cell-density microalgal processes: Achievements, potential and possible limitations. *Appl Microbiol Biotechnol* 91(1): 31–46.

Burns, D., and Dyall-Smith, M. 2006. Cultivation of haloarchaea. In F.A. Rainey and A. Oren (eds.), *Methods in Microbiology*. Vol. 35. Amsterdam: Elsevier Academic Press, pp. 535–552.

Casal, C., Cuaresma, M., Vega, J.M. et al. 2010. Enhanced productivity of a lutein-enriched novel acidophile microalga grown on urea. *Mar Drugs* 9(1): 29–42.

Chen, F. 1996. High cell density culture of microalgae in heterotrophic growth. *Trends Biotechnol* 14(11): 421–426.

Childers, S.E., Vargas, M., and Noll, K.M. 1992. Improved methods for cultivation of the extremely thermophilic bacterium *Thermotoga neapolitana*. *Appl Environ Microbiol* 58(12): 3949–3953.

Chitlaru, E., and Pick, U. 1989. Selection and characterization of *Dunaliella salina* mutants defective in halo-adaptation. *Plant Physiol* 91(2): 788–794.

Christner, B.C., Priscu, J.C., Achberger, A.M. et al. 2014. A microbial ecosystem beneath the West Antarctic ice sheet. *Nature* 512(7514): 310–313.

Coesel, S.N., Baumgartner, A.C., Teles, L.M. et al. 2008. Nutrient limitation is the main regulatory factor for carotenoid accumulation and for Psy and Pds steady state transcript levels in *Dunaliella salina* (Chlorophyta) exposed to high light and salt stress. *Mar Biotechnol (NY)* 10(5): 602–611.

Cuaresma, M., Casal, C., Forján E. et al. 2011. Productivity and selective accumulation of carotenoids of the novel extremophile microalga *Chlamydomonas acidophila* grown with different carbon sources in batch systems. *J Ind Microbiol Biotechnol* 38(1): 167–177.

Deguchi, S., Tsudome, M., Shen, Y. et al. 2007. Preparation and characterisation of nanofibrous cellulose plate as a new solid support for microbial culture. *Soft Matter* 3(9): 1170–1175.

Delgado-García, M., Nicolaus, B., Poli, A. et al. 2015. Isolation and screening of halophilic bacteria from production of hydrolytic enzymes. In D.K. Maheswari and M. Saraf (eds.), *Halophiles: Biodiversity and Sustainable Exploitation*. Vol. 6. Berlin: Springer, pp. 379–401.

Durvasula, R., Hurwitz, I., Fieck, A. et al. 2015. Culture, growth, pigments and lipid content of *Scenedesmus* species, an extremophile microalga from Soda Dam, New Mexico in wastewater. *Algal Res* 10: 128–133.

Durvasula, R.V., Vadrevu, S.H.R., and Subba Rao, D.V. 2013. *Microalgal Biotechnology: Today's (Green) Gold Rush*. In F. Bux (ed.), *Biotechnical Applications of Microalgae: Biodiesel and Value Added Products*. London: Taylor & Francis Publishers, pp. 199–225.

Dyall-Smith, M. 2009. *The Halohandbook: Protocols for Haloarchaeal Genetics*. http://www.haloarchaea.com/resources/halohandbook.

Fan, L., Vonshak, A., Zarka, A. et al. 1998. Does astaxanthin protect *Haematococcus* against light damage? *Z Naturforsch C* 53(1–2): 93–100.

Foflonker, F., Ananyev, G., Qiu, H. et al. 2016. The unexpected extremophile: Tolerance to fluctuating salinity in the green alga *Picochlorum*. *Algal Res* 16: 465–472.

Fox, D. 2014. Lakes under the ice: Antarctica's secret garden. *Nature* 512(7514): 244–246.

Gerwick, W.H., Bernart, M.W., Jiang, Z.D. et al. 1991. Patterns of oxylipin metabolism in marine organisms. In *2nd International Marine Biotechnology Conference*, Baltimore, pp. 369–378.

Ghati, A., Sarkar, K., and Paul, G. 2013. Isolation, characterization and molecular identification of esterolytic thermophilic bacteria from an Indian hot spring. *Curr Res Microbiol Biotechnol* 1: 196–202.

Giddings, L.A., and Newman, D.J. 2015. *Bioactive Compounds from Marine Extremophiles*. Berlin: Springer.

Godfroy, A., Meunier, J.R., Guezennec, J. et al. 1996. *Thermococcus fumicolans* sp. nov., a new hyperthermophilic archaeon isolated from a deep-sea hydrothermal vent in the north Fiji Basin. *Int J Syst Bacteriol* 46(4): 1113–1119.

Godfroy, A., Postec, A., and Raven, N. 2006. Growth of hyperthermophilic microorganisms for physiological and nutritional studies. In F.A. Rainey and A. Oren (eds.), *Methods in Microbiology*. Vol. 35. Amsterdam: Elsevier Academic Press, pp. 93–108.

Godfroy, A., Raven, N.D., and Sharp, R.J. 2000. Physiology and continuous culture of the hyperthermophilic deep-sea vent archaeon *Pyrococcus abyssi* ST549. *FEMS Microbiol Lett* 186(1): 127–132.

González-Toril, E., Gómez, F., Malki, M. et al. 2006. The isolation and study of acidophilic microorganisms. In F.A. Rainey and A. Oren (eds.), *Methods in Microbiology*. Vol. 35. Amsterdam: Elsevier Academic Press, pp. 471–510.

Grammann, K., Volke, A., and Kunte, H.J. 2002. New type of osmoregulated solute transporter identified in halophilic members of the bacteria domain: TRAP transporter TeaABC mediates uptake of ectoine and hydroxyectoine in *Halomonas elongata* DSM 2581(T). *J Bacteriol* 184(11): 3078–3085.

Grant, W.D. 2006. Cultivation of aerobic alkaliphiles. In F.A. Rainey and A. Oren (eds.), *Methods in Microbiology*. Vol. 35. Amsterdam: Elsevier Academic Press, pp. 439–449.

Guillard, R.R., and Ryther, J.H. 1962. Studies of marine planktonic diatoms. I. *Cyclotella nana* Hustedt, and *Detonula confervacea* (cleve) Gran. *Can J Microbiol* 8: 229–239.

Guiry, M.D. 2012. How many species of algae are there? *J Phycol* 48(5): 1057–1063.

Houghton, J.L., Seyfried, W.E., Jr., Banta, A.B. et al. 2007. Continuous enrichment culturing of thermophiles under sulfate and nitrate-reducing conditions and at deep-sea hydrostatic pressures. *Extremophiles* 11(2): 371–382.

Joint, I., Muhling, M., Querellou, J. et al. 2010. Culturing marine bacteria—An essential prerequisite for biodiscovery. *Microb Biotechnol* 3(5): 564–575.

Kaplan, D., Richmond, A.E., Dubinsky, Z. et al. 1986. Algal nutrition. In A. Richmond (ed.), *CRC Handbook of Microalgal Mass Culture*. Boca Raton, FL: CRC Press, pp. 147–198.

Keerthi, S., Koduru, U.D., and Sarma, N.S. 2015. A nutrient medium for development of cell dense inoculum in mixotrophic mode to seed mass culture units of *Dunaliella salina*. *Algol Stud* 147(1): 7–28.

Kevbrin, V.V., and Zavarzin, G.A. 1991. Effect of sulfur compounds on the growth of the halophilic homoacetic bacterium *Acetohalobium arabaticum*. *Mikrobiologiya* 61(5): 812–817.

Kottmeier, S.T., and Sullivan, C.W. 1988. Sea ice microbial communities (SIMCO), 9. Effects of temperature and salinity on rates of metabolism and growth of autotrophs and heterotrophs. *Polar Biol* 8: 293–304.

Krieg, N.R., and Holt, J.C. (eds.). 1984. *Bergey's Manual of Systematic Bacteriology*. 1st ed., vol. 1. Baltimore: Williams and Wilkins.

Kumar, A.G., Vijayakumar, L., Joshi, G. et al. 2014. Biodegradation of complex hydrocarbons in spent engine oil by novel bacterial consortium isolated from deep sea sediment. *Bioresour Technol* 170: 556–554.

Kurosawa, N., Itoh, Y.H., Iwai, T. et al. 1998. *Sulfurisphaera ohwakuensis* gen. nov., sp. nov., a novel extremely thermophilic acidophile of the order Sulfolobales. *Int J Syst Bacteriol* 48: 451–456.

Lam, K.S. 2006. Discovery of novel metabolites from marine actinomycetes. *Curr Opin Microbiol* 9(3): 245–251.

Larsen, P.I., Sydnes, L.K., Landfald, B. et al. 1987. Osmoregulation in *Escherichia coli* by accumulation of organic osmolytes: Betaines, glutamic acid, and trehalose. *Arch Microbiol* 147(1): 1–7.

Lee, H.S., Bae, S.S., Kim, M.S. et al. 2011. Complete genome sequence of hyperthermophilic *Pyrococcus* sp. strain NA2, isolated from a deep-sea hydrothermal vent area. *J Bacteriol* 193(14): 3666–3667.

Lee, Y.K. 2004. *Algal Nutrition: Heterotrophic Carbon Nutrition*. In A. Richmond (ed.), *Handbook of Microalgal Culture: Biotechnology and Applied Phycology*. Oxford: Blackwell, pp. 116–124.

Maheswari, D.K., and Saraf, M. (eds.). 2015. *Halophiles: Biodiversity and Sustainable Exploitation*. Vol. 6. Berlin: Springer.

Meeks, J.C., and Castenholz, R.W. 1971. Growth and photosynthesis in an extreme thermophile, *Synechococcus lividus* (Cyanophyta). *Arch Mikrobiol* 78(1): 25–41.

Mesbah, N.M., and Wiegel, J. 2006. Isolation, cultivation and characterization of alkalithermophiles. In F.A. Rainey and A. Oren (eds.), *Methods in Microbiology*. Vol. 35. Amsterdam: Elsevier Academic Press, pp. 451–468.

Mesbah, N.M., and Wiegel, J. 2012. Life under multiple extreme conditions: Diversity and physiology of the halophilic alkalithermophiles. *Appl Environ Microbiol* 78(12): 4074–4082.

Miquel, P. 1893. De la culture artificielle des diatomées. Introduction. *Le Diatomiste* 1: 73–75.

Mitsuzawa, S., Deguchi, S., and Horikoshi, K. 2006. Cell structure degradation in *Escherichia coli* and *Thermococcus* sp. strain Tc-1-95 associated with thermal death resulting from brief heat treatment. *FEMS Microbiol Lett* 260(1): 100–105.

Mitsuzawa, S., Deguchi, S., Takai, K. et al. 2005. Flow-type apparatus for studying thermotolerance of hyperthermophiles under conditions simulating hydrothermal vent circulation. *Deep Sea Res Pt I* 52(6): 1085–1092.

Morgan-Kiss, R.M., Ivanov, A.G., Modla, S. et al. 2008. Identity and physiology of a new psychrophilic eukaryotic green alga, *Chlorella* sp., strain BI, isolated from a transitory pond near Bratina Island, Antarctica. *Extremophiles* 12(5): 701–711.

Nakagawa, S., and Takai, K. 2006. The isolation of thermophiles from deep-sea hydrothermal environments. In F.A. Rainey and A. Oren (eds.), *Methods in Microbiology*. Vol. 35. Amsterdam: Elsevier Academic Press, pp. 55–91.

Ono, H., Sawada, K., Khunajakr, N. et al. 1999. Characterization of biosynthetic enzymes for ectoine as a compatible solute in a moderately halophilic eubacterium, *Halomonas elongata*. *J Bacteriol* 181(1): 91–99.

Ordonez, O.F., Flores, M.R., Dib, J.R. et al. 2009. Extremophile culture collection from Andean lakes: Extreme pristine environments that host a wide diversity of microorganisms with tolerance to UV radiation. *Microb Ecol* 58(3): 461–473.

Oren, A. 1983. *Halobacterium sodomense* sp. nov., a Dead Sea halobacterium with an extremely high magnesium requirement. *Int J Syst Bacteriol* 33: 381–386.

Park, C.B., Boonyaratanakornkit, B.B., and Clark, D.S. 2006. Toward the large scale cultivation of hyperthermophiles at high-temperature and high-pressure. In F.A. Rainey and A. Oren (eds.), *Methods in Microbiology*. Vol. 35. Amsterdam: Elsevier Academic Press, pp. 109–126.

Pley, U., Schipka, J., Gambacorta, A. et al. 1991. *Pyrodictium abyssi*, new species represents a novel heterotrophic marine archaeal hyperthermophile growing at 110°C. *Syst Appl Microbiol* 14: 245–253.

Postec, A., Pignet, P., Cueff-Gauchard, V. et al. 2005. Optimisation of growth conditions for continuous culture of the hyperthermophilic archaeon *Thermococcus hydrothermalis* and development of sulphur-free defined and minimal media. *Res Microbiol* 156(1): 82–87.

Prokofeva, M.I., Kublanov, I.V., Nercessian, O. et al. 2005. Cultivated anaerobic acidophilic/acidotolerant thermophiles from terrestrial and deep-sea hydrothermal habitats. *Extremophiles* 9(6): 437–448.

Raven, N., Ladwa, N., Cossar, D. et al. 1992. Continuous culture of the hyperthermophilic archaeum *Pyrococcus furiosus*. *Appl Microbiol Biotechnol* 38(2): 263–267.

Raven, N.D.H., and Sharp, R.J. 1997. Development of defined and minimal media for the growth of the hyperthermophilic archaeon *Pyrococcus furiosus* Vc.1. *FEMS Microbiol Lett* 146: 135–141.

Rohban, R., Amoozegar, M.A., and Ventosa, A. 2009. Screening and isolation of halophilic bacteria producing extracellular hydrolyses from Howz Soltan Lake, Iran. *J Ind Microbiol Biotechnol* 36(3): 333–340.

Rosly, N.F., Rabeah A.A.R., Kuppusamy, P. et al. 2013. Induction of bioactive compound composition from marine microalgae (*Lyngbya* sp.) by using different stress condition. *J Coast Life Med* 1: 205–209.

Russell, N.J., and Cowan, D.A. 2006. *Handling of Psychrophilic Microorganisms*. In F.A. Rainey and A. Oren (eds.), *Methods in Microbiology*. Vol. 35. Amsterdam: Elsevier Academic Press, pp. 371–393.

Sakai, H.D., and Kurosawa, N. 2016. Exploration and isolation of novel thermophiles in frozen enrichment cultures derived from a terrestrial acidic hot spring. *Extremophiles* 20(2): 207–214.

Sharp, R.J., and Raven, N.D.H. 1997. Isolation and growth of hyperthermophiles. In P.M. Rhodes and P.F. Stanbury (eds.), *Applied Microbial Physiology: A Practical Approach*. Oxford: IRL Press, pp. 23–51.

Silverman, M.P., and Lundgren, D.G. 1959. Studies on the chemoautotrophic iron bacterium *Ferrobacillus ferrooxidans*. *J Bacteriol* 77: 642–647.

Singh, P., Singh, S.M., and Dhakephalkar, P. 2014. Diversity, cold active enzymes and adaptation strategies of bacteria inhabiting glacier cryoconite holes of High Arctic. *Extremophiles* 18(2): 229–242.

Souza, V., Espinosa-Asuar, L., Escalante, A.E. et al. 2006. An endangered oasis of aquatic microbial biodiversity in the Chihuahuan desert. *Proc Natl Acad Sci USA* 103(17): 6565–6570.

Spijkerman, E., and Wacker, A. 2011. Interactions between P-limitation and different C conditions on the fatty acid composition of an extremophile microalga. *Extremophiles* 15(5): 597–609.

Stephens, E., Ross, I.L., King, Z. et al. 2010. An economic and technical evaluation of microalgal biofuels. *Nat Biotechnol* 28(2): 126–128.

Stetter, K.O. 1982. Ultrathin mycelia-forming organisms from submarine volcanic areas having an optimum growth temperature of 105°C. *Nature* 300: 258–260.

Stetter, K.O., Konig, H., and Stackebrandt, E. 1983. *Pyrodictium* gen. nov., a new genus of submarine disc-shaped sulphur reducing archaebacteria growing optimally at 105 degrees C. *Syst Appl Microbiol* 4(4): 535–551.

Subba Rao, D.V. 2009. Cultivation, growth medium, division rates and applications of *Dunaliella* species. In A. Ben-Amotz, J. E. W. Polle, and D. V. Subba Rao. (eds.), *The Alga Dunaliella: Biodiversity, Physiology, Genomics and Biotechnology*. Enfield, NH: Science Publishers, pp. 45–89.

Subba Rao, D.V., Pan, Y., and Al-Yamani, F. 2005. Growth and photosynthetic rates of *Chalmydomonas plethora* and *Nitzschia frustula* cultures isolated from Kuwait Bay, Arabian Gulf and their potential as live algal food for tropical mariculture. *Mar Ecol* 26: 63–71.

Tang, E.P.Y., Tremblay, R., and Vincent, W.F. 1997a. Cyanobacterial dominance of polar freshwater ecosystems: Are high-latitude mat-formers adapted to low temperature? *J Phycol* 33: 171–181.

Tang, E.P.Y., Vincent, W.F., Proulx, D. et al. 1997b. Polar cyanobacteria versus green algae for tertiary wastewater treatment in cool climates. *J Appl Phycol* 9: 371.

Tsavalos, A.J., and Day, J.G. 1994. Development of media for the mixotrophic/heterotrophic culture of *Brachiomonas submarina*. *J Appl Phycol* 6: 431–433.

Tsudome, M., Deguchi, S., Tsujii, K. et al. 2009. Versatile solidified nanofibrous cellulose-containing media for growth of extremophiles. *Appl Environ Microbiol* 75(13): 4616–4619.

van Wolferen, M., Ajon, M., Driessen, A.J.M. et al. 2013. How hyperthermophiles adapt to change their lives: DNA exchange in extreme conditions. *Extremophiles* 17(4): 545–563.

Varshney, P., Mikulic, P., Vonshak, A. et al. 2015. Extremophilic micro-algae and their potential contribution in biotechnology. *Bioresour Technol* 184: 363–372.

Weber, A.P., Oesterhelt, C., Gross, W. et al. 2004. EST-analysis of the thermo-acidophilic red microalga *Galdieria sulphuraria* reveals potential for lipid A biosynthesis and unveils the pathway of carbon export from rhodoplasts. *Plant Mol Biol* 55(1): 17–32.

White, R.A., III, Grassa, C.J., and Suttle CA. 2013. Draft genome sequence of *Exiguobacterium pavilionensis* strain RW-2, with wide thermal, salinity, and pH tolerance, isolated from modern freshwater microbialites. *Genome Announc* 1(4): e00597-13.

Wilkins, D., Yau, S., Williams, T.J. et al. 2013. Key microbial drivers in Antarctic aquatic environments. *FEMS Microbiol Rev* 37(3): 303–335.

Wolin, E.A., Wolin, M.J., and Wolfe, R.S. 1963. Formation of methane by bacterial extracts. *J Biol Chem* 238: 2882–2886.

Yang, Z., Zhang, Q., and Xu, Z. 2000. Successful culture of *Dunaliella salina* on a solid medium. *Tech Tips Online* 5(1): 23–25.

Zavarzina, D.G., Zhilina, T.N., Tourova, T.P. et al. 2000. *Thermanaerovibrio velox* sp. nov., a new anaerobic, thermophilic, organotrophic bacterium that reduces elemental sulfur, and emended description of the genus *Thermanaerovibrio*. *Int J Syst Evol Microbiol* 50(3): 1287–1295.

Zhilina, T.N., Garnova, E.S., Turova, T.P. et al. 2001. *Halonatronum saccharophilum* gen. nov. sp. nov—A new haloalkalophilic bacteria from the order Haloanaerobiales from Lake Magadi. *Mikrobiologiia* 70(1): 77–85.

Zillig, W., Stetter, K.O., Wunderl, S. et al. 1980. The *Sulfolobus-"Caldariella"* group: Taxonomy on the basis of the structure of DNA-dependent RNA polymerases. *Arch Microbiol* 125: 259–269.

Additional Sources of Information

BOOKS/MONOGRAPHS ETC.

Anitori, R.P., ed. 2012. *Extremophiles, Microbiology and Biotechnology*. Norfolk, UK: Caister Academic Press.

Antranikian, G., ed. 1998. *Biotechnology of extremophiles*. Springer, ISMN978-3-540-63817-9

Barnard, D., Casanueva, A.,Tuffin, M., and Cowan, D. 2010. Extremophiles in biofuel synthesis. *J. Environ Technol* 31 (8-9): 871-888.

Bej, A.K., Aislabie, J., and Atlas, R.M., eds. 2009. *Polar Microbiology: The Ecology, Biodiversity and Bioremediation Potential of Microorganisms in Extremely Cold Environments*. Boca Raton, FL: CRC Press.

Bell, E. 2012. *Life at Extremes: Environments, Organisms, and Strategies for Survival*. Wallingford, UK: CABI.

Bittel, J. 2017. Tardigrade protein helps human DNA withstand radiation. *Nature News and Comment*, May 2.

Boone, D.R., and Calstenholz, R.W., eds. 2001. *Bergey's Manual of Systematic Bacteriology*, vol. 1, *The Archaea and the Deeply Branching and Phototrophic Bacteria*. 2nd ed. Berlin: Springer.

Brock, T.D. 1978. *Thermophilic Microorganisms and Life at High Temperatures*. New York: Springer-Verlag.

Brock, T.D. 2012. *Thermophilic Microorganisms and Life at High Temperatures*. New York: Springer.

Caplan, S.R., and Ginzburg, M., eds. 1978. *Energetics and Structure of Halophilic Microorganisms*. Amsterdam: Elsevier–North Holland Biomedical Press.

Christner, B.C., Priscu, J.C., Achberger, A.M., et al. 2014. A microbial ecosystem beneath the west Antarctic ice sheet. *Nature* 512: 310–313.

Ebel, C., P. Faou, B. Franzetti, B. Kernel, D. Madern, M. Pascu, C. Pfister, S. Richard, and G. Zaccai. 1998. Molecular interactions in extreme halophiles: The solvation-stabilization hypothesis for halophilic proteins. In *Microbiology and Biogeochemistry of Hypersaline Environments*, A. Oren (ed), pp. 227–237. Boca Raton, FL: CRC Press.

Elanor, B.M., ed. 2012. *Life at Extremes: Environments, Organisms and Strategies for Survival*. Wallingford, UK: CABI.

Elleuche, S., Schroder, C., Sahm, K., and Antranikian, G. 2014. Extremozymes-biocatalysts with unique properties from extremophilic microorganisms. *Curr Opin Biotechnol* 29: 116–123.

Hashimoto, T., Horikawa, D.D., and Kunieda, T. 2016. Extremotolerant tardigrade genome and improved radiotolerance of human cultured cells by tardigrade-unique protein. *Nat Commun* 7: 12808.

Hedlund, B.P., Dodsworth, J.A., Murugapiran, S.K., Rinke, C., Woyke, T. 2014. Impact of single-cell genomics and metagenomics on the emerging view of extremophile "microbial dark matter." *Extremophiles* 18 (5): 865–875.

Gerday, C., and Glansdorff, N., eds. 2007. *Physiology and Biochemistry of Extremophiles*. Washington, DC: ASM Press.

Giddings, L.A., and Newman, D.J. 2015. *Bioactive Compounds from Marine Extremophiles*. Berlin: Springer.

Gunde-Cimerman, N., A. Oren, and A. Plemenitaš (ed.). 2005.Adaptation to Life at High Salt Concentrations in *Archaea*, *Bacteria*, and *Eukarya*. Dordrecht, Netherlands: Springer.

Herbert, R.A., and Sharp, R.J., eds. 1992. *Molecular Biology and Biotechnology of Extremophiles*. New York: Blackie.

Horikoshi, K., ed. 1999. *Alkaliphiles*. Tokyo: Harwood Academic Publishers.

Horikoshi, K., Antranikian, G., Bull, A.T., Robb, F.T., and Stetter, K.O., eds. 2011. *Extremophiles Handbook*. Berlin: Springer.

Horikoshi, K., and Grant, W.D., eds. 1998. *Extremophiles: Microbial Life in Extreme Environments*. New York: Wiley-Liss.

Horikoshi, K., and Tsujii, K., eds. *Extremophiles in Deep-Sea Environments*. Tokyo: Springer-Verlag.

Javor, B., ed. 1989. *Hypersaline Environments: Microbiology and Biogeochemistry*. Berlin: Springer-Verlag.

Johnson, D.B., ed. 2009. Extremophiles: Acidic environments. In M. Schaechter (ed.), *Encyclopedia of Microbiology*, 107–126. 3rd ed. Oxford, United Kingdom: Elsevier.

Johri, B.N., Satyanarayana, T., and Olsen, J., eds. 1999. *Thermophilic Moulds in Biotechnology.* Dordrecht: Kluwer Academic.

Kristjansson, J.K., ed. *Thermophilic Bacteria.* Boca Raton, FL: CRC Press.

Kumar, L., Awasthi, G., and Sing, B. 2011. Extremophiles: A novel source of industrially important enzymes. *Biotechnology* 10: 121–135.

Litchfield, C.D., Palme, R., and Piasecki, P., eds. 2001. *Le monde du sel. Melanges offerts a Jean-Claude Hocquet.* Tirol: Berenkamp.

Maheshwari, D.K., and Saraf, M., eds. 2015. *Halophiles Biodiversity and Sustainable Exploitation.* Berlin: Springer.

Margensin, R., and Schinner, F., eds. 1999. *Biotechnological Applications of Cold-Adapted Organisms.* Berlin: Springer-Verlag.

Margensin, R., Schinner, F., Marx, J.C., and Gerday, C., eds. 2008. *Psychrophiles: From Biodiversity to Biotechnology.* Heidelberg: Springer Academic Press.

Oren, A., ed. 1999. *Microbiology and Biogeochemistry of Hypersaline Environments.* Boca Raton, FL: CRC Press.

Oren, A. 2002. Molecular ecology of extremely halophilic Archaea and Bacteria. *FEMS Microbiol Ecol* 39: 1–7.

Oren, A. 2014. Taxonomy of halophilic archaea: Current status and future challenges. *Extremophiles* 18 (5): 825–834.

Oren, A., and Rainey, F., eds. 2006.*Extremophiles.* London: Elsevier Academic Press.

Rainey, F., and Oren, A., eds. 2006. *Extremophiles: Methods in Microbiology.* Vol. 14. Amsterdam: Elsevier.

Rampelotto, P.H., ed. 2016. *Biotechnology of Extremophiles: Grand Challenges in Biology and Biotechnology.* Berlin: Springer.

Rawlings, D.E. 1997. *Biomining: Theory, Microbes, and Industrial Process.* Heidelberg: Springer Academic Press.

Robb, F., Antranikian, G., Grogan, D., and Driessen, A., eds. 2007. *Thermophiles: Biology and Technology at High Temperatures.* Boca Raton, FL: CRC Press.

Satyanarayana, T., Littlechild, J., and Kawarabayasi, Y., eds. 2013. *Thermophilic Microbes in Environmental and Industrial Biotechnology.* Amsterdam: Springer.

Seckbach, J., ed. 2007. *Algae and Cyanobacteria in Extreme Environments.* Berlin: Springer. Singh, O.V., ed. 2012. *Extremophiles: Sustainable Resources and Biotechnological Implications.* Oxford: Wiley-Blackwell.

Stan-Lotter, H., and Fendrihan, S., eds. 2017. *Adaption of Microbial Life to Environmental Extremes: Novel Research Results and Application.* Berlin: Springer.

Thomas, T., and Siddiqui, K.S., eds. 2007. *Protein Adaptation in Extremophiles.* Molecular Anatomy and Physiology of Proteins Series. New York: Nova Science Publisher.

Ventosa, A., and Nieto, J.J. 1995. Biotechnological applications and potentialities of halophilic microorganisms. *World J Microbiol Biotechnol* 11 (1): 85–94.

Ventosa, A., ed. 2004. *Halophilic Microorganisms.* Berlin: Springer-Verlag.

Ventosa, A., Oren, A., and Ma, Y., eds. 2011. *Halophiles and Hypersaline Environments: Current Research and Future Trends.* Berlin: Springer.

Ventosa, A., Fernandez, A.B., Leon, M.J., Sanchez-Porro, C., and Rodriguez-Valera, F. 2014. The Santa Pola saltern as a model for studying the microbiota of hypersaline environments. *Extremophiles* 18 (5): 811–824.

Vincent, W.F., ed. 1989. *Microbial Ecosystems of Antarctica.* Cambridge: Cambridge University Press.

Wiegel, J., and Adams, M.W.W., eds. 1998. *Thermophiles: The Keys to Molecular Evolution and Origin of Life?* Philadelphia: Taylor and Francis Ltd.

CONFERENCE PROCEEDINGS

Adaptation to Life at High Salt Concentrations in Archaea, Bacteria, and Eukarya
Extremophiles 2014 St. Petersburg 10th International Congress on Extremophiles

Physiology and biotechnology of oxygenic phototrophic microorganisms: Looking into the future, presented at International Scientific Conference in Memoriam of the 80th Anniversary of Professor Mikhail V. Gusev, Moscow, Russia, May 27–30, 2014

R. Thane Papke, Preface to the Proceedings of Halophiles 2013 [International Congress on Halophilic Microorganisms], *Front. Microbiol.* 2015, doi: 10.3389/fmicb.2015.00341

Physiology and Biotechnology of Microalgae, presented at international conference devoted to the 80th anniversary of Victor E. Semenenko, K. A. Timiryazev Institute of Plant Physiology, Moscow, Russia, October 16–19, 2012

5th International Congress on Biocatalysis: Biocat 2010; Hamburg, Germany, August 29–September 2, 2010, www.biocat2010.de

6th International Congress on Biocatalysis (biocat2012), Hamburg, Germany, September 2–6, 2012

8th International Congress on Extremophiles (Extremophiles 2010), Ponta Delgada, Azores, Portugal, September 12–16, 2010

9th International Congress on Extremophiles (Extremophiles 2012), Sevilla, Spain, September 10–13, 2012

12th International Congress on Extremophiles, Naples, Italy, September 12–16, 2018

Halophiles 2010, Beijing, China June 29–July 3, 2010: http://www.halophiles2010.org

Thermophiles 2009, Beijing, China, August 16–21, 2009

Thermophiles 2007, Bergen, Norway, September 24–27, 2007: http://sites.web123.no/AtlanticReiser/uib/Thermophiles2007

International Thermophiles Conference: From Evolution to Revolution, Gold Coast, Queensland, Australia, September 2005: http://www.griffith.edu.au/conference/thermophile05

Frontiers in Chemical Biology: Mechanistic Enzymology and Biocatalysis, Exeter, United Kingdom, March 27–31, 2006

International Conference on Alpine and Polar Microbioloyg, Innsbruck, Austria, March 27–30, 2006: http://www.alpine-polar-microbiology2006.at

3rd International Congress on Biocatalysis 2006 (biocat2006), September 3–7, 2006

International Symposium on Extremophiles and Their Applications, Tokyo, Japan, November 29–December 2, 2005: http://www.xbr.jp/isea

International Congress on Extremophiles 2006, Brest, France, September 17–23, 2006

LINKS

http://cccryo.fraunhofer.de/web/infos/scope
http://www.dsmz.de (ncma@bigelow.or)
Journal of Japanese Society for Extremophiles: **J-Stage**: https://www.jstage.jst.go.jp/browse/jjse
International Society for Extremophiles (ISE): News: extremophiles.org/index.php/page/news-2010-11-15
http://serc.carleton.edu/microbelife/extreme/extremeheat/
NASA Science, your gateway to U.S. federal science reports, such as Psychrophilic and Psychrotolerant Microbial Extremophiles in Polar Environments: https://science.nasa.gov
NASA Technical Reports Server (NTRS) https://www.sti.nasa.gov
Proceedings from Halophiles 2013, the International Congress on Halophilic Microorganisms: https://books.google.com/books?isbn=2889195708
htps://books.google.com/books?isbn=3642201989

Glossary

ABC transporters: ATP binding cassette transporters (ABC transporters) are members of a transport system superfamily that is one of the largest and is possibly one of the oldest families, with representatives in all extant phyla, from prokaryotes to humans.

Abiotic: Nonliving components, such as water, sunlight, climate, temperature, rocks, and minerals.

Acidophiles: Organisms able to thrive in highly acid conditions with a growth optimum around pH 2.

Actinobacteria: A phylum of Gram-positive bacteria with high guanine and cytosine content in their DNA.

Adaptation: A process occurring over generations that involve changes in the genetic makeup of cells and selection, ultimately helping offspring of organisms to survive in changing habitats.

Aerobic: Requiring oxygen; processes that use oxygen.

Alga: A nontaxonomic term describing a diverse group of oxygen-evolving photosynthetic organisms lacking true stems, roots, or leaves.

Alkalophiles: Organisms adapted to live in environments where the range of pH is between 8.5 and 11.

Alpha diversity: The difference in terms of number and abundance of species within a habitat unit or in a specific site.

Amplified ribosomal DNA restriction analysis (ARDRA): A restriction analysis often termed as an extension of the restriction fragment length polymorphism.

Anaerobic digestion: The microbial conversion of organic matter to biogas, a mixture of methane, carbon dioxide, water vapor, and small amounts of hydrogen sulfide, and sometimes hydrogen and hydrogen atoms.

Antagonistic: Chemicals exhibit less than additive toxicity in mixture.

Antioxidant: A reactive compound that delays or prevents reactions with oxygen.

Antiporters: Also called an exchanger or countertransporter, this is a cotransporter and integral membrane protein involved in the secondary active transport of two or more different molecules or ions (i.e., solutes) across a phospholipid membrane, such as the plasma membrane, in opposite directions.

Archaea: A major division of microorganisms. Like bacteria, Archaea are single-celled organisms lacking nuclei and are therefore prokaryotes. They are considered to be an ancient form of life that evolved separately from the bacteria and blue-green algae and are classified as belonging to the kingdom Monera in the traditional five-kingdom taxonomy.

Archaeosomes: A type of liposome, or spherical vesicle, with at least one lipid bilayer, which is made from membrane lipids and/or synthetic lipid analogues from Archaea.

Arid: Characterized by lack of rain. A region that receives too little water to support agriculture without irrigation. Rainfall is less than 200 mm/year.

Aridification: The process by which a humid region becomes increasingly dry in a long or medium time frame, as by climatic change.

Asexual reproduction: A process involving mitosis in which a cell first doubles its DNA content and then divides into two identical daughter cells.

Astaxanthin: A ketocarotenoid belonging to a larger class of chemical compounds known as terpenes.

Autochthonous: Indigenous, formed from its present position.

Autotroph: An organism capable of self-nourishment by using inorganic material as a source of nutrients and using photosynthesis or chemosynthesis as a source of energy.

Axenic culture: A culture that contains only one species, free from all contaminants and other organisms, including bacteria.

Barophile (also known as piezophile): Microorganisms that are found exclusively at high barometric pressure, in particular in deep oceans.

Basic Local Alignment Search Tool (BLAST): An algorithm for comparing primary biological sequence information, such as the amino acids of different proteins or the nucleotides of DNA sequences.

Batch culture: A closed-system culture produced by inoculating a sterile medium. In advanced growth stages, shortage of nutrients, light, and carbon dioxide limit algal growth.

Beta-carotene (β-carotene): A plant pigment belonging to the family of isoprenoids that is made up of C5 isoprene units. Derived from phytoene through a sequence of dehydrogenation, isomerization, and cyclization reactions.

Biomass: Quantitative amount of living material; for example, chlorophyll measurement is an estimate of phytoplankton biomass.

Bioremediation: A waste management technique that involves the use of organisms to remove or neutralize pollutants from a contaminated site.

Biosurfactants: Amphiphilic compounds produced on living surfaces, mostly microbial cell surfaces, or excreted extracellularly and containing hydrophobic and hydrophilic moieties that reduce surface tension (ST) and interfacial tensions between individual molecules at the surface and interface, respectively.

Biotic: Relating to or resulting from living organisms.

Bioweathering: Organic weathering, also called biological weathering. The general name for biological processes of weathering that break down rocks.

Brine–seawater interface: Transition layer between the water column and the beginning of the brine body. The interface is usually characterized by steep increases of salinity and temperature and by a sharp decrease of oxygen level.

Canthaxanthin: A red and pink pigment naturally present in both plants and animals.

Carbon concentrating mechanism (CCM): Any process that results in the occurrence of a higher steady-state concentration of CO_2 that is available to RUBISCO (qv) in steady-state photosynthesis than the concentration in the bulk medium. CCMs always involve an energy input.

Carbonic anhydrase: An enzyme that catalyzes the equilibrium between CO_2 and HCO_3^-.

Carotenes: Nonpolar polyene carotenoids (simple hydrocarbons) consisting of only carbon.

Carotenogenesis: The process of developmentally or stress-induced overaccumulation of carotenoids in cells, leading to the production of carotene.

Carotenoid: A class of pigments synthesized by plants, algae, and photosynthetic bacteria, and found within the chloroplasts. A group of yellow to red lipid-soluble polyene pigments derived from a C40 backbone that is made from the building blocks isopentenyl pyrophosphate and dimethylallyl phyrophosphate. Carotenoids are also called tetraterpenoids, and are organic pigments that are found in the chloroplasts and chromoplasts of plants and some other photosynthetic organisms, including some bacteria and fungi.

Chaperones: Proteins that assist the covalent folding or unfolding and the assembly or disassembly of other macromolecular structures.

Chemocline: Chemistry gradient along the water column. In a meromictic lake, the transition between the upper mixolimnion and lower monimolimnion layers, marked by a change from aerobic to anaerobic conditions.

Chemolithoautotroph: An autotrophic microorganism that obtains energy by oxidizing inorganic compounds and uses carbon dioxide as its sole source of carbon for growth.

Chlorophylls: Green lipid-soluble photosynthetic pigments of algae and higher plants, essential in photosynthesis, consisting of closed tetrapyrrole rings with magnesium.

Chlorophyta: A taxonomic division containing the green algae, which are characterized by having photosynthetic pigments similar to those in higher plants.

Chott (also Shott): A depression surrounding a salt marsh or lake, usually dry during the summer and becomes flooded during the winter months.

Coastal zone: The land–sea–air interface zone around continents and islands extending from the landward edge of a barrier beach or shoreline of a coastal bay to the outer extent of the continental shelf.

Codon usage bias: The phenomenon where specific codons are used more often than other synonymous codons during translation of genes, the extent of which varies within and among species.

Compatible solutes: Small molecules (osmoprotectants or compatible solutes) that act as osmolytes and help organisms survive extreme osmotic stress.

Convective layer: In brine pools, a water layer characterized by specific ranges of density, salinity, and temperature, forming a water body well separated from the rest of the water column.

Cryoconite: Water-filled cylindrical melt holes on a glacier ice surface. The bottom of these holes often has dark-colored material, cryoconite, which absorbs solar radiation and promotes melting of the ice beneath it to maintain the hole.

Cyanobacteria: A phylum of Bacteria that obtain their energy through photosynthesis. They are often referred to as blue-green algae, although they are in fact prokaryotes, not algae.

Database: In life sciences, libraries of information, collected from scientific experiments and published literature. Information contained in biological databases includes gene function, structure, localization, and biological sequences and structure.

Deep-sea hydrothermal vent (DSHV): Marine environment where geothermally hot water springs from the seafloor.

Denaturing gradient gel electrophoresis (DGGE): A molecular technique by which DNA is denatured and later subjected to electrophoresis so that DNA fragments of the same size can be separated accordingly.

Dendrogram: A graphically summarized diagram used to represent the similarities and relationships between organisms.

De novo **biosynthesis:** Process wherein molecules are synthesized as per need of cell from raw materials through various pathways.

Deposit: An accumulation of specific minerals within a defined area.

Desert: Arid region receiving less than 200 mm of precipitation annually; tends to evolve into desert.

Diaminopimelic acid (DAP): An amino acid representing an epsilon-carboxy derivative of lysine. DAP is a characteristic of certain cell walls of some bacteria.

Differential gene expression: Gene expression that responds to signals or triggers; a means of gene regulation, and effects of certain hormones on protein biosynthesis.

Diphosphatidylglycerol: Glycerol linked to two molecules of phosphatidic acid; 1,3-diphosphatidylglycerol is cardiolipin.

Diversity: In biology, the degree of variation of living things present in a particular ecosystem or habitat.

DnaK: A bacterial enzyme that couples the cycle of ATP binding, hydrolysis, and ADP unfolded proteins by a C-terminal substrate binding domain.

Domains (super kingdom): In biological taxonomy, the highest taxonomic rank of organisms.

Enantioselectivity: In a chemical reaction, the preferential formation of an enantiomer over another. Enantiomers are molecules that share the same chemical formula and sequence of bonded atoms but are nonsuperimposable.

Ephemeral lake: A lake that is usually dry, but that fills with water for brief periods during and after rainfall or other precipitation.

Epiphyte: Something that grows harmlessly upon another organism or plant and derives its moisture and nutrients from the air, rain, or debris accumulating around it.

Esterase: Hydrolase enzyme that splits esters into an acid and an alcohol.

Estuarine: Of the estuary, habitat where tidal mixing of fresh and salt water occurs.

Eukaryote: An organism with membrane-bound organelles (e.g., nucleus, mitochondria, golgi apparatus, and a 9 + 2 flagellum structure), as distinct from a more primitive prokaryotic organism, lacking membrane-bound organelles.

Euryhaline organism: An organism capable of withstanding a wide range of environmental salinities.

Eury-psychrophile: True psychrophile whose optimal growth temperature is at or below 2°C.

Eutrophic: Having waters rich in phosphates, nitrates, and organic nutrients that promote proliferation of life.

Evaporites: Marine evaporites are sedimentary rocks formed by the precipitation of minerals due to the evaporation of water within ocean basins.

Evolution: Change in the heritable traits of biological populations over successive generations. These processes give rise to diversity at every level of organization, including the levels of species, individual organisms, and molecules.

Exopolysaccharide: Polysaccharides (polymeric carbohydrate molecules composed of long chains of monosaccharide units bound by glycosidic linkages) that are secreted into the external environment.

Extracellular polymeric substances (EPSs): Natural polymers of high molecular weight secreted by microorganisms into their environment.

Extremolytes: Protective organic biomolecules that convey characteristics for survival in extreme environmental conditions.

Extremophile: An organism that can live in extreme environments, such as high pressure (barophile), low temperature (psychrophile), high temperature (thermophile and hyperthermophile), high salinity (halophile), and chemical concentration (acidophile and basophile).

Fluorescence *in situ* hybridization (FISH): A technique for identifying certain genes or organisms in which specific DNA fragments are labeled with fluorescent dye.

Fluorometric: Measurements based on the intensity and wavelength distribution of the emission spectrum after excitation by a certain spectrum of light.

Fluorometry: An analytical method for the measurement of the fluorescence of a compound induced by ultraviolet light.

Fosmid library: A collection of DNA fragments cloned in a vector based on the bacterial F-plasmid.

Fucoxanthin: A xanthophyll, with formula $C_{42}H_{58}O_6$.

Growth rate: Describes the increase of growth in a culture of cells. It is generally measured as an increase in the number of cells, biomass, or optical density.

Habitat: The type of location with resources where an organism or population of organisms occurs.

Haloadaptation: An adaptation strategy continually followed by organisms living under high salinity.

Halobacteria: Type of bacteria capable of living only in hypersaline water bodies.

Halocline: A vertical gradient of salinity in the water column.

Halophile: An organism that requires high salt concentrations of salt for survival.

Halophilic: Term to describe an organism that loves to grow in high salt concentrations.

Halotolerant: Term to describe an organism that can tolerate high saline conditions but prefers to grow at nonsaline conditions.

Heterotroph: An organism requiring organic compounds for its principal source of food, or those that feed on other organisms.

Hofmeister stabilizing ions: The relative stabilizing or destabilizing action of different cations on proteins.

Horizontal gene transfer (HGT): The transfer of genes between organisms without the occurrence of a reproductive event. HGT is known to occur between different species, such as between prokaryotes and eukaryotes, and between the three DNA-containing organelles of eukaryotes: the nucleus, the mitochondrion, and the chloroplast.

Hydrolase: Enzyme that catalyzes the cleavage of chemical bonds by the addition of a water molecule.

Hydrophilic: Having an affinity for water; readily absorbing or dissolving in water.

Hydrophobic: Repelling, tending not to combine with, or incapable of dissolving in water.

Hyperextreme: At the outer end of the range.

Hypersaline: Waters that contain higher concentrations of dissolved mineral salts than seawater.

Hypolimnion: The lower layer of water in a stratified lake, typically cooler than the water above during the summer and warmer than the water above during the winter, and often isolated from surface wind mixing.

Inoculum: Material containing organisms to begin new cultures; usually consists of actively dividing or algal cells in the exponential phase.

Internal transcribed spacer (ITS) region: The spacer DNA situated between the small-subunit ribosomal RNA (rRNA) and large-subunit rRNA genes in the chromosome.

Isosmotic: A state of having the same osmolarity.

Jukes and Cantor: Method used to estimate the amount of change in a nucleotide sequence.

Ketocarotenoid: Any carotenoid that contains a ketone (carbonyl) group.

Ketoreductase activity: Catalysis of the reduction of a ketone group to form the corresponding alcohol.

Lateral gene transfer: A type of horizontal gene transfer that occurs between eukaryotic organisms.

Life cycle: The progression of an organism through a series of developmental changes.

Lipases: Enzymes that catalyze the hydrolysis of lipids.

Lipid: A group of organic molecules that are insoluble in water (also called fats). Examples are oils, waxes, sterols, and triglycerides.

Lutein: A xanthophyll and 1 of 600 known naturally occurring carotenoids.

Macromolecules: Substances that are larger in size than what are normally synthesized.

Maximum likelihood algorithm: A method used for estimating the parameters of a statistical model for given data.

Maximum parsimony: In phylogenetics, the distance method.

Media: To grow marine microalgae seawater, or artificial seawater is enriched with nutrients, sterilized, and seeded with algal cells.

Menaquinone: Form of vitamin K synthesized by bacteria in the large intestine or in putrefying organic matter and essential for the blood-clotting process. It is an isoprenoid derivative of menadione. Menaquinones are abbreviated MK-n, where M stands for menaquinone, K stands for vitamin K, and n represents the number of isoprenoid side-chain residues. For example, menaquinone-4 (MK-4) has four isoprene residues in its side chain.

Meromictic: A lake that has layers of water that do not intermix.

Meso-DAP: The *meso*-isomer of 2,6-diaminopimelic acid. It is a key constituent of bacterial peptidoglycan and is often found in human urine due to the breakdown of the gut microbes.

Metabolic engineering: A procedure that uses recombinant DNA techniques to purposely modify the metabolism of an organism so that it will produce specific desired molecules.

Metabolic homeostasis: A biochemically steady state of a cell.

Metabolic profiling: Chemical fingerprint methods for the description of the secondary metabolites produced under a given, for example, environmental situation.

Metabolite: A chemical molecule made by chemical reactions in cells.

Metalophiles: Organisms that colonize environments where the concentration of metals is very high.

Metatranscriptomics: A branch of transcriptomics that studies, and correlates, the transcriptomes of a group of interacting organisms or species.

Methanogenesis: The biological production of methane by methanogens, a group of anaerobic Archaea.

Methanotroph: Prokaryotes that are able to metabolize methane as their only source of carbon and energy. They can grow aerobically or anaerobically and require single-carbon compounds for survival.

Microalga: An alga of microscopic size with an ability to conduct photosynthesis.

Microbially induced sedimentary structure (MISS): Primary sedimentary structure formed by the interaction of microbes with sediment and the physical agents of erosion, deposition, and transportation.

Microbiome: All the microorganisms that live in a specific habitat.

Mixotroph: An organism that can use a mix of different sources of energy and carbon. Usually, this means that it may be either autotrophic or heterotrophic at different times in its life.

Molecular chaperones: Proteins that assist the covalent folding or unfolding and the assembly or disassembly of other macromolecular structures.

Molecular diversity: Process of studying degree of variation in species on the basis of molecular traits.

Molecular signatures: Sets of genes, proteins, genetic variants, or other variables that can be used as a marker for a particular phenotype.

Multilocus enzyme electrophoresis (MLEEC): A traditional technique that measures the electrophoretic mobility of metabolic enzymes. Nonamplified strain typing. Proteins are isolated from the strain of interest and separated in a gel.

Mycorrhiza: A symbiotic association of a fungus and roots of a vascular plant.

Nanoparticles: Particles measuring between 1 and 100 nm in diameter. These behave as entire units with respect to their transport and other properties. In medicine, they are used for drug and gene delivery, labeling structures, diagnostic imaging, and tumor targeting.

Neighbor-joining: In bioinformatics, a clustering method for the creation of phylogenetic trees based on DNA or protein sequence data.

Next-generation sequencing (NGS): A term used to describe methods of high-throughput DNA and RNA sequencing, allowing rapid determination of sequences of entire microbial and other genomes.

Niches: A term describing the relational position of a species or population in an ecosystem.

Oligotrophic: Lacking in nutrients, such as phosphates, nitrates, and organic matter.

Omics techniques: Comprehensive approaches for the description of the different biological levels of a cell (DNA, genome; RNA, transcriptome; proteins, proteome; metabolites, metabolome). They all share novel and rapid methods for the broad analysis of the given level (the aim is to describe the totality of, e.g., all genes, all proteins, etc.) and the generation of big datasets. A combination of these approaches will lead to a deep understanding of the metabolic potential and regulation on the cellular level.

Open reading frame (ORF): A DNA sequence made by nucleotide triplets, which can code for a protein or a peptide.

Operational taxonomic unit (OTU): An operational definition of a species adopted in microbiology based on the level of sequence similarity of a given gene, usually the 16S rRNA gene.

Osmolyte: The biological molecules synthesized or accumulated by cell that work as osmoprotectants.

Osmophobic effect: Substances that repel or avoid osmosis.

Osmoprotectants: Small molecules, soluble inside or outside the cells, that help organisms survive in highly saline environments.

Osmoregulation: Control and maintenance of the osmotic balance within a cell.

Osmoregulator: A substance that helps maintain osmotic balance.

Osmotic down shock: A state of loss of osmolarity, or plasmolysis.

Osmotic pressure: The pressure that would have to be applied to a pure solvent to prevent it from passing into a given solution by osmosis; often used to express the concentration of the solution.

Osmotic solute: The minimum pressure that needs to be applied to a solution to prevent the inward flow of water across a semipermeable membrane.

Oxycline: Vertical gradient of oxygen concentration in the water column.

Permafrost: A thick subsurface layer of soil that remains frozen throughout the year, occurring chiefly in polar regions.

Photoacclimation: Changes to molecular, physiological, and biochemical components of a cell in response to changes to the redox state of the photosynthetic electron transport chain.

Photobioreactor: In general terms, the description of a photobioreactor system for cultivation of phototroph organisms.

Photoinhibition: The inhibition of a process by high light irradiance. More specifically, a condition when the rate of photodamage exceeds the capacity of photoprotection and of the repair process, causing the decline of the overall photosynthesis rate.

Photosynthetically active radiation (PAR): The visible portion of the spectrum (400–700 nm) that drives photosynthesis.

Photosynthetic photon flux density (PPFD): The rate of flow of photons (400–700 nm) through a standard area. Usually expressed as micromoles of photons per second through an area of 1 m^2. A mole of photons was formerly called an Einstein.

Phycobiliosome: Protein complexes anchored to thylakoid membranes and made of stacks of chromophorylated proteins, the phycobilproteins, and their associated linker polypeptides.

Phycobiliprotein: Water-soluble proteins present in cyanobacteria cryptomonads and rhodophycean algae.

Phylogenetic analysis: In biology, the study of the evolutionary history and relationships among individuals or groups of organisms.

Phylogeny: The evolution of a genetically related group of organisms.

Phylum: A taxonomic rank below kingdom and above class.

Phytoplankton: Photosynthetic microalgae (usually 2–200 μm), living free or suspended (swimming feebly) in the water column.

Piezophiles: Organisms growing optimally at high hydrostatic pressure, such as in the deep-sea environments.

Polyhydroxyalkanoates (PHAs): Linear polyesters produced in nature by bacterial fermentation of sugar or lipids.

Primary carotenoid: A carotenoid that is a structural or functional component of the photosynthetic apparatus. In general, carotenoids produced by cells grown under normal conditions.

Primary productivity: Rate of production of organic compounds from carbon dioxide, principally through the process of photosynthesis.

Productivity: The rate of output (e.g., oxygen during photosynthesis) per given time.

Prokaryote: A primitive organism lacking membrane-bound organelles (e.g., bacteria or blue-green algae).

Protist: Single-celled eukaryotes, traditionally protozoa, algae, and lower fungi.

Psychrophile: Organism (usually bacteria or archaea) that grows and is capable of reproduction, preferentially at low temperatures ranging from −20°C to +10°C. Their optimal growth temperature is about 15°C or lower, and they have a maximal growth temperature of approximately 20°C. Psychrotrophs are cold tolerant, but have optimal and maximal growth temperatures above these values.

Pycnocline: A water layer where a density gradient is observed.

Pyrosequencing: A DNA sequencing technique based on the detection of pyrophosphate released during DNA synthesis.

Radiophiles: Organisms able to thrive in environments characterized by a high level of ionizing radiations.

Rare biosphere: Part of the microbial diversity that is composed of low-abundant taxa.

Rhizosphere: The narrow region of soil that is directly influenced by root secretions and associated soil microorganisms.

Scirroco: A hot, dry, dusty wind blowing from North Africa to the North Mediterranean coast.

Sebkha: Used by the Arabs of North Africa and Arabia to denote flat areas of clay, silt, and sand that are often encrusted with salt.

Secondary metabolites: Set of metabolites produced by an organism for nonprimary purposes (i.e., not for catabolism and anabolism), mainly small molecules (approximately 200–1000 Daltons molecular weight), broad chemodiversity. Another commonly used term is *natural products.*

Siderophores: Greek: "iron carrier." Small, high-affinity iron-chelating compounds secreted by microorganisms such as bacteria, fungi, and grasses. Siderophores are among the strongest soluble Fe^{3+} binding agents known.

Species: A taxonomic unit that today has several different approaches to its definition. An example for the biological species concept is a group of organisms whose members can interbred.

Spectrophotometric: Measurements based on the absorbance of light by a compound at a particular wavelength.

Steno-psychrophile: Psychrotolerant. Those organisms that can grow at 0°C but grow optimally at 20°C–30°C.

Strain: A group of organisms within a species associated with a geographic location.

Structural elucidation: Determination of the chemical structure of a secondary metabolite by means of nuclear magnetic resonance, exact mass measurements, polarimetry, and so forth. In many contexts, this includes the determination of the stereostructure of the molecule, which often determines biological activity.

Sulfate reduction: An anaerobic respiration process catalyzed by sulfate-reducing bacteria (SRB), where sulfate is used as the terminal electron acceptor.

Terminal-restriction fragment length polymorphism (T-RFLP): A molecular biology technique used for the profiling of microbial communities based on a restriction site closest to a labeled end of an amplified gene.

Thermokarst: A form of periglacial topography resembling karst, with hollows produced by the selective melting of permafrost.

Thermophiles: Bacteria or archaea that grow at temperatures up to approximately 70°C, with an optimum of about 50°C. Hyperthermophiles have temperature optima of >75°C, and some can grow at temperatures of >122°C.

Thermotolerant: Organisms able to withstand high temperatures ranging between 41°C and 122°C.

Transposase: An enzyme that binds to the end of a transposon and catalyzes the movement of the transposon to another part of the genome by a cut-and-paste mechanism or a replicative transposition mechanism.

Turgor pressure: The pressure of water pushing the plasma membrane against the cell wall of a plant cell.

Unialgal culture: A culture that contains only one species of alga but may contain contaminants such as bacteria.

Xanthophyll: Yellow pigment that forms one of two major divisions of the carotenoid group. Consists of three carotenoids, violaxanthin, antheraxanthin, and zeaxanthin, which are interconverted by enzymes that are sensitive to the pH of the lumen. While under low-light conditions, violaxanthin is the dominant member of the cycle and aids light harvesting. The pathway of interconversion of these pigments is under high pressure and follows from

Violaxanthin (orange pigments) into Antheraxanthin (a bright yellow accessory pigment) and then to light-yellow pigment zeaxanthin. The carotenoids, in addition to the hydrocarbon backbone, also contain oxygen atoms within the molecule. These are polar oxygenated carotenoids in the chloroplasts of all plant and algal cells that act as accessory light-harvesting pigments, or have a photoprotective function.

Zeaxanthin: One of the most common carotenoid alcohols found in nature.

Zwitterion: Electrically neutral or carrying both positive and negative charges in equal proportion.

Index

Page numbers followed by f and t indicate figures and tables, respectively.